S0-ATY-568

Fundamentals of MATHEMATICS

EIGHTH EDITION

JAMES VAN DYKE

JAMES ROGERS ▪ HOLLIS ADAMS

Portland Community College

THOMSON

BROOKS/COLE™

Australia • Canada • Mexico • Singapore • Spain • United Kingdom • United States

THOMSON
BROOKS/COLE

Sponsoring Editor: *Jennifer Huber*
Editorial Assistant: *Jonathan Wegner*
Technology Project Manager: *Earl Perry*
Marketing Manager: *Leah Thomson*
Marketing Assistant: *Maria Salinas*
Advertising Project Manager: *Bryan Vann*
Project Editor: *Keith Faivre*
Print/Media Buyer: *Vena Dyer*
Permissions Editor: *Sue Ewing*
Production Service: *Nancy Shammas, New Leaf Publishing Services*

Photo Credits: Chapter 1: ©AFP/Corbis; Chapter 2: Reprinted with permission of the U.S. Department of theTreasury; Chapter 3: John Kelly/Stone/Getty Images; Chapter 4: David Madison/Stone/Getty Images; Chapter 5: SuperStock; Chapter 6: J.Silver/SuperStock; Chapter 7: Ping Amranand/SuperStock; Chapter 8: James Balog/Stone/Getty Images.

Printed in the United States of America
1 2 3 4 5 6 7 06 05 04 03 02

For more information about our products, contact us at:
Thomson Learning Academic Resource Center
1-800-423-0563

For permission to use material from this text, contact us by:
Phone: 1-800-730-2214 **Fax:** 1-800-730-2215
Web: http://www.thomsonrights.com

Library of Congress Control Number: 2002106058

ISBN 0-534-39816-2

Text Designer: *Lisa Henry Design*
Photo Researcher: *Kathleen Olson, Kofoto*
Copy Editor: *Carol Dondrea*
Indexer: *Kay Banning*
Illustrator: *Lori Heckelman Illustration*
Cover Designer: *Lisa Henry Design*
Cover Image: *Daryl Benson/Masterfile*
Cover Printer: *Phoenix Color Corporation*
Compositor: *G&S Typesetters, Inc.*
Printer: *Quebecor World, Dubuque*

Brooks/Cole–Thomson Learning
511 Forest Lodge Road
Pacific Grove, CA 93950
USA

Asia
Thomson Learning
5 Shenton Way #01-01
UIC Building
Singapore 068808

Australia
Nelson Thomson Learning
102 Dodds Street
South Melbourne, Victoria 3205
Australia

Canada
Nelson Thomson Learning
1120 Birchmount Road
Toronto, Ontario M1K 5G4
Canada

Europe/Middle East/Africa
Thomson Learning
High Holborn House
50/51 Bedford Row
London WC1R 4LR
United Kingdom

Latin America
Thomson Learning
Seneca, 53
Colonia Polanco
11560 Mexico D.F.
Mexico

Spain
Paraninfo Thomson Learning
Calle/Magallanes, 25
28015 Madrid, Spain

To our spouses:

Carol Van Dyke ✦ Elinore Rogers ✦ Douglas Adams

CONTENTS

CHAPTER 3

GOOD ADVICE FOR STUDYING

FRACTIONS AND MIXED NUMBERS 199

CHAPTER 4

GOOD ADVICE FOR STUDYING

DECIMALS 327

CHAPTER 8

GOOD ADVICE FOR STUDYING

Evaluating Your Performance 658

ALGEBRA PREVIEW: SIGNED NUMBERS 659

TO THE STUDENT

"It looks so easy when you do it, but when I get home . . ." is a popular lament of many students studying mathematics.

The process of learning mathematics evolves in stages. For most students, the first stage is listening to and watching others. In the middle stage, students experiment, discover, and practice. In the final stage, students analyze and summarize what they have learned. Many students try to do only the middle stage because they do not realize how important the entire process is.

Here are some steps that will help you to work through all the learning stages:

1. Go to class every day. Be prepared, take notes, and most of all, think actively about what is happening. Ask questions and keep yourself focused. This is prime study time.
2. Begin your homework as soon after class as possible. Start by reviewing your class notes and then read the text. Each section is organized in the same manner to help you find information easily. The objectives tell you what concepts will be covered, and the vocabulary lists all the new technical words. There is a **How and Why** section for each objective that explains the basic concept, followed by worked sample problems. As you read each example, make sure you understand every step. Then work the corresponding **Warm-Up** problem to reinforce what you have learned. You can check your answer at the bottom of the page. Continue through the whole section in this manner.
3. Now work the exercises at the end of the section. The A group of exercises can usually be done in your head. The B group is harder and will probably require pencil and paper. The C group problems are more difficult, and the objectives are mixed to give you practice at distinguishing the different solving strategies. As a general rule, do not spend more than 15 minutes on any one problem. If you cannot do a problem, mark it and ask someone (your teacher, a tutor, or a study buddy) to help you with it later. Do not skip the **Maintain Your Skills** problems. They are for review and will help you practice earlier procedures so you do not become "rusty." The answers to the odd exercises are in the back of the text so you can check your progress.
4. In this text, you will find **State Your Understanding** exercises in every section. Taken as a whole, these exercises cover *all* the basic concepts in the text. You may do these orally or in writing. Their purpose is to encourage you to analyze or summarize a skill and put it into words. We suggest that you do these in writing and keep them all together in a journal. Then they are readily available as a review for chapter tests and exams.
5. When preparing for a test, work the material at the end of the chapter. The **True/False Concept Review** and the **Chapter Test** give you a chance to review the concepts you have learned. You may want to use the chapter test as a practice test.

If you have never had to write in a math class, the idea can be intimidating. Write as if you are explaining to a classmate who was absent the day the concept was discussed. Use your own words—*do not copy out of the text*. The goal is that you understand the concept, not that you can quote what the authors have said. Always use complete sentences, correct spelling, and proper punctuation. Like everything else, writing about math is a learned skill. Be patient with yourself and you will catch on.

Since we have many students who do not have a happy history with math, we have included **Good Advice for Studying**—a series of eight essays that address various problems that are common for students. They include advice on time organization, test taking, and reducing math anxiety. We talk about these things with our own students, and hope that you will find some useful tips.

We really want you to succeed in this course. If you go through each stage of learning and follow all the steps, you will have an excellent chance for success. But remember, you are in control of your learning. The effort that you put into this course is the single biggest factor in determining the outcome. Good luck!

James Van Dyke
James Rogers
Hollis Adams

CLAST SKILLS AND THEIR LOCATIONS IN THE BOOK

Arithmetic Skills

Arithmetic Skills	Location in Book
Add, subtract, multiply, and divide rational numbers in fractional form	Sections 3.3, 3.4, 3.6–3.10
Add, subtract, multiply, and divide rational numbers in decimal form	Sections 4.3–4.6, 4.8
Solve the sentence "a% of b is c," where the values of the variables are given	Section 6.7
Calculate percent increase and percent decrease	Section 6.8
Recognize the meaning of exponents	Sections 2.5, 2.6, 4.5
Recognize the role of the base number in determining place value in the base ten numeration system	Sections 1.1, 4.1
Identify equivalent forms of decimals, percents, and fractions	Sections 4.2, 4.7, 6.2–6.6
Determine the order relation between real numbers	Sections 1.1, 3.5, 4.2
Identify a reasonable estimate of a sum, average, or product of numbers	Sections 1.2–1.4, 4.3, 4.4, 4.6
Infer relations between numbers in general by examining particular number pairs	Sections 1.1, 2.2, 2.3
Solve real-world problems which do not involve the use of percent	Chapters 1–6, 8
Solve real world problems which involve the use of percent	Sections 6.8, 6.9
Solve problems that involve the structure and logic of arithmetic	Throughout

Geometry and Measurement Skills

Geometry and Measurement Skills	Location in Book
Round measurements	Sections 1.1, 4.1, Chapter 7
Calculate distance, area, and volume	Sections 1.2, 1.3, 7.3–7.5
Classify simple plane figures by recognizing their properties	Sections 1.2, 1.3, 7.3, 7.4
Identify appropriate units of measurement (linear, square, cubic) for geometric objects	Sections 1.2, 1.3, 7.3–7.5
Infer formulas for measuring geometric figures	Sections 1.2, 1.3, Chapter 7
Select applicable formulas for computing measures of geometric figures	Chapter 7
Solve real-world problems involving perimeters, areas, and volumes of geometric figures	Sections 1.2, 1.3, 7.3–7.5
Solve real-world problems involving the Pythagorean property	Section 7.6

Algebra Skills

Algebra Skills	Location in Book
Add, subtract, multiply, and divide real numbers	Sections 8.2–8.5
Apply the order-of-operations agreement to numerical and algebraic expressions	Sections 1.6, 3.10, 4.8, 8.6
Use scientific notation	Section 4.5
Solve linear equations and inequalities	Sections 1.2, 1.4, 1.6, 3.4, 3.9, 4.3, 4.6, 4.8, 8.7
Use formulas to compute results	Throughout
Recognize statements and conditions of proportionality and variation	Chapter 5
Solve real-world problems involving the use of variables	Sections 1.2, 1.4, 1.6, 3.4, 3.9, 4.3, 4.6, 4.8, 8.7

Statistics Skills, Including Probability

Statistics Skills, Including Probability	Location in Book
Identify information contained in bar, line, and circle graphs	Sections 1.8, 6.8
Determine the mean, median, and mode of a set of numbers	Sections 1.7, 3.10, 4.8
Recognize properties and interrelationships among the mean, median, and mode	Sections 1.7, 3.10, 4.8

ELM MATHEMATICAL SKILLS

The following table lists the California ELM MATHEMATICAL SKILLS and where coverage of these skills can be found in the text. Location of the skills is indicated by chapter section or chapter.

Numbers and Data Skills	Location in Book
Carry out basic arithmetic calculations	Chapters 1, 3–4
Understand and use percent in context	Chapter 7
Compare and order rational numbers expressed as fractions and/or decimals	Sections 3.5, 4.2
Solve problems involving fractions and/or decimals in context	Chapters 3, 4
Interpret and use ratio and proportion in context	Chapter 5
Use estimation appropriately	Sections 1.2, 1.3, 1.4, 4.3, 4.4, 4.6
Evaluate reasonableness of a solution to a problem	Sections 1.2, 1.3, 1.4, 4.3, 4.4, 4.6
Evaluate and estimate square roots	Section 7.6
Represent and understand data presented graphically (including pie charts, bar and line graphs, histograms, and other formats for presenting data visually used in print and electronic media)	Sections 1.8, 6.8
Calculate and understand the arithmetic mean	Sections 1.7, 3.10, 4.8
Calculate and understand the median	Sections 1.7, 4.8

Algebra Skills

Use properties of exponents	Sections 1.5, 2.5, 2.6, 4.5
Solve linear equations (with both numerical and literal coefficients)	Sections 1.2, 1.4, 1.6, 3.4, 3.9, 4.3, 4.6, 4.8, 8.7

Geometry Skills

Find the perimeter, area, or volume of geometric figures (including triangles, quadrilaterals, rectangular parallelepipeds, circles, cylinders, and combinations of these figures)	Sections 1.2, 1.3, 7.3. 7.4, 7.5
Use the Pythagorean Theorem	Section 7.6
Solve geometric problems using the properties of basic geometric figures (including triangles, and in problem sets throughout quadrilaterals, polygons, and circles.	Sections 1.2, 1.3, 7.3, 7.4, 7.5

TASP SKILLS AND THEIR LOCATION IN THE BOOK

Fundamental Skills of Mathematics	Location in Book
Solve word problems involving integers, fractions, decimals, and units of measurement.	Chapters 3–8
Solve problems involving data interpretation and analysis	Sections 1.2, 1.8, 6.8, and in problem sets throughout

Algebra Skills

Solve one- and two- variable equations	Sections 1.2, 1.4, 1.6, 3.4, 3.9, 4.3, 4.6, 4.8, 8.7
Solve word problems involving one and two variables	Sections 1.2, 1.4, 1.6, 3.4, 3.9, 4.3, 4.6, 4.8, 8.7

Geometry Skills

Solve problems involving geometric figures	Sections 1.2, 1.3, 7.3, 7.4, 7.5, and in problem sets throughout

TO THE INSTRUCTOR

Fundamentals of Mathematics, eighth edition, is a work text for college students who need to review the basic skills and concepts of arithmetic in order to pass competency or placement exams, or to prepare for courses such as business mathematics or elementary algebra. The text is accompanied by a complete system of ancillaries in a variety of media, affording great flexibility for individual instructors and students.

A TEXTBOOK FOR ADULT STUDENTS

Though the mathematical content of *Fundamentals of Mathematics* is elementary, students using the text are most often mature adults, bringing with them adult attitudes and experiences and a broad range of abilities. Teaching elementary content to these students, therefore, is effective when it accounts for their distinct and diverse adult needs. As you read about and examine the features of *Fundamentals of Mathematics* and its ancillaries, you will see how they especially meet three needs of your students:

- Students must establish good study habits and overcome math anxiety.
- Students must see connections between mathematics and the modern, day-to-day world of adult activities.
- Students must be paced and challenged according to their individual level of understanding.

A TEXTBOOK OF MANY COURSE FORMATS

Fundamentals of Mathematics is suitable for individual study or for a variety of course formats: lab, both supervised and self-paced; lecture; group; or combined formats. For a lecture-based course, for example, each section is designed to be covered in a standard 50-minute class. The lecture can be interrupted periodically so that students individually can work the **Warm-Up Exercises** or work in small groups on the group work. In a self-paced lab course, **Warm-Up Exercises** give students a chance to practice while they learn, and get immediate feedback since warm-up answers are printed on the same page. Using the text's ancillaries, instructors and students have even more options available to them. Computer users, for example, can take advantage of complete electronic tutorial and testing systems that are fully coordinated with the text. A detailed description of each print, video, and computer ancillary is found in the preceding the Resource Integration Guide.

TEACHING METHODOLOGY

As you examine the eighth edition of *Fundamentals of Mathematics*, you will see distinctive format and pedagogy that reflects these aspects of teaching methodology:

Teaching by Objective Each section focuses on a short list of objectives, stated at the beginning of the section. The objectives correspond to the sequence of exposition and tie together other pedagogy, including the highlighted content, the examples, and exercises.

1.1 Whole Numbers and Tables: Writing, Rounding, and Inequalities

OBJECTIVES

1. Write word names from place value names and place value names from word names.
2. Write an inequality statement about two numbers.
3. Round a whole number.
4. Read tables.

VOCABULARY

The **digits** are 0, 1, 2, 3, 4, 5, 6, 7, 8, and 9.

The **natural numbers (counting numbers)** are 1, 2, 3, 4, 5, and so on.

The **whole numbers** are 0, 1, 2, 3, 4, 5, and so on. Numbers larger than 9 are written in **place value name** by writing the digits in positions having standard **place value**.

Word names are written words that represent numerals. The word name of 125 is one hundred twenty-five.

The symbols **less than**, $<$, and **greater than**, $>$, are used to compare two whole numbers that are not equal. So, $2 < 9$, and $13 > 6$.

To **round** a whole number means to give it an approximate value. The sym-

Teaching by Application Each chapter leads off with an application that uses the content of the chapter. Exercise sets have applications that use this material or that are closely related to it. Applications are included in the examples for most objectives. Other applications appear in exercise sets. These cover a diverse range of fields, demonstrating the utility of the content in business, environment, personal health, sports, and daily life.

1 WHOLE NUMBERS

S E C T I O N S

APPLICATION

Professional athletes' salaries have been increasing rapidly, to the concern of many. Table 1.1 shows the top eight National Football League team salaries for the 1999–2000 and 2000–2001 seasons, respectively.

1. How many teams were in the top eight for both seasons?
2. What do you notice about the top salaries in the 1999–2000 season?
3. What do you notice about the top salaries in the 2000–2001 season?

In 1999–2000, the St. Louis Rams won the Superbowl. Their payroll was $52,119,900, which ranked them 16th. In 2000–2001, the Baltimore Ravens won the Superbowl.

4. Does there seem to be a pattern between total payroll and winning the Superbowl?

TABLE 1.1 TOP NFL PAYROLLS

Team	1999–2000 Payroll	Team	2000–2001 Payroll
New England Patriots	56,566,800	Washington Redskins	92,407,000
Tennessee Titans	56,129,500	Baltimore Ravens	90,703,000
Tampa Bay Buccaneers	55,767,700	Detroit Lions	78,376,000
Arizona Cardinals	54,902,000	St. Louis Rams	77,648,000
Dallas Cowboys	54,858,500	Jacksonville Jaguars	76,999,000
Pittsburgh Steelers	54,613,000	Tampa Bay Buccaneers	76,849,000
Philadelphia Eagles	54,155,000	Indianapolis Colts	76,108,000
Jacksonville Jaguars	53,655,100	Chicago Bears	73,095,000

Emphasis on Language New words of each section are explained in the vocabulary segment that precedes the exposition. Exercise sets include questions requiring responses written in the students' own words

1.7 Average, Median, and Mode

VOCABULARY

The **average**, or **mean**, of a set of numbers is the sum of the set of numbers divided by the total number of numbers in the set.

The **median** of a set of numbers, ordered from smallest to largest, is either the middle number of the set or the average of the two middle numbers in the set.

The **mode** of a set of numbers is the number or numbers that appear the most often in the set.

OBJECTIVES

1. Find the average of a set of whole numbers.
2. Find the median of a set of whole numbers.
3. Find the mode of a set of whole numbers.

Emphasis on Skill, Concept, and Problem Solving Each section covers concepts and skills that are fully explained and demonstrated in the exposition for each objective.

HOW & WHY

OBJECTIVE 1 Find the average of a set of whole numbers.

The *average*, or *mean*, of a set of numbers is used in statistics. It is one of the ways to find the middle of a set of numbers (like the average of a set of test grades). Mathematicians call the average or mean a "measure of central tendency." The average of a set of numbers is found by adding the numbers in the set and then dividing that sum by the number of numbers in the set. For example, to find the average of 8, 18, and 25:

$8 + 18 + 25 = 51$
$51 \div 3 = 17$

The average is 17.

The "central" number, or average, does not need to be one of the members of the set. The average, 17, is not a member of the set.

To find the average of a set of whole numbers

1. Add the numbers.
2. Divide the sum by the number of numbers in the set.

Carefully constructed examples for each objective are connected by a common strategy that reinforces both the skill and the underlying concepts. Skills are not treated as isolated feats of memorization but as the practical result of conceptual understanding: Skills are strategies for solving related problems. Students see the connections between problems that require similar strategies.

EXAMPLES A–E

DIRECTIONS: Find the average.

STRATEGY: Add the numbers in the set. Divide the sum by the number of numbers in the set.

A. Find the average of 112, 89, and 153.

$112 + 89 + 153 = 354$
$354 \div 3 = 118$

The average is 118.

B. Find the average of 13, 87, 252, and 124.

$13 + 87 + 252 + 124 = 476$
$476 \div 4 = 119$

The average is 119.

WARM-UPS A–E

A. Find the average of 301, 142, and 499.

B. Find the average of 7, 46, 95, 38, and 89.

Answers to Warm-Ups
A. 314 **B.** 55

EMPHASIS ON SUCCESS AND PREPARATION

Integrated throughout the text, the following features focus on study skills and math anxiety, calculators, and simple algebraic equations.

Good Advice for Studying is continued from the seventh edition. Originally written by the team of Dorette Long and Sylvia Thomas of Rogue Community College, these essays address the unique study problems that students of *Fundamentals of Mathematics* experience. Students learn general study skills and study skills specific to mathematics and to the pedagogy and ancillaries of *Fundamentals of Mathematics*. Special techniques are described to overcome the pervasive problems of math anxiety. Though an essay begins each chapter, students may profit by reading all the essays at once returning to them as the need arises. A description of how to learn from these essays appears in To the Student.

GOOD ADVICE FOR STUDYING

Learning to Learn Math

Learning mathematics is a building process. For example, if you have not mastered fractions, rational expressions are difficult to learn because they require an understanding of the rules for fractions. Therefore, if you are having difficulty with the current topic, you may not have mastered a previous skill that you need. It will be necessary to go back and learn/relearn this skill before you can continue.

Learning math also means learning not just skills, but how and where and when to **apply** the skills. For example, if it takes 16 gallons of gas to travel 320 miles, how many miles to the gallon are you getting? What skill would you use to solve this problem? (Answer: dividing) Reading the application problems and thinking of situations where you have used or could use these concepts helps integrate the concept into your experience.

Learning mathematics is learning something basic to daily life and to virtually every field of science and business. The examples in the book state the problem and a **strategy** for solving the problem. This strategy applies to several related problems. Learning mathematics is learning strategies to solve related problems.

The more you begin to appreciate mathematics as relevant to your life, the more you will see mathematics as worthy of your time, and the more committed you will feel to studying mathematics. Here is an activity that may give you fresh opportunities to see how much mathematics relates to your life. First, create a simple "web" or "map" with "math" at the center and spokes out from this center naming areas in life where math comes up—areas where math is useful. Capture as many areas as you can in a 5-minute period.

Second, turn the page over and construct a second "web," but this time choose one of your "math areas" for the center and create spokes that capture subtopics or subheadings in this particular math area. Take another 5 minutes for this second web. Next, create a problem that you believe to be solvable, from your own experience, and that might be enticing for someone else to solve.

If you are working with a partner, trade problems and see if you can solve each others' problems. Talk about how you might approach the problems and whether they are stated clearly and believably. Here are some criteria for a good problem:

- Enough data and information to solve the problem
- Clear statement of what you need to find
- Not too many questions included
- Appropriate reading level and clearly written
- Appears solvable and not too scary
- Makes the reader care about wanting to solve the problem

Going through this activity may help you become more aware of mathematics in your daily life and give you a greater understanding of problems and problem solving.

Calculator examples, marked by the symbol, demonstrate how a calculator may be used, though the use of a calculator is left to the discretion of the instructor. Nowhere is the use of a calculator required. Appendix A reviews the basics of operating a scientific calculator.

CALCULATOR EXAMPLE

C. Find the average of 777, 888. 914, and 505.

C. Find the average of 673, 821, 415, and 763.

Enter the sum, in parentheses, and divide by 4.

$(673 + 821 + 415 + 763) \div 4$

The average is 668.

D. The average of 23, 35, 11, 19, and ? is 20. Find the missing number.

D. The average of 32, 13, 15, and ? is 21. Find the missing number.

STRATEGY: Because the average of the four numbers is 21, we know that the sum of the four numbers is 4(21) or 84. To find the missing number, subtract the sum of the three given numbers from 84.

$$84 - (32 + 13 + 15) = 84 - (60)$$
$$= 24$$

So the missing number is 24.

E. The local cold storage company ships the following number of cases of frozen peas: Monday, 4018; Tuesday, 1475; Wednesday, 3650; Thursday, 2151; and Friday, 3216. What is the average number of cases shipped each day?

E. In order to help Pete lose weight, the dietician has him record his caloric intake for a week. He records the following: Monday, 3120; Tuesday, 1885; Wednesday, 1600; Thursday, 2466; Friday, 1434; Saturday, 1955; and Sunday, 2016. What is Pete's average caloric intake per day?

STRATEGY: Add the calories for each day and then divide by 7, the number of days.

```
 3120          2068
 1885      7)14476
 1600        14
 2466         4
 1434         0
 1955         47
 2016         42
14,476         56
               56
                0
```

Pete's average caloric intake is 2068 calories per day.

Getting Ready for Algebra segments follow sections 1.2, 1.4, 1.6, 3.4, 3.9, 4.3, 4.6, and 4.8. The operations from these sections lend themselves to solving simple algebraic equations. Though entirely optional, each of these segments includes its own exposition, examples with warm-ups and exercises. Instructors may cover these segments as part of the normal curriculum or assign them to individual students.

Getting Ready for Algebra

HOW & WHY

OBJECTIVE

Solve an equation of the form $\frac{ax}{b} = \frac{c}{d}$, where *a*, *b*, *c*, and *d* are whole numbers.

We have previously solved equations in which variables (letters) are either multiplied or divided by whole numbers. We performed the inverse operations to solve for the variable. To eliminate multiplication, we divide by the number being multiplied. To eliminate division, we multiply by the number that is the divisor. Now we solve equations in which variables are multiplied by fractions. Recall from Chapter 1 that if a number is multiplied times a variable, there is usually no multiplication sign between them. That is, $2x$ is understood to mean 2 times x, and $\frac{2}{3}x$ means $\frac{2}{3}$ times x. However, we usually do not write $\frac{2}{3}x$. Instead, we write this as $\frac{2x}{3}$. We can do this because

$$\frac{2}{3}x = \frac{2}{3} \cdot x = \frac{2}{3} \cdot \frac{x}{1} = \frac{2x}{3}$$

Although we usually write $\frac{2x}{3}$ instead of $\frac{2}{3}x$, for convenience we may use either of these forms. Recall that $\frac{2x}{3}$ means the *product* of 2 and x *divided* by 3.

To solve an equation of the form $\frac{ax}{b} = \frac{c}{d}$

1. Multiply both sides of the equation by *b* to eliminate the division on the left side.
2. Divide both sides by *a* to isolate the variable.

PEDAGOGY

The pedagogical system of *Fundamentals of Mathematics* meets two important criteria: coordinated purpose and consistency of presentation.

Each section begins with numbered **Objectives,** followed by definitions of new **Vocabulary** to be encountered in the section. Following the vocabulary, **How and Why** segments, numbered to correspond to the objectives, explain and demonstrate concepts and skills. Throughout the **How and Why** segments, **skill boxes** clearly summarize and outline the skills in step-by-step form. Also throughout the segments, **concept boxes** highlight appropriate properties, formulas, and theoretical facts underlying the skills. Following each **How and Why** segment are **Examples** and **Warm-Ups.** Each example of an objective is paired with a warm-up, with workspace provided. Solutions to the warm-ups are given at the bottom of the page, affording immediate feedback. The examples also include, where suitable, a relevant application of the objective. Examples similar to each other are linked by common **Directions** and a common **Strategy** for solution. Directions and strategies are closely related to the skill boxes. Connecting examples by a common solution method helps students recognize the similarity of problems and their solutions, despite their specific differences. In this way, students may improve their problem-solving skills. In both **How and Why** segments and in the example, **Caution** remarks help to forestall common mistakes.

EXERCISES, REVIEWS, TESTS

Thorough, varied, properly paced, and well-chosen exercises are a hallmark of *Fundamentals of Mathematics*. Exercise sets are provided at the end of each section and a review set at the end of each chapter. Workspace is provided for all exercises and each exercise set can be torn out and handed in without disturbing other parts of the book.

Section exercises are paired so that virtually each odd-numbered exercise, in Sections A and B, is paired with an even-numbered exercise that is equivalent in type and difficulty. Since answers for odd-numbered exercises are in the back of the book, students can be assigned odd-numbered exercises for practice, and instructors can assign even-numbered exercises for homework.

Section exercises are categorized to satisfy teaching and learning aims. Exercises for estimation, mental computation, pencil and paper computation, application, and calculator skills are provided, as well as opportunities for students to challenge their abilities, master communications skills, and to participate in group problem solving.

- **Category A** exercises, organized by section objective, are those that most students should be able to solve mentally, without pencil, paper, or calculator. Mentally working problems improves students' estimating abilities. These can often be used in class as oral exercises.
- **Category B** exercises, also organized by objective, are similar except for level of difficulty. All students should be able to master Category B.

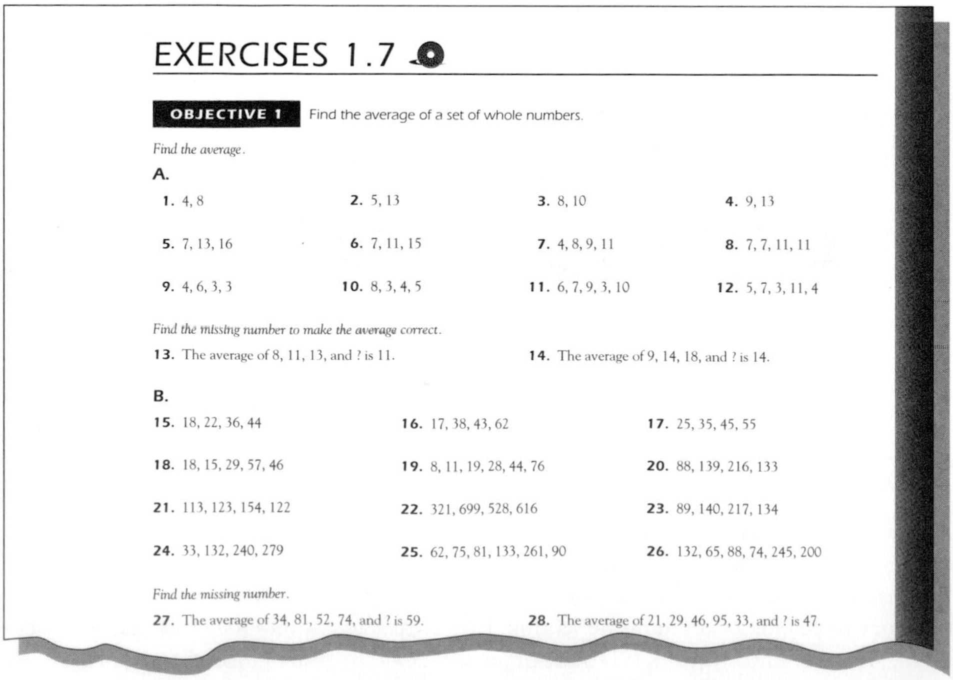

EXERCISES 1.7

OBJECTIVE 1 Find the average of a set of whole numbers.

Find the average.

A.

1. 4, 8	**2.** 5, 13	**3.** 8, 10	**4.** 9, 13
5. 7, 13, 16	**6.** 7, 11, 15	**7.** 4, 8, 9, 11	**8.** 7, 7, 11, 11
9. 4, 6, 3, 3	**10.** 8, 3, 4, 5	**11.** 6, 7, 9, 3, 10	**12.** 5, 7, 3, 11, 4

Find the missing number to make the average correct.

13. The average of 8, 11, 13, and ? is 11.

14. The average of 9, 14, 18, and ? is 14.

B.

15. 18, 22, 36, 44	**16.** 17, 38, 43, 62	**17.** 25, 35, 45, 55
18. 18, 15, 29, 57, 46	**19.** 8, 11, 19, 28, 44, 76	**20.** 88, 139, 216, 133
21. 113, 123, 154, 122	**22.** 321, 699, 528, 616	**23.** 89, 140, 217, 134
24. 33, 132, 240, 279	**25.** 62, 75, 81, 133, 261, 90	**26.** 132, 65, 88, 74, 245, 200

Find the missing number.

27. The average of 34, 81, 52, 74, and ? is 59.

28. The average of 21, 29, 46, 95, 33, and ? is 47.

- **Category C** exercises contain applications and more difficult exercises. Since these are not categorized by objective, the student must decide on the strategy needed to set up and solve the problem. These applications are drawn from business, health and nutrition, environment, consumer, sports, and science fields. Both professional and daily-life uses of mathematics are incorporated.

State Your Understanding exercises require a written response, usually no more than two or three sentences. These responses may be kept in a journal by the student. Maintaining a journal allows students to review concepts as they have written them. These writing opportunities facilitate student writing in accordance with standards endorsed by AMATYC and NCTM.

STATE YOUR UNDERSTANDING

65. Write out the order of operations for fractions. How is it different from the order of operations for whole numbers?

66. Must the average of a group of numbers be larger than the smallest number and smaller than the largest number? Why?

Challenge exercises stretch the content and are more demanding computationally and conceptually.

CHALLENGE

Perform the indicated operations.

67. $2\frac{5}{8}\left(4\frac{1}{5} - 3\frac{5}{6}\right) \div 2\frac{1}{2}\left(3\frac{1}{7} + 2\frac{1}{5}\right)$

68. $1\frac{2}{5}\left(5\frac{1}{5} - 4\frac{3}{4}\right) \div 4\frac{1}{2}\left(3\frac{1}{7} - 2\frac{1}{3}\right)$

69. The Acme Fish Company pays \$1500 per ton for crab. Jerry catches $3\frac{2}{5}$ tons, his brother Joshua catches $1\frac{1}{2}$ times as many as Jerry. Their sister Salicita catches $\frac{7}{8}$ the amount that Joshua does. What is the total amount paid to the three people by Acme Fish Company to the nearest dollar?

Group Work exercises and **Group Projects** provide opportunities for small groups of students to work together to solve problems and create reports. While the use of these is optional, the authors suggest the assignment of two or three of these per semester or term to furnish students with an environment for exchanging ideas. Group Work exercises encourage cooperative learning as recommended by AMATYC and NCTM guidelines.

GROUP WORK

70. Prepare for the chapter exam by having each member of the group make up ten exercises, one from each section of the chapter. Exchange these with another member, work each other's exercises, and check all of your answers. Discuss with the whole group the most common errors and how to avoid them.

GROUP PROJECT (1–2 weeks) OPTIONAL

Much research has been conducted in the area of human nutrition, and yet there are still many unanswered questions. Most nutritionists agree, however, that the American diet has too much fat and refined carbohydrates (primarily sugar).

You are about to become a nutritional consultant to John and Jane Doe. John is a 32-year-old male, 5′10″, 165 lb, who consumes 2800 calories per day. Jane is a 31-year-old female, 5′5″, 130 lb, who consumes 2000 calories per day. Both exercise regularly and try to be careful about their diet. However, neither of them likes to cook, so they eat out a lot.

Experts disagree about the exact ratio of carbohydrates, protein, and fat in an optimal diet. Some recommend a 40% carbohydrate, 30% protein, 30% fat ratio. Others would shift the amounts to 50%, 25%, and 25%, respectively. Almost no one recommends more than 30%

Maintain Your Skills exercises continually reinforce mastery of skills and concepts from previous sections. In the eighth edition the problems are specially chosen to review topics that will be needed in the next section.

MAINTAIN YOUR SKILLS

Perform the indicated operations.

71. $5\frac{1}{3} + \frac{7}{9}$ **72.** $5\frac{1}{3} - \frac{7}{9}$ **73.** $5\frac{1}{3}\left(\frac{7}{9}\right)$

74. $5\frac{1}{3} \div \frac{7}{9}$ **75.** $\frac{15}{28} \cdot \frac{21}{45} \cdot \frac{20}{35}$ **76.** $\frac{9}{15} \cdot \frac{3}{4} \cdot \frac{35}{6}$

77. Find the prime factorization of 650.

78. Find the prime factorization of 975.

79. A coffee table is made of a piece of maple that is $\frac{3}{4}$ in. thick, a piece of chipboard that is $\frac{3}{8}$ in. thick, and a veneer that is $\frac{1}{8}$ in. thick. How thick is the tabletop?

80. Felicia works a 5-day week with the following hours: $6\frac{3}{4}$ hr, $7\frac{1}{3}$ hr, $6\frac{2}{3}$ hr, $9\frac{3}{4}$ hr, and $7\frac{1}{2}$ hr. How many hours does she work for the week? What is her pay if the rate is $\$10\frac{1}{2}$ per hour?

Key Concepts, new to the eighth edition, recap the important concepts and skills covered in the chapter. The **Key Concepts** can serve as a quick review of the chapter material.

KEY CONCEPTS

SECTION 3.1 PROPER AND IMPROPER FRACTIONS; MIXED NUMBERS

Definitions and Concepts	Examples
A fraction has the form $\frac{\text{numerator}}{\text{denominator}}$.	$\frac{4}{81}, \frac{20}{3}, \frac{25}{25}$
A proper fraction has a smaller numerator than denominator.	$\frac{4}{81}$ is a proper fraction.
An improper fraction has a numerator that is not smaller than the denominator.	$\frac{20}{3}$ and $\frac{25}{25}$ are improper fractions.
A mixed number is the sum of a whole number and a fraction.	$4\frac{1}{6}$
To change a mixed number to an improper fraction: Multiply the whole number by the denominator and add the numerator. Place the sum over the denominator.	$4\frac{1}{6} = \frac{4 \cdot 6 + 1}{6} = \frac{25}{6}$
To change an improper fraction to a mixed number: • Divide the numerator by the denominator. • The mixed number is the whole number plus the remainder over the divisor.	$\frac{38}{7} = 5\frac{3}{7}$ because $7\overline{)38}$ = 5; 35; 3

SECTION 3.2 SIMPLIFYING FRACTIONS

Definitions and Concepts	Examples
A fraction is completely simplified when its numerator and denominator have no common factors.	$\frac{6}{7}$ is completely simplified. $\frac{10}{12}$ is not completely simplified because both 10 and 12 have a factor of 2.
To simplify a fraction, eliminate all common factors of the numerator and denominator.	$\frac{36}{72} = \frac{\cancel{9} \cdot 4}{\cancel{9} \cdot 8} = \frac{4}{8} = \frac{\cancel{4} \cdot 1}{\cancel{4} \cdot 2} = \frac{1}{2}$

SECTION 3.3 MULTIPLYING AND DIVIDING FRACTIONS

Definitions and Concepts	Examples
To multiply fractions, simplify if possible and then multiply numerators and multiply denominators.	$\frac{8}{\cancel{15}_3} \cdot \frac{\cancel{5}^1}{7} = \frac{8}{21}$
Two fractions are reciprocals if their product is 1.	$\frac{3}{4}$ and $\frac{4}{3}$ are reciprocals because $\frac{3}{4} \cdot \frac{4}{3} = 1$.
To divide fractions, multiply the first fraction by the reciprocal of the divisor.	$\frac{4}{9} \div \frac{5}{3} = \frac{4}{\cancel{9}_3} \cdot \frac{\cancel{3}^1}{5} = \frac{4}{15}$

Chapter Review Exercises provide a student with a set of exercises, usually 8–10 per section, to verify mastery of the material in the chapter prior to taking an exam.

REVIEW EXERCISES

Section 3.1

Write the fraction represented by the figure.

1. 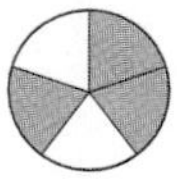

2. 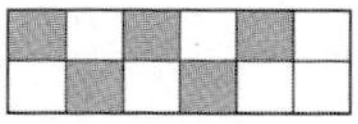

3. 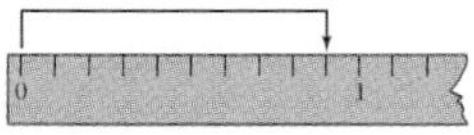

4. 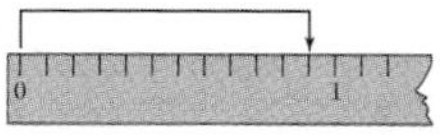

Identify the proper fractions from each list.

5. $\frac{8}{3}, \frac{11}{12}, \frac{9}{9}, \frac{22}{19}, \frac{3}{20}$

6. $\frac{1}{13}, \frac{11}{15}, \frac{8}{10}, \frac{5}{2}, \frac{12}{18}$

Change to a mixed number.

7. $\frac{77}{12}$ 8. $\frac{77}{9}$ 9. $\frac{344}{3}$ 10. $\frac{344}{7}$

Change to an improper fraction

Cumulative Reviews are included at the end of Chapters 3, 5, and 7. Each review covers all of the material in the text that precedes it, allowing students to maintain their skills as the term progresses.

CUMULATIVE REVIEW

Write the word name for each of the following.

1. 6091
2. 110,532

Write the place value name.

3. One million, three hundred ten
4. Sixty thousand, two hundred fifty-seven

Round to the indicated place value.

5. 654,785 (hundred)
6. 43,949 (ten thousand)

Insert > or < between the numbers to make a true statement.

7. 6745 6739
8. 11,899 11,901

Add or subtract.

9. $\begin{array}{r} 76{,}843 \\ 34{,}812 \\ 12{,}833 \\ +\ 9{,}711 \\ \hline \end{array}$

10. $\begin{array}{r} 55{,}304 \\ -\ 37{,}478 \\ \hline \end{array}$

11. $54 + 87 + 124 + 784 + 490 + 54$
12. $70{,}016 - 54{,}942$

13. Find the perimeter of the rectangle.

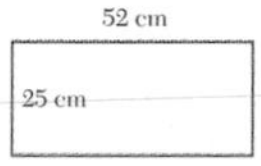

Multiply.

14. $\begin{array}{r} 14{,}654 \\ \times\ 251 \\ \end{array}$

15. (341)(73)

16. Find the area.

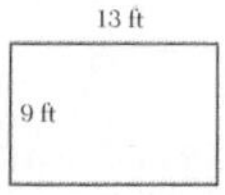

Chapter True/False Concept Review exercises require students to judge whether a statement is true or false and, if false, to rewrite the sentence to make it true. Students evaluate their understanding of concepts and also gain experience using the vocabulary of mathematics.

TRUE/FALSE CONCEPT REVIEW

ANSWERS

Check your understanding of the language of basic mathematics. Tell whether each of the following statements is true (always true) or false (not always true). For each statement you judge to be false, revise it to make a statement that is true.

1. It is not possible to picture an improper fraction using unit regions. **1.** ______

2. The fraction $\frac{7}{8}$ written as a mixed number is $0\frac{7}{8}$. **2.** ______

3. The whole number 1 can also be written as a proper fraction. **3.** ______

4. A fraction is another way of writing a division problem. **4.** ______

5. When a fraction is completely simplified, its value remains the same. **5.** ______

6. Every improper fraction can be simplified. **6.** ______

7. Some fractions with large numerators and denominators cannot be further simplified. **7.** ______

8. Two mixed numbers can be subtracted without first changing them to improper fractions. **8.** ______

9. The reciprocal of a proper fraction is greater than 1. **9.** ______

10. Th

Chapter Test exercises end the chapter. Written to imitate a 50-minute exam, each chapter test covers all of the chapter content. Students can use the chapter test as a self-test before the classroom test.

TEST

ANSWERS

1. Change $\frac{61}{4}$ to a mixed number. **1.** ______

2. Add: $\frac{7}{12} + \frac{5}{8}$ **2.** ______

3. Change $8\frac{7}{9}$ to an improper fraction **3.** ______

4. List these fractions from the smallest to the largest: $\frac{2}{5}, \frac{3}{8}, \frac{3}{7}$ **4.** ______

5. Change 17 to an improper fraction. **5.** ______

6. Find the missing numerator: $\frac{5}{8} = \frac{?}{56}$ **6.** ______

7. Add: $5\frac{3}{10} + 3\frac{5}{6}$ **7.** ______

8. Multiply: $\left(4\frac{5}{8}\right)\left(3\frac{4}{5}\right)$. Write the result as a mixed number. **8.** ______

9. Perform the indicated operations: $\frac{3}{8} \div \frac{3}{4} + \frac{1}{4}$ **9.** ______

10. Simplify $\frac{68}{102}$ completely. **10.** ______

11. Subtract: $17\frac{4}{5} - 11$ **11.** ______

12. Multiply: $\frac{3}{4} \cdot \frac{8}{9} \cdot \frac{15}{16}$ **12.** ______

13. Subtract: $\frac{4}{5} - \frac{3}{10}$ **13.** ______

CHANGES IN THE EIGHTH EDITION

Instructors who have used a previous edition of *Fundamentals of Mathematics* will see changes and improvements in format, pedagogy, exercises, and sectioning of content. Many of these changes are in response to comments and suggestions offered by users and reviewers of the manuscript. Others have been made to bring the text in line with math reform standards and to give the instructor the chance to follow educational guidelines recommended by AMATYC and NTCM.

- In Chapter 1, reading tables has been moved to section 1.2 so that the skill can be used immediately in the exercises. Perimeter and area of common polygons have been added to Chapter 1 so the topics can be used in the exercise sets. Medians and modes have been included with averages to give a more complete coverage of central tendency.
- Measurement and Geometry, old Chapter 2, has been moved to Chapter 7, where it was in editions prior to the seventh edition.
- Finding the Greatest Common Factor (GCF) has been added to Chapter 2, Primes and Multiples.
- Estimating has been rewritten for all operations and is included in whole numbers and decimals.
- Percent applications have been split into two sections, one covering general applications and the other covering business applications.
- Square roots and the Pythagorean Theorem have been added to Chapter 7 along with topics of time and temperature.
- Applications have been updated and several new ones have been added to each section. These are in keeping with the emphasis on real-world data.
- Icons point students to material combined on the Brooks/Cole Web site and on the Interactive Video Skillbuilder CD-ROM that accompanies the text.

ACKNOWLEDGMENTS

The authors appreciate the unfailing and continuous support of their spouses—Carol Van Dyke, Elinore Rogers, and Doug Adams—who made the completion of this work possible. We are also grateful to Jennifer Huber of Brooks/Cole and Nancy Shammas of New Leaf Publishing Services for their suggestions during the preparation and production of the text. We also want to thank the following professors and reviewers for their many excellent contributions to the development of the text: Kinley Alston, Trident Technical College; Carol Barner, Glendale Community College; Karen Driskell, Calhoun Community College; Beverlee Drucker, Northern Virginia Community College; Dale Grussing, Miami-Dade Community College, North Campus; Eric Kaljumagi, Mt. San Antonio College; Joanne Kendall, College of the Mainland; Christopher McNally, Tallahassee Community College; Michael Montano, Riverside Community College; Ellen Sawyer, College of Dupage; Leonard Smiglewski, Penn Valley Community College; Brian Sucevic, Valencia Community College; Stephen Zona, Quinsigamond Community College.

Special thanks to Pauline Siekas of Portland Community College for her careful reading of the text and for the accuracy review of all the problems and exercises in the text.

James Van Dyke
James Rogers
Hollis Adams

Strategies for Success

Are you afraid of math? Do you panic on tests or "blank out" and forget what you have studied, only to recall the material after the test? Then you are just like many other students. In fact, research studies estimate that as many as 50% of you have some degree of math anxiety.

What is math anxiety? It is a learned fear response to math that causes disruptive, debilitating reactions to tests. It can be so encompassing that it becomes a dread of doing **anything** that involves numbers. Although some anxiety at test time is beneficial—it can motivate and energize you, for example—numerous studies show that too much anxiety results in poorer test scores. Besides performing poorly on tests, you may be distracted by worrisome thoughts and be unable to concentrate and recall what you've learned. You may also set unrealistic performance standards for yourself and imagine catastrophic consequences for your failure to be successful in math. Your physical signs could be muscle tightness, stomach upset, sweating, headache, shortness of breath, shaking, or rapid heartbeat.

The good news is that anxiety is a learned behavior and therefore can be unlearned. If you want to stop feeling anxious, the choice is up to you. You can choose to learn behaviors that are more useful to achieve success in math. You can learn and choose the ways that work best for you.

To achieve success, you can focus on two broad strategies. First, you can study math in ways **proven** to be effective in learning mathematics and taking tests. Second, you can learn to physically and mentally **relax**, to manage your anxious feelings, and to think rationally and positively. Make a time commitment to practice relaxation techniques, study math, and record your thought patterns. A commitment of 1 or 2 hours a day may be necessary in the beginning. Remember, it took time to learn your present study habits and to be anxious. It will take time to unlearn these behaviors. After you become proficient with these methods, you can devote less time to them.

Begin now to learn your strategies for success. Be sure you have read the Preface to the Student at the beginning of this book. The purpose of this Preface is to introduce you to the authors' plan for this text. The Preface will help you to understand the authors' organization or "game plan" for your math experience in this course.

At the beginning of each chapter, you will find more Good Advice for Studying sections, which will help you study and take tests more effectively, as well as help you manage your anxiety. You may want to read ahead so that you can improve even more quickly. Good luck!

1 WHOLE NUMBERS

SECTIONS

APPLICATION

Professional athletes' salaries have been increasing rapidly, to the concern of many. Table 1.1 shows the top eight National Football League team salaries for the 1999–2000 and 2000–2001 seasons, respectively.

1. How many teams were in the top eight for both seasons?
2. What do you notice about the top salaries in the 1999–2000 season?
3. What do you notice about the top salaries in the 2000–2001 season?

In 1999–2000, the St. Louis Rams won the Superbowl. Their payroll was $52,119,900, which ranked them 16th. In 2000–2001, the Baltimore Ravens won the Superbowl.

4. Does there seem to be a pattern between total payroll and winning the Superbowl?

TABLE 1.1 TOP NFL PAYROLLS

Team	1999–2000 Payroll	Team	2000–2001 Payroll
New England Patriots	56,566,800	Washington Redskins	92,407,000
Tennessee Titans	56,129,500	Baltimore Ravens	90,703,000
Tampa Bay Buccaneers	55,767,700	Detroit Lions	78,376,000
Arizona Cardinals	54,902,000	St. Louis Rams	77,648,000
Dallas Cowboys	54,858,500	Jacksonville Jaguars	76,999,000
Pittsburgh Steelers	54,613,000	Tampa Bay Buccaneers	76,849,000
Philadelphia Eagles	54,155,000	Indianapolis Colts	76,108,000
Jacksonville Jaguars	53,655,100	Chicago Bears	73,095,000

1.1 Whole Numbers and Tables: Writing, Rounding, and Inequalities

OBJECTIVES

1. Write word names from place value names and place value names from word names.
2. Write an inequality statement about two numbers.
3. Round a whole number.
4. Read tables.

VOCABULARY

The **digits** are 0, 1, 2, 3, 4, 5, 6, 7, 8, and 9.

The **natural numbers (counting numbers)** are 1, 2, 3, 4, 5, and so on.

The **whole numbers** are 0, 1, 2, 3, 4, 5, and so on. Numbers larger than 9 are written in **place value name** by writing the digits in positions having standard **place value**.

Word names are written words that represent numerals. The word name of 125 is one hundred twenty-five.

The symbols **less than**, $<$, and **greater than**, $>$, are used to compare two whole numbers that are not equal. So, $2 < 9$, and $13 > 6$.

To **round** a whole number means to give it an approximate value. The symbol $\approx$ means "approximately equal to."

A **table** is a method of displaying data in an **array**—that is, using a horizontal and vertical arrangement to distinguish the type of data. A **row** is a horizontal line of a table; it reads left to right across the page. A **column** is a vertical line of a table; it reads up and down the page. For example, in Table 1.2 the number 45 is in row 3 and column 2.

		Column 2		
	134	56	89	102
	14	116	7	98
Row 3	65	**45**	12	67
	23	56	7	213

TABLE 1.2

HOW & WHY

OBJECTIVE 1 Write word names from place value names and place value names from word names.

In our written whole number system (called the Hindu-Arabic system), digits and commas are the only symbols used. This system is a positional base 10 (decimal) system. The location of the digit determines its value, from right to left. The first three place value names are one, ten, and hundred. See Figure 1.1.

hundred	ten	one

FIGURE 1.1

For the number 647:

7 is in the ones place, so it contributes 7 ones, or 7, to the value of the number.

4 is in the tens place, so it contributes 4 tens, or 40, to the value of the number.

6 is in the hundreds place, so it contributes 6 hundreds, or 600, to the value of the number.

So 647 is 6 hundreds + 4 tens + 7 ones or 600 + 40 + 7. These are called **expanded forms** of the number. The word name is six hundred forty-seven.

For numbers larger than 999, we use commas to separate groups of three digits. The first four groups are unit, thousand, million, and billion (Figure 1.2). The group on the far left may have one, two, or three digits. All other groups must have three digits. Within each group, the names are the same (hundred, ten, and one).

hundred	ten	one	hundred	ten	one	hundred	ten	one	hundred	ten	one
	billion			million			thousand			(unit)	

FIGURE 1.2

For 82,906,425,777 the group names are:

82	906	425	777
billion	million	thousand	unit

The number is read "82 billion, 906 million, 425 thousand, 777." The word *units* for the units group is not read. The complete word name is eighty-two billion, nine hundred six million, four hundred twenty-five thousand, seven hundred seventy-seven.

To write the word name from a place value name

1. From left to right, write the word name for each set of three digits followed by the group name (except units).
2. Insert a comma after each group name.

CAUTION

The word "and" is not used when writing names of whole numbers. So write: three hundred ten, **NOT** three hundred and ten, or one thousand, two hundred twenty-three, **NOT** one thousand and two hundred twenty-three.

To write the place value name from the word name of a number, we reverse the previous process. First identify the group names and then write each group name in the place value name. Remember to write a 0 for each missing place value. Consider

three billion, two hundred thirty-five million, nine thousand, four hundred thirteen

three *billion*, two hundred thirty-five *million*, nine *thousand*, four hundred thirteen — Identify the group names. (Hint: Look for the commas.)

3 billion, 235 million, 9 thousand, 413 — Write the place value name for each group.

3,235,009,413 — Drop the group names. Keep all commas. Zeros must be inserted to show that there are no hundreds or tens in the thousands group.

To write a place value name from a word name

1. Identify the group names.
2. Write the 3-digit number before each group name, followed by a comma. (The first group, on the left, may have fewer than three digits.) It is common to omit the comma in 4-digit numerals.

Numbers like 43,000,000,000, with all zeros following a single group of digits, are often written in a combination of place value notation and word name. The first set of digits on the left is written in place value notation followed by the group name. So 43,000,000,000 is written 43 billion.

WARM-UPS A–B

A. Write the word name for 27,311,085.

B. Write the word name for 8,431,619.

Answers to Warm-Ups

A. twenty-seven million, three hundred eleven thousand, eighty-five
B. eight million, four hundred thirty-one thousand, six hundred nineteen

EXAMPLES A–B

DIRECTIONS: Write the word name.

STRATEGY: Write the word name of each set of three digits, from left to right, followed by the group name.

A. Write the word name for 19,817,583.

19 nineteen million,

817 eight hundred seventeen thousand,

583 five hundred eighty-three

Write the word name for each group followed by the group name.

CAUTION

Do not write the word "and" when reading or writing a whole number.

The word name is nineteen million, eight hundred seventeen thousand, five hundred eighty-three.

B. Write the word name for 4,781,037.

* four million, seven hundred eighty-one thousand, thirty-seven

WARM-UPS C–F

C. Write the place value name for twenty-two million, seventy-seven thousand, four hundred eleven.

D. Write the place value name for 37 thousand.

Answers to Warm-Ups

C. 22,077,411 **D.** 37,000

EXAMPLES C–F

DIRECTIONS: Write the place value name.

STRATEGY: Write the 3-digit number for each group, followed by a comma.

C. Write the place value name for five million, twenty-six thousand, nine hundred sixty-nine.

5,	Millions group
026,	Thousands group (Note that a zero is inserted on the left to fill out the three digits in the group.)
969	Units group

The place value name is 5,026,969.

D. Write the place value name for 346 million.

The place value name is 346,000,000. Replace the word "million" with six zeros.

E. Write the place value name for four thousand fifty-three. Note the comma is omitted.

The place value name is 4053.

E. Write the place value name for six thousand twelve.

F. The purchasing agent for the Russet Corporation received a telephone bid of forty-three thousand, fifty-one dollars as the price of a new printing press. What is the place value name of the bid that she will include in her report to her superior?

forty-three thousand, fifty-one

43,   051

The place value name she reports is \$43,051.

F. The purchasing agent for the Russet Corporation also received a bid of seventeen thousand, two hundred eighteen dollars for a supply of paper. What is the place value form of the bid that she will include in her report to her superior?

HOW & WHY

OBJECTIVE 2 Write an inequality statement about two numbers.

If two whole numbers are not equal, then the first is either *less than* or *greater than* the second. Look at this number line (or ruler) in Figure 1.3:

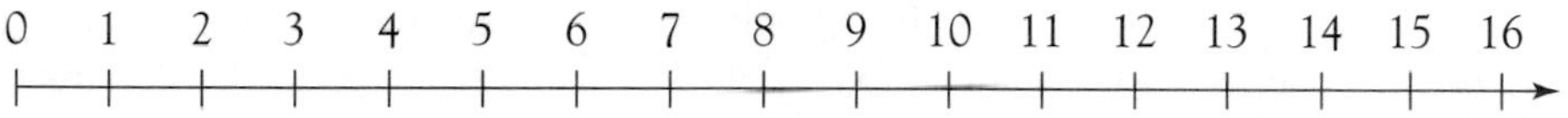

FIGURE 1.3

Given two numbers on a number line or ruler, the number on the right is the larger. For example,

$9 > 7$ 9 is to the right of 7, so 9 is greater than 7.
$11 > 1$ 11 is to the right of 1, so 11 is greater than 1.
$14 > 8$ 14 is to the right of 8, so 14 is greater than 8.
$13 > 0$ 13 is to the right of 0, so 13 is greater than 0.

Given two numbers on a number line or ruler, the number on the left is the smaller. For example,

$3 < 9$ 3 is to the left of 9, so 3 is less than 9.
$5 < 12$ 5 is to the left of 12, so 5 is less than 12.
$1 < 9$ 1 is to the left of 9, so 1 is less than 9.
$10 < 14$ 10 is to the left of 14, so 10 is less than 14.

For larger numbers, imagine a longer number line. Notice how the points in the symbols $<$ and $>$ always point to the smaller of the two numbers. For example,

$239 < 605$
$94 > 56$
$7008 > 7006$

To write an inequality statement about two numbers

1. Insert $<$ between the numbers if the number on the left is smaller.
2. Insert $>$ between the numbers if the number on the left is larger.

Answers to Warm-Ups

E. 6012 **F.** \$17,218

WARM-UPS G–H

G. Insert the appropriate inequality symbol: 135 157

H. Insert the appropriate inequality symbol: 7578 7399

EXAMPLES G–H

DIRECTIONS: Insert $<$ or $>$ to make a true statement.

STRATEGY: Imagine a number line. The smaller number is on the left. Insert the symbol that points to the smaller number.

G. Insert the appropriate inequality symbol: 78 95

$78 < 95$

H. Insert the appropriate inequality symbol: 2714 2044

$2714 > 2044$

HOW & WHY

OBJECTIVE 3 Round a whole number.

Many numbers that we see in daily life are approximations. Such numbers are used to indicate the approximate value when it is believed that the exact value is not important to the discussion. So attendance at a political rally may be stated at 15,000 when it was actually 14,783. The amount of a deficit in the budget may be stated as $2,000,000 instead of $2,067,973. In this chapter, we use approximations to estimate the outcome of operations with whole numbers. The symbol ≈, read "approximately equal to," is used to show approximation. So $2,067,973 ≈ $2,000,000.

We approximate numbers by **rounding**. The number line can be used to see how whole numbers are rounded. Suppose we wish to round 57 to the nearest ten. See Figure 1.4

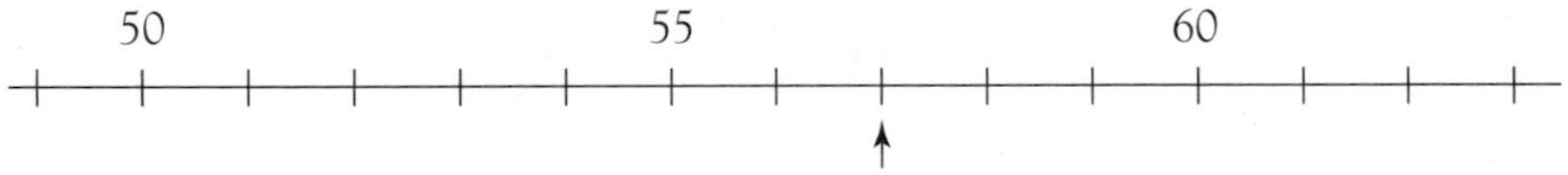

FIGURE 1.4

The arrow under the 57 is closer to 60 than to 50. We say, "To the nearest ten, 57 rounds to 60."

We use the same idea to round any number, although we usually make only a mental image of the number line. The key question is: Is this number closer to the smaller rounded number or closer to the larger one? Practically, we need to determine only if the number is more or less than half the distance between the rounded numbers.

To round 48,598 to the nearest thousand without a number line, draw an arrow under the digit in the thousands place.

48,598
↑

Because 48,598 is between 48,000 and 49,000, we must decide which number it is closer to. Because 48,500 is halfway between 48,000 and 49,000 and because $48{,}598 > 48{,}500$, we conclude that 48,598 is more than halfway to 49,000.

Whenever the number is halfway or closer to the larger number, we choose the larger number.

48,598 ≈ 49,000   48,598 is closer to 49,000 than to 48,000.

Answers to Warm-Ups

G. $<$ H. $>$

To round a number to a given place value

1. Draw an arrow under the given place value.
2. If the digit to the right of the arrow is 5, 6, 7, 8, or 9, add one to the digit above the arrow. (Round to the larger number.)
3. If the digit to the right of the arrow is 0, 1, 2, 3, or 4, do not change the digit above the arrow. (Round to the smaller number.)
4. Replace all the digits to the right of the arrow with zeros.

EXAMPLES I–J

DIRECTIONS: Round to the indicated place value.

STRATEGY: Choose the larger number if the digit to the right of the round-off place is 5 or greater; otherwise, choose the smaller number.

I. Round 347,366 to the nearest ten thousand.

347,366 — Draw an arrow under the ten-thousands place.
↑

350,000 — The digit to the right of the arrow is 7. Because 347,366 > 345,000, choose the larger number.

So 347,366 ≈ 350,000.

J. Round the numbers to the indicated place value.

Number	Ten	Hundred	Thousand
475,833	475,830	475,800	476,000
68,419	68,420	68,400	68,000

WARM-UPS I–J

I. Round 67,327 to the nearest ten.

J. Round the numbers to the indicated place value.

Number	Ten	Hundred	Thousand
581,468			
89,655			

HOW & WHY

OBJECTIVE 4 Read tables.

Data are often displayed in the form of a *table*. We see tables in the print media, in advertisements, and in business presentations. Reading a table involves finding the correct *column* and *row* that describes the needed information, and then reading the data at the intersection of that column and that row.

TABLE 1.3 STUDENT COURSE ENROLLMENT

Class	Mathematics	English	Science	Humanities
Freshman	950	1500	500	1200
Sophomore	600	700	650	1000
Junior	450	200	950	1550
Senior	400	250	700	950

For example, in Table 1.3, to find the number of sophomores who take English, we find the column headed English and the row headed Sophomore and read the number at the intersection.

The number of sophomores taking English is 700.

We can use the table to compare enrollments by class. For instance, are more seniors

Answers to Warm-Ups

I. 67,330

J.

Number	Ten	Hundred	Thousand
581,468	581,470	581,500	581,000
89,655	89,660	89,700	90,000

or sophomores taking science? From the table, we see that 650 sophomores are taking science and 700 seniors are taking science. Because 700 > 650, more seniors than sophomores are taking science.

WARM-UP K

K. Use the table in Example K to answer the questions.

1. Which location has the lowest-priced home sold?

2. Round the average price of a home sold in S.E. Portland to the nearest thousand.

3. Which location has the highest price for a home sold, N.E. Portland or S.E. Portland?

EXAMPLE K

DIRECTIONS: Answer the questions associated with the table.

STRATEGY: Examine the rows and columns of the table to determine the values that are related.

K. This table shows the value of homes sold in the Portland metropolitan area for a given month.

Values of Homes Sold

Location	Lowest	Highest	Average
N. Portland	\$16,000	\$ 58,500	\$ 34,833
N.E. Portland	\$18,000	\$120,000	\$ 47,091
S.E. Portland	\$18,000	\$114,000	\$ 51,490
Lake Oswego	\$40,000	\$339,000	\$121,080
West Portland	\$29,500	\$399,000	\$112,994
Beaverton	\$20,940	\$165,000	\$ 78,737

1. In which location was the highest-priced home sold?

2. Which area has the highest average sale price?

3. Round the highest-priced home in Beaverton to the nearest hundred thousand.

1. We look at the Highest column for the largest entry. It is \$399,000, which is the fifth row. So West Portland is the location of the highest-priced home sold.
2. Looking at the Average column for the largest entry, we find \$121,080 in the fourth row. So Lake Oswego has the largest average sale price.
3. Looking at the sixth row and Highest column, we find \$165,000. So the highest price of a house in Beaverton is \$200,000, rounded to the nearest hundred thousand.

*

Answer to Warm-Up

K. 1. N. Portland has the lowest-priced home sold.
2. The rounded price is \$51,000.
3. N.E. Portland has the higher price.

Name Loretta La Salle Class ____________ Date 9/26/02

EXERCISES 1.1

OBJECTIVE 1 Write word names from place value names and place value names from word names.

A. *Write the word name of each of these numbers.*

1. 781 **2.** 391 **3.** 890

4. 710 **5.** 7020 **6.** 40,051

Write the place value name.

7. Fifty-seven **8.** Ninety-two **9.** Seven thousand, five hundred

10. Seven thousand, five **11.** 100 million **12.** 651 thousand

B. *Write the word name of each of these numbers.*

13. 36,450 **14.** 36,045 **15.** 306,450

16. 360,045 **17.** 45,000,000 **18.** 870,000

Write the place value name.

19. Two hundred forty-three thousand, seven hundred

20. Two hundred forty-three thousand, seven

21. Twenty-two thousand, five hundred seventy

22. Twenty-three thousand, four hundred seventy-seven

23. Seventeen billion

24. Seven hundred thousand, seven

OBJECTIVE 2 Write an inequality statement about two numbers.

Insert $<$ or $>$ between the numbers to make a true statement.

A.

25. 18 $<$ 21 **26.** 43 39 **27.** 61 54 **28.** 62 71

B.

29. 246 251 **30.** 212 208 **31.** 7470 7850 **32.** 275 269

OBJECTIVE 3 Round a whole number.

Round to the indicated place value.

A.

33. 742 (ten) **34.** 577 (ten) **35.** 1658 (hundred) **36.** 3450 (hundred)

B.

	Number	Ten	Hundred	Thousand	Ten Thousand
37.	304,285				
38.	689,377				
39.	8,657,896				
40.	4,309,498				

OBJECTIVE 4 Read tables.

A. *Exercises 41–45. The percentage of people who do and don't exercise regularly, broken down by income levels, is shown in the following table.* Source: *U.S. Centers for Disease Control and Prevention.*

Regular Exercise by Income Level

Income	Does Exercise	Does Not Exercise
$0–$14,999	35%	65%
$15–$24,999	40%	60%
$25–$50,000	45%	55%
Over $50,000	52%	48%

41. What percentage of people in the income level of $15–$24,999 exercise regularly?

42. Which income level has the highest percentage of regular exercisers?

43. Which income level has the highest percentage of nonexercisers?

44. Which income level(s) has (have) more than 50% nonexercisers?

45. Use words to describe the trend indicated in the table.

B. *Exercises 46–50. A profile of the homeless in 26 selected cities, according to data presented to the U.S. Conference of Mayors in 1997, is given in the following table.*

Number of Homeless in Select Cities

	% of Homeless in 1988	% of Homeless in 1997
Single Men	49	47
Substance Abusers	34	43
Families with Children	32	36
Military Veterans	26	22
Children	25	26
Mentally Ill	25	27
Employed	23	17
Single Women	13	14

46. What was the increase in the percent of homeless for families with children between 1988 and 1997?

47. Which of the categories increased over the 9-year period and which decreased?

48. Explain why each column does not add up to 100%.

49. What percentage of the homeless was made up of single men, single women, and children in 1988? Did this percentage increase or decrease in 1997?

50. Did the total number of homeless increase or decrease over the 9-year period?

C. *Write the place value name.*

51. Five hundred sixty million, three hundred fifty-three thousand, seven hundred thirty

52. Six hundred two million, three hundred thirty-three

Exercises 53–54. The average income of the top 20% of the families and the bottom 20% of the families in Iowa is shown in the following figure.

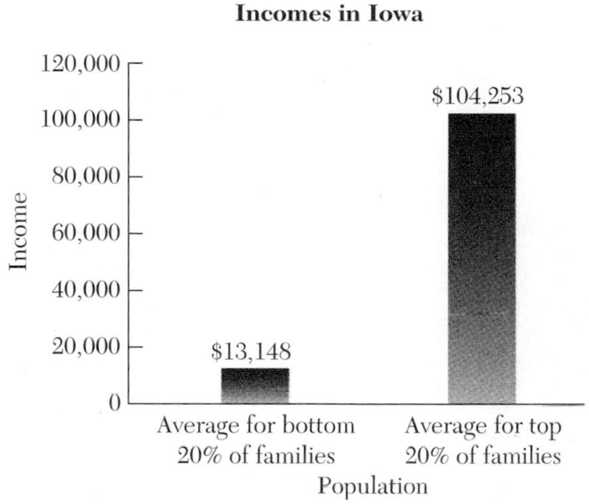

53. Write the word name for the average income of the bottom 20% of families in Iowa.

54. Write the word name for the average income of the top 20% of families in Iowa.

Insert < or > between the numbers to make a true statement.

55. 7523 7525

56. 11,386 11,299

57. What is the smallest 4-digit number?

58. What is the largest 6-digit number?

Round to the indicated place value.

59. 81,634,981 (ten thousand)

60. 62,078,991 (hundred thousand)

61. Round 63,749 to the nearest hundred. Round 63,749 to the nearest ten and then round your result to the nearest hundred. Why did you get a different result the second time? Which method is correct?

62. Hazel bought a jet ski boat for $2075. She wrote a check to pay for it. What word name does she write on the check?

63. Kimo bought a used Infinity for $18,465 and wrote a check to pay for it. What word name does he write on the check?

64. The U.S. Fish and Wildlife Department estimates that salmon runs could be as high as 154,320 fish by 2004 on the Rogue River if new management practices are used in logging along the river. Write the word name for the number of fish.

65. Ducks Unlimited estimated that 389,500 ducks spent the winter at a Wisconsin refuge. Write the word name for the number of ducks.

66. The world population during 1990 exceeded 5 billion, 3 hundred thousand. Write the place value name for the world population.

67. The purchasing agent for Print-It-Right received a telephone bid of thirty-six thousand, four hundred seven dollars as the price for a new printing press. What is the place value name for the bid?

68. An office building in downtown Birmingham sold for \$2,458,950. Give the purchase price of the building to the nearest hundred thousand dollars.

69. Ten thousand shares of Intel Corporation sold for \$846,560. What is the value of the sale, to the nearest thousand dollars?

Exercises 70–72. The figure shows the quantity of the top three industrial releases of toxic materials a few years ago. These values have been decreasing in recent years.

Chemical Industries 1482 million pounds
Metals Industries 327 million pounds
Paper Industries 219 million pounds

70. Write the place value name for the number of pounds of toxic material released by chemical industries.

71. Write the place value name for the number of pounds of toxic material released by metal industries.

72. Write the place value name for the number of pounds of toxic material released by paper industries.

Exercises 73–76. The following table gives the per capita personal income in the New England states.

Per Capita Income in New England

Massachusetts	\$26,994
New Hampshire	\$25,151
Maine	\$20,527
Connecticut	\$30,303
Rhode Island	\$23,310
Vermont	\$20,927

73. Write the word name for the per capita personal income in Maine.

74. Round the per capita personal income in Massachusetts to the nearest thousand.

75. Which state has the smallest per capita personal income?

76. Of Massachusetts and New Hampshire, which has the larger per capita personal income?

77. The distance from Earth to the sun was measured and determined to be 92,875,328 miles. To the nearest million miles, what is the distance?

78. The National Petroleum News reports the following number of brand name retail gasoline outlets in 1995.

Brand Name Gasoline Outlets

Brand	Number of Outlets
Citgo	14,054
Texaco	13,023
Amoco	9600
Shell	8667
Exxon	8250
Chevron	7988
Mobil	7689
Phillips	7106
BP America	6800
Conoco	5017

Have any of these numbers been rounded? If so, which ones and explain how you know. Revise the table, rounding all the figures to the nearest hundred.

Exercises 79–80. The number of marriages each month for a recent year, according to data from the U.S. Census Bureau, is given in the table.

Number of Marriages Per Month

Month	Number of Marriages (in thousands)	Month	Number of Marriages (in thousands)
January	110	July	235
February	155	August	239
March	118	September	225
April	172	October	231
May	241	November	171
June	242	December	184

79. Rewrite the information in the table, ordering the months from most number of marriages to least number of marriages. Use place value notation when writing the number of marriages.

80. Do you think the number of marriages has been rounded? If so, to what place value?

Exercises 81–82. The six longest rivers in the United States are as follows:

Arkansas	1459 miles
Colorado	1450 miles
Mississippi	2340 miles
Missouri	2315 miles
Rio Grande	1900 miles
Yukon	1079 miles

81. List the rivers in order of increasing length.

82. Do you think any of the river lengths have been rounded? If so, which ones?

83. The state motor vehicle department estimated the number of licensed automobiles in the state to be 2,376,000, to the nearest thousand. A check of the records indicated that there were actually 2,376,499. Was their estimate correct?

84. The total land area of the earth is approximately 52,425,000 square miles. What is the land area to the nearest million square miles?

 Name ______________________ Class ______________________ Date ______________

Exercises 85–86. The following figure lists some nutritional facts about two brands of peanut butter.

Skippy® Super Chunk

Nutrition Facts
Serving Size 2 tbsp (32g)
Servings Per Container about 15

Amount Per Serving	
Calories 190 Calories from Fat 140	
	% Daily Values
Total Fat 17g	26%
Saturated Fat 3.5g	17%
Cholesterol 0mg	0%
Sodium 140mg	6%
Total Carbohydrate 7g	2%
Dietary Fiber 2g	8%
Sugars 3g	
Protein 7g	

Jif® Creamy
Simply Jif contains 2g sugar per serving.
Regular Jif contains 3g sugar per serving

Nutrition Facts
Serving Size 2 tbsp (31g)
Servings Per Container about 16

Amount Per Serving	
Calories 190 Calories from Fat 130	
	% Daily Values
Total Fat 16g	25%
Saturated Fat 3g	16%
Cholesterol 0mg	0%
Sodium 65mg	3%
Total Carbohydrate 6g	2%
Dietary Fiber 2g	9%
Sugars 2g	
Protein 8g	

85. List the categories of nutrients where Jif® has fewer of the nutrients than Skippy®.

86. Round the sodium content in each brand to the nearest hundred. Do the rounded numbers give a fair comparison of the amount of sodium in the brands?

Exercises 87–90 relate to the chapter application. See Table 1.1, page 1.

87. To what place do the payroll figures for the 1999–2000 season seem to be rounded?

88. To what place do the payroll figures for the 2000–2001 season seem to be rounded?

89. Write an inequality statement about the highest payrolls for the two seasons.

90. Write the word name for the Dallas Cowboys' payroll in the 1999–2000 season.

STATE YOUR UNDERSTANDING

91. Explain why base 10 is a good name for our number system.

92. Explain what the digit 7 means in 175,892.

93. What is rounding? Explain how to round 87,452 to the nearest thousand and to the nearest hundred.

CHALLENGE

94. What is the place value for the digit 4 in 3,456,709,230,000?

95. Write the word name for 3,456,709,230,000.

96. Arrange the following numbers from smallest to largest: 1234, 1342, 1432, 1145, 1243, 1324, and 1229.

97. What is the largest value of X that makes $2X56 > 2649$ false?

98. Round 8275 to the nearest ten thousand.

99. Round 37,254 to the nearest hundred thousand.

GROUP WORK

100. Two other methods of rounding are called the "odd-even method" and "truncating." Find these methods and be prepared to explain them in class. (*Hint*: Try the library or talk to science and business instructors.)

1.2 Adding and Subtracting Whole Numbers

VOCABULARY

Addends are the numbers that are added. In 9 + 20 + 3 = 32, the addends are 9, 20, and 3.

The result of adding is called the **sum**. In 9 + 20 + 3 = 32, the sum is 32.

The result of subtracting is called the **difference**. So in 62 − 34 = 28, 28 is the difference.

A **polygon** is any closed figure whose sides are line segments.

The **perimeter** of a polygon is the distance around the outside of the polygon.

OBJECTIVES

1. Find the sum of two or more whole numbers.
2. Find the difference of two whole numbers.
3. Estimate the sum or difference of whole numbers.
4. Find the perimeter of a polygon.

HOW & WHY

OBJECTIVE 1 Find the sum of two or more whole numbers.

When Jose graduated from high school, he received cash gifts of \$45, \$20, and \$12. The total number of dollars received is found by adding the individual gifts. The total number of dollars he received is \$77. In this section, we review the procedure for adding and subtracting whole numbers.

The addition facts and place value are used to add whole numbers written with more than one digit. Let's use this to find the sum of the cash gifts that Jose received. We need to find the sum of

45 + 20 + 12

By writing the numbers in expanded form and putting the same place values in columns, it is easy to add.

$$\begin{array}{rl} 45 = & 4 \text{ tens} + 5 \text{ ones} \\ 20 = & 2 \text{ tens} + 0 \text{ ones} \\ +12 = & 1 \text{ ten} + 2 \text{ ones} \\ \hline & 7 \text{ tens} + 7 \text{ ones} = 77 \end{array}$$

So, 45 + 20 + 12 = 77. Jose received \$77 in cash gifts.

Because each place can contain only a single digit, it is often necessary to rewrite the sum of a column.

$$\begin{array}{rl} 68 = & 6 \text{ tens} + 8 \text{ ones} \\ +19 = & 1 \text{ ten} + 9 \text{ ones} \\ \hline & 7 \text{ tens} + 17 \text{ ones} \end{array}$$

Because 17 ones is a 2-digit number, it must be renamed:

$$\begin{aligned} 7 \text{ tens} + 17 \text{ ones} &= 7 \text{ tens} + 1 \text{ ten} + 7 \text{ ones} \\ &= 8 \text{ tens} + 7 \text{ ones} \\ &= 87 \end{aligned}$$

So, the sum of 68 and 19 is 87.

A common shortcut is shown in the following sum. To add 497 + 307 + 135, write the numbers in a column.

$$\begin{array}{r} 497 \\ 307 \\ +135 \\ \hline \end{array}$$

Written this way, the digits in the ones, tens, and hundreds places are aligned vertically.

$$\begin{array}{r} {\scriptstyle 1} \\ 497 \\ 307 \\ +135 \\ \hline 9 \end{array}$$

Add the digits in the ones column: 7 + 7 + 5 = 19.
Write 9 and "carry" the 1 (1 ten) to the tens column.

$$\begin{array}{r} {\scriptstyle 11} \\ 497 \\ 307 \\ +135 \\ \hline 39 \end{array}$$

Add the digits in the tens column: 1 + 9 + 0 + 3 = 13.
Write 3 and carry the 1 (10 tens = 1 hundred) to the hundreds column.

$$\begin{array}{r} {\scriptstyle 11} \\ 497 \\ 307 \\ +135 \\ \hline 939 \end{array}$$

Add the digits in the hundreds column: 1 + 4 + 3 + 1 = 9

To add whole numbers

1. Write the numbers in a column so that the place values are aligned.
2. Add each column, starting with the ones (or units) column.
3. If the sum of any column is greater than 9, write the ones digit and carry the tens digit to the next column.

WARM-UPS A–C

A. Add: 976 + 427

B. Add 76, 3418, 51, and 9. Round the sum to the nearest ten.

C. Add: 9461 + 8377 + 915 + 4132 + 6599

Answers to Warm-Ups

A. 1403 **B.** 3550 **C.** 29,484

EXAMPLES A–C

DIRECTIONS: Add.

STRATEGY: Write the numbers in a column. Add the digits in the columns starting on the right. If the sum is greater than 9, "carry" the tens digit to the next column.

A. Add: 546 + 875

$$\begin{array}{r} {\scriptstyle 11} \\ 546 \\ +\ 875 \\ \hline 1421 \end{array}$$

Add the numbers in the ones column: 6 + 5 = 11. Because the sum is greater than 9, write 1 in the ones column and carry the 1 to the tens column. Add the numbers in the tens column: 1 + 4 + 7 = 12. Write 2 in the tens column and carry the 1 to the hundreds column. Add the numbers in the hundreds column: 1 + 5 + 8 = 14. Because all columns have been added, there is no need to carry.

B. Add 68, 714, 7, and 1309. Round the sum to the nearest ten.

$$\begin{array}{r} {\scriptstyle 1\ 2\ \ } \\ 68 \\ 714 \\ 7 \\ +1309 \\ \hline 2098 \end{array}$$

When writing in a column, make sure the place values are aligned properly.

$2098 \approx 2100$ Round to the nearest ten.

CALCULATOR EXAMPLE

C. Add: 5059 + 798 + 8325 + 3923 + 7089

Calculators are programmed to add numbers just as we have been doing by hand. Simply enter the exercise as it is written horizontally and the calculator will do the rest.

The sum is 25,194.

HOW & WHY

OBJECTIVE 2 Find the difference of two whole numbers.

Felicia went shopping with \$87. She made purchases totaling \$64. How much money does she have left? Finding the difference between two quantities is called subtraction. When we subtract \$64 from \$87, we get \$23.

Subtraction can be thought of as finding the missing addend in an addition exercise. For instance, $9 - 5 = ?$ asks $5 + ? = 9$. Because $5 + 4 = 9$, we know that $9 - 5 = 4$. Similarly, $47 - 15 = ?$ asks $15 + ? = 47$. Because $15 + 32 = 47$, we know that $47 - 15 = 32$.

For larger numbers, such as $875 - 643$, we take advantage of the column form and expanded notation to find the missing addend in each column.

$$\begin{array}{rl} 875 &= 8 \text{ hundreds} + 7 \text{ tens} + 5 \text{ ones} \\ -643 &= 6 \text{ hundreds} + 4 \text{ tens} + 3 \text{ ones} \\ \hline 232 &= 2 \text{ hundreds} + 3 \text{ tens} + 2 \text{ ones} = 232 \end{array}$$

Check by adding:

$$\begin{array}{r} 643 \\ +232 \\ \hline 875 \end{array}$$

So, $875 - 643 = 232$.

Now consider the difference $672 - 438$. Write the numbers in column form.

$$\begin{array}{rl} 672 &= 6 \text{ hundreds} + 7 \text{ tens} + 2 \text{ ones} \\ -438 &= 4 \text{ hundreds} + 3 \text{ tens} + 8 \text{ ones} \\ \hline \end{array}$$

Here we cannot subtract 8 ones from 2 ones, so we rename by "borrowing" one of the tens from the 7 tens (1 ten = 10 ones) and adding the 10 ones to the 2 ones.

$$\begin{array}{rl} & \qquad\qquad\quad \textit{6 tens} \quad\;\; \textit{12 ones} \\ 672 &= 6 \text{ hundreds} + 7 \text{ tens} + 2 \text{ ones} \\ -438 &= 4 \text{ hundreds} + 3 \text{ tens} + 8 \text{ ones} \\ \hline &= 2 \text{ hundreds} + 3 \text{ tens} + 4 \text{ ones} = 234 \end{array}$$

Check by adding:

$$\begin{array}{r} \textit{1}\;\; \\ 438 \\ +234 \\ \hline 672 \end{array}$$

We generally don't bother to write the expanded form when we subtract. We show the shortcut for borrowing in the examples.

To subtract whole numbers

1. Write the numbers in a column so that the place values are aligned.
2. Subtract in each column, starting with the ones (or units) column.
3. When the numbers in a column cannot be subtracted, borrow 1 from the next column and rename by adding 10 to the upper digit in the current column and then subtract.

EXAMPLES D–H

DIRECTIONS: Subtract and check.

STRATEGY: Write the numbers in columns. Subtract in each column. Rename by borrowing when the numbers in a column cannot be subtracted.

WARM-UPS D–H

D. Subtract: 78 − 36

D. Subtract: 96 − 51

$$\begin{array}{r} 96 \\ -51 \\ \hline 45 \end{array}$$

Subtract the ones column: 6 − 1 = 5.
Subtract the tens column: 9 − 5 = 4.

CHECK:

$$\begin{array}{r} 51 \\ +45 \\ \hline 96 \end{array}$$

So 96 − 51 = 45.

E. Find the difference: 746 − 378.

E. Find the difference: 836 − 379.

$$\begin{array}{r} {\scriptstyle 2\ 16} \\ 8\,\cancel{3}\,6 \\ -3\,7\,9 \\ \hline 7 \end{array}$$

In order to subtract in the ones column, we borrow 1 ten (10 ones) from the tens column and rename the ones (10 + 6 = 16).

$$\begin{array}{r} {\scriptstyle 7\ 12} \\ {\scriptstyle 2\ 16} \\ \cancel{8}\,\cancel{3}\,6 \\ -3\,7\,9 \\ \hline 4\,5\,7 \end{array}$$

Now in order to subtract in the tens column, we must borrow 1 hundred (10 tens) from the hundreds column and rename the tens (10 + 2 = 12).

CHECK:

$$\begin{array}{r} 379 \\ +457 \\ \hline 836 \end{array}$$

So 836 − 379 = 457.

F. Subtract 3800 and 759.

F. Subtract 6200 and 874.

$$\begin{array}{r} 6200 \\ -\ \ 874 \\ \hline \end{array}$$

We cannot subtract in the ones column, and because there are 0 tens, we cannot borrow from the tens column.

$$\begin{array}{r} {\scriptstyle 1\ 10\ } \\ 6\,2\,\cancel{0}\,0 \\ -\ \ 8\,7\,4 \\ \hline \end{array}$$

We borrow 1 hundred (1 hundred = 10 tens) from the hundreds place.

$$\begin{array}{r} {\scriptstyle 9\ 10} \\ {\scriptstyle 1\ \cancel{10}\ } \\ 6\,2\,\cancel{0}\,\cancel{0} \\ -\ \ 8\,7\,4 \\ \hline \end{array}$$

Now borrow 1 ten (1 ten = 10 ones). We can now subtract in the ones and tens column but not in the hundreds column.

$$\begin{array}{r} {\scriptstyle 5\,11\ 9\ 10} \\ {\scriptstyle \cancel{1}\ \cancel{10}\ } \\ \cancel{6}\,2\,\cancel{0}\,\cancel{0} \\ -\ \ 8\,7\,4 \\ \hline 5\,3\,2\,6 \end{array}$$

Now borrow 1 thousand (1 thousand = 10 hundreds) and we can subtract in every column.

CHECK:

$$\begin{array}{r} 5326 \\ +\ \ 874 \\ \hline 6200 \end{array}$$

Let's try Example F again, using a technique called "reverse adding." Just ask yourself, "What do I have to add to 874 to get 6200?"

$$\begin{array}{r} 6200 \\ -\ \ 874 \\ \hline 6 \end{array}$$

Begin with the ones column. 4 is larger than 0, so ask "What do I add to 4 to make 10?"

$$\begin{array}{r} 6200 \\ -\ \ 874 \\ \hline 26 \end{array}$$

Because 4 + 6 = 10, we write the 6 in the ones column and carry the 1 over to the 7 to make 8. Now ask, "What do I add to 8 to make 10?"

Answers to Warm-Ups

D. 42 **E.** 368 **F.** 3041

$$\begin{array}{r} 6200 \\ -\ 874 \\ \hline 326 \end{array}$$

Write 2 in the tens column and carry the 1 to the 8 in the hundreds column. Now ask, "What do I add to 9 to make 12?"

$$\begin{array}{r} 6200 \\ -\ 874 \\ \hline 5326 \end{array}$$

Finally, ask, "What do I add to the carried 1 to make 6?"

The advantage of this method is that 1 is the largest value carried. Most people can do this mentally.

So 6200 − 874 = 5326.

CALCULATOR EXAMPLE

G. Subtract 34,985 from 78,459.

G. Subtract 33,966 from 81,088.

CAUTION

When a subtraction exercise is worded "Subtract A from B," it is necessary to reverse the order of the numbers. The difference is B − A.

Enter 78,459 − 34,985.

The difference is 43,474.

H. Maxwell Auto is advertising a $986 rebate on all new cars priced above $15,000. What is the cost after rebate of a car originally priced at $16,798?

H. Maxwell Auto is also advertising a $2138 rebate on all new cars priced above $32,000. What is the cost after rebate of a car originally priced at $38,971?

STRATEGY: Because the price of the car is over $15,000, we subtract the amount of the rebate to find the cost.

$$\begin{array}{r} 16{,}798 \\ -\ \ \ 986 \\ \hline 15{,}812 \end{array} \qquad \textbf{CHECK:} \qquad \begin{array}{r} 15{,}812 \\ +\ \ \ 986 \\ \hline 16{,}798 \end{array}$$

The car costs $15,812.

*

HOW & WHY

OBJECTIVE 3 Estimate the sum or difference of whole numbers.

The sum or difference of whole numbers can be estimated by rounding each number to a specified place value and then adding or subtracting the rounded values. For instance, estimate the sum by rounding to the nearest thousand.

$$\begin{array}{r} 6359 \\ 3790 \\ 9023 \\ 4825 \\ +\ \ 899 \\ \hline \end{array} \qquad \begin{array}{r} 6000 \\ 4000 \\ 9000 \\ 5000 \\ +\ 1000 \\ \hline 25{,}000 \end{array}$$

Round each number to the nearest thousand.

The estimate of the sum is 25,000.

Answers to Warm-Ups

G. 47,122

H. The cost of the car is $36,833.

Another estimate can be found by rounding each number to the nearest hundred.

6359	6400	Round each number to the nearest hundred.
3790	3800	
9023	9000	
4825	4800	
+ 899	+ 900	
	24,900	

One use of the estimate is to see if the sum of a group of numbers is correct. If the calculated sum is not close to the estimated sum, you should check the addition by readding. In this case, the calculated sum, 24,896, is close to the estimated sums of 25,000 and 24,900.

Estimate the difference by rounding each number to the hundreds place and then subtract the rounded numbers.

8967	9000	Round each number to the nearest hundred.
−5141	−5100	Subtract.
	3900	

The estimate of the difference is 3900. One use of the estimate is to see if the difference is correct. If the calculated difference is not close to 3900, you should check the subtraction. In this case, the difference is 3826, which is close to the estimate.

To estimate the sum or difference of whole numbers

1. Round each number to the same place value.
2. Add or subtract the rounded numbers.

WARM-UPS I–J

I. Round each number to the nearest hundred, then add and compare to the actual sum:
643 + 72 + 422 + 875 + 32 + 91

Answers to Warm-Ups

I. The sum, 2135, is close to the estimation, 2100.

EXAMPLES I–J

DIRECTIONS: Estimate the sum or difference. Then add or subtract and compare.

STRATEGY: Round each number to the specified place value. Then add or subtract and compare.

I. Round each number to the nearest hundred:

475 + 8795 + 976 + 6745 + 5288 + 12

475	500	Round each number to the nearest hundred. With practice, this can be done mentally for a quick check
8795	8800	
976	1000	
6745	6700	
5288	5300	
+ 12	+ 0	
	22,300	

Now add and compare.

475
8795
976
6745
5288
+ 12
22,291

The sum, 22,291, is close to the estimation, 22,300.

J. Estimate the difference of 87,456 and 37,921 by rounding to the nearest thousand. Then subtract and compare.

$$\begin{array}{r} 87{,}000 \\ -38{,}000 \\ \hline 49{,}000 \end{array}$$ Round each number to the nearest thousand.

Now subtract.

$$\begin{array}{r} 87{,}456 \\ -37{,}921 \\ \hline 49{,}535 \end{array}$$

The estimated difference and the calculated difference are not close.

Check to see if the subtraction was done correctly.

$$\begin{array}{r} 87{,}456 \\ -37{,}921 \\ \hline 49{,}535 \end{array}$$ In the first subtraction, the fact that 1 was borrowed from the 8 was ignored.

Now the estimated difference, 49,000, is close to the actual difference, 49,535.

J. Estimate the difference of 43,981 and 11,765 by rounding to the nearest thousand. Then subtract and compare.

HOW & WHY

OBJECTIVE 4 Find the perimeter of a polygon.

A polygon is a closed figure whose sides are line segments, such as rectangles, squares, and triangles (Figure 1.5). An expanded discussion of polygons can be found in Section 7.3.

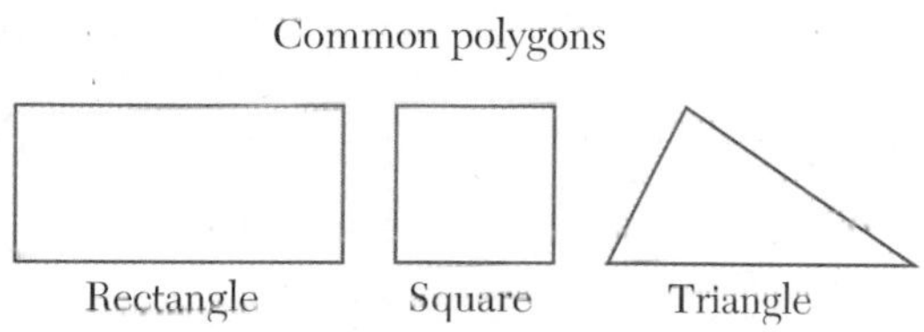

FIGURE 1.5

The perimeter of a polygon is the distance around the outside. To find the perimeter, we add the lengths of the sides.

Answers to Warm-Ups

J. The estimated difference, 32,000, is close to the difference, 32,216.

EXAMPLES K–L

DIRECTIONS: Find the perimeter of the polygon.

STRATEGY: Add up the lengths of all the sides.

K. Find the perimeter of the triangle.

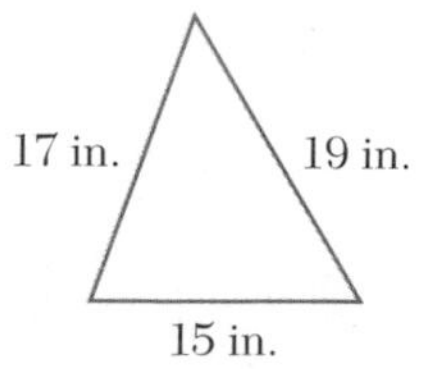

17 in. + 19 in. + 15 in. = 51 in. Add the lengths of the sides.

The perimeter is 51 in.

WARM-UPS K–L

K. Find the perimeter of the triangle.

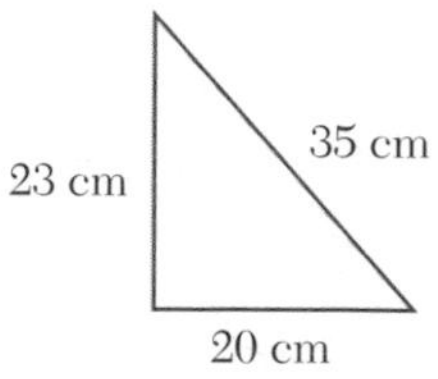

L. Find the perimeter of the polygon.

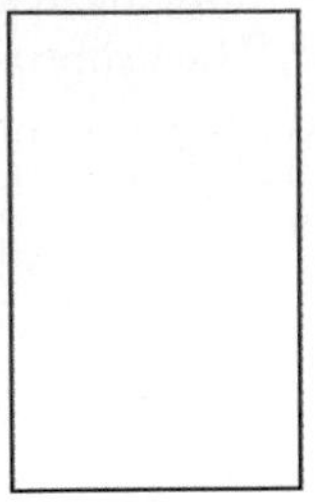

L. Find the perimeter of the polygon.

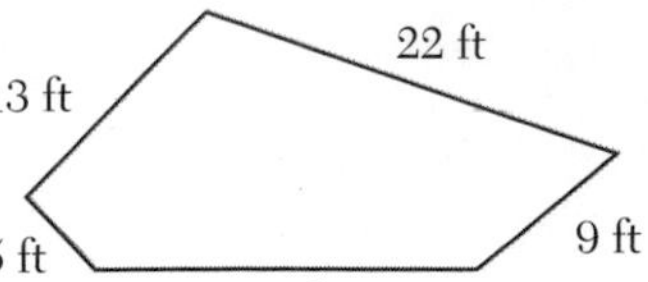

13 ft + 22 ft + 9 ft + 19 ft + 5 ft = 68 ft

The perimeter is 68 ft.

Answers to Warm-Ups

K. The perimeter is 78 cm.
L. The perimeter is 96 in.

EXERCISES 1.2

OBJECTIVE 1 Find the sum of two or more whole numbers.

Add.

A.

1. 65 + 32

2. 57 + 22

3. 748 + 231

4. 341 + 547

5. $\begin{array}{r} 212 \\ +495 \\ \hline \end{array}$

6. $\begin{array}{r} 468 \\ +324 \\ \hline \end{array}$

7. When you add 36 and 48, the sum of the ones column is 14. You must carry the _____ to the tens column.

8. In 463 + 385, the sum is X48. The value of X is _____.

B.

9. 624 + 4815 + 298

10. 783 + 5703 + 529

11. 8 + 90 + 403 + 6070

12. 7 + 90 + 709 + 3090

13. 2795 + 3643 + 7055 + 4004 (Round sum to the nearest hundred.)

14. 5962 + 7633 + 8049 + 6222 (Round sum to the nearest hundred.)

OBJECTIVE 2 Find the difference of two whole numbers.

Subtract.

A.

15. $\begin{array}{r} 5 \text{ hundreds} + 9 \text{ tens} + 6 \text{ ones} \\ -3 \text{ hundreds} + 9 \text{ tens} + 5 \text{ ones} \\ \hline \end{array}$

16. $\begin{array}{r} 6 \text{ hundreds} + 3 \text{ tens} + 5 \text{ ones} \\ -4 \text{ hundreds} + 2 \text{ tens} + 4 \text{ ones} \\ \hline \end{array}$

17. 509 − 72

18. 642 − 80

19. 768 − 345

20. 848 − 611

21. When subtracting 54 − 19, you must "borrow" 1 from the 5. The value of the "borrowed 1" is _____ ones.

22. To subtract 526 − 462, you must borrow from the _____ column to subtract in the _____ column.

B.

23. 632 − 457

24. 932 − 857

25. 600 − 437

26. 600 − 338

27. 9641 − 5098 (Round difference to the nearest hundred.)

28. 9045 − 5786 (Round difference to the nearest hundred.)

OBJECTIVE 3 Estimate the sum or difference of whole numbers.

A. *Estimate the sum by rounding each number to the nearest hundred.*

29. 345 + 782

30. 495 + 912

31.
3411
2001
+4561

32.
5467
3811
+2199

Estimate the difference by rounding each number to the nearest hundred.

33. 773 − 523

34. 854 − 392

35.
8678
−3914

36.
6235
−5991

B. *Estimate the sum by rounding each number to the nearest thousand and then add.*

37.
3209
7095
4444
2004
+3166

38.
5038
4193
2121
5339
+6560

39.
45,902
33,333
57,700
+23,563

40.
12,841
29,671
21,951
+73,846

Estimate the difference by rounding each number to the nearest thousand and then subtract.

41.
8753
−4067

42.
7661
−3089

43.
65,808
−32,175

44.
92,150
−67,498

45. Estimate the sum of 23,706, 34, 7561, 9346, and 236, by rounding each number to the nearest thousand and then find the sum.

46. Estimate the difference of 567,234 and 315,732 by rounding to the nearest ten thousand and then find the difference.

OBJECTIVE 4 Find the perimeter of a polygon.

Find the perimeter of the following polygons.

A.

47.

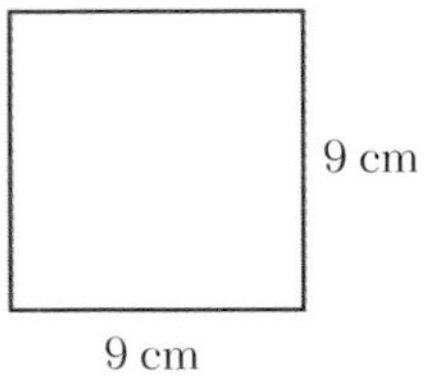

48.

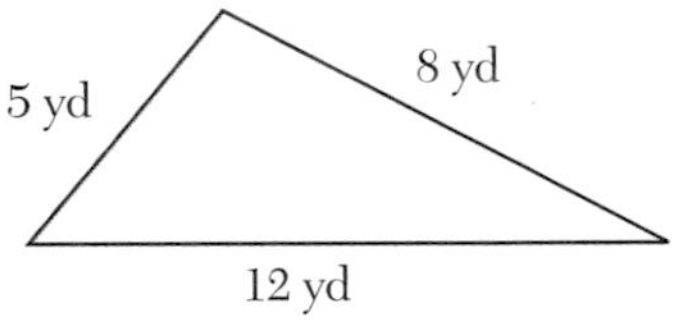

49.

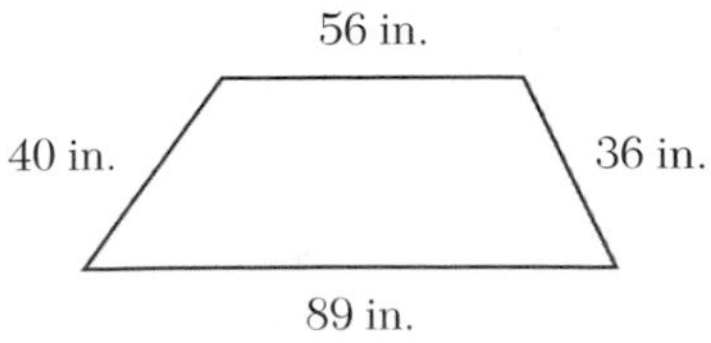

50.

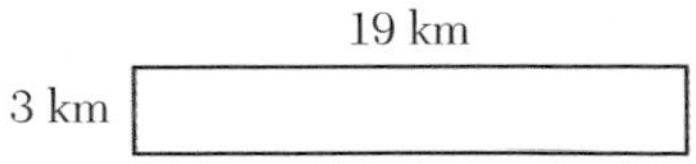

B.

51.

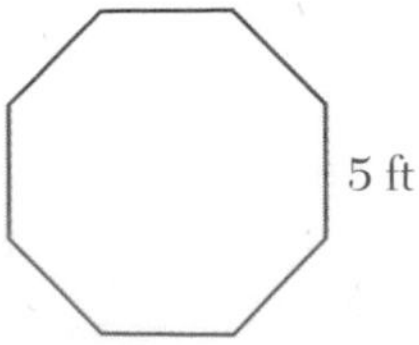

52.

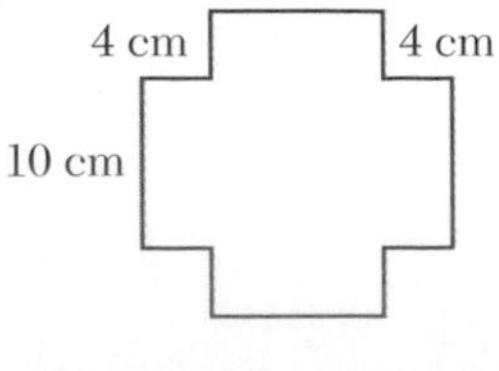

53.

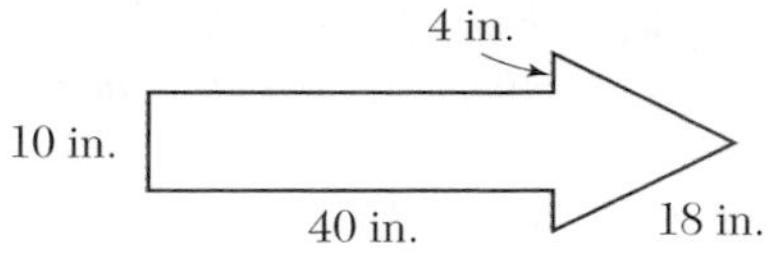

54.

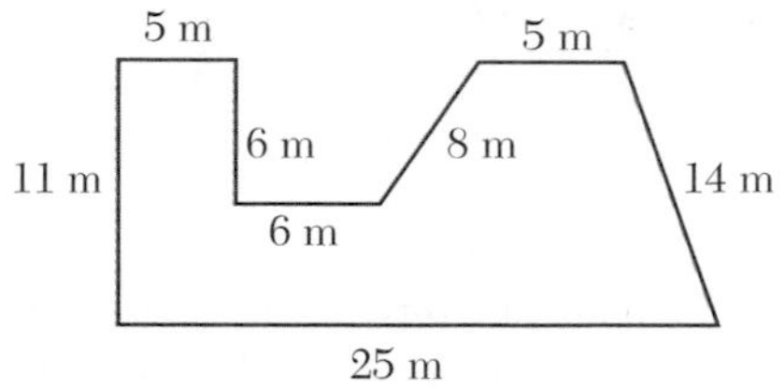

C. *Exercises 55–60 refer to the following sales chart, which gives the distribution of car sales among dealers in Wisconsin.*

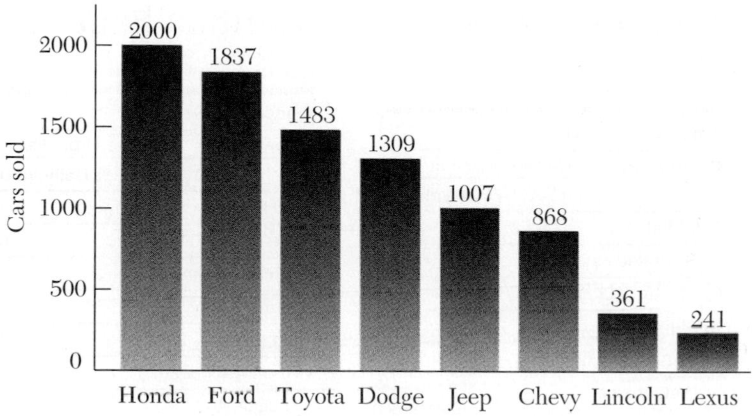

55. What is the total number of Fords, Toyotas, and Lexuses sold?

56. What is the total number of Chevys, Lincolns, Dodges, and Hondas sold?

57. How many more Hondas are sold than Fords?

58. How many more Toyotas are sold than Jeeps?

59. What is the total number of the three best-selling cars sold?

60. What is the difference in cars sold between the best-selling car and the least-selling car?

61. The biologist at the Bonneville fish ladder counted the following number of coho salmon during a 1-week period: Monday, 895; Tuesday, 675; Wednesday, 124; Thursday, 1056; Friday, 308; Saturday, 312; and Sunday, 219. How many salmon went through the ladder that week? How many more salmon went through the ladder on Tuesday than on Saturday?

Exercises 62–64. The estimated number of arrests in a recent year in the United States, according to the FBI, is given in the table. Source: *Uniform Crime Reports.*

Estimated Number of Arrests

Murder/manslaughter	18,856
Forcible rape	29,432
Robbery	143,877
Aggravated assault	408,148
Burglary	308,849
Larceny-theft	1,131,768
Motor vehicle theft	156,711
Arson	14,504

62. Find the total number of estimated arrests for violent crimes (murder/manslaughter, forcible rape, robbery, and aggravated assault).

63. Find the total number of estimated arrests for property crimes (burglary, larceny-theft, motor vehicle theft, and arson).

64. How many more aggravated assaults than robberies were there?

 Name ______________________ Class ______________ Date __________

Exercises 65–67. A home furnace uses natural gas, oil, or electricity for the energy needed to heat the house. We humans get our energy for body heat and physical activity from calories in our food. Even when resting we use energy for muscle actions such as breathing, heartbeat, digestion, and other functions. If we consume more calories than we use up, we gain weight. If we consume less, we lose weight. Some nutritionists recommend about 2270 calories per day for women aged 18–30 who are reasonably active.

Sasha, who is 22 years old, sets 2250 calories per day as her goal. She plans to have pasta with marinara sauce for dinner. The product labels shown here give the number of calories in each food.

Pasta

Nutrition Facts
Serving Size 2 oz (56g) dry
(1/8 of the package)
Servings Per Container 8

Amount Per Serving	
Calories 200	Calories from Fat 10
	% Daily Value*
Total Fat 1 g	2%
Saturated Fat 0g	0%
Cholesterol 0mg	0%
Sodium 0mg	0%
Total Carbohydrate 41g	14%
Dietary Fiber 2g	8%
Sugars 3g	
Protein 7g	
Vitamin A 0% •	Vitamin C 0%
Calcium 0% •	Iron 10%
Thiamin 35% •	Riboflavin 15%
Niacin 15% •	

*Percent Daily Values are based on a 2,000 calorie diet. Your daily values may be higher or lower depending on your calorie needs.

Marinara Sauce

Nutrition Facts
Serving Size 1/2 cup (125g)
Servings per Container approx 6

Amount Per Serving	
Calories 60	Calories from Fat 20
	% Daily Value*
Total Fat 2g	3%
Saturated Fat 3g	0%
Cholesterol 0mg	0%
Sodium 370mg	15%
Total Carbohydrate 7g	2%
Dietary Fiber 2g	8%
Sugars 4g	
Protein 3g	
Vitamin A 15% •	Vitamin C 40%
Calcium 0% •	Iron 4%

65. If she eats two servings each of pasta and sauce for dinner, how many calories does she consume?

66. If Sasha has 550 more calories in bread, butter, salad, drink, and desert for dinner, how many total calories does she consume at dinner?

67. If Sasha keeps to her goal, how many calories could she have eaten at breakfast and lunch?

68. The attendance at three consecutive Super Bowls is 78,943, 85,782, and 103,456. What is the total attendance at the three games? How many more fans were at the game with the largest attendance as opposed to the one with the least attendance?

69. A forester counted 23,679 trees that are ready for harvest on a certain acreage. If Forestry Service rules require that 8543 mature trees must be left on the acreage, how many trees can be harvested?

70. The new sewer line being installed in downtown Kearney will handle 345,760 gallons of refuse per minute. The old line handled 178,550 gallons per minute. How many more gallons per minute will the new line handle?

71. Fong's Grocery owes a supplier $25,875. During the month, Fong's makes payments of $460, $983, $565, and $10,730. How much does Fong's still owe, to the nearest hundred dollars?

72. In the spring of 1989, an oil tanker hit a reef and spilled 10,100,000 gallons of oil off the coast of Alaska. The tanker carried a total of 45,700,000 gallons of oil. The oil that did not spill was pumped into another tanker. How many gallons of oil were pumped into the second tanker? Round to the nearest million gallons.

73. The median family income of a region is a way of estimating the middle income. Half the families in the region make more than the median income, and the other half of the families make less than the median income. In 1997, the median family income for San Francisco was $64,400, and the median for Seattle was $55,100. What place value were these figures rounded to and how much higher was San Francisco's median income than Seattle's.

74. Bryce Canyon, the Grand Canyon, and Zion national parks are found in the southwestern United States. Geologic changes over a billion years have created these formations and canyons. The table shows the highest and lowest elevations in each of these parks. Find the change in elevation in each park. In which park is the change greatest? How much greater is this change than that in the park with the smallest change?

Elevations at National Parks

	Highest Elevation	Lowest Elevation
Bryce Canyon	8500 ft	6600 ft
Grand Canyon	8300 ft	2500 ft
Zion	7500 ft	4000 ft

Exercises 75–78. The average number of murder victims per year in the United States who are related to the murderer, according to statistics from the FBI is given in the table.

Murder Victims Related to the Murderer

Wives	Husbands	Sons	Daughters	Fathers	Brothers	Mothers	Sisters
913	383	325	235	169	167	121	42

75. In an average year, how many more husbands killed their wives than wives killed their husbands?

76. In an average year, how many people killed their child?

77. In a average year, how many people killed a sibling?

78. In an average year, did more people kill their child or their parents?

Exercises 79–81. A sales report prepared by General Motors contains the following information for sales in June 2001.

	June Sales
General Motors	456,855
Ford	392,433
Daimler-Chrysler	244,551

79. How many more General Motors vehicles were sold than Ford vehicles?

80. Did General Motors outsell the other two companies combined?

81. Rewrite the table, rounding the sales figures to the nearest thousand.

82. In the National Football League, the salary cap is the absolute maximum that a club can spend on player salaries. For the 2000 season, the maximum salary cap per club was $62,172,000. The minimum salary cap per team was $51,561,000. What is the difference between the maximum and minimum salary caps?

Exercises 83–84. The World Health Organization's estimates of the number of HIV infections in adults in 1999 are given in the table.

Number of HIV Infections—1999

Region	HIV Infections
Sub-Saharan Africa	25 million
South and Southeast Asia	6 million
Latin America/Caribbean	1 million
North America	1 million
Western Europe	500 thousand
East Asia/Pacific	500 thousand
Eastern Europe/Central Asia	400 thousand
Caribbean	400 thousand
North Africa/Middle East	200 thousand
Australia	15 thousand

83. What is the total estimated number of HIV infections for Asian regions?

84. How many more estimated cases of HIV infections are there in Western Europe than in Eastern Europe/Central Asia?

85. Find the perimeter of a rectangular house that is 52 ft long and 28 ft wide.

86. A farmer wants to put a fence around a triangular plot of land that measures 3 km by 8 km by 4 km. How much fence does he need?

87. Blanche wants to sew lace around the edge of a rectangular tablecloth that measures 56 in. by 36 in. How much lace does she need, ignoring the corners and seam allowances?

88. Annisa wants to trim a picture frame in ribbon. The outside of the rectangular frame is 25 cm by 30 cm. How much ribbon does she need, ignoring the corners?

STATE YOUR UNDERSTANDING

89. Explain to a 6-year-old child why $15 - 9 = 6$.

90. Explain to a 6-year-old child why $8 + 7 = 15$.

91. Define and give an example of a sum.

92. Define and give an example of a difference.

CHALLENGE

93. Add the following numbers, round the sum to the nearest hundred, and write the word name for the rounded sum: one hundred sixty; eighty thousand, three hundred twelve; four hundred seventy-two thousand, nine hundred fifty-two; and one hundred forty-seven thousand, five hundred twenty-three.

94. How much greater is seven million, two hundred forty-seven thousand, one hundred ninety-five than two million, eight hundred four thousand, fifty-three? Write the word name for the difference.

95. Peter sells three Honda Civics for $14,385 each, four Accords for $17,435 each, and two Acuras for $26,548 each. What is the total dollar sales for the nine cars? How many more dollars were paid for the four Accords than for the three Civics?

Complete the sum or difference by writing in the correct digit whenever you see a letter.

96.

```
  5A68
   241
 +10A9
 -----
  B64C
```

97.

```
  4A6B
 - C251
 ------
  15D1
```

GROUP WORK

98. Add and round to the nearest hundred.

```
   14,657
    3,766
  123,900
      569
   54,861
 +346,780
```

Now round each addend to the nearest hundred and then add. Discuss why the answers are different. Be prepared to explain why this happens.

99. If Ramon delivers 112 loaves of bread to each store on his delivery route, how many stores are on the route if he delivers a total of 4368 loaves? (*Hint*: Subtract 112 loaves for each stop from the total number of loaves.) What operation does this perform? Make up three more examples and be prepared to demonstrate them in class.

Getting Ready for Algebra

VOCABULARY

An **equation** is a statement about numbers that says that two expressions are equal. Letters, called **variables** or **unknowns**, are often used to represent numbers.

OBJECTIVE

Solve an equation of the form $x + a = b$ or $x - a = b$, where a, b, and x are whole numbers.

HOW & WHY

OBJECTIVE Solve an equation of the form $x + a = b$ or $x - a = b$, where a, b, and x are whole numbers.

Examples of equations are

$9 = 9$ $\quad 13 = 13$ $\quad 123 = 123$ $\quad 30 + 4 = 34$ $\quad 52 - 7 = 45$

When variables are used, an equation can look like this:

$x = 7$ $\quad x = 10$ $\quad y = 18$ $\quad x + 5 = 13$ $\quad y - 8 = 23$

An equation containing a variable can be true only when the variable is replaced by a specific number. For example,

$x = 7$ is true only when x is replaced by 7.
$x = 10$ is true only when x is replaced by 10.
$y = 18$ is true only when y is replaced by 18.
$x + 5 = 13$ is true only when x is replaced by 8, so that $8 + 5 = 13$.
$y - 8 = 23$ is true only when y is replaced by 31, so that $31 - 8 = 23$.

The numbers that make equations true are called *solutions*. Solutions of equations, such as $x - 7 = 12$, can be found by trial and error, but let's develop a more practical way.

Addition and subtraction are inverse, or opposite, operations. For example, if 14 is added to a number and then 14 is subtracted from that sum, the difference is the original number. So

$23 + 14 = 37$	Add 14 to 23.
$37 - 14 = 23$	Subtract 14 from the sum, 37. The difference is the original number, 23.

We use this idea to solve the following equation:

$x + 21 = 35$	21 is added to the number represented by x.
$x + 21 - 21 = 35 - 21$	To remove the addition and have only x on the left side
$x = 14$	of the equal sign, we subtract 21. To keep a true equation, we must subtract 21 from both sides.

To check, replace x in the original equation with 14 and see if the result is a true statement.

$x + 21 = 35$	
$14 + 21 = 35$	
$35 = 35$	The statement is true, so the solution is 14.

We can also use the idea of inverses to solve an equation in which a number is subtracted from a variable (letter):

$b - 17 = 12$	Because 17 is subtracted from the variable, we eliminate
$b - 17 + 17 = 12 + 17$	the subtraction by adding 17 to both sides of the equation. Recall that addition is the inverse of subtraction.
$b = 29$	The equation will be true when b is replaced by 29.

CHECK:

$$b - 17 = 12$$
$$29 - 17 = 12 \quad \text{Substitute 29 for } b.$$
$$12 = 12 \quad \text{True.}$$

So the solution is $b = 29$.

> **To solve an equation using addition or subtraction**
>
> 1. Add the same number to each side of the equation to isolate the variable.
> 2. Or subtract the same number from each side of the equation to isolate the variable.
> 3. Check the solution by substituting it for the variable in the original equation.

WARM-UPS A–E

A. $x + 15 = 32$

B. $y - 20 = 46$

C. $56 = z + 25$

EXAMPLES A–E

DIRECTIONS: Solve and check.

STRATEGY: Isolate the variable by adding or subtracting the same number to or from each side.

A. $x + 7 = 23$

$$x + 7 = 23$$
$$x + 7 - 7 = 23 - 7$$

Because 7 is added to the variable, eliminate the addition by subtracting 7 from both sides of the equation.

$$x = 16$$

Simplify.

CHECK:

$$x + 7 = 23$$

Check by substituting 16 for x in the original equation.

$$16 + 7 = 23$$
$$23 = 23$$

The statement is true.

The solution is $x = 16$.

B. $a - 24 = 50$

$$a - 24 = 50$$
$$a - 24 + 24 = 50 + 24$$

Because 24 is subtracted from the variable, eliminate the subtraction by adding 24 to both sides of the equation.

$$a = 74$$

Simplify.

CHECK:

$$a - 24 = 50$$

Check by substituting 74 for a in the original equation.

$$74 - 24 = 50$$
$$50 = 50$$

The statement is true.

The solution is $a = 74$.

C. $45 = b + 22$

In this example, we do the subtraction vertically.

$$\begin{array}{rcl} 45 & = & b + 22 \\ -22 & & -22 \\ \hline 23 & = & b \end{array}$$

Subtract 22 from both sides to eliminate the addition of 22.

Simplify.

CHECK:

$$45 = b + 22$$
$$45 = 23 + 22 \quad \text{Substitute 23 for } b.$$
$$45 = 45 \quad \text{The statement is true.}$$

The solution is $b = 23$.

Answers to Warm-Ups

A. $x = 17$ **B.** $y = 66$ **C.** $z = 31$

D. $z - 33 = 41$

$$z - 33 = 41$$
$$z - 33 + 33 = 41 + 33 \quad \text{Add 33 to both sides.}$$
$$z = 74 \quad \text{Simplify.}$$

CHECK: $z - 33 = 41$

$$74 - 33 = 41 \quad \text{Substitute 74 for } z.$$
$$41 = 41 \quad \text{The statement is true.}$$

The solution is $z = 74$.

E. The selling price for a pair of Nike AirMax® Deluxe shoes is \$139. If the markup on the shoes is \$43, what is the cost to the store?

Cost + markup = selling price.

$$C + M = S \quad \text{Cost + Markup = Selling price}$$
$$C + 43 = 139 \quad \text{Substitute 43 for the markup and 139 for the selling price.}$$
$$C + 43 - 43 = 139 - 43 \quad \text{Subtract 43 from both sides.}$$
$$C = 96$$

CHECK: Does the cost + the markup equal \$139?

\$96	Cost
+ 43	Markup
\$139	Selling price

So the cost of the shoes to the store is \$96.

D. $b - 43 = 51$

E. The selling price of a set of golf clubs is \$576. If the markup is \$138, what is the cost to the store?

Answers to Warm-Ups

D. $b = 94$ **E.** The golf clubs cost the store \$438.

EXERCISES

OBJECTIVE Solve an equation of the form $x + a = b$ or $x - a = b$, where a, b, and x are whole numbers.

Solve and check.

1. $x + 12 = 24$
2. $x - 11 = 14$
3. $x - 6 = 17$
4. $x + 10 = 34$

5. $z + 13 = 27$
6. $b - 21 = 8$
7. $c + 24 = 63$
8. $y - 33 = 47$

9. $a - 40 = 111$
10. $x + 75 = 93$
11. $x + 91 = 105$
12. $x - 76 = 43$

13. $y + 67 = 125$
14. $z - 81 = 164$
15. $k - 56 = 112$
16. $c + 34 = 34$

17. $73 = x + 62$
18. $534 = a + 495$
19. $87 = w - 29$
20. $373 = d - 112$

21. The selling price for a computer is $1265. If the cost to the store is $917, what is the markup?

22. The selling price of a trombone is $675. If the markup is $235, what is the cost to the store?

23. The length of a rectangular garage is 2 meters more than the width. If the width is 7 meters, what is the length?

24. The width of a rectangular fish pond is 6 feet shorter than the length. If the length is 27 feet, what is the width?

25. A Saturn with manual transmission has an EPA City rating of 4 miles per gallon more than the EPA rating of a Subaru® Impreza®. Write an equation that describes this relationship. Be sure to define all variables in your equation. If the Saturn has an EPA City rating of 28 mpg, find the rating of the Impreza.

26. In a recent year in the United States, the number of deaths by drowning was 1700 less than the number of deaths by fire. Write an equation that describes this relationship. Be sure to define all variables in your equation. If there were approximately 4800 deaths by drowning that year, how many deaths by fire were there?

Exercises 27–28. A city treasurer made the following report to the city council regarding monies allotted and dispersed from a city parks bond.

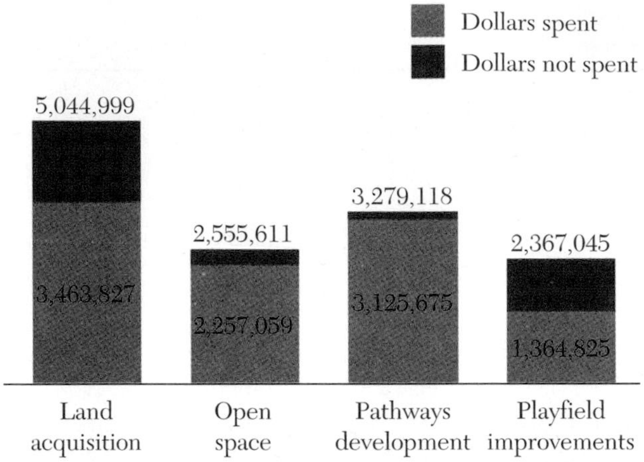

27. Write an equation that relates the total money budgeted per category to the amount of money spent and the amount of money not yet spent. Define all the variables.

28. Use your equation (Exercise 27) to calculate the amount of money not yet spent in each of the four categories.

1.3 Multiplying Whole Numbers

VOCABULARY

There are several ways to indicate multiplication. Here are examples of most of them, using 28 and 41.

28×41	$28 \cdot 41$	$\begin{array}{r} 28 \\ \times\ 41 \\ \hline \end{array}$
$(28)(41)$	$28(41)$	$(28)41$

The **factors** of a multiplication exercise are the numbers being multiplied. In $7(9) = 63$, 7 and 9 are the factors.

The **product** is the answer to a multiplication exercise. In $7(9) = 63$, the product is 63.

The **area** of a rectangle is the measure of the surface inside the rectangle.

OBJECTIVES

1. Multiply whole numbers.
2. Estimate the product of whole numbers.
3. Find the area of a rectangle.

HOW & WHY

OBJECTIVE 1 Multiply whole numbers.

Multiplying whole numbers is a shortcut for repeated addition:

$$\underbrace{8 + 8 + 8 + 8 + 8 + 8}_{\text{6 eights}} = 48 \qquad \text{or} \qquad 6 \cdot 8 = 48$$

As numbers get larger, the shortcut saves time. Imagine adding 152 eights.

$$\underbrace{8 + 8 + 8 + 8 + 8 + \cdots + 8}_{\text{152 eights}} = ?$$

We multiply 8 times 152 using the expanded form of 152.

$$\begin{array}{rl} 152 = 100 + \ 50 + \ 2 & \text{Write 152 in expanded form.} \\ \times\ \ 8 = \qquad\qquad\quad 8 & \\ \hline = 800 + 400 + 16 & \text{Multiply 8 times each addend.} \\ = 1216 & \text{Add.} \end{array}$$

The exercise can also be performed in column form without expanding the factors.

$$\begin{array}{rl} 152 & \\ \times\ 8 & \\ \hline 16 & 8(2) = 16 \\ 400 & 8(50) = 400 \\ 800 & 8(100) = 800 \\ \hline 1216 & \end{array} \qquad\qquad \begin{array}{r} {\scriptstyle 4\,1\ } \\ 152 \\ \times\ 8 \\ \hline 1216 \end{array}$$

The form on the right shows the usual shortcut. The carried digit is added to the product of each column. Study this example.

$$\begin{array}{r} 738 \\ \times\ 53 \\ \hline \end{array}$$

First multiply 738 by 3.

$$\begin{array}{r} {\scriptstyle 1\,2} \\ 738 \\ \times\ 53 \\ \hline 2214 \end{array}$$

3(8) = 24. Carry the 2 to the tens column.
3(3 tens) = 9 tens. Add the 2 tens that were carried:
(9 + 2) tens = 11 tens. Carry the 1 to the hundreds column.
3(7 hundreds) = 21 hundreds. Add the 1 hundred that was carried: (21 + 1) hundreds = 22 hundreds.

Now multiply 738 by 50.

$$\begin{array}{r} {\scriptstyle 1\,4} \\ {\scriptstyle \not{1}\,\not{2}} \\ 738 \\ \times\ 53 \\ \hline 2214 \\ 36900 \end{array}$$

50(8) = 400 or 40 tens. Carry the 4 to the hundreds column.
50(30) = 1500 or 15 hundreds. Add the 4 hundreds that were carried.
(15 + 4) hundreds = 19 hundreds. Carry the 1 to the thousands column.
50(700) = 35,000 or 35 thousands. Add the 1 thousand that was carried: (35 + 1) thousands = 36 thousands. Write the 6 in the thousands column and the 3 in the ten-thousands column.

$$\begin{array}{r} {\scriptstyle 1\,4} \\ {\scriptstyle \not{1}\,\not{2}} \\ 738 \\ \times\ 53 \\ \hline 2214 \\ 36900 \\ \hline 39114 \end{array}$$

Add the products.

Multiplication Property

Multiplication property of zero:

$a \cdot 0 = 0 \cdot a = 0$

Any number times 0 is 0.

Multiplication Property

Multiplication property of one:

$a \cdot 1 = 1 \cdot a = a$

Any number times 1 is that number.

Two important properties of arithmetic and higher mathematics are the *multiplication property of zero* and the *multiplication property of one*.

As a result of the multiplication property of zero, we know that

$0 \cdot 23 = 23 \cdot 0 = 0$ and $0(215) = 215(0) = 0$

As a result of the multiplication property of one, we know that

$1 \cdot 47 = 47 \cdot 1 = 47$ and $1(698) = 698(1) = 698$

WARM-UPS A–F

A. Multiply: (382)(0)

B. Find the product: 9(4892)

Answers to Warm-Ups

A. 0 **B.** 44,028

EXAMPLES A–F

DIRECTIONS: Multiply.

STRATEGY: Write the factors in columns. Start multiplying with the ones digit. If the product is 10 or more, carry the tens digit to the next column and add it to the product in that column. Repeat the process for every digit in the second factor. When the multiplication is complete, add to find the product.

A. Multiply: 1(777)

1(777) = 777 Multiplication property of one.

B. Find the product: 7(6497)

$$\begin{array}{r} {\scriptstyle 364} \\ 6497 \\ \times\quad 7 \\ \hline 45{,}479 \end{array}$$

Multiply 7 times each digit, carry when necessary, and add the number carried to the next product.

C. Multiply: $54 \cdot 49$

$$\begin{array}{r} 4 \\ 3 \\ 49 \\ \times\ 54 \\ \hline 196 \\ 2450 \\ \hline 2646 \end{array}$$

When multiplying by the 5 in the tens place, write a 0 in the ones column to keep the places lined up.

C. Multiply: $76 \cdot 63$

D. Find the product of 477 and 308.

STRATEGY: When multiplying by zero in the tens place, rather than showing a row of zeros, just put a zero in the tens column. Then multiply by the 2 in the hundreds place.

$$\begin{array}{r} 477 \\ \times\ \ 308 \\ \hline 3816 \\ 143100 \\ \hline 146{,}916 \end{array}$$

D. Find the product of 663 and 409.

CALCULATOR EXAMPLE

E. 786(43)

Enter the multiplication.

The product is 33,798.

E. 411(579)

F. The Sweet & Sour Company ships 72 cartons of packaged candy to the Chewum Candy Store. Each carton contains 48 packages of candy. What is the total number of packages of candy shipped?

STRATEGY: To find the total number of packages, multiply the number of cartons by the number of packages.

$$\begin{array}{r} 72 \\ \times\ 48 \\ \hline 576 \\ 2880 \\ \hline 3456 \end{array}$$

The company shipped 3456 packages of candy.

F. The Sweet & Sour Company ships 125 cartons of chewing gum to the Sweet Shop. If each carton contains 72 packages of gum, how many packages of gum are shipped to the Sweet Shop?

HOW & WHY

OBJECTIVE 2 Estimate the product of whole numbers.

The product of two numbers can be estimated by rounding each factor and then multiplying the rounded factors. For instance,

$$\begin{array}{r} 6110 \\ \times\ \ 87 \\ \hline \end{array} \qquad \begin{array}{r} 6000 \\ \times\ \ \ 90 \\ \hline 540{,}000 \end{array}$$

Round to the nearest thousand.
Round to the nearest ten.
Multiply.

or

$$\begin{array}{r} 6110 \\ \times\ \ 87 \\ \hline \end{array} \qquad \begin{array}{r} 6100 \\ \times\ \ \ 90 \\ \hline 549{,}000 \end{array}$$

Round to the nearest hundred.
Round to the nearest ten.
Multiply.

The estimates of the products are 540,000 and 549,000. One use of the estimate is to see if the product is correct. If the calculated product is not close to the estimate, you

Answers to Warm-Ups

C. 4788 **D.** 271,167 **E.** 237,969
F. The company shipped 9000 packages of gum.

should check the multiplication. In this case, the actual product is 531,570, which is close to the estimates.

To estimate the product of two whole numbers

1. Round each number to a specified place value.
2. Multiply the rounded numbers.

WARM-UP G

G. 913 and 489 Round both numbers to the nearest hundred.

EXAMPLE G

DIRECTIONS: Estimate the product. Then find the product and compare.

STRATEGY: Round each number to the specified place value. Multiply the rounded numbers. Multiply the original numbers.

G. 879 and 411 Round both numbers to the the nearest hundred.

$$\begin{array}{r} 900 \\ \times\ \ 400 \\ \hline 360{,}000 \end{array}$$

900 Round to the nearest hundred.
400 Round to the nearest hundred.

$$\begin{array}{r} 879 \\ \times\ \ 411 \\ \hline 879 \\ 8790 \\ 351600 \\ \hline 361{,}269 \end{array}$$

The product is 361,269 and is close to the estimated product of 360,000.

HOW & WHY

OBJECTIVE 3 Find the area of a rectangle.

The area of a polygon is the space inside the polygon. We use area when describing the size of a plot of land, the living space in a house, or an amount of carpet. Area is measured in square units such as square feet or square meters. A square foot is literally a square with sides of 1 foot. The measure of the surface inside the square is 1 square foot.

When measuring the space inside a polygon, we divide the space into squares and count them. For example, consider an area rug that is 2 ft by 3 ft (Figure 1.6).

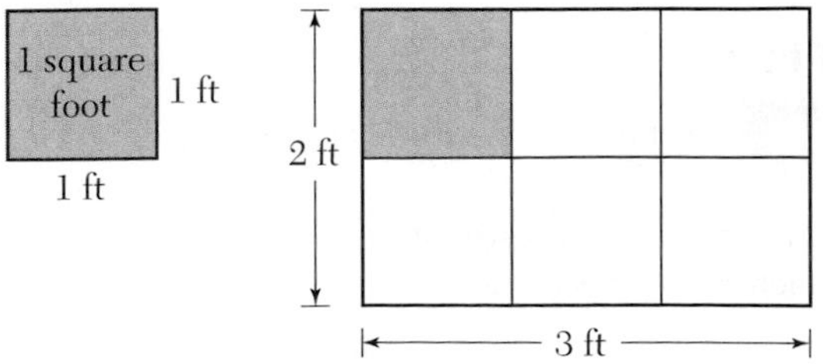

FIGURE 1.6

There are six squares in the subdivided rug so the area of the rug is 6 square feet.

Finding the area of a rectangle, such as the area rug in the example, is relatively easy because a rectangle has straight sides, and it is easy to fit squares inside it. The length of the rectangle determines how many squares will be in each row, and the width of the rectangle determines the number of rows. In the area rug shown in Figure 1.6, there are two rows of three squares each because the width is 2 ft and the length is 3 ft. The product of the length and width gives the number of squares inside the rectangle.

Area of a rectangle = length • width

Answer to Warm-Up

G. The product is 446,457 and is close to the estimated product of 450,000.

Finding the area of other shapes is a little more complicated and is discussed in Section 7.4.

EXAMPLE H

DIRECTIONS: Find the area of the rectangle.

STRATEGY: Multiply the length and width.

H. Find the area of the rectangle.

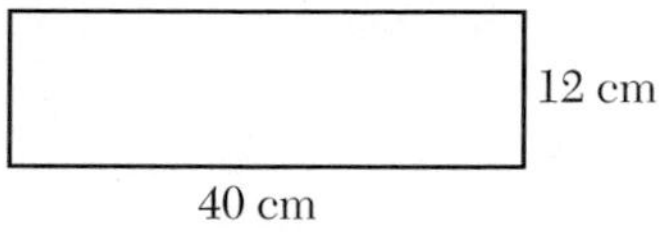

$$\begin{aligned} \text{Area} &= \text{length} \cdot \text{width} \\ &= 40 \cdot 12 \\ &= 480 \end{aligned}$$

The area is measured in square centimeters because the sides are measured in centimeters, and so each square is a square centimeter.

The area is 480 square centimeters.

WARM-UP H

H. Find the area of the rectangle.

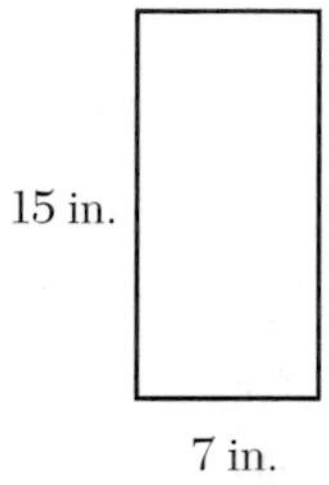

Answer to Warm-Up

H. 105 square inches

Name ______________________ Class ______________________ Date ____________

EXERCISES 1.3

OBJECTIVE 1 Multiply whole numbers.

Multiply.

A.

1. 62×5

2. 54×3

3. 44×3

4. 23×6

5. 67×4

6. 46×6

7. 73×7

8. 94×8

9. $9 \cdot 53$

10. $6 \cdot 55$

11. (81)(0)

12. (0)(215)

13. 66×40

14. 38×60

15. In 326×52, the place value of the product of 5 and 3 is ____________.

16. In 326×52, the product of 5 and 6 is 30 and you must carry the 3 to the ____________ column.

B.

17. 353×9

18. 652×6

19. 703×6

20. 505×8

21. (42)(38)

22. (36)(55)

23. (78)(46)

24. (98)(23)

25. 513×300

26. 487×700

27. 509×76

28. 509×58

29. 646×45

30. 328×52

31. (83)(254) Round product to the nearest hundred.

32. (78)(246) Round product to the nearest thousand.

OBJECTIVE 2 Estimate the product of whole numbers.

A. *Estimate the product by rounding both numbers to the nearest ten.*

33. 58(33)

34. 82(59)

35. 213(81)

36. 365(24)

37. 362×19

38. 756×37

39. 733×81

40. 448×49

B. *Estimate the product and multiply. Round each number to its largest place value.*

41. $\begin{array}{r} 312 \\ \times\ 50 \\ \hline \end{array}$

42. $\begin{array}{r} 675 \\ \times\ 40 \\ \hline \end{array}$

43. $\begin{array}{r} 527 \\ \times\ 73 \\ \hline \end{array}$

44. $\begin{array}{r} 265 \\ \times\ 57 \\ \hline \end{array}$

45. $\begin{array}{r} 475 \\ \times\ 39 \\ \hline \end{array}$

46. $\begin{array}{r} 823 \\ \times\ 65 \\ \hline \end{array}$

47. (4321)(76)

48. (3510)(83)

OBJECTIVE 3 Find the area of a rectangle.

A. *Find the area of the following rectangles.*

49.
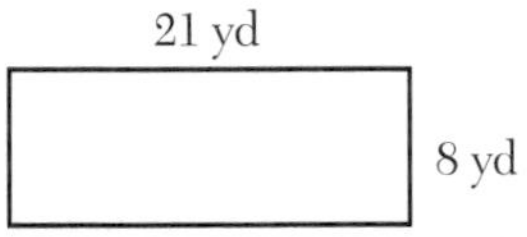

50.
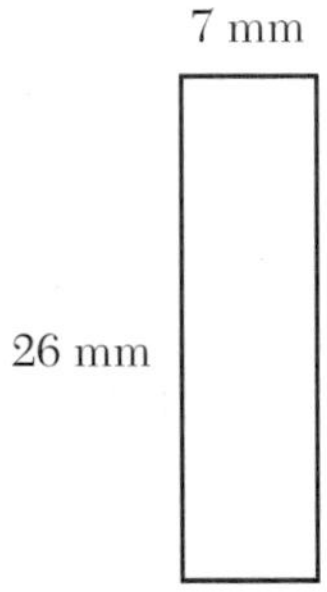

51.
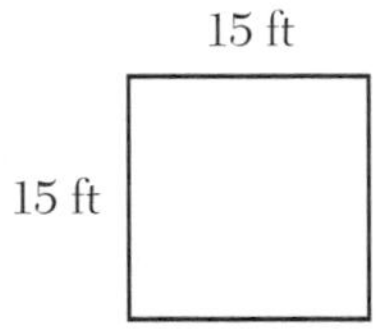

52.
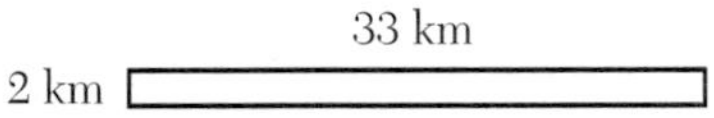

53.
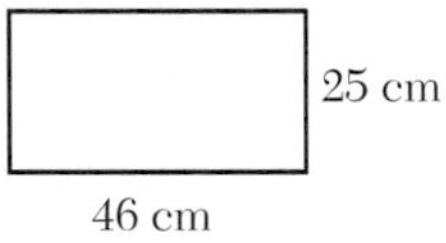

54.
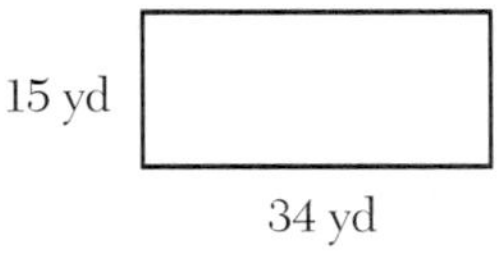

55. What is the area of a rectangle that has a length of 14 ft and a width of 4 ft?

56. What is the area of a rectangle that measures 20 cm by 50 cm?

B. *Find the area of the following.*

57.
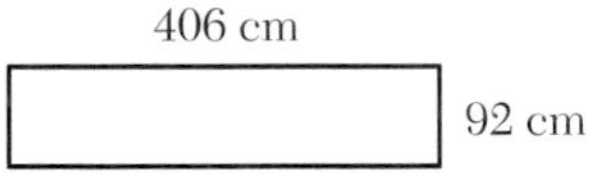

58.
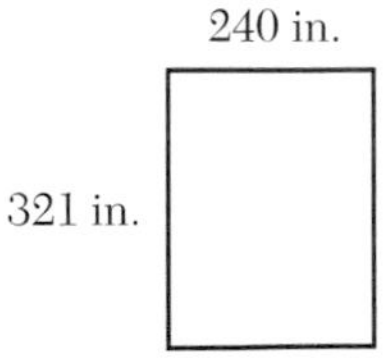

59.
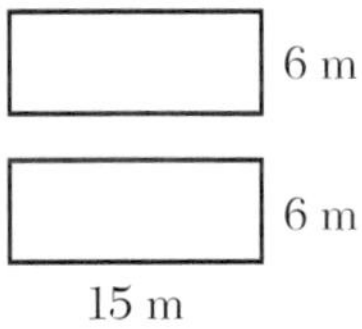

60.
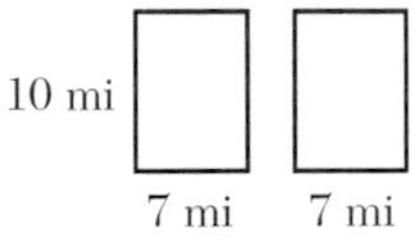

61.
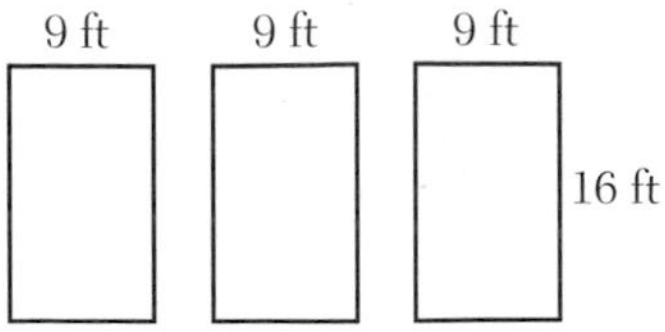

62.
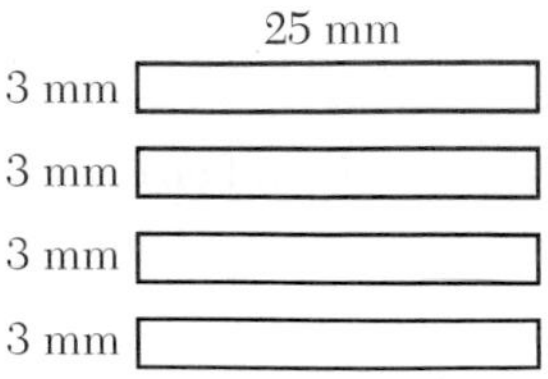

63. What is the area of three goat pens that each measure 5 ft by 8 ft?

64. What is the area of five bath towels that measure 70 cm by 150 cm?

C.

65. Find the product of 707 and 395.

66. Find the product of 707 and 526.

67. Estimate the product and multiply: (744)(3193).

68. Estimate the product and multiply: (6004)(405).

69. During the first week of the Rotary Club rose sale, 235 dozen roses are sold. The club estimates that a total of 18 times that number will be sold during the sale. What is the estimated number of dozens of roses that will be sold?

Exercises 70–73. Use the information in the table on monthly sales at Dick's Country Cars.

Monthly Sales—Dick's Country Cars

Car Model	Number of Cars Sold	Average Price per Sale
Escort ZX2	27	$10,115
Mustang Coupe	42	$14,988
Windstar LX	17	$21,795

70. Estimate the gross receipts from the sale of the Escort ZX2s.

71. What are the gross receipts from the sale of the Mustang Coupes?

72. Estimate the gross receipts from the sale of Windstar LXs and find the actual gross receipts.

73. Estimate the gross receipts for the month (the sum of the estimates for each model) and then find the actual gross receipts, rounded to the nearest thousand dollars.
Round off to Ten thousand

74. An average of 134 salmon per day are counted at the Bonneville fish ladder during a 17-day period. How many total salmon are counted during the 17-day period?

75. During 1997, the population of Washington County grew at a pace of 2043 people per month. What was the total growth in population during 1997?

76. The CEO of Apex Corporation exercised his option to purchase 2355 shares of Apex stock at $13 per share. He immediately sold the shares for $47 per share. If broker fees came to $3000, how much money did he realize from the purchase and sale of the shares?

77. The comptroller of Apex Corporation exercised her option to purchase 895 shares of Apex stock at $18 per share. She immediately sold the shares for $51 per share. If broker fees came to $950, how much money did she realize from the purchase and sale of the shares?

78. A certain bacteria culture triples its size every hour. If the culture has a count of 375 at 10 A.M., what will the count be at 2 P.M. the same day?

Exercises 79–80. The depth of water is often measured in fathoms. There are 3 feet in a yard and 2 yards in a fathom.

79. How many feet in a fathom?

80. How many feet are there in 20 fathoms?

Exercises 81–83. A league is an old measure of about 3 nautical miles. A nautical mile is about 6076 feet.

81. How many feet in a league?

82. There is famous book by Jules Verne titled *20,000 Leagues Under the Sea*. How many feet are in 20,000 leagues?

83. The Mariana Trench in the Pacific Ocean is the deepest point of all the world's oceans. It is 35,840 ft deep. Is it physically possible to be 20,000 leagues under the sea?

Exercises 84–86. Because distances between bodies in the universe are so large, scientists use large units. One such unit is the light-year, which is the distance traveled by light in 1 year, or about 5880 billion miles.

84. Write the place value notation for the number of miles in a light-year.

85. The star Sirius is recognized as the brightest star in the sky (other than the sun). It is 8 light-years from Earth. How many miles is Sirius from Earth?

86. The star Rigel in the Orion constellation is 545 light-years from Earth. How many miles away is Rigel from Earth?

Exercises 87–88. An inkjet computer printer has three resolutions: 400 dpi (dots per square inch), 720 dpi, and 1440 dpi.

87. Gabe prints a draft of his term paper on the lowest setting. His paper is three pages long, each page with a printing surface of 6 by 9 in. What is the maximum number of dots used by the printer to print the term paper?

88. Max is printing out two 4×6 in. photographs using the highest setting. How many dots were used?

Exercises 89–90. In computers, a byte is the amount of space needed to store one character. Knowing something about the metric system, one might think that a kilobyte is 1000 bytes, but actually it is 1024 bytes.

89. A computer has 128 Kb (kilobytes) of RAM. How many bytes is this?

90. A megabyte is 1024 Kb. A zip disk used for backing up files is 100 Mb (megabytes). How many bytes will the zip disk hold?

Exercises 91–92. A gram of fat contains about 9 calories, as does a gram of protein. A gram of carbohydrate contains about 4 calories.

91. A tablespoon of olive oil has 14 g of fat. How many calories is this?

92. One ounce of cream cheese contains 2 g of protein and 10 grams of fat. How many calories from fat and protein are in the cream cheese?

93. The water consumption in Hebo averages 320,450 gallons per day. How many gallons of water are consumed in a 31-day month, rounded to the nearest thousand gallons?

94. Ms. Munos orders four hundred fifty radios to sell in her discount store. If she pays \$78 per radio and sells them for \$112, how much do the radios cost her and what is the net income from their sale? How much are her profits from the sale of the radios?

95. Mr. Garcia orders three hundred twenty-five snow shovels for his hardware store. He pays \$12 per shovel and plans to sell them for \$21 each. What do the shovels cost Mr. Garcia and what will be his gross income from their sale? What net income does he receive from the sale of the shovels?

96. In 1997, Bill Gates of Microsoft® was the richest person in the United States, with an estimated net worth of \$36 billion. Write the place value name for this number. A financial analyst made the observation that the average person has a hard time understanding such large amounts. She gave the example that in order to spend \$1 billion, one would have to spend \$40,000 per day for 69 years. Exactly how much money would you spend if you did this?

Exercises 97–98 relate to the chapter application. See page 1.

97. If the St. Louis Rams doubled their 1999–2000 payroll, where would they rank in the 2000–2001 season?

98. Suppose the Jaguars tripled their 1999–2000 payroll and the Bears doubled their 2000–2001 payroll. Which payroll would be larger?

STATE YOUR UNDERSTANDING

99. Explain to an 8-year-old child that $3(8) = 24$.

100. When 74 is multiplied by 8, we carry 3 to the tens column. Explain why this is necessary.

101. Define and give an example of a product.

CHALLENGE

102. Find the product of twenty-four thousand, fifty-five and two hundred thirteen thousand, two hundred seventy-six. Write the word name for the product.

103. Tesfay harvests 75 bushels of grain per acre from his 11,575 acres of grain. If Tesfay can sell the grain for \$27 a bushel, what is the crop worth, to the nearest thousand dollars?

Complete the problems by writing in the correct digit wherever you see a letter.

104.
```
  51A
×  B2
 10B2
 154C
1A5D2
```

105.
```
 1A57
×  42
 B71C
 D428
569E4
```

GROUP WORK

106. Multiply 23, 56, 789, 214, and 1345 each by 10, 100, and 1000. What do you observe? Can you devise a rule for multiplying by 10, 100, and 1000?

1.4 Dividing Whole Numbers

VOCABULARY

There are a variety of ways to indicate division. These are the most commonly used:

$63 \div 3 \qquad 3\overline{)63} \qquad \frac{63}{3}$

The **dividend** is the number being divided, so in $48 \div 8 = 6$, the dividend is 48.

The **divisor** is the number that we are dividing by, so in $48 \div 8 = 6$, the divisor is 8.

The **quotient** is the answer to a division exercise, so in $48 \div 8 = 6$, the quotient is 6.

When a division exercise does not come out even, as in $53 \div 3$, the quotient is not a whole number.

$$\begin{array}{r} 13 \\ 5\overline{)67} \\ \underline{65} \\ 2 \end{array}$$

We call 13 the **partial quotient** and 2 the **remainder**. The quotient is written 13 R 2.

OBJECTIVES

1. Divide whole numbers.
2. Estimate the quotient of whole numbers.

HOW & WHY

OBJECTIVE 1 Divide whole numbers.

The division exercise $135 \div 27 = ?$ (read "135 divided by 27") can be interpreted in one of two ways.

How many times can 27 be subtracted from 135?	This is called the "repeated subtraction" version.
What number times 27 is equal to 135?	This is called the "missing factor" version.

All division problems can be done using repeated subtraction.

In $135 \div 27 = ?$, we can find the missing factor by repeatedly subtracting 27 from 135:

$$\begin{array}{r} 135 \\ \underline{-\ 27} \\ 108 \\ \underline{-\ 27} \\ 81 \\ \underline{-\ 27} \\ 54 \\ \underline{-\ 27} \\ 27 \\ \underline{-\ 27} \\ 0 \end{array}$$

Five subtractions, so $135 \div 27 = 5$.

The process can be shortened using the traditional method of guessing the number of 27s and subtracting from 135:

$$\begin{array}{r|l} 27\overline{)135} & \\ \underline{81} & \text{3 twenty-sevens} \\ 54 & \\ \underline{54} & \underline{\text{2 twenty-sevens}} \\ 0 & 5 \end{array} \qquad \text{or} \qquad \begin{array}{r|l} 27\overline{)135} & \\ \underline{108} & \text{4 twenty-sevens} \\ 27 & \\ \underline{27} & \underline{\text{1 twenty-seven}} \\ 0 & 5 \end{array}$$

$$\text{or} \qquad \begin{array}{r|l} 27\overline{)135} & \\ \underline{135} & \underline{\text{5 twenty-sevens}} \\ 0 & 5 \end{array}$$

In each case, $135 \div 27 = 5$.

We see that the missing factor in $(27)(?) = 135$ is 5. So $27(5) = 135$ or $135 \div 27 = 5$.

This leads to a method for checking division. If we multiply the divisor times the quotient, we will get the dividend. To check $135 \div 27 = 5$, we multiply 27 and 5.

$(27)(5) = 135$

So 5 is correct.

This process works regardless of the size of the numbers. If the divisor is considerably smaller than the dividend, you will want to guess a rather large number.

$$\begin{array}{r|r} 63\overline{)19{,}593} & \\ \underline{6\ 300} & 100 \\ 13\ 293 & \\ \underline{6\ 300} & 100 \\ 6\ 993 & \\ \underline{6\ 300} & 100 \\ 693 & \\ \underline{630} & 10 \\ 63 & \\ \underline{63} & \underline{1} \\ 0 & 311 \end{array}$$

So, $19{,}593 \div 63 = 311$.

All divisions can be done by this method. However, the process can be shortened by finding the number of groups, starting with the largest place value on the left, in the dividend, and then working toward the right. Study the following example. Note that the answer is written above the problem for convenience.

$31\overline{)17{,}422}$ Working from left to right, we note that 31 does not divide 1, and it does not divide 17. However, 31 does divide 174 five times. Write the 5 above the 4 in the dividend.

$$\begin{array}{r} 562 \\ 31\overline{)17{,}422} \\ \underline{155} \\ 192 \\ \underline{186} \\ 62 \\ \underline{62} \\ 0 \end{array}$$

$5(31) = 155$. Subtract 155 from 174. Because the difference is less than the divisor, no adjustment is necessary. Bring down the next digit, which is 2. Next, 31 divides 192 six times. The 6 is placed above the first 2 in the dividend. $6(31) = 186$. Subtract 186 from 192. Again, no adjustment is necessary since $6 < 31$. Bring down the next digit, which is 2. Finally, 31 divides 62 two times. Place the 2 above the second 2 in the dividend. $2(31) = 62$. Subtract 62 from 62; the remainder is zero. The division is complete.

CHECK:

$$\begin{array}{r} 562 \\ \underline{\times\ \ 31} \\ 562 \\ \underline{16{,}860} \\ 17{,}422 \end{array}$$

Check by multiplying the quotient by the divisor.

So $17{,}422 \div 31 = 562$.

Not all division problems come out even (have a zero remainder). In

$$\begin{array}{r} 4 \\ 17\overline{)81} \\ \underline{68} \\ 13 \end{array}$$

we see that 81 contains 4 seventeens and 13 toward the next group of seventeen. The answer is written as 4 remainder 13. The word *remainder* is abbreviated "R" and the result is 4 R 13.

Check by multiplying (17)(4) and adding the remainder.

$(17)(4) = 68$
$68 + 13 = 81$

So $81 \div 17 = 4 \text{ R } 13$.

The division, $45 \div 0 = ?$, can be restated: What number times 0 is 45? $0 \times ? = 45$. According to the multiplication property of zero, we know that $0 \times (\text{any number}) = 0$, so it cannot equal 45.

CAUTION

Division by zero is not defined. It is an operation that cannot be performed.

When dividing by a single-digit number the division can be done mentally using "short division."

$$\begin{array}{r} 423 \\ 3\overline{)1269} \end{array}$$

Divide 3 into 12. Write the answer, 4, above the 2 in the dividend. Now divide the 6 by 3 and write the answer, 2, above the 6. Finally, divide the 9 by 3 and write the answer, 3, above the 9.

The quotient is 423.

If the "mental" division does not come out even, each remainder is used in the next division.

$$\begin{array}{r} 4\ 5\ 2 \text{ R } 2 \\ 3\overline{)1\ 3\ {}^{1}5\ 8} \end{array}$$

$13 \div 3 = 4 \text{ R } 1$. Write the 4 above the 3 in the dividend. Now form a new number, 15, using the remainder 1 and the next digit 5. Divide 3 into 15. Write the answer, 5, above 5 in the dividend. Because there is no remainder, divide the next digit, 8, by 3. The result is 2 R 2. Write this above the 8.

The quotient is 452 R 2.

EXAMPLES A–E

DIRECTIONS: Divide and check.

STRATEGY: Divide from left to right. Use short division for single-digit divisors.

A. $8\overline{)5624}$

STRATEGY: Because there is a single-digit divisor, we use short division.

$$\begin{array}{r} 703 \\ 8\overline{)5624} \end{array}$$

8 divides 56 seven times. 8 divides 2 zero times with a remainder of 2. Now form a new number, 24, using the remainder and the next number, 4. 8 divides 24 three times.

WARM-UPS A–E

A. $9\overline{)6318}$

Answers to Warm-Ups

A. 702

CAUTION

A zero must be placed in the quotient so that the 7 and the 3 have the correct place values.

The quotient is 703.

B. Divide: $13\overline{)3133}$

B. Divide: $23\overline{)2898}$

STRATEGY: Write the partial quotients above the dividend with the place values aligned.

```
     126
23)2898
   23        23(1) = 23
    59
    46       23(2) = 46
    138
    138      23(6) = 138
      0
```

CHECK:

```
   126
×   23
   378
  2520
  2898
```

The quotient is 126.

C. Find the quotient: $\frac{233{,}781}{482}$

C. Find the quotient: $\frac{127{,}257}{482}$

STRATEGY: When a division is written as a fraction, the dividend is above the fraction bar and the divisor is below.

```
          264
482)127,257     482 does not divide 1.
     96 4       482 does not divide 12.
     30 85      482 does not divide 127.
     28 92      482 divides 1272 two times.
      1 937     482 divides 3085 six times.
      1 928     482 divides 1937 four times.
          9     The remainder is 9.
```

CHECK: Multiply the divisor by the partial quotient and add the remainder.

$$\begin{aligned} 264(482) + 9 &= 127{,}248 + 9 \\ &= 127{,}257 \end{aligned}$$

The answer is 264 with a remainder of 9 or 264 R 9.

You may recall other ways to write a remainder using fractions or decimals. These are covered in a later chapter.

CALCULATOR EXAMPLE

D. Divide 45,682 by 415.

D. Divide 54,762 by 213.

Enter the division: 54,762 ÷ 213

54,762 ÷ 213 ≈ 257.09859

The quotient is not a whole number. This means that 257 is the partial quotient and there is a remainder. To find the remainder, multiply 257 times 213.

Answers to Warm-Ups

B. 241 **C.** 485 R 11 **D.** 110 R 32

Subtract the product from 54,762. The result is the remainder.

54,762 − 257(213) = 21

So 54,762 ÷ 213 = 257 R 21.

E. When planting Christmas trees, the Greenfir Tree Farm allows 64 square feet per tree. If there are 43,520 square feet in 1 acre, how many trees will the tree farm plant per acre?

STRATEGY: Because each tree is allowed 64 square feet, we divide the number of square feet in 1 acre by 64 to find out how many trees will be planted in 1 acre.

```
     680
64)43520
   384
    512
    512
     00
      0
      0
```

A total of 680 trees will be planted per acre.

E. The Greenfir Tree Farm allows 256 square feet per large spruce tree. If there are 43,520 square feet in 1 acre, how many trees will the tree farm plant per acre?

HOW & WHY

OBJECTIVE 2 Estimate the quotient of whole numbers.

The quotient of two numbers can be estimated by rounding the divisor and the dividend to specified place values and then dividing the rounded numbers. For instance,

27)6345   30)6300   Round the divisor to the nearest ten and the dividend to the nearest hundred.

```
   210
30)6300
   6000
    300
    300
```

Divide.

The estimate of the quotient is 210. One use of the estimate is to see if the quotient is correct. If the calculated quotient is not close to 210, you should check the division. In this case, the actual quotient is 235, which is close to the estimate.

The estimated quotient will not always come out even. When this happens, divide to find the first two digits of the quotient and then write zeros to complete the estimated quotient. For instance,

628)247678

600)250000   Round the divisor to the nearest hundred and the dividend to the nearest ten thousand.

```
    410
600)250000
    2400
     1000
      600
```

Divide to find the first two digits in the quotient.
Write a zero to complete the estimated quotient.

The estimated quotient is 410.
The estimate, 410, is close to the actual quotient, 394 R 246.

Answers to Warm-Ups

E. The tree farm will plant 170 trees per acre.

To estimate the quotient of two whole numbers

1. Round the divisor to one digit and the dividend to two digits.
2. Divide the rounded numbers.
3. If the first partial quotient has a remainder, continue the division so the quotient has two digits and write zeros to complete the estimated quotient.

WARM-UP F

F. 78,624 ÷ 416

EXAMPLE F

DIRECTIONS: Estimate the quotient and divide.

STRATEGY: Round the divisor to one digit and the dividend to two digits and then divide the rounded numbers. If the first partial quotient has a remainder, continue the division so the quotient has two digits, and write zeros to complete the estimated quotient.

F. 58,590 ÷ 188

$188\overline{)58590}$ $\quad 200\overline{)59000}$ Round the divisor to the nearest hundred and dividend to the nearest thousand.

```
   290
200)59000
    400
    1900
    1800
```

Divide to two places and write a zero to complete the estimate.

```
     311
188)58590
    564
     219
     188
      310
      188
      122
```

The quotient is estimated at 290 and is 311 R 122.

Answer to Warm-Up

F. 190; 189

EXERCISES 1.4

OBJECTIVE 1 Divide whole numbers.

Divide.

A.

1. $8\overline{)56}$ **2.** $12\overline{)96}$ **3.** $9\overline{)45}$ **4.** $5\overline{)60}$

5. $5\overline{)535}$ **6.** $3\overline{)126}$ **7.** $5\overline{)305}$ **8.** $9\overline{)189}$

9. $144 \div 9$ **10.** $252 \div 6$ **11.** $680 \div 17$ **12.** $900 \div 18$

13. $497 \div 7$ **14.** $639 \div 3$ **15.** $21 \div 8$ **16.** $23 \div 5$

17. The division has a remainder when the last difference in the division is smaller than the ____________ and is not zero.

18. For $2600 \div 13$, in the partial division $26 \div 13 = 2$, 2 has place value ____________.

B.

19. $16{,}248 \div 4$ **20.** $12{,}324 \div 6$ **21.** $\frac{768}{24}$ **22.** $\frac{664}{83}$

23. $34\overline{)884}$ **24.** $54\overline{)4212}$ **25.** $26\overline{)2496}$ **26.** $37\overline{)2664}$

27. $234\overline{)11{,}934}$ **28.** $322\overline{)24{,}794}$ **29.** $47\overline{)8143}$ **30.** $34\overline{)9822}$

31. $47\overline{)675}$ **32.** $31\overline{)456}$ **33.** $(62)(?) = 3596$ **34.** $(?)(73) = 2555$

35. $57{,}224 \div 56$ Round quotient to the nearest ten.

36. $31{,}000 \div 55$ Round quotient to the nearest hundred.

OBJECTIVE 2 Estimate the quotient of whole numbers.

A. *Estimate the quotient.*

37. $625 \div 57$ **38.** $789 \div 29$ **39.** $4500 \div 53$ **40.** $6328 \div 57$

41. $610\overline{)34{,}560}$ **42.** $459\overline{)55{,}923}$ **43.** $\frac{44{,}842}{815}$ **44.** $\frac{83{,}695}{169}$

B. *Estimate the quotient and divide.*

45. $83\overline{)30{,}046}$ **46.** $34\overline{)6290}$ **47.** $17{,}121 \div 39$

48. $52{,}812 \div 81$ **49.** $187\overline{)81{,}012}$ **50.** $115\overline{)64{,}891}$

51. $780{,}854 \div 436$ **52.** $560{,}999 \div 356$

C.

53. Estimate the quotient and divide: $7{,}224{,}135 \div 382$.

54. Estimate the quotient and divide: $6{,}089{,}477 \div 219$.

Exercises 55–58. The revenue department of a central state had the following collection data for the first 3 weeks of March.

Taxes Collected

Number of Returns	Total Taxes Paid
Week 1: 4563	$24,986,988
Week 2: 3981	$19,315,812
Week 3: 11,765	$48,660,040

55. Estimate the taxes paid per return during week 2.

56. Find the actual taxes paid per return during week 1.

57. Find the taxes paid per return during week 3, rounded to the nearest hundred dollars.

58. Find the taxes paid per return during the 3 weeks, rounded to the nearest hundred dollars.

59. Estimate the quotient and divide: $\frac{209{,}458}{1195}$

60. Estimate the quotient and divide: $\frac{1{,}893{,}451}{4016}$

61. A forestry survey finds that 1664 trees are ready to harvest on a 13-acre plot. On average, how many trees are ready to harvest per acre?

62. Rosebud Lumber Company replants 5696 seedling fir trees on a 16-acre plot of logged-over land. What is the average number of seedlings planted per acre?

63. It takes the Morris Packing Plant 8 hours to process 12 tons of lima beans. How long does it take the plant to process 780 tons of lima beans?

64. It takes the Pacific Packing Plant 6 hours to process 4 tons of Dungeness crab. How long does it take the plant to process 392 tons of Dungeness crab?

65. In 2001, Bill Gates of Microsoft was the richest person in the United States, with an estimated net worth of $54 billion. How much would you have to spend per day in order to spend all of Bill Gates's $54 billion in 72 years, rounded to the nearest thousand?

66. How much money would you have to spend per day in order to spend Bill Gates's $54 billion in 50 years, rounded to the nearest hundred? In 20 years, rounded to the nearest hundred?

Exercises 67–69. Use the estimated population in 1999 and the area of the country as shown in the table.

Population in 1999

Country	Estimated Population	Area (in square miles)
China	1,246,871,951	3,691,521
Republic of Mali	10,429,124	478,819
United States	272,171,813	3,761,363

67. What was the population density (people per square mile) of China in 1999, rounded to the nearest whole person?

68. What was the population density (people per square mile) of the Republic of Mali, rounded to the nearest whole person?

69. What was the population density (people per square mile) of the United States, rounded to the nearest whole person?

Exercises 70–71. According to Consumer Reports, *Americans have 59 million cats, 56 million fish, 53 million dogs, 13 million birds, 6 million rabbits and ferrets, 5 million rodents, and 4 million reptiles.*

70. Americans spend about $23 billion on their pets per year. How much do they spend on average per pet, rounded to the nearest dollar?

71. There are about 70,880,000 households in the United States. What is the average number of pets per household, rounded to the nearest whole pet?

Exercises 72–74. The 2000 census population and the number of House of Representative seats in the United States and in two states is given in the table.

Population and House Representation

	Population	Number of House Seats
United States	272,171,813	435
California	33,145,121	53
Montana	882,779	2

72. How many people does each House member represent in the United States?

73. How many people does each representative from California represent?

74. How many people does each representative from Montana represent?

75. In 2000, the estimated population of Florida was 15,982,378 and the gross state product was $418,900,000,000. What was the state product per person, rounded to the nearest dollar?

76. In 2000, the estimated population of Kansas was 2,688,418 and the total personal income for the state was about $71,600,000,000. What was the per capita income, rounded to the nearest dollar?

Exercises 77–78. A bag of rice crackers contains 8 servings, a total of 900 calories, and 1500 mg of sodium.

77. How many calories are there per serving?

78. How many milligrams of sodium are there per serving?

79. Juan is advised by his doctor not to exceed 2700 mg of aspirin per day for his arthritic pain. If he takes capsules containing 325 mg of aspirin, how many capsules can he take without exceeding the doctor's orders?

Exercises 80–82 refer to the chapter application. See Table 1.1, page 1.

80. There are 16 games in a season. How much did the Washington Redskins spend per game for their payroll, rounded to the nearest thousand dollars?

81. In the 1999–2000 season, the St. Louis Rams played a total of 19 games. What was their payroll per game, rounded to the nearest thousand dollars?

82. In 2000–2001, the Baltimore Ravens had a roster of 60 players. What was the average salary, rounded to the nearest thousand dollars?

83. Jerry Rice of the San Francisco 49ers holds the Superbowl record for most pass receptions. In a 1989 game, he had 11 receptions, for a total of 215 yards. What was the average yardage per reception, rounded to the nearest whole yard?

STATE YOUR UNDERSTANDING

84. Explain to an 8-year-old child that $45 \div 9 = 5$.

85. Explain the concept of remainder.

86. Define and give an example of a quotient.

CHALLENGE

87. The Belgium Bulb Company has 171,000 tulip bulbs to market. Eight bulbs are put in a package when shipping to the United States and sold for $3 per package. Twelve bulbs are put in a package when shipping to France and sold for $5 per package. In which country will the Belgium Bulb Company get the greatest gross return? What is the difference in gross receipts?

Complete the problems by writing in the correct digit wherever you see a letter.

88. $3\overline{)1653C}$ with quotient 5AB2

89. $A3\overline{)4CC1}$ with quotient 21B

GROUP WORK

90. Divide 23,000,000 and 140,000,000 by 10, 100, 1000, 10,000, and 100,000. What do you observe? Can you devise a rule for dividing by 10, 100, 1000, 10,000, and 100,000?

Getting Ready for Algebra

HOW & WHY

OBJECTIVE

Solve an equation of the form $ax = b$ or $\frac{x}{a} = b$, where x, a, and b are whole numbers.

In Section 1.2, the equations involved the inverse operations addition and subtraction. Multiplication and division are also inverse operations. We can use this idea to solve equations containing those operations.

For example, if 4 is multiplied by 2, $4 \cdot 2 = 8$, the product is 8. If the product is divided by 2, $8 \div 2$, the result is 4, the original number. In the same manner, if 12 is divided by 3, $12 \div 3 = 4$, the quotient is 4. If the quotient is multiplied by 3, $4 \cdot 3 = 12$, the original number. We use this idea to solve equations in which the variable is either multiplied or divided by a number.

When a variable is multiplied or divided by a number, the multiplication symbols ($\cdot$ or $\times$) and the division symbol ($\div$) normally are not written. We write $3x$ for 3 times x and $\frac{x}{3}$ for x divided by 3.

Consider the following:

$$5x = 30$$

$$\frac{5x}{5} = \frac{30}{5} \quad \text{Division will eliminate multiplication.}$$

$$x = 6$$

If x in the original equation is replaced by 6, we have

$$5x = 30$$

$$5 \cdot 6 = 30$$

$$30 = 30 \quad \text{A true statement}$$

Therefore, the solution is 6.

If the variable is divided by a number

$$\frac{x}{7} = 21$$

$$7 \cdot \frac{x}{7} = 7 \cdot 21 \quad \text{Multiplication will eliminate division.}$$

$$x = 147$$

If x in the original equation is replaced by 147, we have

$$\frac{147}{7} = 21$$

$$21 = 21 \quad \text{A true statement}$$

Therefore, the solution is 147.

To solve an equation using multiplication or division

1. Divide both sides by the same number to isolate the variable.
2. Or multiply both sides by the same number to isolate the variable.
3. Check the solution by substituting it for the variable in the original equation.

WARM-UPS A–E

A. $6y = 18$

B. $\frac{a}{5} = 10$

C. $\frac{b}{3} = 33$

D. $8t = 96$

Answers to Warm-Ups

A. $y = 3$ **B.** $a = 50$ **C.** $b = 99$ **D.** $t = 12$

EXAMPLES A–E

DIRECTIONS: Solve and check.

STRATEGY: Isolate the variable by multiplying or dividing both sides of the equation by the same number. Check the solution by substituting it for the variable in the original equation.

A. $3x = 24$

$3x = 24$

$\frac{3x}{3} = \frac{24}{3}$ Isolate the variable by dividing both sides of the equation by 3.

$x = 8$ Simplify.

CHECK: $3x = 24$

$3(8) = 24$ Substitute 8 for x in the original equation.

$24 = 24$ The statement is true.

The solution is $x = 8$.

B. $\frac{x}{4} = 9$

$\frac{x}{4} = 9$

$4 \cdot \frac{x}{4} = 4 \cdot 9$ Isolate the variable by multiplying both sides by 4.

$x = 36$ Simplify.

CHECK: $\frac{x}{4} = 9$

$\frac{36}{4} = 9$ Substitute 36 for x in the original equation.

$9 = 9$ The statement is true.

The solution is $x = 36$.

C. $\frac{c}{7} = 12$

$\frac{c}{7} = 12$

$7 \cdot \frac{c}{7} = 7 \cdot 12$ Isolate the variable by multiplying both sides of the equation by 7.

$c = 84$ Simplify.

CHECK: $\frac{c}{7} = 12$

$\frac{84}{7} = 12$ Substitute 84 for c in the original equation.

$12 = 12$ The statement is true.

The solution is $c = 84$.

D. $9y = 117$

$9y = 117$

$\frac{9y}{9} = \frac{117}{9}$ Isolate the variable by dividing both sides of the equation by 9.

$y = 13$ Simplify.

CHECK:

$$9y = 117$$
$$9(13) = 117 \quad \text{Substitute 13 for } y \text{ in the original equation.}$$
$$117 = 117 \quad \text{The statement is true.}$$

The solution is $y = 13$.

E. What is the width (w) of a rectangular lot in a subdivision if the length (ℓ) is 125 feet and and the area (A) is 9375 square feet? See the figure. Use the formula $A = \ell w$.

STRATEGY: To find the width of the lot, substitute the area, $A = 9375$, and the length, $\ell = 125$, into the formula and solve.

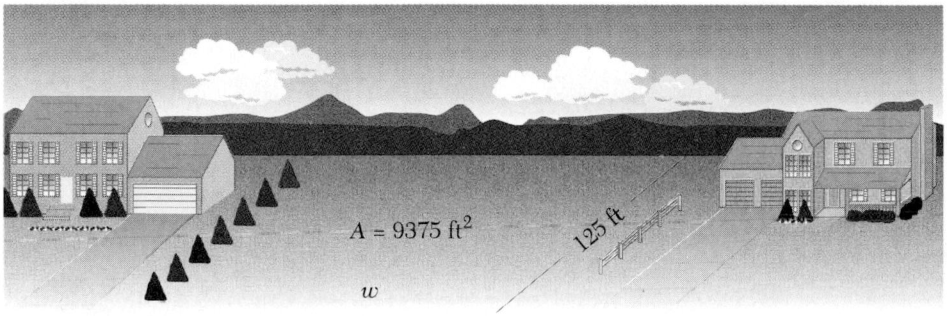

$$A = \ell w$$
$$9375 = 125w \quad A = 9375, \ell = 125$$
$$\frac{9375}{125} = \frac{125w}{125} \quad \text{Divide both sides by 125.}$$
$$75 = w$$

CHECK: If the width is 75 feet and the length is 125 feet, is the area 9375 square feet?

$A = (125 \text{ ft})(75 \text{ ft}) = 9375 \text{ sq ft}$ True.

The width of the lot is 75 feet.

E. What is the length (ℓ) of a second lot in the subdivision if the width (w) is 90 feet and the area (A) is 10,350 square feet? Use the formula $A = \ell w$.

Answers to Warm-Ups

E. The length of the lot is 115 ft.

EXERCISES

OBJECTIVE Solve an equation of the form $ax = b$ or $\frac{x}{a} = b$, where x, a, and b are whole numbers.

Solve and check.

1. $3x = 15$

2. $\frac{z}{4} = 5$

3. $\frac{c}{3} = 6$

4. $8x = 32$

5. $13x = 52$

6. $\frac{y}{4} = 14$

7. $\frac{b}{2} = 23$

8. $15a = 135$

9. $12x = 144$

10. $\frac{x}{14} = 12$

11. $\frac{y}{13} = 24$

12. $23c = 184$

13. $27x = 648$

14. $\frac{a}{32} = 1216$

15. $\frac{b}{12} = 2034$

16. $57z = 2451$

17. $1098 = 18x$

18. $616 = 11y$

19. $34 = \frac{w}{23}$

20. $64 = \frac{c}{33}$

21. Find the width of a rectangular garden plot that has a length of 35 feet and an area of 595 square feet. Use the formula $A = \ell w$.

22. Find the length of a room that has an area of 391 square feet and a width of 17 feet.

23. Crab sells at the dock for \$2 per pound. A fisherman sells his catch and receives \$4680. How many pounds of crab does he sell?

24. Felicia earns \$7 an hour. Last week she earned \$231. How many hours did she work last week?

25. If the wholesale cost of 18 stereo sets is \$5580, what is the wholesale cost of one set? Use the formula $C = np$, where C is the total cost, n is the number of units purchased, and p is the price per unit.

26. Using the formula in Exercise 25, if the wholesale cost of 24 personal computers is \$18,864, what is the wholesale cost of one computer?

27. The average daily low temperature in Toronto in July is twice the average high temperature in January. Write an equation that describes this relationship. Be sure to define all variables in your equation. If the average daily low temperature in July is 60°F, what is the average daily high temperature in January?

28. Car manufactures recommend that the fuel filter in a car be replaced when the mileage is ten times the recommended mileage for an oil change. Write an equation that describes this relationship. Be sure to define all variables in your equation. If a fuel filter should be replaced every 30,000 miles, how often should the oil be changed?

1.5 Whole Number Exponents and Powers of 10

VOCABULARY

A **base** is a number used as a repeated factor. An **exponent** indicates the number of times the base is used as a factor and is always written as a superscript to the base. In 2^3, 2 is the base and 3 is the exponent.

The **value** of 2^3 is 8.

An exponent of 2 is often read **"squared"** and an exponent of 3 is often read **"cubed."**

A **power of 10** is the value obtained when 10 is written with an exponent.

OBJECTIVES

1. Find the value of an expression written in exponential form.
2. Multiply or divide a whole number by a power of 10.

HOW & WHY

OBJECTIVE 1 Find the value of an expression written in exponential form.

Exponents show repeated multiplication. Whole number *exponents* greater than 1 are used to write repeated multiplications in shorter form. For example,

5^4 means $5 \cdot 5 \cdot 5 \cdot 5$

and because $5 \cdot 5 \cdot 5 \cdot 5 = 625$, we write $5^4 = 625$. The number 625 is sometimes called the "fourth power of five" or "the *value* of 5^4."

EXPONENT
↓
BASE → $5^4 = 625$ ← VALUE

Similarly, the value of 7^6 is

$7^6 = 7 \cdot 7 \cdot 7 \cdot 7 \cdot 7 \cdot 7 = 117{,}649$

The base, the repeated factor, is 7. The exponent, which indicates the number of times the base is used as a factor, is 6.

The exponent 1 is a special case.

In general, $x^1 = x$. So $2^1 = 2$, $13^1 = 13$, $7^1 = 7$, and $(413)^1 = 413$.

We can see a reason for the meaning of 6^1 ($6^1 = 6$) by studying the following pattern:

$6^4 = 6 \cdot 6 \cdot 6 \cdot 6$
$6^3 = 6 \cdot 6 \cdot 6$
$6^2 = 6 \cdot 6$
$6^1 = 6$

Exponential Property

If 1 is used as an exponent, the value is equal to the base.

To find the value of an expression with a whole number exponent

1. If the exponent is 1, the value is the same as the base.
2. If the exponent is greater than 1, use the base number as a factor as many times as shown by the exponent. Multiply.

Exponents give us a second way to write an area measurement. Using exponents we can write 74 square inches as 74 in.^{2}. The symbol 74 in.^{2} is still read "seventy-four square inches." Also, 65 square feet is written as 65 ft^2.

WARM-UPS A–F

A. Find the value of 12^2.

B. Simplify: 56^1

C. Find the value of 10^5.

D. Evaluate: 7^3

E. Find the value of 4^{11}.

F. Herman's vegetable garden covers a total of 700 square feet. Write the area of his garden, using an exponent to show the measure.

Answers to Warm-Ups

A. 144 **B.** 56 **C.** 100,000 **D.** 343 **E.** 4,194,304 **F.** The area of Herman's garden is 700 ft^2.

EXAMPLES A–F

DIRECTIONS: Find the value.

STRATEGY: Identify the exponent. If it is 1, the value is the base number. If it is greater than 1, use it to tell how many times the base is used as a factor and then multiply.

A. Find the value of 8^3.

$8^3 = 8 \cdot 8 \cdot 8$ Use 8 as a factor three times.
$= 512$

The value is 512.

B. Simplify: 32^1

$32^1 = 32$ If the exponent is 1, the value is the base number.

The value is 32.

C. Find the value of 10^8.

$10^8 = (10)(10)(10)(10)(10)(10)(10)(10)$
$= 100{,}000{,}000$ One hundred million
Note that the value has eight zeros.

The value is 100,000,000.

D. Evaluate: 6^4

$6^4 = (6)(6)(6)(6) = 1296$

The value is 1296.

CALCULATOR EXAMPLE

E. Find the value of 8^7.

Calculators usually have an exponent key, marked [y^x] or [^]. If your calculator doesn't have such a key, you can multiply the repeated factors.

The value is 2,097,152.

F. Mabel needs 125 square yards of carpet to carpet her house wall to wall. Write the amount of carpet needed, using an exponent to show the measure.

1254 square yards = 125 yd^2 Square yards is written yd^2.

Mabel needs 125 yd^2 of carpet.

HOW & WHY

OBJECTIVE 2 Multiply or divide a whole number by a power of 10.

It is particularly easy to multiply or divide a whole number by a power of 10. Consider the following and their products when multiplied by 10.

$3 \times 10 = 30$ $8 \times 10 = 80$ $15 \times 10 = 150$ $77 \times 10 = 770$

The place value of every digit becomes ten times larger when the number is multiplied by 10. So, to multiply by 10, we need to merely write a zero on the right of the whole number. If a whole number is multiplied by 10 more than once, a zero is written on the right for each 10. So

$37 \times 10^3 = 37{,}000$ Three zeros are written on the right, one for each 10.

Because division is the inverse of multiplication, dividing by 10 will eliminate the last zero on the right of a whole number. So,

590,000 ÷ 10 = 59,000 Eliminate the final zero on the right.

If we divide by 10 more than once, one zero is eliminated for each 10. So,

590,000 ÷ 10^3 = 590 Eliminate three zeros.

To multiply a whole number by a power of 10

1. Identify the exponent of 10.
2. Write as many zeros to the right of the whole number as the exponent of 10.

To divide a whole number by a power of 10

1. Identify the exponent of 10.
2. Eliminate the same number of zeros on the right of the whole number as the exponent of 10.

Using powers of 10, we have a third way of writing a whole number in expanded form.

$$\begin{aligned}7623 &= 7000 + 600 + 20 + 3, \text{ or}\\ &= 7 \text{ thousands} + 6 \text{ hundreds} + 2 \text{ tens} + 3 \text{ ones, or}\\ &= (7 \times 10^3) + (6 \times 10^2) + (2 \times 10^1) + (3 \times 1)\end{aligned}$$

EXAMPLES G–K

DIRECTIONS: Multiply or divide.

STRATEGY: Identify the exponent of 10. For multiplication, write the same number of zeros on the right of the whole number as the exponent of 10. For division, eliminate the same number of zeros on the right of the whole number as the exponent of 10.

G. Multiply: 33,456 × 10^4

33,456 × 10^4 = 334,560,000 The exponent of 10 is 4. To multiply, write 4 zeros on the right of the whole number.

The product is 334,560,000.

H. Simplify: 834 × 10^3

834 × 10^3 = 834,000

The product is 834,000.

I. Divide: $\frac{68,000}{10^3}$

$\frac{68,000}{10^3} = 68$ The exponent of 10 is 3. To divide, eliminate 3 zeros on the right of the whole number.

The quotient is 68.

J. Simplify: 803,200,000 ÷ 10^5

803,200,000 ÷ 10^5 = 8032 Eliminate 5 zeros on the right.

The quotient is 8032.

WARM-UPS G–K

G. Multiply: 4712 × 10^6

H. Simplify: 98 × 10^5

I. Divide: $\frac{15,770,000}{10^4}$

J. Simplify: 1,450,000 ÷ 10^2

Answers to Warm-Ups

G. 4,172,000,000 **H.** 9,800,000
I. 1577 **J.** 14,500

K. A survey of 10,000 (10^4) people indicated that they pay an average of \$5781 in federal taxes. What was the total paid in taxes?

K. A recent fund-raising campaign raised an average of \$236 per donor. How much was raised if there were 100,000 (10^5) donors?

STRATEGY: To find the total raised, multiply the average donation by the number of donors.

$236 \times 10^5 = 23{,}600{,}000$ The exponent is 5. To multiply, write 5 zeros on the right of the whole number.

The campaign raised \$23,600,000.

Answers to Warm-Ups

K. The total paid in taxes was \$57,810,000.

EXERCISES 1.5

OBJECTIVE 1 Find the value of an expression written in exponential form.

A. *Write in exponential form.*

1. (12)(12)(12)(12)(12)(12)(12)(12)

2. $43 \times 43 \times 43 \times 43 \times 43 \times 43 \times 43$

Find the value.

3. 9^2

4. 8^2

5. 2^3

6. 3^3

7. 11^2

8. 17^1

9. In $5^3 = 125$, 5 is the ________, 3 is the ________, and 125 is the ________.

10. In $6^4 = 1296$, 1296 is the ________, 6 is the ________, and 4 is the ________.

B. *Find the value.*

11. 6^3

12. 2^7

13. 18^2

14. 23^2

15. 10^4

16. 10^6

17. 9^3

18. 8^4

19. 3^8

20. 5^6

OBJECTIVE 2 Multiply or divide a whole number by a power of 10.

Multiply or divide.

A.

21. 37×10^2

22. 65×10^1

23. 11×10^4

24. 13×10^3

25. $1700 \div 10^2$

26. $3100 \div 10^2$

27. $730{,}000 \div 10^4$

28. $14{,}000 \div 10^1$

29. To multiply a number by a power of 10, write as many zeros to the right of the number as the ________ of 10.

30. To divide a number by a power of 10, eliminate as many ________ on the right of the number as the exponent of 10.

B.

31. 363×10^5

32. 419×10^3

33. $2{,}310{,}000 \div 10^4$

34. $72{,}000{,}000 \div 10^3$

35. 3591×10^4

36. 6711×10^3

37. $\dfrac{30{,}200}{100}$

38. $\dfrac{95{,}500}{10^2}$

39. 1310×10^8

40. 215×10^5

41. $820{,}000{,}000 \div 10^6$

42. $4{,}613{,}000{,}000 \div 10^5$

C.

43. Write in exponent form:
(10)(10)(10)(10)(10)(10)(10)(10)(10)(10)(10)(10)

44. Write in exponent form:
(4)(4)(4)(4)(4)(4)(4)(4)(4)(4)(4)(4)(4)(4)(4)(4)(4)

Find the value.

45. 15^4

46. 17^4

47. 9^9

48. 8^8

Multiply or divide.

49. $47{,}160 \times 10^9$

50. 630×10^{13}

51. $\frac{79{,}000{,}000{,}000}{10^7}$

52. $\frac{3{,}120{,}000{,}000{,}000{,}000}{10^8}$

53. Salvador bought a lot for his new house in a suburban subdivision. The lot consists of 12,000 square feet. Write the size of Salvador's lot, using an exponent to express the measure.

54. To make her niece's wedding gown, Marlene needs 25 square meters of material. Write the amount of material she needs, using an exponent to express the measure.

55. The operating budget of a community college is approximately 73×10^6 dollars. Write this amount in place value form.

56. A congressional committee proposes to decrease the national debt by 125×10^5 dollars. Write this amount in place value form.

57. The distance from Earth to the nearest star outside our solar system (Alpha Centauri) is approximately 255×10^{11} miles. Write this distance in place value form.

58. A high roller in Atlantic City places nine consecutive bets at the "Twenty-one" table. The first bet is \$5, and each succeeding bet is five times the one before. How much does she wager on the ninth bet? Express the answer as a power of 5 and as a whole number.

59. The distance that light travels in a year is called a light-year. This distance is almost 6 trillion miles. Write the place value name for this number. Write the number as 6 times a power of 10.

60. The average distance from Earth to the sun is approximately 93 million miles. Write the place value name for this distance and write it as 93 times a power of 10.

Exercises 61–63. A bacteria culture is doubling in size every hour. There is one bacterium at the start of the experiment.

61. How many bacteria are there after 4 hours? Write your answer in exponential form and also in place value notation.

62. How many bacteria are there after 24 hours? Write your answer in exponential form and also in place value notation.

63. How many hours will it take for the number of bacteria to exceed 1000?

64. The distance around Earth at the equator is about 25×10^3 miles. Write this number in place value notation.

65. The surface area of the Pacific Ocean is about 642×10^5 square miles. Write this number in place value notation.

66. The area of the three largest deserts in the world is given in the table.

Three Largest Deserts in the World

Desert	Sahara (N. Africa)	Arabian (Arabian Peninsula)	Gobi (Mongolia/ China)
Approximate Area (in square miles)	3,320,000	900,000	450,000

Write each area as the product of a whole number and a power of 10.

67. Disk space in a personal computer is commonly measured in gigabytes, which are billions of bytes. Write the number of bytes in 1 gigabyte as a power of 10.

68. A googol is 1 followed by 100 zeros. Legend has it that the name for this number was chosen by a 9-year-old nephew of mathematician Edward Kasner. Write a googol as a power of 10.

Exercises 69–72 refer to the chapter application. See Table 1.1, page 1.

69. Express the Pittsburgh Steelers' payroll as the product of a whole number and a power of 10.

70. Round the New England Patriots' payroll to the nearest million; then express this number as the product of a whole number and a power of 10.

71. What is the smallest power of 25 that is larger than the payroll for the Arizona Cardinals?

72. In the 2000–2001 season, the Kansas City Chiefs had a payroll of \$69,331,000. What power of 37 is approximately equal to their payroll?

STATE YOUR UNDERSTANDING

73. Explain what is meant by 4^{10}.

74. Explain how to multiply a whole number by a power of 10. Give at least two examples.

CHALLENGE

75. Mitchell's grandparents deposit \$3 on his first birthday and triple that amount on each succeeding birthday until he is 12. What amount did Mitchell's grandparents deposit on his twelfth birthday? What is the total amount they have deposited in the account?

76. Find the sum of the cubes of the digits.

77. Find the difference between the sum of the cubes of 4, 8, 11, and 23 and the sum of the fourth powers of 2, 5, 6 and 7.

78. The sun is estimated to weigh 2 octillion tons. Write the place value for this number. Write it as two times a power of 10.

GROUP WORK

79. Research your local newspaper and find at least five numbers that could be written using a power of 10. Write these as a product using a power of 10.

80. Have each member of your group work the following problem independently:

$$19 - 2 \cdot 6 - 5 + 7$$

Compare your answers. If they are different, be prepared to discuss in class why they are different.

1.6 Order of Operations

VOCABULARY

Parentheses ( ) and **brackets** [ ] are used in mathematics as **grouping** symbols. These symbols indicate that the operations inside are to be performed first. Other grouping symbols that are often used are **braces** { } and the **fraction bar** —.

OBJECTIVE

1. Perform any combination of operations on whole numbers.

HOW & WHY

OBJECTIVE 1 Perform any combination of operations on whole numbers.

Without a rule, it is possible to interpret $6 + 4 \cdot 11$ in two ways:

$$\begin{aligned} 6 + 4 \cdot 11 &= 6 + 44 \\ &= 50 \end{aligned}$$

or

$$\begin{aligned} 6 + 4 \cdot 11 &= 10 \cdot 11 \\ &= 110 \end{aligned}$$

In order to decide which answer to use, we agree to use a standard set of rules. Among these is the rule that we multiply before adding. So

$$6 + 4 \cdot 11 = 50$$

The order in which the operations are performed is important because the order often determines the answer. Therefore, there is an established *order of operations*. This established order was agreed upon many years ago and is built into most of today's calculators and computers.

Order of Operations

To evaluate an expression with more than one operation

1. Parentheses: Do the operations within grouping symbols first (parentheses, fraction bar, and so on) in the order given in steps 2, 3, and 4.
2. Exponents: Do the operations indicated by exponents.
3. Multiply and divide: Do only multiplication and division as they appear from left to right.
4. Add and subtract: Do addition and subtraction as they appear from left to right.

So we see that

$$\begin{aligned} 12 - 9 \div 3 &= 12 - 3 \\ &= 9 \end{aligned}$$ Divide first. Subtract.

$$\begin{aligned} (11 - 8)(6) &= 3(6) \\ &= 18 \end{aligned}$$ Subtract in parentheses first. Multiply.

$$\begin{aligned} 72 \div 8 \cdot 4 &= 9 \cdot 4 \\ &= 36 \end{aligned}$$ Neither multiplication nor division takes preference over the other. So do them from left to right.

As you can see, the rules for the order of operations are fairly complicated, and it is important that you learn them all. A standard memory trick is to use the first letters to make an easy-to-remember phrase:

Parentheses
Exponents
Multiplication/**D**ivision
Addition/**S**ubtraction

Consider the phrase: **P**lease **E**xcuse **M**y **D**ear **A**unt **S**ally. Note the first letters of the words in this phrase are exactly the same (and in the same order) as the first letters for the order of operations.

Many students use "Please excuse my dear Aunt Sally" to help them remember the order for operations. Why not give it a try?

CAUTION

Brackets [ ] and braces { } are often used differently in calculators. Consult your calculator manual.

Exercises involving all of the operations are shown in the examples.

WARM-UPS A–G

A. Simplify: $5 \cdot 3 + 6 \cdot 7$

B. Simplify: $4 \cdot 11 - 8 \div 4 + 5 \cdot 2$

C. Simplify: $36 \div 6 + 9 - 2(7 - 4)$

D. Simplify: $6 \cdot 4^2 - 3 \cdot 5^2 + 47 - 8 \cdot 4$

E. Simplify: $(4^3 - 120 \div 2)^2 + 8^2$

Answers to Warm-Ups

A. 57 **B.** 52 **C.** 9 **D.** 36 **E.** 80

EXAMPLES A–G

DIRECTIONS: Simplify.

STRATEGY: The operations are done in this order: operations in parentheses first, exponents next, then multiplication and division, and finally addition and subtraction.

A. Simplify: $7 \cdot 9 + 6 \cdot 2$

$7 \cdot 9 + 6 \cdot 2 = 63 + 12$ Multiply first.
$= 75$ Add.

B. Simplify: $25 - 6 \div 3 + 8 \cdot 4$

$25 - 6 \div 3 + 8 \cdot 4 = 25 - 2 + 32$ Divide and multiply.
$= 23 + 32$ Subtract.
$= 55$ Add.

C. Simplify: $5 \cdot 8 + 11 - 5(6 + 1)$

$5 \cdot 8 + 11 - 5(6 + 1) = 5 \cdot 8 + 11 - 5(7)$ Add in parentheses first.
$= 40 + 11 - 35$ Multiply.
$= 51 - 35$ Add.
$= 16$ Subtract.

D. Simplify: $2 \cdot 3^4 - 5 \cdot 2^2 + 31$

$2 \cdot 3^4 - 5 \cdot 2^2 + 31 = 2 \cdot 81 - 5 \cdot 4 + 31$ Do exponents first.
$= 162 - 20 + 31$ Multiply.
$= 142 + 31$ Subtract.
$= 173$ Add.

E. Simplify: $(5^2 - 7 \cdot 3)^2 + 7^2$

STRATEGY: First do the operations in the parentheses, following the proper order.

$(5^2 - 7 \cdot 3)^2 + 7^2 = (25 - 7 \cdot 3)^2 + 7^2$ Do the exponents first.
$= (25 - 21)^2 + 7^2$ Multiply.
$= (4)^2 + 7^2$ Subtract.

Now that the operations inside the parentheses are complete, continue, using the order of operations.

$= 16 + 49$ Do the exponents.
$= 65$ Add.

CALCULATOR EXAMPLE

F. Simplify: $2469 + 281 \cdot 11 - 3041$

Enter the numbers and operations as they appear from left to right.

The answer is 2519.

G. The Lend A Helping Hand Association prepares two types of food baskets for distribution to the needy. The family pack contains 12 cans of vegetables and the elderly pack contains 5 cans of vegetables. How many cans of vegetables are needed for 215 family packs and 65 elderly packs?

STRATEGY: To find the number of cans of vegetables needed for the packs, multiply the number of packs by the number of cans per pack. Then add the two amounts.

$215(12) + 65(5) = 2580 + 325$ Multiply.
$= 2905$ Add.

The Lend A Helping Hand Association needs 2905 cans of vegetables.

F. Simplify:
$4236 + 17 \cdot 584 \div 8$

G. The Fruit-of-the Month Club prepares two types of boxes for shipment. Box A contains 8 apples and Box B contains 18 apples. How many apples are needed for 85 orders of Box A and 115 orders of Box B?

Answers to Warm-Ups

F. 5477 **G.** The order requires 2750 apples.

EXERCISES 1.6

OBJECTIVE 1 Perform any combination of operations on whole numbers.

Simplify.

A.

1. $6 \cdot 7 + 11$ **2.** $19 + 6 \cdot 9$ **3.** $15 - 5 \cdot 3$ **4.** $28 \cdot 4 - 7$

5. $32 + 8 \div 4$ **6.** $48 \div 6 - 4$ **7.** $41 - (17 + 1)$ **8.** $(23 - 5) - 7$

9. $42 \div 7 \times 6$ **10.** $60 \div 6 \times 2$ **11.** $61 - 5 \cdot 4 - 2$ **12.** $25 - 3 \cdot 7 + 4$

13. $3^3 - 5 + 2^3$ **14.** $6^2 \div 9 + 4^2$ **15.** $4 \cdot 7 + 3 \cdot 5$ **16.** $6 \cdot 7 - 5 \cdot 4$

B.

17. $3^2 - 4 \cdot 2 + 5 \cdot 6$ **18.** $6^2 + 20 \div 5 + 2 \cdot 3$ **19.** $36 \div 9 + 8 - 5$

20. $56 \cdot 3 \div 14 + 4 - 6$ **21.** $(21 + 13) - (42 - 37)$ **22.** $(49 - 18) - (12 + 9)$

23. $64 \div 8 \cdot 2^3 + 6 \cdot 7$ **24.** $90 \div 15 \cdot 5^2 + 4 \cdot 9$ **25.** $96 \div 12 \cdot 3$

26. $100 \div 4 \cdot 5$ **27.** $93 - 3(21 - 12) + 15 - 23$

28. $35(41 - 33) \div 8 - 11 + 17$ **29.** $3^4 + 4^3$

30. $3^5 - 2^4 + 7^2$ **31.** $5 \cdot 3^3 - 72 \div 6 + 38$ **32.** $7 \cdot 2^3 - 45 \div 9 + 21$

C.

33. $50 - 12 \div 6 - 36 \div 6 + 3$ **34.** $80 - 24 \div 4 + 30 \div 6 + 4$

Exercises 35–38. The graph shows the count of ducks, by species, at a lake in northern Idaho, as taken by a chapter of Ducks Unlimited.

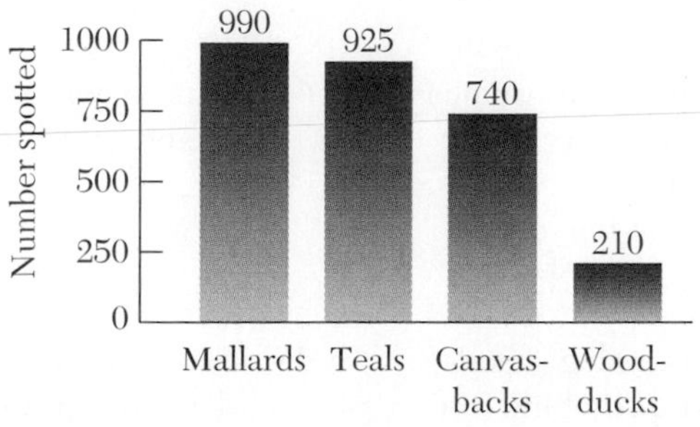

35. How many more mallards and canvasbacks were counted than teals and woodducks?

36. If twice as many canvasbacks had been counted, how many more canvasbacks would there have been than teals?

37. If four times the number of woodducks had been counted, how many more woodducks and mallards would there have been than teals and canvasbacks?

38. If twice the number of teals had been counted, how many more teals and woodducks would there have been compared with mallards and canvasbacks?

Simplify.

39. $9(2^3 \cdot 3 - 4) \div 4 + 13$

40. $13(3^3 \cdot 5 - 51) \div 7 - 42$

41. $4(8 - 3)^3 - 4^3$

42. $3(7 - 3)^3 - 8(3 - 1)^2$

43. At the beginning of a month, Mario's Hair Barn has an inventory of 76 bottles of shampoo at \$5 each, 65 bottles of conditioner at \$4 each, and 35 perms at \$22 each. At the end of the month, the shop has 18 bottles of shampoo, 9 bottles of conditioner, and 7 perms left. What is the cost of the supplies used for the month?

44. Juanita's Camera Shop ordered 40 cameras that cost \$346 each and 32 lenses at \$212 each. She sells the cameras for \$575 each and the lenses for \$316 each. If the shop sells 27 cameras and 19 lenses, how much net income is realized on the sale of the cameras and lenses?

45. A long-haul trucker is paid \$25 for every 100 miles driven and \$15 per stop. If he averages 2500 miles and 40 stops per week, what is his average weekly income? Assuming he takes 2 weeks off each year for vacation, what is his average yearly income?

Exercises 46–47. The labels for bread, cereal with milk, orange juice, and jam show their nutrition facts.

Grape Nuts

Nutrition Facts
Serving Size 1/2 cup (58g)
Servings per Container about 12

Amount Per Serving	Cereal	Cereal with 1/2 cup Skim milk
Calories	200	240
Calories from Fat	10	10
	% Daily Values**	
Total Fat 1g*	2%	2%
Saturated Fat 3g	0%	0%
Cholesterol 0mg	0%	0%
Sodium 350mg	15%	17%
Potassium 160mg	5%	10%
Total Carbohydrate 47g	16%	18%
Dietary Fiber 5g	21%	21%
Sugars 7g		
Other Carbohydrates 35g		
Protein 6g		

Orange Juice

Nutrition Facts
Serving Size 8 fl oz (240 mL)
Servings Per Container 8

Amount Per Serving	
Calories 110	Calories from Fat 0
	% Daily Value*
Total Fat 0 g	0%
Sodium 0mg	0%
Potassium 450mg	13%
Total Carbohydrate 26g	9%
Sugars 22g	
Protein 2g	
Vitamin C 120% •	Calcium 2%
Thiamin 10% •	Niacin 4%
Vitamin B6 6% •	Folate 15%

Not a significant source of saturated fat, cholesterol, dietary fiber, vitamin A and iron

*Percent Daily Values are based on a 2,000 calorie diet.

Jam

Nutrition Facts
Serving Size 1 Tbsp. (20g)
Servings per Container about 25

Amount Per Serving	
Calories 50	
	% Daily Value*
Total Fat 0g	0%
Sodium 15mg	1%
Total Carbohydrate 13g	4%
Sugars 13g	
Protein 0g	

*Percent Daily Values are based on a 2,000 calorie diet.

Bread

Nutrition Facts
Serving Size 1 slice (49g)
Servings Per Container 14

Amount Per Serving	
Calories 120	Calories from Fat 10
	% Daily Value*
Total Fat 1 g	2%
Saturated Fat 0g	0%
Cholesterol 0mg	0%
Sodium 360mg	15%
Total Carbohydrate 24g	8%
Dietary Fiber 1g	5%
Sugars 1g	
Protein 4g	
Vitamin A 0% •	Vitamin C 0%
Thiamin 10% •	Riboflavin 8%
Calcium 2% •	Iron 10%
Niacin 8% •	Folic Acid 10%

*Percent Daily Values are based on a 2,000 calorie diet. Your daily values may be higher or lower depending on your calorie needs.

46. How many milligrams (mg) of sodium are consumed if Marla has 3 servings of orange juice, 2 servings of cereal, and 2 slices of bread with jam for breakfast? Milk contains 62 mg of sodium per one-half cup serving.

47. How many calories does Marla consume when she eats the breakfast listed in Exercise 46?

48. Ruth orders flowers from her garden catalog. She orders two shrub roses for \$18 each, one climbing rose for \$15, an iron arbor for \$80, and three boxes of rose food for \$12 each. The shipping and handling charge for her order is \$20. What is the total charge for Ruth's order?

49. Sally orders birthday presents for her twin granddaughters from a catalog. She orders each girl a stuffed bear for \$19 each and a T-shirt for \$16 each. She also orders one of them a picture frame for \$14 and the other a puzzle for \$12. The shipping and handling charges for the order are \$12. What is the total charge for Sally's order?

50. Ron makes the following purchases at his local auto supply store: 4 spark plugs at \$1 each, a can of tire sealant for \$4, two heavy-duty shocks at \$10 each, a case (12 quarts) of Quaker State motor oil at \$1 per quart, and a gallon of antifreeze coolant for \$5. The sales tax for his purchases is \$3. Ron has a store credit of \$25. How much does he owe the store for this transaction?

51. Clay and Connie have a \$100 gift certificate to the Olive Garden. One night they order an appetizer for \$6, veal Parmesan for \$17, and fettuccine Alfredo for \$15. They each have a glass of wine at \$4 each, and they order desserts for \$6 each. How much do they have left on their gift certificate after the meal.

Exercises 52–55. To determine the intensity of cardiovascular training, a person needs to know her maximal heart rate (MHR) and resting heart rate (RHR). The MHR may be approximated by subtracting your age from 220. The RHR can be measured by sitting quietly for about 15 minutes, taking your pulse for 15 sec, and multiplying by 4. The difference between the MHR and the RHR is the heart rate reserve (HRR), the number of beats available to go from rest to all-out effort. To determine a target training rate, divide the HRR by 2 and add the results to the RHR.

52. Jessie is 24 years old and has a RHR of 68 beats per minute (bpm). Determine her cardiovascular target training rate.

53. June is 52 years old. She measures her resting heart rate as 18 beats in 15 sec. Determine her cardiovascular target training rate.

54. Determine your own cardiovascular target training rate.

55. As you age, does your target training rate increase or decrease? Why?

Exercises 56–58 refer to the chapter application. See Table 1.1, page 1.

56. Using figures for the 1999–2000 season, calculate the sum of the Patriots' and Titans' payrolls and divide by 2.

57. Again using the figures from the 1999–2000 season, calculate the sum of the Patriots' payroll and the Titans' payroll divided by 2.

58. Use the order of operations to explain why the answers to Exercise 56 and Exercise 57 are not the same.

STATE YOUR UNDERSTANDING

59. Which of the following is correct? Explain.

$$3 + 5 \cdot 4 = 8 \cdot 4 = 32 \quad \text{or} \quad 3 + 5 \cdot 4 = 3 + 20 = 23$$

60. Explain how to simplify $2(1 + 36 \div 3^2) - 3$ using order of operations.

CHALLENGE

61. Simplify: $(6 \cdot 3 - 8)^2 - 50 + 2 \cdot 3^2 + 2(9 - 5)^3$

62. USA Video buys three first-run movies: #1 for \$185, #2 for \$143, and #3 for \$198. During the first 2 months, #1 is rented 10 times at the weekend rate of \$5 and 26 times at the weekday rate of \$3; #2 is rented 12 times at the weekend rate and 30 times at the weekday rate; and #3 is rented 8 times at the weekend rate and 18 times at the weekday rate. How much money must still be raised to pay for the cost of the three videos?

GROUP WORK

63. It is estimated that hot water heaters need to be big enough to accommodate the water usage for an entire hour. To figure what size heater you need, first identify the single hour of the day in which water usage is highest. Next, identify the types of water usage during this hour. The estimates of water usage for various activities is shown in the table.

Water Usage

Activity	Gallons Used
Shower	20
Bath	20
Washing hair	4
Shaving	2
Washing hands/face	4
Dishwasher	14

Now calculate the total number of gallons of water used in your hour. Your water heater must have this capacity.

Getting Ready for Algebra

HOW & WHY

OBJECTIVE

Solve equations of the form $ax + b = c$, $ax - b = c$, $\frac{x}{a} + b = c$, and $\frac{x}{a} - b = c$, in which x, a, b, and c are whole numbers.

Recall that we have solved equations involving only one operation. Let's look at some equations that involve two operations.

To solve $x - 5 = 7$, we add 5 to both sides of the equation. To solve $4x = 16$, we divide both sides of the equation by 4. The following equation requires both steps.

$4x - 5 = 7$

$4x - 5 + 5 = 7 + 5$ First, eliminate the subtraction by adding 5 to both sides.

$4x = 12$ Simplify both sides.

$\frac{4x}{4} = \frac{12}{4}$ Eliminate the multiplication. Divide both sides by 4.

$x = 3$ Simplify.

CHECK: $4x - 5 = 7$

$4(3) - 5 = 7$ Substitute 3 for x.

$12 - 5 = 7$ Multiply.

$7 = 7$ Subtract.

Thus, if x is replaced by 3 in the original equation, the statement is true. So the solution is 3. Now solve

$5x + 11 = 46$

$5x + 11 - 11 = 46 - 11$ First, eliminate the addition by subtracting 11 from both sides.

$5x = 35$ Simplify.

$\frac{5x}{5} = \frac{35}{5}$ Eliminate the multiplication by dividing both sides by 5.

$x = 7$ Simplify.

CHECK: $5x + 11 = 46$

$5(7) + 11 = 46$ Substitute 7 for x in the original equation.

$35 + 11 = 46$ Multiply.

$46 = 46$ Add.

Thus, if x is replaced by 7 in the original equation, the statement is true. So the solution is 7.

Note that in each of the previous examples the operations are eliminated in the opposite order in which they are performed. That is, addition and subtraction were eliminated first and then the multiplication and division.

To solve an equation of the form
$ax + b = c$, $ax - b = c$, $\frac{x}{a} + b = c$, or $\frac{x}{a} - b = c$

1. Eliminate the addition or subtraction by subtracting or adding the same number to both sides.
2. Eliminate the multiplication or division by dividing or multiplying by the same number on both sides.
3. Check the solution by substituting it into the original equation.

WARM-UPS A–E

A. $6x - 11 = 55$

B. $\frac{a}{7} + 11 = 15$

C. $\frac{t}{5} - 11 = 14$

Answers to Warm-Ups

A. $x = 11$ **B.** $a = 28$ **C.** $t = 125$

EXAMPLES A–E

DIRECTIONS: Solve and check.

STRATEGY: Isolate the variable by first adding or subtracting the same number from both sides. Second, multiply or divide both sides by the same number.

A. $3x - 23 = 7$

$$3x - 23 = 7$$

$$3x - 23 + 23 = 7 + 23$$ Add 23 to both sides to eliminate the subtraction.

$$3x = 30$$ Simplify.

$$\frac{3x}{3} = \frac{30}{3}$$ Divide both sides by 3 to eliminate the multiplication.

$$x = 10$$ Simplify.

CHECK:

$$3x - 23 = 7$$

$$3(10) - 23 = 7$$ Substitute 10 for x in the original equation.

$$30 - 23 = 7$$ Simplify.

$$7 = 7$$ The statement is true.

The solution is $x = 10$.

B. $\frac{y}{9} + 3 = 13$

$$\frac{y}{9} + 3 = 13$$

$$\frac{y}{9} + 3 - 3 = 13 - 3$$ Subtract 3 from both sides to eliminate the addition.

$$\frac{y}{9} = 10$$ Simplify.

$$9\left(\frac{y}{9}\right) = 9(10)$$ Multiply both sides by 9 to eliminate the division.

$$y = 90$$

CHECK:

$$\frac{y}{9} + 3 = 13$$

$$\frac{90}{9} + 3 = 13$$ Substitute 90 for y in the original equation.

$$10 + 3 = 13$$ Simplify.

$$13 = 13$$ The statement is true.

The solution is $y = 90$.

C. $\frac{z}{4} - 8 = 14$

$$\frac{z}{4} - 8 + 8 = 14 + 8$$ Add 8 to both sides to eliminate the subtraction.

$$\frac{z}{4} = 22$$ Simplify.

$$4\left(\frac{z}{4}\right) = 4(22)$$ Multiply both sides by 4 to eliminate the division.

$$z = 88$$ Simplify.

CHECK: $\frac{z}{4} - 8 = 14$

$\frac{88}{4} - 8 = 14$ Substitute 88 for z in the original equation.

$22 - 8 = 14$ Simplify.

$14 = 14$ The statement is true.

The solution is $z = 88$.

D. $6b + 14 = 26$

$6b + 14 = 26$

$6b + 14 - 14 = 26 - 14$ Subtract 14 from both sides to to eliminate the addition.

$6b = 12$ Simplify.

$\frac{6b}{6} = \frac{12}{6}$ Divide both sides by 6 to eliminate the multiplication.

$b = 2$ Simplify.

CHECK: $6b + 14 = 26$

$6(2) + 14 = 26$ Substitute 2 for b in the original equation.

$12 + 14 = 26$ Simplify.

$26 = 26$ The statement is true.

The solution is $b = 2$.

E. The formula for the balance of a loan (D) is $D + NP = B$, where P represents the monthly payment, N represents the number of payments, and B represents the amount of money borrowed. Find the number of payments that have been made on an original loan of $875 with a current balance of $425 if the payment is $25 per month.

STRATEGY: Substitute the given values into the formula and solve.

$D + NP = B$

$425 + N(25) = 875$ Substitute 425 for D, 25 for P, and 875 for B.

$25N = 875 - 425$ Subtract 425 from each side.

$25N = 450$

$N = \frac{450}{25}$ Divide both sides by 25.

$N = 18$

CHECK: If 18 payments have been made, is the balance $425?

$875
− 450 18 payments of $25 is $450.
$425 True

Eighteen payments have been made.

D. $8c + 11 = 43$

E. Find the monthly payment on an original loan of $1155 if the balance after 14 payments is $385.

Answers to Warm-Ups

D. $c = 4$ **E.** The monthly payment is $55.

EXERCISES

OBJECTIVE Solve equations of the form $ax + b = c$, $ax - b = c$, $\frac{x}{a} + b = c$, and $\frac{x}{a} - b = c$, in which x, a, b, and c are whole numbers.

Solve and check.

1. $4x - 16 = 12$

2. $\frac{a}{4} + 9 = 13$

3. $\frac{y}{3} - 7 = 4$

4. $36 = 5x + 6$

5. $45 = 6x + 9$

6. $\frac{a}{9} + 5 = 10$

7. $\frac{c}{8} + 23 = 27$

8. $12x - 10 = 38$

9. $11x + 32 = 54$

10. $7y + 53 = 123$

11. $15c - 63 = 117$

12. $6 = \frac{w}{3} - 33$

13. $81 = \frac{a}{14} + 67$

14. $\frac{x}{21} + 92 = 115$

15. $673 = 45b - 272$

16. $804 = 43c + 30$

17. Fast-Tix charges \$43 per ticket for a rock concert plus an \$8 service charge. How many tickets did Remy buy if he was charged \$309? Use the formula $C = PN + S$, where C is the total cost, P is the price per ticket, N is the number of tickets purchased, and S is the service charge.

18. Ticket Master charges Jose \$253 for nine tickets to the Festival of Jazz. If the service charge is \$10, what is the price per ticket? Use the formula in Exercise 17.

19. Rana is paid \$40 per day plus \$8 per artificial flower arrangement she designs and completes. How many arrangements did she complete if she earned \$88 for the day? Use the formula $S = B + PN$, where S is the total salary earned, B is the base pay for the day, P is the pay per unit, and N is the number of units completed.

20. Rana's sister works at a drapery firm where the pay is \$50 per day plus \$12 per unit completed. How many units did she complete if she earned \$122 for the day?

Exercises 21–24. Several different long-distance calling plans are shown in the table.

Long-Distance Calling Plans

Company	Monthly Fee	Charge per Minute
AT&T	none	15¢
Tone	\$4.90	10¢
Pace	\$6.96	9¢

21. Jessica has \$30 budgeted for long-distance calls each month. Write an equation for the number of minutes she would get from AT&T. Let m be the number of minutes of long-distance calls per month. Let C represent the monthly bill in cents. Find the number of minutes that Jessica can purchase from AT&T each month.

22. Jessica has \$30 budgeted for long-distance calls each month. Write an equation for the number of minutes she would get from Tone. Let m be the number of minutes of long-distance calls per month. Let C represent the monthly bill in cents. Find the number of minutes that Jessica can purchase from Tone each month.

23. Jessica has \$30 budgeted for long-distance calls each month. Write an equation for the number of minutes she would get from Pace. Let *m* be the number of minutes of long-distance calls per month. Let C represent the monthly bill in cents. Find the number of minutes that Jessica can purchase from Pace each month.

24. Based on the results of Exercises 21–23, which company will give Jessica the most minutes for her \$30?

1.7 Average, Median, and Mode

VOCABULARY

The **average**, or **mean**, of a set of numbers is the sum of the set of numbers divided by the total number of numbers in the set.

The **median** of a set of numbers, ordered from smallest to largest, is either the middle number of the set or the average of the two middle numbers in the set.

The **mode** of a set of numbers is the number or numbers that appear the most often in the set.

OBJECTIVES

1. Find the average of a set of whole numbers.
2. Find the median of a set of whole numbers.
3. Find the mode of a set of whole numbers.

HOW & WHY

OBJECTIVE 1 Find the average of a set of whole numbers.

The *average*, or *mean*, of a set of numbers is used in statistics. It is one of the ways to find the middle of a set of numbers (like the average of a set of test grades). Mathematicians call the average or mean a "measure of central tendency." The average of a set of numbers is found by adding the numbers in the set and then dividing that sum by the number of numbers in the set. For example, to find the average of 8, 18, and 25:

$8 + 18 + 25 = 51$ Find the sum of the numbers in the set.
$51 \div 3 = 17$ Divide the sum by the number of numbers, 3.

The average is 17.

The "central" number, or average, does not need to be one of the members of the set. The average, 17, is not a member of the set.

To find the average of a set of whole numbers

1. Add the numbers.
2. Divide the sum by the number of numbers in the set.

EXAMPLES A–E

DIRECTIONS: Find the average.

STRATEGY: Add the numbers in the set. Divide the sum by the number of numbers in the set.

A. Find the average of 112, 89, and 153.

$112 + 89 + 153 = 354$ Add the numbers in the group.
$354 \div 3 = 118$ Divide the sum by the number of numbers.

The average is 118.

B. Find the average of 13, 87, 252, and 124.

$13 + 87 + 252 + 124 = 476$ Add the numbers in the group.
$476 \div 4 = 119$ Divide the sum by the number of numbers.

The average is 119.

WARM-UPS A–E

A. Find the average of 301, 142, and 499.

B. Find the average of 7, 46, 95, 38, and 89.

Answers to Warm-Ups
A. 314 **B.** 55

C. Find the average of 777, 888. 914, and 505.

CALCULATOR EXAMPLE

C. Find the average of 673, 821, 415, and 763.

Enter the sum, in parentheses, and divide by 4.

$(673 + 821 + 415 + 763) \div 4$

The average is 668.

D. The average of 23, 35, 11, 19, and ? is 20. Find the missing number.

D. The average of 32, 13, 15, and ? is 21. Find the missing number.

STRATEGY: Because the average of the four numbers is 21, we know that the sum of the four numbers is 4(21) or 84. To find the missing number, subtract the sum of the three given numbers from 84.

$$84 - (32 + 13 + 15) = 84 - (60) \\ = 24$$

So the missing number is 24.

E. The local cold storage company ships the following number of cases of frozen peas: Monday, 4018; Tuesday, 1475; Wednesday, 3650; Thursday, 2151; and Friday, 3216. What is the average number of cases shipped each day?

E. In order to help Pete lose weight, the dietician has him record his caloric intake for a week. He records the following: Monday, 3120; Tuesday, 1885; Wednesday, 1600; Thursday, 2466; Friday, 1434; Saturday, 1955; and Sunday, 2016. What is Pete's average caloric intake per day?

STRATEGY: Add the calories for each day and then divide by 7, the number of days.

```
  3120        2068
  1885     7)14476
  1600       14
  2466        4
  1434        0
  1955        47
  2016        42
------         56
14,476         56
                0
```

Pete's average caloric intake is 2068 calories per day.

HOW & WHY

OBJECTIVE 2 Find the median of a set of whole numbers.

Another measure of central tendency is called the *median*. The median of a set of numbers is a number such that half the numbers in the set are smaller than the median and half the numbers in the set are larger than the median. If the set is ordered from smallest to largest and there is an odd number of numbers in the set, the median is literally the middle number. Consider the set of numbers

3, 45, 47, 49, 51

The median is the middle number, 47.

If the set is ordered from smallest to largest and there is an even number of numbers in the set, the median is the average (mean) of the two middle numbers. Consider the set of numbers

4, 6, 9, 13, 15, 30

The median is the average of the middle numbers, 9 and 13.

$(9 + 13) \div 2 = 11$

The median is 11.

Answers to Warm-Ups

C. 771 **D.** 12

E. The company shipped an average of 2902 cases of frozen peas each day.

The median is used when a set of numbers has one or two numbers that are much larger or smaller than the others. In the set, 3, 45, 47, 49, and 51, the average of the set is 39, which is near the low end of the set. The median is the middle number, which in this case is 47, and may give us a better centerpoint.

To find the median of a set of numbers

1. List the numbers in order from smallest to largest.
2. If there is an odd number of numbers in the set, the median is the middle number.
3. If there is an even number of numbers in the set, the median is the average (mean) of the two middle numbers.

EXAMPLES F–G

DIRECTIONS: Find the median of the set of whole numbers.

STRATEGY: List the numbers from smallest to largest. If there is an odd number of numbers in the set, choose the middle number. If there is an even number of numbers in the set, find the average of the two middle numbers.

F. Find the median of 23, 45, 36, 17, 46, 64, and 70.

17, 23, 36, 45, 46, 64, 70 — List the numbers from smallest to largest.

The median is 45. — Because there is an odd number of numbers in the set, the median is the middle number.

G. Find the median of 38, 46, 65, 38, 19, 137, 71, and 3.

3, 19, 38, 38, 46, 65, 71, 137 — List the numbers from smallest to largest.

$$\frac{38 + 46}{2} = \frac{84}{2} = 42$$

Because there is an even number of numbers in the set, the median is the average of the two middle numbers, 38 and 46.

The median is 42.

WARM-UPS F–G

F. Find the median of 93, 56, 102, 54, and 20.

G. Find the median of 123, 140, 101, 95, 117, and 341.

HOW & WHY

OBJECTIVE 3 Find the mode of a set of whole numbers.

A third measure of central tendency is the *mode*. The mode is the number that occurs most often in the set. For example, consider

45, 67, 88, 88, 92, 100

The number that occurs most often is 88, so 88 is the mode.

Unlike the average and median, where there is just one per set of numbers, there can be more than one mode or no mode at all. See Examples I and J.

To find the mode of a set of numbers

1. Find the number or numbers that occur most often.
2. If all the numbers occur the same number of times, there is no mode.

Answers to Warm-Ups

F. 56 **G.** 120

WARM-UPS H–K

H. Find the mode of 64, 76, 80, 58, 80, and 91.

I. Find the mode of 6, 7, 11, 7, 9, 13, 11, and 14.

J. Find the mode of 1, 4, 6, 3, 11, 13, and 9.

K. In a class of 30 seniors, the following weights are recorded on Health Day:

Number of Students	Weight per Student
1	123 lb
3	132 lb
7	153 lb
4	175 lb
5	183 lb
5	191 lb
3	210 lb
2	215 lb

What are the average, median, and mode of the weight of a student in the class?

Answers to Warm-Ups

H. 80 **I.** 7 and 11

J. There is no mode.

K. The average weight of a student is 174 lb. The median weight is 179 lb. The mode is 153 lb.

EXAMPLES H–K

DIRECTIONS: Find the mode of the set of numbers.

STRATEGY: Find the number or numbers that occur most often. If all the numbers occur the same number of times, there is no mode.

H. Find the mode of 21, 13, 17, 13, 15, 21, and 13.

The mode is 13. The number 13 appears three times. No other number appears three times.

I. Find the mode of 22, 56, 72, 22, 81, 72, 93, and 105.

The modes are 22 and 72. Both 22 and 72 occur twice and the other numbers appear just once.

J. Find the mode of 13, 35, 16, 35, 16, and 13.

There is no mode. All the numbers occur the same number of times, 2. So no number appears most often.

K. During the annual Salmon Fishing Derby, 35 fish are entered. The weights of the fish are recorded as shown:

Number of Fish	Weight per Fish
2	6 lb
4	7 lb
8	10 lb
10	12 lb
5	15 lb
4	21 lb
1	22 lb
1	34 lb

What are the average, median, and mode of the weight of a fish entered in the Derby?

STRATEGY: To find the average, first find the total weight of all 35 fish.

$2(6) = 12$ Multiply 2 times 6 because there are 2 fish that weigh 6 lb, for a total of 12 lb, and so on.

$$\begin{aligned} 4(7) &= 28 \\ 8(10) &= 80 \\ 10(12) &= 120 \\ 5(15) &= 75 \\ 4(21) &= 84 \\ 1(22) &= 22 \\ 1(34) &= \underline{34} \\ & \quad 455 \end{aligned}$$

Now divide the total weight by the number of salmon, 35.

$455 \div 35 = 13$

The average weight per fish is 13 lb.

The median of the weight of fish is 12 lb. Because there are 35 fish, the middle, or 18th, weight is the median.

The mode of the weight of the fish is 12 lb. 12 lb occurs most often in the list.

EXERCISES 1.7

OBJECTIVE 1 Find the average of a set of whole numbers.

Find the average.

A.

1. 4, 8

2. 5, 13

3. 8, 10

4. 9, 13

5. 7, 13, 16

6. 7, 11, 15

7. 4, 8, 9, 11

8. 7, 7, 11, 11

9. 4, 6, 3, 3

10. 8, 3, 4, 5

11. 6, 7, 9, 3, 10

12. 5, 7, 3, 11, 4

Find the missing number to make the average correct.

13. The average of 8, 11, 13, and ? is 11.

14. The average of 9, 14, 18, and ? is 14.

B.

15. 18, 22, 36, 44

16. 17, 38, 43, 62

17. 25, 35, 45, 55

18. 18, 15, 29, 57, 46

19. 8, 11, 19, 28, 44, 76

20. 88, 139, 216, 133

21. 113, 123, 154, 122

22. 321, 699, 528, 616

23. 89, 140, 217, 134

24. 33, 132, 240, 279

25. 62, 75, 81, 133, 261, 90

26. 132, 65, 88, 74, 245, 200

Find the missing number.

27. The average of 34, 81, 52, 74, and ? is 59.

28. The average of 21, 29, 46, 95, 33, and ? is 47.

OBJECTIVE 2 Find the median of a set of whole numbers.

Find the median.

A.

29. 13, 56, 102

30. 34, 43, 57

31. 14, 20, 32, 40

32. 17, 21, 51, 82

33. 35, 72, 18, 47

34. 53, 12, 29, 80

B.

35. 36, 72, 91, 103, 112

36. 15, 23, 45, 67, 91

37. 102, 105, 113, 119, 123, 131

38. 14, 22, 14, 18, 25, 36, 55

39. 93, 55, 93, 55, 71, 106

40. 203, 214, 203, 233, 261, 275, 261, 280

OBJECTIVE 3 Find the mode of a set of whole numbers.

Find the mode.

A.

41. 12, 17, 17, 34, 41

42. 2, 4, 5, 5, 7, 8

43. 1, 1, 2, 2, 3, 3, 3, 6, 7, 7

44. 12, 20, 34, 56, 20, 14

45. 23, 45, 19, 36, 22

46. 17, 32, 63, 17, 18, 34, 12

B.

47. 25, 38, 56, 34, 38, 19, 38, 56

48. 76, 54, 82, 76, 54, 82, 91, 54

49. 32, 72, 47, 32, 55, 102, 47

50. 12, 16, 18, 13, 12, 17, 18, 19

51. 19, 23, 14, 14, 19, 23

52. 35, 27, 27, 55, 55, 35, 27, 19, 21

C.

53. A chapter of Ducks Unlimited counts the following number of ducks at a lake in northern Idaho: mallards, 990; teals, 924; canvasbacks, 740; woodducks, 210; and widgeons, 511. Find the average and the median number of ducks per species.

54. The following number of fish were counted at the Bonneville fish ladder during 1 week in July: coho, 148; shad, 356; silver salmon, 214; and sturgeon, 94. Find the average and median number of fish per species.

55. A golfer shoots the following scores for eleven rounds of golf: 84, 90, 103, 78, 91, 87, 75, 80, 78, 81, 77. What are the average, median, and mode scores per round?

56. A bowler has the following scores for 9 games: 251, 200, 210, 300, 195, 211, 271, 200, 205. What are the average, median, and mode scores per game?

57. A consumer magazine tests 18 makes of cars for gas mileage. The results are shown in the table.

New Car Gas Mileage

Number of Makes	Gas Mileage Based on 200 Miles
2	14 mpg
3	20 mpg
2	25 mpg
4	31 mpg
3	36 mpg
2	40 mpg
2	45 mpg

What are the average, median, and mode of the gas mileage of the cars?

58. A home economist lists the costs of 15 brands of canned fruit drinks.

Cost of Canned Fruit Drinks

Number of Brands	Price per Can
3	84¢
1	82¢
4	78¢
3	74¢
2	71¢
2	65¢

What are the average, median, and mode prices per can?

59. Twenty-five football players are weighed in on the first day of practice.

Weights of Football Players

Number of Players	Weight
1	167 lb
4	172 lb
3	187 lb
5	195 lb
4	206 lb
3	215 lb
3	225 lb
2	245 lb

What is the average weight of the players?

60. Eighty-two ladies at the Rock Creek Country Club ladies championship tournament recorded the scores shown in the table.

Ladies Championship Scores

Number of Golfers	Score
1	66
3	68
7	69
8	70
12	71
16	72
15	74
10	76
6	78
3	82
1	85

What are the average, median, and mode scores for the tournament?

61. A West Coast city is expanding its mass transit system. It is building a 15-mile east-west light rail line for $780 million and an 11-mile north-south light rail line for $850 million. What is the average cost per mile, to the nearest million dollars, of the new lines?

62. A school district in Iowa has a budget of $7,880,000 and serves 1600 students. A second school district has a budget of $13,850,000 and serves 2500 students. What is the average cost per student in each district? What is the average cost per student in the combined districts?

63. The table lists information on the top five coffee makers, as rated by the *Consumer Reports Buying Guide 2000*.

Top Five Coffee Makers

Rank	1	2	3	4	5
Brand	Braun FlavorSelect KF187	Krups ProAroma 12 Time 45342	Mr. Coffee Accel PRX23	Mr. Coffee Accel PR 17	Braun FlavorSelect KF140
Price	$78	$100	$33	$21	$30

What was the average price (rounded to the nearest dollar) of the top five coffee makers? How many of the top five retail for less than the average price?

64. The table lists information on the top six telephone answering machines, as rated in the *Consumer Reports Buying Guide 2000*.

Top Six Telephone Answering Machines

Rank	1	2	3	4	5	6
Brand	AT&T 1725	Casio TA-117	RadioShack TAD-798	Casio TA-105	AT&T 1717	GE 2-9992
Price	$55	$30	$60	$25	$30	$70

What is the average cost of the top six rated machines? Which of the answering machines are more expensive than the average price?

Exercises 65–68. According to the U.S. Census Bureau, the population of South Carolina in 1990 was 3,486,703, and the population in 2000 was 3,885,736.

65. What was the average population of South Carolina over the 10-year period?

66. What was the average increase per year over the 10-year period?

67. Use your answer to Exercise 66 to complete the table.

Population by Year in South Carolina

Year	1990	1991	1992	1993	1994	1995	1996	1997	1998	1999	2000
Population	3,486,703										

68. Is it true that the actual population of South Carolina is given by the table in Exercise 67?

69. The attendance at the Epcot Center at Walt Disney World in Florida was 9,700,000 in 1994, 10,700,000 in 1995, and 11,200,000 in 1996. What was the average yearly attendance for the 3-year period?

70. According to the U.S. Environmental Protection Agency, the three facilities that released the most toxic chemicals in 1995 were: Magnesium Corp. of America in Rowly, UT, 64,339,080 lb; Asarco Inc. of East Helena, MT, 39,517,514 lb; and Courtlands Fibers Inc. of Axis, AL, 34,018,200 lb. What was the average release of toxic chemicals for the top three plants?

Exercises 71–74. The information on the assets of the top five U.S. commercial banks in 1997 is given in the table.

Assets of the Top Five U.S. Commercial Banks

Bank	Manhattan Corp.	Citicorp	NationsBank Corp.	BankAmerica Corp.	J.P. Morgan & Co.
Assets (in billions)	\$366	\$311	\$265	\$262	\$260

71. What were the average assets for the top three banks?

72. What were the average assets for the top four banks?

73. What were the average assets for the top five banks?

74. Compare your answers for Exercises 71, 72, and 73. Are the averages increasing or decreasing? Explain.

Exercises 75–80. Information on the top shopping web sites as of May 1999 is given in the table.

Top Shopping Web Sites, May 1999

Rank	Web Site	Unique Visitors (thousands)	Rank	Web Site	Unique Visitors (thousands)
1	AOL Shopping	10,297	6	Barnesandnoble.com	4330
2	Bluemountainarts.com	10,089	7	Mypoints.com	3491
3	Amazon.com	9933	8	Cdnow.com	3345
4	Ebay.com	8227	9	Valupage.com	3244
5	Cnet software download svcs	4709	10	Freeshop.com	2719

75. What is the mean number of unique visitors for the top five shopping web sites?

76. How many web sites are above the mean for the top five web sites and how many are below it?

77. What is the median number of unique visitors for the top five web shopping sites?

78. What is the mean number of unique visitors for the top ten shopping web sites, rounded to the nearest thousand?

79. How many web sites are above the mean for the top ten web sites and how many are below it?

80. What is the median number of unique visitors for the top ten web shopping sites, rounded to the nearest thousand?

81. On Jupiter Island in Florida, the median home price is about $4 million. State in words what this means.

82. According to U.S. Census figures for 1990, the city of Austin, Texas, had a population of 465,622 and a median age of 29. Approximately how many people in the city of Austin were 29 years old or older, rounded to the nearest thousand?

83. In 1900, the median age of women in the United States at their first marriage was 22, whereas in 1996 the median age was 25. Explain in words what this means.

Exercises 84–87 refer to the chapter application See Table 1.1, page 1.

84. Calculate the average payroll of the top three teams in the 1999–2000 season.

85. Calculate the average payroll of the top three teams in the 2000–2001 season.

86. Which season had the largest average of the top three teams? Explain why this should be obvious from the table, even without doing the calculations.

87. Estimate the average payroll of the Patriots and Jaguars for the 1999–2000 season and explain your reasoning.

STATE YOUR UNDERSTANDING

88. Explain what is meant by the average of two or more numbers.

89. Explain how to find the average (mean) of 2, 4, 5, 5, and 9. What does the average of a set of numbers tell you about the set?

CHALLENGE

90. A patron of the arts estimates that the average donation to the fund drive will be $72. She will donate $150 for each dollar by which she misses the average. The 150 donors made the following contributions:

Contributions to the Arts

Number of Donors	Donation
5	$153
13	$125
24	$110
30	$100
30	$ 75
24	$ 50
14	$ 25
10	$ 17

How much does the patron donate to the fund drive?

GROUP WORK

91. Divide each of the following numbers by 2 and by 5: 35, 68, 120, 44, 56, 75, 82, 170, 92. Which of these numbers are divisible by 2 (the division has no remainder)? Which ones are divisible by 5? See if your group can find simple rules for looking at a number and telling whether or not it is divisible by 2 and/or 5.

92. Using the new car ads in the newspaper, find four advertised prices for the same model of a car. What is the average price, to the nearest ten dollars?

1.8 Drawing and Interpreting Graphs

VOCABULARY

Graphs are used to illustrate sets of numerical information.

A **bar graph** uses solid lines or heavy bars of fixed length to represent numbers from a set. Bar graphs contain two **scales**, a vertical scale and a horizontal scale. The **vertical scale** represents one set of values and the **horizontal scale** represents a second set of values. These depend on the information to be presented. The bar graph in Figure 1.7 illustrates four types of cars (first set of values) and the number of each type of car sold (second set of values).

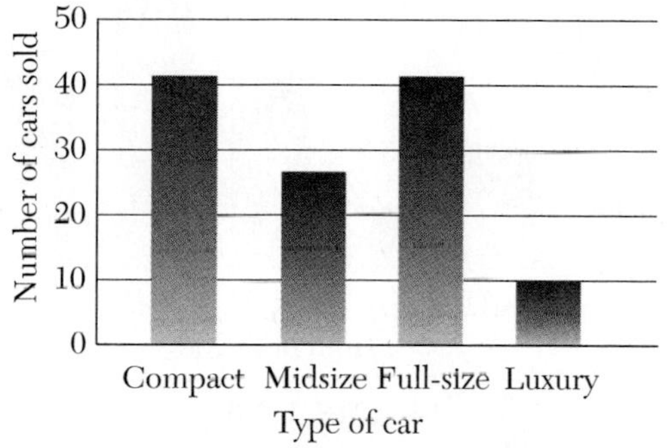

FIGURE 1.7

A **line graph** uses lines connecting points to represent numbers from a set. A line graph has a vertical and a horizontal scale, like a bar graph. A line graph showing the percentage of women in real estate by age is shown in Figure 1.8.

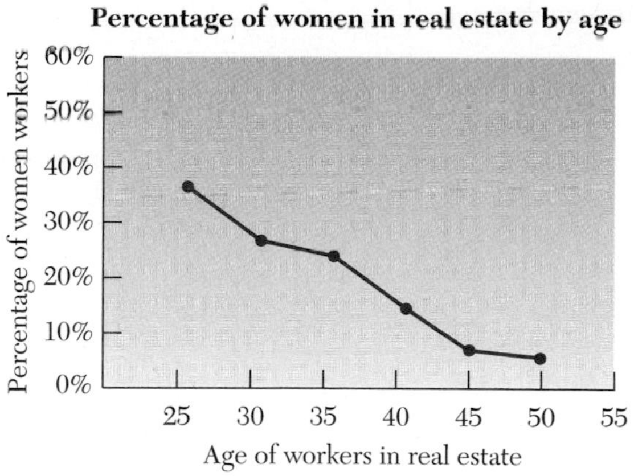

FIGURE 1.8

A **pictograph** uses symbols or simple drawings to represent numbers from a set. The pictograph in Figure 1.9 shows the distribution of mathematics students at a community college.

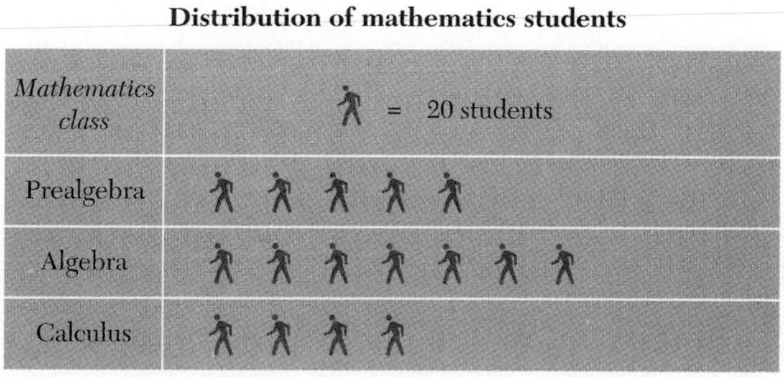

FIGURE 1.9

OBJECTIVES

1. Read data from bar, pictorial, and line graphs.
2. Construct a bar, pictorial, or line graph.

HOW & WHY

OBJECTIVE 1 Read data from bar, pictorial, and line graphs.

A graph or chart is a picture used to present data for the purpose of comparison. To "read a graph" means to find values from the graph.

Examine the bar graph in Figure 1.10.

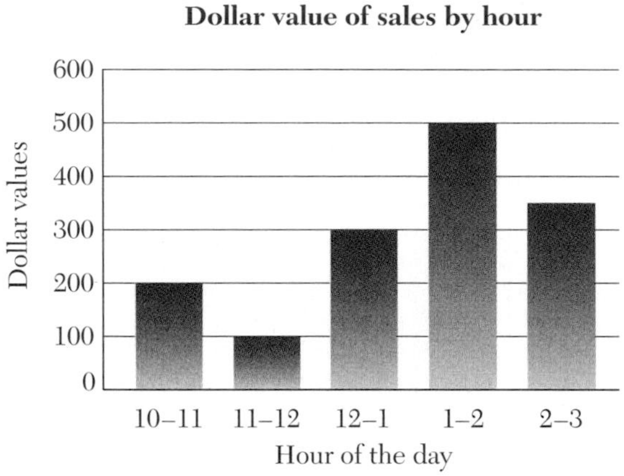

FIGURE 1.10

The vertical scale shows dollar values and is divided into units of \$100. The horizontal scale shows time values and is divided into units of 1-hour periods from 10:00 A.M. to 3:00 P.M. From the graph we see that

1. The hour of greatest sales (highest bar) is 1 P.M. to 2 P.M., with \$500 in sales.
2. The hour of least sales (lowest bar) is 11:00 A.M. to 12 noon, with sales of \$100.
3. The total sales during the 2 P.M. to 3 P.M. time period is estimated to be \$350 because the top of the bar falls between the scale divisions.
4. The total of the morning sales is \$300.
5. The total of the afternoon sales is \$1150.

Some advantages of displaying data with a graph:

1. Each person can easily find the data most useful to him or her.
2. The visual display is easy for most people to read.
3. Some questions can be answered by a quick look at the graph—for example, What time does the store need the most sales clerks?

EXAMPLES A–B

DIRECTIONS: Answer the questions associated with the graph.

STRATEGY: Examine the graph to determine the values that are related.

A. The graph shows the number of cars that used Highway 37 during a 1-week period.

Cars on Highway 37

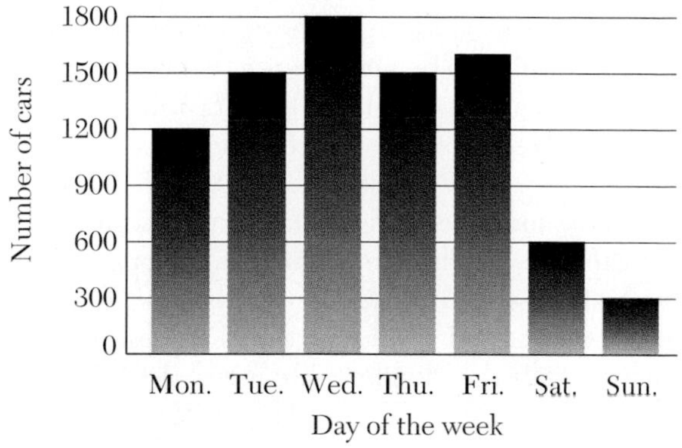

1. What day had the most traffic?
2. What day had the least traffic?
3. How many cars used the highway on Tuesday?
4. How many cars used the highway on the weekend?

1. Wednesday	The tallest bar shows the largest number of cars.
2. Sunday	The shortest bar shows the least number of cars.
3. 1500	Read the vertical scale at the top of the bar for Tuesday.
4. 900	Add the number of cars for Saturday and Sunday.

B. The total sales from hot dogs, soda, T-shirts, and buttons during an air show are shown in the pictorial graph.

Sales at air show

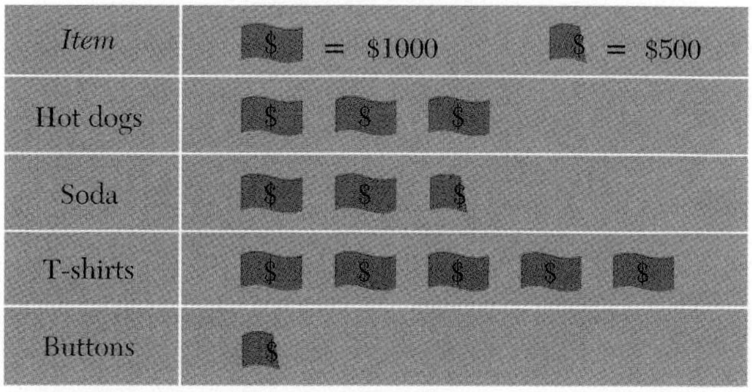

1. What item has the largest dollar sales?
2. What were the total sales from hot dogs and buttons?
3. How many more dollars were realized from the sale of T-shirts than from buttons?

1. T-shirts — T-shirts have the greatest number of symbols.
2. Hot Dogs: $3000
   Buttons: +$ 500
   Total: $3500
3. T-shirts: $5000 — Subtract the sales of buttons from the sales of T-shirts.
   Buttons: −$ 500
   Difference: $4500

The sales from T-shirts were $4500 more than from buttons.

WARM-UPS A–B

A. The number of bus riders for each day of the week is shown in the bar graph.

Bus riders

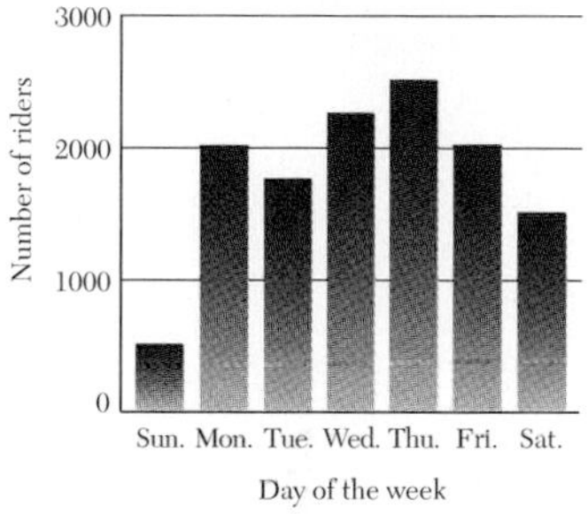

1. What day had the greatest ridership?
2. What day had the least?
3. How many people rode the bus on Friday?
4. How many people rode the bus during the seven days?

B. The number of birds spotted during a a recent expedition of the Huntsville Bird Society is shown in the pictorial graph.

Birds spotted

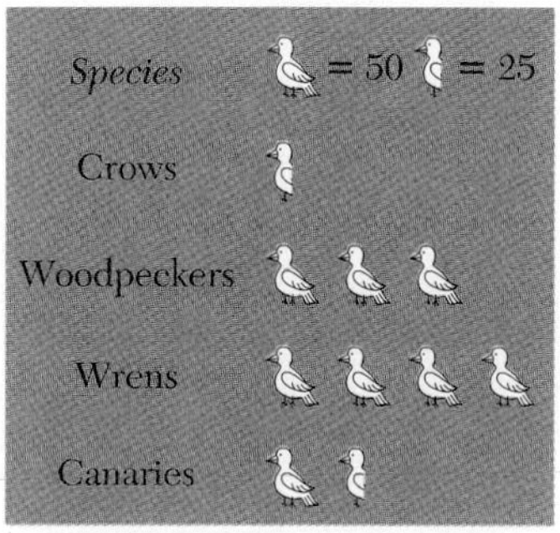

1. Which species of bird was spotted most often?
2. How many woodpeckers and wrens were spotted?
3. How many more canaries were spotted than crows?

Answers to Warm-Ups

A. **1.** Thursday **2.** Sunday **3.** 2000 **4.** 12,500

* **B.** **1.** wrens **2.** 350 **3.** 50

HOW & WHY

OBJECTIVE 2 Construct a bar, pictorial, or line graph.

TABLE 1.4
USED CAR SALES

Month	Cars Sold
January	30
February	15
March	25
April	20
May	15
June	35

Let us construct a bar graph to show the variation in used car sales at Oldies but Goldies used car lot. The data are shown in Table 1.4

To draw and label the bar graph for these data, we show the number of cars sold on the vertical scale and the months on the horizontal scale. This is a logical display because we will most likely be asked to find the highest and lowest month of car sales, and a vertical display of numbers is easier to read than a horizontal display of numbers. This is the typical way bar graphs are displayed. Be sure to write the names on the vertical and horizontal scales as soon as you have chosen how the data will be displayed. Now title the graph so that the reader will recognize the data it contains.

The next step is to construct the two scales of the graph. Because each monthly total is divisible by 5, we choose multiples of 5 for the vertical scale. We could have chosen units of 1 for the vertical scale, but the bars would be very long and the graph would take up a lot of space. If we had chosen a larger scale—for instance, 10—then the graph might be too compact and we would need to find fractional values on the scale. It is easier to draw the graph if we use a scale that divides each unit of data. The months are displayed on the horizontal scale. Be sure to draw the bars with uniform width because each of them represents a month of scales. A vertical display of between 5 and 12 units is typical. The vertical display should start with zero. See Figure 1.11.

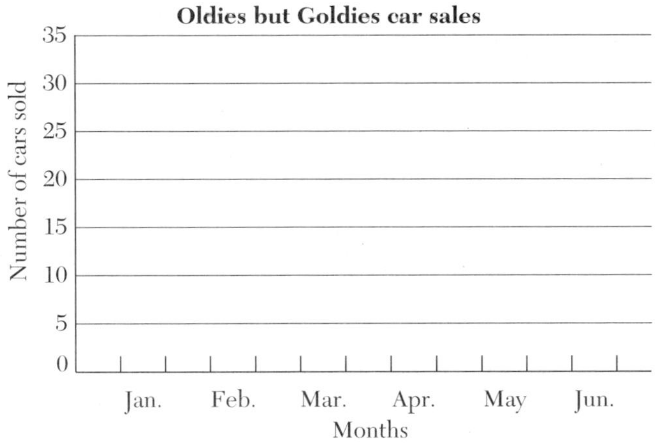

FIGURE 1.11

We stop the vertical scale at 35 because that is the maximum number of cars to be displayed. The next step is to draw the bars. Start by finding the number of cars for January. Thirty cars were sold in January, so we draw the bar for January until the height of 30 is reached. This is the top of the bar. Now draw the solid bar for January. See Figure 1.12.

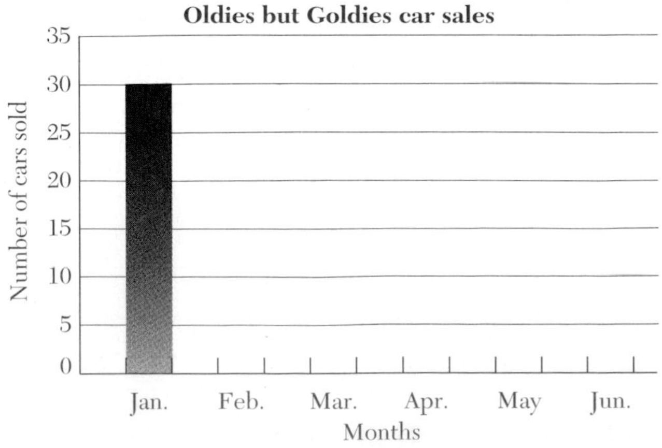

FIGURE 1.12

Complete the graph by drawing the bars for the other months. See Figure 1.13.

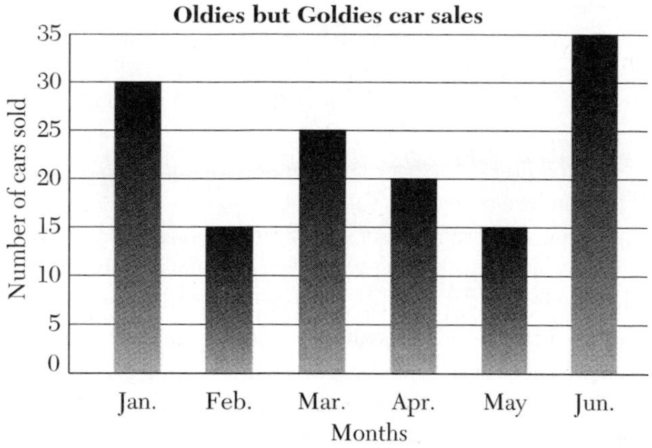

FIGURE 1.13

A line graph is similar to a bar graph in that it has vertical and horizontal scales. The data are represented by points rather than bars, and the points are connected by line segments. We use a line graph to display the data in Table 1.5.

The vertical scale represents the tax rate and each unit represents $2. This requires using a half space for the $15.00 rate.

Another possibility is to use a vertical scale in which each unit represents $1, but this would require 20 units on the vertical scale and would make the graph much taller. We opt to save space by using $2 units on the vertical scale. See Figure 1.14.

TABLE 1.5
PROPERTY TAXES

Year	Tax Rate (per $1000)
1970	$12.00
1975	$15.00
1980	$14.00
1985	$16.00
1990	$20.00

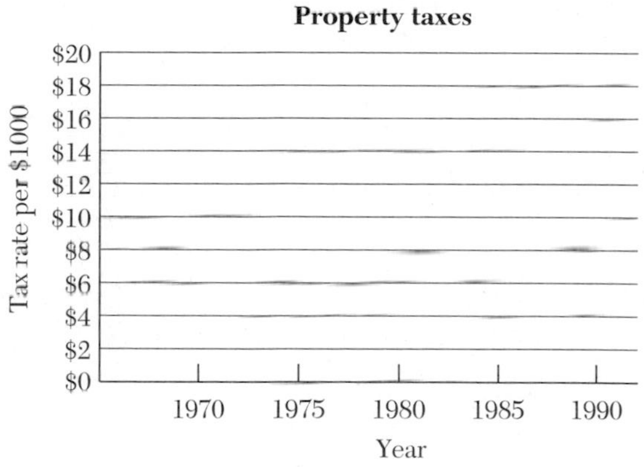

FIGURE 1.14

To find the points that represent the data, locate the points that are at the intersection of the horizontal line through the tax rate and the vertical line through the corresponding year. Once all the points have been located, connect them with line segments. See Figure 1.15.

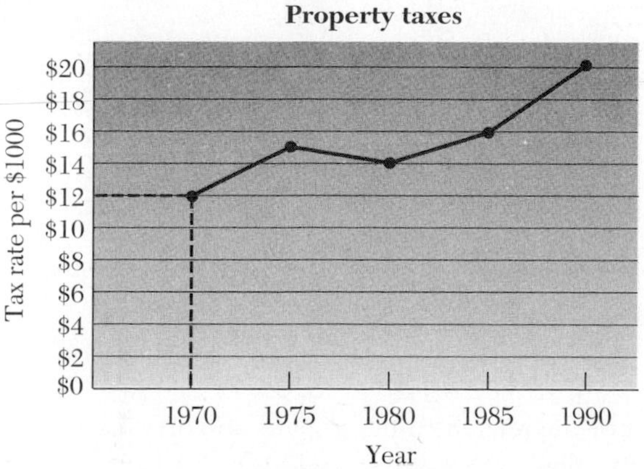

FIGURE 1.15

From the graph, we can conclude the following:

1. Only during one 5-year period (1975–1980) did the tax rate decline.
2. The largest tax increase in the tax rate took place from 1985 to 1990.
3. The tax rate has increased $8 per thousand from 1970 to 1990.

TABLE 1.6 CAR SALES AT WESTERN CAR CORRAL

Make	Number Sold
Ford	50
Chrysler	35
Honda	60
Toyota	25
Pontiac	20

Construct a pictorial graph to show the number of cars sold, by model, at the Western Car Corral. The data are shown in Table 1.6.

First, select a symbol and the number of cars it represents. Here we use a picture of a car, with each symbol representing 10 cars. By letting each symbol represent 10 cars and a half symbol represent 5, we can save space. We could have chosen 5 cars per symbol, but then we would need to display 12 symbols to represent the number of Hondas, and 12 symbols would make the graph quite large. Next, determine the number of symbols we need for each model. The number of symbols can be found by dividing each number of cars by 10. See Table 1.7.

TABLE 1.7 CAR SALES AT WESTERN CAR CORRAL

Make	Number Sold	Symbols Needed
Ford	50	5 tens
Chrysler	35	3 tens and 1 five
Honda	60	6 tens
Toyota	25	2 tens and 1 five
Pontiac	20	2 tens

Now draw the graph using the symbols (pictures) to represent the data. See Figure 1.16.

Car sales at Western Car Corral

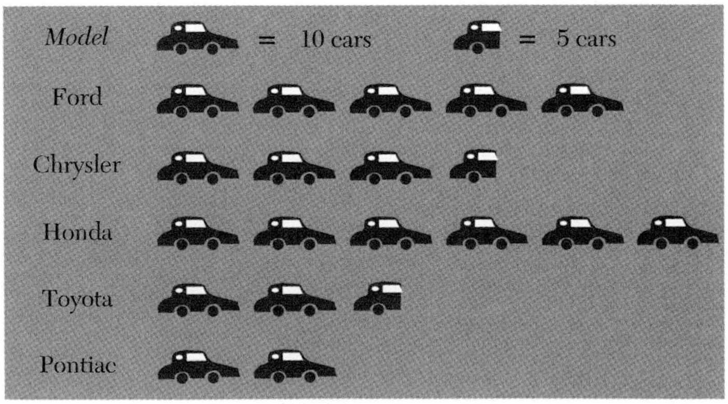

FIGURE 1.16

WARM-UP C

EXAMPLE C

DIRECTIONS: Construct a bar graph.

STRATEGY: List the related values in pairs and draw two scales to show the pairs of values.

C. The number of babies born during the first six months at Tuality Hospital were: January, 15; February, 9; March, 7; April, 18; May, 10; June, 6.

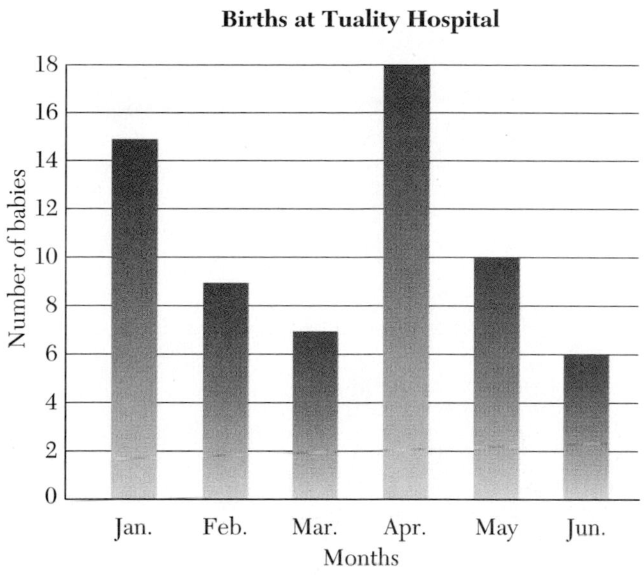

Choose a scale of 1 unit = 2 babies for the vertical scale. Divide the horizontal scale so that it will accommodate 6 months with a common space between them. Construct the graph, label the scales, and give the graph a title.

C. The number of plants sold at the Pick-a-Posey Nursery are: geraniums, 45; fuchsias, 60; marigolds, 150; impatiens, 90; daisies, 120.

Answer to Warm-Up

C.

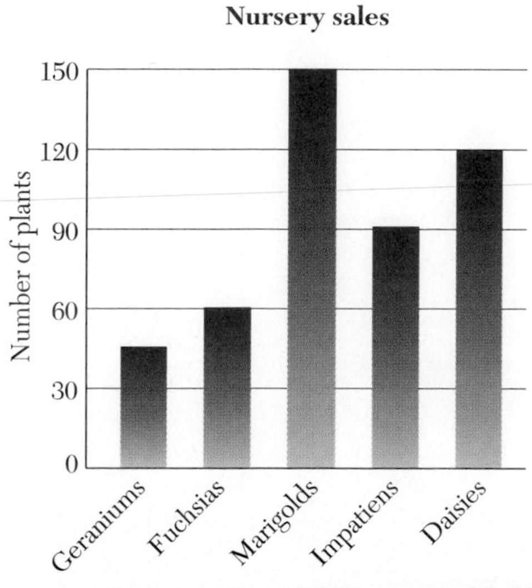

EXERCISES 1.8

OBJECTIVE 1 Read data from bar, pictorial, and line graphs.

A.

Exercises 1–6. The graph shows the variation in the number of phone calls during normal business hours:

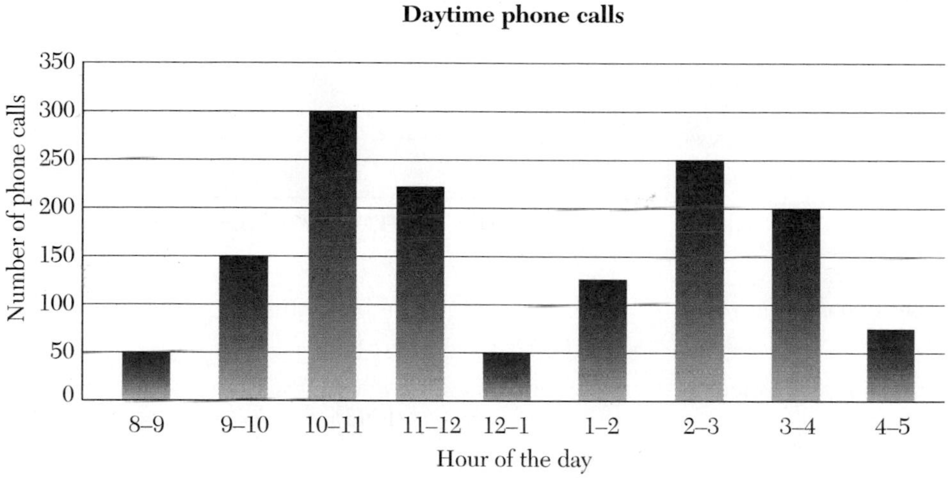

1. At what hour of the day is the number of phone calls greatest?

2. At what hour of the day is the number of phone calls least?

3. What is the number of phone calls made between 2 and 3?

4. What is the number of phone calls made between 8 and 12?

5. What is the total number of phone calls made during the times listed?

6. Are there more phone calls in the morning, 8 to 12, or the afternoon, 12 to 5?

Exercises 7–12. The graph shows the number of cars in the shop for repair during a given year:

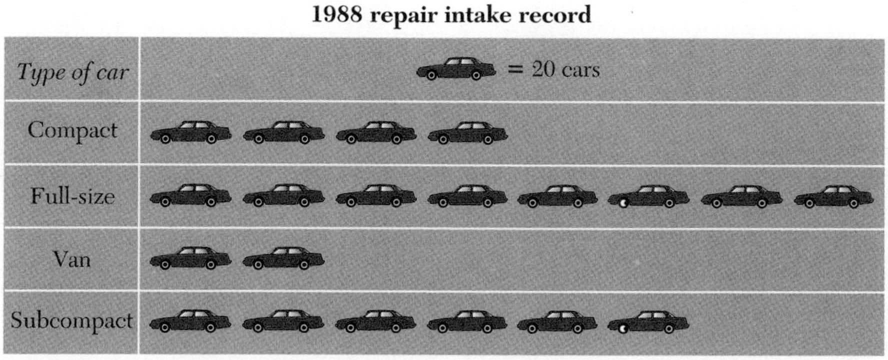

7. How many vans are in the shop for repair during the year?

8. How many compacts and subcompacts are in the shop for repair during the year?

9. What type of car has the most cars in for repair?

10. Are more subcompacts or compacts in for repair during the year?

11. How many cars are in for repair during the year?

12. If the average repair costs for compacts is $210, what is the gross income on compact repairs for the year?

B.

Exercises 13–18. The graph shows the number of production units at NERCO during the period 1992 to 1996.

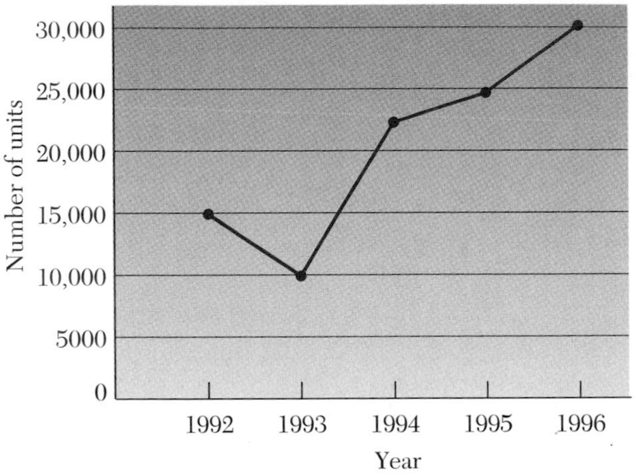

13. What is the greatest production year?

14. What is the year of least production?

15. What is the increase in production between 1993 and 1994?

16. What is the decrease in production between 1992 and 1993?

17. What is the average production per year?

18. If the cost of producing a unit in 1995 is $2750 and the unit is sold for $4560, what is the net income for the year?

Exercises 19–24. The graph shows the amounts paid for raw materials at Southern Corporation during a production period.

Raw material costs

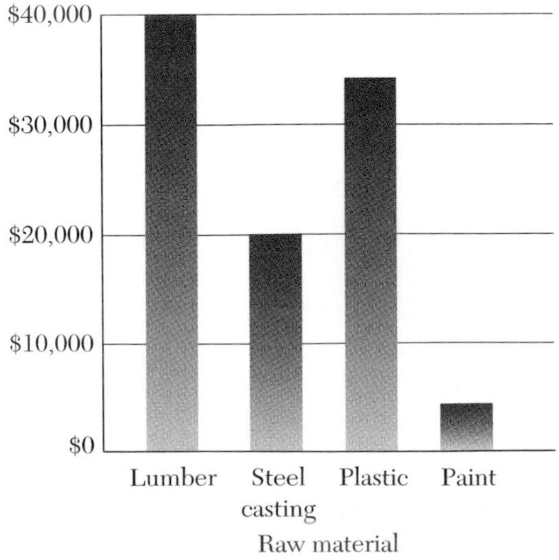

19. What is the total paid for paint and lumber?

20. What is the total paid for raw materials?

21. How much less is paid for steel casting than for plastic?

22. How much more is paid for plastic than for paint?

23. If Southern Corporation decides to double its production during the next period, what will it pay for steel casting?

24. If Southern Corporation decides to double its production during the next period, how much more will it pay for lumber and steel casting than plastic and paint?

OBJECTIVE 2 Construct a bar, pictorial, or line graph.

A.

In Exercises 25–28, draw bar graphs to display the data. Be sure to title the graph and label the parts.

25. Distribution of grades in an algebra class: A, 8; B, 6; C, 15; D, 8; F, 4

26. The distribution of monthly income: rent, $450; automobile, $300; taxes, $250; clothes, $50; food, $300; miscellaneous, $100

27. Career preference as expressed by a senior class: business, 120; law, 20; medicine, 40; science, 100; engineering, 50; public service, 80; armed service, 10

28. According to statistics from Financial World, the average value of a team in the NFL is $153 million, the average value of a major league baseball (MLB) team is $107 million, and the average value of an NBA team is $99 million.

In Exercises 29–32, draw line graphs to display the data. Be sure to title the graph and label the parts.

29. Daily sales at the local men's store: Monday, \$1500; Tuesday, \$2500; Wednesday, \$1500; Thursday, \$3500; Friday, \$4000; Saturday, \$6000; Sunday, \$4500

30. The gallons of water used each quarter of the year by a small city in New Mexico:

Jan.–Mar. 20,000,000
Apr.–June 30,000,000
July–Sept. 45,000,000
Oct.–Dec. 25,000,000

31. Income from various sources for a given year for the Smith family: wages, \$36,000; interest, \$2000; dividends, \$4000; sale of property, \$24,000

32. Jobs in the electronics industry in a western state: 1990, 15,000; 1991, 21,000; 1992, 18,000; 1993, 21,000; 1994, 27,000; 1995, 30,000

B.

In Exercises 33–36, draw pictorial graphs to display the data. Be sure to title the graph and label the parts.

33. The cost of an average three-bedroom house in a rural city:

Year	Cost
1970	$60,000
1975	$70,000
1980	$75,000
1985	$70,000
1990	$65,000
1995	$80,000

34. The population of Wilsonville that is over 20 years of age: 1975, 22,500; 1980, 30,000; 1985, 32,500; 1990, 37,500; 1995, 40,000

35. The oil production from a local well over a 5-year period:

Year	Barrels Produced
1991	15,000
1992	22,500
1993	35,000
1994	32,500
1995	40,000

36. The estimated per capita income in the Great Lakes region in 1995:

State	Income
Illinois	$25,000
Michigan	$22,500
Ohio	$22,500
Wisconsin	$20,000
Indiana	$20,000

37. Draw a line graph to display the maximum monthly benefit at retirement for individuals who retire at age 65, according to the Social Security Administration.

Monthly Social Security Benefits

1970	1980	1990	1999
\$190	\$572	\$975	\$1373

38. Draw a bar graph to display the five most frequently prescribed drugs in physicians' offices for 1998, according to the National Center for Health Statistics.

Most Frequently Prescribed Drugs

Name of Drug	Therapeutic Use	Times Prescribed
Amoxicillin	Antibiotic	16,744
Tylenol	Pain relief	15,840
Claritin	Antihistamine	14,218
Lasix	Diuretic	13,850
Premarin	Estrogen replacement	12,735

39. Draw a bar graph to display the median price of existing single-family houses in Miami, Florida, as follows: \$121,500 in 1998; \$134,600 in 1999; and \$138,200 in 2000, according to data from the National Association of Realtors.

40. The number of thousands of veterans in the United States over the last century is given in the table.

Veterans in the United States (in thousands)

1900	1910	1920	1930	1940	1950	1960	1970	1980	1990	1998
1100	1000	5100	4700	4300	19,100	22,500	27,600	28,600	27,000	25,100

a. Make a line graph for this information.

b. What historical event caused the dramatic increase in 1950?

c. What historic event caused the increase in 1920?

d. What year had the maximum number of veterans?

e. What has been happening to the number of veterans since 1980? Why?

C.

Exercises 41–44 refer to the bar graph, which shows U.S. and worldwide employment in the auto industry in a recent year, according to the American Automobile Manufacturers Association.

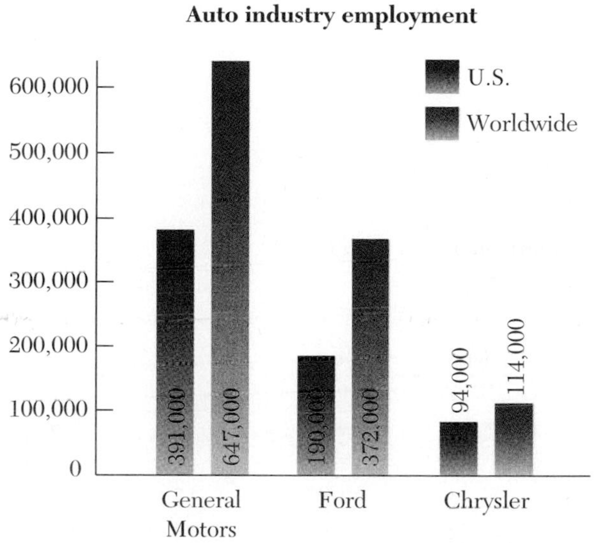

41. Which company has the largest number of employees worldwide? Which has the smallest number of employees worldwide?

42. How many employees does Ford have who work outside the United States?

43. What is the total number of car workers in the United States employed by the three car makers?

44. Do the numbers in the graph appear to be rounded? Why or why not? If so, what place value were they rounded to?

Exercises 45–46. Use the information on zoo attendance given in the table.

Zoo Attendance

	1997	1998	1999	2000	2001
Fisher Zoo	2,367,246	2,356,890	2,713,455	2,745,111	2,720,567
Delaney Zoo	1,067,893	1,119,875	1,317,992	1,350,675	1,398,745
Shefford Garden	2,198,560	2,250,700	2,277,300	2,278,345	2,311,321
Utaki Park	359,541	390,876	476,200	527,893	654,345

45. Draw a line graph to display the attendance at Utaki Park for the 5-year period.

46. Draw a bar graph to display the data on attendance at the zoos in 2001.

Exercises 47–48. The table gives visitor information for Lizard Lake State Park.

Visitors at Lizard Lake State Park

	May	June	July	August	September
Overnight camping	231	378	1104	1219	861
Picnics	57	265	2371	2873	1329
Boat rental	29	147	147	183	109
Hiking/climbing	48	178	178	192	56
Horse rental	22	43	43	58	27

47. Draw a line graph to display the data on overnight camping for the 5-month period.

48. Draw a bar graph to display the data on hiking/climbing for the 5-month period.

Exercise 49 refers to the chapter application. See Table 1.1, page 1.

49. Draw a line graph that shows the top NFL payrolls. Use 1–8 on the horizontal axis to indicate the spending rank, and put the payroll amounts on the vertical axis. Put the data for both seasons on the same graph, using different colors to indicate different seasons.

STATE YOUR UNDERSTANDING

50. Explain the advantages of each type of graph. Which do you prefer? Why?

CHALLENGE

51. The figures for United States casualties in the four declared wars of the 20th century are: World War I, 321,000; World War II, 1,076,000; Korean War, 158,000; Vietnam War, 211,000. Draw a bar graph, a line graph, and a pictorial graph to illustrate the information. Which of your graphs do you think does the best job of displaying the data? (For the pictorial graph, you may want to round to the nearest hundred thousand.)

GROUP WORK

52. Have each member select a country and find the most recent population statistics for that country. Put the numbers together and have each member draw a different kind of graph of the populations.

CHAPTER 1

KEY CONCEPTS

SECTION 1.1 WHOLE NUMBERS AND TABLES: WRITING, ROUNDING, AND INEQUALITIES

Definitions and Concepts	Examples
The whole numbers are 0, 1, 2, 3, and so on.	238 two hundred thirty-eight 6,198,349 six million, one hundred ninety-eight thousand, three hundred forty-nine
One whole number is smaller than another if it is to the left on the number line.	$3 < 6$
One whole number is larger than another if it is to the right on the number line.	$14 > 2$
To round a whole number: • Round to the larger number if the digit to the right is 5 or more. • Round to the smaller number if the digit to the right is 4 or less.	 $6745 \approx 7000$ (nearest thousand) $6745 \approx 6700$ (nearest hundred)
Tables are a method of organizing information or data in rows and columns.	See table below.

Enrollment by Gender at River CC

	Males	Females
English	52	67
Math	71	64
Science	69	75
History	63	59

There are 71 males taking math and 75 females taking science.

SECTION 1.2 ADDING AND SUBTRACTING WHOLE NUMBERS

Definitions and Concepts	Examples
To add whole numbers, write the numbers in columns so the place values are aligned. Add each column, starting with the ones. Carry as necessary.	$\begin{array}{r} {\scriptstyle 1} \\ 372 \\ +594 \\ \hline 966 \end{array}$ $\quad \begin{array}{r} {\scriptstyle 11} \\ 36 \\ +785 \\ \hline 821 \end{array}$
addend + *addend* = *sum*	
To subtract whole numbers, write the numbers in columns so the place values are aligned. Subtract, starting with the ones column. Borrow if necessary. The answer to a subtraction problem is called the *difference*.	$\begin{array}{r} 4597 \\ -\ 362 \\ \hline 4235 \end{array}$ $\quad \begin{array}{r} {\scriptstyle 2\,14\,4\,12} \\ \not{3}\,\not{4}\,\not{5}\,\not{2} \\ -\ \ 7\ 3\ 5 \\ \hline 2\ 7\ 1\ 7 \end{array}$

SECTION 1.2 ADDING AND SUBTRACTING WHOLE NUMBERS (continued)

Definitions and Concepts	Examples
The perimeter of a polygon is the distance around the outside. To calculate the perimeter, add the lengths of the sides.	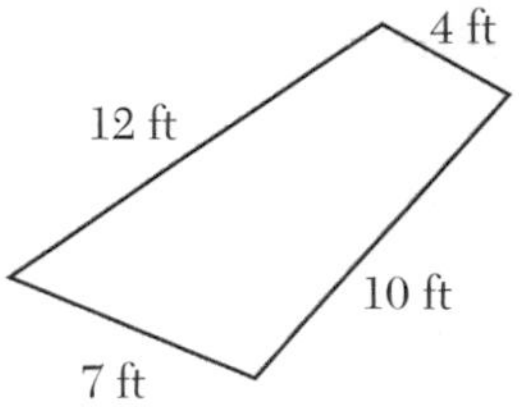 Perimeter $= 12 + 4 + 10 + 7$ $= 33$ $P = 33$ ft

SECTION 1.3 MULTIPLYING WHOLE NUMBERS

Definitions and Concepts	Examples
To multiply whole numbers, multiply the first factor by each digit in the second factor, keeping alignment. Add the partial products. *(factor)(factor) = product*	482 × 12 964 2 × 482 482 1 × 482 5784
The area of a rectangle is the space inside it: $A = \ell w$	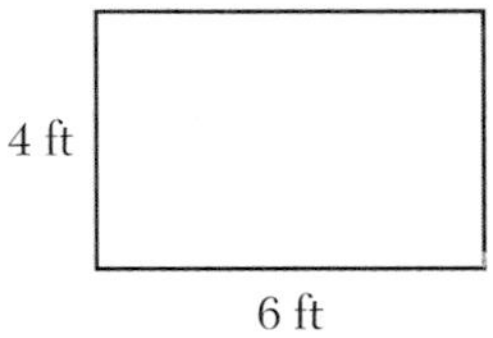 Area $= \ell w$ $= 6 \cdot 4$ $= 24$ The area is 24 square feet.

SECTION 1.4 DIVIDING WHOLE NUMBERS

Definitions and Concepts	Examples
To divide whole numbers, use long division as shown. $\text{divisor}\overline{)\text{dividend}}$ with *quotient* above	205 32)6578 64 32(2) = 64 17 Subtract and bring down the 7. 0 32(0) = 0 178 Subtract and bring down the 8. 160 32(5) = 160 18 Subtract. The remainder is 18.

SECTION 1.5 WHOLE NUMBER EXPONENTS AND POWERS OF 10

Definitions and Concepts	Examples
An exponent indicates how many times a number is used as a factor. $base^{exponent} = value$	$2^3 = (2)(2)(2) = 8$
A power of 10 is the value of 10 with some exponent.	$10^4 = (10)(10)(10)(10) = 10{,}000$

SECTION 1.6 ORDER OF OPERATIONS

Definitions and Concepts	Examples
The order of operations is: Parentheses Exponents Multiplication and Division Addition and Subtraction	$2(7 + 4) - 36 \div 3^2 + 5$ $2(11) - 36 \div 3^2 + 5$ $2(11) - 36 \div 9 + 5$ $22 - 4 + 5$ $18 + 5$ 23

SECTION 1.7 AVERAGE, MEDIAN, AND MODE

Definitions and Concepts	Examples
	{4, 13, 26, 51, 51}
To find the average of a set of numbers: • Add the numbers. • Divide by the number of numbers.	$4 + 13 + 26 + 51 + 51 = 145$ $145 \div 5 = 29$ The average is 29.
To find the median of a set of numbers: • List the numbers in order from smallest to largest. • If there is an odd number of numbers in the set, the median is the middle number. • If there is an even number of numbers in the set, the median is the average (mean) of the middle two.	The median is 26.
To find the mode of a set of numbers: • Find the number or numbers that occur most often • If all the numbers occur the same number of times, there is no mode.	The mode is 51.

SECTION 1.8 DRAWING AND INTERPRETING GRAPHS

Definitions and Concepts	Examples
A line graph uses a line to connect data points.	Items received

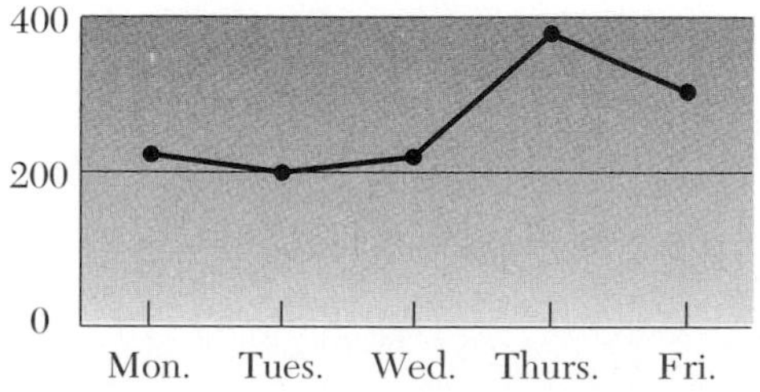

SECTION 1.8 DRAWING AND INTERPRETING GRAPHS (continued)

Definitions and Concepts	Examples
A bar graph uses bars to represent data values.	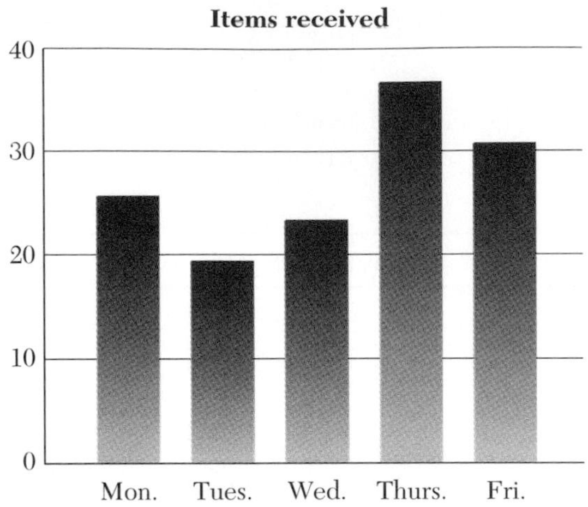
A pictograph uses pictures to represent data values.	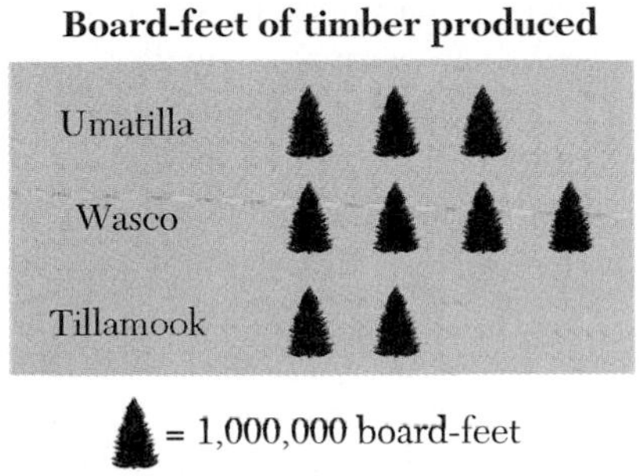

CHAPTER 1

REVIEW EXERCISES

Section 1.1

Write the word name for each of these numbers.

1. 451,810 **2.** 7,005,200

Write the place value name for each of these numbers.

3. Ninety thousand, three hundred thirty-seven
4. Six million, three hundred forty-four thousand, fifty-nine

Insert > or < between the numbers to make a true statement.

5. 347 351 **6.** 56 69 **7.** 809 811

Round to the nearest ten, hundred, thousand, and ten thousand.

8. 34,529
9. 183,659

Exercises 10–14. The table displays the population of Hepner by age group.

Population of Hepner by Age Group

Age	Number of Residents	Age	Number of Residents
Under 15	472	36–50	1098
15–25	398	51–70	602
26–35	612	Over 70	89

10. Which age group is the largest?

11. How many more residents are under 15 years old than are in the 15–25 age group?

12. What is the total population in the 26–50 age group?

13. How many more people are there in the under-15 age group as opposed to the over-70 age group?

14. What is the population of Hepner?

Section 1.2

Estimate the sum by rounding to the largest place value in each group and then add.

15. 116 + 49 + 312

16. 1734 + 210 + 819

17. 34 + 455 + 17 + 377 + 881

18. 6,891 + 12,055 + 51,932 + 4,772 + 492

Estimate the difference by rounding to the largest place value in each group and then subtract.

19. 756 − 325

20. 803 − 738

21. 7415 − 5923

22. 246,892 − 149,558

23. Find the perimeter of the polygon.

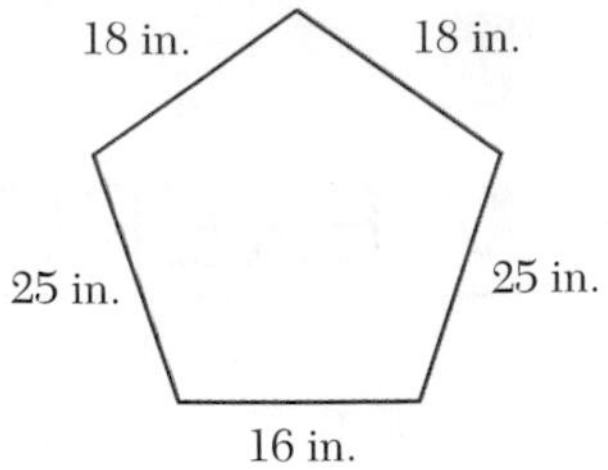

Section 1.3

Estimate the product by rounding each number to its largest place value and then multiply.

24. $\begin{array}{r} 45 \\ \times\ 12 \\ \hline \end{array}$

25. $\begin{array}{r} 406 \\ \times\ 29 \\ \hline \end{array}$

26. $\begin{array}{r} 197 \\ \times\ 98 \\ \hline \end{array}$

27. (449)(171)

28. The local Girl Scout Troop sold 34 cases of cookies during the recent sale. Each case contained 24 boxes of cookies. If the cookies sold for $4 a box, how much did they gross from the sale?

29. Find the area of the rectangle.

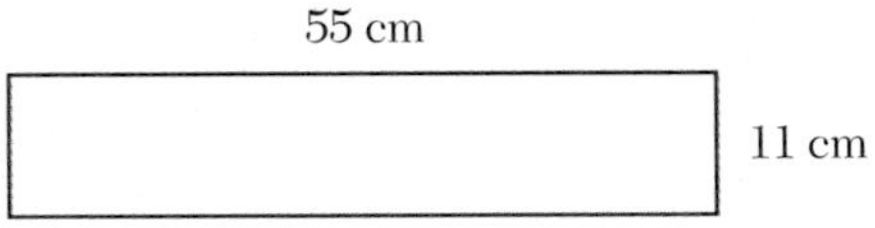

Section 1.4

Estimate the quotient by rounding the divisor to one digit and the dividend to two digits, and then divide.

30. $13\overline{)117}$

31. $18\overline{)576}$

32. $192\overline{)14{,}980}$

33. $675{,}200 \div 63$

34. The Candy Basket packs chocolates in boxes containing 32 pieces of assorted chocolates. How many boxes can be made from 4675 chocolates? Will any pieces of candy be left over? If so, how many?

Section 1.5

Find the value.

35. 11^3

36. 4^5

Multiply or divide.

37. 18×10^3

38. $134{,}000{,}000 \div 10^5$

39. 516×10^6

40. $71{,}500{,}000 \div 10^4$

41. In 2001, President Bush proposed a $\$16 \times 10^8$ cut in federal taxes. Write this amount in place value form.

Section 1.6

Simplify.

42. $8 \cdot 11 + 5$

43. $14 \div 2 - 3 \cdot 2$

44. $94 \div 47 + 47 - 16 \cdot 2 + 6$

45. $35 - (25 - 17) + (12 - 10)^2 + 5 \cdot 2$

Section 1.7

Find the average, median, and mode.

46. 41, 64, 23, 70, 87

47. 89, 106, 212, 313

48. 61, 72, 53, 61, 37, 82

49. 1086, 4008, 3136, 8312, 8312, 1474

50. The six children of Jack and Mary Barker had annual incomes of \$54,500, \$45,674, \$87,420, \$110,675, \$63,785, and \$163,782. Find the average salary of the six siblings to the nearest hundred dollars.

Section 1.8

Exercises 51–52. The graph displays the temperature readings for a 24-hour period.

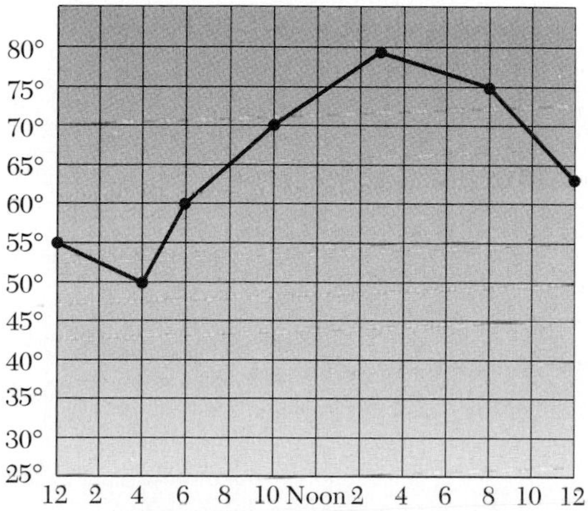

51. What time do the highest and lowest temperatures occur?

52. What is the difference in temperature from 6 A.M. to 8 P.M.?

53. The grade distribution in an algebra class is displayed in the table.

Algebra Grades

Grade	Number of Students
A	5
B	10
C	12
D	4
F	2

Construct a bar graph to display the data.

CHAPTER 1

TRUE / FALSE CONCEPT REVIEW

ANSWERS

Check your understanding of the language of basic mathematics. Tell whether each of the following statements is true (always true) or false (not always true). For each statement you judge to be false, revise it to make a statement that is true.

1. All whole numbers can be written using nine digits. **1.** ____________

2. In the number 8425, the digit 4 represents 400. **2.** ____________

3. The word *and* is not used when writing the word names of whole numbers. **3.** ____________

4. The inequality $7 < 23$ can be read "seven is greater than twenty-three." **4.** ____________

5. $2567 < 2566$ **5.** ____________

6. To the nearest thousand, 7398 rounds to 7000. **6.** ____________

7. It is possible for the rounded value of a number to be equal to the original number **7.** ____________

8. The expanded form of a whole number shows the plus signs that are usually not written. **8.** ____________

9. The sum of 80 and 7 is 807. **9.** ____________

10. The process of "carrying" when doing an addition problem with pencil and paper is based on the place value of the numbers. **10.** ____________

11. A line graph has at least two scales. **11.** ____________

12. The product of 15 and 3 is 18. **12.** ____________

13. It is possible to subtract 47 from 65 without "borrowing." **13.** ____________

14. The number 8 is a factor of 72. **14.** ____________

15. The multiplication sign is sometimes omitted when writing a multiplication problem. **15.** ____________

16. Whenever a number is multiplied by zero, the value remains unchanged. **16.** ____________

17. There is more than one method for doing division problems. **17.** ____________

18. In $104 \div 4 = 26$, the quotient is 104. **18.** ____________

19. If a division exercise has a remainder, then we know there is no whole number quotient. 19. ______

20. When zero is divided by any whole number from 25 to 91, the result is 0. 20. ______

21. The result of zero divided by zero can be either 1 or 0. 21. ______

22. The value of 7^2 is 14. 22. ______

23. The value of 222^1 is 222. 23. ______

24. One billion is a power of 10. 24. ______

25. The product of 450×10^3 is equal to 45,000. 25. ______

26. The quotient of 9000 and 10 is 900. 26. ______

27. In the order of operations, exponents always take precedence over addition. 27. ______

28. In the order of operations, multiplication always takes precedence over subtraction. 28. ______

29. The value of $2^3 + 2^3$ is the same as the value of 2^4. 29. ______

30. The average of three different numbers is smaller than the largest of the three numbers. 30. ______

31. The word *mean* sometimes has the same meaning as "average." 31. ______

32. A table is a method of displaying data in an array using a horizontal and vertical arrangement to distinguish the type of data. 32. ______

33. The median of 34, 54, 14, 44, 67, 81, and 90 is 44. 33. ______

34. The distance around a polygon is called the perimeter. 34. ______

CHAPTER 1

TEST

ANSWERS

1. Divide: $63\overline{)7812}$ — 1. ______

2. Subtract: $8741 - 5976$ — 2. ______

3. Simplify: $45 \div 5 + 3 \cdot 4 - 8$ — 3. ______

4. Multiply: $47(538)$ — 4. ______

5. Insert $<$ or $>$ to make the statement true: $468 \quad 459$ — 5. ______

6. Multiply: 83×10^5 — 6. ______

7. Multiply: $803(452)$ — 7. ______

8. Write the place value name for five hundred sixty thousand, seventy-two. — 8. ______

9. Find the average of 2450, 472, 2966, 389, and 783. — 9. ______

10. Multiply: $45(3074)$. Round the product to the nearest hundred. — 10. ______

11. Round 19,524 to the nearest thousand. — 11. ______

12. Estimate the sum of 93,914, 27,348, 57,699, and 20,341 by rounding to the nearest ten thousand. — 12. ______

13. Find the value of 9^3. — 13. ______

14. Add: $56 + 877 + 24 + 5882$ — 14. ______

15. Find the perimeter of the rectangle. — 15. ______

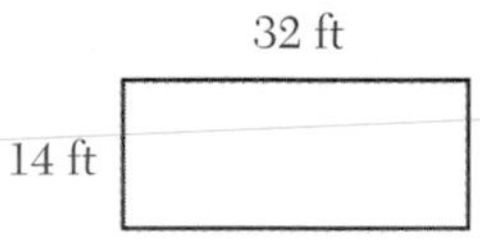

16. Estimate the product: $687(28)$ — 16. ______

17. Subtract:
$$\begin{array}{r} 9050 \\ -\ 794 \\ \hline \end{array}$$
— 17. ______

18. Write the word name for 3006. — 18. ______

19. Simplify: $45 + 3^2 - 55 \div 11$ **19.** ______

20. Add:

$$\begin{array}{r} 63{,}872 \\ 42{,}773 \\ 89{,}032 \\ 19{,}088 \\ +58{,}447 \\ \hline \end{array}$$

20. ______

21. Divide: $8{,}060{,}000{,}000 \div 10^7$ **21.** ______

22. Round 675,937,558 to the nearest ten thousand. **22.** ______

23. Estimate the quotient and then divide: $63{,}527 \div 42$ **23.** ______

24. Simplify: $84 - 8^2 + 36 \div 4$ **24.** ______

25. Find the area of a rectangle that measures 12 cm by 17 cm. **25.** ______

26. Simplify: $(5 \cdot 3)^2 + (4^2)^2 + 11 \cdot 3$ **26.** ______

27. Find the average, median, and mode of 674, 416, 592, 674, 313, and 805. **27.** ______

28. A secretary can type an average of 75 words per minute. If there are approximately 700 words per page, how long will it take the secretary to type 15 pages? **28.** ______

29. Twelve people share in the Nationwide lottery jackpot. If the jackpot is worth \$37,758,000, how much will each person receive? If each person's share is to be distributed evenly over a 20-year period, how much will each person receive per year? **29.** ______

Exercises 30–32. The graph shows the auto sales distribution for a local dealer.

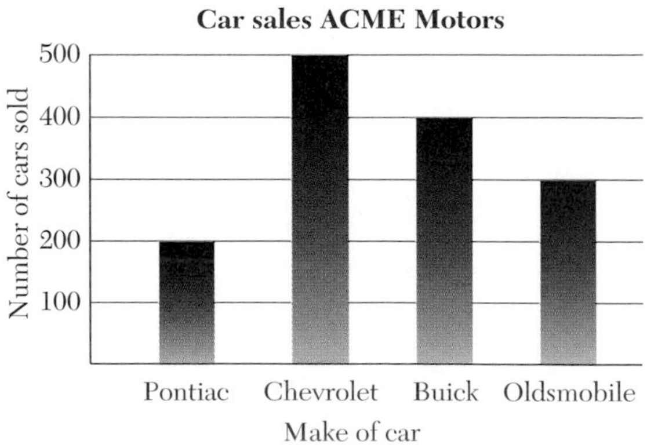

30. What make of auto had the greatest sales? **30.** ______

31. What was the total number of Pontiac and Chevrolets sold? **31.** ______

32. How many more Buicks were sold than Oldsmobiles? **32.** ______

Exercises 33–35. The table shows the number of employees by division and shift for Beaver Horseradish.

Employees by Division: Beaver Horseradish

Division	Day Shift	Night Shift
A	380	205
B	430	155
C	155	55

33. Which division has the greatest number of employees? **33.** ______________

34. How many more employees are in the day shift in Division A as compared to the day shift in Division C? **34.** ______________

35. How many employees are in the three divisions? **35.** ______________

CHAPTER 1

GROUP PROJECT (1–2 weeks)

OPTIONAL

All tables, graphs, and charts should be clearly labeled and computer generated if possible. Written responses should be typed and checked for spelling and grammar.

a. Go to the library and find the population and area for each state in the United States. Organize your information by geographic region. Record your information in a table.

b. Calculate the total population and the total area for each region. Calculate the population density (number of people per square mile, rounded to the nearest whole person) for each region, and put this and the other regional totals in a regional summary table. Then make three separate graphs, one for regional population, one for regional area, and the third for regional population density.

c. Calculate the average population per state for each region, rounding as necessary. Put this information in a bar graph. What does this information tell you about the regions? How is it different from the population density of the region?

d. How did your group decide on the makeup of the regions? Explain your reasoning.

e. Are your results what you expected? Explain. What surprised you?

GOOD ADVICE FOR STUDYING

New Habits from Old

If you are in the habit of studying math by only reading the examples to learn how to do the exercises, stop now! Instead, read the assigned section—all of it—before class. It is important that you read more than the examples so that you fully understand the concepts. Knowing how to do a problem isn't all that needs to be learned. Knowing where and when to use specific skills is also essential.

When you read, read interactively. This means that you should be both writing and thinking about what you are reading. Write down new vocabulary; perhaps start a list of new terms paraphrased in words that are clear to you. Take notes on the How & Why segments, jotting down questions you may have, for example. As you read examples, work the Warm-Up problems in the margin. Begin the exercise set only when you understand what you have read in the section. This process should make your study sessions go much faster and be more effective.

If you have written down questions during your study session, be sure to ask them at the next class session, seek help from a tutor, or discuss them with a classmate. Don't leave these questions unanswered.

Pay particular attention to the objectives at the beginning of each section. Read these at least twice: first, when you do your reading before class and again, after attending class. Ask yourself, "Do I understand what the purpose of this section is?" Read the objectives again before test time to see if you feel that you have met these objectives.

During your study session, if you notice yourself becoming tense and your breathing shallow (light and from your throat or upper part of your lungs), follow this simple coping strategy. Say to yourself, "I'm in control. Relax and take a deep breath." Breathe deeply and properly by relaxing your stomach muscle (that's right, you have permission to let your stomach protrude!) and inhaling so that the air reaches the bottom of your lungs. Hold the air in for a few seconds; then slowly exhale, pulling your stomach muscle in as you exhale. This easy exercise not only strengthens your stomach muscle but gives your body and brain the oxygen you need to perform free from physical stress and anxiety. This deep breathing relaxation method can be done in 1 to 5 minutes. You may want to use it several times a day, especially during an exam.

These techniques can help you to start studying math more effectively and to begin managing your anxiety. Begin today.

2 PRIMES AND MULTIPLES

SECTIONS

APPLICATION

Mathematicians have always been fascinated by numbers—their structure and their uses. All the other chapters in this text explain ways to use numbers. This chapter explores the structure of counting numbers. The structure of the counting numbers—their basic building blocks—is similar in nature to the concept of all molecules being made up of atoms of the basic elements. As a molecule of water is formed from two atoms of hydrogen and one atom of oxygen, H_2O, so the number 12 is made up of two factors of 2 and one factor of 3—that is, $12 = 2 \cdot 2 \cdot 3$.

Throughout the history of humanity, people have studied different sets of numbers. Some groups have even attributed magical powers to certain numbers ("lucky seven") or groups of numbers because of some special properties. We investigate here the notion of a "magic square." Magic squares are arrangements of the numbers 1–4, or 1–9, or 1–16, or 1–25, or 1–any number squared. These numbers are written in square arrays that are 2×2, or 3×3, or 4×4, and so on. The numbers are placed so that the sum of each row and of each column and of each diagonal is the same. Figure 2.1 shows an example of a 3-×-3 magic square, which uses the numbers from 1 to 9 (3^2).

2	9	4
7	5	3
6	1	8

FIGURE 2.1

Row 1	$2 + 9 + 4 = 15$	Column 1	$2 + 7 + 6 = 15$
Row 2	$7 + 5 + 3 = 15$	Column 2	$9 + 5 + 1 = 15$
Row 3	$6 + 1 + 8 = 15$	Column 3	$4 + 3 + 8 = 15$
Diagonal	$2 + 5 + 8 = 15$	Diagonal	$4 + 5 + 6 = 15$

In medieval times, some people wore magic squares and used them as talismans. The talisman wearers considered them powerful enough to furnish protection from evil spirits.

Benjamin Franklin was a big fan of magic squares. As clerk to the Pennsylvania Assembly, he admitted to creating them when he was bored with the proceedings. He created more and more complex squares as time passed. He even experimented with magic circles. Figure 2.2 shows an 8-×-8 square, using the numbers from 1 to 64 (8^2), that Franklin created. This square has several interesting features. Note the sum of the diagonals is not the same as the sum of the rows and the columns.

52	61	4	13	20	29	36	45
14	3	62	51	46	35	30	19
53	60	5	12	21	28	37	44
11	6	59	54	43	38	27	22
55	58	7	10	23	26	39	42
9	8	57	56	41	40	25	24
50	63	2	15	18	31	34	47
16	1	64	49	48	33	32	17

FIGURE 2.2

GROUP ACTIVITY

1. What is the sum of each row and each column?
2. What is the sum of the four corners? Of the four middle squares?
3. Starting at the lower left corner, 16, move diagonally up three times, move right once, move down and right diagonally to the lower right corner, 17. What is the sum of this path?
4. Start at 50 and trace the path "parallel" to the one in Exercise 3. What is the sum of this path?
5. Find at least six other paths of eight numbers through the square that have the same sum.
6. What is the sum of the first four numbers in each row? In each column? Of the last four numbers in each row or column? What is the sum of these half rows and half columns?

2.1 Divisibility Tests

VOCABULARY

A whole number is **divisible** by another whole number if the quotient of these numbers is a natural number greater than 1 and the remainder is 0. The second number is said to be a **divisor** of the first. Thus, 7 is a divisor of 42 because 42 ÷ 7 = 6.

We also say 42 is **divisible** by 7.

The **even digits** are 0, 2, 4, 6, and 8.

The **odd digits** are 1, 3, 5, 7, and 9.

OBJECTIVES

1. Determine whether a natural number is divisible by 2, 3, or 5.
2. Determine whether a natural number is divisible by 6, 9, or 10.

HOW & WHY

OBJECTIVE 1 Determine whether a natural number is divisible by 2, 3, or 5.

To ask if a number is divisible by 3 is to ask if the division of the number by 3 comes out even (has no remainder). We can answer this question by doing the division and checking to see that there is no remainder. Or, we can bypass the division by using divisibility tests. For many numbers, we can answer questions about divisibility mentally. Table 2.1 provides clues for some of these tests.

TABLE 2.1 PATTERNS OF MULTIPLYING

Some Natural Numbers	Multiply by 2	Multiply by 3	Multiply by 5	Multiply by 9
1	2	3	5	9
2	4	6	10	18
3	6	9	15	27
4	8	12	20	36
5	10	15	25	45
6	12	18	30	54
10	20	30	50	90
16	32	48	80	144
24	48	72	120	216
44	88	132	220	396
66	132	198	330	594

In the second column (Multiply by 2), the ones digit of each number is an even digit that is either 0, 2, 4, 6, or 8. Because the ones place is even, the number itself is also even.

In the third column (Multiply by 3), the sum of the digits of each number in the column is divisible by 3. For example, the sum of the digits of 48 (16 · 3) is 4 + 8 or 12, and 12 is divisible by 3. Likewise, the sum of the digits of 198 is divisible by 3 because 1 + 9 + 8 is 18.

In the fourth column (Multiply by 5), the ones digit of each number is 0 or 5.

To test for divisibility of a natural number by 2, 3, or 5

1. If the ones-place digit of the number is even (0, 2, 4, 6, or 8), the number is divisible by 2.
2. If the sum of the digits of the number is divisible by 3, then the number is divisible by 3.
3. If the ones-place digit of the number is 0 or 5, the natural number is divisible by 5.

WARM-UPS A–D

A. Is 63 divisible by 2, 3, or 5?

B. Is 390 divisible by 2, 3, or 5?

C. Is 215 divisible by 2, 3, or 5?

D. If Anna has ¥ 626 to divide among the children, will each child receive the same number of yen in whole numbers? Why or why not?

Answers to Warm-Ups

A. 63 is divisible by 3; 63 is not divisible by 2 or 5.
B. 390 is divisible by 2, 3, and 5.
C. 215 is divisible by 5; 215 is not divisible by 2 or 3.
D. No, each child will not receive the same amount because 626 is not divisible by 3.

EXAMPLES A–D

DIRECTIONS: Determine whether the natural number is divisible by 2, 3, or 5.

STRATEGY: First check the ones-place digit. If it is even, the number is divisible by 2. If it is 0 or 5, the number is divisible by 5. Next find the sum of the digits. If the sum is divisible by 3, the number is divisible by 3.

A. Is 42 divisible by 2, 3, or 5?

42 is divisible by 2. — The ones-place digit is 2.
42 is divisible by 3. — $4 + 2 = 6$, which is divisible by 3.
42 is not divisible by 5. — The ones-place digit is neither 0 nor 5.

B. Is 210 divisible by 2, 3, or 5?

210 is divisible by 2. — The ones-place digit is 0.
210 is divisible by 3. — $2 + 1 + 0 = 3$, which is divisible by 3.
210 is divisible by 5. — The ones-place digit is 0.

C. Is 523 divisible by 2, 3, or 5?

523 is not divisible by 2. — The ones-place digit is not even.
523 is not divisible by 3. — $5 + 2 + 3 = 10$, which is not divisible by 3.
523 is not divisible by 5. — The ones-place digit is neither 0 nor 5.

D. Georgio and Anna and their three children are on a trip to Okinawa. Anna has a total of ¥ 546 to divide among the children. She wants each child to receive the same number of yen in whole numbers. Is this possible? Why or why not?

Yes, each child will receive the same number of yen in whole numbers (¥ 182 each) because 546 is divisible by 3.

*

HOW & WHY

OBJECTIVE 2 Determine whether a natural number is divisible by 6, 9, or 10.

In Table 2.1, some numbers appear in both the Multiply by 2 column and the Multiply by 3 column. These numbers are divisible by 6. Because $6 = 2 \cdot 3$, every number divisible by 6 must also be divisible by 2 and 3. For example, 132 is divisible by both 2 and 3. Therefore, it is also divisible by 6.

In the Multiply by 9 column, the sum of the digits of each number is divisible by 9. For example, the sum of the digits in 36 is $3 + 6$, or 9, which is divisible by 9. Also, the sum of the digits in 981 is $9 + 8 + 1$, or 18, which is divisible by 9.

Notice that, because 0 is even, all natural numbers ending in 0 are divisible by 2. They are also divisible by 5 because they end in 0. The numbers are also divisible by 10. They appear in both the Multiply by 2 column and the Multiply by 5 column. The numbers 20, 90, and 130 are all divisible by 10.

To test for divisibility of a natural number by 6, 9, or 10

1. If the number is divisible by both 2 and 3, then the number is divisible by 6.
2. If the sum of the digits of the number is divisible by 9, then the number is divisible by 9.
3. If the ones-place digit of the number is 0, then the number is divisible by 10.

EXAMPLES E–F

DIRECTIONS: Determine whether a natural number is divisible by 6, 9, or 10.

STRATEGY: First, check whether the number is divisible by both 2 and 3. If so, the number is divisible by 6. Second, find the sum of the digits. If the sum is divisible by 9, then the number is divisible by 9. Finally, check the ones-place digit. If the digit is 0, the number is divisible by 10.

E. Is 720 divisible by 6, 9, or 10?

720 is divisible by 6.	720 is divisible by both 2 and 3.
720 is divisible by 9.	$7 + 2 + 0 = 9$, which is divisible by 9.
720 is divisible by 10.	The ones-place digit is 0.

F. Is 1770 divisible by 6, 9, or 10?

1770 is divisible by 6.	1770 is divisible by both 2 and 3.
1770 is not divisible by 9.	$1 + 7 + 7 + 0 = 15$, which is not divisible by 9.
1770 is divisible by 10.	The ones-place digit is 0.

WARM-UPS E–F

E. Is 900 divisible by 6, 9, or 10?

F. Is 1152 divisible by 6, 9, or 10?

Answers to Warm-Ups

E. 900 is divisible by 6, 9, and 10.

F. 1152 is divisible by 6 and 9; 1152 is not divisible by 10.

EXERCISES 2.1

OBJECTIVE 1 Determine whether a natural number is divisible by 2, 3, or 5.

A. *Is each number divisible by 2?*

1. 25 **2.** 22 **3.** 20

4. 73 **5.** 38 **6.** 58

Is each number divisible by 5?

7. 52 **8.** 23 **9.** 45

10. 95 **11.** 551 **12.** 507

Is each number divisible by 3?

13. 75 **14.** 30 **15.** 36

16. 35 **17.** 43 **18.** 53

B. *Determine whether the natural number is divisible by 2, 3, or 5.*

19. 3660 **20.** 2670 **21.** 3998 **22.** 4998

23. 4815 **24.** 1845 **25.** 7275 **26.** 5280

27. 11,115 **28.** 11,116

OBJECTIVE 2 Determine whether a natural number is divisible by 6, 9, or 10.

A. *Is each number divisible by 6?*

29. 108 **30.** 801 **31.** 254

32. 333 **33.** 444 **34.** 452

Is each number divisible by 9?

35. 108 **36.** 801 **37.** 376

38. 333 **39.** 414 **40.** 765

Is each number divisible by 10?

41. 233 **42.** 330 **43.** 875

44. 607 **45.** 1920 **46.** 9210

B. *Determine whether the natural number is divisible by 6, 9, or 10.*

47. 4920 **48.** 4896 **49.** 6993 **50.** 3780

51. 5555 **52.** 8888 **53.** 5700 **54.** 7880

55. 7290 **56.** 9066

C.

57. Pedro and two friends plan to buy a used car that is priced at $1221. Is it possible for each of them to spend the same whole number of dollars? Explain.

58. Janna and four of her friends plan to run 82 miles in relays. Is it possible for each runner to run the same whole number of miles? Explain.

59. A marching band has 175 members. Can the band march in rows of 3 without any member being left over? Rows of 5? Rows of 10?

60. Joe has a bookshelf that is 36 in. wide. He wants to use it for a set of encyclopedia. Each volume is 2 in. thick. Is it possible for him to completely fill the shelf with volumes? Could he completely fill the shelf with 3-in. volumes?

61. Lucia has 120 students in a single class at a community college. She wants to divide the students into small, equal-sized groups to work on a group project. Is it possible to have groups of 5 students? Of 6 students? Of 9 students? Explain.

62. Six merchants combine to build one new store in each of 318 cities. Is it possible for each merchant to oversee the building of the same number of new stores? Explain.

63. Allen has an extensive collection of 864 recordings. He plans to divide the collection evenly among his five nieces and four nephews. Will each receive the same number of recordings?

64. A movie theater is to have 250 seats. The manager plans to arrange them in rows, with the same number of seats in each row. Is it possible to have rows of 10 seats each? Rows of 15 seats? Determine the answers by using divisibility tests.

Exercises 65–68 relate to the chapter application.

65. Mark and Barbara attend the zoo and see a pen with peacocks and water buffalo. On the way out, Barbara remarks that the pen has 30 eyes and 44 feet. How many animals are in the pen?

66. In Exercise 65, how many peacocks and how many water buffalo are in the pen?

67. A small circus has an act with elephants and riders. In all, there are 48 eyes and 68 feet. What is the total number of elephants plus riders?

68. In Exercise 67, how many elephants and how many riders are in the act?

STATE YOUR UNDERSTANDING

69. Explain what it means to say, "This number is divisible by 5." Give an example of a number that is divisible by 5 and one that is not divisible by 5.

70. Explain why a number that is divisible by both 2 and 3 must also be divisible by 6.

71. Explain the difference between the divisibility tests for 2 and 3.

72. Write a short statement to explain why every number divisible by 9 is also divisible by 3.

CHALLENGE

73. Is 23,904 divisible by 6?

74. Is 11,370 divisible by 15? Write a divisibility test for 15.

75. Is 11,370 divisible by 30? Write a divisibility test for 30.

76. Is 99,000,111,370 divisible by:

a. 2?________ because the ________________

b. 3?________ because the ________________

c. 5?________ because the ________________

d. 6?________ because the ________________

e. 9?________ because the ________________

f. 10?________ because the ________________

GROUP WORK

77. As a group, find divisibility tests for 4, 8, 20, and 25. Report to the class and compare your tests with the other groups.

78. Call a local recycling center and ask for the number of pounds of newspaper the center collects for a complete week. Use the figure given and round to the nearest whole number. Is this amount divisible into 2-, 3-, 5-, 6-, 9-, or 10-lb bins? Which size is the most efficient? Why?

79. Make up a puzzle that is similar to those in Exercises 65–68. Trade puzzles with another group and solve each others' puzzle. One student group made up a puzzle with a group of animals having 25 eyes and 37 feet. Barring deformed animals, explain why this is not possible.

MAINTAIN YOUR SKILLS

80. Round 56,857 to the nearest thousand and nearest ten thousand.

81. Round 5,056,857 to the nearest ten thousand and nearest hundred thousand.

82. Divide: $792 \div 66$

83. Divide: $1386 \div 66$

84. Find the perimeter of a square that is 14 cm on a side.

85. Find the area of a square that is 14 cm on a side.

86. Multiply 12 by 1, 2, 3, 4, 5, and 6.

87. Multiply 13 by 1, 2, 3, 4, 5, and 6.

88. Multiply 123 by 1, 2, 3, 4, 5, and 6.

89. Multiply 1231 by 1, 2, 3, 4, 5, and 6

2.2 Multiples

VOCABULARY

A **multiple** of a whole number is the product of that number and a natural number. For instance,

21 is a multiple of 7 because 7(3) = 21.

77 is a multiple of 7 because 7(11) = 77.

98 is a multiple of 7 because 7(14) = 98.

147 is a multiple of 7 because 7(21) = 147.

OBJECTIVES

1. List multiples of a whole number.
2. Determine whether a given whole number is a multiple of another whole number.

HOW & WHY

OBJECTIVE 1 List multiples of a whole number.

To list the multiples of 7, we multiply 7 by each natural number. See Table 2.2.

TABLE 2.2 MULTIPLES OF 7

Natural Number	Multiple of 7
1	7
2	14
3	21
4	28
5	35
6	42
...	...
15	105
...	...
47	329
...	...

Table 2.2 can be continued without end. We say that the 1st multiple of 7 is 7, the 2nd multiple of 7 is 14, the 15th multiple of 7 is 105, the 47th multiple of 7 is 329, and so on. To find a particular multiple of 7, say the 23rd, we multiply by 23. The 23rd multiple of 7 is 161.

EXAMPLES A–E

DIRECTIONS: List the designated multiples.

STRATEGY: Multiply the natural number by the given value.

A. List the first five multiples of 8.

$1 \cdot 8$ *or* 8 Multiply 8 by 1, 2, 3, 4, and 5.
$2 \cdot 8$ *or* 16
$3 \cdot 8$ *or* 24
$4 \cdot 8$ *or* 32
$5 \cdot 8$ *or* 40

The first five multiples of 8 are 8, 16, 24, 32, and 40.

WARM-UPS A–E

A. List the first five multiples of 4.

Answers to Warm-Ups

A. 4, 8, 12, 16, 20

B. List the first five multiples of 16.

B. List the first five multiples of 15.

$1 \cdot 15$ *or* 15 — Multiply 15 by 1, 2, 3, 4, and 5.
$2 \cdot 15$ *or* 30
$3 \cdot 15$ *or* 45
$4 \cdot 15$ *or* 60
$5 \cdot 15$ *or* 75

The first five multiples of 15 are 15, 30, 45, 60, and 75.

CALCULATOR EXAMPLE

C. Find the 5th, 23rd, 28th, and 452nd multiples of 9.

C. Find the 5th, 23rd 28th, and 452nd multiples of 14.

STRATEGY: Use a calculator to multiply 5, 23, 28, and 452 by 14. The 5th multiple of 14 is 70, the 23rd multiple of 14 is 322, the 28th multiple of 14 is 392, and the 452nd multiple of 14 is 6328.

CALCULATOR EXAMPLE

D. Find all of the multiples of 8 between 300 and 340.

D. Find all of the multiples of 7 between 300 and 340.

STRATEGY: Use a calculator to make a quick estimate. Start with, say, the 42nd multiple. Multiply 42 times 7.

$42(7) = 294$ — The product, 294, is less than 300.
$43(7) = 301$ — This is the first multiple of 7 larger than 300.

The 43rd multiple is 301, so keep on multiplying.

$43(7) = 301$ $\quad 44(7) = 308$ $\quad 45(7) = 315$
$46(7) = 322$ $\quad 47(7) = 329$ $\quad 48(7) = 336$

The multiples of 7 between 300 and 340 are 301, 308, 315, 322, 329, and 336.

E. Maria's mathematics teacher assigns homework problems from 1 to 70 that are multiples of 4. Which problems should she work?

E. Jordan's mathematics teacher assigns homework problems from 1 to 60 that are multiples of 6. Which problems should he work?

STRATEGY: Find the multiples of 6 from 1 to 60 by multiplying by 1,2, 3, 4, and so on, until the product is 60 or larger.

$1(6) = 6$ $\quad 2(6) = 12$ $\quad 3(6) = 18$ $\quad 4(6) = 24$ $\quad 5(6) = 30$
$6(6) = 36$ $\quad 7(6) = 42$ $\quad 8(6) = 48$ $\quad 9(6) = 54$ $\quad 10(6) = 60$

Jordan should work problems 6, 12, 18, 24, 30, 36, 42, 48, 54, and 60.

HOW & WHY

OBJECTIVE 2 Determine whether one whole number is a multiple of another whole number.

If one number is a multiple of another number, the first must be divisible by the second. To determine whether 870 is a multiple of 6, we check to see if 870 is divisible by 6. That is, we check to see whether 870 is divisible by both 2 and 3.

852 — Divisible by 2 because the ones-place digit is even (2)
852 — Divisible by 3 because $8 + 5 + 2 = 15$, which is divisible by 3

So 852 is a multiple of 6.

Answers to Warm-Ups

B. 16, 32, 48, 64, 80
C. 45, 207, 252, 4068
D. 304, 312, 320, 328, 336
E. Maria should work problems 4, 8, 12, 16, 20, 24, 28, 32, 36, 40, 44, 48, 52, 56, 60, 64, and 68.

If there is no divisibility test, use long division. For example, is 299 a multiple of 13? Divide 299 by 13.

$$\begin{array}{r} 23 \\ 13\overline{)299} \\ \underline{26} \\ 39 \\ \underline{39} \\ 0 \end{array}$$

Because 299 ÷ 13 = 21 with no remainder, 299 = 13 × 23. Therefore, 299 is a multiple of 13.

EXAMPLES F–J

DIRECTIONS: Determine whether a given whole number is a multiple of another whole number.

STRATEGY: Use divisibility tests or long division to determine whether the first number is divisible by the second.

F. Is 891 a multiple of 9?

Use the divisibility test for 9.

8 + 9 + 1 = 18 The sum of the digits is divisible by 9.

So 891 is a multiple of 9.

G. Is 6114 a multiple of 6?

Use the divisibility tests for 2 and 3.

6114 is divisible by 2. The ones-place digit is even (2).
6114 is divisible by 3. The sum of the digits, 6 + 1 + 1 + 4 = 12, is divisible by 3.

So 6114 is a multiple of 6.

H. Is 2103 a multiple of 19?

Use long division.

$$\begin{array}{r} 110 \\ 19\overline{)2103} \\ \underline{19} \\ 20 \\ \underline{19} \\ 13 \end{array}$$

The remainder is not 0.

So 2103 is not a multiple of 19.

I. Is 765 a multiple of 15?

Use the divisibility tests for 3 and 5 or use long division.

765 is divisible by 3. The sum of the digits, 7 + 6 + 5 = 18 , is divisible by 3.
765 is divisible by 5. The ones-place digit is 5.

Long division gives the same result because 765 ÷ 15 = 51.
So 765 is a multiple of 15.

J. Is 4 a multiple of 20?

The number 4 is not divisible by 20. The multiples of 20 are 20, 40, 60, and so on. The smallest multiple of 20 is 20 · 1 = 20.
No, 4 is not a multiple of 20.

WARM-UPS F–J

F. Is 711 divisible by 9?

G. Is 4132 a multiple of 6?

H. Is 481 a multiple of 13?

I. Is 675 a multiple of 45?

J. Is 10 a multiple of 70?

Answers to Warm-Ups

F. yes **G.** no **H.** yes
* **I.** yes **J.** no

EXERCISES 2.2

OBJECTIVE 1 List multiples of a whole number.

List the first five multiples of the whole number.

A.

1. 3 **2.** 14 **3.** 17 **4.** 16

5. 12 **6.** 24 **7.** 25 **8.** 38

9. 50 **10.** 45

B.

11. 56 **12.** 49 **13.** 68 **14.** 72

15. 83 **16.** 108 **17.** 157 **18.** 234

19. 345 **20.** 409

OBJECTIVE 2 Determine whether a given whole number is a multiple of another whole number.

A. *Is each number a multiple of 6?*

21. 42 **22.** 38 **23.** 78

24. 90 **25.** 89 **26.** 114

Is each number a multiple of 9?

27. 39 **28.** 63 **29.** 93

30. 108 **31.** 306 **32.** 199

B. *Is each number a multiple of 7?*

33. 56 **34.** 65 **35.** 84

36. 105 **37.** 119 **38.** 167

Is each number a multiple of 6? of 9? of 15?

39. 648 **40.** 480 **41.** 570

42. 675 **43.** 690 **44.** 768

Is each number a multiple of 13? of 19?

45. 286 **46.** 323 **47.** 494 **48.** 741

C.

49. Jean is driving home from work one evening when she comes across a police safety inspection team stopping cars. She knows that the team chooses every fifth car to inspect. She counts and determines that she is 16th in line. Will she be selected to have her car safety-checked?

50. In Exercise 49, would Jean be selected if she were 35th in line?

51. A teacher assigns problems numbered from 1 to 52 that are multiples of 4. Which problems should the students work?

52. A teacher assigns problems numbered from 1 to 105 that are multiples of 8. Which problems should the students work?

53. Katy reported that she counted 26 goats' feet in a pen at a petting zoo. Explain how you know that she made a mistake.

54. Vance reported that he counted 45 duck feet in a pen. Can he be correct?

55. Joaquim supervises the quality-control team at a bottling plant. His team is responsible for checking the quality of a line of soft drinks. The team is assigned to check bottles numbered 500 to 700 in each batch. Jaoquim decides to have his team check every bottle that is a multiple of 15. List the bottle numbers the team should check.

56. Minh was recently appointed to a supervisory position at a candy manufacturing plant. She oversees two employees who count the daily total number of bonbons. They count one at a time and find that the daily average produced is 1800. To increase efficiency, what numbers less than 10 could Minh recommend that the employees use? Explain why you chose these numbers.

57. A gear has 20 teeth. The gear is rotated through 260 teeth. Is it in the original position after the rotation?

58. A gear has 24 teeth. The gear is rotated through 204 teeth. Is it in its original position after the rotation?

59. According to recent nationwide estimates by the Consumer Products Safety Commission, approximately 105,000 injuries treated in hospital emergency rooms were related to in-line skating. During the same period, there were approximately 35,000 injuries due to skateboarding. Is the number of in-line skating injuries a multiple of the number of skateboarding injuries? If you were reporting a comparison of these injuries, how might you write the comparison so that it is easily understood?

Exercises 60–63 relate to the chapter application. The Pythagorean theorem states that the sides of a right triangle satisfy the equation $a^2 + b^2 = c^2$, where a, b, and c are the sides.

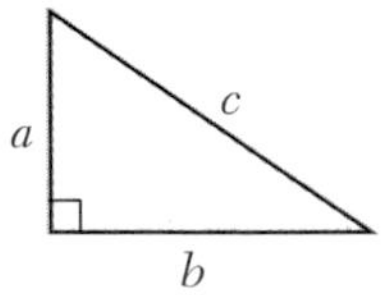

Any set of three numbers that satisfies the equation is called a Pythagorean triple. A famous Pythagorean triple is 3, 4, 5.

60. Verify that 3, 4, 5 is a Pythagorean triple.

61. A branch of mathematics called "number theory" informs us that all multiples of a Pythagorean triple are also Pythagorean triples. Find three different Pythagorean triples that are multiples of 3, 4, 5, and verify that they are also Pythagorean triples.

62. Another Pythagorean triple includes $a = 5$ and $b = 12$. What is the third number in this triple?

63. Find three triples that are multiples of the triple 5, 12, 13, and verify that they are Pythagorean triples.

STATE YOUR UNDERSTANDING

64. Write a short statement to explain why *every* multiple of 12 is also a multiple of 6.

65. One of the factors of 135 is 9 because $9 \cdot 15 = 135$. The number 9 is also a divisor of 135 because $135 \div 9 = 15$. Every factor of a number is also a divisor of that number. Explain in your own words why you think we have two different words, *factor* and *divisor*, for such numbers.

66. Suppose you are asked to list all of the multiples of 7 from 126 to 175. Describe a method you could use to be sure all the multiples are listed.

CHALLENGE

67. Is 3645 a multiple of 27?

68. Is 8008 a multiple of 91?

69. Find the largest number less than 6000 that is a multiple of 6, 9, and 17.

70. How many multiples of 3 are there between 1000 and 5000?

GROUP WORK

71. A high school marching band is practicing. When they march in rows of 2, Elmo, the sousaphone player, has no one to march with. The band director rearranges them into rows of 3, but Elmo is still left marching alone.

a. Will anyone be left over if the band marches in rows of 4? Why or why not?

b. When the band marches by 5s, Elmo is in one of the rows. Find two possible values for number of band members.

72. Draw a large circle on a sheet of paper. Mark nine points that are approximately the same distance apart on the circle and number them from 1 to 9. Next list ten consecutive multiples of your age. The first multiple and the last multiple will be associated with point number 1. Repeatedly add the digits together until you get a single digit. For example, for 75: $7 + 5 = 12$, then $1 + 2 = 3$. Connect the points on the circle in the order of the sums of the digits. Make one circle for each member of the group. Compare the designs.

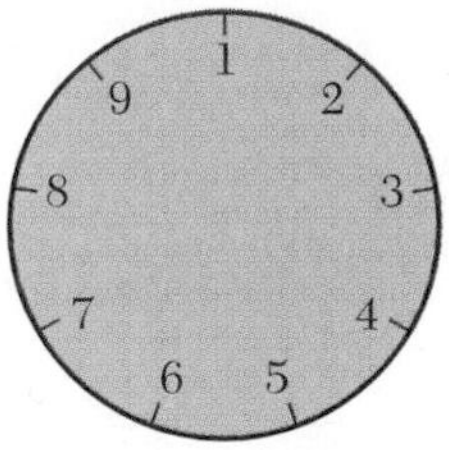

 Name ______________ Class ______________ Date ______________

MAINTAIN YOUR SKILLS

List all of the numbers from 1 to 18 that divide the given number evenly (0 remainder).

73. 12

74. 14

75. 15

76. 16

77. 17

78. 18

Find the smallest number whose square is greater than the given number.

79. 33

80. 87

81. 279

82. 500

83. Is every number divisible by both 3 and 6 also divisible by 18? Explain.

84. Is every number divisible by both 2 and 8 also divisible by 16? Explain.

2.3 DIVISORS, FACTORS, AND GREATEST COMMON FACTOR

VOCABULARY

The product of two numbers is a multiple of each. The two numbers are called **factors**. Thus, 8 and 5 are factors of 40 because $8 \cdot 5 = 40$.

Recall that $12^2 = 12 \cdot 12 = 144$, so the **square** of 12 is 144. The number 144 is called a **perfect square**.

When two or more numbers are multiplied, each number is a factor of the product. If a number is a factor of a second number, it is also a **divisor** of the second number.

The **greatest common factor (GCF)** of two or more counting numbers is the largest number that is a factor of each. The GCF of 16 and 20 is 4.

OBJECTIVES

1. Write a counting number as the product of two factors in all possible ways.
2. List all of the factors (divisors) of a counting number.
3. Find the greatest common factor (GCF) of two or more counting numbers.

HOW & WHY

OBJECTIVE 1 Write a counting number as the product of two factors in all possible ways.

Finding the factors of a number can be visualized using blocks, pennies, or other small objects. To illustrate, let's examine the factors of 12. Arrange 12 blocks or squares in a rectangle. The rectangle will have an area of 12 square units. The length and width of the rectangle are factors of 12. For instance, Figure 2.3 shows a rectangle with 3 rows of 4 squares.

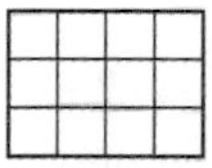

FIGURE 2.3

Because $3 \times 4 = 12$, 3 and 4 are factors of 12. Figure 2.4 shows the same 12 squares arranged into a different rectangle.

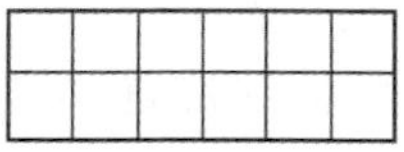

FIGURE 2.4

Figure 2.4 shows that $2 \times 6 = 12$, so 2 and 6 are also factors of 12.

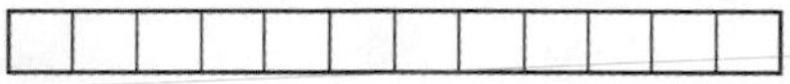

FIGURE 2.5

Figure 2.5 shows a third arrangement: $1 \times 12 = 12$, so 1 and 12 are factors of 12. Figure 2.6 shows that the result of using rows of 5, 7, 8, 9, 10, or 11 do not form rectangles.

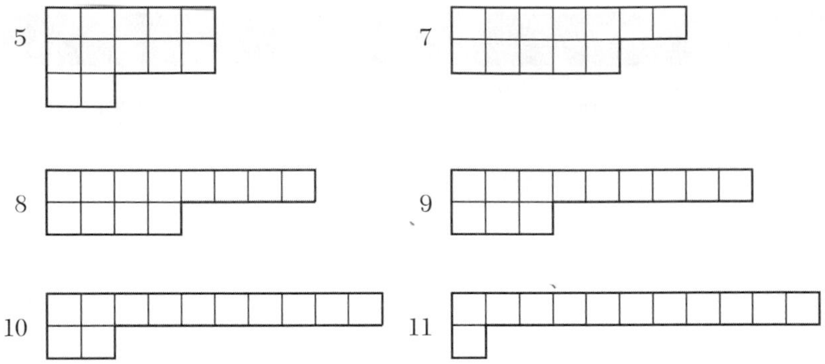

FIGURE 2.6

We conclude that the *only* pairs of factors of 12 are $1 \cdot 12$, $2 \cdot 6$, and $3 \cdot 4$.

To write a larger number, say 250, as the product of two factors in all possible ways, we could again draw rectangles, or divide 250 by every number smaller than 250. Either method takes too long. The following steps save time.

1. List all the counting numbers from 1 to the first number whose square is larger than 250. Because $15 \times 15 = 225$ and $16 \times 16 = 256$, we stop at 16.

1	6	11	16
2	7	12	
3	8	13	
4	9	14	
5	10	15	

We can stop at 16 because 250 divided by any number larger than 16 gives a quotient that is less than 16. But all the possible factors less than 16 are already in the chart.

2. Divide each of the listed numbers into 250. List the factors of the numbers that divide evenly. Otherwise, cross out the number.

$1 \cdot 250$	~~6~~	~~11~~	~~16~~
$2 \cdot 125$	~~7~~	~~12~~	
~~3~~	~~8~~	~~13~~	
~~4~~	~~9~~	~~14~~	
$5 \cdot 50$	$10 \cdot 25$	~~15~~	

When you find a number that is not a factor, you can also eliminate all the multiples of that number. For example, because 3 is not a factor of 250, we can also eliminate 6, 9, 12, and 15.

These steps give us a list of all the 2-factor products. Hence, 250 written as a product in all possible ways is

$1 \cdot 250 \qquad 2 \cdot 125 \qquad 5 \cdot 50 \qquad 10 \cdot 25$

To use the square method to write a counting number as the product of two factors in all possible ways

1. List all the counting numbers from 1 to the first number whose square is larger than the given number.
2. For each number on the list, test whether the number is a divisor of the given number.
3. If the number is not a divisor, cross it and all of its multiples off the list.
4. If the number is a divisor, write the indicated product of the two factors. The first factor is the tested number; the second factor is the quotient.

EXAMPLES A–C

DIRECTIONS: Write the counting number as the product of two factors in all possible ways.

STRATEGY: Use the square method. Begin by testing all of the counting numbers from 1 to the first number whose square is larger than the given number.

A. Write 102 as the product of two factors in all possible ways.

$1 \cdot 102$	$6 \cdot 17$	~~11~~
$2 \cdot 51$	~~7~~	
$3 \cdot 34$	~~8~~	
~~4~~	~~9~~	
~~5~~	~~10~~	

We can stop at 11 because $11^2 = 121$, which is larger than 102.

The pairs of factors whose product is 102 are $1 \cdot 102$, $2 \cdot 51$, $3 \cdot 34$, and $6 \cdot 17$.

B. Write 198 as the product of two factors in all possible ways.

$1 \cdot 198$	$6 \cdot 33$	$11 \cdot 18$
$2 \cdot 99$	~~7~~	~~12~~
$3 \cdot 66$	~~8~~	~~13~~
~~4~~	$9 \cdot 22$	~~14~~
~~5~~	~~10~~	~~15~~

We can stop at 15 because $15^2 = 225$, which is larger than 198.

The pairs of factors whose product is 198 are $1 \cdot 198$, $2 \cdot 99$, $3 \cdot 66$, $6 \cdot 33$, $9 \cdot 22$, and $11 \cdot 18$.

C. A television station has 130 minutes of late-night programming to fill. In what ways can the time be scheduled if each program must last a whole number of minutes and if each schedule must include programs all the same length?

STRATEGY: List the pairs of factors of 130.

$1 \cdot 130$
[1 program that is 130 minutes long or 130 programs that are 1 minute long (probably too short)]
$2 \cdot 65$
[2 programs that are 65 minutes long or 65 programs that are 2 minutes long]
~~3~~
~~4~~
$5 \cdot 26$
[5 programs that are 26 minutes long or 26 programs that are 5 minutes long]
~~6~~
~~7~~
~~8~~
~~9~~
$10 \cdot 13$
[10 programs that are 13 minutes long or 13 programs that are 10 minutes long]
~~11~~
~~12~~ Stop here because $12^2 = 144$, which is larger than 130.

WARM-UPS A–C

A. Write 56 as the product of two factors in all possible ways.

B. Write 150 as the product of two factors in all possible ways.

C. A television station has 90 minutes of programming to fill. In what ways can the time be scheduled if each program must last a whole number of minutes and if each schedule must include programs all the same length?

Answers to Warm-Ups

A. $1 \cdot 56$, $2 \cdot 28$, $4 \cdot 14$, and $7 \cdot 8$

B. $1 \cdot 150$, $2 \cdot 75$, $3 \cdot 50$, $5 \cdot 30$, $6 \cdot 25$, and $10 \cdot 15$

C. The time can be filled with 1 program of 90 minutes or 90 programs 1 minute long, 2 programs 45 minutes long or 45 programs 2 minutes long; 3 programs 30 minutes long or 30 programs 3 minutes long; 5 programs 18 minutes long or 18 programs 5 minutes long; 6 programs 15 minutes long or 15 programs 6 minutes long; 9 programs 10 minutes long or 10 programs 9 minutes long.

HOW & WHY

OBJECTIVE 2 List all of the factors (divisors) of a counting number.

The square method to find pairs of factors also gives us a list of *all* factors or divisors of a given whole number. To make a list of all factors of 250, in order, we can use the chart for all pairs of factors of 250.

↓
$1 \cdot 250$ Read down the left column of factors and then up the right column of factors.
$2 \cdot 125$
$5 \cdot 50$
$10 \cdot 25$
└─↑

Reading in the direction of the arrows, we see that the ordered list of all factors of 250 is: 1, 2, 5, 10, 25, 50, 125, and 250.

To list, in order, all the factors or divisors of a number

1. List all pairs of factors of the number in vertical form.
2. Read down the left column of factors and up the right column.

WARM-UPS D–E

D. List all the factors of 124.

E. List all the factors of 53.

Answers to Warm-Ups

D. 1, 2, 4, 31, 62, and 124
E. 1 and 53

EXAMPLES D–E

DIRECTIONS: List, in order, all factors of a given whole number.

STRATEGY: Use the square method to find all the pairs of factors. List the factors in order, and read down the left column of factors and up the right column.

D. List all the factors of 126.

$1 \cdot 126$	$6 \cdot 21$	~~11~~	We can stop at 12 because $12^2 = 144$, which is larger than 126.
$2 \cdot 63$	$7 \cdot 18$	~~12~~	
$3 \cdot 42$	~~8~~		
~~4~~	$9 \cdot 14$		
~~5~~	~~10~~		

↓
$1 \cdot 126$ List the pairs, and read the factors, following the arrow.
$2 \cdot 63$
$3 \cdot 42$
$6 \cdot 21$
$7 \cdot 18$
$9 \cdot 14$
└─↑

In order, all the factors of 126 are 1, 2, 3, 6, 7, 9, 14, 18, 21, 42, 63, and 126.

E. List all the factors of 41.

$1 \cdot 41$	~~4~~	~~7~~	Stop at 7 because $7^2 = 49$.
~~2~~	~~5~~		
~~3~~	~~6~~		

All the factors of 41 are 1 and 41.

HOW & WHY

OBJECTIVE 3 *Find the greatest common factor (GCF) of two or more counting numbers.*

To find the greatest common factor of 45 and 120, we first list all the factors of each number.

The factors of 45 are 1, 3, 5, 9, 15, and 45.
The factors of 120 are 1, 2, 3, 4, 5, 6, 8, 10, 12, 15, 20, 24, 30, 40, 60, and 120.

Now, circle the factors that are common to both lists.

The factors of 45 are ①, ③, ⑤, 9, ⑮, and 45.
The factors of 120 are ①, 2, ③, 4, ⑤, 6, 8, 10, 12, ⑮, 20, 24, 30, 40, 60, and 120.

The common factors are 1, 3, 5, and 15. The greatest common factor (GCF) is 15.

To find the greatest common factor (GCF) of two or more counting numbers

1. List all the factors of each number.
2. Circle the factors common to each list.
3. Write the largest (greatest) common factor that is circled.

EXAMPLES F–G

DIRECTIONS: Find the GCF, greatest common factor, of the given counting numbers.

STRATEGY: List the factors of each number. Circle the common factors. Write the greatest common factor.

F. Find the GCF of 75 and 100.

The factors of 75 are ①, 3, ⑤, 15, ㉕, and 75.
The factors of 100 are ①, 2, 4, ⑤, 10, 20, ㉕, 50, and 100.

List the factors of each number and circle the common factors.

The GCF of 75 and 100 is 25.

The largest common factor is 25.

G. Find the GCF of 72, 120, and 144.

The factors of 72 are ①, ②, ③, ④, ⑥, ⑧, 9, ⑫, 18, ㉔, 36, and 72.
The factors of 120 are ①, ②, ③, ④, 5, ⑥, ⑧, 10, ⑫, 20, ㉔, 30, 40, 60, and 120.
The factors of 144 are ①, ②, ③, ④, 6, ⑧, 9, ⑫, 16, 18, ㉔, 36, 72, and 144.

List the factors of each number and circle the common factors.

The GCF of 72, 120, and 144 is 24.

WARM-UPS F–G

F. Find the GCF of 60 and 80.

G. Find the GCF of 90, 150, and 180.

Answers to Warm-Ups
F. 20 **G.** 30

EXERCISES 2.3

OBJECTIVE 1 Write a counting number as the product of two factors in all possible ways.

Write the counting number as the product of two factors in all possible ways.

A.

1. 16 **2.** 18 **3.** 19 **4.** 31

5. 20 **6.** 27 **7.** 46 **8.** 50

9. 49 **10.** 71

B.

11. 95 **12.** 98 **13.** 100 **14.** 104

15. 112 **16.** 105 **17.** 115 **18.** 128

19. 444 **20.** 445

OBJECTIVE 2 List all of the factors (divisors) of a counting number.

List all of the factors (divisors) of the counting number.

A.

21. 16 **22.** 18 **23.** 23 **24.** 29

25. 30 **26.** 36 **27.** 51 **28.** 56

29. 65 **30.** 68

B.

31. 72 **32.** 76 **33.** 84 **34.** 90

35. 102 **36.** 116 **37.** 122 **38.** 142

39. 162 **40.** 182

OBJECTIVE 3 Find the greatest common factor (GCF) of two or more counting numbers.

Find the GCF of each set of numbers.

A.

41. 8, 16

42. 9, 21

43. 10, 28

44. 10, 35

45. 30, 105

46. 28, 126

47. 48, 120

48. 90, 108

49. 130, 208

50. 168, 216

B.

51. 8, 16, 88

52. 9, 21, 51

53. 12, 84, 114

54. 14, 70, 126

55. 28, 56, 98

56. 105, 120, 210

57. 112, 144, 192

58. 36, 117, 207

59. 28, 60, 108, 180

60. 36, 54, 108, 198

C. *Write the counting number as the product of two factors in all possible ways.*

61. 445

62. 650

63. 652

64. 480

65. 680

66. 720

67. In what ways can a television station schedule 120 minutes of time if each program must last a whole number of minutes, and each schedule must include programs of the same length?

Number of Programs	Length of Each

68. In what ways can a television station schedule 75 minutes of time if each program must last a whole number of minutes, and each schedule must include programs of the same length?

Number of Programs	Length of Each

69. Child care experts recommend that child care facilities have 1 adult for every 3 or 4 infants. The Bee Fore School Daycare has 24 infants. If the school provides staff according to the low end of the recommendation, how many adults are needed? If it staffs according to the high end of the recommendation, how many adults are needed?

70. Child care experts recommend that child care facilities have 1 caregiver for 7 to 10 preschoolers. KinderCare has 65 preschoolers. If KinderCare has 5 caregivers, does the facility meet the recommendations?

71. A marching band has 50 members. Excluding single-file marchers, list all the rectangular configurations possible for the band.

72. Montereigh High School has a marching band with 72 members. Excluding single-file marchers, list all of the rectangular configurations possible for the band.

73. The director of the Forefront Marching Band arranged the band members in rows of 4 but had one person left over. He then arranged the band in rows of 5 but still had one person left over. What do you know about the number of members in the band? What is the smallest possible number of band members?

74. Jack has 36 poker chips. He is arranging them in the shape of a rectangle. How many different ways can he do this?

Exercises 75–78 relate to the chapter application.

The ancient Greeks were fascinated by numbers, in part because every Greek letter has a number associated with it. In particular, each Greek name had its own number and everyone was very interested in the properties of the number for his or her name. An exceptional individual was one whose number was a **perfect number**. A **perfect number** is a number that is the sum of all its divisors, excluding the number itself. *Historical note*: The Greeks did not consider a number to be a divisor of itself, although today we do.

75. List all the divisors of 6. Find the sum of the divisors that are less than 6. Is 6 a perfect number?

76. Find another perfect number less than 20.

77. Find a perfect number between 20 and 30.

78. The third perfect number is 496. Verify that it is a perfect number.

STATE YOUR UNDERSTANDING

79. Explain the difference between a factor and a divisor of a number.

80. Describe how multiples, factors, and divisors are related to each other.

CHALLENGE

81. Find the largest factor of 2973 that is less than 2973.

82. Find the largest factor of 3381 that is less than 3381.

GROUP WORK

83. Have each person in the group choose a whole number, making sure each person chooses a different whole number. Have each person with an even number divide by 2. Have each person with an odd number, multiply by 3 and add 1. Record the results. Repeat the same process with the new number—that is, divide each even number by 2 and multiply each odd number by 3 and add 1. Do it again a third time. Continue until you know you have gone as far as possible. Now have each person compare her or his final result with that of other members of the group. Everyone should end with the same number, regardless of the number she or he started with. Mathematicians believe that this will always happen, yet they have been unable to prove that it is true for *all* whole numbers. This is called the "Syracuse conjecture."

84. In a recent year, the average American consumed approximately 42 gallons of soft drinks, 25 gallons of milk, and 40 gallons of alcoholic beverages. Determine the number of ounces of soft drinks, milk, and alcoholic beverages each member of your group consumes in 1 week. Multiply these amounts by 52 to get an estimate of the annual consumption. Divide by 128 to find the number of gallons per category per person. Determine a group average for each category. Compare these with the given national average in a chart or graph.

MAINTAIN YOUR SKILLS

Multiply.

85. 49(51)

86. 68(404)

87. 88(432)

88. 407(702)

Divide.

89. $78\overline{)2574}$

90. $82\overline{)24,682}$

91. $306\overline{)8265}$

92. $306\overline{)92,110}$

93. How many speakers can be wired from a spool of wire containing 1000 feet if each speaker requires 24 feet of wire? How much wire is left?

94. A consumer magazine tested 15 brands of tires to determine the number of miles each brand could travel before the tread would be worn away. The results are shown in the graph. What was the average mileage of the 15 brands?

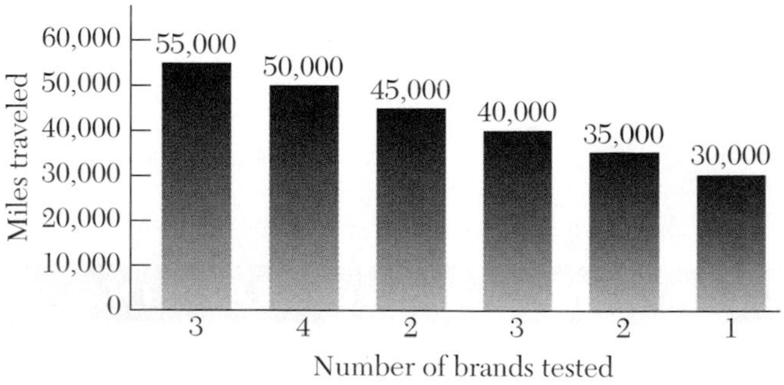

2.4 PRIMES AND COMPOSITES

VOCABULARY

A **prime number** is a whole number greater than 1 with exactly two different factors (divisors).

The two factors are the number 1 and the number itself.

A **composite number** is a whole number greater than 1 with more than two different factors (divisors).

OBJECTIVE

1. Determine whether a whole number is prime or composite.

HOW & WHY

OBJECTIVE 1 Determine whether a whole number is prime or composite.

The whole numbers zero (0) and one (1) are neither prime nor composite. The number 2 is the first prime number ($2 = 1 \cdot 2$), because 1 and 2 are the only factors of 2. The number 3 ($3 = 3 \cdot 1$) is also prime because 1 and 3 are its only factors. The number 4 is a composite number ($4 = 1 \cdot 4$ and $4 = 2 \cdot 2$) because 4 has more than two factors.

To determine whether a number is prime or composite, list its factors or divisors in a chart like those in Section 2.3; then count the number of factors. For instance, the chart for 247 is

1 · 247	~~7~~	13 · 19
~~2~~	~~8~~	14
~~3~~	~~9~~	15
~~4~~	~~10~~	16
~~5~~	~~11~~	
~~6~~	~~12~~	

We would have to test up to 16 in order to find all the factors of 247, as $16^2 = 256$ and $256 > 247$. Because we do not need all the factors, we stop at 13 because we know that 247 has at least four factors.

Therefore, 247 is a composite number.

The chart for 311 is

1 · 311	~~6~~	~~11~~	~~16~~
~~2~~	~~7~~	~~12~~	~~17~~
~~3~~	~~8~~	~~13~~	~~18~~
~~4~~	~~9~~	~~14~~	
~~5~~	~~10~~	~~15~~	

Stop here because $18 \cdot 18 = 324$.

The number 311 is a prime number because it has exactly two factors, 1 and itself.

All prime numbers up to any given number may be found by a method called the Sieve of Eratosthenes. Eratosthenes (born around 230 B.C.) is remembered for both the prime Sieve and his method of measuring the circumference of Earth. The accuracy of his measurement of Earth, compared with modern methods, is within 50 miles, or six-tenths of 1%.

To use the famous Sieve to find all the primes from 2 to 35, list the numbers from 2 to 35.

(2)	3	~~4~~	5	~~6~~	7	~~8~~	9
~~10~~	11	~~12~~	13	~~14~~	15	~~16~~	17
~~18~~	19	~~20~~	21	~~22~~	23	~~24~~	25
~~26~~	27	~~28~~	29	~~30~~	31	~~32~~	33
~~34~~	35						

The number 2 is prime, but all other multiples of 2 are not prime. They are crossed off.

The next number is 3, which is prime.

(2)	(3)	~~4~~	5	~~6~~	7	~~8~~	~~9~~	All remaining multiples of 3 are not prime, so they are crossed off—that is, 9, 15, 21, and so on.
~~10~~	11	~~12~~	13	~~14~~	~~15~~	~~16~~	17	
~~18~~	19	~~20~~	~~21~~	~~22~~	23	~~24~~	25	
~~26~~	~~27~~	~~28~~	29	~~30~~	31	~~32~~	~~33~~	
~~34~~	35							

The number 4 has already been eliminated. The next number, 5, is prime.

(2)	(3)	~~4~~	(5)	~~6~~	(7)	~~8~~	~~9~~	All remaining multiples of 5 are not prime, so they are crossed off.
~~10~~	(11)	~~12~~	(13)	~~14~~	~~15~~	~~16~~	(17)	
~~18~~	(19)	~~20~~	~~21~~	~~22~~	(23)	~~24~~	~~25~~	
~~26~~	~~27~~	~~28~~	(29)	~~30~~	(31)	~~32~~	~~33~~	
~~34~~	~~35~~							

The multiples of the remaining numbers, except themselves, have been crossed off. We only need to test divisors up to the first number whose square is larger than 30 $(6 \cdot 6 = 36)$.

The prime numbers less than 35 are 2, 3, 5, 7, 11, 13, 17, 19, 23, 29, and 31.

Looking at the Sieve, we see that we can shorten the factor chart by omitting all the numbers except those that are prime. For example, is 413 prime or composite?

$1 \cdot 413$
~~2~~
~~3~~
~~5~~
$7 \cdot 59$ Stop here because we do not need all the factors.

The number 413 is composite because it has *at least* four factors: 1, 7, 59, and 413. We know that a number is prime if no smaller prime number divides it evenly.

Keep the divisibility tests for 2, 3, and 5 in mind because they are prime numbers.

To determine whether a given number is prime or composite (0 and 1 are neither prime nor composite)

Test every prime number whose square is less than the given number.

a. If the given number has exactly two divisors (factors), the number 1 and itself, it is prime.
b. If the given number has more than two divisors (factors), it is composite.

WARM-UPS A–F

A. Is 79 prime or composite?

Answers to Warm-Ups

A. prime

EXAMPLES A–F

DIRECTIONS: Determine whether a given number is prime or composite.

STRATEGY: Test all possible prime factors of the given number. If there are exactly two factors, the number 1 and itself, the number is prime.

A. Is 107 prime or composite?

Test 1 and the prime numbers from 2 to 11. We stop at 11 because $11^2 > 107$.

$1 \cdot 107$ ~~7~~
~~2~~ ~~11~~
~~3~~
~~5~~

The numbers 2, 3, and 5 can be crossed out using the divisibility tests. Eliminate 7 and 11 by division.

The number 107 is a prime number.

B. Is 187 prime or composite?

Test 1 and the prime numbers from 2 to 17.

$1 \cdot 187$	~~7~~	
~~2~~	$11 \cdot 17$	Stop testing at 11 because we have at least four factors.
~~3~~		
~~5~~		

The number 187 is a composite number.

B. Is 123 prime or composite?

C. Is 257 prime or composite?

Test 1 and the prime numbers from 2 to 17.

$1 \cdot 257$	~~7~~	
~~2~~	~~11~~	
~~3~~	~~13~~	
~~5~~	~~17~~	Stop at 17 because $17^2 > 251$.

After testing all prime numbers in the list, we see that 257 has only two factors, 1 and itself. So 257 is a prime number.

C. Is 263 prime or composite?

D. Is 481 prime or composite?

Test 1 and the prime numbers from 2 to 23.

$1 \cdot 257$	~~7~~	
~~2~~	~~11~~	
~~3~~	$13 \cdot 37$	We stop testing at 13 because we have at least four factors (1, 13, 37, and 481).
~~5~~		

So 481 is a composite number.

D. Is 493 prime or composite?

E. Is 124,647 prime or composite?

124,647 is divisible by 3. The sum of the digits $1 + 2 + 4 + 6 + 4 + 7 = 24$, which is divisible by 3.

We have at least three factors, 1, 3, and 124,647.
So 124,647 is a composite number.

E. Is 301,767 prime or composite?

F. Christina won a math contest at her school. One of the questions in the contest was: Is 234,423 prime or composite? What should Christina have answered?

234,423 is not divisible by 2. The ones-place digit is 3 (not even)
234,423 is divisible by 3. $2 + 3 + 4 + 4 + 2 + 3 = 18$, which is divisible by 3.

We have at least three factors, 1, 3, and 234,423 .
Christina should have answered, "Composite."

F. In the same contest Mikey was asked: Is the number 234,425 prime or composite? What should he have answered.

Answers to Warm-Ups

B. composite **C.** prime
D. composite **E.** composite
F. composite

EXERCISES 2.4

OBJECTIVE 1 Determine whether a whole number is prime or composite.

A. *Tell whether the number is prime or composite.*

1. 4 **2.** 8 **3.** 11 **4.** 7

5. 12 **6.** 16 **7.** 17 **8.** 18

9. 24 **10.** 29 **11.** 31 **12.** 37

13. 38 **14.** 39 **15.** 41 **16.** 43

17. 40 **18.** 50 **19.** 53 **20.** 57

21. 58 **22.** 59

B. *Tell whether the number is prime or composite.*

23. 60 **24.** 63 **25.** 61 **26.** 67

27. 72 **28.** 77 **29.** 79 **30.** 75

31. 87 **32.** 89 **33.** 97 **34.** 99

35. 109 **36.** 110 **37.** 123 **38.** 133

39. 149 **40.** 153 **41.** 201 **42.** 203

43. 237 **44.** 253 **45.** 305 **46.** 321

C. *Tell whether the number is prime or composite.*

47. 401 **48.** 421 **49.** 433 **50.** 437

51. 575 **52.** 929 **53.** 3456 **54.** 3457

55. What was the last year in the 20th century that was a prime number? What is the first year in the 21st century that is a prime number?

56. How many prime numbers are there between 1 and 100? How many between 100 and 200?

57. The Wright brothers made the first sustained airplane flight at Kitty Hawk, North Carolina, in December 1903. Is 1903 a prime number?

58. Tell whether the year of your birth is a prime or composite number.

59. In the motion picture *Contact*, a scientist, played by Jodie Foster, intercepts a message from outer space. The message begins with a series of impulses grouped according to successive prime numbers. How many impulses came in the group that followed 53?

60. Vigorous physical activity can lead to accidental injury. The estimates by the Consumer Product Safety Commission of nationwide basketball injuries in recent years are given in the table. These estimates are the number of related hospital emergency room visits.

Year	1	2	3	4
Injuries	761,171	716,182	692,396	653,675

Three of the four entries are clearly not prime numbers. Identify them and tell how you know they are not prime. (*Optional*: Is the fourth number prime?)

61. How many rectangular arrangements can be made with 19 blocks?

62. One year Mr. Tyson had 73 members in his marching band. Explain why he could not arrange the band members in a rectangular shape.

Exercises 63–66 relate to the chapter application.

Many mathematicians have been fascinated with prime numbers. A French mathematician named Marin Mersenne (1588–1648) identified a special form of prime numbers, which bears his name. A Mersenne prime number has the form $M_p = 2^p - 1$, where p is a prime number. The smallest Mersenne prime is 3 because when $p = 2$, the smallest prime, $M_2 = 2^2 - 1 = 4 - 1 = 3$.

63. Find the value of the next three Mersenne prime numbers by substituting the prime numbers 3, 5, and 7 for p.

64. Mersenne was originally hoping that all numbers with his special form were prime. He discovered that this was not so when he came to M_{11}. Calculate M_{11} by substituting 11 for p and show that M_{11} is not prime.

65. Every Mersenne prime has a companion perfect number. See Exercises 75–78 in Section 2.3. The perfect number is expressed by $P = M_p \times (2^{p-1})$. Find the perfect number companion to M_3.

66. What perfect number is companion to M_5?

STATE YOUR UNDERSTANDING

67. Explain the difference between prime numbers and composite numbers. Give an example of each.

68. What is the minimum number of factors that a composite number can have?

69. What is the maximum number of factors that a composite number can have?

CHALLENGE

70. Is 23,341 a prime or composite number?

71. Is 37,789 a prime or composite number?

72. A **sphenic number** is a number that is a product of three unequal prime numbers. The smallest sphenic number is $2 \cdot 3 \cdot 5 = 30$. Is 4199 a sphenic number?

GROUP WORK

73. Mathematicians have established that every prime number greater than 3 is either one more than or one less than a multiple of 6. Use the Sieve of Eratosthenes to identify all the prime numbers less than 100. Divide up the list of these primes and verify that each of them is one away from a multiple of 6. Here is a step-by-step confirmation of this fact. As a group, supply the reasons for each step of the confirmation.

Consider this following portion of the number line. The first point, a, is at an even number.

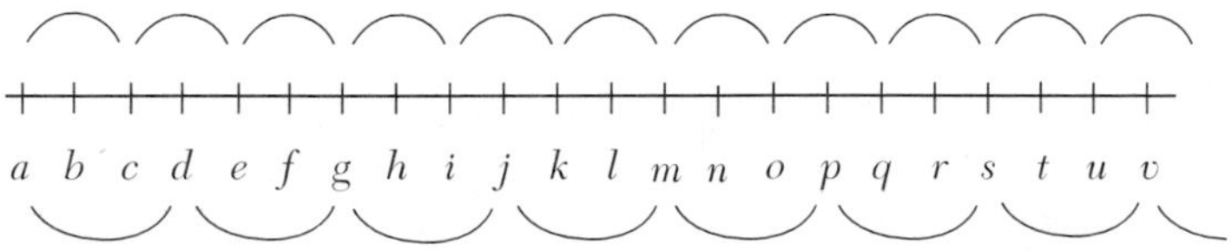

Statement

a. If a is even, then so are c, e, g, l, k, m, o, q, s, and u.
b. If a is a multiple of 3, then so are d, g, j, m, p, s, and v.
c. a, g, m, and s are multiples of 6.
d. None of the points in steps 1, 2, or 3 can be prime.
e. The only possible places for prime numbers to occur are at b, f, h, l, n, r, and t.
f. All primes are one unit from a multiple of 6.

Reason

a.
b.
c.
d.
e.
f.

74. Have each person in your group determine the total of the ages of the people now living in their household. Are these prime numbers? Report back to your class and compute the class total. Is this a prime number?

MAINTAIN YOUR SKILLS

75. Because the last digit of the number 160 is 0, 160 is divisible by 2. Is the quotient of 160 and 2 divisible by 2? Continue dividing by 2 until the quotient is not divisible by 2. What is this quotient?

76. Because the last digit of the number 416 is 6, 416 is divisible by 2. Is the quotient of 416 and 2 divisible by 2? Continue dividing by 2 until the quotient is not divisible by 2. What is this quotient?

77. Because the last digit of the number 928 is 8, 928 is divisible by 2. Is the quotient of 928 and 2 divisible by 2? Continue dividing by 2 until the quotient is not divisible by 2. What is this quotient?

78. Because the sum of the digits of 1701 is 9, 1701 is divisible by 3. Is the quotient of 1701 and 3 divisible by 3? Continue dividing by 3 until the quotient is not divisible by 3. What is this quotient?

79. Because the sum of the digits of 1029 is 12, 1029 is divisible by 3. Is the quotient of 1029 and 3 divisible by 3? Continue dividing by 3 until the quotient is not divisible by 3. What is this quotient?

80. Because the last digit of the number 2875 is 5, 2875 is divisible by 5. Is the quotient of 2875 and 5 divisible by 5? Continue dividing by 5 until the quotient is not divisible by 5. What is this quotient?

81. Because the last digit of the number 2880 is 0, 2880 is divisible by 5. Is the quotient of 2880 and 5 divisible by 5? Continue dividing by 5 until the quotient is not divisible by 5. What is this quotient?

82. The number 1029 is divisible by 7. Is the quotient of 1029 and 7 divisible by 7? Continue dividing by 7 until the quotient is not divisible by 7. What is this quotient?

83. The number 1859 is divisible by 13. Is the quotient of 1859 and 13 divisible by 13? Continue dividing by 13 until the quotient is not divisible by 13. What is this quotient?

84. Because the last digit of the number 17,408 is 8, 17,408 is divisible by 2. Is the quotient of 17,408 and 2 divisible by 2? Continue dividing by 2 until the quotient is not divisible by 2. What is this quotient?

2.5 Prime Factorization and Greatest Common Factor

VOCABULARY

The **prime factorization** of a counting number is the indicated product of prime numbers. There are two ways of asking the same question:

1. What is the prime factorization of this number?
2. Write this number in prime-factored form.

$35 = 5 \cdot 7$ and $42 = 2 \cdot 3 \cdot 7$ are prime factorizations.

$30 = 3 \cdot 10$ is not a prime factorization because 10 is not a prime number.

Recall that **exponents** show repeated factors. Using them can save space in writing.

$2 \cdot 2 \cdot 2 = 2^3$ and $3 \cdot 3 \cdot 3 \cdot 3 \cdot 7 \cdot 7 \cdot 7 = 3^4 \cdot 7^3$

Recall that the **GCF**, greatest common factor, of two or more counting numbers is the largest number that is a factor of each.

OBJECTIVES

1. Write the prime factorization of a counting number by repeated division.
2. Write the prime factorization of a counting number using the Tree Method.
3. Find the greatest common factor (GCF) of two or more counting numbers using prime factorization.

HOW & WHY

OBJECTIVE 1 Write the prime factorization of a counting number by repeated division.

In chemistry, we learn that every compound in the world is made of a particular combination of basic elements. For instance, salt is NaCl and water is H_2O. This means that every molecule of salt contains one atom of sodium (Na) and one atom of chlorine (Cl). Sodium and chlorine are two of the basic elements. Similarly, every molecule of water is made up of two atoms of hydrogen (H) and one of oxygen (O), which are also basic elements. The chemical formula tells how many units of each basic element are needed to make the compound.

In mathematics, the basic elements are the prime numbers. Every counting number (except 1) is either prime or a unique combination of prime factors. Finding the prime factorization of a composite number is comparable to finding the chemical formula for a compound. The prime factorization simply allows us to see the basic elements of a number.

To find a prime factorization, repeatedly divide by prime numbers until the quotient is 1. The prime factorization is the product of the prime divisors. We begin by dividing a given number by 2 repeatedly, until the quotient is odd. Then we divide by 3, 5, 7, 11, and so on to check for other prime factors. To save time and space, we do not rewrite the division problem each time. Instead, divide each quotient, starting at the top, and divide down until the quotient is 1. The division is shown for the number 72.

$$\begin{array}{r} 2\underline{)72} \\ 2\underline{)36} \\ 2\underline{)18} \\ 3\underline{)\ 9} \\ 3\underline{)\ 3} \\ 1 \end{array} \qquad 72 = 2 \cdot 2 \cdot 2 \cdot 3 \cdot 3 = 2^3 \cdot 3^2$$

If there is a large prime factor, you will find it when you have divided by each prime whose square is smaller than the number. Consider the number 970.

$$\begin{array}{r|l} 2\overline{)970} & \\ 5\overline{)485} & \\ 97\overline{)\ 97} & \text{97 is not divisible by 2, 3, 5, 7, or 11; also } 11^2 = 121 > 97. \\ 1 & \text{Therefore, 97 is prime.} \end{array}$$

So $970 = 2 \cdot 5 \cdot 97$.

In each case, the goal is to keep dividing until the quotient is 1.

To write the prime factorization of a counting number using repeated division

1. Divide the counting number and each following quotient by a prime number until the quotient is 1.
2. Write the indicated product of all of the divisors.

WARM-UPS A–D

A. Find the prime factorization of 60.

B. Find the prime factorization of 150.

C. Find the prime factorization of 540.

Answers to Warm-Ups

A. $2^2 \cdot 3 \cdot 5$ **B.** $2 \cdot 3 \cdot 5^2$ **C.** $2^2 \cdot 3^3 \cdot 5$

EXAMPLES A–D

DIRECTIONS: Write the prime factorization of the counting number.

STRATEGY: Use repeated division by prime numbers.

A. Find the prime factorization of 105.

STRATEGY: The last digit is not even, so start dividing by 3.

$$\begin{array}{r|l} 3\overline{)105} & \text{105 is divisible by 3.} \\ 5\overline{)\ 35} & \text{35 is divisible by 5.} \\ 7\overline{)\ \ 7} & \text{The number 7 is prime and the quotient is 1.} \\ 1 & \end{array}$$

CHECK: Multiply all of the divisors.

$3 \cdot 5 \cdot 7 = 105$

The prime factorization of 105 is $3 \cdot 5 \cdot 7$.

B. Find the prime factorization of 984.

STRATEGY: 984 is even, so start dividing by 2. Continue until 2 is no longer a divisor. Then try another prime number as a divisor.

$$\begin{array}{r|l} 2\overline{)984} & \\ 2\overline{)492} & \\ 2\overline{)246} & \\ 3\overline{)123} & \\ 41\overline{)\ 41} & \text{The number 41 is not divisible by 3, 5, or 7.} \\ 1 & \end{array}$$

CHECK: $2 \cdot 2 \cdot 2 \cdot 3 \cdot 41 = 984$

The prime factorization of 984 is $2^3 \cdot 3 \cdot 41$.

CALCULATOR EXAMPLE

C. Find the prime factorization of 5280.

STRATEGY: Record the number of times you divide by each prime number.

$2 \cdot 2 \cdot 2 \cdot 2 \cdot 2 \cdot 3 \cdot 5 \cdot 11$

The prime factorization of 5280 is $2^5 \cdot 3 \cdot 5 \cdot 11$.

D. When she was adding two fractions, Christine needed to find the least common denominator (LCD) of $\frac{5}{18}$ and $\frac{7}{24}$; therefore, she needed to write the prime factorization of 18 and 24. What are the prime factorizations of 18 and 24?

STRATEGY: Because 18 and 24 are both divisible by 2, start dividing by 2.

$$\begin{array}{r|r} 2 & 18 \\ \hline 3 & 9 \\ \hline 3 & 3 \\ \hline & 1 \end{array} \qquad \begin{array}{r|r} 2 & 24 \\ \hline 2 & 12 \\ \hline 2 & 6 \\ \hline 3 & 3 \\ \hline & 1 \end{array}$$

The prime factors of 18 and 24 are

$18 = 2 \cdot 3 \cdot 3$ or $2 \cdot 3^2$

$24 = 2 \cdot 2 \cdot 2 \cdot 3$ or $2^3 \cdot 3$

D. Chris needs to know the least common denominator (LCD) of $\frac{7}{54}$ and $\frac{11}{36}$. What are the prime factorizations of 36 and 54?

HOW & WHY

OBJECTIVE 2 *Write the prime factorization of a counting number using the Tree Method.*

Another method for prime factoring is the Tree Method. We draw "factor branches" from a given number using *any* two factors of the number. We then draw additional branches from the end of each original branch by using factors of the number at the end of the branch. A branch stops splitting when it ends in a prime number. Figure 2.7 shows the Tree Method used to find the prime factorization of 140.

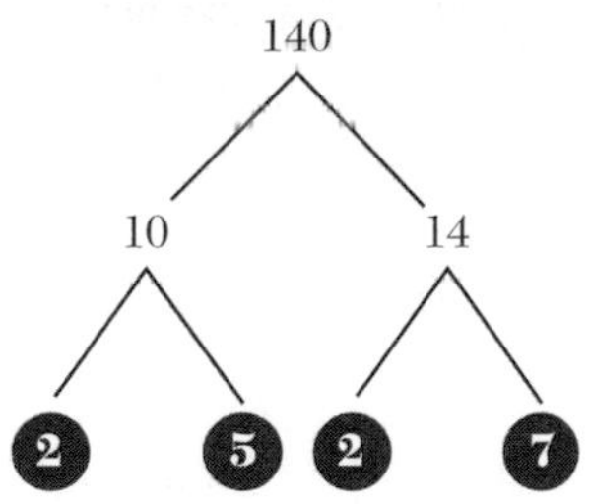

FIGURE 2.7

Often, different trees are created for the same number. Examine the two trees in Figure 2.8 used to find the prime factorization of 140.

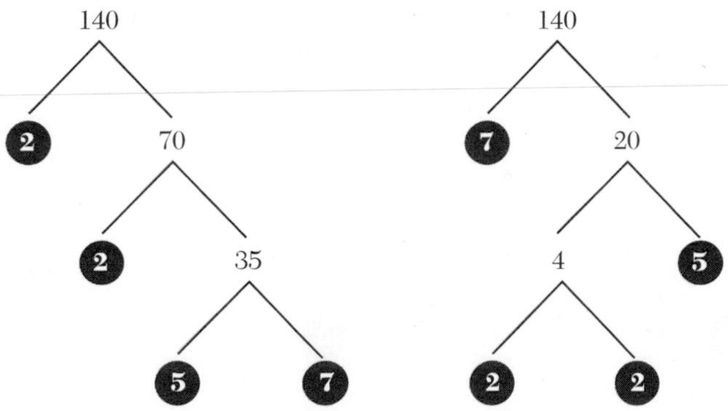

FIGURE 2.8

Answers to Warm-Ups

D. The prime factors of 36 are $2^2 \cdot 3^2$ and the prime factors of 54 are $2 \cdot 3^3$

In each case, we can see that $140 = 2^2 \cdot 5 \cdot 7$. It does not matter which tree we use. As long as we keep branching until we come to a prime number, we get the same prime factorization. In fact, *any* tree results in the same prime factorization as repeated division.

To write the prime factorization of a counting number using the Tree Method

Draw factor branches starting with any two factors of the given number. Form additional branches by using factors of the number at the end of each branch. The factoring is complete when the number at the end of each branch is a prime number.

WARM-UPS E–F

E. Find the prime factorization of 252.

F. Find the prime factorization of 468.

EXAMPLES E–F

DIRECTIONS: Write the prime factorization of the counting number.

STRATEGY: Use the Tree Method to write the given counting number as a product of primes.

E. Find the prime factorization of 198.

STRATEGY: We start with $3 \cdot 66$ as the pair of factors of 198.

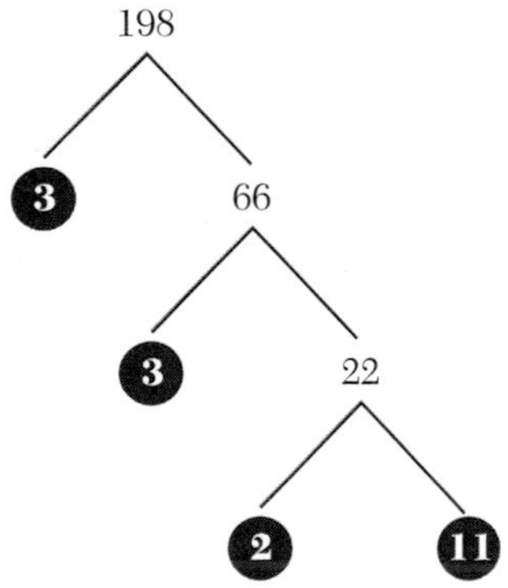

We could have used 2 and 99, or 6 and 33, or 11 and 12 as the first pairs of factors. The same prime factors would have been discovered.

The circled number at the end of each branch is a prime number.
The prime factorization of 198 is $2 \cdot 3^2 \cdot 11$.

F. Find the prime factorization of 612.

STRATEGY: Start with any two factors whose product is 612. In this case, we develop two trees, one showing factors $9 \cdot 68$ and the other, $4 \cdot 153$. The tree is shorter if larger factors are used first.

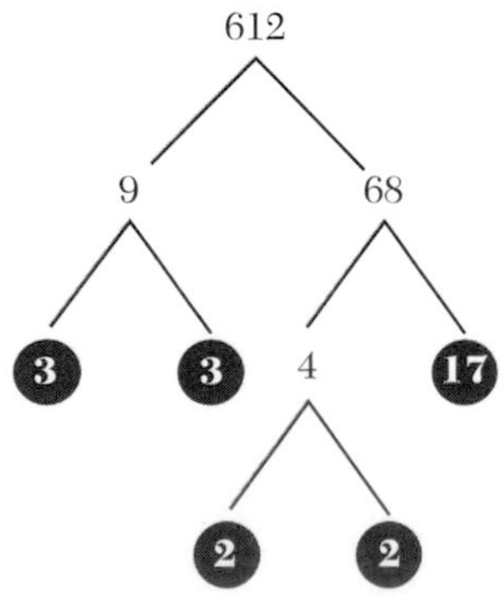

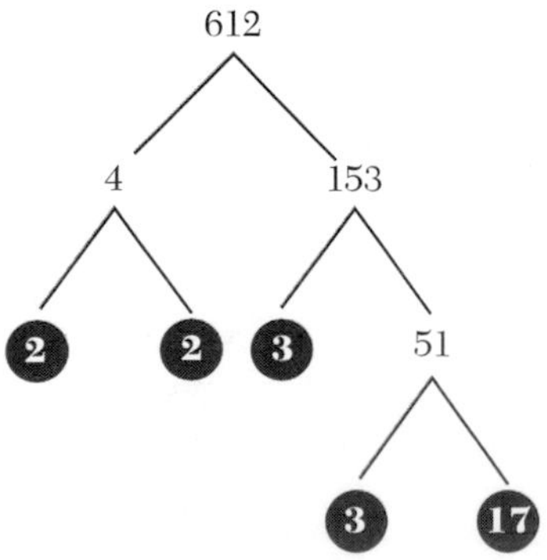

The same set of factors is shown in each tree. The factors just appear in a different order.

The prime factorization of 612 is $2^2 \cdot 3^2 \cdot 17$.

Answers to Warm-Ups

E. $2^2 \cdot 3^2 \cdot 7$ **F.** $2^2 \cdot 3^2 \cdot 13$

HOW & WHY

OBJECTIVE 3 Find the greatest common factor (GCF) of two or more counting numbers using prime factorization.

In Section 2.3, we found the GCF of 72, 120, and 144 by listing all the factors of each number and looking for the largest of them. The GCF of two or more numbers can be found more quickly using prime factorization. For example, to find the GCF of 300 and 450, find the prime factorization of each.

$300 = 2^2 \cdot 3 \cdot 5^2$

$450 = 2 \cdot 3^2 \cdot 5^2$

The GCF is the product of *all* the common factors of each number. The common prime factors are 2, 3, and 5^2.

The GCF of 300 and 450 is $2 \cdot 3 \cdot 5^2 = 150$.

To find the greatest common factor (GCF) of two or more counting numbers using prime factorization

1. Write each number in prime factorization form.
2. Identify the common prime factors. A given prime number may be a common factor more than once.
3. Find the product of the common prime factors.

EXAMPLES G–H

DIRECTIONS: Find the GCF, greatest common factor, of the given counting numbers.

STRATEGY: Find the prime factorization of each number. List all the common prime factors and multiply.

G. Find the GCF of 68 and 102.

$68 = 2^2 \cdot 17$ Find the prime factorization of each number.

$102 = 2 \cdot 3 \cdot 17$ Identify the common factors.

The common prime factors are 2 and 17.

The GCF is $2 \cdot 17 = 34$. Find the product of the common prime factors.

H. Find the GCF of 90, 252, and 630.

$90 = 2 \cdot 3^2 \cdot 5$ Find the prime factorization of each number.

$252 = 2^2 \cdot 3^2 \cdot 7$ Identify the common factors.

$630 = 2 \cdot 3^2 \cdot 5 \cdot 7$

The common prime factors are 2 and 3^2.

The GCF is $2 \cdot 3^2 = 18$. Find the product of the common prime factors.

WARM-UPS G–H

G. Find the GCF of 66 and 88.

H. Find the GCF of 80, 100, and 280.

Answers to Warm-Ups

G. 22 **H.** 20

Name ______________________ Class ______________________ Date ______________

EXERCISES 2.5

OBJECTIVE 1 Write the prime factorization of a counting number by repeated division.

OBJECTIVE 2 Write the prime factorization of a counting number using the Tree Method.

Write the prime factorization of the whole number using either method.

A.

1. 12	**2.** 14	**3.** 18	**4.** 20
5. 22	**6.** 24	**7.** 25	**8.** 26
9. 28	**10.** 30	**11.** 32	**12.** 34
13. 40	**14.** 48	**15.** 68	**16.** 72
17. 76	**18.** 84	**19.** 88	**20.** 90

B.

21. 91	**22.** 96	**23.** 102	**24.** 104
25. 105	**26.** 106	**27.** 127	**28.** 149
29. 154	**30.** 160	**31.** 162	**32.** 170
33. 180	**34.** 190	**35.** 194	**36.** 208
37. 318	**38.** 325	**39.** 396	**40.** 399

OBJECTIVE 3 Find the greatest common factor (GCF) of two or more counting numbers using prime factorization.

Find the GCF of each set of numbers.

A.

41. 40, 56	**42.** 54, 81	**43.** 45, 105	**44.** 42, 54
45. 52, 56	**46.** 54, 57	**47.** 54, 55	**48.** 55, 56
49. 56, 64	**50.** 56, 76		

B.

51. 92, 115
52. 98, 126
53. 30, 42, 66
54. 42, 60, 84

55. 56, 126, 140
56. 80, 150, 350
57. 70, 140, 350
58. 70, 105, 210

59. 116, 145, 174
60. 148, 185, 222

C. *Write the prime factorization of the counting number.*

61. 303
62. 515
63. 323
64. 437

65. 456
66. 400
67. 401
68. 409

69. 625
70. 1024
71. 1218
72. 1302

73. Determine the prime factorization of the highest temperature in the United States for today.

74. Determine the prime factorization for the sum of the digits in today's numerical date (month, day, and year).

75. Determine the prime factorization of your birth year.

Exercises 76–79 relate to the chapter application.

76. Two numbers are said to be **relatively prime** if they do not have any common factors. The numbers 14 and 15 are relatively prime because $14 = 2 \cdot 7$ and $15 = 3 \cdot 5$. Are the numbers 12 and 15 relatively prime? Explain.

77. Are the numbers 119 and 143 relatively prime? Explain.

78. Find five different numbers that are relatively prime to 12.

79. What is the largest number less than 100 that is relatively prime to 180?

STATE YOUR UNDERSTANDING

80. Explain how to write the prime factorization of 990.

81. Explain how to determine whether a number is written in prime-factored form.

Name ______________________ Class ______________________ Date ______________

CHALLENGE

82. Find the prime factorization of 1492.

83. Find the prime factorization of 2190.

84. Find the prime factorization of 1547.

85. Find the prime factorization of 1792.

GROUP WORK

86. The Goldbach conjecture states that every even number greater than 4 can be written as the sum of two prime numbers. Mathematicians believe this to be true, but as yet no one has proved that it is true for *all* even numbers. Complete the table.

Even Number	Sum of Primes	Even Number	Sum of Primes
6	3 + 3	26	
8	3 + 5	28	
10		30	
12		32	
14		34	
16		36	
18		38	
20		40	
22		42	
24		44	

Do you believe the Goldbach conjecture?

MAINTAIN YOUR SKILLS

87. Find the perimeter of the figure.

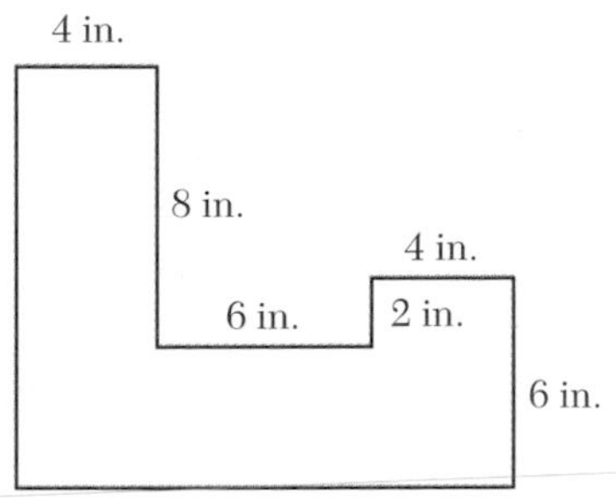

88. Find the area of the figure in Exercise 87.

89. Is 3003 a multiple of *both* 3 and 13?

90. Is 4004 a multiple of *both* 2 and 13?

91. Is 1001 a multiple of *both* 3 and 13?

92. Is 5005 a multiple of *both* 5 and 13?

93. Is 3003 a multiple of *both* 3 and 7?

94. Is 4004 a multiple of *both* 2 and 7?

95. Find two numbers that are multiples of both 3 and 5.

96. Find two numbers that are multiples of both 6 and 10.

2.6 LEAST COMMON MULTIPLE

VOCABULARY

The **least common multiple** of two or more whole numbers is

1. the smallest natural number that is a multiple of each whole number, and
2. the smallest natural number that has each whole number as a factor, and
3. the smallest natural number that has each whole number as a divisor, and
4. the smallest natural number that each whole number will divide evenly.

The abbreviation of least common multiple is **LCM**.

OBJECTIVES

1. Find the least common multiple (LCM) of two or more whole numbers using the Individual Prime-Factoring Method.
2. Find the least common multiple (LCM) of two or more whole numbers using the Group Prime-Factoring Method.

HOW & WHY

OBJECTIVE 1 Find the least common multiple (LCM) of two or more whole numbers using the Individual Prime-Factoring Method.

The factorizations of this chapter are most often used to simplify fractions and find LCMs. LCMs are used to compare, add, and subtract fractions. In algebra, LCMs are useful in equation solving.

The LCM of 21 and 35 can be discovered by listing the multiples of both and finding the smallest value common to both lists:

Multiples of 21: 21, 42, 63, 84, (105), 126, 147, 168, 189, (210), 231, . . .

Multiples of 35: 35, 70, (105), 140, 175, (210), 245, 280, 315, 650, . . .

The LCM of 21 and 35 is 105 because it is the smallest multiple in both lists. This fact can be stated in four equivalent ways.

1. 105 is the smallest natural number that is a multiple of both 21 and 35.
2. 105 is the smallest natural number that has both 21 and 35 as factors.
3. 105 is the smallest natural number that has both 21 and 35 as divisors.
4. 105 is the smallest natural number that both 21 and 35 will divide evenly.

Finding the LCM by this method has a big drawback: You might have to list hundreds of multiples to find it. For this reason, we look for a shortcut.

To find the LCM of 12 and 18, write the prime factorization of each. Write these prime factors in columns so that the prime factors of 18 are under the prime factors of 12. Leave blank spaces for prime factors that do not match.

		Primes with the Largest Exponents
$12 = 2 \cdot 2 \cdot 3$	$= (2^2) \cdot 3^1$	
$18 = 2 \quad \cdot 3 \cdot 3$	$= 2^1 \quad \cdot (3^2)$	2^2 and 3^2

Because 12 must divide the LCM, the LCM must have $2 \cdot 2 \cdot 3$ as part of its factorization; and because 18 must divide the LCM, the LCM must also have $2 \cdot 3 \cdot 3$ as part of its factorization. Thus, the LCM is $2 \cdot 2 \cdot 3 \cdot 3$, or 36. Observe that the LCM is the product of the highest power of each prime factor.

Find the LCM of 16, 10, and 24.

		Primes with the Largest Exponents
$16 = 2 \cdot 2 \cdot 2 \cdot 2$	$= (2^4)$	
$10 = 2 \qquad \cdot 5$	$= 2^1 \cdot (5^1)$	2^4, 3^1, and 5^1
$24 = 2 \cdot 2 \cdot 2 \quad \cdot 3$	$= 2^3 \cdot (3^1)$	

The LCM $= 2 \cdot 2 \cdot 2 \cdot 2 \cdot 3 \cdot 5 = 2^4 \cdot 3^1 \cdot 5^1$
$= 16 \cdot 15 = 240$

To write the least common multiple of two or more whole numbers using the Individual Prime-Factoring Method

1. Write each number in prime-factored form using exponents.
2. Find the product of the highest power of each prime factor.

WARM-UPS A–C

A. Find the LCM of 16 and 20.

B. Find the LCM of 10, 15, and 18.

C. Find the LCM of 9, 12, 18, and 24.

Answers to Warm-Ups

A. 80 **B.** 90 **C.** 72

EXAMPLES A–C

DIRECTIONS: Write the least common multiple (LCM) of two or more whole numbers.

STRATEGY: Use the Individual Prime-Factoring Method.

A. Find the LCM of 21 and 35.

STRATEGY: Write the prime factors of 21 and 35 in exponent form.

$21 = 3 \cdot 7 = 3^1 \cdot 7^1$
$35 = 5 \cdot 7 = 5^1 \cdot 7^1$

The different prime factors are 3, 5, and 7. The largest exponent of each is 1. Multiply the powers.

The LCM of 21 and 35 is $3 \cdot 5 \cdot 7 = 105$.

B. Find the LCM of 12, 15, and 20.

STRATEGY: Write the prime factors of 12, 15, and 20 in exponent form.

$12 = 2 \cdot 2 \cdot 3 = 2^2 \cdot 3^1$
$15 = 3 \cdot 5 = 3^1 \cdot 5^1$
$20 = 2 \cdot 2 \cdot 5 = 2^2 \cdot 5^1$

The different prime factors are 2, 3, and 5. The largest exponent of 2 is 2, of 3 is 1, and of 5 is 1. Multiply the powers.

The LCM of 12, 15, and 20 is $2^2 \cdot 3 \cdot 5 = 60$.

C. Find the LCM of 9, 16, 18, and 36.

STRATEGY: Write the prime factors of each in exponent form.

$9 = 3 \cdot 3 = 3^2$
$16 = 2 \cdot 2 \cdot 2 \cdot 2 = 2^4$
$18 = 2 \cdot 3 \cdot 3 = 2^1 \cdot 3^2$
$36 = 2 \cdot 2 \cdot 3 \cdot 3 = 2^2 \cdot 3^2$

The different prime factors are 2 and 3. The largest exponent of 2 is 4 and of 3 is 2. Multiply the powers.

The LCM of 9, 16, 18, and 36 is $2^4 \cdot 3^2 = 144$.

HOW & WHY

OBJECTIVE 2 Find the least common multiple (LCM) of two or more whole numbers using the Group Prime-Factoring Method.

A second method for finding the LCM is sometimes referred to as the Group Prime-Factoring Method. To use this method with two numbers, find a prime number that will divide both. If there is none, multiply the two numbers. The product is the LCM of the two numbers. Continue dividing by primes until no prime will divide the quotients. The product of all the divisors and the remaining quotients is the LCM. For example, find the LCM of 18 and 24.

2)	18	24	Divide both numbers by 2.
3)	9	12	Divide both numbers by 3.
	3	4	The quotients 3 and 4 have no common factors.

The LCM is the product of the divisors and remaining quotients. The LCM of 18 and 24 is $2 \cdot 3 \cdot 3 \cdot 4 = 72$.

To find the LCM of three or more numbers, find a prime number that will divide *at least* two of the numbers. Divide all the numbers if possible. Bring down the numbers that are not multiples of the prime divisor. Continue until no common factors remain. For example, find the LCM of 12, 18, and 45.

2)	12	18	45	Divide 12 and 18 by 2. Bring down 45.
3)	6	9	45	Divide each number by 3.
3)	2	3	15	Bring down 2. Divide 3 and 15 by 3.
	2	1	5	The remaining quotients have no common factors.

The LCM is the product of the divisors and the remaining quotients. The LCM of 12, 18, and 45 is $2 \cdot 3 \cdot 3 \cdot 2 \cdot 5 = 180$.

To find the LCM of two or more numbers using the Group Prime-Factoring Method

1. Divide at least two of the numbers by any common prime number. Continue dividing the remaining quotients in the same manner until no two quotients have a common divisor. When a number cannot be divided, bring it down as a remaining quotient.
2. Write the product of the divisors and the remaining quotients.

EXAMPLES D–G

DIRECTIONS: Find the LCM of two or more whole numbers.

STRATEGY: Use the Group Prime-Factoring Method.

D. Find the LCM of 32 and 40.

2)	32	40	Divide both numbers by 2.
2)	16	20	Divide both numbers by 2 again.
2)	8	10	Divide both numbers by 2 again.
	4	5	The quotients 4 and 5 have no common factors.

$2 \cdot 2 \cdot 2 \cdot 4 \cdot 5 = 160$ Multiply the divisors and remaining quotients.

The LCM of 32 and 40 is 160.

E. Find the LCM of 12, 20, 25, and 48.

2)	12	20	25	48	Divide 12, 20, and 48 by 2. Bring down 25.
2)	6	10	25	24	Divide 6, 10, and 24 by 2. Bring down 25.
3)	3	5	25	12	Divide 3 and 12 by 3. Bring down 5 and 25.
5)	1	5	25	4	Divide 5 and 25 by 5. Bring down 1 and 4.
	1	1	5	4	The quotients 5 and 4 have no common factors.

$2 \cdot 2 \cdot 3 \cdot 5 \cdot 5 \cdot 4 = 1200$ Multiply the divisors and remaining quotients.

The LCM of 12, 20, 25, and 48 is 1200.

WARM-UPS D–G

D. Find the LCM of 30 and 42.

E. Find the LCM of 12, 20, 24, and 50.

Answers to Warm-Ups

D. 210 **E.** 600

F. Find the LCM of the denominators of $\frac{5}{6}, \frac{5}{9}, \frac{7}{12}$, and $\frac{5}{18}$.

The denominators are 6, 9, 12, and 18.

2) 6	9	12	18	Divide 6, 12, and 18 by 2. Bring down 9.
3) 3	9	6	9	Divide 3, 9, 6, and 9 by 3.
3) 1	3	2	3	Divide both 3s by 3. Bring down 2.
1	1	2	1	The remaining quotients have no common factors other than 1.

$2 \cdot 3 \cdot 3 \cdot 2 = 36$ Multiply the divisors and remaining quotient.

The LCM of 6, 9, 12, and 18 is 36, so the common denominator of the fractions is 36.

G. Jane and Robin have each been saving coins. Jane saved dimes and Robin saved quarters. The girls went shopping together and each bought the same item. What is the least amount the item could cost if neither girl receives change?

STRATEGY: The least amount the item could cost is the smallest number that is divisible by both 10 and 25. Find the LCM of 10 and 25.

5) 10	25	Divide both numbers by 5.
2	5	The quotients 2 and 5 have no common factors.

$5 \cdot 2 \cdot 5 = 50$ Multiply the divisors and remaining quotients.

The LCM of 10 and 25 is 50.

* The least cost of the item is 50¢—five dimes or two quarters.

F. Find the LCM of the denominators of $\frac{5}{2}, \frac{5}{3}, \frac{5}{12}$, and $\frac{7}{10}$.

G. If Jane had saved nickels and Robin had saved quarters, what is the least each could pay for the same item if neither girl receives change?

Answers to Warm-Ups

F. 60

G. The least each could have paid is 25¢.

EXERCISES 2.6

OBJECTIVE 1 Find the least common multiple of two or more whole numbers using the Individual Prime-Factoring Method.

OBJECTIVE 2 Find the least common multiple of two or more whole numbers using the Group Prime-Factoring Method.

A. *Find the LCM of each group of whole numbers using either method.*

1. 5, 10 **2.** 6, 12 **3.** 3, 9 **4.** 4, 8

5. 2, 14 **6.** 3, 15 **7.** 7, 14 **8.** 8, 16

9. 6, 9 **10.** 4, 6 **11.** 9, 12 **12.** 8, 12

13. 3, 6, 12 **14.** 4, 6, 8 **15.** 4, 6, 9 **16.** 3, 4, 9

17. 2, 8, 12 **18.** 2, 6, 10 **19.** 2, 5, 10 **20.** 3, 8, 12

B. *Find the LCM of each group of whole numbers using either method.*

21. 12, 15 **22.** 12, 18 **23.** 16, 24 **24.** 22, 33

25. 8, 20 **26.** 12, 20 **27.** 12, 18 **28.** 16, 20

29. 18, 24 **30.** 28, 42 **31.** 8, 12, 16 **32.** 6, 8, 10

33. 9, 12, 15 **34.** 2, 6, 12, 24 **35.** 4, 12, 10, 15 **36.** 3, 6, 12, 24

37. 16, 24, 48 **38.** 20, 25, 40 **39.** 4, 10, 24 **40.** 8, 9, 12

C. *Find the LCM of each group of whole numbers using either method.*

41. 12, 18, 24, 36 **42.** 7, 14, 28, 32 **43.** 12, 17, 51, 68

44. 14, 35, 49, 56 **45.** 35, 40, 70, 80, 140 **46.** 15, 20, 25, 30, 50, 75

Find the least common denominator for each set of fractions.

47. $\frac{2}{3}, \frac{1}{4}, \frac{5}{8}$ **48.** $\frac{1}{6}, \frac{3}{5}, \frac{7}{9}$ **49.** $\frac{4}{15}, \frac{15}{16}, \frac{1}{12}$ **50.** $\frac{21}{30}, \frac{14}{15}, \frac{5}{12}$

51. A gear with 36 teeth is engaged with another gear that has 9 teeth. How many turns of the first gear are necessary in order for the two gears to return to their original positions?

52. A gear with 36 teeth is engaged with another gear that has 23 teeth. How many turns of the first gear are necessary in order for the two gears to return to their original positions?

53. A gear with 36 teeth is engaged with another gear that has 24 teeth. How many turns of the first gear are necessary in order for the two gears to return to their original positions?

54. A gear with 36 teeth is engaged with another gear that has 45 teeth. How many turns of the first gear are necessary in order for the two gears to return to their original positions?

55. Melissa withdraws some 20 dollar bills from a bank. Sean withdraws some 50 dollar bills from the same bank. Luigi, too, goes to the bank and withdraws some 5 dollar bills. If they all intend to buy the same item with their money, what is the least price they could pay without receiving change?

56. Find the least common multiple for the highest and lowest temperatures in the United States yesterday.

STATE YOUR UNDERSTANDING

57. Explain how to find the LCM of 20, 24, and 45.

58. Identify the error in the following problem and correct it. Find the prime factorization of 100, 75, and 45 and determine the LCM of the three numbers.

$100 = 2 \cdot 2 \cdot 5 \cdot 5$
$75 = 3 \cdot 5 \cdot 5$
$45 = 3^2 \cdot 5$

The LCM is $2 \cdot 3 \cdot 3 \cdot 5 \cdot 5 = 450$

CHALLENGE

59. Find the LCM of 144, 240, and 360.

60. Find the LCM of 64, 128, and 192.

61. Find the LCM of 144, 216, and 324.

62. Find the LCM of 1200, 1500, and 1800.

GROUP WORK

63. Here is a classic Hindu puzzle from the 7th century: A woman carrying a basket of eggs is frightened by a horse galloping by. She drops the basket and breaks all the eggs. Concerned passersby ask how many eggs she lost, but she can't remember. She does remember that there was one egg left over when she counted by 2s, two eggs left over when she counted by 3s, three eggs left over when she counted by 4s, and four eggs left over when she counted by 5s. How many eggs did she have? Show calculations that justify your answer by showing it fits all the conditions of the puzzle.

MAINTAIN YOUR SKILLS

64. List all the factors of 425.

65. List all the factors of 460.

66. List all the divisors of 530.

67. List all the divisors of 560.

68. Is 3080 divisible by 35?

69. Is 3080 divisible by 56?

70. Is 3080 divisible by 77?

71. Is 3080 divisible by 21?

72. The Won-Stawp General Store made a profit of \$112 so far this week. If their profit on each appliance sold is \$14, how many more appliances must they sell so that the profit for the week will be more than \$198?

73. A marketing researcher checked the weekly attendance at 14 multiplex theaters. The results are shown in the table. What was the average attendance at the 14 theaters?

Theater Attendance

Number of Theaters	Attendance
1	900
2	1000
2	1200
3	1300
2	1400
2	1600
1	1800
1	1900

CHAPTER 2

KEY CONCEPTS

SECTION 2.1 DIVISIBILITY TESTS

Definitions and Concepts	Examples
A whole number is divisible by 2 if it ends in 0, 2, 4, 6, or 8.	34, 876, and 9078 are divisible by 2. 35, 879, and 9073 are not divisible by 2.
A whole number is divisible by 3 if the sum of the digits is divisible by 3.	876 is divisible by 3 because $8 + 7 + 6 = 21$. 116 is not divisible by 3 because $1 + 1 + 6 = 8$
A whole number is divisible by 5 if it ends in 0 or 5.	9345 is divisible by 5 because it ends in a 5. 8762 is not divisible by 5 because it ends in a 2.

SECTION 2.2 MULTIPLES

Definitions and Concepts	Examples
A multiple of a number is the product of that number and a natural number.	Multiples of 7 are: $(7)(1) = 7$ $(7)(2) = 14$ $(7)(3) = 21$ and so on

SECTION 2.3 DIVISORS, FACTORS, AND GREATEST COMMON FACTOR

Definitions and Concepts	Examples
When one number is a multiple of a second number, the second number is a divisor or factor of the first number.	$3(8) = 24$ so 3 and 8 are factors and divisors of 24.
To find all the factors of a number: • List all counting numbers to the first number whose square is larger than the number. • Test each number to see if it is a factor and record each pair.	Factors of 24: $1 \cdot 24$ $4 \cdot 6$ $2 \cdot 12$ ~~5~~ $3 \cdot 8$ The factors of 24 are 1, 2, 3, 4, 6, 8, 12, 24.
To find the greatest common factor of two numbers: • Find all the factors of each number. • The greatest common factor is the largest number that is on both lists.	Find the GCF of 24 and 18. The factors of 24 are 1, 2, 3, 4, 6, 8, 12, 24. The factors of 18 are 1, 2, 3, 6, 9, 18. The GCF of 24 and 18 is 6.

SECTION 2.4 PRIMES AND COMPOSITES

Definitions and Concepts	Examples
A prime number is a counting number greater than one with exactly two factors, itself and 1.	2, 3, 5, 7, 11, 13, 17, and 19 are the primes less than 20.
A composite number is a counting number with more than two different factors.	4, 6, 8, 9, 10, 12, 14, 15, 16, and 18 are the composite numbers less than 20.
To determine if a number is prime: • Systematically search for factor pairs up to the number whose square is larger than the given number. • If none are found, then the given number is prime.	Is 53 prime? ~~1~~ ~~6~~ ~~2~~ ~~7~~ ~~3~~ ~~8~~ ~~4~~ Stop because $8^2 = 64 > 53$. ~~5~~ 53 is a prime number.

SECTION 2.5 PRIME FACTORIZATION AND GREATEST COMMON FACTOR

Definitions and Concepts	Examples
The prime factorization of a number is the number written as the product of primes.	$40 = 2^3 \cdot 5$
To find the prime factorization of a number: • Divide the number and each succeeding quotient by a prime number until the quotient is 1.	2) 60 2) 30 3) 15 5) 5 1
• Write the indicated product of all the primes.	$60 = 2^2 \cdot 3 \cdot 5$

SECTION 2.6 LEAST COMMON MULTIPLE

Definitions and Concepts	Examples
The least common multiple (LCM) of two or more numbers is the smallest natural number that is a multiple of each of the numbers.	The LCM of 6 and 8 is 24.
To find the least common multiple: • Write each number in prime-factored form. • The LCM is the product of the highest power of each prime factor.	Find the LCM of 20 and 150. $20 = 2^2 \cdot 5$ $150 = 2 \cdot 3 \cdot 5^2$ $\text{LCM} = 2^2 \cdot 3 \cdot 5^2 = 300$

CHAPTER 2

REVIEW EXERCISES

Section 2.1

1. Which of these numbers is divisible by 2: 6, 36, 63, 636, 663?

2. Which of these numbers is divisible by 3: 6, 36, 63, 636, 663?

3. Which of these numbers is divisible by 5: 15, 51, 255, 525, 552?

4. Which of these numbers is divisible by 6: 6, 36, 63, 636, 663?

5. Which of these numbers is divisible by 9: 6, 36, 63, 636, 663?

6. Which of these numbers is divisible by 10: 50, 55, 505, 550, 555?

7. Which of these numbers is divisible by both 2 and 3: 444, 555, 666, 777, 888, 999?

8. Which of these numbers is divisible by both 2 and 9: 444, 555, 666, 777, 888, 999?

9. Which of these numbers is divisible by both 3 and 5: 445, 545, 645, 745, 845?

10. Which of these numbers is divisible by both 3 and 10: 440, 550, 660, 770, 880?

Determine whether the number is divisible by 6, 7, or 9.

11. 567 **12.** 576 **13.** 756 **14.** 765

Determine whether the number is divisible by 4, 5, or 10.

15. 560 **16.** 575

Determine whether the number is divisible by 6 or 15.

17. 690 **18.** 975

19. Ed and his four partners made a profit of $2060 in their stereo installation business. Can the profits be divided evenly in whole dollars among them? Explain.

20. Bobbie and her five partners incurred expenses of $3060 in their satellite dish installation enterprise. Can the expenses be divided evenly in whole dollars among them? Explain.

Section 2.2

List the first five multiplies of the whole number.

21. 8 **22.** 19 **23.** 51

24. 64 **25.** 85 **26.** 97

27. 122 **28.** 125 **29.** 141

30. 252

Is each number a multiple of 6? of 9? of 15?

31. 72 **32.** 135 **33.** 150

34. 290 **35.** 465 **36.** 540

Is each number a multiple of 12? of 16?

37. 288 **38.** 348

39. The designer of new production-line equipment at the Bredworthy Jelly plant predicts that 55 small jars of jelly can be produced per minute. The next day a total of 605 jars is produced in 11 minutes. Is the designer's prediction accurate?

40. In Exercise 39, if the plant produced 880 jars in 16 minutes, is the prediction accurate?

Section 2.3

Write the whole number as the product of two factors in all possible ways.

41. 15 **42.** 24 **43.** 38

44. 136 **45.** 236 **46.** 336

List all the factors (divisors) of the whole number.

47. 33 **48.** 39 **49.** 42

50. 60 **51.** 78 **52.** 97

Find the GCF of each set of numbers.

53. 33, 36 **54.** 36, 42 **55.** 42, 63

56. 63, 72, 90 **57.** 90, 120, 135 **58.** 135, 180, 270

59. In a recent year, the average single-family detached residence consumed $1590 in energy costs. Find the possible costs of energy consumption (whole number amounts) per person. Assume that each household has from 1 to 10 people.

60. A rectangular floor area requires 180 square-foot tiles. List all the possible whole-number dimensions that this floor could measure that would require using all the tiles.

Section 2.4

Tell whether the number is prime or composite.

61. 13 **62.** 25 **63.** 47

64. 49 **65.** 51 **66.** 61

67. 71 **68.** 73 **69.** 77

70. 81 **71.** 83 **72.** 91

73. 337 **74.** 339 **75.** 341

76. 343 **77.** 347 **78.** 479

79. The year 2003 is a prime number? What is the next year that is a prime number?

80. What was the last year before 1997 that was a prime number?

Section 2.5

Write the prime factorization of each number.

81. 27 **82.** 36 **83.** 38

84. 42 **85.** 52 **86.** 64

87. 222 **88.** 232 **89.** 252

Find the greatest common factor (GCF) of each set of numbers.

90. 75, 120 **91.** 120, 174 **92.** 120, 144

93. 120, 210 **94.** 120, 600 **95.** 32, 40, 48

96. 32, 40, 60 **97.** 40, 60, 105 **98.** 92, 115, 207

99. Write the prime factorization of the next several years.

100. Write the prime factorization of the last year.

Section 2.6

Find the LCM of each group of numbers.

101. 6, 8

102. 8, 10

103. 10, 12

104. 12, 16

105. 16, 18

106. 18, 27

107. 20, 25

108. 28, 35

109. 30, 35

110. 3, 6, 8

111. 6, 8, 12

112. 8, 10, 12

113. 8, 12, 18

114. 18, 24, 36

115. 15, 45, 60

116. 32, 40, 60

117. 40, 60, 105

118. 30, 35, 40

119. Find two numbers that have an LCM of 98.

120. Find two numbers that have an LCM of 102.

CHAPTER 2

TRUE / FALSE CONCEPT REVIEW

ANSWERS

Check your understanding of the language of basic mathematics. Tell whether each of the following statements is true (always true) or false (no always true). For those statements you judge to be false, revise them to make them true.

1. Every multiples of 3 ends with the digit 3. **1.** ____________

2. Every multiple of 10 ends with the digit 0. **2.** ____________

3. Every multiple of 11 is divisible by 11. **3.** ____________

4. Every multiple of 5 is the product of 5 and some natural number. **4.** ____________

5. Every whole number, except the number 1, has at least two different factors. **5.** ____________

6. Every factor of 300 is also a divisor of 300. **6.** ____________

7. Every multiple of 300 is also a factor of 300. **7.** ____________

8. The square of 25 is 50. **8.** ____________

9. Every natural number ending in 6 is divisible by 6. **9.** ____________

10. Every natural number ending in 6 is divisible by 2. **10.** ____________

11. Every natural number ending in 9 is divisible by 3. **11.** ____________

12. The GCF of 6 and 8 is 24. **12.** ____________

13. The GCF of three numbers is a multiple of each of the numbers. **13.** ____________

14. The number 123,321,231 is divisible by 9. **14.** ____________

15. The number 111,111,115 is divisible by 5. **15.** ____________

16. All prime numbers are odd. **16.** ____________

17. All numbers that end with the digit 8 are composite. **17.** ____________

18. Every prime number has exactly two multiples. **18.** ____________

19. It is possible for a composite number to have exactly 5 factors. **19.** ____________

 Name ______________________ Class ______________________ Date ____________

20. All of the prime factors of a composite number are smaller than the number. **20.** ____________

21. The least common multiple (LCM) of three different numbers is the product of the three numbers. **21.** ____________

22. The greatest common factor (GCF) of three different prime numbers is the smallest of the three numbers. **22.** ____________

23. The largest divisor of the least common multiple (LCM) of three numbers is the largest of the three numbers. **23.** ____________

24. It is possible for the GCF of three natural numbers to be one of the three numbers. **24.** ____________

25. If all numbers in a group of numbers are even numbers, then the greatest common factor (GCF) of the group is an even number. **25.** ____________

CHAPTER 2

TEST

ANSWERS

1. Is 411,234 divisible by 6? 1. ____________
2. List all of the factors (divisors) of 112. 2. ____________
3. Is 8617 divisible by 7? 3. ____________
4. What is the GCF of 12 and 32? 4. ____________
5. What is the LCM of 12 and 32? 5. ____________
6. Write 75 as the product of two factors in as many ways as possible. 6. ____________
7. Write the prime factorization of 280. 7. ____________
8. Find the LCM of 18, 42, and 84. 8. ____________
9. Find the GCF of 18, 42, and 84. 9. ____________
10. Write all multiples of 13 between 200 and 250. 10. ____________
11. Is 200 a multiple of 400? 11. ____________
12. Is 109 a prime or a composite number? 12. ____________
13. Is 111 a prime or a composite number? 13. ____________
14. Write the prime factorization of 605. 14. ____________
15. What is the LCM of 18, 21, and 56? 15. ____________
16. What is the smallest prime number? 16. ____________
17. What is the largest composite number that is less than 300? 17. ____________
18. What is the smallest natural number that 6, 18, 24, and 30 will divide evenly? 18. ____________
19. Can two different numbers have the same prime factorization? Explain. 19. ____________
20. List two sets of three different numbers whose LCM is 42. 20. ____________

CHAPTER 2

GROUP PROJECT (2–3 weeks)

OPTIONAL

In the early 1200s, the Italian mathematician Leonardo de Pisa, also known as Fibonacci, was interested in a set of numbers that bear his name. The set begins {1, 1, 2, 3, 5, . . .} and continues infinitely using the rule that the next number in the set is the sum of the previous two.

a. Mathematicians often name numbers of a set according to the order in which they occur. For the Fibonacci numbers, it is customary to designate F_n as the nth element of the set. So $F_3 = 2$ and $F_4 = 3$. Calculate all the Fibonacci numbers less than 500. Make a table that lists them in order, along with their name.

b. The Fibonacci numbers are strongly related to items in nature that spiral, such as the cells on pineapples or the petals on pinecones. The number of spirals in normal specimens is always a Fibonacci number, although the number changes from species to species. Find two whole pineapples and two pinecones from different kinds of trees. Look at each item from the top and count the spirals that move clockwise. Then count the spirals that move counterclockwise. Record your results in the following table. Write a paragraph that summarizes your findings.

Item	Number of Clockwise Spirals	Number of Counterclockwise Spirals
Pineapple 1		
Pineapple 2		
Pinecone 1		
Pinecone 2		

c. Just as each number has a unique prime factorization, so each number can also be written as the sum of Fibonacci numbers. Always start with the largest Fibonacci number less than the given number, then go to the largest Fibonacci number less than the difference, and continue until the whole number is accounted for. For example,

$$\begin{aligned} 46 &= 34 + 12 \\ &= 34 + 8 + 4 \\ &= 34 + 8 + 3 + 1 \end{aligned}$$

Make a table that gives each number from 1 to 50 as the sum of Fibonacci numbers.

d. The game of Fibonacci Nim is an easy game where players take turns removing chips from a central pot. You may start with any number of chips. Each player must take at least one chip out of the pile but not more than twice the number just removed by his opponent. (The first player may not remove all the chips on the first move.) The player who takes the last chips wins. Start with 20 chips and play Nim 5 times. What do you notice about the number of chips remaining just before the game is won?

e. A winning strategy is to take the smallest Fibonacci number in the sum of Fibonacci numbers that makes the number of chips remaining. For example, if 46 chips are remaining, then one chip should be taken because $46 = 34 + 8 + 3 + 1$. Play Nim for 5 more times starting with 20 chips and using this strategy. Did it work? If both players employ the strategy, which player will win? Write a paragraph summarizing how to win at Nim.

Managing Anxiety

For those of you who become anxious as you begin to study, we recommend that you devote a section in your math notebook to record your thoughts and feelings each time you study. Record your thoughts in the first person, as "I-statements." For example, "Nobody ever uses this stuff in real life," rephrased as an I-statement, it would be: "I would like to know where I would use this in my life." Research shows that a positive attitude is the single most important key to your success in mathematics. By recognizing negative thoughts and replacing them with positive thoughts, you are beginning to work on changing your attitude. The first step is to become aware of your self-talk.

Negative self-talk falls into three categories of irrational beliefs that you have about yourself and how you view the world. You think that (1) Worrying helps. Wrong. Worrying leads to excessive anxiety, which is distracting and hinders performance. (2) Your worth as a person is determined by your performance. Wrong. Not being able to solve math problems doesn't mean you won't amount to anything. If you think it does, you are a victim of "catastrophizing." (3) You are the only one who feels anxious. Wrong. By thinking that other students have some magical coping skills that allows them to avoid anxiety, you are comparing yourself to some irrational mythical norm.

As you begin your first section, ask yourself such questions as What am I saying to myself now? What is triggering these thoughts? How do I feel physically? and What emotions am I feeling now? Your answers will likely reveal a pattern to your thoughts and feelings. You need to analyze your statements and change them into more positive and rational self-talk. You can use a simple technique called "rational emotive therapy."[1] This method, if practiced regularly, can quickly and effectively change the way you think and feel about math. When you find yourself getting upset, watch for words such as "should," "must," "ought to," "never," "always," and "I can't." They are clues to negative self-talk and signals for you to direct your attention back to math. Use the following ABCD model.

A. Triggering event: You start to do your math homework and your mind goes blank.

B. Negative self-talk in response to the trigger: "I can't do math. I'll never pass. My life is ruined!"

C. Anxiety caused by the negative self-talk: panic, anger, tight neck, etc.

D. Positive self-talk to cope with anxiety: "This negative self-talk is distracting. It doesn't help me solve these problems. Focus on the problems."; or say: "I may be uncomfortable, but I'm working on it."

Recognize that negative self-talk (B) is the culprit. It causes the anxiety (C) and must be restructured to positive rational statements (D). Practice with the model using your own self-talk statements (B) and then complete the remaining steps.

[1] R.E.T. was created by Albert Ellis.

3 FRACTIONS AND MIXED NUMBERS

SECTIONS

APPLICATION

Sandy and Hannah have decided to start a gourmet cookie business. They each have several family recipes that they consider excellent. Since they are just starting their business, they bake only when they have orders to fill. The women need to decide how to price their cookies. This will depend on their costs and on the local market conditions.

They begin by estimating the cost of making their cookies. They determine that they will use only the finest and freshest ingredients, including real butter, fresh eggs, and high-quality chocolate chips and pecans. Hannah's pecan chocolate chip recipe makes 2 dozen 4-in. cookies, which cost about 20¢ each. Sandy's dipped butterfingers recipe makes 3 dozen 3-in. cookies, which cost 18¢ each. The women buy small boxes for 38¢ each. The boxes hold 20 pecan chocolate chip cookies or 30 dipped butterfingers each. Each box is lined with a sheet of tissue paper that costs 2¢.

Calculate the total cost of a single box of each type of cookie. Contact a bakery in your area and find the price per cookie of comparable cookies. Determine the cost of boxes of bakery cookies that are the same number and type of cookie as Hannah's and Sandy's boxes of cookies. Make a recommendation for the selling price of each type of box of cookies. How much profit will the women make on each type of box of cookies? Fill in Table 3.1 with your information.

TABLE 3.1

	Pecan chocolate chip	Dipped butterfinger
Cost per cookie		
Cost of box and tissue		
Total cost per box		
Bakery cost per cookie		
Total cost per bakery cookie box		
Selling price per box		
Profit per box		

3.1 Proper and Improper Fractions; Mixed Numbers

OBJECTIVES

1. Write a fraction to describe the parts of a unit.
2. Select proper or improper fractions from a list of fractions.
3. Change improper fractions to mixed numbers.
4. Change mixed numbers to improper fractions.

VOCABULARY

A **fraction** $\left(\text{for example, } \frac{5}{9}\right)$ is a name for a number. The upper numeral (5) is the **numerator**. The lower numeral (9) is the **denominator**.

A **proper fraction** is one in which the numerator is less than the denominator $\left(\text{for example, } \frac{5}{9}\right)$.

An **improper fraction** is one in which the numerator is not less than the denominator $\left(\text{for example, } \frac{9}{5} \text{ or } \frac{7}{7}\right)$.

A **mixed number** is the sum of a whole number and a fraction $\left(4 + \frac{3}{5}\right)$ with the plus sign left out $\left(4\frac{3}{5}\right)$. The fraction part is usually a proper fraction.

HOW & WHY

OBJECTIVE 1 Write a fraction to describe the parts of a unit.

A unit (here we use a rectangle) may be divided into smaller parts of equal size to help us picture a fraction. The rectangle in Figure 3.1 is divided into 7 parts, 6 of which are shaded. The fraction $\frac{6}{7}$ represents the shaded part. The denominator (7) tells the number of parts in the unit. The numerator (6) tells the number of shaded parts. The fraction $\frac{1}{7}$ represents the part that is not shaded.

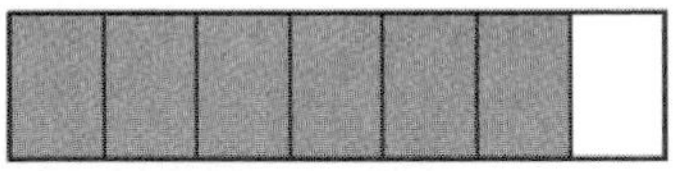

FIGURE 3.1

Because fractions are another way of writing a division problem and because division by zero is not defined, the denominator can never be zero. There will always be at least one part in a unit.

The unit may also be shown on a ruler. The fraction $\frac{6}{10}$ represents the distance from 0 to the arrowhead in Figure 3.2.

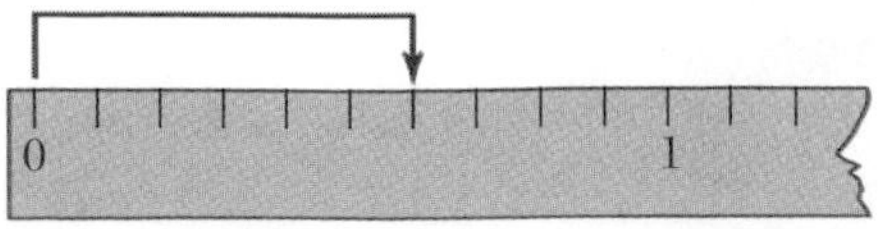

FIGURE 3.2

To write a fraction to describe the parts of a unit

Write the fraction:

$$\frac{\text{numerator}}{\text{denominator}} = \frac{\text{number of shaded parts}}{\text{total number of parts in one unit}}$$

To write a fraction from a ruler or a number line

Write the fraction:

$$\frac{\text{numerator}}{\text{denominator}} = \frac{\text{number of spaces between zero and end of arrow}}{\text{number of spaces between zero and one}}$$

EXAMPLES A–E

DIRECTIONS: Write the fraction represented by the figure.

STRATEGY: First count the number of parts that are shaded or marked. This number is the numerator. Now count the total number of parts in the unit. This number is the denominator.

A. Write the fraction represented by

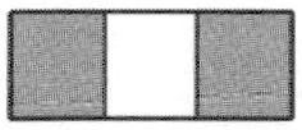

The figure represents $\frac{2}{3}$. The count of shaded parts is 2. The total count is 3.

B. Write the fraction represented by

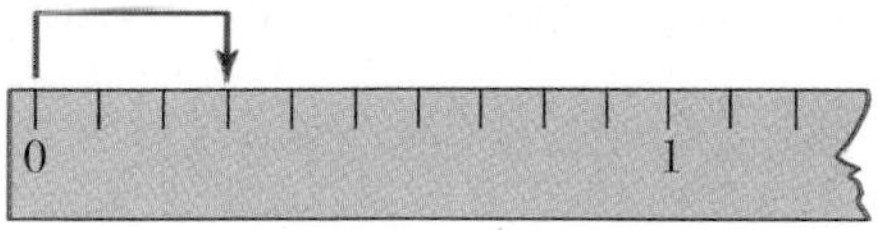

The figure represents $\frac{3}{10}$. There are 3 spaces between 0 and the arrowhead. There are 10 spaces between 0 and 1.

C. Write the fraction represented by

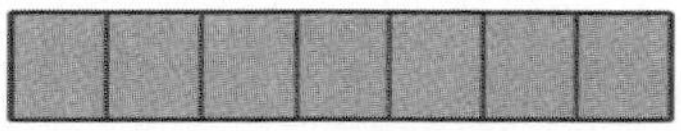

The figure represents $\frac{7}{7}$ or 1. The number of shaded parts is the same as the total number of parts. The whole unit, or 1, is shaded.

D. Write the fraction represented by

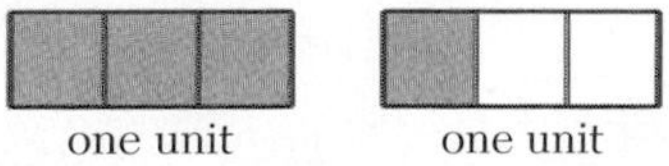

We have two units. The denominator is the number of parts in one unit.

The figure represents $\frac{4}{3}$. Four parts are shaded and there are 3 parts in each unit.

WARM-UPS A–E

A. Write the fraction represented by

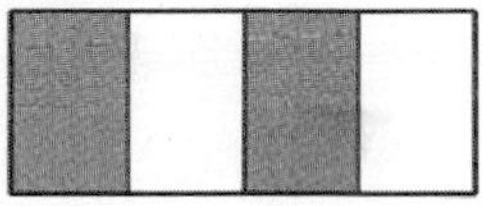

B. Write the fraction represented by

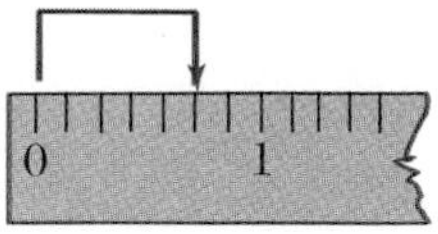

C. Write the fraction represented by

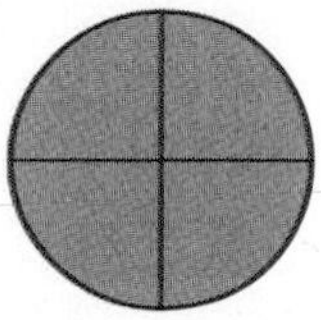

D. Write the fraction represented by

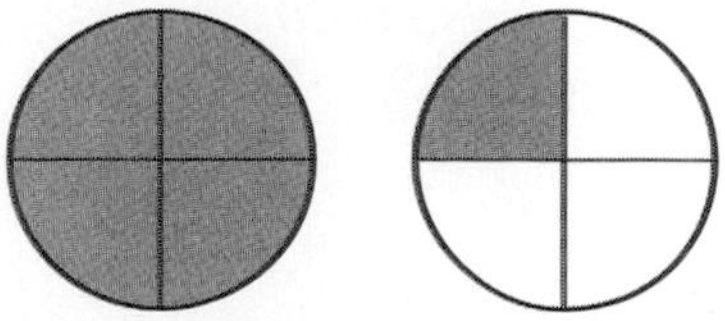

Answers to Warm-Ups

A. $\frac{2}{4}$ **B.** $\frac{5}{7}$ **C.** $\frac{4}{4}$ **D.** $\frac{5}{4}$

E. In a second low-cholesterol margarine spread, 1 tablespoon contains 10 grams of fat. Each fat gram has 9 calories. The total number of calories in a tablespoon is 97. Write a fraction that represents the portion of calories from fat.

E. One tablespoon of a low-cholesterol margarine spread contains 11 grams of fat. Each fat gram has 9 calories, so there are 99 calories from fat in a tablespoon of margarine. The total number of calories in a tablespoon of margarine is 100. Write a fraction that represents the portion of calories from fat.

The fraction is the number of calories from fat over the total number of calories.

The fraction is $\frac{99}{100}$. 99 calories are from fat, and the total number of calories is 100.

HOW & WHY

OBJECTIVE 2 Select proper or improper fractions from a list of fractions.

Fractions are called "proper" if the numerator is smaller than the denominator. If the numerator is equal to or greater than the denominator, the fractions are called "improper." So in the list $\frac{5}{6}, \frac{11}{9}, \frac{19}{19}, \frac{10}{11}, \frac{14}{16}, \frac{18}{11}, \frac{24}{29}$ the proper fractions are $\frac{5}{6}, \frac{10}{11}, \frac{14}{16}$, and $\frac{24}{29}$. The improper fractions are $\frac{11}{9}, \frac{19}{19}$, and $\frac{18}{11}$.

If the numerator and the denominator are equal, as in $\frac{19}{19}$, the value of the fraction is 1. This is easy to see from a picture because the entire unit is shaded. Improper fractions have a value that is greater than or equal to 1. Proper fractions have a value that is less than 1 because some part of the unit is not shaded. See Table 3.2.

To determine if a fraction is proper or improper

1. Compare the size of the numerator and the denominator.
2. If the numerator is smaller, the fraction is proper. Otherwise, the fraction is improper.

TABLE 3.2 REGIONS THAT SHOW PROPER AND IMPROPER FRACTIONS

Proper fractions	Improper fractions	
Value less than 1	Value equal to 1	Value greater than 1

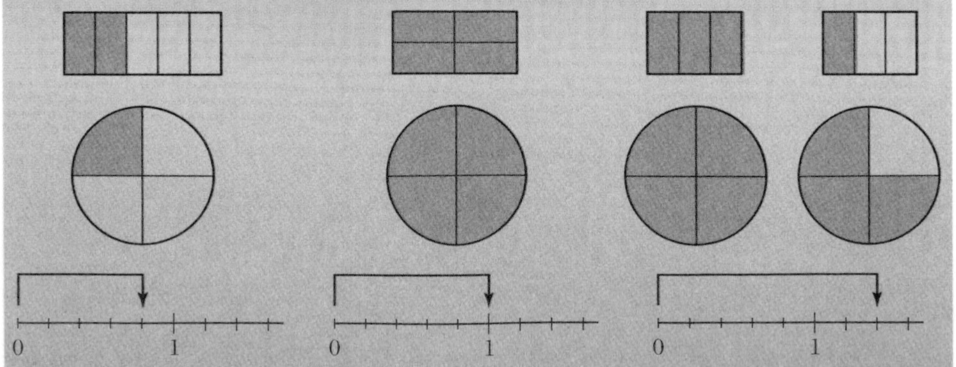

Answers to Warm-Ups

E. The fraction of calories from fat is $\frac{90}{97}$.

EXAMPLE F

DIRECTIONS: Identify the proper and improper fractions in the list.

STRATEGY: Compare the numerator and the denominator. If the numerator is smaller, the fraction is proper. If not, the fraction is improper.

F. Identify the proper and improper fractions: $\frac{5}{6}, \frac{4}{5}, \frac{5}{5}, \frac{7}{6}, \frac{19}{20}, \frac{21}{21}, \frac{25}{23}, \frac{25}{30}$

The proper fractions are:

$\frac{5}{6}, \frac{4}{5}, \frac{19}{20}, \frac{25}{30}$ The numerators are smaller than the denominators.

The improper fractions are

$\frac{5}{5}, \frac{7}{6}, \frac{21}{21}, \frac{25}{23}$ The numerators are not smaller than the denominators.

WARM-UP F

F. Identify the proper and improper fractions:

$\frac{8}{8}, \frac{13}{16}, \frac{5}{10}, \frac{11}{8}, \frac{8}{4}, \frac{1}{21}, \frac{25}{1}, \frac{11}{11}$

HOW & WHY

OBJECTIVE 3 Change improper fractions to mixed numbers.

An improper fraction is equal to a whole number or to a mixed number. A mixed number is the sum of a whole number and a fraction. Figures 3.3 and 3.4 show the conversions.

An improper fraction changed to a whole number:

$$\frac{6}{3} = 2$$

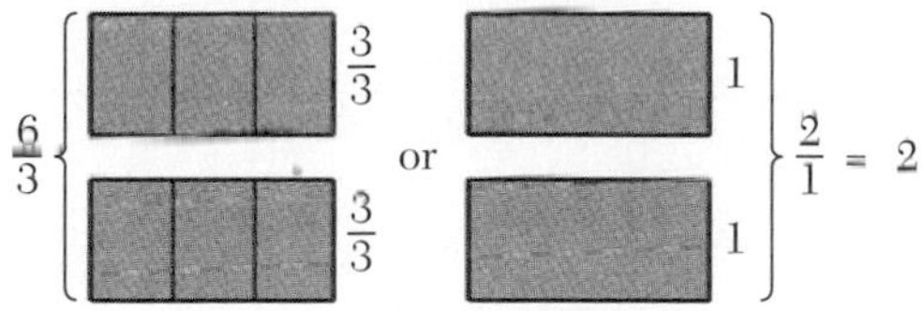

FIGURE 3.3

An improper fraction changed to a mixed number:

$$\frac{9}{7} = 1\frac{2}{7}$$

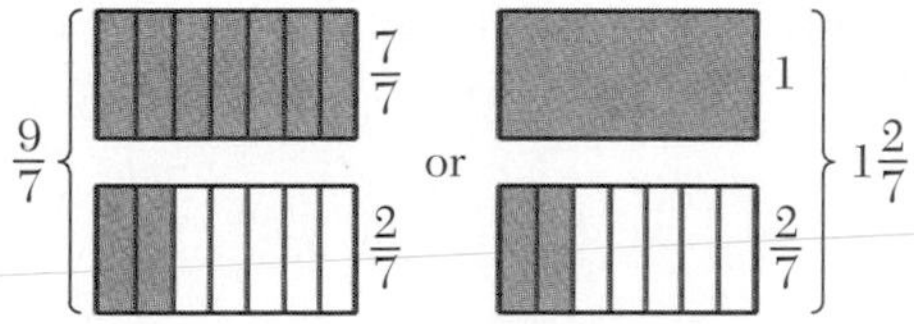

FIGURE 3.4

The shortcut for changing an improper fraction to a mixed number is to divide.

$$\frac{24}{4} = 24 \div 4 = 6$$

$$\frac{11}{5} = 11 \div 5 = \begin{array}{r} 2 \\ 5\overline{)11} \\ \underline{10} \\ 1 \end{array} = 2\frac{1}{5}$$

Answer to Warm-Up

F. The proper fractions are $\frac{13}{16}, \frac{5}{10}, \frac{1}{21}$.
The improper fractions are $\frac{8}{8}, \frac{11}{8}, \frac{8}{4}, \frac{25}{1}, \frac{11}{11}$.

To change an improper fraction to a mixed number

1. Divide the numerator by the denominator.
2. If there is a remainder, write the whole number and then write the fraction: $\dfrac{\text{remainder}}{\text{divisor}}$.

CAUTION

Do not confuse the process of changing an improper fraction to a mixed number with "simplifying." Simplifying fractions is a totally different procedure. See Section 3.2.

WARM-UPS G–I

G. Change $\frac{20}{7}$ to a mixed number.

H. Change $\frac{117}{3}$ to a mixed number.

I. Change $\frac{319}{14}$ to a mixed number.

Answers to Warm-Ups

G. $2\frac{6}{7}$ **H.** 39 **I.** $22\frac{11}{14}$

EXAMPLES G–I

DIRECTIONS: Change the improper fraction to a mixed number.

STRATEGY: Divide the numerator by the denominator to find the whole number. If there is a remainder, write it over the denominator to form the fraction part.

G. Change $\frac{11}{3}$ to a mixed number.

$$\frac{11}{3} = 3\overline{)11} = 3\frac{2}{3} \qquad \begin{array}{r} 3 \\ 3\overline{)11} \\ \underline{9} \\ 2 \end{array}$$

Divide 11 by 3.

So $\frac{11}{3} = 3\frac{2}{3}$.

H. Change $\frac{114}{6}$ to a mixed number.

$$\frac{114}{6} = \begin{array}{r} 19 \\ 6\overline{)114} \\ \underline{6} \\ 54 \\ \underline{54} \end{array}$$

Divide 114 by 6. Because there is no remainder, the fraction is equal to a whole number.

So $\frac{114}{6} = 19$.

CALCULATOR EXAMPLE

I. Change $\frac{348}{7}$ to a mixed number.

If your calculator has a key for fractions, refer to the manual to see how you can use it to change fractions to mixed numbers. If your calculator does not have a fraction key, divide 348 by 7. The quotient $\frac{348}{7} \approx 49.714$, so the whole number part is 49. Now subtract to find the remainder: $348 - 7(49) = 5$.

$$\begin{array}{r} 49 \\ 7\overline{)348} \\ \underline{343} \\ 5 \end{array}$$

$348 \div 7 = 49 \text{ R } 5$, so $\frac{348}{7} = 49\frac{5}{7}$.

*

HOW & WHY

OBJECTIVE 4 *Change mixed numbers to improper fractions.*

Despite the value judgment attached to the name "improper," in many cases improper fractions are a more convenient and useful form than mixed numbers. Thus, it is important to be able to convert from mixed number to improper fractions.

Every mixed number can be changed to an improper fraction. See Figure 3.5.

$$1\frac{3}{7} = \frac{?}{7}$$

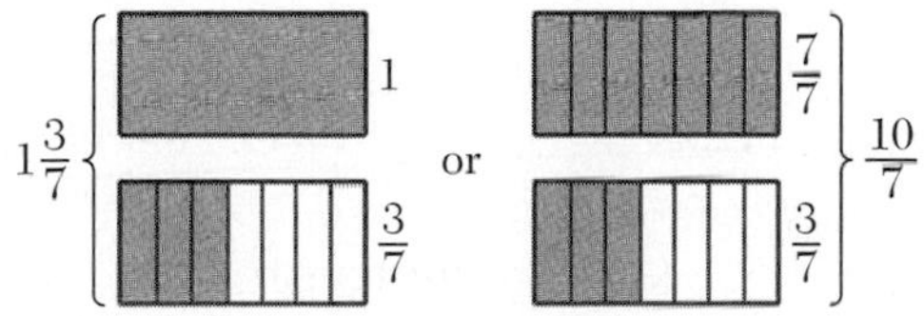

FIGURE 3.5

$$1\frac{3}{7} = \frac{10}{7}$$

The shortcut uses multiplication and addition:

$$1\frac{3}{7} = \frac{1(7) + 3}{7} = \frac{7 + 3}{7} = \frac{10}{7}$$

Similarly,

$$4\frac{5}{7} = \frac{4(7) + 5}{7} = \frac{28 + 5}{7} = \frac{33}{7}$$

To change a mixed number to an improper fraction

1. Multiply the denominator times the whole number.
2. Add the numerator to the product in step 1.
3. Write the sum from step 2 over the denominator.

WARM-UPS J–L

J. Change $4\frac{2}{5}$ to an improper fraction.

K. Change $1\frac{14}{15}$ to an improper fraction.

L. Change 6 to an improper fraction.

EXAMPLES J–L

DIRECTIONS: Change each mixed number to an improper fraction.

STRATEGY: Multiply the whole number times the denominator. Add the numerator. Write the sum over the denominator.

J. Change $2\frac{4}{7}$ to an improper fraction.

$$2\frac{4}{7} = \frac{2(7) + 4}{7} = \frac{18}{7}$$

Multiply the whole number times the denominator and add the numerator. Write the sum over the denominator.

K. Change $6\frac{6}{11}$ to an improper fraction.

$$6\frac{6}{11} = \frac{6(11) + 6}{11} = \frac{72}{11}$$

Multiply the whole number times the denominator and add the numerator. Write the sum over the denominator.

L. Change 12 to an improper fraction.

First, rewrite the whole number as a mixed number. Use the fraction $\frac{0}{1}$.

$$12 = 12\frac{0}{1} = \frac{12(1) + 0}{1} = \frac{12}{1}$$

Note: Any fraction that equals 0 can be used.

As example L illustrates, any whole number can be written as an improper fraction by using a denominator of 1.

Answers to Warm-Ups

J. $\frac{22}{5}$ **K.** $\frac{29}{15}$ **L.** $\frac{6}{1}$

EXERCISES 3.1

OBJECTIVE 1 Write a fraction to describe the parts of a unit.

Write the fraction represented by the figure.

A.

1. 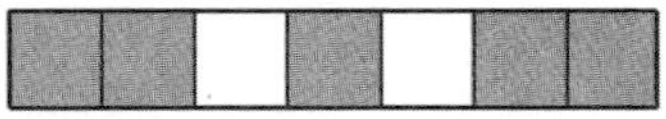

2. 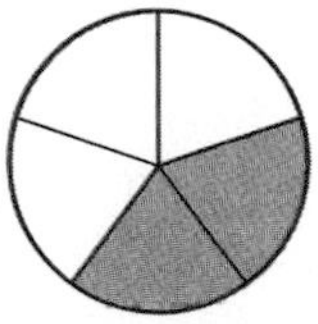

3. 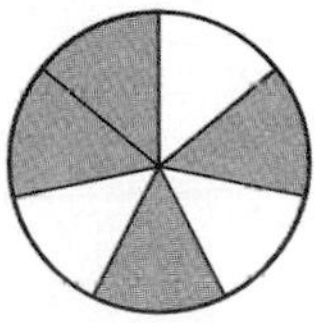

4. 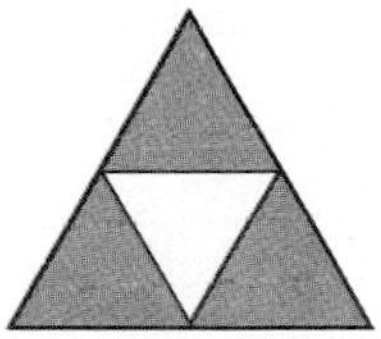

5. 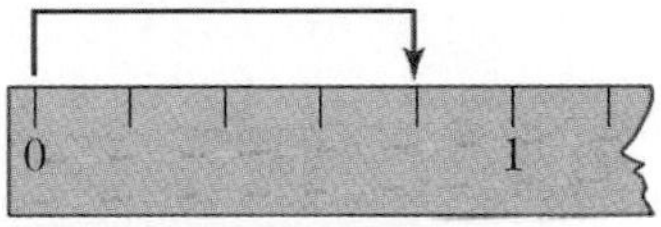

6. 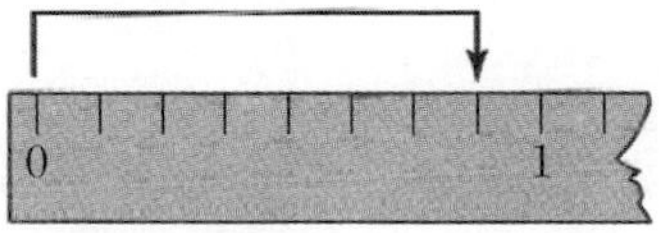

B.

7. 

8. 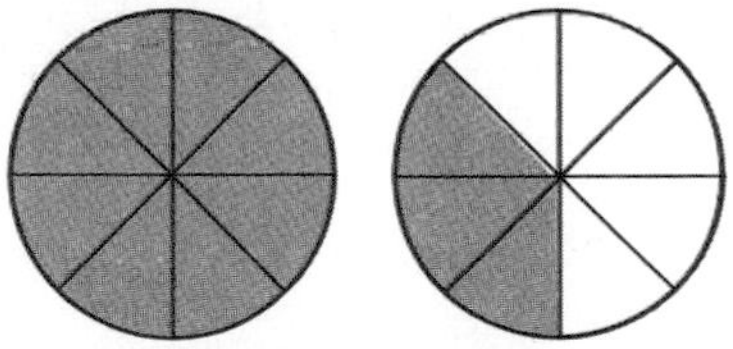

9. 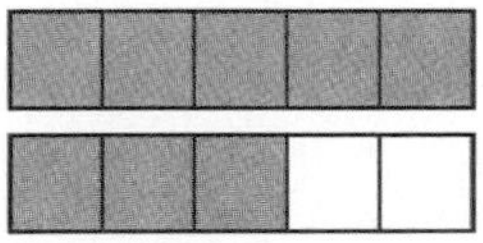

10. 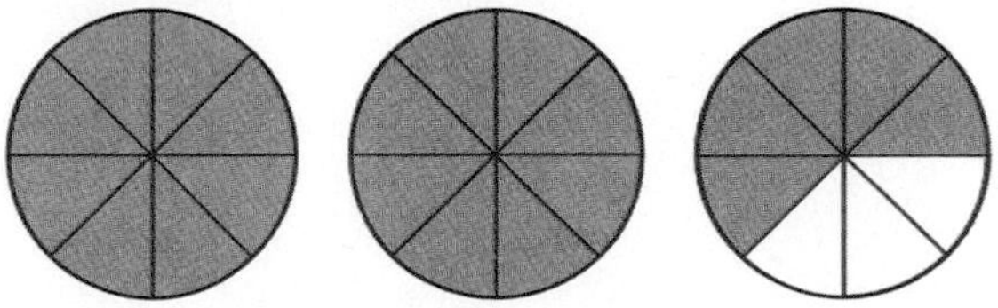

11. 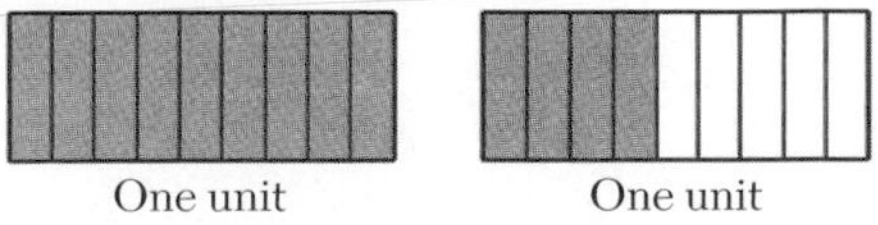

12. 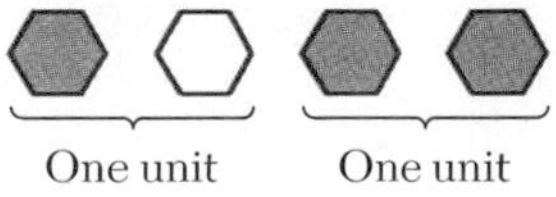

 Name ______________________ Class ______________ Date ________

OBJECTIVE 2 Select proper or improper fractions from a list of fractions.

Identify the proper and improper fractions from the list.

A.

13. $\frac{3}{7}, \frac{4}{7}, \frac{5}{7}, \frac{6}{7}, \frac{7}{7}, \frac{8}{7}, \frac{9}{7}$

14. $\frac{5}{6}, \frac{8}{7}, \frac{14}{15}, \frac{16}{18}, \frac{17}{17}, \frac{23}{25}$

15. $\frac{17}{16}, \frac{18}{17}, \frac{29}{19}, \frac{30}{21}, \frac{23}{23}$

16. $\frac{7}{12}, \frac{11}{12}, \frac{13}{12}, \frac{12}{13}, \frac{7}{13}$

17. $\frac{7}{4}, \frac{10}{11}, \frac{13}{13}, \frac{20}{19}, \frac{3}{5}$

18. $\frac{6}{11}, \frac{10}{8}, \frac{11}{6}, \frac{10}{12}, \frac{9}{9}$

B.

19. $\frac{5}{5}, \frac{9}{10}, \frac{18}{18}, \frac{102}{103}, \frac{147}{147}$

20. $\frac{5}{6}, \frac{6}{7}, \frac{7}{8}, \frac{8}{9}, \frac{10}{10}, \frac{11}{12}, \frac{12}{13}, \frac{13}{13}, \frac{15}{15}$

OBJECTIVE 3 Change improper fractions to mixed numbers.

Change to a mixed number.

A.

21. $\frac{213}{5}$

22. $\frac{213}{4}$

23. $\frac{211}{7}$

24. $\frac{111}{10}$

25. $\frac{112}{9}$

26. $\frac{103}{6}$

B.

27. $\frac{98}{13}$

28. $\frac{83}{21}$

29. $\frac{329}{22}$

30. $\frac{400}{13}$

31. $\frac{421}{12}$

32. $\frac{421}{15}$

OBJECTIVE 4 Change mixed numbers to improper fractions.

Change to an improper fraction.

A.

33. $4\frac{5}{9}$

34. $5\frac{9}{10}$

35. 12

36. 19

37. $7\frac{3}{4}$

38. $6\frac{5}{6}$

B.

39. $29\frac{2}{3}$ **40.** $37\frac{1}{2}$ **41.** $15\frac{1}{5}$

42. $18\frac{5}{8}$ **43.** $19\frac{3}{8}$ **44.** $21\frac{7}{9}$

C.

Exercises 45–46. Fill in the boxes so the statement is true. Explain your answer.

45. The fraction $\frac{0}{\square}$ is a proper fraction.

46. $30\frac{\square}{9} = \frac{278}{9}$

47. Find the error(s) in the statement: $2\frac{2}{5} = \frac{4}{10}$. Correct the statement.

48. Find the error(s) in the statement: $\frac{37}{8} = 3\frac{7}{8}$. Correct the statement.

Write the fraction represented by the figure.

49. 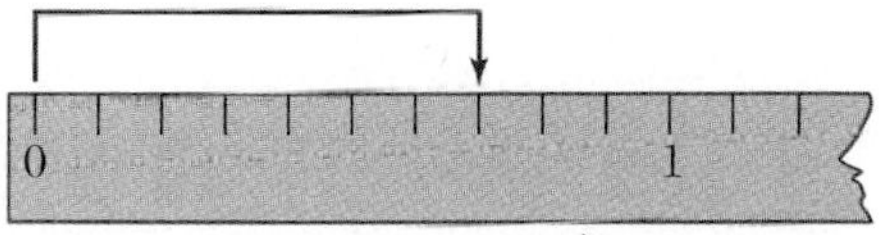

50. 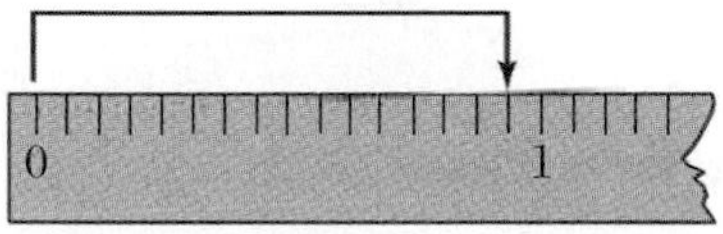

51. 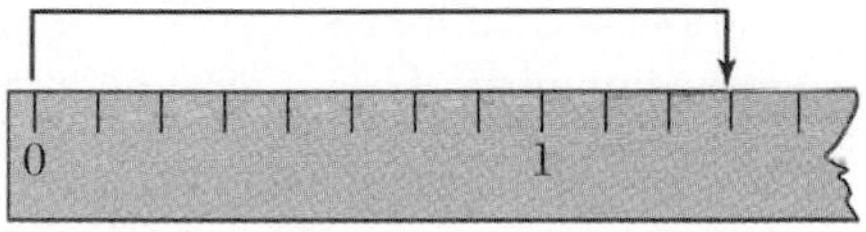

52. 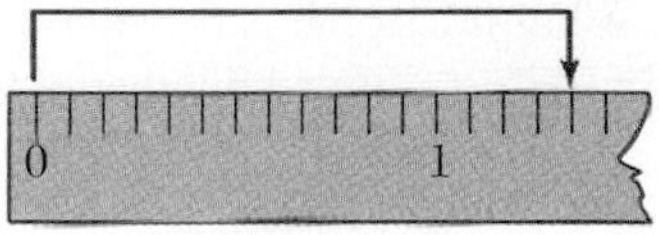

53. Draw a rectangular unit divided into equal parts that shows $\frac{4}{7}$.

54. Draw rectangular units divided into equal parts that shows $\frac{7}{4}$.

55. In a class of 21 students there are 13 women. What fraction represents the part of the class that is female?

56. In a history class there are 45 students; 23 of them are men. What fraction represents the part of the class that is male?

57. The U.S. Postal Service defines nonstandard mail as anything that is longer than $11\frac{1}{2}$ in., taller than $6\frac{1}{8}$ in., and/or thicker than $\frac{1}{4}$ in. Which of these dimensions can be changed to improper fractions and which cannot? Rewrite the appropriate dimensions as improper fractions.

58. The Adams family budgets $1155 for food and housing. They spend $772 per month for housing. What fractional part of their food and housing budget is spent for food?

59. What fraction of a full tank of gas is indicated by the gas gauge?

60. What fraction of a full tank of gas is indicated by the gas gauge?

61. A weight scale is marked with a whole number at each pound. What whole number mark is closest to a weight of $\frac{61}{16}$ lb?

62. A ruler is marked with a whole number at each centimeter. What whole number mark is closest to a length of $\frac{83}{10}$ cm?

63. Sue's construction company must place concrete section barriers, each $\frac{1}{352}$ mile long, between two sides of a freeway. How many such sections are needed for $7\frac{31}{352}$ miles of freeway?

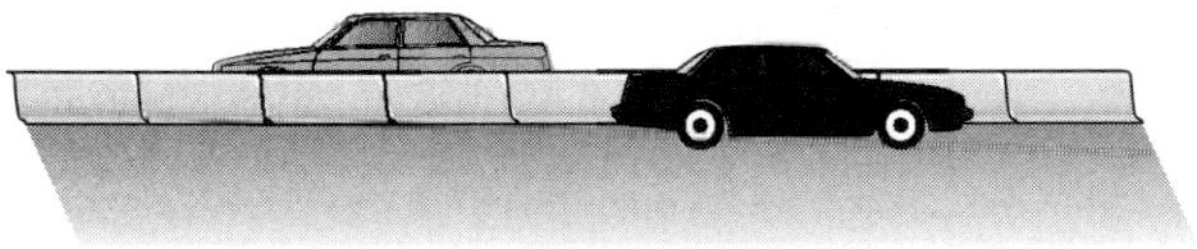

64. How many sections from Exercise 63 are needed for $37\frac{197}{352}$ miles of freeway?

65. In clothing, junior sizes are designed for young women who are 5′7″ or less. What fractional part of a foot is 7″? Express the height 5′7″ as a mixed number of feet.

66. Linh awoke from a bad dream at 4:23 A.M. What fractional part of an hour is 23 minutes? Express the time 4:23 A.M. as a mixed number of hours after midnight.

67. The figure shows a measuring cup that contains oil for a zucchini bread recipe. The oil is what fractional part of a cup?

Exercises 68–70 relate to the chapter application.

68. Sandy doesn't like to dirty a lot of dishes when she bakes, so she often uses a $\frac{1}{2}$-c measure for all of her flour and sugar. The dipped butterfingers recipe calls for $2\frac{1}{2}$ c of flour. How many $\frac{1}{2}$-c measures can she use?

69. The recipe in Exercise 68 also calls for 2 c of sugar. How many $\frac{1}{2}$-c measures can she use?

70. The recipe in Exercise 68 also calls for $1\frac{3}{4}$ tsp of baking soda. If Sandy has only a $\frac{1}{4}$-tsp measuring spoon, how many $\frac{1}{4}$-tsp measures of baking soda will she use in the cookie dough?

STATE YOUR UNDERSTANDING

71. Explain how to change $\frac{34}{5}$ to a mixed number. Explain how to change $7\frac{3}{8}$ to an improper fraction.

72. Tell why mixed numbers and improper fractions are both useful. Give examples of the use of each.

73. Explain why a proper fraction cannot be changed into a mixed number.

CHALLENGE

74. Write the whole number 13 as an improper fraction with (a) the numerator 117 and (b) the denominator 117.

75. Write the whole number 16 as an improper fraction with (a) the numerator 144 and (b) the denominator 144.

76. The Swete Tuth candy company packs 30 pieces of candy in its special Mother's Day box. Write, as a mixed number, the number of special boxes that can be made from 67,892 pieces of candy. The company then packs 25 of the special boxes in a carton for shipping. Write, as a mixed number, the number of cartons that can be filled. If it costs Swete Tuth \$45 per carton to ship the candy, what is the shipping cost for the number of full cartons that can be shipped?

77. Jose has $21\frac{3}{4}$ yd of rope to use in a day care class. If the rope is cut into $\frac{1}{4}$-yd pieces, how many pieces will there be? If there are 15 children in the class, how many pieces of rope will each child get? How many pieces will be left over?

GROUP WORK

78. Often, data about a population are presented as a pie chart. Pie charts are based on fractions. If the fractions are easy to draw, a pie chart can be drawn quickly. Sometimes a drawing that is a reasonable estimate is adequate. A survey revealed that half of a class preferred pepperoni pizza, one-quarter preferred cheese pizza, one-eighth preferred Canadian bacon pizza, and one-eighth preferred sausage pizza. Draw a pie chart that illustrates this survey. Explain your strategy.

79. According to a radio advertising survey, 77 of 100 people in the United States say they listen to the radio daily. Make a pie chart that illustrates this survey. Explain your strategy.

80. Using the following figure, write a fraction to name each of the division marks. Now divide each section in half and write new names for each mark. Divide the new sections in half and again write new fraction names for the marks. You now have three names for each of the original marks. What conclusion can you make about these names?

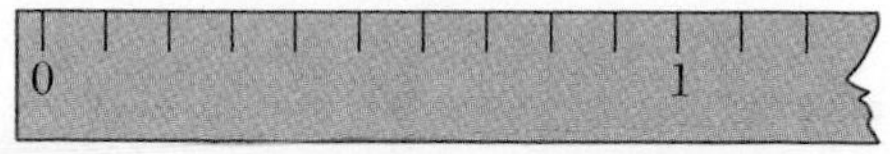

81. Write $\frac{45678}{37}$ as a mixed number with the aid of a calculator. Use the calculator to find the whole number part and the numerator of the fraction part of the mixed number. Be prepared to show the class how to do this.

MAINTAIN YOUR SKILLS

82. Divide: $78 \div 3$

83. Divide: $78 \div 2$

84. Divide: $390 \div 6$

85. Divide: $390 \div 26$

86. Is 271 prime or composite?

87. Is 273 prime or composite?

88. Find the prime factorization of 275.

89. Find the prime factorization of 279.

90. Bonnie averages 40 miles per gallon of gasoline on her vintage motorcycle. On a recent trip she traveled 680 miles. How many gallons of gas did she use?

91. After a tune-up, Bonnie traveled 221 miles, 203 miles, 192 miles, and 188 miles on four tanks of gas. If her tank holds 4 gallons of gas, what was her average number of miles per tank? To the nearest whole number, what was her average number of miles per gallon?

3.2 Simplifying Fractions

VOCABULARY

Equivalent fractions are fractions that are the different names for the same number.

Simplifying a fraction is the process of renaming it by using a smaller numerator and denominator.

A fraction is **completely simplified** when its numerator and denominator have no common factors other than 1. For instance, $\frac{12}{18} = \frac{2}{3}$.

OBJECTIVE

Simplify a fraction.

HOW & WHY

OBJECTIVE Simplify a fraction.

Fractions are *equivalent* if they represent the same quantity. When we compare the two units in Figure 3.6, we see that each is divided into four parts. The shaded part on the left is named $\frac{2}{4}$, whereas the shaded part on the right is labeled $\frac{1}{2}$. It is clear that the two are the same size, and therefore we say $\frac{2}{4} = \frac{1}{2}$.

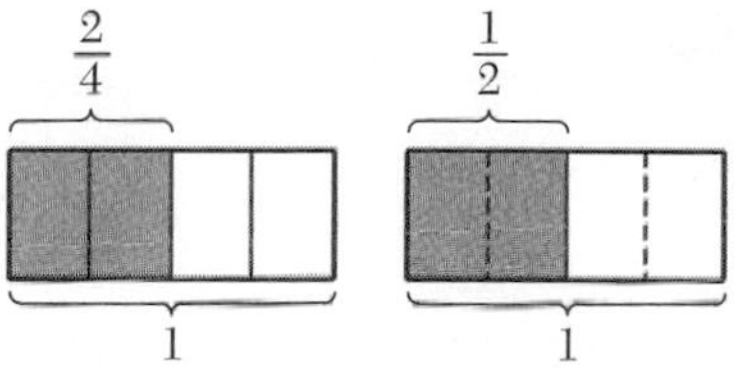

FIGURE 3.6

The arithmetical way of showing that $\frac{2}{4} = \frac{1}{2}$ is to eliminate the common factors by dividing:

$$\frac{2}{4} = \frac{2 \div 2}{4 \div 2} = \frac{1}{2}$$

The division can also be shown by eliminating the common factors.

$$\frac{2}{4} = \frac{1 \cdot \cancel{2}^{1}}{2 \cdot \cancel{2}_{1}} = \frac{1}{2}$$

This method works for all fractions. To simplify $\frac{28}{21}$:

$$\frac{28}{21} = \frac{4 \cdot \cancel{7}^{1}}{3 \cdot \cancel{7}_{1}} = \frac{4}{3}$$ Eliminate the common factors.

or

$$\frac{28}{21} = \frac{28 \div 7}{21 \div 7} = \frac{4}{3}$$ Divide out the common factors.

When all common factors have been eliminated (divided out), the fraction is completely simplified.

$$\frac{24}{40} = \frac{12}{20} = \frac{6}{10} = \frac{3}{5}$$ Completely simplified.

If the common factors are not discovered easily, they can be found by writing the numerator and denominator in prime-factored form. See Example E.

To simplify a fraction completely

Eliminate all common factors other than 1 in the numerator and the denominator.

WARM-UPS A–H

A. Simplify: $\frac{22}{33}$

B. Simplify: $\frac{18}{45}$

C. Simplify: $\frac{48}{32}$

Answers to Warm-Ups

A. $\frac{2}{3}$ **B.** $\frac{2}{5}$ **C.** $\frac{3}{2}$

EXAMPLES A–H

DIRECTIONS: Simplify completely.

STRATEGY: Eliminate the common factors in the numerator and the denominator.

A. Simplify: $\frac{27}{45}$

$$\frac{27}{45} = \frac{3 \cdot \overset{1}{\cancel{9}}}{5 \cdot \underset{1}{\cancel{9}}} = \frac{3}{5}$$

The common factor is 9. Eliminate the common factor by dividing.

B. Simplify: $\frac{16}{24}$

$$\frac{16}{24} = \frac{\overset{1}{\cancel{2}} \cdot 8}{\underset{1}{\cancel{2}} \cdot 12} = \frac{8}{12}$$

A common factor of 2 is eliminated.

$$= \frac{\overset{1}{\cancel{2}} \cdot 4}{\underset{1}{\cancel{2}} \cdot 6} = \frac{4}{6}$$

There is still a factor of 2 in the numerator and the denominator.

$$= \frac{\overset{1}{\cancel{2}} \cdot 2}{\underset{1}{\cancel{2}} \cdot 3} = \frac{2}{3}$$

Again, a common factor of 2 is eliminated by division.

or

$$\frac{16}{24} = \frac{2 \cdot \overset{1}{\cancel{8}}}{3 \cdot \underset{1}{\cancel{8}}} = \frac{2}{3}$$

Rather than divide by 2 three times, divide by 8 once, and the fraction is simplified completely.

C. Simplify: $\frac{51}{34}$

$$\frac{51}{34} = \frac{\overset{3}{\cancel{51}}}{\underset{2}{\cancel{34}}} = \frac{3}{2}$$

Divide both numerator and denominator by 17.

or

$$\frac{51}{34} = \frac{\overset{3}{\cancel{51}}}{\underset{2}{\cancel{34}}} = \frac{3}{2}$$

D. Simplify: $\frac{100}{600}$

$$\frac{100}{600} = \frac{100 \div 100}{600 \div 100} = \frac{1}{6}$$ Divide both numerator and denominator by 100.

or

$$\frac{100}{600} = \frac{1}{6}$$ Divide by 100 mentally, using the shortcut for dividing by powers of 10: $100 = 10^2$.

E. Simplify: $\frac{126}{144}$

STRATEGY: Because the numbers are large, write them in prime-factored form.

$$\frac{126}{144} = \frac{\overset{1}{\cancel{2}} \cdot \overset{1}{\cancel{3}} \cdot \overset{1}{\cancel{3}} \cdot 7}{\underset{1}{\cancel{2}} \cdot 2 \cdot 2 \cdot 2 \cdot \underset{1}{\cancel{3}} \cdot \underset{1}{\cancel{3}}}$$ Eliminate the common factors.

$$= \frac{7}{2 \cdot 2 \cdot 2}$$

$$= \frac{7}{8}$$ Multiply.

F. Simplify: $\frac{18}{35}$

$$\frac{18}{35} = \frac{2 \cdot 3 \cdot 3}{5 \cdot 7} = \frac{18}{35}$$ There are no common factors. The fraction is already completely simplified.

CALCULATOR EXAMPLE

G. Simplify $\frac{493}{551}$.

$$\frac{493}{551} = \frac{17}{19}$$ Use the fraction key on your calculator.

H. Mel washes cars on Saturday to earn extra money. On a particular Saturday, he has 12 cars to wash. After he has washed 8 of them, what fraction of the total has he washed? Simplify the fraction completely.

STRATEGY: Form the fraction: $\frac{\text{number of cars washed}}{\text{total number of cars}}$.

$\frac{8}{12}$ Eight cars are washed out of 12.

$$\frac{8}{12} = \frac{2 \cdot \overset{1}{\cancel{4}}}{3 \cdot \underset{1}{\cancel{4}}} = \frac{2}{3}$$ Simplify.

So $\frac{2}{3}$ of the cars are washed.

D. Simplify: $\frac{4000}{14{,}000}$

E. Simplify: $\frac{160}{256}$

F. Simplify: $\frac{25}{81}$

G. Simplify $\frac{703}{851}$

H. In Example H, if Mel washes only 4 cars, what fraction of the total is washed?

Answers to Warm-Ups

D. $\frac{2}{7}$ **E.** $\frac{5}{8}$ **F.** $\frac{25}{81}$ **G.** $\frac{19}{23}$

* **H.** He has washed $\frac{1}{3}$ of the cars.

EXERCISES 3.2

OBJECTIVE Simplify a fraction.

Simplify completely.

A.

1. $\frac{6}{12}$ **2.** $\frac{6}{15}$ **3.** $\frac{6}{9}$ **4.** $\frac{8}{12}$

5. $\frac{10}{25}$ **6.** $\frac{16}{18}$ **7.** $\frac{40}{50}$ **8.** $\frac{30}{80}$

9. $\frac{15}{20}$ **10.** $\frac{21}{28}$ **11.** $\frac{12}{20}$ **12.** $\frac{20}{22}$

13. $\frac{32}{40}$ **14.** $\frac{30}{40}$ **15.** $\frac{60}{36}$ **16.** $\frac{55}{22}$

17. $\frac{14}{18}$ **18.** $\frac{28}{36}$ **19.** $\frac{28}{35}$ **20.** $\frac{18}{45}$

21. $\frac{130}{10}$ **22.** $\frac{180}{60}$

B.

23. $\frac{32}{12}$ **24.** $\frac{50}{35}$ **25.** $\frac{13}{39}$ **26.** $\frac{26}{39}$

27. $\frac{12}{36}$ **28.** $\frac{20}{36}$ **29.** $\frac{27}{36}$ **30.** $\frac{32}{36}$

31. $\frac{29}{36}$ **32.** $\frac{23}{26}$ **33.** $\frac{50}{75}$ **34.** $\frac{30}{75}$

35. $\frac{55}{75}$ **36.** $\frac{15}{75}$ **37.** $\frac{600}{800}$ **38.** $\frac{500}{900}$

39. $\frac{45}{80}$ **40.** $\frac{65}{80}$ **41.** $\frac{72}{96}$ **42.** $\frac{36}{39}$

43. $\frac{38}{19}$ **44.** $\frac{96}{24}$ **45.** $\frac{75}{105}$ **46.** $\frac{64}{120}$

C.

51. Which of these fractions is completely simplified: $\frac{14}{35}$, $\frac{26}{33}$, $\frac{34}{51}$, $\frac{28}{42}$?

52. Which of these fractions is completely simplified: $\frac{39}{65}$, $\frac{44}{76}$, $\frac{102}{119}$, $\frac{76}{135}$?

Simplify completely.

53. $\frac{120}{144}$

54. $\frac{96}{144}$

55. $\frac{496}{620}$

56. $\frac{372}{620}$

57. $\frac{728}{910}$

58. $\frac{805}{910}$

59. In clothing, petite sizes are designed for women who are 5′4″ or less. What fractional part of a foot is 4″? Express the height 5′4″ as a mixed number of feet.

60. In a recent playoff game, Karl Malone of the Utah Jazz made 6 free throws out of 8 attempts. What fraction of free throw attempts did he make? What fraction did he miss?

61. In a recent NHL game, the goalie for Toronto had 32 shots on goal and made 30 saves. What fraction of the total shots did the goalie save? How many points did the opposing team make?

62. In 1980 in the United States, 122,000,000 cars, 34,000,000 trucks, 500,000 buses, and 5,700,000 motorcycles were registered. What fraction of the total number of vehicles registered was motorcycles?

63. Maria performs a tune-up on her automobile. She finds that 4 of the 6 spark plugs are fouled. What fraction represents the number of fouled plugs? Simplify.

64. The float on a tank registers 12 feet. If the tank is full when it registers 28 feet, what fraction of the tank is full? Simplify.

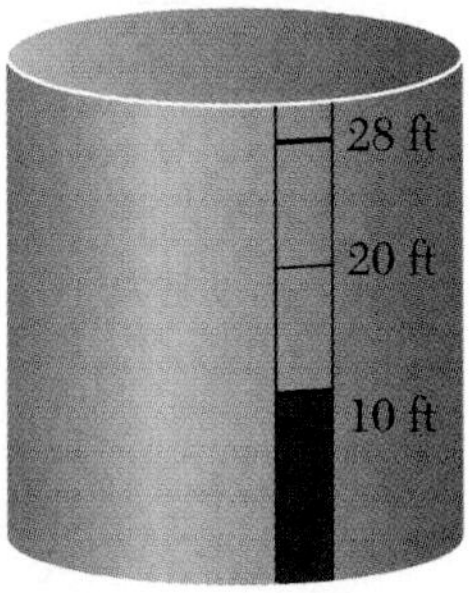

65. Gyrid completes 32 hours out of her weekly shift of 40 hours. What fraction of her weekly shift remains?

66. The Bonneville Dam fish counter shows 4320 salmon passing over the dam. Of this number, 220 are cohos. What fraction of the salmon is cohos? Simplify.

67. A district attorney successfully prosecutes 44 cases and is unsuccessful on another 20 cases. What fraction of the attorney's cases is successfully prosecuted? Simplify.

68. The energy used to produce 1 lb of virgin rubber is 15,800 BTUs. Producing 1 lb of recycled rubber requires only 4600 BTUs. What fraction of the BTUs needed to produce 1 lb of virgin rubber is used to produce 1 lb of recycled rubber? Simplify.

69. One hundred thirty elk are tallied at the Florence Refuge. Forty-five of the elk are bulls. What fraction of the elk are cows? Simplify.

Exercises 70–71 relate to the chapter application.

70. Hannah and Sandy intend to sell a box of pecan chocolate chip cookies for \$6.00. Write a simplified fraction that represents the portion of the selling price that is profit.

71. The women also agree to price a box of dipped butterfingers at \$7.20. Write a simplified fraction that represents the portion of the selling price that is profit.

STATE YOUR UNDERSTANDING

72. Draw a picture that illustrates that $\frac{9}{12} = \frac{6}{8} = \frac{3}{4}$.

73. Explain how to simplify $\frac{525}{1125}$.

74. Explain why $\frac{12}{16}$ and $\frac{15}{20}$ are equivalent fractions.

CHALLENGE

75. Are these four fractions equivalent? Justify your answer.

$\frac{495}{1188}$ $\frac{665}{1596}$ $\frac{1095}{2628}$ $\frac{890}{2136}$

76. Are these four fractions equivalent? Justify your answer.

$\frac{1170}{2925}$ $\frac{864}{2160}$ $\frac{672}{1440}$ $\frac{1134}{2430}$

GROUP WORK

77. In the figure, the circle is divided into halves. Divide each of the halves in half. Now use the figure to answer the question: What is $\frac{1}{2}$ of $\frac{1}{2}$? Devise a rule for finding the product without using the circle.

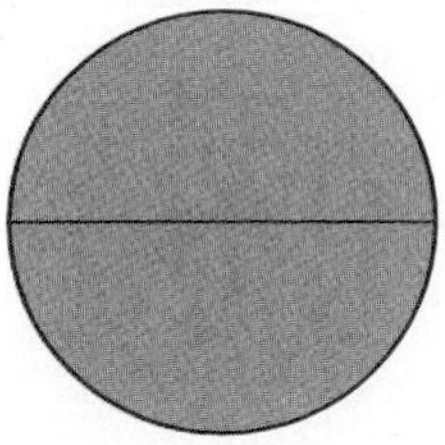

78. In the figure, the circle is divided into thirds. Divide each of the thirds in half. Now use the figure to answer the question: What is $\frac{1}{2}$ of $\frac{1}{3}$? Devise a rule for finding the product without using the circle.

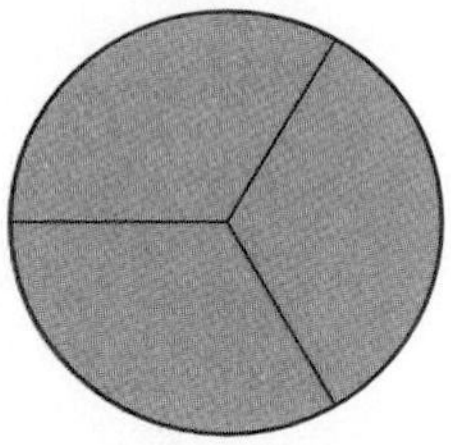

MAINTAIN YOUR SKILLS

Multiply.

79. 8(77)

80. 9(88)

81. 10(99)

82. 7(333)

83. 8(444)

Divide.

84. $270 \div 6$

85. $270 \div 15$

86. $270 \div 45$

87. $336 \div 21$

88. $336 \div 28$

3.3 Multiplying and Dividing Fractions

VOCABULARY

A **product** is the answer to a multiplication problem. If two fractions have a product of 1, each fraction is called the **reciprocal** of the other. For example, $\frac{2}{3}$ is the reciprocal of $\frac{3}{2}$. A **quotient** is the answer to a division exercise. The quotient of 33 and 3 is 11.

OBJECTIVES

1. Multiply fractions.
2. Find the reciprocal of a number.
3. Divide fractions.

HOW & WHY

OBJECTIVE 1 Multiply fractions.

The word *of* often indicates multiplication. For example, what is $\frac{1}{2}$ of $\frac{1}{3}$? That is, $\frac{1}{2} \cdot \frac{1}{3} = ?$

See Figure 3.7. The rectangle is divided into three parts. One part, $\frac{1}{3}$, is shaded yellow. To find $\frac{1}{2}$ of the yellow shaded region, divide each of the thirds into two parts (halves). Figure 3.8 shows the rectangle divided into six parts. So, $\frac{1}{2}$ of the shaded third is $\frac{1}{6}$ of the rectangle, which is shaded blue.

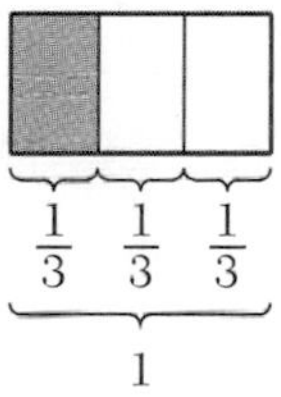

FIGURE 3.7

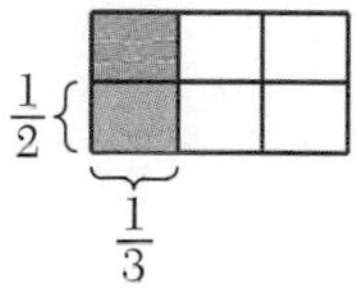

FIGURE 3.8

$$\frac{1}{2} \text{ of } \frac{1}{3} = \frac{1}{2} \cdot \frac{1}{3} = \frac{1}{6} = \frac{\text{number of parts shaded yellow}}{\text{total number of parts}}$$

What is $\frac{1}{4}$ of $\frac{3}{4}$? $\left(\frac{1}{4} \cdot \frac{3}{4} = ?\right)$ In Figure 3.9, the rectangle has been divided into 4 parts, and $\frac{3}{4}$ is represented by the parts that are shaded yellow. To find $\frac{1}{4}$ of the yellow shaded regions, divide each of the fourths into 4 parts. The rectangle is now divided into 16 parts, and $\frac{1}{4}$ of each of the three original yellow regions is shaded blue. The blue shaded region represents $\frac{3}{16}$. See Figure 3.10.

$$\frac{1}{4} \text{ of } \frac{3}{4} = \frac{1}{4} \cdot \frac{3}{4} = \frac{3}{16}$$

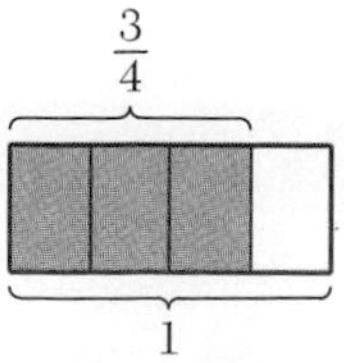

FIGURE 3.9

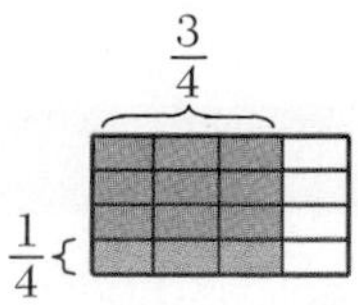

FIGURE 3.10

We have seen that $\frac{1}{2}\cdot\frac{1}{3}=\frac{1}{6}$ and that $\frac{1}{4}\cdot\frac{3}{4}=\frac{3}{16}$. The shortcut is to multiply the numerators and multiply the denominators.

$$\begin{aligned}
\frac{6}{25}\cdot\frac{35}{24} &= \frac{210}{600} \\
&= \frac{210 \div 10}{600 \div 10} && \text{Simplify by dividing both numerator and denominator by 10.} \\
&= \frac{21}{60} \\
&= \frac{21 \div 3}{60 \div 3} && \text{Simplify by dividing both numerator and denominator by 3.} \\
&= \frac{7}{20} && \text{The fraction is completely simplified.}
\end{aligned}$$

Multiplying two or more fractions like those just given can often be done more quickly by simplifying before multiplying.

$$\begin{aligned}
\frac{6}{25}\cdot\frac{35}{24} &= \frac{\overset{1}{\cancel{6}}}{25}\cdot\frac{35}{\underset{4}{\cancel{24}}} && \text{Simplify by dividing 6 and 24 by 6.} \\
&= \frac{\overset{1}{\cancel{6}}}{\underset{5}{\cancel{25}}}\cdot\frac{\overset{7}{\cancel{35}}}{\underset{4}{\cancel{24}}} && \text{Divide 25 and 35 by 5.} \\
&= \frac{7}{20} && \text{Multiply.}
\end{aligned}$$

The next example shows all of the simplifying done in one step.

$$\begin{aligned}
\frac{20}{36}\cdot\frac{18}{25} &= \frac{\overset{\overset{1}{\cancel{5}}}{\cancel{20}}}{\underset{\underset{1}{\cancel{9}}}{\cancel{36}}}\cdot\frac{\overset{2}{\cancel{18}}}{\underset{5}{\cancel{25}}} && \text{Divide 20 and 36 by 4; then divide 5 and 25 by 5.} \\
& && \text{Next, divide 18 and 9 by 9. Then multiply.} \\
&= \frac{2}{5}
\end{aligned}$$

If the numbers are large, find the prime factorization of each numerator and denominator. See Example E.

To multiply fractions

1. Simplify.
2. Write the product of the numerators over the product of the denominators.

CAUTION

Simplifying before doing the operation works only for multiplication because it is based on multiplying by 1. It does **not** work for addition, subtraction, or division.

EXAMPLES A–G

DIRECTIONS: Multiply. Simplify completely.

STRATEGY: Simplify and then multiply.

A. Multiply and simplify: $\frac{1}{3}\cdot\frac{2}{5}\cdot\frac{4}{7}$

$\frac{1}{3}\cdot\frac{2}{5}\cdot\frac{4}{7}=\frac{8}{105}$ There are no common factors. Multiply the numerators and multiply the denominators.

B. Multiply and simplify: $\frac{7}{3}\cdot 2$

STRATEGY: First change 2 to an improper fraction.

$\frac{7}{3}\cdot 2=\frac{7}{3}\cdot\frac{2}{1}$ $\quad 2=\frac{2}{1}$

$=\frac{14}{3}$ or $4\frac{2}{3}$ Write the product of the numerators over the product of the denominators.

C. Multiply and simplify: $\frac{5}{6}\cdot\frac{3}{4}$

$\frac{5}{6}\cdot\frac{3}{4}=\frac{5}{\underset{2}{\cancel{6}}}\cdot\frac{\overset{1}{\cancel{3}}}{4}$ Eliminate the common factor of 3 in 6 and 3.

$=\frac{5}{8}$ Multiply.

D. Multiply and simplify: $\frac{7}{9}\cdot\frac{18}{5}\cdot\frac{10}{21}$

$\frac{7}{9}\cdot\frac{18}{5}\cdot\frac{10}{21}=\frac{\overset{1}{\cancel{7}}}{\underset{1}{\cancel{9}}}\cdot\frac{\overset{2}{\cancel{18}}}{\underset{1}{\cancel{5}}}\cdot\frac{\overset{2}{\cancel{10}}}{\underset{3}{\cancel{21}}}$ Eliminate the common factors of 9, 5, and 7.

$=\frac{4}{3}$ Multiply.

E. Multiply and simplify: $\frac{20}{30}\cdot\frac{15}{88}$

STRATEGY: Finding the prime factorization makes the common factors easier to detect.

$\frac{20}{30}\cdot\frac{15}{88}=\frac{2\cdot 2\cdot 5}{2\cdot 3\cdot 5}\cdot\frac{3\cdot 5}{2\cdot 2\cdot 2\cdot 11}$

$=\frac{\cancel{2}\cdot\cancel{2}\cdot\cancel{5}}{\cancel{2}\cdot\cancel{3}\cdot\cancel{5}}\cdot\frac{\cancel{3}\cdot 5}{\cancel{2}\cdot 2\cdot 2\cdot 11}$ Eliminate the common factors.

$=\frac{5}{44}$ Multiply.

WARM-UPS A–G

A. Multiply and simplify: $\frac{1}{2}\cdot\frac{3}{4}\cdot\frac{5}{7}$

B. Multiply and simplify: $3\cdot\frac{5}{7}$

C. Multiply and simplify: $\frac{7}{10}\cdot\frac{4}{9}$

D. Multiply and simplify: $\frac{8}{16}\cdot\frac{12}{9}\cdot\frac{15}{18}$

E. Multiply and simplify: $\frac{48}{55}\cdot\frac{33}{30}$

Answers to Warm-Ups

A. $\frac{15}{56}$ **B.** $\frac{15}{7}$ or $2\frac{1}{7}$ **C.** $\frac{14}{45}$ **D.** $\frac{5}{9}$ **E.** $\frac{24}{25}$

F. Multiply and simplify:
$\frac{18}{21} \cdot \frac{28}{30}$

G. During 1 year, $\frac{4}{5}$ of all the tires sold by Tredway Tires were highway tread tires. If $\frac{1}{50}$ of the highway tread tires were recalled, what fraction of all tires sold was highway treads that were recalled?

CALCULATOR EXAMPLE

F. Multiply and simplify: $\frac{16}{75} \cdot \frac{45}{56}$

$\frac{16}{75} \cdot \frac{45}{56} = \frac{6}{35}$ Use the fraction key. The calculator will automatically simplify the product.

G. In a 1-year period, $\frac{9}{10}$ of all cars sold by Trustum Used Cars had automatic transmissions. Of the cars sold with automatic transmissions, $\frac{1}{18}$ had rebuilt transmissions. What fraction of all cars sold had rebuilt automatic transmissions?

$$\frac{9}{10} \cdot \frac{1}{18} = \frac{\overset{1}{\cancel{9}}}{10} \cdot \frac{1}{\underset{2}{\cancel{18}}}$$

$$= \frac{1}{20}$$

So $\frac{1}{20}$ of all cars sold had rebuilt automatic transmissions.

HOW & WHY

OBJECTIVE 2 Find the reciprocal of a number.

Finding the reciprocal of a fraction is often called "inverting" the fraction. For instance, the reciprocal of $\frac{3}{8}$ is $\frac{8}{3}$. We check by showing that the product of the fraction and its reciprocal is 1.

$$\frac{3}{8} \cdot \frac{8}{3} = \frac{24}{24} = 1$$

To find the reciprocal of a fraction

Interchange the numerator and denominator.

CAUTION

The number zero (0) does not have a reciprocal.

Answers to Warm-Ups

F. $\frac{4}{5}$ **G.** $\frac{2}{125}$ of the highway treads were recalled.

WARM-UPS H–I

EXAMPLES H–I

DIRECTIONS: Find the reciprocal.

STRATEGY: Interchange the numerator and the denominator. That is "invert" the fraction.

H. Find the reciprocal of $\frac{5}{12}$.

$\frac{12}{5}$ Exchange the numerator and denominator.

CHECK: $\frac{5}{12} \cdot \frac{12}{5} = \frac{60}{60} = 1$

The reciprocal of $\frac{5}{12}$ is $\frac{12}{5}$ or $2\frac{2}{5}$.

I. Find the reciprocal of $1\frac{4}{9}$.

STRATEGY: First write $1\frac{4}{9}$ as an improper fraction.

$1\frac{4}{9} = \frac{13}{9}$

$\frac{9}{13}$ Invert the fraction.

CHECK: $\frac{13}{9} \cdot \frac{9}{13} = \frac{117}{117} = 1$

The reciprocal of $1\frac{4}{9}$ is $\frac{9}{13}$.

H. Find the reciprocal of $\frac{14}{5}$.

I. Find the reciprocal of $2\frac{5}{6}$.

HOW & WHY

OBJECTIVE 3 Divide fractions.

In Chapter 1, you learned that division is the inverse of multiplication. That is, the quotient of a division problem is the number that is multiplied times the divisor (second number) to get the dividend. For example, the quotient of $21 \div 3$ is 7. When 7 is multiplied by the divisor, 3, the product is 21, the dividend. Another way of thinking of division is to ask: How many groups of a given size are contained in a number?

	Think	Answer
$21 \div 3$	How many 3s in 21?	7
$\frac{4}{5} \div \frac{1}{10}$	How many $\frac{1}{10}$s in $\frac{4}{5}$?	See Figure 3.11.

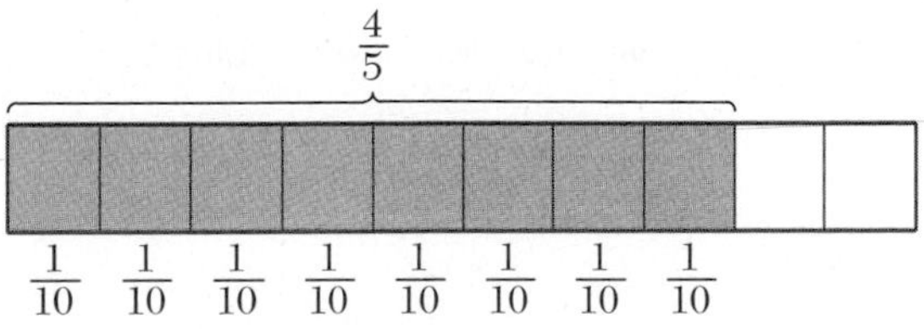

FIGURE 3.11

Figure 3.11 shows that there are eight $\frac{1}{10}$s in $\frac{4}{5}$. Therefore, we write $\frac{4}{5} \div \frac{1}{10} = 8$.

Check by multiplying. Because $8 \cdot \frac{1}{10} = \frac{8}{10} = \frac{4}{5}$, we know that the quotient is correct.

Answers to Warm-Ups

H. $\frac{5}{14}$ **I.** $\frac{6}{17}$

The quotient can also be found by multiplying $\frac{4}{5}$ by the reciprocal of $\frac{1}{10}$. This is the customary shortcut.

$$\frac{4}{5} \div \frac{1}{10} = \frac{4}{5} \cdot \frac{10}{1} = \frac{40}{5} = 8$$

To divide fractions

Multiply the first fraction by the reciprocal of the divisor; that is, invert and multiply.

CAUTION

Do not simplify the fractions before changing the division to multiplication; that is, invert before simplifying.

WARM-UPS J–N

J. Divide: $\frac{5}{12} \div \frac{7}{12}$

K. Divide: $\frac{5}{3} \div \frac{10}{7}$

L. Divide $\frac{5}{11} \div \frac{2}{3}$

Answers to Warm-Ups

J. $\frac{5}{7}$ **K.** $\frac{7}{6}$ or $1\frac{1}{6}$ **L.** $\frac{15}{22}$

EXAMPLES J–N

DIRECTIONS: Divide. Simplify completely.

STRATEGY: Multiply by the reciprocal of the divisor.

J. Divide: $\frac{8}{17} \div \frac{11}{17}$

$$\frac{8}{17} \div \frac{11}{17} = \frac{8}{\cancel{17}} \cdot \frac{\cancel{17}}{11}$$ Multiply by the reciprocal of the divisor.

$$= \frac{8}{11}$$

K. Divide: $\frac{9}{4} \div \frac{3}{7}$

$$\frac{9}{4} \div \frac{3}{7} = \frac{\overset{3}{\cancel{9}}}{4} \cdot \frac{7}{\underset{1}{\cancel{3}}}$$ Invert the divisor and multiply.

$$= \frac{21}{4} \text{ or } 5\frac{1}{4}$$

L. Divide: $\frac{1}{12} \div \frac{3}{5}$

$$\frac{1}{12} \div \frac{3}{5} = \frac{1}{12} \cdot \frac{5}{3}$$ Invert the divisor and multiply.

$$= \frac{5}{36}$$

CALCULATOR EXAMPLE

M. Divide and simplify: $\frac{8}{21} \div \frac{24}{77}$

STRATEGY: Use the fraction key. The calculator will automatically invert the divisor and simplify the quotient.

$$\frac{8}{21} \div \frac{24}{77} = \frac{11}{9}$$

N. The distance a nut moves one turn on a bolt is $\frac{3}{16}$ inch. How many turns will it take to move the nut $\frac{3}{4}$ inch?

STRATEGY: To find the number of turns required to move the nut $\frac{3}{4}$ inch, divide $\frac{3}{4}$ by the distance the nut moves in one turn.

$$\frac{3}{4} \div \frac{3}{16} = \frac{\overset{1}{\cancel{3}}}{\underset{1}{\cancel{4}}} \cdot \frac{\overset{4}{\cancel{16}}}{\underset{1}{\cancel{3}}} = 4$$ Invert the divisor and multiply.

It takes four turns of the nut to move it $\frac{3}{4}$ inch.

M. Divide and simplify: $\frac{9}{64} \div \frac{27}{80}$

N. Suppose in Example N that the distance the nut moves on the bolt with one turn is $\frac{3}{32}$ inch. How many turns will it take to move the nut $\frac{3}{4}$ inch?

Answers to Warm-Ups

M. $\frac{5}{12}$ **N.** Eight turns are needed to move the nut $\frac{3}{4}$ inch.

Name ______________________ Class ______________________ Date ____________

EXERCISES 3.3

OBJECTIVE 1 Multiply fractions.

Multiply. Simplify completely.

A.

1. $\frac{1}{4} \cdot \frac{3}{4}$

2. $\frac{7}{6} \cdot \frac{1}{8}$

3. $\frac{3}{2} \cdot \frac{5}{14}$

4. $\frac{3}{7} \cdot \frac{6}{5}$

5. $\frac{5}{2} \cdot \frac{7}{10}$

6. $\frac{5}{4} \cdot \frac{7}{15}$

7. $\frac{2}{3} \cdot \frac{3}{10}$

8. $\frac{5}{2} \cdot \frac{2}{15}$

9. $\frac{5}{3} \cdot \frac{3}{20}$

10. $\frac{7}{9} \cdot \frac{3}{14}$

B.

11. $\frac{4}{6} \cdot \frac{9}{30} \cdot \frac{10}{6}$

12. $\frac{2}{3} \cdot \frac{4}{15} \cdot \frac{18}{7}$

13. $8 \cdot \frac{1}{5} \cdot \frac{15}{16}$

14. $\frac{14}{3} \cdot 4 \cdot \frac{2}{7}$

15. $\frac{11}{3} \cdot \frac{3}{4} \cdot \frac{4}{11}$

16. $\frac{21}{16} \cdot 2 \cdot \frac{8}{7}$

17. $\frac{5}{6} \cdot \frac{6}{7} \cdot \frac{7}{8} \cdot \frac{8}{9}$

18. $\frac{11}{12} \cdot \frac{12}{13} \cdot \frac{13}{14} \cdot \frac{14}{15}$

19. $\frac{6}{5} \cdot \frac{7}{23} \cdot \frac{5}{8} \cdot \frac{0}{6}$

20. $\frac{5}{3} \cdot \frac{3}{0} \cdot \frac{20}{7} \cdot \frac{4}{3}$

OBJECTIVE 2 Find the reciprocal of a number.

Find the reciprocal.

A.

21. $\frac{3}{10}$

22. $\frac{11}{5}$

23. 8

24. 0

25. $\frac{2}{0}$

B.

26. $3\frac{7}{11}$

27. $2\frac{5}{14}$

28. 23

29. $\frac{1}{8}$

30. 10

OBJECTIVE 3 Divide fractions.

Divide. Simplify completely.

A.

31. $\frac{3}{7} \div \frac{4}{9}$

32. $\frac{9}{8} \div \frac{5}{7}$

33. $\frac{7}{15} \div \frac{7}{10}$

34. $\frac{9}{11} \div \frac{10}{11}$

35. $\frac{8}{3} \div \frac{8}{9}$

36. $\frac{5}{6} \div \frac{5}{3}$

37. $\frac{6}{55} \div \frac{10}{33}$

38. $\frac{5}{12} \div \frac{25}{6}$

39. $\frac{12}{16} \div \frac{3}{4}$

40. $\frac{9}{16} \div \frac{3}{4}$

B.

41. $\frac{28}{25} \div \frac{21}{80}$

42. $\frac{30}{49} \div \frac{10}{21}$

43. $\frac{20}{24} \div \frac{20}{18}$

44. $\frac{7}{12} \div \frac{12}{5}$

45. $\frac{9}{50} \div \frac{6}{5}$

46. $\frac{35}{32} \div \frac{10}{3}$

47. $\frac{90}{55} \div \frac{9}{5}$

48. $\frac{28}{35} \div \frac{7}{5}$

49. $\frac{51}{85} \div \frac{3}{5}$

50. $\frac{42}{98} \div \frac{3}{7}$

C.

Multiply. Simplify completely.

51. $\frac{36}{55} \cdot \frac{33}{54} \cdot \frac{35}{24}$

52. $\frac{21}{15} \cdot \frac{18}{14} \cdot \frac{15}{42}$

53. $\frac{40}{44} \cdot \frac{24}{160} \cdot \frac{77}{35}$

54. $\frac{32}{675} \cdot \frac{250}{16} \cdot \frac{135}{50}$

Divide. Simplify completely.

55. $\frac{150}{54} \div \frac{100}{81}$

56. $\frac{13}{75} \div \frac{39}{35}$

57. $\frac{16}{81} \div \frac{8}{108}$

58. $\frac{75}{90} \div \frac{50}{72}$

Fill in the boxes with a single number so the statement is true. Explain your answer.

59. $\frac{2}{3} \cdot \frac{\square}{16} = \frac{5}{24}$

60. $\frac{3}{\square} \cdot \frac{8}{15} = \frac{8}{45}$

61. Find the error(s) in the statement: $\frac{3}{8} \cdot \frac{4}{5} = \frac{15}{32}$. Correct the statement. Explain how you would avoid this error.

62. Find the error(s) in the statement: $\frac{3}{8} \div \frac{4}{5} = \frac{32}{15}$. Correct the statement. Explain how you would avoid this error.

63. For families with four children, there are 16 possible combinations of gender and birth order. Half of these have a boy as firstborn. How many of the possible combinations have a boy as firstborn? How many have a girl as firstborn?

64. For families with four children, there are 16 possible combinations of gender and birth order. One-fourth of these have three girls and one boy. How many of the possible combinations have three girls?

65. For families with four children, there are 16 possible combinations of gender and birth order. One-fourth of the possible combinations have three girls and one boy, and one-fourth of these (3 girl–1 boy combinations) have a firstborn boy followed by three girls. How many of the possible combinations have a firstborn boy followed by three girls?

66. In 1998, the average family income of the poorest $\frac{2}{5}$ of the population in the United States was about $21,510. The average family income of the poorest $\frac{2}{5}$ of the population in 1929 was one-third of the 1998 figure. What was the average family income of the poorest $\frac{2}{5}$ of the population in 1929?

67. A container for a household cleaner holds $\frac{7}{5}$ gallons. How much does it hold when it is three-fourths full?

68. The National Zoo has T-shirts that regularly sell for $27. The shirts are marked $\frac{1}{3}$ off during a Labor Day sale. What is the sale price of a T-shirt?

69. Underinflation of car tires can waste up to $\frac{1}{20}$ of a car's fuel due to an increase in "rolling resistance." If Becky uses 820 gallons of gas in a year, how many gallons can she save with proper tire inflation?

70. Underinflation of truck tires can waste up to $\frac{1}{15}$ of a truck's fuel due to an increase in "rolling resistance." If Melvin uses 500 gallons of gas while unaware that the tires are low, how many gallons might be wasted?

Exercises 71–73 relate to the chapter application.

71. The pecan chocolate chip recipe calls for $\frac{1}{2}$ tsp of salt and $\frac{3}{4}$ c of brown sugar. Hannah has an order for seven boxes.

a. How many cookies does she need?

b. How many times must she make the recipe to fill the order?

c. How much salt does she need for the entire order?

d. How much brown sugar does she need for the entire order?

72. Sandy has an order of six boxes of dipped butterfingers. Her recipe calls for $\frac{3}{8}$ c of bittersweet chocolate.

a. How many cookies does she need?

b. How many times must she make the recipe to fill the order?

c. How much bittersweet chocolate does she need for the entire order?

73. Sally is a friend of Sandy's who runs a catering business. She has been hired to cater a dinner party for 15 people. She is serving lasagna, but her recipe yields 40 servings.

a. Assuming that each guest will get one serving, write a simplified fraction that represents the part of the recipe Sally needs to make for the party.

b. The recipe calls for 3 lb of ricotta cheese. How many pounds does she need for the party?

STATE YOUR UNDERSTANDING

74. Explain how to find the product of $\frac{35}{24}$ and $\frac{40}{14}$.

75. Explain how to find the quotient of $\frac{35}{24}$ and $\frac{40}{14}$.

76. Evalynne's supervisor tells her that her salary is to be divided by one-half. Should she quit her job? Explain.

CHALLENGE

77. Simplify: $\left(\frac{81}{75} \cdot \frac{96}{99} \cdot \frac{55}{125}\right) \div \frac{128}{250}$

78. The In-n-Out Grocery has a standard workweek of 40 hours. Jan works $\frac{3}{4}$ of a standard week, Jose works $\frac{5}{8}$ of a standard week, Aria works $\frac{9}{8}$ of a standard week, and Bill works $\frac{6}{5}$ of a standard week. How many hours did each employee work? In-n-Out pays an average salary of $11 an hour. What is the total week's payroll for these four employees?

GROUP WORK

79. Divide your group into two subgroups. Have the two subgroups take opposite sides to discuss the following statement: Since computers and calculators use decimals more often than they use fractions, the use of fractions will eventually disappear. Have each subgroup write down their arguments for or against the statement.

To make accurate pie charts, it is necessary to use a measuring instrument for angles. One such instrument is called a "protractor." It is also necessary to convert the fractions of the components into equivalent fractions with denominators of 360 because there are 360° in a complete circle. Starting with any radius, use the protractor to measure the correct angle for each component.

80. An aggressive investment strategy allocates $\frac{3}{4}$ of a portfolio to stocks, $\frac{1}{5}$ to bonds, and $\frac{1}{20}$ to money market funds. Make an accurate pie chart for this strategy.

81. A moderate investment strategy allocates $\frac{3}{5}$ of a portfolio to stocks, $\frac{3}{10}$ to bonds, and $\frac{1}{10}$ to money market funds. Make an accurate pie chart for this strategy.

82. A conservative investment strategy allocates $\frac{2}{5}$ of a portfolio to stocks, $\frac{9}{20}$ to bonds, and $\frac{3}{20}$ to money market funds. Make an accurate pie chart for this strategy.

MAINTAIN YOUR SKILLS

Change each improper fraction to a mixed number.

83. $\frac{18}{5}$ **84.** $\frac{18}{7}$ **85.** $\frac{29}{16}$ **86.** $\frac{29}{15}$

Change each mixed number to an improper fraction.

87. $4\frac{1}{2}$ **88.** $3\frac{2}{3}$ **89.** $3\frac{1}{5}$ **90.** $2\frac{5}{8}$

91. Kevin needs a load of gravel for a drainage field. His truck can safely haul a load of $1\frac{3}{4}$ tons (3500 lb). If the gravel is sold in 222-lb scoops, how many scoops can Kevin haul?

3.4 Multiplying and Dividing Mixed Numbers

OBJECTIVES

1. Multiply mixed numbers.
2. Divide mixed numbers.

HOW & WHY

OBJECTIVE 1 Multiply mixed numbers.

In Section 3.1, we changed mixed numbers to improper fractions. For instance,

$$4\frac{3}{7} = \frac{4(7) + 3}{7} = \frac{31}{7}$$

To multiply mixed numbers, we change them to improper fractions and then multiply.

$$\left(3\frac{1}{2}\right)\left(4\frac{2}{3}\right) = \left(\frac{7}{2}\right)\left(\frac{14}{3}\right) \quad \text{Change to improper fractions.}$$

$$= \frac{98}{6} \quad \text{Multiply.}$$

$$= \frac{49}{3} = 16\frac{1}{3} \quad \text{Simplify and write as a mixed number.}$$

Products may be left as improper fractions or as mixed numbers; either is acceptable. In this text, we write mixed numbers. In algebra, improper fractions are often preferred.

To multiply mixed numbers and/or whole numbers

1. Change to improper fractions.
2. Simplify and multiply.

EXAMPLES A–D

DIRECTIONS: Multiply and simplify. Write as a mixed number.

STRATEGY: Change the mixed numbers and whole numbers to improper fractions. Multiply and simplify. Write the answer as a mixed number.

A. Multiply: $\frac{3}{5}\left(2\frac{1}{4}\right)$

$$\frac{3}{5}\left(2\frac{1}{4}\right) = \frac{3}{5}\cdot\frac{9}{4} \quad 2\frac{1}{4} = \frac{9}{4}$$

$$= \frac{27}{20} \quad \text{Multiply.}$$

$$= 1\frac{7}{20} \quad \text{Write as a mixed number.}$$

B. Multiply: $\left(2\frac{1}{3}\right)\left(2\frac{4}{7}\right)$

$$\left(2\frac{1}{3}\right)\left(2\frac{4}{7}\right) = \frac{7}{3}\cdot\frac{18}{7}$$

$$= \frac{\overset{1}{\cancel{7}}}{\underset{1}{\cancel{3}}}\cdot\frac{\overset{6}{\cancel{18}}}{\underset{1}{\cancel{7}}} \quad \text{Simplify.}$$

$$= 6 \quad \text{Multiply and write as a whole number.}$$

WARM-UPS A–D

A. Multiply: $\frac{3}{5}\left(1\frac{1}{2}\right)$

B. Multiply: $\left(1\frac{1}{2}\right)\left(5\frac{1}{3}\right)$

Answers to Warm-Ups

A. $\frac{9}{10}$ **B.** 8

C. Multiply: $4\left(3\frac{3}{5}\right)\left(\frac{5}{9}\right)$

C. Multiply: $6\left(\frac{3}{4}\right)\left(2\frac{1}{3}\right)$

CAUTION

$6\left(\frac{3}{4}\right)$ means $6 \cdot \frac{3}{4}$, but $6\frac{3}{4}$ means $6 + \frac{3}{4}$.

$$6\left(\frac{3}{4}\right)\left(2\frac{1}{3}\right) = \frac{\overset{3}{\cancel{6}}}{1} \cdot \frac{\overset{1}{\cancel{3}}}{\underset{2}{\cancel{4}}} \cdot \frac{7}{\underset{1}{\cancel{3}}}$$

$$= \frac{21}{2} = 10\frac{1}{2}$$ Multiply and write as a mixed number.

CALCULATOR EXAMPLE

D. Multiply: $5\frac{3}{8}\left(6\frac{1}{2}\right)$

D. Multiply: $4\frac{2}{3}\left(8\frac{5}{6}\right)$

$$4\frac{2}{3}\left(8\frac{5}{6}\right) = 41\frac{2}{9}$$

On a calculator with fraction keys, it is not necessary to change the mixed numbers to improper fractions. The calculator is programmed to operate with simple fractions or with mixed numbers.

*

HOW & WHY

OBJECTIVE 2 Divide mixed numbers.

Division of mixed numbers is also done by changing to improper fractions first.

$$\left(6\frac{1}{6}\right) \div \left(2\frac{1}{2}\right) = \left(\frac{37}{6}\right) \div \left(\frac{5}{2}\right)$$ Change to improper fractions.

$$= \left(\frac{37}{6}\right)\left(\frac{2}{5}\right)$$ Multiply by the reciprocal of the divisor.

$$= \frac{37}{15}$$ Simplify and multiply.

$$= 2\frac{7}{15}$$ Write as a mixed number.

To divide mixed numbers and/or whole numbers

1. Change to improper fractions.
2. Divide.

Answers to Warm-Ups

C. 8 **D.** $34\frac{15}{16}$

WARM-UPS E–I

EXAMPLES E–I

DIRECTIONS: Divide. Write as a mixed number.

STRATEGY: Change the mixed numbers and whole numbers to improper fractions. Divide and simplify completely. Write the answer as a mixed number.

E. Divide: $2\frac{1}{2} \div 4\frac{7}{8}$

$$2\frac{1}{2} \div 4\frac{7}{8} = \frac{5}{2} \div \frac{39}{8}$$

$2\frac{1}{2} = \frac{5}{2}$ and $4\frac{7}{8} = \frac{39}{8}$

$$= \frac{5}{\cancel{2}_{1}} \cdot \frac{\cancel{8}^{4}}{39}$$

Invert the divisor and multiply.

$$= \frac{20}{39}$$

F. Divide: $10\frac{5}{16} \div 16\frac{1}{2}$

$$10\frac{5}{16} \div 16\frac{1}{2} = \frac{165}{16} \div \frac{33}{2}$$

$$= \frac{\cancel{165}^{5}}{\cancel{16}_{8}} \cdot \frac{\cancel{2}^{1}}{\cancel{33}_{1}}$$

Invert the divisor and multiply.

$$= \frac{5}{8}$$

G. Divide: $6 \div 4\frac{2}{3}$

$$6 \div 4\frac{2}{3} = \frac{6}{1} \div \frac{14}{3}$$

$$= \frac{\cancel{6}^{3}}{1} \cdot \frac{3}{\cancel{14}_{7}}$$

Invert the divisor and multiply.

$$= \frac{9}{7} = 1\frac{2}{7}$$

CALCULATOR EXAMPLE

H. Divide: $25\frac{1}{4} \div 30\frac{3}{4}$

$$25\frac{1}{4} \div 30\frac{3}{4} = \frac{101}{123}$$

On a calculator with fraction keys, it is not necessary to change the mixed numbers to improper fractions. The calculator is programmed to operate with simple fractions or with mixed numbers.

I. The Sillvary Aluminum Works produces ingots that are $6\frac{3}{4}$ in. thick. What is the height in feet of a stack of 18 ingots?

$$18\left(6\frac{3}{4}\right) = \frac{18}{1} \cdot \frac{27}{4}$$

$$= \frac{\cancel{18}^{9}}{1} \cdot \frac{27}{\cancel{4}_{2}}$$

$$= \frac{243}{2} = 121\frac{1}{2}$$

E. Divide: $3\frac{1}{6} \div 2\frac{5}{12}$

F. Divide: $12\frac{1}{2} \div 15\frac{5}{8}$

G. Divide: $6\frac{3}{5} \div 6$

H. Divide: $13\frac{4}{5} \div 12\frac{3}{10}$

I. The Sillvary Aluminum Works also produces ingots that are $7\frac{1}{2}$ in. thick. What is the height in feet of a stack of 18 ingots?

Answers to Warm-Ups

E. $1\frac{9}{29}$ **F.** $\frac{4}{5}$ **G.** $1\frac{1}{10}$ **H.** $1\frac{5}{41}$

I. The height is $11\frac{1}{4}$ ft.

The height of the stack is $121\frac{1}{2}$ in. To find the height in feet, divide by 12 because there are 12 in. in 1 ft.

$$
\begin{aligned}
121\frac{1}{2} \div 12 &= \frac{243}{2} \div \frac{12}{1} \\
&= \frac{\overset{81}{\cancel{243}}}{2} \cdot \frac{1}{\underset{4}{\cancel{12}}} \\
&= \frac{81}{8} = 10\frac{1}{8}
\end{aligned}
$$

The stack of ingots is $10\frac{1}{8}$ ft high.

Name ________________ Class ________________ Date ________________

EXERCISES 3.4

OBJECTIVE 1 Multiply mixed numbers.

Multiply. Simplify completely and write as a mixed number, if possible.

A.

1. $\left(\frac{2}{3}\right)\left(1\frac{1}{4}\right)$

2. $\left(\frac{2}{7}\right)\left(1\frac{3}{7}\right)$

3. $\left(1\frac{3}{4}\right)\left(1\frac{1}{2}\right)$

4. $\left(2\frac{1}{4}\right)\left(1\frac{3}{5}\right)$

5. $3\left(2\frac{2}{3}\right)$

6. $4\left(1\frac{1}{4}\right)$

7. $\left(3\frac{3}{5}\right)\left(1\frac{1}{4}\right)(5)$

8. $\left(\frac{3}{5}\right)\left(2\frac{1}{3}\right)(3)$

9. $\left(7\frac{1}{2}\right)\left(3\frac{1}{5}\right)$

10. $\left(4\frac{1}{6}\right)\left(4\frac{1}{5}\right)$

11. $\left(3\frac{7}{17}\right)\left(1\frac{3}{14}\right)$

12. $\left(2\frac{2}{9}\right)\left(\frac{1}{15}\right)$

B.

13. $\left(2\frac{1}{6}\right)\left(2\frac{5}{8}\right)$

14. $\left(4\frac{1}{5}\right)\left(2\frac{5}{6}\right)$

15. $\left(7\frac{3}{4}\right)\left(\frac{2}{3}\right)(0)$

16. $\left(3\frac{5}{9}\right)\left(6\frac{5}{7}\right)(0)$

17. $\left(2\frac{1}{2}\right)(4)\left(3\frac{3}{4}\right)$

18. $\left(3\frac{1}{4}\right)(3)\left(2\frac{1}{3}\right)$

19. $\left(2\frac{2}{3}\right)\left(\frac{11}{12}\right)\left(6\frac{3}{4}\right)$

20. $\left(2\frac{2}{5}\right)\left(\frac{1}{4}\right)\left(5\frac{5}{8}\right)$

21. $\left(4\frac{1}{5}\right)\left(1\frac{1}{3}\right)\left(6\frac{2}{7}\right)$

22. $\left(12\frac{1}{4}\right)\left(1\frac{1}{7}\right)\left(2\frac{1}{3}\right)$

23. $(14)\left(6\frac{1}{2}\right)\left(1\frac{2}{13}\right)$

24. $12\left(1\frac{4}{15}\right)\left(6\frac{1}{4}\right)$

OBJECTIVE 2 Divide mixed numbers.

Divide. Simplify completely and write as a mixed number, if possible.

A.

25. $6 \div 1\frac{1}{3}$

26. $3 \div 1\frac{1}{6}$

27. $2\frac{7}{8} \div 3\frac{5}{6}$

28. $4\frac{2}{3} \div 8\frac{2}{3}$

29. $10\frac{1}{2} \div 5\frac{1}{4}$

30. $6\frac{7}{8} \div 3\frac{7}{16}$

31. $1\frac{5}{8} \div 2\frac{1}{4}$

32. $1\frac{5}{8} \div 1\frac{3}{4}$

33. $3\frac{1}{3} \div 2\frac{1}{2}$

34. $5\frac{1}{3} \div 1\frac{1}{7}$

35. $4\frac{4}{15} \div 6\frac{2}{5}$

36. $6\frac{1}{4} \div 7\frac{1}{2}$

B.

37. $\frac{7}{8} \div 3\frac{3}{4}$

38. $\frac{11}{15} \div 2\frac{4}{5}$

39. $6\frac{1}{4} \div 5$

40. $3\frac{5}{7} \div 13$

41. $4\frac{5}{6} \div \frac{1}{3}$

42. $8\frac{3}{4} \div 2\frac{1}{3}$

43. $3\frac{3}{4} \div \frac{7}{15}$

44. $5\frac{5}{6} \div \frac{20}{9}$

45. $23\frac{1}{4} \div 1\frac{1}{3}$

46. $26\frac{7}{8} \div 3\frac{3}{4}$

47. $10\frac{2}{3} \div 2\frac{2}{7}$

48. $22\frac{2}{3} \div 6\frac{6}{7}$

C.

Multiply. Simplify completely and write as a mixed number, if possible.

49. $\left(5\frac{1}{3}\right)\left(2\frac{3}{14}\right)(7)\left(2\frac{1}{4}\right)$

50. $\left(3\frac{3}{5}\right)\left(4\frac{1}{3}\right)(3)\left(2\frac{7}{9}\right)$

Divide. Simplify completely and write as a mixed number, if possible.

51. $\left(1\frac{1}{2}\right) \div \left(2\frac{1}{3}\right) \div \left(1\frac{2}{7}\right)$

52. $\left(2\frac{1}{2}\right) \div \left(4\frac{1}{3}\right) \div \left(1\frac{1}{4}\right)$

53. Find the error(s) in the statement: $1\frac{2}{3} \cdot 1\frac{1}{2} = 1\frac{1}{3}$. Correct the statement. Explain how you would avoid this error.

54. Find the error(s) in the statement: $6\frac{2}{9} \div 2\frac{2}{3} = 3\frac{1}{3}$. Correct the statement. Explain how you would avoid this error.

55. A 6-ft by 8-ft ready-made storage shed has interior dimensions of $94\frac{3}{4}$ in. wide by 66 in. deep. How many square inches are in the interior of the shed?

56. The recommended foundation size for the storage shed in Exercise 55 is $97\frac{1}{2}$ in. by $68\frac{3}{4}$ in. What is the area of the recommended slab?

57. During a recent fish count at an Idaho dam, 3568 fish were counted. Five-sixteenths of these were steelhead. How many steelhead were counted?

58. Joe has an 8-ft board that he wants to cut into $20\frac{1}{2}$ in. lengths to make shelves. How many shelves can Joe get from the board?

59. The iron content in a water sample at Lake Hieda is 8 parts per million. The iron content in Swan Lake is $2\frac{3}{4}$ times greater than the content in Lake Hieda. What is the iron content in Swan Lake in parts per million?

60. The water pressure during a bad neighborhood grass fire is reduced to $\frac{5}{9}$ its original pressure at the hydrant. What is the reduced pressure if the original pressure was $70\frac{1}{5}$ lb per square inch?

61. A glass of reconstituted juice contains $\frac{3}{16}$ oz of concentrate. How many glasses of juice can Karla make from 36 oz of the concentrate?

62. The Top Notch Candy Company packages jellybeans in $1\frac{3}{8}$-lb bags. How many bags can be made from 660 lb of jellybeans?

63. The amount of CO_2 a car emits is directly related to the amount of gas it uses. Cars give off 20 lb of CO_2 for every gallon of gas used. A car averaging 27 mpg will emit 2000 lb of CO_2 in 2700 mi. A car averaging 18 mpg will emit $1\frac{1}{2}$ times more CO_2 in the same distance. How many pounds of CO_2 does the less-efficient car emit in the 2700 mi?

64. Nutritionists recommend that not more than $\frac{3}{10}$ of your daily intake in calories come from fat grams. If each gram of fat is 10 calories, what is the recommended upper limit on fat from a diet of 2400 calories?

Exercises 65–67 relate to the chapter application.

65. Hannah is making 10 recipes of her pecan chocolate chip cookies.

a. The recipe calls for $2\frac{1}{4}$ c of flour. How much flour does she need for 10 recipes?

b. The recipe also calls for $1\frac{1}{2}$ c of sugar. How much sugar does she need for 10 recipes?

66. Sandy has an order of 3 boxes of dipped butterfingers.

a. How many cookies does she need?

b. How many times does she need to increase her recipe in order to get exactly the correct number of cookies?

c. The recipe calls for $1\frac{1}{2}$ c of flour. How much flour does she need for the order?

d. The recipe calls for 2 tsp of vanilla. How much vanilla does she need for the order?

e. The recipe calls for $\frac{1}{4}$ tsp baking powder. How much baking powder does she need for the order?

67. Quan really wants to make brownies, but his recipe calls for three eggs and he only has two.

a. Write a fraction that represents the part of the recipe Quan can make with only two eggs.

$\frac{2}{3}$

b. His recipe calls for 6 tbs of butter. How much butter should Quan use?

c. His recipe also calls for $1\frac{1}{2}$ tsp of vanilla. How much vanilla should he use?

STATE YOUR UNDERSTANDING

68. Explain how to simplify $5\frac{1}{4} \div 1\frac{7}{8}$.

69. When a number is multiplied by $1\frac{1}{2}$, the result is larger than the original number. But when you divide by $1\frac{1}{2}$, the result is smaller. Explain why.

70. Why is it helpful to change mixed numbers to improper fractions before multiplying or dividing?

CHALLENGE

71. Multiply and simplify completely:

$$\left(\frac{2}{15}\right)\left(1\frac{6}{7}\right)\left(2\frac{2}{49}\right)\left(16\frac{2}{5}\right)\left(8\frac{3}{4}\right)\left(5\frac{1}{4}\right)\left(3\frac{3}{13}\right)$$

72. The Celtic Candy Company has two packs of mints that it sells in discount stores. One pack contains $1\frac{1}{4}$ lb of mints and the other contains $3\frac{1}{3}$ lb of mints. If the smaller pack sells for \$2 and the larger pack for \$5, which size should the company use to get the most income from 3000 lb of mints? How much more is the income?

GROUP WORK

73. Some people say that mixed numbers are no longer useful because of the ease of working with calculator approximations. See how many examples of the use of mixed numbers your group can find. Share the results with the class.

74. With your group members, find a way to multiply two mixed numbers without changing them to improper fractions. Describe your method to the class.

MAINTAIN YOUR SKILLS

Change to a mixed number.

75. $\frac{65}{8}$

76. $\frac{553}{15}$

77. $\frac{17}{15}$

78. $\frac{37}{15}$

79. $\frac{66}{5}$

Multiply.

80. $7200\left(\frac{1}{60}\right)\left(\frac{1}{60}\right)$

81. $60\left(\frac{1}{60}\right)\left(\frac{1}{60}\right)\left(\frac{5280}{1}\right)$

82. $25\left(\frac{1}{12}\right)\left(\frac{1}{12}\right)\left(\frac{15}{1}\right)$

83. To be eligible for a drawing at the Flick Film Mall, your ticket stub number must be a multiple of 3. If Jean's ticket number is 234572, is she eligible for the drawing?

84. Sales for the Fireyear Company totaled \$954,000 last year. During the first 6 months of last year, monthly sales were \$72,400, \$68,200, \$85,000, \$89,500, \$92,700, and \$87,200. What were the average monthly sales for the rest of the year?

Getting Ready for Algebra

HOW & WHY

OBJECTIVE

Solve an equation of the form $\frac{ax}{b} = \frac{c}{d}$, where a, b, c, and d are whole numbers.

We have previously solved equations in which variables (letters) are either multiplied or divided by whole numbers. We performed the inverse operations to solve for the variable. To eliminate multiplication, we divide by the number being multiplied. To eliminate division, we multiply by the number that is the divisor. Now we solve equations in which variables are multiplied by fractions. Recall from Chapter 1 that if a number is multiplied times a variable, there is usually no multiplication sign between them. That is, $2x$ is understood to mean 2 times x, and $\frac{2}{3}x$ means $\frac{2}{3}$ times x. However, we usually do not write $\frac{2}{3}x$. Instead, we write this as $\frac{2x}{3}$. We can do this because

$$\frac{2}{3}x = \frac{2}{3} \cdot x = \frac{2}{3} \cdot \frac{x}{1} = \frac{2x}{3}$$

Although we usually write $\frac{2x}{3}$ instead of $\frac{2}{3}x$, for convenience we may use either of these forms. Recall that $\frac{2x}{3}$ means the *product* of 2 and x *divided* by 3.

To solve an equation of the form $\frac{ax}{b} = \frac{c}{d}$

1. Multiply both sides of the equation by b to eliminate the division on the left side.
2. Divide both sides by a to isolate the variable.

EXAMPLES A–C

DIRECTIONS: Solve.

STRATEGY: Multiply both sides by the denominator of the fraction containing the variable. Solve as before.

A. Solve: $\frac{3x}{4} = 2$

$4\left(\frac{3x}{4}\right) = 4(2)$ To eliminate the division, multiply both sides by 4.

$3x = 8$ Simplify.

$\frac{3x}{3} = \frac{8}{3}$ To eliminate the multiplication, divide both sides by 3.

$x = \frac{8}{3}$

CHECK: $\frac{3}{4} \cdot \frac{8}{3} = 2$ Substitute $\frac{8}{3}$ for x in the original equation. Recall that $\frac{3x}{4} = \frac{3}{4}x$.

$2 = 2$

The solution is $x = \frac{8}{3}$ or $x = 2\frac{2}{3}$.

WARM-UPS A–C

A. Solve: $\frac{6x}{7} = 12$

Answers to Warm-Ups

A. $x = 14$

B. Solve: $\frac{5x}{8} = \frac{3}{2}$

B. Solve: $\frac{2}{3} = \frac{4x}{5}$

$5\left(\frac{2}{3}\right) = 5\left(\frac{4x}{5}\right)$ To eliminate the division by 5, multiply both sides by 5.

$\frac{10}{3} = 4x$ Simplify.

$\frac{10}{3(4)} = \frac{4x}{4}$ To eliminate the multiplication, divide both sides by 4.

$\frac{10}{12} = x$

The solution is $x = \frac{5}{6}$. The check is left for the student.

C. It requires about two-fifths the amount of energy to make "new" paper from recycled paper as it does to make paper from trees. If the amount of energy needed to make a given amount of paper from recycled paper is equivalent to 1500 BTUs, how much energy is needed to make the same amount of paper from trees?

C. Commuting to work accounts for one-third of the mileage put on a private car. If Nancy averages 510 miles per month in commuting to work, how many miles does she put on her car in a month?

STRATEGY: First write the English statement of the equation.

One-third times total number of miles = miles of commute

$\frac{1}{3}x = 510$ Let x represent the number of total miles. Miles of commute is 510.

$3\left(\frac{1}{3}x\right) = 3(510)$ Multiply both sides by 3 to eliminate the division.

$x = 1530$

CHECK: Is one-third of 1530 equal to 510?

$\frac{1}{3}(1530) = 510$ Yes

* Nancy puts 1530 miles on her car per month.

Answers to Warm-Ups

B. $x = \frac{12}{5}$ or $x = 2\frac{2}{5}$

C. It would take 3750 BTUs.

EXERCISES

Solve

1. $\frac{2x}{3} = \frac{1}{4}$

2. $\frac{2x}{5} = \frac{1}{3}$

3. $\frac{3y}{5} = \frac{2}{3}$

4. $\frac{3y}{4} = \frac{5}{8}$

5. $\frac{4z}{5} = \frac{3}{4}$

6. $\frac{5z}{4} = \frac{8}{9}$

7. $\frac{17}{9} = \frac{8x}{9}$

8. $\frac{29}{10} = \frac{9x}{5}$

9. $\frac{6a}{5} = \frac{3}{10}$

10. $\frac{12b}{5} = 8$

11. $\frac{47}{4} = \frac{47b}{6}$

12. $\frac{13}{3} = \frac{52w}{9}$

13. $\frac{41z}{6} = \frac{41}{3}$

14. $\frac{3b}{20} = \frac{7}{10}$

15. $\frac{15a}{2} = \frac{11}{4}$

16. $\frac{119x}{12} = \frac{119}{8}$

17. Merle walks one-third of the distance from his home to school. If he walks $\frac{1}{2}$ mi, what is the distance from his home to school?

18. Mona cuts a board into nine pieces of equal length. If each piece is $1\frac{4}{9}$ ft long, what was the length of the board?

19. In a southern city, $3\frac{1}{2}$ times more pounds of glass are recycled than tin. How many pounds of tin are recycled when 630 lb of glass are recycled?

20. Washing machines use about $\frac{7}{50}$ of all the water consumed in the home. If Matthew uses 280 gallons of water per month to operate his washing machine, how many gallons of water does he use in a month?

3.5 Building Fractions; Listing in Order; Inequalities

VOCABULARY

Recall that **equivalent fractions** are fractions that are different names for the same number. For instance, $\frac{4}{8}$ and $\frac{11}{22}$ are equivalent because both represent one-half $\left(\frac{1}{2}\right)$ of a unit. Two or more fractions have a **common denominator** when they have the same denominator.

OBJECTIVES

1. Rename fractions by multiplying by 1 in the form $\frac{a}{a}$.
2. Build a fraction by finding the missing numerator.
3. List a group of fractions from smallest to largest.

HOW & WHY

OBJECTIVE 1 Rename fractions by multiplying by 1 in the form $\frac{a}{a}$.

The process of renaming fractions is often referred to as "building fractions." Building a fraction means renaming the fraction by multiplying both numerator and denominator by a common factor. This process is often necessary when we add and subtract fractions. We "build fractions" to a common denominator so they can be compared, added, or subtracted. Building fractions is the opposite of "simplifying" fractions.

Simplifying a fraction	Building a fraction
$\frac{8}{10} = \frac{4\cdot\cancel{2}}{5\cdot\cancel{2}} = \frac{4}{5}$	$\frac{4}{5} = \frac{4\cdot 2}{5\cdot 2} = \frac{8}{10}$

We can see this visually in Figure 3.12.

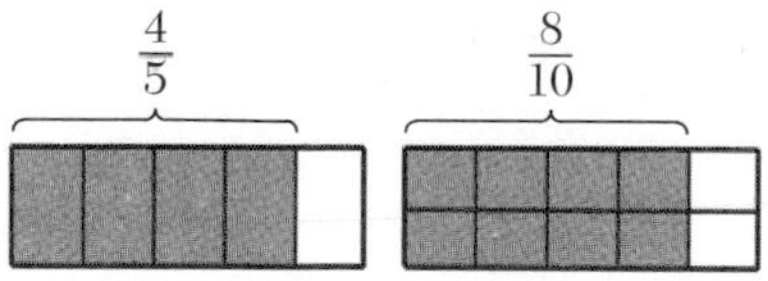

FIGURE 3.12

Table 3.3 shows five fractions built to equivalent fractions.

TABLE 3.3 EQUIVALENT FRACTIONS

	Multiply numerator and denominator by					
	2	3	4	6	10	15
$\frac{2}{5}$	$\frac{4}{10} =$	$\frac{6}{15} =$	$\frac{8}{20} =$	$\frac{12}{30} =$	$\frac{20}{50} =$	$\frac{30}{75}$
$\frac{1}{2}$	$\frac{2}{4} =$	$\frac{3}{6} =$	$\frac{4}{8} =$	$\frac{6}{12} =$	$\frac{10}{20} =$	$\frac{15}{30}$
$\frac{5}{9}$	$\frac{10}{18} =$	$\frac{15}{27} =$	$\frac{20}{36} =$	$\frac{30}{54} =$	$\frac{50}{90} =$	$\frac{75}{135}$
$\frac{7}{8}$	$\frac{14}{16} =$	$\frac{21}{24} =$	$\frac{28}{32} =$	$\frac{42}{48} =$	$\frac{70}{80} =$	$\frac{105}{120}$
$\frac{7}{4}$	$\frac{14}{8} =$	$\frac{21}{12} =$	$\frac{28}{16} =$	$\frac{42}{24} =$	$\frac{70}{40} =$	$\frac{105}{60}$

To rename a fraction

Multiply both the numerator and the denominator of the fraction by a common factor; that is, multiply the fraction by 1 in the form $\frac{a}{a}$, $a \neq 1$ and $a \neq 0$.

WARM-UPS A–B

A. Rename $\frac{5}{12}$ using $\frac{8}{8}$ for 1.

B. Write three fractions equivalent to $\frac{6}{7}$ using $\frac{2}{2}$, $\frac{3}{3}$, and $\frac{4}{4}$.

EXAMPLES A–B

DIRECTIONS: Rename the fraction.

STRATEGY: Multiply the fraction by 1 in the form $\frac{a}{a}$.

A. Rename $\frac{7}{10}$ using $\frac{5}{5}$ for 1.

$\frac{7}{10} \cdot \frac{5}{5} = \frac{35}{50}$ The new fraction, $\frac{35}{50}$, is equivalent to $\frac{7}{10}$.

B. Write three fractions equivalent to $\frac{5}{6}$ using $\frac{3}{3}$, $\frac{4}{4}$, and $\frac{5}{5}$.

$\frac{5}{6} \cdot \frac{3}{3} = \frac{15}{18}$

$\frac{5}{6} \cdot \frac{4}{4} = \frac{20}{24}$

$\frac{5}{6} \cdot \frac{5}{5} = \frac{25}{30}$

*

HOW & WHY

OBJECTIVE 2 Build a fraction by finding the missing numerator.

To find the missing numerator in

$$\frac{4}{5} = \frac{?}{30}$$

divide 30 by 5 to find out what form of 1 to multiply by.

$30 \div 5 = 6$

The correct factor (multiplier) is $\frac{6}{6}$. So

$$\frac{4}{5} = \frac{4}{5} \cdot \frac{6}{6} = \frac{24}{30}$$

The shortcut is to write 30, the target denominator. Then multiply the original numerator, 4, by 6, the quotient of the target denominator and the original denominator.

$$\frac{4}{5} = \frac{4 \cdot 6}{30} = \frac{24}{30}$$

The fractions $\frac{4}{5}$ and $\frac{24}{30}$ are equivalent. Either fraction can be used in place of the other.

Answers to Warm-Ups

A. $\frac{40}{96}$ **B.** $\frac{12}{14}, \frac{18}{21}, \frac{24}{28}$

To find the missing numerator when building fractions

1. Divide the target denominator by the original denominator.
2. Multiply this quotient by the original numerator.

EXAMPLES C–E

DIRECTIONS: Find the missing numerator.

STRATEGY: Multiply the given numerator by the same factor used in the denominator.

C. Find the missing numerator: $\frac{3}{5} = \frac{?}{80}$

$80 \div 5 = 16$ Divide the denominators.

$\frac{3}{5} = \frac{3(16)}{80} = \frac{48}{80}$ Multiply the quotient, 16, by the given numerator.

D. Find the missing numerator: $\frac{7}{9} = \frac{?}{54}$

$54 \div 9 = 6$ Divide the denominators.

$\frac{7}{9} = \frac{7(6)}{54} = \frac{42}{54}$ Multiply the quotient, 6, by the given numerator.

E. Find the missing numerator: $\frac{8}{5} = \frac{?}{140}$

$140 \div 5 = 28$

$\frac{8}{5} = \frac{8(28)}{140} = \frac{224}{140}$

WARM-UPS C–E

C. Find the missing numerator: $\frac{3}{4} = \frac{?}{80}$

D. Find the missing numerator: $\frac{3}{7} = \frac{?}{56}$

E. Find the missing numerator: $\frac{19}{15} = \frac{?}{135}$

HOW & WHY

OBJECTIVE 3 List a group of fractions from smallest to largest.

If two fractions have the same denominator, the one with the smaller numerator has the smaller value. Figure 3.13 shows that $\frac{2}{5}$ is smaller than $\frac{4}{5}$—that is, $\frac{2}{5} < \frac{4}{5}$.

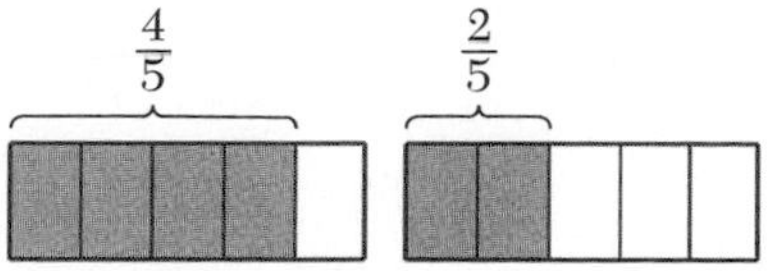

FIGURE 3.13

$\frac{2}{5} < \frac{4}{5}$ means "two-fifths is less than four-fifths."

$\frac{4}{5} > \frac{2}{5}$ means "four-fifths is greater than two-fifths."

Answers to Warm-Ups
C. 60 **D.** 24 **E.** 171

If fractions to be compared do not have a common denominator, then one or more must be renamed so that all have a common denominator. The preferred common denominator is the least common multiple (LCM) of all the denominators.

To list $\frac{3}{8}, \frac{5}{16}, \frac{1}{2}$, and $\frac{9}{16}$ from smallest to largest, we write each with a common denominator. Then we compare the numerators. The LCM of the denominators is 16. We build each fraction so that it has a denominator of 16.

$$\frac{3}{8} = \frac{6}{16} \qquad \frac{5}{16} = \frac{5}{16} \qquad \frac{1}{2} = \frac{8}{16} \qquad \frac{9}{16} = \frac{9}{16}$$

Each fraction now has a denominator of 16.

We arrange the fractions with denominator 16 in order from smallest to largest according to the values of the numerators.

$$\frac{5}{16} < \frac{6}{16} < \frac{8}{16} < \frac{9}{16}$$

The fractions are listed in order from the smallest to the largest with common denominator 16.

Next, replace each fraction by the original, so

$$\frac{5}{16} < \frac{3}{8} < \frac{1}{2} < \frac{9}{16}$$

The original fractions are listed in order from smallest to largest.

To list fractions from smallest to largest

1. Build the fractions so they have a common denominator. Use the LCM of the denominators.
2. List the fractions with common denominators so the numerators range from smallest to largest.
3. Simplify.

WARM-UP F

F. Which is larger: $\frac{7}{13}$ or $\frac{3}{5}$?

Answer to Warm-Up

F. $\frac{3}{5}$

EXAMPLE F

DIRECTIONS: Tell which fraction is larger.

STRATEGY: Write the fractions with a common denominator. The fraction with the larger numerator is the larger.

F. Which is larger: $\frac{5}{11}$ or $\frac{1}{2}$?

STRATEGY: The LCM of 11 and 2 is 22. Build each fraction so it has 22 for a denominator.

$$\frac{5}{11} = \frac{10}{22} \quad \text{and} \quad \frac{1}{2} = \frac{11}{22}$$

$\frac{1}{2}$ is larger. 11 > 10, so $\frac{1}{2} > \frac{5}{11}$.

WARM-UPS G–I

EXAMPLES G–I

DIRECTIONS: List the group of fractions from smallest to largest.

STRATEGY: Build each of the fractions to a common denominator. List the fractions from smallest to largest by the value of the numerator. Simplify.

G. List from smallest to largest: $\frac{5}{12}, \frac{1}{3}, \frac{3}{8}$

$\frac{5}{12} = \frac{10}{24}$ $\frac{1}{3} = \frac{8}{24}$ $\frac{3}{8} = \frac{9}{24}$ The LCM of 3, 8, and 12 is 24. Build the fractions to the denominator 24.

$\frac{8}{24}, \frac{9}{24}, \frac{10}{24}$ List the fractions in the order of the numerators: $8 < 9 < 10$.

The list is $\frac{1}{3}, \frac{3}{8}, \frac{5}{12}$. Simplify.

H. List from smallest to largest: $2\frac{3}{8}, 2\frac{2}{5}, 2\frac{3}{10}$

STRATEGY: Each mixed number has the same whole number part. Compare the fraction parts.

$2\frac{3}{8} = 2\frac{15}{40}$ $2\frac{2}{5} = 2\frac{16}{40}$ $2\frac{3}{10} = 2\frac{12}{40}$ The LCM of 8, 5, and 10 is 40.

$2\frac{12}{40}, 2\frac{15}{40}, 2\frac{16}{40}$ List the numbers in the order of the numerators from smallest to largest.

The list is $2\frac{3}{10}, 2\frac{3}{8}, 2\frac{2}{5}$. Simplify.

I. The Acme Hardware Store sells bolts with diameters of $\frac{5}{16}, \frac{3}{8}, \frac{1}{2}, \frac{5}{8}, \frac{1}{4}$, and $\frac{7}{16}$ in. List the diameters from smallest to largest.

$\frac{5}{16} = \frac{5}{16}$ $\frac{3}{8} = \frac{6}{16}$ $\frac{1}{2} = \frac{8}{16}$ Write each diameter using the common denominator 16.

$\frac{5}{8} = \frac{10}{16}$ $\frac{1}{4} = \frac{4}{16}$ $\frac{7}{16} = \frac{7}{16}$

$\frac{4}{16}, \frac{5}{16}, \frac{6}{16}, \frac{7}{16}, \frac{8}{16}, \frac{10}{16}$ List the diameters in order of the numerators from smallest to largest.

From smallest to largest, the diameters are

$\frac{1}{4}, \frac{5}{16}, \frac{3}{8}, \frac{7}{16}, \frac{1}{2}$, and $\frac{5}{8}$ in. Simplify.

G. List from smallest to largest: $\frac{2}{3}, \frac{5}{7}, \frac{9}{14}$

H. List from smallest to largest: $3\frac{4}{9}, 3\frac{2}{3}, 3\frac{3}{5}$

I. The Acme Hardware Store sells "rebar" with diameters of $\frac{7}{8}, \frac{5}{16}, \frac{17}{32}, \frac{17}{64}$, and $\frac{5}{12}$ in. List the diameters from smallest to largest.

Answers to Warm-Ups

G. $\frac{9}{14}, \frac{2}{3}, \frac{5}{7}$ **H.** $3\frac{4}{9}, 3\frac{3}{5}, 3\frac{2}{3}$

I. $\frac{17}{64}, \frac{5}{16}, \frac{5}{12}, \frac{17}{32}, \frac{7}{8}$

EXAMPLES J–K

DIRECTIONS: Tell whether the statement is true or false.

STRATEGY: Build both fractions to a common denominator and compare the numerators.

J. True or false: $\frac{4}{5} > \frac{7}{9}$?

$\frac{4}{5} = \frac{36}{45}$ $\frac{7}{9} = \frac{35}{45}$ The common denominator is 45.

The statement is true. Compare the numerators.

WARM-UPS J–K

J. True or false: $\frac{2}{15} > \frac{3}{20}$?

Answers to Warm-Ups

J. false

K. True or false: $\frac{13}{16} < \frac{23}{27}$?

K. True or false: $\frac{18}{35} < \frac{12}{25}$?

$\frac{18}{35} = \frac{90}{175}$ $\frac{12}{25} = \frac{84}{175}$ The common denominator is 175.

* The statement is false. Compare the numerators.

Answers to Warm-Ups

K. true

EXERCISES 3.5

OBJECTIVE 1 Rename fractions by multiplying by 1 in the form $\frac{a}{a}$.

Write four fractions equivalent to each of the given fractions by multiplying by $\frac{2}{2}, \frac{3}{3}, \frac{4}{4}$, *and* $\frac{5}{5}$.

A.

1. $\frac{3}{4}$

2. $\frac{1}{5}$

3. $\frac{7}{8}$

4. $\frac{5}{6}$

5. $\frac{4}{9}$

6. $\frac{3}{10}$

B.

7. $\frac{9}{11}$

8. $\frac{9}{13}$

9. $\frac{7}{3}$

10. $\frac{6}{5}$

OBJECTIVE 2 Build a fraction by finding the missing numerator.

Find the missing numerator.

A.

11. $\frac{1}{2} = \frac{?}{16}$

12. $\frac{3}{4} = \frac{?}{36}$

13. $\frac{2}{3} = \frac{?}{42}$

14. $\frac{3}{8} = \frac{?}{96}$

15. $\frac{4}{5} = \frac{?}{35}$

16. $\frac{7}{8} = \frac{?}{32}$

17. $\frac{3}{4} = \frac{?}{36}$

18. $\frac{2}{3} = \frac{?}{24}$

19. $\frac{1}{5} = \frac{?}{75}$

20. $\frac{5}{9} = \frac{?}{45}$

B.

21. $\frac{?}{12} = \frac{2}{3}$

22. $\frac{?}{66} = \frac{3}{11}$

23. $\frac{19}{6} = \frac{?}{12}$

24. $\frac{11}{5} = \frac{?}{100}$

25. $\frac{?}{300} = \frac{4}{75}$

26. $\frac{2}{3} = \frac{?}{108}$

27. $\frac{?}{126} = \frac{19}{42}$

28. $\frac{?}{147} = \frac{16}{7}$

29. $\frac{13}{18} = \frac{?}{144}$

30. $\frac{13}{16} = \frac{?}{144}$

OBJECTIVE 3 List a group of fractions from smallest to largest.

A.

List the fractions from smallest to largest.

31. $\frac{4}{7}, \frac{5}{7}, \frac{3}{7}$

32. $\frac{5}{11}, \frac{4}{11}, \frac{2}{11}$

33. $\frac{1}{2}, \frac{3}{4}, \frac{3}{8}$

34. $\frac{4}{5}, \frac{1}{2}, \frac{3}{10}$

35. $\frac{1}{2}, \frac{3}{8}, \frac{1}{3}$

36. $\frac{2}{3}, \frac{8}{15}, \frac{3}{5}$

Are the following statements true or false?

37. $\frac{3}{5} < \frac{4}{5}$

38. $\frac{7}{8} > \frac{5}{8}$

39. $\frac{11}{16} > \frac{7}{8}$

40. $\frac{9}{16} > \frac{5}{8}$

41. $\frac{5}{12} < \frac{2}{3}$

42. $\frac{8}{9} < \frac{7}{8}$

B.

List the fractions from smallest to largest.

43. $\frac{4}{5}, \frac{2}{3}, \frac{3}{4}$

44. $\frac{5}{8}, \frac{7}{10}, \frac{3}{4}$

45. $\frac{14}{15}, \frac{2}{3}, \frac{17}{30}, \frac{3}{5}$

46. $\frac{7}{9}, \frac{2}{3}, \frac{5}{6}, \frac{13}{18}$

47. $\frac{11}{24}, \frac{17}{36}, \frac{35}{72}$

48. $\frac{8}{25}, \frac{31}{50}, \frac{59}{100}$

49. $\frac{13}{28}, \frac{17}{35}, \frac{3}{7}$

50. $\frac{11}{15}, \frac{17}{20}, \frac{3}{4}$

51. $3\frac{7}{9}, 3\frac{5}{6}, 3\frac{2}{3}$

52. $6\frac{3}{16}, 6\frac{1}{4}, 6\frac{1}{8}$

Are the following statements true or false?

53. $\frac{3}{5} < \frac{5}{8}$

54. $\frac{7}{25} < \frac{17}{50}$

55. $\frac{3}{4} > \frac{11}{15}$

56. $\frac{19}{40} > \frac{31}{60}$

57. $\frac{11}{30} < \frac{17}{18}$

58. $\frac{11}{27} > \frac{29}{36}$

C.

59. Find the LCM of the denominators of $\frac{1}{2}, \frac{2}{3}, \frac{1}{6}$, and $\frac{5}{8}$. Build the four fractions so that each has the LCM as the denominator.

60. Find the LCM of the denominators of $\frac{1}{4}, \frac{4}{13}$, and $\frac{5}{26}$. Build the three fractions so that each has the LCM as the denominator.

61. The night nurse at Malcom X Community Hospital finds bottles containing codeine tablets out of the usual order. The bottles contain tablets having the following strengths of codeine: $\frac{1}{8}, \frac{3}{32}, \frac{5}{16}, \frac{3}{8}, \frac{9}{16}, \frac{1}{2}$, and $\frac{1}{4}$ grain, respectively. Arrange the bottles in order of the strength of codeine from the smallest to the largest.

62. Joe, an apprentice, is given the task of sorting a bin of bolts according to their diameters. The bolts have the following diameters: $\frac{11}{16}, \frac{7}{8}, 1\frac{1}{16}, \frac{3}{4}, 1\frac{1}{8}$, and $1\frac{3}{32}$ in. How should he list the diameters from the smallest to the largest?

63. In the Republic of Niger, $\frac{2}{25}$ of the population is from the ethnic group Tuareg and $\frac{9}{100}$ of the population is from the ethnic group Fula. Is more of the population from the Tuareg or from the Fula?

64. According to the General Services Administration, the federal government owns about $\frac{1}{30}$ of the acreage in Alaska and about $\frac{6}{125}$ of the acreage of Kentucky. Which state has the larger portion of federally owned land?

65. The state of New Mexico has African Americans as $\frac{13}{500}$ of its population and Asian/Pacific Islanders as $\frac{3}{200}$ of its population. Which group is the larger portion of the state's population?

66. In 1999, $\frac{1}{250}$ of the world's Buddhists lived in Europe and $\frac{7}{1000}$ of them lived in North America. Which continent had the smaller portion of the world's Buddhists?

67. Three chemistry students weigh a container of a chemical. Mary records the weight as $3\frac{1}{8}$ lb. George reads the weight as $3\frac{3}{16}$ lb. Chang reads the weight as $3\frac{1}{4}$ lb. Whose measurement is heaviest?

68. Three rulers are marked in inches. On the first ruler, the spaces are divided into tenths. On the second, they are divided into sixteenths, and on the third, they are divided into eighths. All are used to measure a line on a scale drawing. The nearest mark on the first ruler is $5\frac{7}{10}$, the nearest mark on the second is $5\frac{11}{16}$, and the nearest mark on the third is $5\frac{6}{8}$. Which is the largest (longest) measurement?

Exercise 69 relates to the chapter application.

69. a. If Sandy and Hannah sell a box of pecan chocolate chip cookies for \$6.00, then $\frac{4}{15}$ of the selling price is profit. If they sell a box of dipped butterfingers for \$7.20, then $\frac{1}{6}$ of the selling price is profit. Which kind of cookie gives them a bigger profit?

b. Hannah is doubling a recipe that calls for 5 tbs of oil. Rather than taking the time to measure out 10 single tablespoons, she pours $\frac{3}{4}$ c of oil. Is this more or less than she needs? (*Hint*: 1 c = 16 tbs)

STATE YOUR UNDERSTANDING

70. Explain why it is easier to compare the size of two fractions if they have common denominators.

71. What is the difference between simplifying fractions and building fractions?

CHALLENGE

72. List $\frac{12}{25}, \frac{14}{29}, \frac{29}{60}, \frac{35}{71}, \frac{39}{81}$, and $\frac{43}{98}$ from smallest to largest

73. Build $\frac{5}{7}$ so that is has denominators 70, 91, 161, 784, and 4067.

74. Fernando and Filipe are hired to sell tickets for the holiday raffle. Fernando sells $\frac{14}{17}$ of his quota of 765 tickets. Filipe sells $\frac{19}{23}$ of his quota of 759 tickets. Who sells the most of his quota? Who sells the most tickets?

GROUP WORK

75. We can see that $\frac{2}{5}$ is less than $\frac{4}{5}$ by looking at a rectangle representing each fraction. Show the sum, $\frac{2}{5} + \frac{4}{5}$, visually using a rectangle divided into five parts. Similarly, show $\frac{2}{7} + \frac{3}{7}$.

MAINTAIN YOUR SKILLS

Find the LCM of the denominators of each set of fractions.

76. $\frac{1}{2}, \frac{1}{5}$

77. $\frac{1}{8}, \frac{3}{4}$

78. $\frac{5}{8}, \frac{3}{10}$

79. $\frac{5}{6}, \frac{1}{10}$

80. $\frac{11}{96}, \frac{35}{72}$

81. $\frac{1}{32}, \frac{3}{16}, \frac{3}{4}$

82. Find the prime factorization of 96.

83. Find the prime factorization of 72.

84. The Forest Service rents a two-engine plane at $625 per hour and a single-engine plane at $365 per hour to drop fire retardant. During a forest fire, the two-engine plane was used for 4 hr and the single-engine plane was used for 2 hr. What was the cost of using the two planes?

85. Ms. Wallington is taking one capsule containing 250 mg of a drug every 8 hr. Beginning next Wednesday, her doctor's instructions are to increase the dosage to 500 mg every 6 hr. How many 250-mg capsules should the pharmacist give her for the following week (7 days)?

3.6 Adding Fractions

VOCABULARY

Like fractions are fractions with common denominators. **Unlike fractions** are fractions with different denominators.

OBJECTIVES

1. Add like fractions.
2. Add unlike fractions.

HOW & WHY

OBJECTIVE 1 Add like fractions.

What is the sum of $\frac{1}{5}$ and $\frac{2}{5}$? The denominators tell us the number of parts in the unit. The numerators tell us how many of these parts are shaded. By adding the numerators, we find the total number of shaded parts. The common denominator keeps track of the size of the parts. See Figure 3.14.

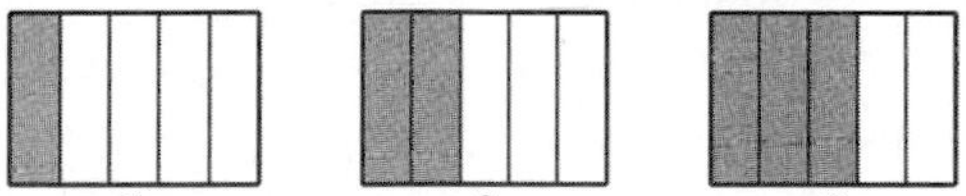

$$\frac{1}{5} + \frac{2}{5} = \frac{3}{5}$$

FIGURE 3.14

To add like fractions

1. Add the numerators.
2. Write the sum over the common denominator.

CAUTION

Do not add the denominators.

$$\frac{2}{5} + \frac{2}{5} \neq \frac{4}{10}$$

EXAMPLES A–D

DIRECTIONS: Add and simplify.

STRATEGY: Add the numerators and write the sum over the common denominator. Simplify.

A. Add: $\frac{3}{7} + \frac{2}{7}$

$\frac{3}{7} + \frac{2}{7} = \frac{5}{7}$ Add the numerators. Keep the common denominator

WARM-UPS A–D

A. Add: $\frac{3}{11} + \frac{5}{11}$

Answers to Warm-Ups

A. $\frac{8}{11}$

B. Add: $\frac{5}{6} + \frac{5}{6} + \frac{1}{6}$

B. Add: $\frac{3}{4} + \frac{3}{4} + \frac{1}{4}$

$\frac{3}{4} + \frac{3}{4} + \frac{1}{4} = \frac{7}{4}$ Add.

$= 1\frac{3}{4}$ Write as a mixed number.

C. Add: $\frac{1}{10} + \frac{1}{10} + \frac{3}{10}$

C. Add: $\frac{1}{9} + \frac{1}{9} + \frac{4}{9}$

$\frac{1}{9} + \frac{1}{9} + \frac{4}{9} = \frac{6}{9}$ Add.

$= \frac{2}{3}$ Simplify.

D. According to a report from Leed Community College, $\frac{3}{14}$ of the college's total revenues were from alumni contributions and $\frac{9}{14}$ were from tuition. What portion of the total revenues was from these two sources?

D. According to the annual report from Radd Community College, $\frac{11}{20}$ of the college's total revenues were from state taxes and $\frac{7}{20}$ were from tuition and fees. What portion of the total revenues was from these two sources?

$\frac{11}{20} + \frac{7}{20} = \frac{18}{20}$ Add.

$= \frac{9}{10}$ Simplify.

Taxes and tuition and fees were $\frac{9}{10}$ of the revenue for the school.

HOW & WHY

OBJECTIVE 2 Add unlike fractions.

The sum $\frac{1}{2} + \frac{1}{5}$ cannot be found in this form. A look at Figure 3.15 shows that the parts are not the same size.

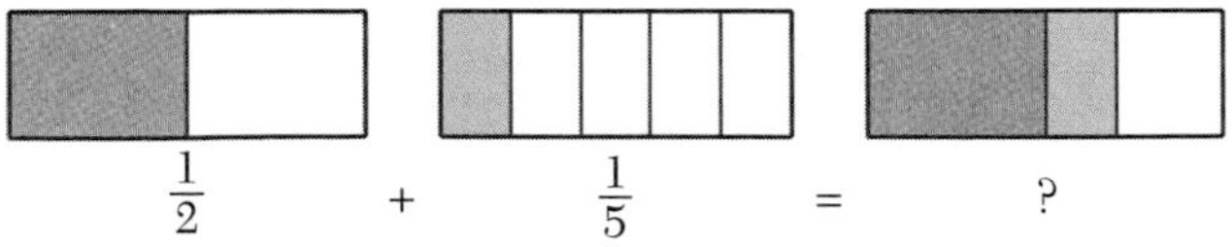

FIGURE 3.15

To add, rename $\frac{1}{2}$ and $\frac{1}{5}$ as like fractions. The LCM (least common multiple) of the two denominators serves as the least common denominator. The LCM of 2 and 5 is 10. Renaming the fractions we write

$$\frac{1}{2} = \left(\frac{1}{2}\right)\left(\frac{5}{5}\right) = \frac{5}{10} \quad \text{and} \quad \frac{1}{5} = \left(\frac{1}{5}\right)\left(\frac{2}{2}\right) = \frac{2}{10}$$

Figure 3.16 shows the regions with equal size parts.

Answers to Warm-Ups

B. $\frac{11}{6}$ or $1\frac{5}{6}$ **C.** $\frac{1}{2}$

D. Contributions and tuition were $\frac{6}{7}$ of the revenue for the school.

$$\frac{1}{2} + \frac{1}{5} = ?$$

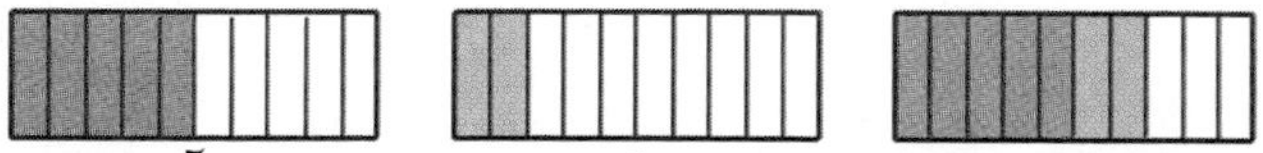

$$\frac{5}{10} + \frac{2}{10} = \frac{7}{10}$$

FIGURE 3.16

To add unlike fractions

1. Build the fractions so they have a common denominator.
2. Add and simplify.

EXAMPLES E–J

DIRECTIONS: Add and simplify.

STRATEGY: Build each of the fractions to a common denominator, add, and simplify.

E. Add: $\frac{1}{8} + \frac{3}{4}$

$\frac{1}{8} + \frac{3}{4} = \frac{1}{8} + \frac{6}{8}$ The LCM of 8 and 4 is 8. Build the fractions.

$= \frac{7}{8}$ Add.

F. Add: $\frac{5}{8} + \frac{3}{10}$

$\frac{5}{8} + \frac{3}{10} = \frac{25}{40} + \frac{12}{40}$ The LCM of 8 and 10 is 40: $\frac{5}{8}\cdot\frac{5}{5} = \frac{25}{40}$ and $\frac{3}{10}\cdot\frac{4}{4} = \frac{12}{40}$.

$= \frac{37}{40}$ Add.

G. Add: $\frac{5}{6} + \frac{1}{10}$

$\frac{5}{6} + \frac{1}{10} = \frac{25}{30} + \frac{3}{30}$ The LCM of 6 and 10 is 30.

$= \frac{28}{30}$ Add.

$= \frac{14}{15}$ Simplify.

H. Add: $\frac{11}{96} + \frac{35}{72}$

STRATEGY: Find the prime factorization of the denominators to help find the LCM.

$\frac{11}{96} + \frac{35}{72} = \frac{11(3)}{288} + \frac{35(4)}{288}$ $96 = 2^5 \cdot 3$; $72 = 2^3 \cdot 3^2$, so the LCM is $2^5 \cdot 3^2 = 288$.

$= \frac{33}{288} + \frac{140}{288}$

$= \frac{173}{288}$ Add.

WARM-UPS E–J

E. Add: $\frac{1}{6} + \frac{5}{12}$

F. Add: $\frac{5}{12} + \frac{2}{9}$

G. Add: $\frac{3}{5} + \frac{3}{8}$

H. Add: $\frac{13}{45} + \frac{28}{75}$

Answers to Warm-Ups

E. $\frac{7}{12}$ **F.** $\frac{23}{36}$ **G.** $\frac{39}{40}$ **H.** $\frac{149}{225}$

I. Add: $\frac{8}{35} + \frac{9}{28}$

J. A nail must be long enough to reach through three thicknesses of wood and penetrate a fourth piece $\frac{1}{4}$ in. If the first piece of wood is $\frac{5}{16}$ in., the second is $\frac{3}{8}$ in., and the third is $\frac{9}{16}$ in., how long must the nail be?

CALCULATOR EXAMPLE

I. Add: $\frac{41}{60} + \frac{13}{84}$

$\frac{41}{60} + \frac{13}{84} = \frac{88}{105}$ On a calculator with fraction keys, it is not necessary to find the common denominator. The calculator is programmed to add and simplify.

J. Sheila is assembling a composting bin for her lawn and garden debris. She needs a bolt that will reach through a $\frac{1}{32}$-in.-thick washer, a $\frac{3}{16}$-in.-thick plastic bushing, a $\frac{3}{4}$-in.-thick piece of steel tubing, a second $\frac{1}{32}$-in.-thick washer, and a $\frac{1}{4}$-in.-thick nut. How long a bolt does she need?

STRATEGY: Add the thicknesses of each part to find the total length needed.

$\frac{1}{32} + \frac{3}{16} + \frac{3}{4} + \frac{1}{32} + \frac{1}{4}$ The LCM of 32, 16, and 4 is 32.

$\frac{1}{32} + \frac{6}{32} + \frac{24}{32} + \frac{1}{32} + \frac{8}{32}$ Build the fractions to a common denominator.

$= \frac{40}{32}$ Add.

$= \frac{5}{4} = 1\frac{1}{4}$ Simplify and write as a mixed number.

The bolt must be $1\frac{1}{4}$ in. long.

Answers to Warm-Ups

I. $\frac{11}{20}$ **J.** The nail must be at least $1\frac{1}{2}$ in. long.

Name ______________________ Class ______________________ Date ______________

EXERCISES 3.6

OBJECTIVE 1 Add like fractions.

Add. Simplify completely.

A.

1. $\frac{3}{7} + \frac{3}{7}$

2. $\frac{5}{11} + \frac{5}{11}$

3. $\frac{1}{6} + \frac{1}{6} + \frac{1}{6}$

4. $\frac{3}{8} + \frac{1}{8} + \frac{2}{8}$

5. $\frac{3}{4} + \frac{5}{4}$

6. $\frac{6}{7} + \frac{8}{7}$

7. $\frac{2}{10} + \frac{5}{10} + \frac{1}{10}$

8. $\frac{4}{12} + \frac{3}{12} + \frac{3}{12}$

9. $\frac{4}{13} + \frac{5}{13} + \frac{1}{13}$

10. $\frac{5}{11} + \frac{2}{11} + \frac{1}{11}$

11. $\frac{5}{12} + \frac{5}{12} + \frac{5}{12}$

12. $\frac{9}{16} + \frac{7}{16} + \frac{4}{16}$

B.

13. $\frac{1}{16} + \frac{2}{16} + \frac{5}{16}$

14. $\frac{7}{20} + \frac{1}{20} + \frac{7}{20}$

15. $\frac{5}{48} + \frac{7}{48} + \frac{3}{48}$

16. $\frac{3}{16} + \frac{2}{16} + \frac{5}{16}$

17. $\frac{8}{30} + \frac{9}{30} + \frac{1}{30}$

18. $\frac{13}{50} + \frac{7}{50} + \frac{2}{50}$

19. $\frac{3}{20} + \frac{3}{20} + \frac{9}{20}$

20. $\frac{7}{18} + \frac{2}{18} + \frac{7}{18}$

OBJECTIVE 2 Add unlike fractions.

Add. Simplify completely.

A.

21. $\frac{5}{6} + \frac{1}{8}$

22. $\frac{5}{8} + \frac{1}{6}$

23. $\frac{3}{8} + \frac{1}{24}$

24. $\frac{5}{8} + \frac{3}{24}$

25. $\frac{7}{16} + \frac{3}{8}$

26. $\frac{4}{9} + \frac{5}{18}$

27. $\frac{1}{10} + \frac{1}{5} + \frac{1}{3}$

28. $\frac{1}{20} + \frac{1}{5} + \frac{1}{4}$

29. $\frac{1}{5} + \frac{1}{10} + \frac{1}{2}$

30. $\frac{2}{15} + \frac{2}{5} + \frac{1}{3}$

B.

31. $\frac{6}{35}+\frac{16}{21}$

32. $\frac{1}{21}+\frac{1}{28}$

33. $\frac{7}{10}+\frac{3}{20}+\frac{7}{30}$

34. $\frac{5}{8}+\frac{5}{12}+\frac{1}{3}$

35. $\frac{1}{10}+\frac{2}{5}+\frac{5}{6}+\frac{1}{15}$

36. $\frac{1}{2}+\frac{3}{10}+\frac{3}{5}+\frac{1}{4}$

37. $\frac{3}{4}+\frac{3}{8}+\frac{3}{16}$

38. $\frac{15}{36}+\frac{1}{6}+\frac{5}{12}$

39. $\frac{5}{6}+\frac{5}{8}+\frac{3}{4}+\frac{5}{12}$

40. $\frac{7}{10}+\frac{4}{5}+\frac{1}{15}+\frac{31}{35}$

41. $\frac{4}{9}+\frac{2}{3}+\frac{32}{81}+\frac{5}{9}$

42. $\frac{8}{13}+\frac{13}{15}+\frac{5}{13}+\frac{14}{25}$

43. $\frac{7}{9}+\frac{4}{5}+\frac{4}{15}+\frac{11}{30}$

44. $\frac{11}{15}+\frac{7}{12}+\frac{9}{10}+\frac{17}{20}$

C.

45. $\frac{17}{100}+\frac{31}{100}+\frac{9}{100}+\frac{3}{100}$

46. $\frac{29}{50}+\frac{3}{50}+\frac{3}{50}+\frac{11}{50}$

47. $\frac{7}{120}+\frac{9}{120}+\frac{25}{120}+\frac{15}{120}$

48. $\frac{29}{144}+\frac{17}{144}+\frac{3}{144}+\frac{25}{144}$

49. $\frac{25}{36}+\frac{19}{48}$

50. $\frac{72}{85}+\frac{69}{102}$

51. According to the annual report from George Fox University, $\frac{1}{8}$ of the university's total revenues were from contributions and $\frac{31}{50}$ were from tuition and fees. What portion of the total revenues was from these two sources?

52. Ten different church organizations in the United States identify themselves as Lutheran. The largest two are the Evangelical Lutheran Church in America (E.L.C.A.), which has $\frac{5}{8}$ of the total members, and the Lutheran Church–Missouri Synod (L.C.M.S.), which has $\frac{3}{10}$ of the total members. What portion of the total number of Lutherans in the United States are E.L.C.A. or L.C.M.S.?

53. Of the worldwide adherents to the Muslim faith, slightly more than $\frac{1}{4}$ of them are in Africa, whereas $\frac{7}{10}$ of them reside in Asia. What portion of the world's Muslims are on these two continents?

54. The Republic of Panama has four major ethnic groups. The largest is Mestizo. About $\frac{7}{50}$ of the population is West Indian, $\frac{1}{10}$ of the population is white, and $\frac{6}{100}$ of the population is Amerindian. What portion of Panama's population is not Mestizo?

55. Find the error(s) in the statement: $\frac{1}{5}+\frac{2}{5}=\frac{3}{10}$. Correct the statement. Explain how you would avoid this error.

56. Find the error(s) in the statement: $\frac{1}{2}+\frac{4}{7}=\frac{5}{14}$. Correct the statement. Explain how you would avoid this error.

57. Chef Ramon prepares a punch for the stockholders' meeting of the Northern Corporation. The punch calls for $\frac{1}{4}$ gal lemon juice, $\frac{3}{4}$ gal raspberry juice, $\frac{2}{4}$ gal cranberry juice, $\frac{1}{4}$ gal lime juice, $\frac{5}{4}$ gal 7-Up, and $\frac{3}{4}$ gal vodka. How many gallons of punch does the recipe make?

58. A physical therapist advises Belinda to swim $\frac{1}{4}$ mi on Monday and increase this distance by $\frac{1}{16}$ mi each day from Tuesday through Friday. What is the total number of miles she advises?

59. Jonnie is assembling a rocking horse for his granddaughter. He needs a bolt to reach through a $\frac{7}{8}$-in. piece of steel tubing, a $\frac{1}{16}$-in. bushing, a $\frac{1}{2}$-in. piece of tubing, a $\frac{1}{8}$-in.-thick washer, and a $\frac{1}{4}$-in.-thick nut. How long a bolt does he need?

60. What is the perimeter of this triangle?

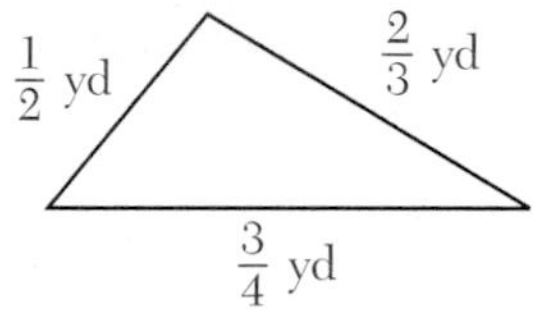

61. Find the length of this pin.

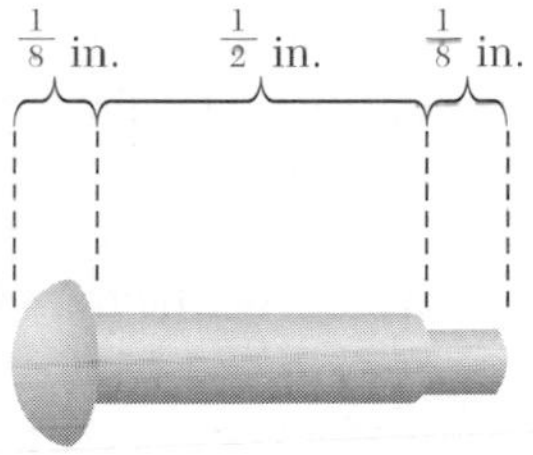

62. The Sandoz family spends $\frac{2}{15}$ of their income on rent, $\frac{1}{4}$ on food, $\frac{1}{20}$ on clothes, $\frac{1}{10}$ on transportation, and $\frac{5}{24}$ on taxes. What fraction of their income is spent on these costs?

63. Find the length of the rod in the figure. Assume that the grooves and teeth are uniform in length.

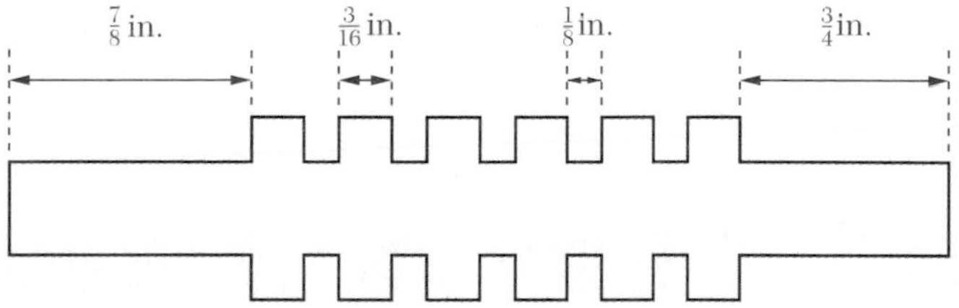

64. An electronics mogul leaves $\frac{2}{9}$, $\frac{1}{5}$, and $\frac{4}{15}$ shares of her estate to her three children. What share of the estate will the children receive?

Exercise 65 relates to the chapter application.

65. Sandy and Hannah propose to make a Christmas sampler box with nut balls, thumbprints, and lemon squares. One recipe of each type of cookie makes enough to fill four boxes.

a. Nutballs require $\frac{1}{4}$ tsp of baking soda, thumbprints require $\frac{1}{2}$ tsp, and lemon squares require $\frac{1}{2}$ tsp. How much baking soda is needed for all three cookie recipes?

b. How much baking soda is required to fill an order for 12 sampler boxes?

c. Nutballs require $\frac{3}{4}$ tsp of salt, thumbprints require $\frac{1}{2}$ tsp, and lemon squares need $\frac{3}{4}$ tsp. How much salt is needed for all three recipes?

d. How much salt is needed to fill an order for 6 sampler boxes?

STATE YOUR UNDERSTANDING

66. Explain how to find the sum of $\frac{5}{12}$ and $\frac{3}{20}$.

67. Why is it important to write fractions with a common denominator before adding?

CHALLENGE

68. Find the sum of $\frac{107}{372}$ and $\frac{41}{558}$.

69. Find the sum of $\frac{67}{124} + \frac{27}{868}$.

70. Janet left $\frac{1}{7}$, $\frac{3}{14}$, and $\frac{1}{6}$ of her estate to Bob, Greta, and Joe Guerra. She also left $\frac{1}{8}$, $\frac{5}{16}$, and $\frac{1}{9}$ of the estate to Pele, Rhonda, and Shauna Contreras. Which family received the greater share of the estate?

71. Jim is advised by his doctor to limit fat intake in his diet. For breakfast his fat intake is: a bagel, $\frac{3}{4}$ g; a banana, $\frac{13}{16}$ g; cereal, $\frac{17}{10}$ g; milk, $\frac{9}{8}$ g; jelly, 0 g; and coffee, 0 g. Rounded to the nearest whole number, how many grams of fat does Jim consume at breakfast? If each gram of fat represents 9 calories, and the total calories for breakfast is 330, what fraction represents the number of calories from fat? Use the rounded whole number of grams of fat.

GROUP WORK

72. In the next section, we add mixed numbers. Formulate a procedure for adding $3\frac{3}{4} + 5\frac{4}{5}$. Also, find two or three applications for which the sum of mixed numbers is necessary. Be prepared to share the results with the class.

MAINTAIN YOUR SKILLS

Add:

73. $2 + 8 + \frac{1}{4} + \frac{1}{6}$

74. $3 + 5 + \frac{2}{9} + \frac{7}{18}$

75. $1 + 7 + 10 + \frac{1}{6} + \frac{7}{10} + \frac{1}{5}$

76. $2 + 9 + 5 + \frac{2}{9} + \frac{5}{9} + \frac{2}{3}$

77. $3 + \frac{3}{8} + 2 + \frac{1}{8} + 1 + \frac{3}{16}$

78. $4 + \frac{1}{12} + 1 + \frac{3}{4} + 3 + \frac{1}{8}$

Perform the indicated operations.

79. $(21 - 4)6 - 3(19 - 11)$

80. $(13 - 2^3)^3 - 7(11)$

81. Mrs. Teech has five classes to teach this term. The enrollment in the classes is 42, 36, 56, 32, and 34. What is the average class size?

82. In a metal benchwork class that has 36 students, each student is allowed $11\frac{5}{8}$ in. of wire solder. How many inches of wire solder must the instructor provide for the class?

3.7 Adding Mixed Numbers

HOW & WHY

OBJECTIVE Add mixed numbers.

What is the sum of $3\frac{1}{6}$ and $5\frac{1}{4}$? Pictorially, we can show the sum by drawing rectangles such as those in Figure 3.17.

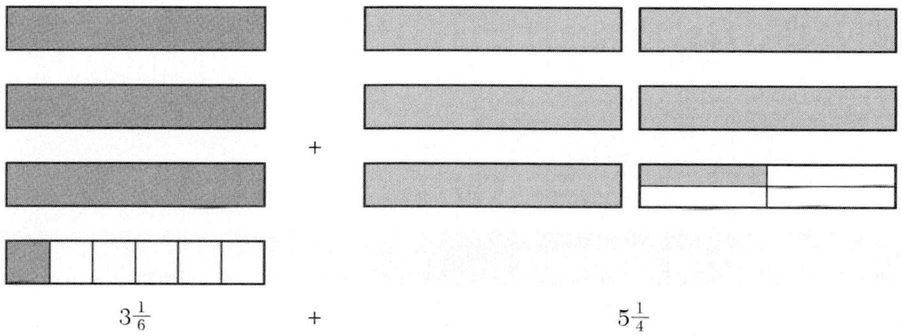

FIGURE 3.17

It is easy to see that the sum contains 8 whole units. The sum of the fractional parts requires finding a common denominator. The LCM of 6 and 4 is 12. Figure 3.18 shows the divided rectangles.

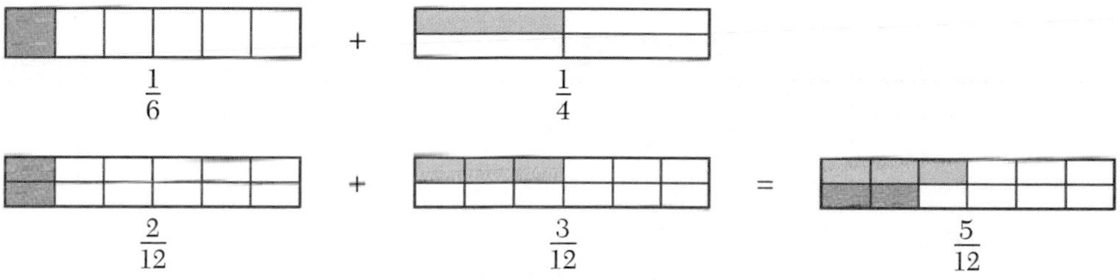

FIGURE 3.18

So the sum is $8\frac{5}{12}$.

Mixed numbers can be added horizontally or in columns. The sum of $3\frac{1}{6}$ and $5\frac{1}{4}$ is shown both ways.

$$\left(3 + \frac{1}{6}\right) + \left(5 + \frac{1}{4}\right) = (3 + 5) + \left(\frac{1}{6} + \frac{1}{4}\right)$$
$$= 8 + \left(\frac{2}{12} + \frac{3}{12}\right)$$
$$= 8\frac{5}{12}$$

When we write the sum vertically, the grouping of the whole numbers and the fractions takes place naturally.

$$\begin{aligned} 3\frac{1}{6} &= 3\frac{2}{12} \\ +5\frac{1}{4} &= 5\frac{3}{12} \\ \hline &= 8\frac{5}{12} \end{aligned}$$

OBJECTIVE

Add mixed numbers.

Sometimes the sum of the fractions is greater than 1. In this case, change the fractional sum to a mixed number and add it to the whole number part.

$$17\frac{7}{10} = 17\frac{21}{30}$$ Write the fractions with a common denominator.

$$+14\frac{8}{15} = 14\frac{16}{30}$$

$$31\frac{37}{30}$$ Add the whole numbers and the fractions.

$$= 31 + 1\frac{7}{30}$$ Rewrite the improper fraction as a mixed number.

$$= 32\frac{7}{30}$$ Add the mixed number to the whole number.

To add mixed numbers

1. Add the whole numbers.
2. Add the fractions. If the sum of the fractions is more than 1, change the fraction to a mixed number and add again.
3. Simplify.

WARM-UPS A–E

A. Add: $4\frac{1}{2} + 9\frac{5}{16}$

Answers to Warm-Ups

A. $13\frac{13}{16}$

EXAMPLES A–E

DIRECTIONS: Add. Write as a mixed number.

STRATEGY: Add the whole numbers and add the fractions. If the sum of the fractions is an improper fraction, rewrite it as a mixed number and add to the sum of the whole numbers. Simplify.

A. Add: $3\frac{3}{4} + 2\frac{5}{12}$

STRATEGY: Write the mixed numbers in a column to group the whole numbers and to group the fractions.

$$3\frac{3}{4} = 3\frac{9}{12}$$ Build the fractions to the common denominator, 12.

$$+2\frac{5}{12} = 2\frac{5}{12}$$

$$= 5\frac{14}{12}$$ Add.

$$5\frac{14}{12} = 5 + 1\frac{2}{12}$$ Write the improper fraction as a mixed number.

$$= 6\frac{2}{12}$$ Add the whole numbers.

$$= 6\frac{1}{6}$$ Simplify.

B. Add: $25\frac{7}{8} + 13\frac{5}{9} + 7\frac{1}{6}$

$$25\frac{7}{8} = 25\frac{63}{72}$$ The LCM of 8, 9, and 6 is 72.

$$13\frac{5}{9} = 13\frac{40}{72}$$

$$+\ 7\frac{1}{6} = \ 7\frac{12}{72}$$

$$45\frac{115}{72}$$ Add.

$$45\frac{115}{72} = 45 + 1\frac{43}{72}$$ Write the improper fraction as a mixed number.

$$= 46\frac{43}{72}$$ Add.

C. Add: $18 + 7\frac{4}{15}$

$$18 + 7\frac{4}{15} = 25\frac{4}{15}$$ Add the whole numbers.

CALCULATOR EXAMPLE

D. Add: $31\frac{3}{8} + 62\frac{13}{15}$

$$31\frac{3}{8} + 62\frac{13}{15} = \frac{11309}{120}$$

$$= 94\frac{29}{120}$$

On a calculator with fraction keys, it is not necessary to find the common denominator. The calculator is programmed to add and simplify. Some calculators may not change the improper sum to a mixed number. See your calculator manual.

E. A report by Environmental Hazards Management lists the following amounts of hazardous material that a city of 100,000 discharges into city drains each month: $3\frac{3}{4}$ tons of toilet bowl cleaner, $13\frac{3}{4}$ tons of liquid household cleaners, and $3\frac{2}{5}$ tons of motor oil. How many tons of these materials are discharged each month?

STRATEGY: Find the sum of the number of tons of hazardous material.

$$3\frac{3}{4} = 3\frac{15}{20}$$ The LCM of 4 and 5 is 20.

$$13\frac{3}{4} = 13\frac{15}{20}$$

$$+\ 3\frac{2}{5} = 3\frac{8}{20}$$

$$19\frac{38}{20}$$ Add.

$$19\frac{38}{20} = 20\frac{9}{10}$$ Change the improper fraction to a mixed number and simplify.

The residents discharge $20\frac{9}{10}$ tons of hazardous material each month.

B. Add: $12\frac{8}{9} + 14\frac{3}{4} + 6\frac{2}{3}$

C. Add: $21\frac{3}{11} + 19$

D. Add: $51\frac{7}{8} + 36\frac{5}{12}$

E. The same report also records that the city recycles $10\frac{1}{3}$ tons of paper, $3\frac{3}{8}$ tons of aluminum, and $4\frac{5}{12}$ tons of glass each month. How many tons of material are recycled each month?

Answers to Warm-Ups

B. $34\frac{11}{36}$ **C.** $40\frac{3}{11}$ **D.** $88\frac{7}{24}$

E. The city recycles $18\frac{1}{8}$ tons of material per month.

EXERCISES 3.7

OBJECTIVE Add mixed numbers.

Add. Write the results as a mixed number where possible.

A.

1. $2\frac{3}{5} + 1\frac{1}{5}$

2. $7\frac{2}{7} + 2\frac{2}{7}$

3. $2\frac{4}{5} + 5\frac{1}{5}$

4. $6\frac{3}{8} + 5\frac{5}{8}$

5. $5\frac{11}{12} + 2\frac{1}{4}$

6. $6\frac{5}{6} + 8\frac{2}{3}$

7. $5\frac{5}{9} + \frac{2}{3}$

8. $8\frac{5}{12} + 7\frac{1}{2}$

9. $1\frac{3}{7} + 4\frac{9}{14}$

10. $7\frac{5}{6} + 4\frac{5}{12}$

11. $8\frac{2}{3} + 6\frac{7}{15}$

12. $6\frac{8}{15} + 3\frac{3}{5}$

13. $9\frac{3}{8} + 5\frac{11}{16}$

14. $8\frac{5}{9} + 2\frac{13}{27}$

15. $3\frac{2}{3}$, $2\frac{1}{2}$, $+4\frac{5}{6}$

16. $7\frac{3}{10}$, $8\frac{1}{5}$, $+6\frac{1}{2}$

17. $4\frac{1}{2} + 8\frac{3}{4} + 6 + 7\frac{3}{8}$

18. $5\frac{2}{5} + 4 + 3\frac{2}{3} + 9\frac{7}{15}$

B.

19. $4\frac{5}{8} + 10\frac{5}{6}$

20. $7\frac{7}{15} + 10\frac{5}{6}$

21. $3\frac{1}{10}$, $2\frac{3}{5}$, $+4\frac{7}{15}$

22. $7\frac{1}{8}$, $3\frac{5}{12}$, $+8\frac{5}{6}$

23. $12\frac{9}{16} + 9\frac{9}{20}$

24. $21\frac{17}{18} + 19\frac{5}{24}$

25. $21\frac{5}{7} + 15\frac{9}{14} + 12\frac{10}{21}$

26. $18\frac{3}{4} + 17\frac{7}{8} + 23\frac{1}{6}$

27. $\begin{array}{r} 213\frac{5}{18} \\ +506\frac{7}{12} \\ \hline \end{array}$

28. $\begin{array}{r} 213\frac{5}{6} \\ +347\frac{3}{10} \\ \hline \end{array}$

29. $47\frac{1}{5} + 23\frac{2}{3} + 15\frac{1}{2}$

30. $47\frac{3}{8} + 23 + 42\frac{5}{12}$

31. $15\frac{3}{4} + 18\frac{2}{3} + 21\frac{1}{2}$

32. $39\frac{11}{12} + 49\frac{7}{8} + 59\frac{3}{4}$

33. $62 + 18\frac{5}{9} + 37\frac{7}{15}$

34. $41 + 29\frac{9}{14} + 3\frac{1}{6}$

35. $37\frac{18}{35} + 29\frac{9}{14} + 36$

36. $12\frac{11}{12} + 22\frac{5}{8} + 8$

37. $\begin{array}{r} 11\frac{1}{6} \\ 12\frac{3}{10} \\ 13\frac{1}{12} \\ +14\frac{1}{20} \\ \hline \end{array}$

38. $\begin{array}{r} 11\frac{5}{18} \\ 21\frac{2}{9} \\ 31\frac{1}{3} \\ +41\frac{1}{6} \\ \hline \end{array}$

C. *Add. Simplify completely.*

39. $2\frac{2}{5} + 7\frac{1}{6} + \frac{4}{15} + 3\frac{1}{10}$

40. $2\frac{1}{5} + 13\frac{1}{10} + 24\frac{1}{3} + \frac{13}{15}$

41. $12\frac{3}{5} + 7\frac{1}{8} + 29\frac{3}{4} + 14\frac{9}{10}$

42. $14\frac{17}{25} + 4\frac{3}{5} + 25\frac{2}{15} + 10$

43. Nancy painted a portrait of her daughters. The canvas measures $18\frac{1}{2}$ by $24\frac{3}{4}$ in. She plans to frame it with molding, which will require an extra inch and a half on each end to make mitered corners, as in the figure. How much molding does Nancy need to frame her portrait?

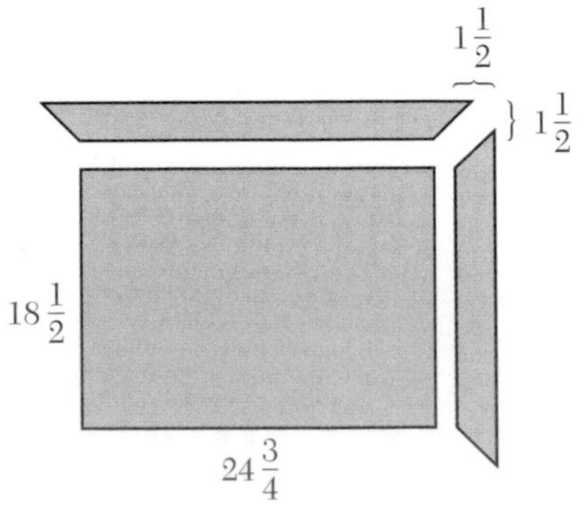

44. Tom is making a bookshelf with six pieces of wood, as shown in the figure. He needs three pieces of wood for the shelves, which are 24 in. long; one piece of wood for the top, which is $25\frac{1}{2}$ in. long; and two pieces of wood for the vertical supports, which are $30\frac{3}{4}$ in. long. What is the total length of wood that Tom needs?

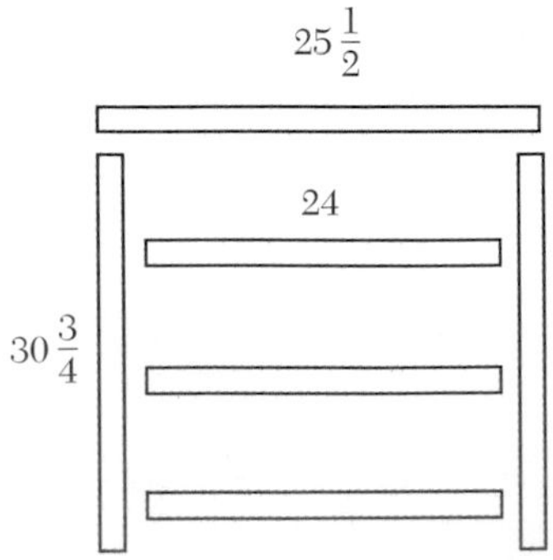

45. Dave is planning to edge one side of his driveway with bricks and mortar, as shown in the figure. Each brick is $3\frac{3}{4}$ in. wide, and Dave will use a $\frac{3}{8}$-in. mortar joint between bricks. Assuming that Dave begins and ends the edging with a brick and that he needs 65 bricks, how long is the driveway?

46. Juanita worked the following hours at her part-time job during the month of October:

Week	Oct. 1–7	Oct. 8–14	Oct. 15–21	Oct. 22–28	Oct. 29–31
Hours	$25\frac{1}{2}$	$19\frac{2}{3}$	10	$16\frac{5}{6}$	$4\frac{3}{4}$

How many hours did she work during October?

47. Michelle worked the following hours at her part-time job during the week:

Day	Mon.	Tues.	Wed.	Thurs.	Fri.	Sat.
Hours	$4\frac{1}{2}$	$4\frac{1}{4}$	$4\frac{1}{2}$	$2\frac{1}{4}$	0	$8\frac{1}{4}$

How many hours did she work during the week?

48. Find the perimeter of a rectangle that has length $24\frac{1}{2}$ ft and width $17\frac{3}{4}$ ft.

49. Find the perimeter of a triangle with sides $12\frac{1}{6}$ in., $16\frac{3}{4}$ in., and $14\frac{7}{12}$ in.

50. What is the overall length of this bolt?

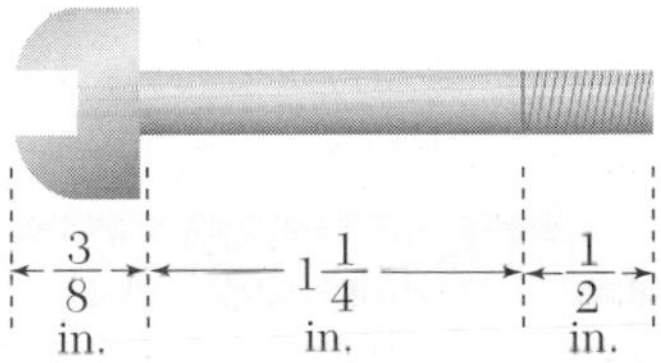

51. The graph displays the average yearly rainfall for five cities.

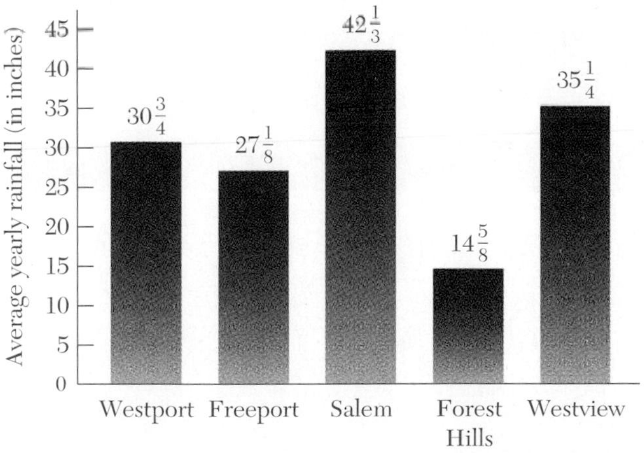

What is the total amount of rain that falls in a given year in Salem, Forest Hills, and Westview?

52. The rain gauge at the water reservoir recorded the following rainfall during a 6-month period:

$1\frac{2}{3}$ in., $4\frac{5}{6}$ in., $3\frac{3}{4}$ in., $\frac{2}{3}$ in., $7\frac{7}{8}$ in., and $2\frac{1}{4}$ in.

What is the total rainfall recorded during the 6 months? If an inch of rain means a gain of 2,400,000 gal of water in the reservoir, what is the water gain during the 6 months?

53. The State Department of Transportation must resurface part of seven roads this summer. The distances to be paved are $6\frac{11}{16}$ mi, $8\frac{3}{5}$ mi, $9\frac{3}{4}$ mi, $17\frac{1}{2}$ mi, $5\frac{1}{8}$ mi, $12\frac{4}{5}$ mi, and $\frac{7}{8}$ mi. How many miles of highway are to be resurfaced? If it costs \$15,000 to resurface 1 mile, what is the cost of the resurfacing project? Round to the nearest thousand dollars.

Exercises 54–55 relate to the chapter application.

54. Sandy and Hannah can fill four Christmas sampler boxes with one recipe each of nutballs, thumbprints, and lemon squares.

a. Nutballs call for $1\frac{1}{2}$ c of flour, thumbprints call for $2\frac{3}{4}$ c of flour, and lemon squares call for 2 c of flour. How much flour is required for all three recipes?

b. How much flour is needed to fill an order for 20 boxes?

c. Nutballs call for $\frac{3}{4}$ c of sugar, thumbprints call for $1\frac{1}{2}$ c, and lemon squares call for $1\frac{1}{4}$ c. How much sugar is needed for all three recipes?

d. How much sugar is needed to fill an order for 20 boxes?

55. a. Hannah's recipe for pecan chocolate chip cookies makes 2 dozen, but a box holds only 20 cookies. Write a simplified fraction that represents the number of boxes one recipe makes.

b. Sandy's dipped butterfinger recipe makes 3 dozen, but a box holds only 30 cookies. Write a simplified fraction that represents the total number of boxes one recipe makes.

STATE YOUR UNDERSTANDING

56. Explain why it is sometimes necessary to rename the sum of two mixed numbers after adding the whole numbers and the fractional parts. Give an example where this happens.

57. Add $8\frac{4}{5} + 7\frac{3}{8}$ by the procedures given in this section. Then change each mixed number to an improper fraction and add. Be sure you get the same result for both. Which method do you prefer? Why?

CHALLENGE

58. Is $3\left(4\frac{1}{4}\right) + 2\frac{2}{3} + 5\left(2\frac{5}{6}\right) = 3\left(3\frac{1}{8}\right) + 7\frac{1}{12}$ a true statement?

59. Is $6\left(6\frac{4}{9}\right) + 5\left(2\frac{5}{6}\right) + 7\left(4\frac{1}{3}\right) = 6\left(5\frac{1}{3}\right) + 8\left(3\frac{2}{3}\right) + 7\left(3\frac{5}{42}\right)$ a true statement?

60. During the month of January, the rangers at Yellowstone National Park record the following snowfall: week 1, $6\frac{9}{10}$ in.; week 2, $8\frac{3}{4}$ in.; week 3, $13\frac{5}{6}$ in.; week 4, $9\frac{2}{3}$ in. How many inches of snow have fallen during the month? If the average snowfall for the month is $38\frac{17}{32}$ in., does this January exceed the average?

GROUP WORK

61. We know that a mixed number can be changed to an improper fraction, so $5\frac{3}{7} = \frac{38}{7}$. How many ways can you find to express $5\frac{3}{7}$ as a mixed number using improper fractions?

MAINTAIN YOUR SKILLS

Subtract.

62. 103 − 77

63. 212 − 128

64. 1111 − 889

65. 2222 − 1798

Find the missing number.

66. $\frac{7}{8} = \frac{?}{40}$

67. $\frac{3}{16} = \frac{?}{80}$

68. Simplify: $\frac{1950}{4095}$

69. Simplify: $\frac{462}{847}$

70. Par on the first nine holes of the Ricochet Golf Course is 36. If Millie records scores of 5, 4, 6, 2, 3, 3, 3, 5, and 4 on the nine holes, what is her total score? Is she under or over par for the first nine holes?

71. Dried prunes weigh one-third the weight of fresh prunes. How many pounds of fresh prunes are required to make 124 half-pound packages of dried prunes?

3.8 Subtracting Fractions

HOW & WHY

OBJECTIVE Subtract fractions.

OBJECTIVE
Subtract fractions.

What is the difference of $\frac{2}{3}$ and $\frac{1}{3}$? In Figure 3.19, we can see that we subtract the numerators and keep the common denominator (subtract the yellow region from the blue region).

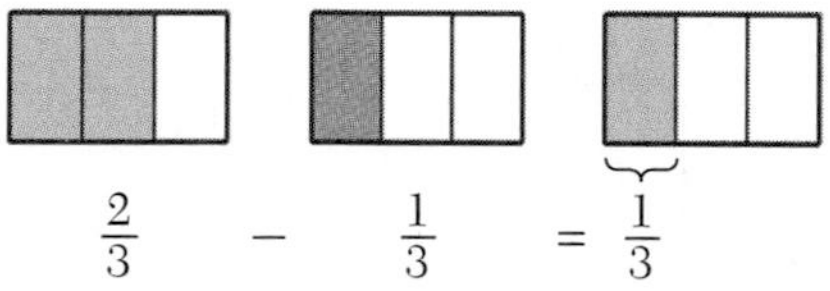

FIGURE 3.19

What is the value of $\frac{3}{4} - \frac{1}{3}$? See Figure 3.20.

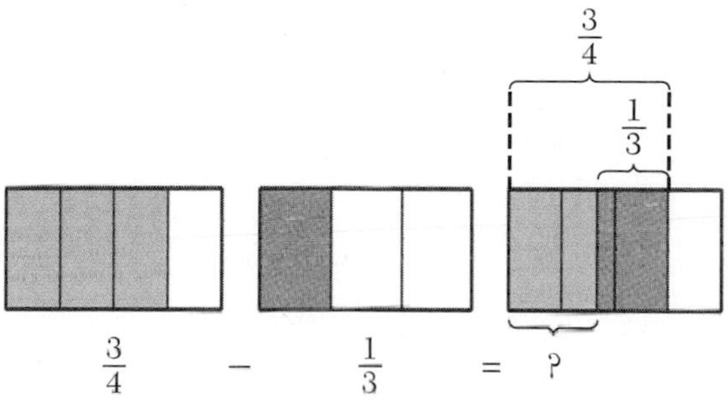

FIGURE 3.20

The region shaded blue with the question mark cannot be named immediately because the original parts are not the same size. If the fractions had a common denominator, we could subtract, as in Figure 3.19. Using the common denominator 12, we see in Figure 3.21 that the difference is $\frac{5}{12}$.

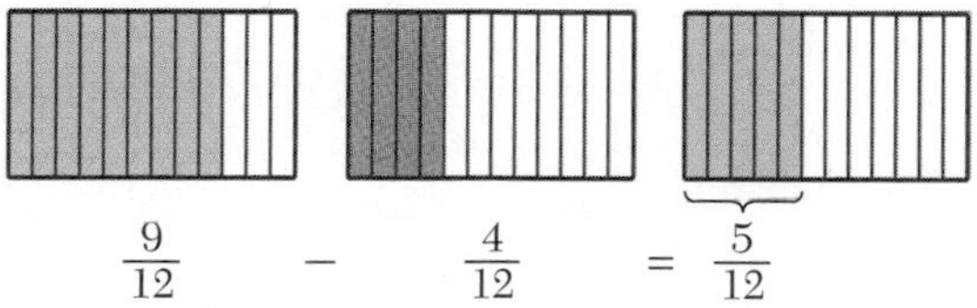

FIGURE 3.21

$$\frac{3}{4} - \frac{1}{3} = \frac{9}{12} - \frac{4}{12} = \frac{5}{12}$$

The method for subtracting fractions is similar to that for adding fractions.

To subtract fractions

1. Build each fraction to a common denominator.
2. Subtract the numerators and write the difference over the common denominator.
3. Simplify.

WARM-UPS A–F

A. Subtract: $\frac{13}{18} - \frac{7}{18}$

B. Subtract: $\frac{4}{5} - \frac{5}{12}$

C. Subtract: $\frac{13}{18} - \frac{9}{24}$

D. Subtract: $\frac{67}{72} - \frac{11}{90}$

E. Subtract: $\frac{43}{48} - \frac{23}{32}$

Answers to Warm-Ups

A. $\frac{1}{3}$ B. $\frac{23}{60}$ C. $\frac{25}{72}$ D. $\frac{97}{120}$ E. $\frac{17}{96}$

EXAMPLES A–F

DIRECTIONS: Subtract and simplify.

STRATEGY: Build each fraction to a common denominator.

A. Subtract: $\frac{11}{20} - \frac{7}{20}$

$\frac{11}{20} - \frac{7}{20} = \frac{4}{20}$ Subtract the numerators.

$= \frac{1}{5}$ Simplify.

B. Subtract: $\frac{5}{8} - \frac{1}{6}$

$\frac{5}{8} - \frac{1}{6} = \frac{15}{24} - \frac{4}{24}$ The LCM of 8 and 6 is 24.

$= \frac{11}{24}$ Subtract the numerators.

C. Subtract: $\frac{13}{18} - \frac{7}{12}$

$\frac{13}{18} - \frac{7}{12} = \frac{26}{36} - \frac{21}{36}$ The LCM of 18 and 12 is 36.

$= \frac{5}{36}$ Subtract the numerators.

D. Subtract: $\frac{43}{60} - \frac{5}{48}$

$\frac{43}{60} - \frac{5}{48} = \frac{172}{240} - \frac{25}{240}$ The LCM of 60 and 48 is 240.

$= \frac{147}{240}$ Subtract the numerators.

$= \frac{49}{80}$ Simplify.

CALCULATOR EXAMPLE

E. Subtract: $\frac{39}{50} - \frac{8}{15}$

$\frac{39}{50} - \frac{8}{15} = \frac{37}{150}$ On a calculator with fraction keys, it is not necessary to find the common denominator. The calculator is programmed to subtract and simplify.

F. Lumber mill operators must plan for the shrinkage of "green" (wet) boards when they cut logs. If the shrinkage for a $\frac{5}{8}$-in.-thick board is expected to be $\frac{1}{16}$ in., what will be the thickness of the dried board?

STRATEGY: To find the thickness of the dried board, subtract the shrinkage from the thickness of the green board.

$$\frac{5}{8} - \frac{1}{16} = \frac{10}{16} - \frac{1}{16}$$ Build $\frac{5}{8}$ to a fraction with denominator 16.

$$= \frac{9}{16}$$

The dried board will be $\frac{9}{16}$ in. thick.

F. Mike must plane $\frac{3}{32}$ in. from the thickness of a board. If the board is now $\frac{3}{8}$ in. thick, how thick will it be after he has planed it?

Answers to Warm-Ups

F. The board will be $\frac{9}{32}$ in. thick.

EXERCISES 3.8

OBJECTIVE Subtract fractions.

Subtract. Simplify completely.

A.

1. $\frac{6}{7} - \frac{4}{7}$
2. $\frac{4}{5} - \frac{3}{5}$
3. $\frac{3}{4} - \frac{2}{4}$
4. $\frac{5}{9} - \frac{1}{9}$
5. $\frac{17}{30} - \frac{7}{30}$
6. $\frac{14}{15} - \frac{11}{15}$
7. $\frac{6}{7} - \frac{5}{14}$
8. $\frac{14}{15} - \frac{3}{5}$
9. $\frac{3}{4} - \frac{5}{16}$
10. $\frac{8}{9} - \frac{5}{18}$
11. $\frac{3}{15} - \frac{2}{45}$
12. $\frac{11}{18} - \frac{4}{9}$
13. $\frac{5}{6} - \frac{2}{3}$
14. $\frac{5}{6} - \frac{1}{2}$
15. $\frac{17}{18} - \frac{2}{3}$
16. $\frac{19}{24} - \frac{3}{8}$
17. $\frac{13}{20} - \frac{2}{5}$
18. $\frac{17}{25} - \frac{2}{5}$
19. $\frac{23}{40} - \frac{1}{8}$
20. $\frac{17}{36} - \frac{1}{4}$

B.

21. $\frac{7}{8} - \frac{5}{6}$
22. $\frac{2}{3} - \frac{3}{8}$
23. $\frac{9}{16} - \frac{5}{12}$
24. $\frac{11}{15} - \frac{11}{20}$
25. $\frac{2}{7} - \frac{5}{21}$
26. $\frac{5}{6} - \frac{7}{15}$
27. $\frac{7}{16} - \frac{3}{8}$
28. $\frac{4}{5} - \frac{3}{4}$
29. $\frac{8}{9} - \frac{5}{6}$
30. $\frac{4}{7} - \frac{5}{14}$
31. $\frac{5}{8} - \frac{1}{12}$
32. $\frac{7}{10} - \frac{7}{15}$
33. $\frac{9}{10} - \frac{1}{2}$
34. $\frac{11}{15} - \frac{7}{12}$
35. $\frac{8}{9} - \frac{3}{4}$
36. $\frac{11}{12} - \frac{11}{15}$
37. $\frac{19}{25} - \frac{11}{15}$
38. $\frac{23}{32} - \frac{21}{48}$
39. $\frac{13}{16} - \frac{11}{24}$
40. $\frac{13}{18} - \frac{7}{12}$

C.

41. $\frac{17}{18} - \frac{11}{12}$
42. $\frac{9}{10} - \frac{7}{8}$
43. $\frac{3}{10} - \frac{1}{15}$
44. $\frac{23}{48} - \frac{21}{80}$
45. $\frac{19}{20} - \frac{8}{15}$
46. $\frac{23}{24} - \frac{23}{60}$

47. If the shrinkage of a $\frac{5}{4}$-in.-thick "green" board is $\frac{1}{8}$ in., what will be the thickness when the board has dried?

48. One month the reservoir at Bull Run watershed was at $\frac{7}{8}$ capacity. During the following month, the reservoir lost $\frac{3}{64}$ of its capacity due to evaporation. What fraction of its capacity does it now hold?

49. The fat content of 1 ounce of hotdog relish is $\frac{29}{35}$ g. The fat content of one slice of smoked chicken breast is $\frac{38}{45}$ g. Which foodstuff has less fat? How much less?

50. A water sample from Lake Tuscumba contains 61 parts per million of phosphate. A sample from Lost Lake contains 6 parts per hundred thousand. Which lake has the greatest phosphate content? By how much?

51. In a recent year, 13 of every 25 Americans owned stock in a publicly held company or mutual fund. What fraction of Americans did not own stock?

52. In a recent year, approximately 80 million people attended a professional ice hockey, basketball, football, or baseball game. If $\frac{1}{8}$ of them attended ice hockey, $\frac{1}{8}$ of them attended basketball, and $\frac{3}{16}$ of them attended football, what fraction attended baseball?

53. In 1999, according to the Recording Industry Association of America, $\frac{1}{10}$ of all the recorded music and music video sales was cassettes. In that same year, $\frac{4}{5}$ of all sales were CDs. What part of the total sales for the year was not CDs or cassettes?

54. In 1820, according to figures from the U.S. Dept. of Agriculture, approximately $\frac{18}{25}$ of the U.S. labor force were farm occupations. By the 1990s, only $\frac{1}{40}$ of the labor force was farm occupations. What part of the labor force has changed from farm to nonfarm occupations over this time period?

55. When a person goes on a reduced calorie diet to lose weight, about $\frac{3}{4}$ of the weight lost is body fat, and the rest is water. How much of the weight lost is water?

56. A population survey finds that $\frac{34}{75}$ of the residents of Ukiah are Caucasian and that $\frac{17}{40}$ are African American. What fraction represents the difference in population between Caucasians and African Americans?

57. The diameter at the large end of a tapered pin is $\frac{7}{8}$ in., and at the smaller end, it is $\frac{3}{16}$ in. What is the difference between the diameters?

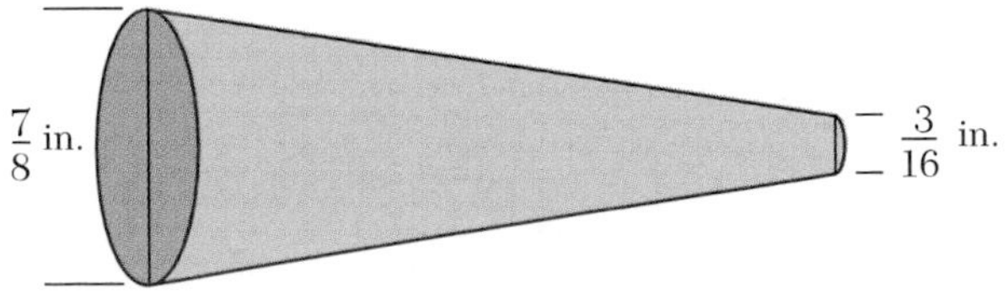

58. A landscaper is building a brick border, one brick wide, around a formal rose garden. The garden is a 10-ft by 6-ft rectangle. Standard bricks are 8 in. by $3\frac{3}{4}$ in. by $2\frac{1}{4}$ in., and the landscaper is planning to use a $\frac{3}{8}$-in.-wide mortar in the joints. How many whole bricks are needed for the project? Explain your reasoning.

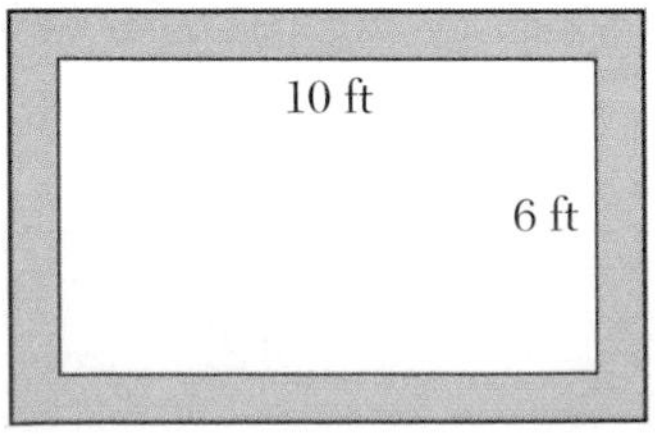

Exercise 59 relates to the chapter application.

59. How can Hannah measure $\frac{3}{8}$ of a cup of nuts with only a $\frac{1}{2}$-c measure and a tablespoon measure? (*Hint*: 1 c = 16 tbs)

STATE YOUR UNDERSTANDING

60. Explain in writing how you would teach a child to subtract fractions.

61. Explain why $\frac{3}{4} - \frac{1}{2}$ is not equal to $\frac{2}{2}$.

CHALLENGE

Subtract.

62. $\frac{213}{560} - \frac{109}{430}$

63. $\frac{93}{125} - \frac{247}{625}$

64. A donor agrees to donate $1000 for each foot that Skola outdistances Sheila in 13 minutes. Skola walks $\frac{19}{24}$ mi. Sheila walks $\frac{47}{60}$ mi. Does Skola outdistance Sheila? By what fraction of a mile? How much does the donor contribute? (A mile equals 5280 ft.)

GROUP WORK

65. With your group, find a way to subtract $4\frac{7}{8}$ from $6\frac{3}{4}$. Can you find more than one way? Write the procedure down and share it with the class.

MAINTAIN YOUR SKILLS

Subtract.

66. 41,695 − 3969

67. 21,596 − 9396

Subtract. Write as a mixed number.

68. $(9 - 5) + \left(\frac{2}{3} - \frac{5}{12}\right)$

69. $(22 - 16) + \left(\frac{2}{3} - \frac{4}{15}\right)$

70. $(92 - 87) + \left(\frac{2}{3} - \frac{7}{18}\right)$

71. $(81 - 76) + \left(\frac{2}{3} - \frac{5}{21}\right)$

72. $(54 - 45) + \left(\frac{3}{4} - \frac{5}{12}\right)$

73. $(63 - 36) + \left(\frac{3}{4} - \frac{1}{20}\right)$

74. Three bricklayers can each lay 795 bricks per day, on the average. How many bricks can they lay in 5 days?

75. If a retaining wall requires 19,080 bricks, how many days will it take the three bricklayers in Exercise 74 to build the wall?

3.9 Subtracting Mixed Numbers

HOW & WHY

OBJECTIVE Subtract mixed numbers.

OBJECTIVE
Subtract mixed numbers.

A subtraction problem may be written in horizontal or vertical form. Horizontally:

$$8\frac{7}{9} - 3\frac{2}{9} = (8 - 3) + \left(\frac{7}{9} - \frac{2}{9}\right)$$
$$= 5 + \frac{5}{9} = 5\frac{5}{9}$$

Because the denominators are the same, subtract the whole-number parts and then subtract the fractional parts.

Vertically:

$$\begin{array}{r} 8\frac{7}{9} \\ -3\frac{2}{9} \\ \hline 5\frac{5}{9} \end{array}$$

The process is similar to that for adding mixed numbers.

It is sometimes necessary to "borrow" from the whole number in order to subtract the fractions. For example,

$$8\frac{2}{5} - 3\frac{3}{4} = ?$$

First, write in columns and build each fraction to the common denominator, 20.

$$\begin{array}{rcr} 8\frac{2}{5} & = & 8\frac{8}{20} \\ -3\frac{3}{4} & = & 3\frac{15}{20} \\ \hline \end{array}$$

Because we cannot subtract $\frac{15}{20}$ from $\frac{8}{20}$, we need to borrow. To do this, we rename $8\frac{8}{20}$ by borrowing 1 from 8.

$$8\frac{8}{20} = 7 + 1\frac{8}{20}$$ Borrow 1 from 8 and add it to the fractional part.

$$= 7 + \frac{28}{20}$$ Change the mixed number, $1\frac{8}{20}$, to an improper fraction.

$$= 7\frac{28}{20}$$ Write as a mixed number.

CAUTION

Do not write $\frac{18}{20}$. If we "borrow" 1 from 8, we must add 1—that is $\frac{20}{20}$—to $\frac{8}{20}$.

The example can now be completed.

$$\begin{array}{rcrcr} 8\frac{2}{5} & = & 8\frac{8}{20} & = & 7\frac{28}{20} \\ -3\frac{3}{4} & = & 3\frac{15}{20} & = & 3\frac{15}{20} \\ \hline & & & & 4\frac{13}{20} \end{array}$$

Rename by borrowing 1.

Subtract the whole numbers. Subtract the fractions.

To subtract mixed numbers

1. Build the fractions so they have a common denominator.
2. Subtract the fractions. If the fractions cannot be subtracted, rename the first mixed number by borrowing 1 from the whole-number part to add to the fractional part. Then subtract the fractions.
3. Subtract the whole numbers.
4. Simplify.

WARM-UPS A–G

A. Subtract: $22\frac{3}{5} - 15\frac{2}{9}$

B. Subtract: $61\frac{9}{10} - 18$

C. Subtract: $25 - 16\frac{5}{7}$

Answers to Warm-Ups

A. $7\frac{17}{45}$ **B.** $43\frac{9}{10}$ **C.** $8\frac{2}{7}$

EXAMPLES A–G

DIRECTIONS: Subtract. Write as a mixed number.

STRATEGY: Subtract the fractions and subtract the whole numbers. If necessary, borrow. Simplify.

A. Subtract: $15\frac{4}{5} - 13\frac{1}{6}$

STRATEGY: Write the mixed numbers in columns to group the whole numbers and group the fractions.

$$\begin{array}{rcl} 15\frac{4}{5} & = & 15\frac{24}{30} \\ -13\frac{1}{6} & = & 13\frac{5}{30} \\ \hline & & 2\frac{19}{30} \end{array}$$

Build the fraction to the common denominator, 30.

Subtract.

B. Subtract: $21\frac{3}{4} - 9$

$$21\frac{3}{4} - 9 = 12\frac{3}{4}$$

Subtract the whole numbers.

C. Subtract: $21 - 9\frac{3}{4}$

STRATEGY: Notice the difference between Examples B and C. Here we must also subtract the fraction. We may think of 21 as $21\frac{0}{4}$ in order to get a common denominator for the improper fraction. Or think: $21 = 20 + 1 = 20 + \frac{4}{4} = 20\frac{4}{4}$

$$\begin{array}{rcl} 21 & = 20 + \frac{4}{4} = & 20\frac{4}{4} \\ -\ 9\frac{3}{4} & = & 9\frac{3}{4} \\ \hline & = & 11\frac{1}{4} \end{array}$$

CAUTION

Do not just bring down the fraction $\frac{3}{4}$ and then subtract the whole numbers:

$21 - 9\frac{3}{4} \neq 12\frac{3}{4}$. The fraction must also be subtracted.

D. Subtract: $15\frac{5}{12} - 7\frac{7}{12}$

STRATEGY: Because $\frac{7}{12}$ cannot be subtracted from $\frac{5}{12}$, we need to borrow.

$$\begin{aligned} 15\frac{5}{12} &= 14 + 1\frac{5}{12} = 14\frac{17}{12} \\ -\ 7\frac{7}{12} &= \phantom{14 + 1\frac{5}{12}} = 7\frac{7}{12} \end{aligned}$$

Borrow 1 from 15 and change the mixed number to an improper fraction.

$$= 7\frac{10}{12}$$ Subtract.

$$= 7\frac{5}{6}$$ Simplify.

E. Subtract. $23\frac{5}{12} - 11\frac{7}{15}$

$$\begin{aligned} 23\frac{5}{12} &= 22\frac{25}{60} = 22 + 1\frac{25}{60} = 22\frac{85}{60} \\ -11\frac{7}{15} &= 11\frac{28}{60} \phantom{= 22 + 1\frac{25}{60}} = 11\frac{28}{60} \end{aligned}$$

The LCM of 12 and 15 is 60. Borrow 1 from 23 and change the mixed number to an improper fraction.

$$11\frac{57}{60}$$ Subtract.

$$= 11\frac{19}{20}$$ Simplify.

CALCULATOR EXAMPLE

F. Subtract: $31\frac{1}{3} - 18\frac{3}{4}$

$$31\frac{1}{3} - 18\frac{3}{4} = \frac{151}{12} = 12\frac{7}{12}$$

On a calculator with fraction keys, it is not necessary to find the common denominator. The calculator is programmed to subtract and simplify. Some calculators may not change the improper result to a mixed number.

D. Subtract: $18\frac{11}{15} - 8\frac{12}{15}$

E. Subtract: $58\frac{3}{8} - 43\frac{14}{15}$

F. Subtract: $72\frac{7}{9} - 25\frac{13}{15}$

Answers to Warm-Ups

D. $9\frac{14}{15}$ **E.** $14\frac{53}{120}$ **F.** $46\frac{41}{45}$

G. Jamie weighs $138\frac{1}{2}$ lb and decides to lose weight. She loses a total of $5\frac{3}{4}$ lb in 2 weeks. What is her weight after the loss?

G. Shawn brings a roast home for Sunday dinner that weighs 7 lb. He cuts off some fat and takes out a bone. The meat left weighs $4\frac{1}{3}$ lb. How many pounds of fat and bone does he trim off?

STRATEGY: Subtract the weight of the remaining meat from the original weight of the roast.

$$\begin{array}{rl} 7 & = 6 + 1\frac{0}{3} = 6\frac{3}{3} \\ -4\frac{1}{3} & = 4\frac{1}{3} \\ \hline & = 2\frac{2}{3} \end{array}$$

Borrow 1 from 7 and rename it as an improper fraction.

* Shawn trims off $2\frac{2}{3}$ lb of fat and bone.

Answers to Warm-Ups

G. Her weight is $132\frac{3}{4}$ lb.

EXERCISES 3.9

OBJECTIVE Subtract mixed numbers.

Subtract. Write the results as mixed numbers where possible.

A.

1. $12\frac{3}{5} - 9\frac{1}{5}$

2. $22\frac{5}{7} - 13\frac{2}{7}$

3. $202\frac{17}{24} - 200\frac{13}{24}$

4. $107\frac{7}{9} - 98\frac{1}{9}$

5. $10\frac{7}{8} - 5\frac{3}{4}$

6. $11\frac{4}{5} - 7\frac{3}{10}$

7. $15 - 8\frac{3}{7}$

8. $32 - 14\frac{1}{6}$

9. $105\frac{7}{8} - 37\frac{3}{4}$

10. $106\frac{4}{5} - 48\frac{7}{10}$

11. $5\frac{1}{4} - 2\frac{3}{4}$

12. $7\frac{3}{8} - 4\frac{5}{8}$

13. $19\frac{3}{8} - 7\frac{5}{8}$

14. $77\frac{5}{12} - 18\frac{7}{12}$

15. $212\frac{1}{9} - 57$

16. $400\frac{3}{8} - 62$

17. $9\frac{1}{3} - 2\frac{5}{12}$

18. $8\frac{1}{2} - 6\frac{5}{12}$

B.

19. $32\frac{13}{15} - 17\frac{7}{15}$

20. $24\frac{19}{24} - 22\frac{11}{24}$

21. $310\frac{23}{24} - 254\frac{5}{8}$

22. $418\frac{7}{12} - 93\frac{1}{4}$

23. $67\frac{7}{12} - 42\frac{2}{15}$

24. $52\frac{7}{12} - 47\frac{3}{16}$

25. $37\frac{2}{3} - 15\frac{11}{12}$

26. $28\frac{1}{3} - 15\frac{7}{9}$

27. $40\frac{1}{6} - 24\frac{3}{16}$

28. $31\frac{3}{10} - 17\frac{5}{18}$

29. $45 - 16\frac{2}{3}$

30. $76 - 26\frac{3}{4}$

31. $3\frac{17}{32} - 1\frac{5}{16}$

32. $9\frac{3}{10} - \frac{17}{20}$

33. $7\frac{13}{18} - \frac{5}{12}$

34. $9\frac{1}{6} - \frac{5}{8}$

35. $93\frac{7}{8} - 19$

36. $55\frac{7}{16} - 28$

37. $64\frac{13}{15} - 28\frac{29}{40}$

38. $70\frac{11}{18} - 58\frac{41}{45}$

39. $75\frac{7}{12} - 47\frac{13}{18}$

40. $82\frac{4}{15} - 56\frac{7}{12}$

C.

41. $16 - 14\frac{17}{20}$

42. $41 - 29\frac{17}{36}$

43. $3\frac{11}{15} - 2\frac{9}{10}$

44. $6\frac{1}{3} - 5\frac{5}{6}$

45. Find the error(s) in the statement: $16 - 13\frac{1}{4} = 3\frac{3}{4}$. Correct the statement. Explain how you would avoid this error.

46. Find the error(s) in the statement: $5\frac{1}{2} - 2\frac{3}{4} = 3\frac{1}{4}$. Correct the statement. Explain how you would avoid this error.

47. Han Kwong trims bone and fat from a $6\frac{3}{4}$-lb roast. The meat left weighs $5\frac{1}{4}$ lb. How many pounds does she trim off?

48. Patti has a piece of lumber that measures $10\frac{7}{12}$ ft, which is to be used in a spot that calls for a length of $8\frac{5}{12}$ ft. How much of the board must she cut off?

49. Dick harvests $30\frac{3}{4}$ tons of wheat. He sells $12\frac{3}{10}$ tons to the Cartwright Flour Mill. How many tons of wheat does he have left?

50. A $14\frac{3}{4}$-in. casting shrinks $\frac{3}{16}$ in. on cooling. Find the size when the casting is cold.

51. Nancy bought a carpet remnant that is 12 ft long and 10 ft wide. Her small bedroom is $9\frac{2}{3}$ ft long and 10 ft wide. How much of the remnant is left over?

52. According to the International Game Fish Association, the largest cubera snapper ever caught weighed $121\frac{5}{8}$ lb and was caught in Louisiana in 1982 by Mike Hebart. The largest red snapper ever caught weighed $50\frac{1}{3}$ lb and was caught in the Gulf of Mexico off Louisiana in 1996 by Doc Kennedy. How much larger was the record cubera snapper than the red snapper?

53. An airline defines overweight luggage as anything over 70 lb. Amber's suitcase weighs $82\frac{7}{8}$ lb. How much extra weight is she charged for?

54. A McDonnell Douglas DC-9 seats 158 and is $147\frac{5}{6}$ ft long. A Boeing 737 seats 128–149 and is $109\frac{7}{12}$ ft long. How much longer is the DC-9?

55. Larry and Greg set out to hike 42 mi in 2 days. At the end of the first day, they have covered $24\frac{3}{10}$ mi. How many miles do they have to go?

56. Frank pours $8\frac{3}{10}$ yd of cement for a fountain. Another fountain takes $5\frac{3}{8}$ yd. How much more cement is needed for the larger fountain?

57. The Williams family recycles an average of $134\frac{2}{5}$ lb of material per month. The Madera family recycles an average of $102\frac{5}{8}$ lb of material per month. How many more pounds of material are recycled by the Williams family in a year?

58. The town of Fredonia averages $35\frac{1}{5}$ in. of rain per year. The town of Wheatland averages $27\frac{13}{15}$ in. of rain per year. Over a 10-yr period, how much more rain does Fredonia get?

59. The town of Hartville averages $38\frac{1}{6}$ in. of rain per year. The town of Lingle averages $25\frac{3}{16}$ in. of rain per year. Over a 6-year period, how much more rain falls on Hartville than on Lingle?

Exercise 60 relates to the chapter application.

60. Some recipes call for regular milk and vinegar as a substitute for buttermilk. A recipe calls for 1 c of regular milk less 1 tbs, and then 1 tbs of vinegar is added, for a total of 1 c of "buttermilk replacement." How much regular milk is needed for this recipe?

STATE YOUR UNDERSTANDING

61. Explain how to simplify $4\frac{1}{3} - 2\frac{5}{8}$.

62. When you "borrow" 1 to subtract mixed numbers, explain the fraction form it is written in and explain why.

CHALLENGE

63. Does $4\left(5\frac{5}{6}\right) - 3\left(2\frac{7}{12}\right) = 6\left(5\frac{1}{2}\right) - 5\frac{3}{4}$?

64. Does $3\left(7\frac{7}{8}\right) - 13\frac{1}{10} = 6\left(3\frac{3}{16}\right) - 7\left(1\frac{8}{35}\right)$?

65. A snail climbs $5\frac{1}{3}$ ft in a day and slips back $1\frac{2}{3}$ ft at night. What is the snail's net distance in 24 hr? How many days will it take the snail to make a net gain of over 20 ft?

GROUP WORK

66. Have each member of your group create an application involving subtraction of mixed numbers. Trade them around and have other members solve them. Share the best two with the rest of the class.

MAINTAIN YOUR SKILLS

Simplify.

67. $82 - 4 \cdot 3 + 10$

68. $8^2 - 4 \cdot 3 + 10$

69. $82 - 4^3 + 10$

70. $6 \cdot 18 \div 3 \cdot 2$

71. $6 \cdot 18 \div 3^2$

72. $6 + 18 \div 3^2$

Multiply.

73. $\frac{16}{25} \cdot \frac{25}{28} \cdot \frac{14}{15}$

74. $\frac{1}{2} \cdot \frac{5}{8} \cdot \frac{3}{5} \cdot 24$

75. Last week, when Karla filled the tank of her car with gasoline, the odometer read 57,832 miles. Yesterday, when she filled the tank with 18 gallons of gasoline, the odometer read 58,336 miles. What is Karla's mileage? That is, how many miles to the gallon did she get?

76. A pet food canning company packs Feelein Cat Food in cans, each containing $7\frac{3}{4}$ oz of cat food. Each empty can weighs $1\frac{1}{2}$ oz. Twenty-four cans are packed in a case that weighs 10 oz when empty. What is the shipping weight of five cases of the cat food?

Getting Ready for Algebra

HOW & WHY

OBJECTIVE

Solve equations of the form $x + \frac{a}{b} = \frac{c}{d}$, where a, b, c, and d are whole numbers.

We have solved equations in which whole numbers are either added to or subtracted from a variable. Now we solve equations in which fractions or mixed numbers are either added to or subtracted from the variable. We use the same procedure as with whole numbers.

EXAMPLES A–C

DIRECTIONS: Solve.

STRATEGY: Add or subtract the same number from each side of the equation to isolate the variable.

A. Solve: $x - \frac{4}{5} = 3\frac{1}{2}$

$x - \frac{4}{5} + \frac{4}{5} = 3\frac{1}{2} + \frac{4}{5}$ To eliminate the subtraction, add $\frac{4}{5}$ to each side of the equation.

$x = 3\frac{1}{2} + \frac{4}{5}$ Simplify the left side.

$x = 3\frac{5}{10} + \frac{8}{10}$ Build each fraction to the denominator 10.

$x = 3\frac{13}{10}$ Add.

$x = 4\frac{3}{10}$ Change the improper fraction to a mixed number and add.

CHECK: $4\frac{3}{10} - \frac{4}{5} = 3\frac{1}{2}$ Substitute $4\frac{3}{10}$ for x in the original equation.

$3\frac{1}{2} = 3\frac{1}{2}$

The solution is $x = 4\frac{3}{10}$.

B. Solve: $4\frac{1}{2} = x + 2\frac{2}{3}$

$4\frac{1}{2} - 2\frac{2}{3} = x + 2\frac{2}{3} - 2\frac{2}{3}$ To eliminate the addition, subtract $2\frac{2}{3}$ from each side of the equation.

$4\frac{1}{2} - 2\frac{2}{3} = x$

$1\frac{5}{6} = x$

CHECK: $4\frac{1}{2} = 1\frac{5}{6} + 2\frac{2}{3}$ Substitute $1\frac{5}{6}$ for x in the original equation.

$4\frac{1}{2} = 4\frac{1}{2}$

The solution is $x = 1\frac{5}{6}$.

WARM-UPS A–C

A. Solve: $x - \frac{5}{8} = 3\frac{7}{8}$

B. Solve: $6\frac{1}{2} = a + 5\frac{5}{8}$

Answers to Warm-Ups

A. $x = 4\frac{1}{2}$ **B.** $a = \frac{7}{8}$

C. When $3\frac{7}{8}$ mi of new freeway opened last month, it brought the total length to $21\frac{3}{5}$ mi. What was the original length of the freeway?

C. On Tuesday, $2\frac{3}{8}$ in. of rain fell on Kansas City. This brought the total for the last five consecutive days to $14\frac{1}{2}$ in. What was the rainfall for the first 4 days?

STRATEGY: First write the English version of the equation.

$$\begin{pmatrix}\text{rain on first}\\ \text{4 days}\end{pmatrix} + \begin{pmatrix}\text{rain on}\\ \text{Tuesday}\end{pmatrix} = \text{total rain}$$

Let x represent the number of inches of rain on the first 4 days.

$$x + 2\frac{3}{8} = 14\frac{1}{2}$$ Translate to algebra.

$$\underline{-2\frac{3}{8} \quad - 2\frac{3}{8}}$$ Subtract $2\frac{3}{8}$ from each side.

$$x = 12\frac{1}{8}$$

Because $12\frac{1}{8} + 2\frac{3}{8} = 14\frac{1}{2}$, $12\frac{1}{8}$ in. of rain fell during the first 4 days.

Answers to Warm-Ups

C. The original freeway was $17\frac{29}{40}$ mi long.

EXERCISES

Solve

1. $a + \frac{3}{8} = \frac{7}{8}$

2. $y - \frac{1}{8} = \frac{5}{8}$

3. $c - \frac{3}{16} = \frac{7}{16}$

4. $w + \frac{5}{12} = \frac{11}{12}$

5. $x + \frac{5}{8} = \frac{7}{9}$

6. $x - \frac{7}{8} = \frac{3}{2}$

7. $y - \frac{5}{7} = \frac{8}{9}$

8. $y + \frac{5}{9} = \frac{9}{10}$

9. $a + \frac{9}{8} = \frac{12}{5}$

10. $a - \frac{7}{4} = \frac{5}{8}$

11. $c - 1\frac{1}{2} = 2\frac{2}{3}$

12. $c + 2\frac{4}{5} = 3\frac{5}{8}$

13. $x + 4\frac{3}{4} = 7\frac{8}{9}$

14. $x - 7\frac{8}{9} = 5\frac{7}{8}$

15. $12 = w + 8\frac{5}{6}$

16. $25 = m + 15\frac{5}{8}$

17. $a - 13\frac{5}{6} = 22\frac{11}{18}$

18. $b + 23\frac{11}{12} = 34\frac{1}{3}$

19. $c + 44\frac{13}{21} = 65\frac{5}{7}$

20. $x - 27\frac{5}{8} = 48\frac{2}{3}$

21. A native pine tree grew $2\frac{7}{8}$ ft in the last 10 years to its present height of $45\frac{1}{2}$ ft. What was the height of the tree 10 years ago?

22. Perez brought in $35\frac{3}{4}$ lb of tin to be recycled. This brings his total for the month to $112\frac{1}{2}$ lb. How many pounds had he already brought in this month?

23. Freeda purchased a supply of nails for a construction project. She has used $18\frac{2}{3}$ lb and has $27\frac{1}{3}$ lb left. How many pounds of nails did she buy?

24. For cross-country race practice, Althea has run $10\frac{7}{10}$ mi. She needs to run an additional $13\frac{3}{10}$ mi to meet the goal set by her coach. How many miles does the coach want her to run?

3.10 Order of Operations; Average

VOCABULARY

Recall that the **mean**, or **average**, of a set of numbers is the sum of the set of numbers divided by the total number of numbers in the set.

OBJECTIVES

1. Do any combination of operations with fractions.
2. Find the average of a set of fractions.

HOW & WHY

OBJECTIVE 1 Do any combination of operations with fractions.

The order of operations for fractions is the same as for whole numbers.

To evaluate an expression with more than one operation

1. Parentheses: Do the operations within grouping symbols first (parentheses, fraction bar, and so on) in the order given in steps 2, 3, and 4.
2. Exponents: Do the operations indicated by exponents.
3. Multiply and divide: Do only multiplication and division as they appear from left to right.
4. Add and subtract: Do addition and subtraction as they appear from left to right.

Table 3.4 summarizes some of the processes that need to be remembered when working with fractions.

TABLE 3.4 OPERATIONS WITH FRACTIONS

Operation	Find the LCM and build	Change mixed numbers to improper fractions	Invert divisor and multiply	Simplify answer
Add	Yes	No	No	Yes
Subtract	Yes	No	No	Yes
Multiply	No	Yes	No	Yes
Divide	No	Yes	Yes	Yes

EXAMPLES A–D

DIRECTIONS: Perform the indicated operations.

STRATEGY: Follow the order of operations that are used for whole numbers.

A. Simplify: $\frac{5}{6} - \frac{1}{2} \cdot \frac{2}{3}$

$\frac{5}{6} - \frac{1}{2} \cdot \frac{2}{3} = \frac{5}{6} - \frac{1}{3}$ Multiplication is performed first.

$= \frac{5}{6} - \frac{2}{6}$ Build $\frac{1}{3}$ to a denominator of 6.

$= \frac{3}{6} = \frac{1}{2}$ Subtract and simplify.

WARM-UPS A–D

A. Simplify: $\frac{7}{8} - \frac{3}{4} \cdot \frac{8}{9}$

Answers to Warm-Ups

A. $\frac{5}{24}$

B. Simplify: $\frac{3}{8} \div \frac{5}{16} \cdot \frac{1}{3}$

B. Simplify: $\frac{1}{3} \div \frac{3}{4} \cdot \frac{1}{2}$

$\frac{1}{3} \div \frac{3}{4} \cdot \frac{1}{2} = \frac{1}{3} \cdot \frac{4}{3} \cdot \frac{1}{2}$ Division is performed first, as it appears from left to right.

$= \frac{4}{18}$ Multiply from left to right.

$= \frac{2}{9}$ Simplify.

C. Simplify: $\left(\frac{3}{4}\right)^2 \cdot \frac{2}{5} - \frac{1}{5}$

C. Simplify: $\left(\frac{2}{3}\right)^2 \cdot \frac{1}{2} - \frac{1}{5}$

$\left(\frac{2}{3}\right)^2 \cdot \frac{1}{2} - \frac{1}{5} = \frac{\overset{2}{\cancel{4}}}{9} \cdot \frac{1}{\underset{1}{\cancel{2}}} - \frac{1}{5}$ Do exponentiation first; then simplify.

$= \frac{2}{9} - \frac{1}{5}$ Multiply.

$= \frac{10}{45} - \frac{9}{45}$ Build to a common denominator.

$= \frac{1}{45}$ Subtract.

D. Jill, Jean, and Joan have equal shares in a gift shop. Jill sells her share. She sells $\frac{1}{4}$ to Jean and the rest to Joan. What share of the gift shop does Joan now own?

D. Gwen, George, Carlos, and Maria have equal shares in a florist shop. George sells $\frac{3}{8}$ of his share to Gwen, $\frac{1}{2}$ to Carlos, and the rest to Maria. What share of the florist shop does Gwen now own?

STRATEGY: Because each of the four had equal shares, each of them owned $\frac{1}{4}$ of the business. To find the share Gwen now owns, add her original share, $\frac{1}{4}$, to $\frac{3}{8}$ of George's share, which was $\frac{1}{4}$.

$\frac{1}{4} + \frac{3}{8} \cdot \frac{1}{4} = \frac{1}{4} + \frac{3}{32}$ Multiply first.

$= \frac{8}{32} + \frac{3}{32}$ Build $\frac{1}{4}$ to the denominator 32.

$= \frac{11}{32}$

Gwen now owns $\frac{11}{32}$ of the florist shop.

HOW & WHY

OBJECTIVE 2 Find the average of a set of fractions.

To find the average of a set of fractions, divide the sum of the fractions by the number of fractions. The procedure is the same for all numbers.

To find the average of a set of numbers

1. Add the numbers.
2. Divide the sum by the number of numbers in the set.

Answers to Warm-Ups

B. $\frac{2}{5}$ **C.** $\frac{1}{40}$

D. Joan owns $\frac{7}{12}$ of the gift shop.

EXAMPLES E–G

DIRECTIONS: Find the mean.

STRATEGY: Find the sum of the set of fractions; then divide by the number of fractions.

E. Find the mean: $\frac{5}{6}, \frac{2}{3}, \frac{1}{2}$

$\frac{5}{6} + \frac{2}{3} + \frac{1}{2} = \frac{5}{6} + \frac{4}{6} + \frac{3}{6}$ Add the three fractions in the set.

$= \frac{12}{6}$ or 2

$2 \div 3 = \frac{2}{3}$ Divide the sum by 3, the number of fractions in the set.

The mean is $\frac{2}{3}$.

F. A class of 10 students takes a 12-problem quiz. The results are listed in the table. What is the average score?

Number of students	Fraction of problems correct
1	$\frac{12}{12}$
2	$\frac{11}{12}$
3	$\frac{10}{12}$
4	$\frac{9}{12}$

STRATEGY: To find the class average, add all the grades and divide by 10. There were two scores of $\frac{11}{12}$, three scores of $\frac{10}{12}$, and four scores of $\frac{9}{12}$, in addition to one perfect score of $\frac{12}{12}$.

$\frac{12}{12} + 2\left(\frac{11}{12}\right) + 3\left(\frac{10}{12}\right) + 4\left(\frac{9}{12}\right)$ Find the sum of the 10 scores.

$\frac{12}{12} + \frac{22}{12} + \frac{30}{12} + \frac{36}{12} = \frac{100}{12}$

$\frac{100}{12} \div 10 = \frac{100}{12} \cdot \frac{1}{10}$ Divide the sum by the number of students in the class.

$= \frac{10}{12}$

CAUTION

Do not simplify the answer because the test scores are based on 12.

The class average is $\frac{10}{12}$ of the problems correct, or 10 problems correct.

WARM-UPS E–G

E. Find the mean: $\frac{1}{6}, \frac{5}{8}, \frac{3}{4}$

F. A class of 12 students takes a 20-problem test. The results are listed in the table. What is the average score?

Number of students	Fraction of problems correct
1	$\frac{20}{20}$
2	$\frac{19}{20}$
4	$\frac{16}{20}$
5	$\frac{14}{20}$

Answers to Warm-Ups

E. $\frac{37}{72}$

F. The class average is $\frac{16}{20}$ of the problems correct.

G. Find the average: $3\frac{5}{6}$, $4\frac{1}{2}$, and $2\frac{2}{3}$

G. Find the average: $7\frac{1}{2}$, $6\frac{1}{4}$, and $4\frac{5}{8}$

$$7\frac{1}{2} + 6\frac{1}{4} + 4\frac{5}{8} = 18\frac{3}{8}$$

$$\begin{aligned} 18\frac{3}{8} \div 3 &= \frac{147}{8} \div 3 \\ &= \frac{147}{8} \cdot \frac{1}{3} \\ &= \frac{49}{8} = 6\frac{1}{8} \end{aligned}$$

The average is $6\frac{1}{8}$.

*

Answers to Warm-Ups

G. $3\frac{2}{3}$

EXERCISES 3.10

OBJECTIVE 1 Do any combination of operations with fractions.

Perform the indicated operations.

A.

1. $\frac{4}{15} + \frac{2}{15} - \frac{1}{15}$

2. $\frac{4}{15} - \frac{2}{15} + \frac{1}{15}$

3. $\frac{4}{15} - \left(\frac{2}{15} + \frac{1}{15}\right)$

4. $\frac{4}{15} - \left(\frac{2}{15} - \frac{1}{15}\right)$

5. $\frac{5}{6} - \frac{1}{2} \cdot \frac{2}{3}$

6. $\frac{1}{6} + \frac{1}{2} \div \frac{3}{2}$

7. $\frac{1}{2} + \frac{1}{8} \div \frac{1}{2}$

8. $\frac{1}{2} \div \frac{1}{8} + \frac{1}{2}$

9. $\frac{3}{4} \cdot \frac{1}{2} - \frac{3}{8}$

10. $\frac{5}{6} \cdot \frac{1}{3} - \frac{5}{18}$

11. $\frac{5}{8} \div \left(\frac{1}{2} + \frac{3}{4}\right)$

12. $\frac{1}{3} \div \left(\frac{1}{6} - \frac{1}{9}\right)$

13. $\frac{2}{3} + \left(\frac{1}{2}\right)^2$

14. $\frac{1}{2} + \left(\frac{1}{3}\right)^2$

B.

15. $\frac{3}{4} \div \frac{1}{3} \cdot \frac{1}{6}$

16. $\frac{3}{4} \div \left(\frac{1}{3} \cdot \frac{1}{6}\right)$

17. $\frac{5}{6} - \left(\frac{1}{6} \cdot \frac{1}{3}\right)$

18. $\frac{2}{3} - \left(\frac{1}{3} \cdot \frac{1}{6}\right)$

19. $\frac{5}{6} - \frac{3}{4} \div \frac{3}{2} + \frac{1}{2}$

20. $\frac{5}{8} - \frac{5}{6} \div \frac{4}{3} + \frac{1}{6}$

21. $\frac{5}{8} \cdot \frac{1}{3} + \frac{1}{2} \div \frac{1}{3} - \frac{7}{8}$

22. $\frac{1}{9} \div \frac{1}{2} - \frac{2}{3} \cdot \frac{1}{4} + \frac{1}{9}$

23. $\frac{7}{12} + \left(\frac{3}{4}\right)^2 - \frac{5}{9} \cdot \frac{33}{16}$

24. $\frac{3}{4} \cdot \frac{4}{5} - \frac{3}{0} + \left(\frac{2}{3}\right)^2$

25. $\frac{15}{16} - \left(\frac{5}{8}\right)^2 + \frac{7}{8} \div \frac{4}{3}$

26. $\frac{19}{25} - \frac{2}{5} + \left(\frac{3}{5}\right)^2 \div \frac{2}{3}$

27. $\frac{3}{4} - \left(\frac{1}{3} \div \frac{2}{5} - \frac{1}{4}\right)$

28. $\frac{2}{3} + \left(\frac{3}{4} \cdot \frac{4}{9} + \frac{1}{2}\right)$

OBJECTIVE 2 Find the average of a set of fractions.

Find the average.

A.

29. $\frac{1}{6}$ and $\frac{5}{6}$

30. $\frac{3}{8}$ and $\frac{5}{8}$

31. $\frac{1}{5}, \frac{2}{5}$, and $\frac{3}{5}$

32. $\frac{2}{7}, \frac{3}{7}$, and $\frac{4}{7}$

33. $\frac{1}{11}, \frac{2}{11}$, and $\frac{9}{11}$

34. $\frac{1}{5}, \frac{3}{5}$, and $\frac{8}{5}$

35. $\frac{1}{5}, \frac{2}{5}$, and $\frac{4}{15}$

36. $\frac{5}{12}, \frac{1}{4}$, and $\frac{7}{12}$

37. $2, 3\frac{1}{2}$, and $6\frac{1}{2}$

38. $1\frac{2}{3}, 3\frac{2}{3}$, and $6\frac{2}{3}$

B.

39. $\frac{3}{4}, \frac{4}{3}, \frac{3}{2}$, and $\frac{2}{3}$

40. $\frac{5}{6}, \frac{7}{12}, \frac{5}{4}$, and $\frac{3}{2}$

41. $\frac{1}{3}, \frac{5}{12}, \frac{1}{2}$, and $\frac{1}{4}$

42. $\frac{2}{5}, \frac{1}{10}, \frac{3}{10}$, and $\frac{3}{5}$

43. $3\frac{1}{3}, 4\frac{5}{6}$, and $2\frac{1}{6}$

44. $6\frac{5}{8}, 7\frac{1}{4}$, and $8\frac{1}{2}$

45. $5\frac{1}{3}, 6\frac{2}{5}$, and $9\frac{13}{15}$

46. $2\frac{1}{2}, 1\frac{1}{6}, 1\frac{7}{9}$, and $2\frac{1}{3}$

C.

Perform the indicated operations.

47. $\frac{8}{9} - \left(\frac{1}{2} + \frac{1}{3} \div \frac{3}{4}\right) \cdot \frac{1}{2}$

48. $\frac{1}{2} - \left(\frac{1}{2} - \frac{3}{4} \div \frac{12}{5}\right) \cdot \frac{1}{6}$

49. $\left(\frac{5}{8} - \frac{1}{2} \cdot \frac{3}{4}\right) \div \frac{1}{2} + \frac{1}{2}$

50. $\left(\frac{29}{12} - \frac{5}{4} \cdot \frac{3}{2}\right) \div \frac{5}{6} + \frac{3}{4}$

Find the average.

51. $3\frac{2}{3}, 5\frac{1}{6}$, and $9\frac{2}{9}$

52. $2\frac{1}{4}, 2\frac{4}{5}$, and $9\frac{9}{10}$

53. $\frac{11}{12}, 6\frac{1}{3}$, and 5

54. $5, \frac{8}{9}$, 12, and $\frac{5}{6}$

55. Wayne catches six salmon. The salmon measure $23\frac{1}{4}$ in., $31\frac{5}{8}$ in., $42\frac{3}{4}$ in., $28\frac{5}{8}$ in., $35\frac{3}{4}$ in., and 40 in. in length. What is the average length of the salmon?

56. Karla, a nurse at General Hospital, weighs five new babies. They weigh $6\frac{1}{2}$ lb, $7\frac{3}{4}$ lb, $9\frac{3}{8}$ lb, $7\frac{1}{2}$ lb, and $8\frac{7}{8}$ lb. What is the average weight of the babies?

57. The mothers of a swim team are making Rice Krispie treats for a big meet for the team. Each recipe calls for 6 cups of cereal, $\frac{1}{4}$ cup of butter, $1\frac{3}{4}$ cups chocolate chips, and $3\frac{1}{2}$ cups of marshmallows. They are also making Gorp, using $2\frac{1}{2}$ cups of cereal, 4 cups of pretzels, 1 cup of marshmallows, 1 cup of raisins, and $2\frac{1}{2}$ cups of chocolate chips. How much of each ingredient do they need if they intend to make 5 recipes of Rice Krispie treats and 10 recipes of Gorp?

58. Nellie is making the trellis shown here out of $\frac{1}{2}$-in. copper pipe, which comes in 10-ft lengths. How many 10-ft lengths does she need and how much will she have left over?

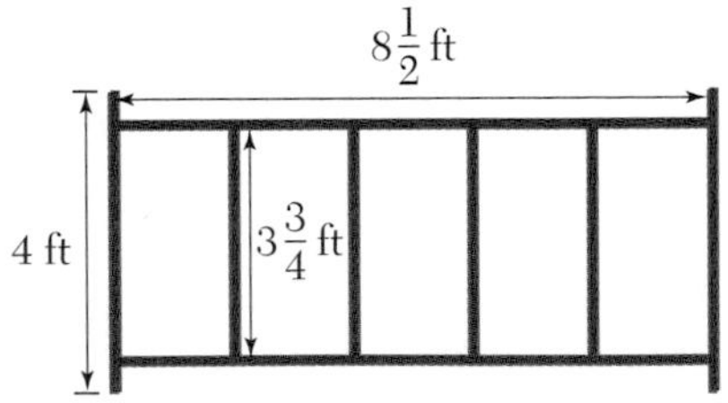

59. A class of 15 students took a 10-problem quiz. Their results were as follows.

Number of students	Fraction of problems correct
1	$\frac{10}{10}$ (all correct)
2	$\frac{9}{10}$
3	$\frac{8}{10}$
5	$\frac{7}{10}$
1	$\frac{6}{10}$
2	$\frac{5}{10}$
1	$\frac{2}{10}$

What was the class average?

60. What is the average of the five highest scores in Exercise 59?

61. Kohough Inc. packs a variety carton of canned seafood. Each carton contains three $3\frac{1}{2}$-oz cans of smoked sturgeon, five $6\frac{3}{4}$-oz cans of tuna, four $5\frac{1}{2}$-oz cans of salmon, and four $10\frac{1}{2}$-oz cans of sardines. How many ounces of seafood are in the carton? If the carton sells for \$52, to the nearest cent what is the average cost per ounce?

62. In a walk for charity, seven people walk $2\frac{7}{8}$ miles, six people walk $3\frac{4}{5}$ miles, nine people walk $4\frac{1}{4}$ miles, and five people walk $5\frac{3}{4}$ miles. What is the total number of miles walked? If the charity raises \$2355, what is the average amount raised per mile, rounded to the nearest dollar?

Exercises 63–64 relate to the chapter application.

63. Sandy and Hannah can fill four Christmas sampler boxes with one recipe each of nutballs, thumbprints, and lemon squares.

a. Nutballs require $1\frac{1}{2}$ c of flour, thumbprints require $2\frac{3}{4}$ c of flour, and lemon squares require 2 c of flour. What is the average amount of flour needed for the three recipes?

b. Nutballs require $\frac{3}{4}$ c of sugar, thumbprints require $1\frac{1}{2}$ c of sugar, and lemon squares require $1\frac{1}{4}$ c of sugar. What is the average amount of sugar needed for the three recipes?

64. Hannah's pecan chocolate chip cookie recipe calls for $2\frac{1}{4}$ c of flour and $1\frac{1}{2}$ c of sugar. Sandy's dipped butterfingers recipe calls for $1\frac{1}{2}$ c of flour and $1\frac{1}{4}$ c of sugar. The women make three pecan chocolate chip recipes and four dipped butterfingers recipes.

a. What is the total amount of flour needed?

b. What is the total amount of sugar needed?

STATE YOUR UNDERSTANDING

65. Write out the order of operations for fractions. How is it different from the order of operations for whole numbers?

66. Must the average of a group of numbers be larger than the smallest number and smaller than the largest number? Why?

CHALLENGE

Perform the indicated operations.

67. $2\frac{5}{8}\left(4\frac{1}{5} - 3\frac{5}{6}\right) \div 2\frac{1}{2}\left(3\frac{1}{7} + 2\frac{1}{5}\right)$

68. $1\frac{2}{5}\left(5\frac{1}{5} - 4\frac{3}{4}\right) \div 4\frac{1}{2}\left(3\frac{1}{7} - 2\frac{1}{3}\right)$

69. The Acme Fish Company pays \$1500 per ton for crab. Jerry catches $3\frac{2}{5}$ tons, his brother Joshua catches $1\frac{1}{2}$ times as many as Jerry. Their sister Salicita catches $\frac{7}{8}$ the amount that Joshua does. What is the total amount paid to the three people by Acme Fish Company to the nearest dollar?

GROUP WORK

70. Prepare for the chapter exam by having each member of the group make up ten exercises, one from each section of the chapter. Exchange these with another member, work each other's exercises, and check all of your answers. Discuss with the whole group the most common errors and how to avoid them.

MAINTAIN YOUR SKILLS

Perform the indicated operations.

71. $5\frac{1}{3} + \frac{7}{9}$

72. $5\frac{1}{3} - \frac{7}{9}$

73. $5\frac{1}{3}\left(\frac{7}{9}\right)$

74. $5\frac{1}{3} \div \frac{7}{9}$

75. $\frac{15}{28} \cdot \frac{21}{45} \cdot \frac{20}{35}$

76. $\frac{9}{15} \cdot \frac{3}{4} \cdot \frac{35}{6}$

77. Find the prime factorization of 650.

78. Find the prime factorization of 975.

79. A coffee table is made of a piece of maple that is $\frac{3}{4}$ in. thick, a piece of chipboard that is $\frac{3}{8}$ in. thick, and a veneer that is $\frac{1}{8}$ in. thick. How thick is the tabletop?

80. Felicia works a 5-day week with the following hours: $6\frac{3}{4}$ hr, $7\frac{1}{3}$ hr, $6\frac{2}{3}$ hr, $9\frac{3}{4}$ hr, and $7\frac{1}{2}$ hr. How many hours does she work for the week? What is her pay if the rate is $\$10\frac{1}{2}$ per hour?

CHAPTER 3

KEY CONCEPTS

SECTION 3.1 PROPER AND IMPROPER FRACTIONS; MIXED NUMBERS

Definitions and Concepts	Examples
A fraction has the form $\frac{\textit{numerator}}{\textit{denominator}}$.	$\frac{4}{81}, \frac{20}{3}, \frac{25}{25}$
A proper fraction has a smaller numerator than denominator.	$\frac{4}{81}$ is a proper fraction.
An improper fraction has a numerator that is not smaller than the denominator.	$\frac{20}{3}$ and $\frac{25}{25}$ are improper fractions.
A mixed number is the sum of a whole number and a fraction.	$4\frac{1}{6}$
To change a mixed number to an improper fraction: Multiply the whole number by the denominator and add the numerator. Place the sum over the denominator.	$4\frac{1}{6} = \frac{4 \cdot 6 + 1}{6} = \frac{25}{6}$
To change an improper fraction to a mixed number: • Divide the numerator by the denominator. • The mixed number is the whole number plus the remainder over the divisor.	$\frac{38}{7} = 5\frac{3}{7}$ because $7\overline{)38}$ = 5, $38 - 35 = 3$

SECTION 3.2 SIMPLIFYING FRACTIONS

Definitions and Concepts	Examples
A fraction is completely simplified when its numerator and denominator have no common factors.	$\frac{6}{7}$ is completely simplified. $\frac{10}{12}$ is not completely simplified because both 10 and 12 have a factor of 2.
To simplify a fraction, eliminate all common factors of the numerator and denominator.	$\frac{36}{72} = \frac{\cancel{9} \cdot 4}{\cancel{9} \cdot 8} = \frac{4}{8} = \frac{\cancel{4} \cdot 1}{\cancel{4} \cdot 2} = \frac{1}{2}$

SECTION 3.3 MULTIPLYING AND DIVIDING FRACTIONS

Definitions and Concepts	Examples
To multiply fractions, simplify if possible and then multiply numerators and multiply denominators.	$\frac{8}{\cancelto{3}{15}} \cdot \frac{\cancelto{1}{5}}{7} = \frac{8}{21}$
Two fractions are reciprocals if their product is 1.	$\frac{3}{4}$ and $\frac{4}{3}$ are reciprocals because $\frac{3}{4} \cdot \frac{4}{3} = 1$.
To divide fractions, multiply the first fraction by the reciprocal of the divisor.	$\frac{4}{9} \div \frac{5}{3} = \frac{4}{\cancelto{3}{9}} \cdot \frac{\cancelto{1}{3}}{5} = \frac{4}{15}$

SECTION 3.4 MULTIPLYING AND DIVIDING MIXED NUMBERS

Definitions and Concepts	Examples
To multiply or divide mixed numbers, change them to improper fractions first. Then multiply or divide.	$3\frac{1}{5}\cdot 2\frac{3}{4} = \frac{\overset{4}{\cancel{16}}}{5}\cdot\frac{11}{\underset{1}{\cancel{4}}} = \frac{44}{5} = 8\frac{4}{5}$

SECTION 3.5 BUILDING FRACTIONS; LISTING IN ORDER; INEQUALITIES

Definitions and Concepts	Examples
Building fractions means writing an equivalent fraction with a different denominator.	$\frac{1}{2} = \frac{15}{30}$
To build a fraction, multiply both its numerator and denominator by the same factor.	$\frac{3}{5} = \frac{3\cdot 6}{5\cdot 6} = \frac{18}{30}$
To list fractions in order: • Rewrite each fraction with a common denominator. • Order the fractions according to their numerators.	List $\frac{3}{5}$, $\frac{7}{10}$, and $\frac{5}{8}$ in order from smallest to largest. $\frac{3}{5} = \frac{24}{40}$, $\frac{7}{10} = \frac{28}{40}$, $\frac{5}{8} = \frac{25}{40}$ $\frac{24}{40} < \frac{25}{40} < \frac{28}{40}$, so $\frac{3}{5} < \frac{5}{8} < \frac{7}{10}$

SECTION 3.6 ADDING FRACTIONS

Definitions and Concepts	Examples
Like fractions have common denominators.	$\frac{3}{54}$ and $\frac{18}{54}$ are like fractions.
Unlike fractions have different denominators.	$\frac{1}{4}$ and $\frac{2}{5}$ are unlike fractions.
To add fractions: • Rewrite with common denominators (if necessary). • Add the numerators and keep the common denominator. • Simplify.	$\frac{1}{2} + \frac{1}{6} = \frac{3}{6} + \frac{1}{6} = \frac{4}{6} = \frac{2}{3}$

SECTION 3.7 ADDING MIXED NUMBERS

Definitions and Concepts	Examples
To add mixed numbers: • Add the whole numbers. • Add the fractions. If this sum is more than 1, change to a mixed number and add again. • Simplify.	$\begin{array}{rcl} 5\frac{1}{2} & = & 5\frac{2}{4} \\ +\,2\frac{3}{4} & = & 2\frac{3}{4} \\ \hline & & 7\frac{5}{4} = 7 + 1\frac{1}{4} = 8\frac{1}{4} \end{array}$

SECTION 3.8 SUBTRACTING FRACTIONS

Definitions and Concepts	Examples
To subtract fractions: • Rewrite with common denominators (if necessary). • Subtract the numerators and keep the common denominator. • Simplify.	$\frac{71}{72} - \frac{31}{90} = \frac{355}{360} - \frac{124}{360}$ $= \frac{231}{360}$ $= \frac{77}{120}$

SECTION 3.9 SUBTRACTING MIXED NUMBERS

Definitions and Concepts	Examples
To subtract mixed numbers: • Subtract the fractions. If the fractions cannot be subtracted, borrow 1 from the whole-number part and add it to the fractional part. Then subtract the fractions. • Subtract the whole numbers. • Simplify.	$17\frac{3}{8} = 17\frac{45}{120} = 16 + 1\frac{45}{120} = 16\frac{165}{120}$ $-12\frac{14}{15} = 12\frac{112}{120} \qquad = 12\frac{112}{120}$ $4\frac{53}{120}$

SECTION 3.10 ORDER OF OPERATIONS; AVERAGE

Definitions and Concepts	Examples
The order of operations for fractions is the same as that for whole numbers: • Parentheses • Exponents • Multiplication / Division • Addition / Subtraction	$\left(\frac{1}{2} + \frac{2}{5}\right) \div \left(\frac{2}{3}\right)^2 = \left(\frac{5}{10} + \frac{4}{10}\right) \div \left(\frac{2}{3}\right)^2$ $= \left(\frac{9}{10}\right) \div \left(\frac{2}{3}\right)^2$ $= \frac{9}{10} \div \left(\frac{4}{9}\right)$ $= \frac{9}{10} \cdot \frac{9}{4}$ $= \frac{81}{40} = 2\frac{1}{40}$
Finding the average of a set of fractions is the same as for whole numbers. • Add the fractions. • Divide by the number of fractions.	Find the average of $\frac{1}{4}$, $\frac{1}{3}$, and $\frac{1}{6}$. $\frac{1}{4} + \frac{1}{3} + \frac{1}{6} = \frac{3}{12} + \frac{4}{12} + \frac{2}{12} = \frac{9}{12} = \frac{3}{4}$ $\frac{3}{4} \div 3 = \frac{3}{4} \cdot \frac{1}{3} = \frac{1}{4}$ The average is $\frac{1}{4}$.

CHAPTER 3

REVIEW EXERCISES

Section 3.1

Write the fraction represented by the figure.

1.

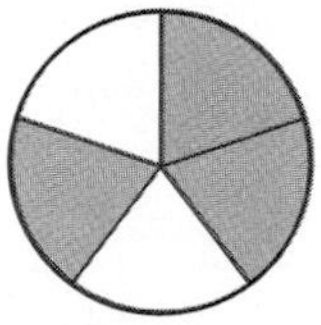

2.

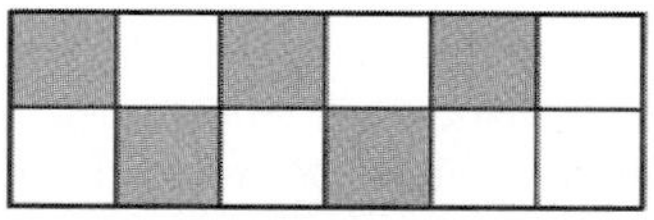

3.

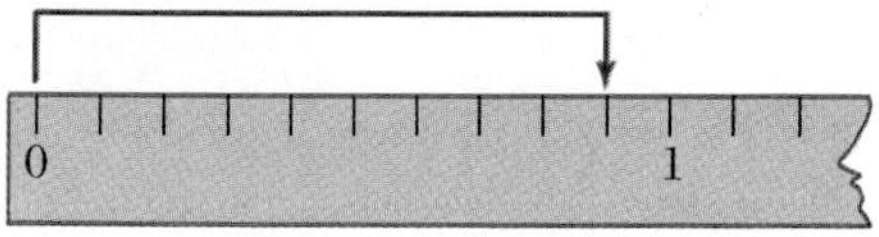

4.

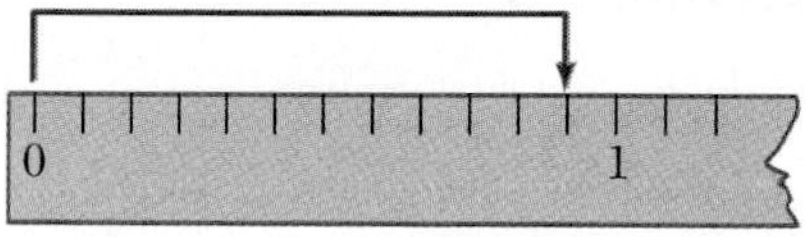

Identify the proper fractions from each list.

5. $\frac{8}{3}, \frac{11}{12}, \frac{9}{9}, \frac{22}{19}, \frac{3}{20}$

6. $\frac{1}{13}, \frac{11}{15}, \frac{8}{10}, \frac{5}{2}, \frac{12}{18}$

Change to a mixed number.

7. $\frac{77}{12}$ **8.** $\frac{77}{9}$ **9.** $\frac{344}{3}$ **10.** $\frac{344}{7}$

Change to an improper fraction.

11. $8\frac{7}{12}$ **12.** $3\frac{2}{9}$ **13.** $12\frac{5}{6}$ **14.** $9\frac{2}{3}$

15. 17 **16.** 21

17. A food wholesaler packs 24 cans of beans in a case for shipping. Write as a mixed number the number of cases that can be packed if she has 64,435 cans of beans.

18. A food wholesaler packs 16 cans of fruit juice in a case for shipping. Write as a mixed number the number of cases that can be packed if he has 22,827 cans of fruit juice.

Section 3.2

Simplify.

19. $\frac{20}{28}$ **20.** $\frac{10}{25}$ **21.** $\frac{40}{70}$ **22.** $\frac{18}{24}$

23. $\frac{21}{35}$ **24.** $\frac{25}{45}$ **25.** $\frac{102}{6}$ **26.** $\frac{132}{33}$

27. $\frac{14}{42}$ **28.** $\frac{30}{45}$ **29.** $\frac{88}{92}$ **30.** $\frac{75}{125}$

31. $\frac{26}{130}$ **32.** $\frac{84}{144}$ **33.** $\frac{268}{402}$ **34.** $\frac{630}{1050}$

35. Cardis completes 12 hours of his weekly part-time job of 30 hours. What fraction of his weekly shift remains?

36. On a math test, a student answers 42 items correctly and 18 incorrectly. What fraction of the items is answered correctly? Simplify.

Section 3.3

Multiply. Simplify completely.

37. $\frac{1}{5} \cdot \frac{2}{5}$ **38.** $\frac{7}{8} \cdot \frac{1}{6}$ **39.** $\frac{3}{5} \cdot \frac{6}{11}$

40. $7 \cdot \frac{1}{4} \cdot \frac{8}{21}$ **41.** $\frac{21}{5} \cdot \frac{5}{4} \cdot \frac{4}{21}$ **42.** $\frac{24}{30} \cdot \frac{3}{8} \cdot \frac{4}{9}$

Find the reciprocal.

43. $\frac{3}{8}$ **44.** 5

Divide. Simplify completely.

45. $\frac{7}{20} \div \frac{14}{15}$ **46.** $\frac{8}{13} \div \frac{2}{13}$ **47.** $\frac{15}{18} \div \frac{30}{27}$

48. $\frac{12}{15} \div \frac{15}{8}$ **49.** $\frac{9}{100} \div \frac{3}{14}$ **50.** $\frac{32}{45} \div \frac{8}{9}$

51. Lois spends half of the family income on rent, utilities, and food. She pays $\frac{2}{7}$ of this amount for rent. What fraction of the family income goes for rent?

52. As part of his job at a pet store, Perry feeds each gerbil $\frac{1}{8}$ cup of seeds each day. If the seeds come in packages of $\frac{5}{4}$ cups, how many gerbils can be fed from one package?

Section 3.4

Multiply. Simplify completely and write as a mixed number, if possible.

53. $\left(\frac{3}{4}\right)\left(1\frac{3}{4}\right)$ **54.** $\left(\frac{3}{7}\right)\left(2\frac{5}{7}\right)$ **55.** $\left(4\frac{1}{2}\right)\left(2\frac{2}{3}\right)$

56. $\left(3\frac{1}{3}\right)\left(1\frac{4}{5}\right)$ **57.** $\left(3\frac{4}{7}\right)\left(\frac{14}{15}\right)$ **58.** $\left(4\frac{4}{9}\right)\left(\frac{12}{25}\right)$

59. $\left(3\frac{2}{3}\right)\left(\frac{15}{22}\right)\left(7\frac{1}{2}\right)$ **60.** $\left(4\frac{3}{4}\right)\left(3\frac{1}{5}\right)\left(5\frac{5}{8}\right)$

Divide. Simplify completely and write as a mixed number, if possible.

61. $3 \div 1\frac{1}{2}$

62. $4 \div 1\frac{1}{4}$

63. $3\frac{3}{8} \div 3$

64. $2\frac{1}{6} \div 4$

65. $8\frac{2}{5} \div 2\frac{1}{3}$

66. $2\frac{1}{2} \div 1\frac{1}{5}$

67. $31\frac{1}{3} \div 1\frac{1}{9}$

68. $21\frac{3}{7} \div 8\frac{1}{3}$

69. A wheat farmer in Iowa averages $55\frac{3}{4}$ bushels of wheat per acre on 150 acres of wheat. How many bushels of wheat does she harvest?

70. A wildlife survey in a water fowl preserve finds that there are $3\frac{1}{3}$ times as many brant geese in the preserve as there are Canada geese. If the survey counts 7740 Canada geese, how many brant geese are there?

Section 3.5

Write four fractions equivalent to each of the given fractions by multiplying by $\frac{2}{2}, \frac{3}{3}, \frac{5}{5},$ *and* $\frac{8}{8}$.

71. $\frac{2}{3}$

72. $\frac{3}{5}$

73. $\frac{3}{14}$

74. $\frac{4}{11}$

Find the missing numerator.

75. $\frac{3}{4} = \frac{?}{24}$

76. $\frac{6}{7} = \frac{?}{56}$

77. $\frac{5}{6} = \frac{?}{144}$

78. $\frac{5}{8} = \frac{?}{224}$

List the fractions from smallest to largest.

79. $\frac{1}{2}, \frac{3}{5}, \frac{7}{10}$

80. $\frac{1}{3}, \frac{5}{12}, \frac{2}{5}$

81. $\frac{2}{9}, \frac{1}{5}, \frac{3}{11}$

82. $\frac{10}{9}, \frac{4}{3}, \frac{7}{6}, \frac{19}{18}$

83. $\frac{3}{5}, \frac{8}{25}, \frac{31}{50}, \frac{59}{100}$

84. $7\frac{3}{4}, 7\frac{7}{8}, 7\frac{5}{6}$

Are the following statements true or false?

85. $\frac{3}{14} < \frac{5}{14}$

86. $\frac{11}{8} > \frac{9}{8}$

87. $\frac{11}{14} > \frac{17}{21}$

88. $\frac{13}{15} < \frac{22}{25}$

89. Four pickup trucks are advertised in the local car ads. The load capacities listed are $\frac{3}{4}$ ton, $\frac{5}{8}$ ton, $\frac{7}{16}$ ton, and $\frac{1}{2}$ ton. Which capacity is the smallest and which is the largest?

90. During 1 week on her diet, Samantha ate five servings of chicken, each containing $\frac{3}{16}$ oz of fat. During the same period, her brother ate four servings of beef, each containing $\frac{6}{25}$ oz of fat. Who ate the greatest amount of fat from these entrees?

Section 3.6

Add. Simplify completely.

91. $\frac{5}{11} + \frac{2}{11}$

92. $\frac{7}{12} + \frac{4}{12}$

93. $\frac{2}{9} + \frac{2}{9} + \frac{2}{9}$

94. $\frac{3}{16} + \frac{2}{16} + \frac{3}{16}$

95. $\frac{7}{32} + \frac{8}{32} + \frac{5}{32}$

96. $\frac{5}{24} + \frac{7}{24} + \frac{9}{24}$

97. $\frac{4}{15} + \frac{1}{3}$

98. $\frac{7}{24} + \frac{3}{8}$

99. $\frac{3}{35} + \frac{8}{21}$

100. $\frac{11}{30} + \frac{9}{20} + \frac{3}{10}$

101. $\frac{1}{6} + \frac{7}{8} + \frac{7}{12}$

102. $\frac{7}{15} + \frac{11}{30} + \frac{5}{6}$

103. An elephant ear bamboo grew $\frac{1}{2}$ in. on Tuesday, $\frac{3}{8}$ in. on Wednesday, and $\frac{1}{4}$ in. on Thursday. How much did the bamboo grow in the three days?

104. In order to complete a project, Preston needs $\frac{1}{10}$ in. of foam, $\frac{3}{10}$ in. of metal, $\frac{4}{10}$ in. of wood, and $\frac{7}{10}$ in. of plexiglass. What will be the total thickness of this project when these materials are piled up?

Section 3.7

Add. Write the results as a mixed number where possible.

105. $2\frac{4}{7} + 3\frac{11}{14}$

106. $17\frac{5}{12} + 1\frac{5}{6}$

107. $2\frac{7}{15} + 13\frac{2}{5}$

108. $4\frac{7}{15} + 6\frac{2}{3}$

109. $7\frac{3}{8} + 5\frac{5}{6}$

110. $15\frac{4}{15} + 5\frac{5}{6}$

111. $14\frac{7}{20} + 11\frac{3}{16}$

112. $11\frac{7}{24} + 32\frac{7}{18}$

113. $28\frac{1}{4} + 39\frac{1}{3} + 12\frac{5}{12}$

114. $18\frac{3}{4} + 19 + 25\frac{7}{12}$

115. $25\frac{2}{3} + 16\frac{1}{6} + 18\frac{3}{4}$

116. $29\frac{7}{8} + 19\frac{5}{12} + 32\frac{3}{4}$

117. Russ rode his bicycle $3\frac{3}{8}$ miles on Monday, $1\frac{2}{3}$ miles on Tuesday, $1\frac{7}{8}$ miles on Wednesday, $\frac{1}{2}$ mile on Thursday, and $4\frac{1}{8}$ miles on Friday. What was his total mileage for the week?

118. On a fishing excursion, Roona caught four fish weighing $6\frac{3}{4}$ lb, $1\frac{3}{5}$ lb, $2\frac{2}{3}$ lb, and $5\frac{1}{2}$ lb. What was the total weight of her catch?

Section 3.8

Subtract.

119. $\frac{5}{7} - \frac{3}{14}$

120. $\frac{5}{18} - \frac{2}{9}$

121. $\frac{5}{6} - \frac{1}{3}$

122. $\frac{17}{20} - \frac{1}{5}$

123. $\frac{19}{30} - \frac{1}{5}$

124. $\frac{7}{15} - \frac{3}{20}$

125. $\frac{5}{6} - \frac{4}{5}$

126. $\frac{7}{10} - \frac{1}{4}$

127. $\frac{18}{25} - \frac{7}{15}$

128. $\frac{21}{32} - \frac{5}{16}$

129. $\frac{47}{60} - \frac{17}{24}$

130. $\frac{13}{15} - \frac{9}{20}$

131. Wanda finds $\frac{3}{4}$ oz of gold during a day of panning along the Snake River. She gives a $\frac{1}{3}$-oz nugget to Jose, her guide. What fraction of an ounce of gold does she have left?

132. A carpenter planes the thickness of a board from $\frac{13}{16}$ in. to $\frac{5}{8}$ in. How much is removed?

Section 3.9

Subtract. Write the results as mixed numbers where possible.

133. $145\frac{2}{3} - 27\frac{1}{2}$

134. $6\frac{5}{6} - 3\frac{3}{10}$

135. $26\frac{1}{10} - 10\frac{9}{10}$

136. $19\frac{3}{8} - 8\frac{5}{8}$

137. $7\frac{1}{4} - 5\frac{5}{8}$

138. $76\frac{7}{15} - 50\frac{1}{12}$

139. $9\frac{9}{16} - 3\frac{5}{12}$

140. $30\frac{7}{16} - 22\frac{5}{6}$

141. $9\frac{6}{16} - 2\frac{5}{6}$

142. $33\frac{17}{30} - 25\frac{7}{9}$

143. $5\frac{31}{32} - \frac{3}{16}$

144. $8\frac{11}{20} - \frac{9}{10}$

145. The graph displays the average yearly rainfall for five cities.

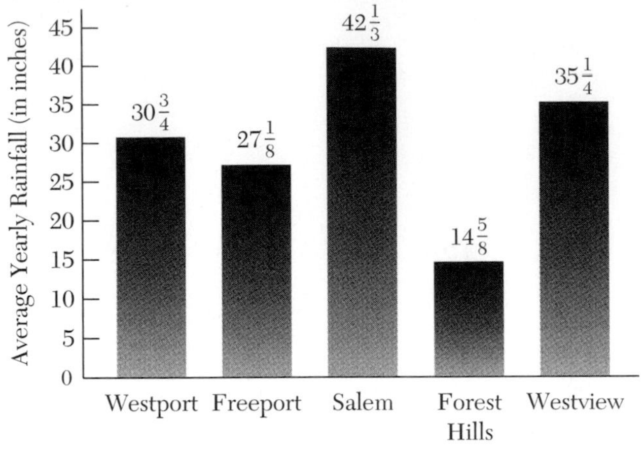

a. How much more rain falls in Westport during a year than in Freeport?

b. In a 10-yr period, how much more rain falls in Salem than in Forest Hills?

146. Use the graph in Exercise 145. If the average rainfall in Westview doubles, how much more rain would it receive than Salem?

Section 3.10

Perform the indicated operations.

147. $\frac{1}{4} + \frac{3}{8} \div \frac{1}{2}$

148. $\frac{1}{4} \div \frac{3}{8} + \frac{1}{2}$

149. $\frac{5}{8} \div \frac{3}{4} + \frac{3}{4}$

150. $\frac{1}{3} \div \frac{1}{6} + \frac{4}{9}$

151. $\frac{3}{4} - \left(\frac{1}{2}\right)^2$

152. $\frac{3}{4} - \left(\frac{1}{4}\right)^2$

153. $\left(\frac{9}{8}\right)^2 - \left(\frac{1}{2} \div \frac{4}{5} - \frac{3}{8}\right)$

154. $\left(\frac{1}{2}\right)^2 + \left(\frac{4}{5} \cdot \frac{5}{8} + \frac{2}{3}\right)$

Find the average.

155. $\frac{3}{8}, \frac{1}{4}, \frac{1}{2},$ and $\frac{3}{4}$

156. $\frac{3}{8}, \frac{3}{4}, \frac{1}{6},$ and $\frac{5}{8}$

157. $1\frac{2}{3}, 1\frac{5}{12}, 2\frac{1}{2}, \frac{3}{4},$ and $\frac{5}{6}$

158. $\frac{2}{3}, \frac{5}{12}, \frac{1}{2}, 1\frac{3}{4},$ and $1\frac{5}{6}$

159. A class of 20 students took a 10-problem quiz. Their results were as follows:

Number of Students	Fraction of Problems Correct
3	$\frac{10}{10}$ (all correct)
1	$\frac{9}{10}$
3	$\frac{8}{10}$
5	$\frac{7}{10}$
4	$\frac{6}{10}$
3	$\frac{5}{10}$
1	$\frac{3}{10}$

What was the class average?

160. What is the average of the top six scores in Exercise 159?

CHAPTER 3

TRUE / FALSE CONCEPT REVIEW

ANSWERS

Check your understanding of the language of basic mathematics. Tell whether each of the following statements is true (always true) or false (not always true). For each statement you judge to be false, revise it to make a statement that is true.

1. It is not possible to picture an improper fraction using unit regions. **1.** ________

2. The fraction $\frac{7}{8}$ written as a mixed number is $0\frac{7}{8}$. **2.** ________

3. The whole number 1 can also be written as a proper fraction. **3.** ________

4. A fraction is another way of writing a division problem. **4.** ________

5. When a fraction is completely simplified, its value remains the same. **5.** ________

6. Every improper fraction can be simplified. **6.** ________

7. Some fractions with large numerators and denominators cannot be further simplified. **7.** ________

8. Two mixed numbers can be subtracted without first changing them to improper fractions. **8.** ________

9. The reciprocal of a proper fraction is greater than 1. **9.** ________

10. The quotient of two nonzero fractions can always be found by multiplication. **10.** ________

11. Simplifying fractions is the opposite of building fractions. **11.** ________

12. The primary reason for building fractions is so they will have a common denominator. **12.** ________

13. Unlike fractions have different numerators. **13.** ________

14. Mixed numbers must be changed to improper fractions before they are added. **14.** ________

15. It is sometimes necessary to use borrowing to subtract mixed numbers, just as we do when subtracting some whole numbers. **15.** ________

16. The order of operations for fractions is the same as the order of operations for whole numbers. **16.** ________

17. The average of three nonequivalent fractions is larger than at least one of the fractions. **17.** ________

18. The product of two fractions is always larger than the two fractions. **18.** ________

TRUE/FALSE CONCEPT REVIEW

CHAPTER 3

TEST

ANSWERS

1. Change $\frac{61}{4}$ to a mixed number. **1.** ________

2. Add: $\frac{7}{12} + \frac{5}{8}$ **2.** ________

3. Change $8\frac{7}{9}$ to an improper fraction **3.** ________

4. List these fractions from the smallest to the largest: $\frac{2}{5}, \frac{3}{8}, \frac{3}{7}$ **4.** ________

5. Change 17 to an improper fraction. **5.** ________

6. Find the missing numerator: $\frac{5}{8} = \frac{?}{56}$ **6.** ________

7. Add:

$$\begin{array}{r} 5\frac{3}{10} \\ +\ 3\frac{5}{6} \\ \hline \end{array}$$

7. ________

8. Multiply: $\left(4\frac{5}{8}\right)\left(3\frac{4}{5}\right)$. Write the result as a mixed number. **8.** ________

9. Perform the indicated operations: $\frac{3}{8} \div \frac{3}{4} + \frac{1}{4}$ **9.** ________

10. Simplify $\frac{68}{102}$ completely. **10.** ________

11. Subtract: $17\frac{4}{5} - 11$ **11.** ________

12. Multiply: $\frac{3}{4} \cdot \frac{8}{9} \cdot \frac{15}{16}$ **12.** ________

13. Subtract: $\frac{4}{5} - \frac{3}{10}$ **13.** ________

14. Divide: $1\frac{2}{9} \div 3\frac{2}{3}$ **14.** ________

15. Multiply: $\frac{3}{7} \cdot \frac{4}{5}$ **15.** ________

16. Subtract:

$$\begin{array}{r} 11\frac{7}{12} \\ -\ 4\frac{14}{15} \\ \hline \end{array}$$

16. ________

17. Simplify $\frac{132}{352}$ completely. **17.** ________

18. Add: $\frac{2}{35} + \frac{5}{14} + \frac{3}{10}$ **18.** ________

19. What is the reciprocal of $2\frac{5}{8}$? **19.** ____________

20. What is the reciprocal of $\frac{5}{16}$? **20.** ____________

21. Which of these fractions are proper: $\frac{7}{8}, \frac{8}{8}, \frac{9}{8}, \frac{7}{9}, \frac{9}{7}, \frac{8}{9}, \frac{9}{9}$? **21.** ____________

22. Divide: $\frac{10}{3} \div \frac{8}{9}$ **22.** ____________

23. Subtract:

$$\begin{array}{r} 11\frac{7}{10} \\ -\ 3\frac{3}{8} \\ \hline \end{array}$$

23. ____________

24. Write the fraction for the shaded part of this figure. **24.** ____________

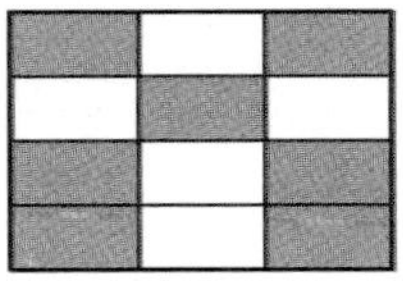

25. Subtract: $13 - 7\frac{7}{12}$ **25.** ____________

26. Add: $\frac{5}{14} + \frac{5}{14}$ **26.** ____________

27. Find the average of $1\frac{3}{8}, \frac{1}{4}, 3\frac{1}{4}$, and $2\frac{3}{8}$. **27.** ____________

28. Multiply: $\left(\frac{8}{25}\right)\left(\frac{9}{16}\right)$ **28.** ____________

29. True or false: $\frac{5}{7} < \frac{11}{16}$? **29.** ____________

30. Which of the fractions represent the number 1: $\frac{6}{5}, \frac{5}{5}, \frac{7}{6}, \frac{6}{6}, \frac{7}{7}, \frac{6}{7}, \frac{5}{7}$? **30.** ____________

31. A rail car contains $126\frac{1}{2}$ tons of baled hay. A truck that is being used to unload the hay can haul $5\frac{3}{4}$ tons in one load. How many truckloads of hay are in the rail car? **31.** ____________

32. Jill wants to make up 20 bags of homemade candy for the local bazaar. Each bag will contain $1\frac{1}{4}$ pounds of candy. How many pounds of candy must she make? **32.** ____________

CHAPTER 3

GROUP PROJECT (2 weeks)

OPTIONAL

One of the major applications of statistics is predicting future occurrences. Before the future can be predicted, statisticians study what has happened in the past and look for patterns. If a pattern can be detected and it is reasonable to assume that nothing will happen to interrupt the pattern, then it is a relatively easy matter to predict the future simply by continuing the pattern. Insurance companies, for instance, study the occurrences of traffic accidents among various groups of people. Once they have identified a pattern, they use this to predict future accident rates, which in turn are used to set insurance rates. When a group, such as teenaged boys, is identified as having a higher incidence of accidents, the insurance rates for the group are set higher.

Dice

Although predicting accident rates is a very complicated endeavor, the patterns of other activities are relatively easy to find. Take, for instance, the act of rolling a die. The die has six sides, marked 1 to 6. Theoretically, each side has an equal chance of ending in the up position after a roll. Fill in the following table by rolling a die 120 times.

Side up	1	2	3	4	5	6
Times rolled						

Theoretically, each side will be rolled the same number of times as the others. Because you rolled the die 120 times and there were six possible outcomes, each side should have come up $120 \div 6 = 20$ times. How close to 20 are your outcomes in the table? What do you suppose are reasons for not getting a perfectly distributed table?

Mathematicians are likely to express the relationships in this situation using the concept of *probability*, which is a measure of the likelihood of a particular event occurring. We describe the probability of an event with a fraction. The numerator of the fraction is the number of different ways the desired event can occur, and the denominator of the fraction is the total number of possible outcomes. So the probability of rolling a 2 on the die is $\frac{1}{6}$ because there is only one way to roll a 2, but there are 6 possible outcomes when rolling a die. What is the probability of rolling a 5? What is the probability of rolling a 6? Nonmathematicians are more likely to express this relationship using the concept of *odds*. They would say that the odds of rolling a 2 are 1 in 6. This means that for every six times you roll a die, you can expect one of them to result in a 2.

Coin Toss

Suppose you and a friend each flip a coin. What are all the possible joint outcomes? What is the probability of getting two heads? What is the probability of getting two tails? What is the probability of getting one head and one tail? What does it mean if the probability of an event is $\frac{3}{3}$? Is it possible for the probability of an event to be $\frac{5}{4}$? Explain.

Cards

Suppose you pick a card at random out of a deck of playing cards. What is the probability that the card will be the queen of hearts? What is the probability that the card will be a queen? What is the probability that the card will be a heart? Fill out the following table and try to discover the relationship among these three probabilities.

Probability of a Queen	Probability of a Heart	Probability of the Queen of Hearts

For a card to be the queen of hearts, two conditions must hold true at the same time. The card must be a queen *and* the card must be a heart. Make a guess about the relationship of the probabilities when two conditions must occur simultaneously. Test your guess by considering the probability of drawing a black 7. What are the two conditions that must be true in order for the card to be a black 7? What are their individual probabilities? Was your guess correct?

What two conditions must be true when you draw a red face card? What is the probability of drawing a red face card?

Suppose you pick a card at random out of a deck of playing cards. What is the probability that the card will be a 3 or a 4? What is the probability that the card will be a 3? A 4? Fill in the following table to try to discover the relationship among these probabilities.

Probability of a 3	Probability of a 4	Probability of a 3 or a 4

A card is a 3 *or* a 4 if either condition holds. Make a guess about the relationship of the probabilities when either of two conditions must be true. Test your guess by calculating the probability that a card will be a heart or a club. Was your guess correct?

Sometimes a complicated probability is easier to calculate using a backdoor approach. For instance, suppose you needed to calculate the probability that a card drawn is an ace or a 2 or a 3 or a 4 or a 5 or a 6 or a 7 or an 8 or a 9 or a 10 or a jack or a queen. You can certainly add the individual probabilities (what do you get?). However, another way to look at the situation is to ask what is the probability of not getting a king. We reason that if you do not get a king, then you do get one of the desired cards. We calculate this by subtracting from 1 the probability of getting a king. This is because 1 must be the sum of all the probabilities that totally define the set (in this case, the sum of the probabilities of getting a king and the probability of getting one of the other cards). Verify that you get the same probability using both methods.

CHAPTERS 1–3

CUMULATIVE REVIEW

Write the word name for each of the following.

1. 6091

2. 110,532

Write the place value name.

3. One million, three hundred ten

4. Sixty thousand, two hundred fifty-seven

Round to the indicated place value.

5. 654,785 (hundred)

6. 43,949 (ten thousand)

Insert > or < between the numbers to make a true statement.

7. 6745   6739

8. 11,899   11,901

Add or subtract.

9.
$$\begin{array}{r} 76{,}843 \\ 34{,}812 \\ 12{,}833 \\ +\ \ 9{,}711 \\ \hline \end{array}$$

10.
$$\begin{array}{r} 55{,}304 \\ -\ 37{,}478 \\ \hline \end{array}$$

11. $54 + 87 + 124 + 784 + 490 + 54$

12. $70{,}016 - 54{,}942$

13. Find the perimeter of the rectangle.

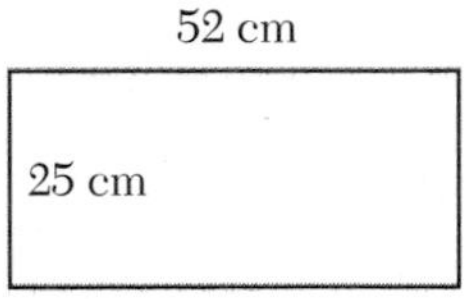

Multiply.

14.
$$\begin{array}{r} 14{,}654 \\ \times\ \ \ \ 251 \\ \hline \end{array}$$

15. (341)(73)

16. Find the area.

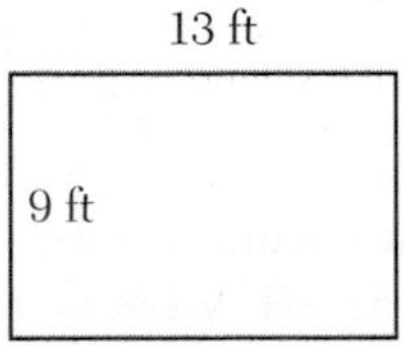

Divide.

17. $76\overline{)19{,}532}$

18. $16{,}702 \div 95$

19. Write in exponential form: (7)(7)(7)(7)(7)(7)(7)(7)

Multiply or divide.

20. $54{,}000{,}000 \div 10^4$

21. 321×10^3

Simplify.

22. $8 \cdot 2^3 - 10 \div 2 + 6 \cdot 10 \div 12$

23. $80 \div 16 \cdot 3^2 - (5 \cdot 4 - 7)$

Find the average and median.

24. 345, 672, 801, 943, 144

25. 10,504, 13,654, 92,230, 65,236

Exercises 26–28 refer to the table.

Golf Scores

	Round 1	Round 2	Round 3	Round 4
Tiger Woods	66	70	68	66
David Duval	70	65	73	69
Phil Mickelson	67	68	70	64

26. Find the average, median, and mode of all the scores.

27. Who scored the lowest score for a single round of golf ? What was his score?

28. Who had the best score for the 72 holes (all four rounds)?

Exercises 29–30. The graph gives the weekly sales at the local convenience mart.

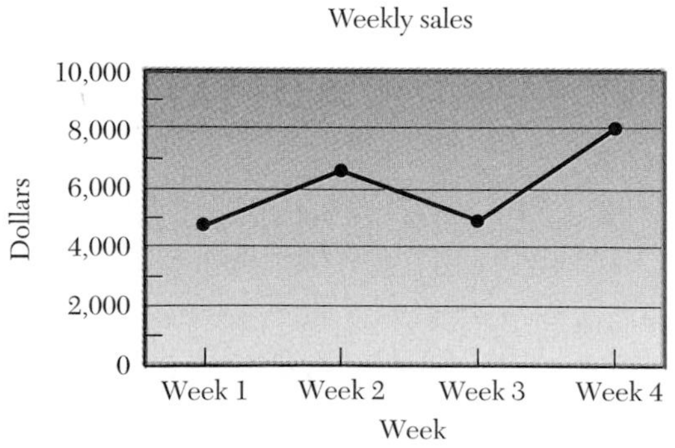

29. Which week had the highest weekly sales?

30. What is the approximate total sales for weeks 1 and 2?

31. Holli decided to redecorate her living room. She bought a new painting for $235, a mirror for $410, a love seat for $895, and new carpeting for $1864. What is the total amount she spent on the redecorating?

32. Russ left an estate valued at \$568,950. He designated that \$150,000 would go to the local community college foundation for scholarships. The remaining money was to be divided equally among his seven nieces and nephews. How much money did each niece and nephew receive?

33. For four performances of *Guys and Dolls, A Semi-Staged Musical*, the Portland Symphony sold 2143 tickets at \$60 per ticket, 1867 tickets at \$47 per ticket, 976 tickets at \$36 per ticket, and 561 tickets at \$19 per ticket. What was the total revenue generated by the ticket sales?

Determine whether the natural number is divisible by 2, 3, or 5.

34. 2850 **35.** 4436 **36.** 4515 **37.** 35,742

List the first five multiples of the whole number.

38. 34 **39.** 140

40. Is 3584 a multiple of 7? **41.** Is 1340 a multiple of 8?

Write the whole number as a product of two factors in all possible ways.

42. 224 **43.** 495

List all the factors (divisors) of the whole number.

44. 120 **45.** 165

Find the greatest common factor (GCF) of the two whole numbers.

46. 84 and 48 **47.** 330 and 465

Tell whether the number is prime or composite.

48. 45,873 **49.** 167

Write the prime factorization of the whole number.

50. 420 **51.** 11,850

52. Find the greatest common factor (GCF) of 420 and 560.

Find the least common multiple of the group of whole numbers.

53. 7, 21, 30 **54.** 24, 28, 35, 45

55. Write the fraction represented by the figure.

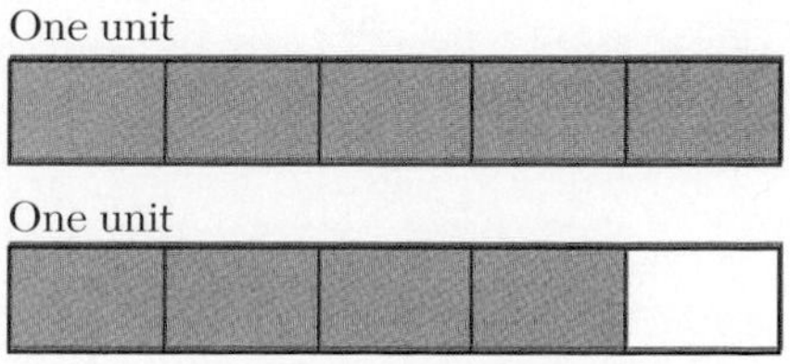

56. Select the proper and improper fractions from the list. $\frac{6}{5}, \frac{8}{9}, \frac{12}{11}, \frac{18}{18}, \frac{19}{20}, \frac{23}{24}, \frac{31}{30}$

Change to a mixed number.

57. $\frac{27}{8}$

58. $\frac{165}{31}$

Change to an improper fraction.

59. $7\frac{1}{6}$

60. $17\frac{9}{11}$

Simplify completely.

61. $\frac{66}{102}$

62. $\frac{216}{360}$

Multiply or divide. Simplify completely.

63. $\frac{12}{35} \cdot \frac{49}{28}$

64. $\frac{40}{57} \cdot \frac{38}{48}$

65. $\frac{15}{16} \div \frac{18}{20}$

66. $\frac{25}{32} \div \frac{75}{40}$

Multiply or divide. Simplify completely and write as a mixed number, if possible.

67. $\left(4\frac{8}{9}\right)\left(3\frac{7}{8}\right)$

68. $\left(11\frac{5}{7}\right)\left(9\frac{5}{8}\right)$

69. $8\frac{3}{5} \div 5\frac{3}{5}$

70. $13\frac{1}{8} \div 15\frac{2}{5}$

71. Write four fractions equivalent to $\frac{6}{7}$ by multiplying by $\frac{2}{2}, \frac{3}{3}, \frac{4}{4}$, and $\frac{5}{5}$.

Find the missing numerator.

72. $\frac{4}{7} = \frac{?}{35}$

73. $\frac{9}{22} = \frac{?}{66}$

74. List the fractions from smallest to largest: $\frac{3}{5}, \frac{6}{25}, \frac{11}{15}, \frac{1}{6}$

75. True or false: $\frac{21}{40} < \frac{18}{35}$

76. True or false: $\frac{9}{25} > \frac{13}{40}$

Add or subtract. Simplify completely and write as a mixed number, if possible.

77. $\frac{4}{11} + \frac{1}{11} + \frac{5}{11}$

78. $\frac{8}{9} + \frac{5}{12} + \frac{2}{3}$

79. $\frac{17}{35} - \frac{8}{21}$

80. $5\frac{2}{3} + 1\frac{5}{6} + 3\frac{3}{4}$

81.
$$\begin{array}{r} 19\frac{4}{15} \\ -\ 14\frac{7}{12} \\ \hline \end{array}$$

82. $\frac{7}{10} + \frac{8}{15} + \frac{9}{20} + \frac{5}{6}$

Perform the indicated operations.

83. $\frac{7}{8} - \frac{5}{6} \cdot \frac{8}{15}$

84. $\frac{3}{5} + \left(\frac{2}{3}\right)^2 - \frac{4}{9} \div \frac{8}{3}$

85. Find the average and median of $\frac{1}{6}, \frac{7}{18}, \frac{7}{12}, \frac{5}{18}, \frac{3}{4}$, and $\frac{8}{9}$.

86. Find the perimeter of the rectangle.

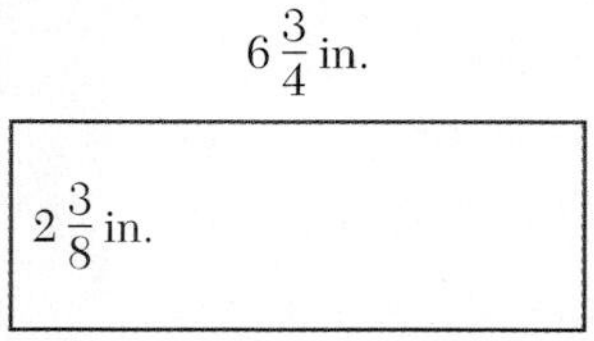

87. Find the area of the rectangle in Exercise 86.

88. Ersula is making a batch of cookies. The recipe calls for $2\frac{2}{3}$ cups of flour. If Ersula doubles the recipe, how many cups of flour will she need?

89. The Perez family spends $\frac{3}{20}$ of their income on rent, $\frac{1}{6}$ on food, $\frac{1}{15}$ on transportation, and $\frac{1}{5}$ on taxes. What fraction of their income is spent on these items?

90. During the last 6 months, the following amounts of rain were recorded: $2\frac{4}{5}$ in., $1\frac{1}{2}$ in., $3\frac{1}{3}$ in., $5\frac{2}{5}$ in., $2\frac{3}{5}$ in., and $4\frac{1}{2}$ in. Find the average monthly rainfall during these 6 months.

Planning Makes Perfect

Now is the time to formalize a study plan. Set aside a time of day, every day, to focus all of your attention on math. For some students, finding a quiet place in the library to study regularly for 1 hour is far more efficient than studying 2 hours at home, where there are constant distractions. For others, forming a study group where you can talk about what you have learned is helpful. Decide which works best for you.

Try to schedule time as close to the class session as possible while the concepts are fresh in your mind. If you wait several hours to practice what seemed clear during class, you may find that what was clear earlier may no longer be meaningful. This may mean planning a schedule of classes that includes an hour after class to study.

If, on some days, you cannot devote 1 or 2 hours to math, find at least a few minutes and review one thing—perhaps read your notes, reread the section objectives, or if you want to do a few problems, do the section Warm-Ups. This helps to keep the concepts fresh in your memory.

If you choose to form or join a study group with other math-anxious students, don't use your time together to gripe. Instead, use it to discuss and recognize the content of your negative self-talk and to write positive coping statements.

Plan, too, for your physical health. Notice how your anxious thought patterns trigger physical tension. When you wrinkle your forehead, squint your eyes, make a trip to the coffee machine, or light up a cigarette, you are looking for a way to release these tensions. Learning relaxation techniques, specifically progressive relaxation, is a healthier alternative to controlling body tension. Briefly, relaxation training involves alternately tensing and relaxing all of the major muscles in the body with the goal of locating your specific muscle tension and being able to relax it away. Use a professionally prepared progressive relaxation tape, or take a stress management class to properly learn this technique. Allow at least 20 to 30 minutes for this exercise daily. The time it takes for you to deeply relax will become briefer as you become more skilled. Soon relaxation will be as automatic as breathing, and when you find yourself feeling math anxious, you can stop, take control, and relax.

4 DECIMALS

SECTIONS

APPLICATION

Sports hold a universal attraction. People all over the world enjoy a good game. For some sports, it is relatively easy to determine which athlete is the best. In track, downhill skiing, and swimming, for instance, each contestant races against the clock and the fastest time wins. In team sports, it is easy to tell which team wins but sometimes difficult to determine how the individual athletes compare with one another. In order to make comparisons more objective, we often use sports statistics.

The simplest kind of statistic is to count how many times an athlete performs a particular feat in a single game. In basketball, for instance, it is usual to count the number of points scored, the number of rebounds made, and the number of assists for each player.

Consider the following statistics for members of the Los Angeles Lakers in their win of the 2000–2001 NBA championship against the Philadelphia 76ers.

Player	Points Scored	Rebounds	Assists
O'Neal	165	79	24
Bryant	123	39	29
Fisher	49	6	10
Fox	49	23	19
Horry	42	25	6

1. Which player had the best overall statistics? Justify your answer.

2. Which is more important in basketball, rebounds or assists? Explain.

3. Among Fisher, Fox, and Horry, who had the best overall performance? Explain.

4.1 Decimals: Reading, Writing, and Rounding

OBJECTIVES

1. Write word names from place value names and place value names from word names.
2. Round a given decimal.

VOCABULARY

Decimal numbers, more commonly referred to as **decimals**, are another way of writing fractions and mixed numbers. The digits used to write whole numbers and a period called a **decimal point** are used to write place value names for these numbers.

The **number of decimal places** is the number of digits to the right of the decimal point. **Exact decimals** are decimals that show exact values. **Approximate decimals** are rounded values.

HOW & WHY

OBJECTIVE 1 Write word names from place value names and place value names from word names.

Decimals are written by using a standard place value in the same way we write whole numbers. Numbers such as 12.65, 0.45, 0.795, 1306.94 and 19.36956 are examples of decimals.

In general, the place value for decimals is

1. The same as whole numbers for digits to the left of the decimal point.
2. A fraction whose denominator is 10, 100, 1000, and so on, for digits to the right of the decimal point.

The digits to the right of the decimal point have place values of

$$\frac{1}{10^1} = \frac{1}{10} = 0.1$$

$$\frac{1}{10^2} = \frac{1}{10 \cdot 10} = \frac{1}{100} = 0.01$$

$$\frac{1}{10^3} = \frac{1}{10 \cdot 10 \cdot 10} = \frac{1}{1000} = 0.001$$

$$\frac{1}{10^4} = \frac{1}{10 \cdot 10 \cdot 10 \cdot 10} = \frac{1}{10{,}000} = 0.0001$$

and so on, in that order from left to right.

Using the ones place as the central position, the place values of a decimal are shown in Figure 4.1.

	Whole-number part					Decimal point	Fractional part					
. . .	Ten thousands	Thousands	Hundreds	Tens	Ones	.	Tenths	Hundredths	Thousandths	Ten-thousandths	Hundred-thousandths	. . .
. . .	10,000	1000	100	10	1		$\frac{1}{10}$	$\frac{1}{100}$	$\frac{1}{1000}$	$\frac{1}{10{,}000}$	$\frac{1}{100{,}000}$	. . .
. . .	10^4	10^3	10^2	10^1	1		$\frac{1}{10^1}$	$\frac{1}{10^2}$	$\frac{1}{10^3}$	$\frac{1}{10^4}$	$\frac{1}{10^5}$	. . .

FIGURE 4.1

Note that the decimal point separates the whole-number part from the fractional part.

$$82.459 = \underbrace{82}_{\text{Whole-number part}} + \underbrace{.459}_{\text{Fractional part}}$$

If the decimal point is not written, as in the case of a whole number, the decimal point is understood to follow the ones place—thus,

$71 = 71. \qquad 7 = 7. \qquad 345 = 345.$

We can write an expanded form of the decimal using fractions with a denominator that is a power of 10. So

$$73.183 = 70 + 3 + \frac{1}{10} + \frac{8}{100} + \frac{3}{1000}$$

The expanded form can also be written:

7 tens + 3 ones + 1 tenth + 8 hundredths + 3 thousandths.

Table 4.1 shows how to write the word names for 448.26 and 0.639.

TABLE 4.1 WORD NAMES FOR DECIMALS

	Number to Left of Decimal Point	Decimal Point	Number to Right of Decimal Point	Place Value of Last Digit
Place value name	448	.	26	$\frac{1}{100}$
Word name of each	Four hundred forty-eight	and	twenty-six	hundredths
Word name of decimal	Four hundred forty-eight and twenty-six hundredths			
Place value name	0	.	639	$\frac{1}{1000}$
Word name of each	Omit	Omit	Six hundred thirty-nine	thousandths
Word name of decimal	Six hundred thirty-nine thousandths			

For numbers greater than zero and less than one (such as 0.639), the digit 0 is written in the ones place.

To write the word name for a decimal

1. Write the name for the whole number to the left of the decimal point.
2. Write the word "and" for the decimal point.
3. Write the whole number name for the number to the right of the decimal point.
4. Write the place value of the digit farthest to the right.

If the decimal has only zero or no digit to the left of the decimal point, omit steps 1 and 2.

Table 4.2 shows some other numbers and their corresponding word names.

TABLE 4.2 WORD NAMES FOR DECIMALS

Number	Word Name
12.75	Twelve and seventy-five hundredths
0.543	Five hundred forty-three thousandths
0.00023	Twenty-three hundred-thousandths
9.001	Nine and one thousandth

To write the place value name for a decimal

1. Write the whole number (the number before the word "and").
2. Write a decimal point for the word and."
3. Ignoring the place value name, write the name for the number following the word "and." Insert zeros, if necessary, between the decimal point and the digits following it to ensure that the place on the far right has the correct (given) place value.

So the place value notation for five hundred eleven and thirty-two thousandths is

511	First write the whole number to the left of the word "and."
511.	Write a decimal point for the word "and."
511.032	The "whole number" after the word "and" is 32, and a zero is inserted to place the 2 in the thousandths place.

WARM-UPS A–D

A. Write the word name for 0.67.

B. Write the word name for 0.0053.

C. Write the word name for 45.091.

Answers to Warm-Ups

A. Sixty-seven hundredths
B. Fifty-three ten-thousandths
C. Forty-five and ninety-one thousandths.

EXAMPLES A–D

DIRECTIONS: Write the word name.

STRATEGY: Write the word name for the whole number to the left of the decimal point. Then write *and* for the decimal point. Finally, write the word name for the number to the right of the decimal point followed by the place value of the digit farthest to the right.

A. Write the word name for 0.97.

Ninety-seven	Write the word name for the number right of the decimal point.
Ninety-seven hundredths	Next, write the place value of the digit 7.The word name for 0 in the ones place may be written or omitted. "Zero and ninety-seven hundredths" is correct but unnecessary.

B. Write the word name for 0.077.

Seventy-seven thousandths

C. Write the word name for 213.17.

Two hundred thirteen	Write the word name for the whole number left of the decimal point.
Two hundred thirteen and	Write "and" for the decimal point.
Two hundred thirteen and seventeen	Write the word name for the number right of the decimal point.
Two hundred thirteen and seventeen hundredths	Write the place value of the digit 7.

D. Janet called an employee to find the measurement of the outside diameter of a new wall clock the company is manufacturing. She asked the employee to check the plans. What is the word name the employee will read to her? The clock is shown here.

9.225 in.

The employee will read "Nine and two hundred twenty-five thousandths inches."

D. The measurement of the outside diameter of another clock is shown in the following diagram. What word name will the employee read?

11.375 in.

Answers to Warm-Ups

D. The employee will read "Eleven and three hundred seventy-five thousandths inches."

EXAMPLES E–F

DIRECTIONS: Write the place value name.

STRATEGY: Write the digit symbols for the corresponding words. Replace the word *and* with a decimal point.

E. Write the place value name for twenty-six hundred-thousandths.

26	First, write the number for twenty-six.
.00026	The place value hundred-thousandths indicates five decimal places, so write three zeros before the numeral twenty-six and then a decimal point. This puts the numeral 6 in the hundred-thousandths place.
0.00026	Because the number is between zero and one, we write a 0 in the ones place.

F. Write the place value name for four hundred five and four hundred five ten-thousandths.

405	The whole number part is 405.
405.	Write the decimal point for "and."
405.0405	The whole number after the "and" is 405. A zero is inserted so the numeral 5 is in the ten-thousandths place.

WARM-UPS E–F

E. Write the place value name for thirty-five thousandths.

F. Write the place value name for seven hundred three and three hundred seven ten-thousandths.

HOW & WHY

OBJECTIVE 2 Round a given decimal.

Decimals can be either *exact* or *approximate*. For example, decimals that count money are exact. The figure $56.35 shows an exact amount. Most decimals that describe measurements are approximations. For example, 6.1 ft shows a person's height to the nearest tenth of a foot, and 1.9 m shows the height to the nearest tenth of a meter, but neither is an exact measure.

Answers to Warm-Ups

E. 0.035 **F.** 703.0307

Decimals are rounded using the same procedure as for whole numbers. Using a ruler (see Figure 4.2), we round 2.563.

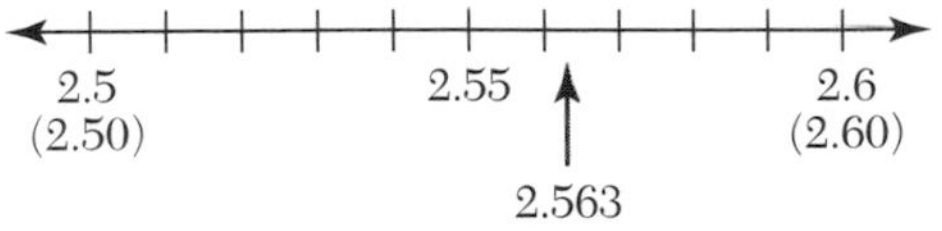

FIGURE 4.2

To the nearest tenth, 2.563 is rounded to 2.6 because it is closer to 2.6 than to 2.5. Rounded to the nearest hundredth, 2.563 is rounded to 2.56 because it is closer to 2.56 than to 2.57.

To round 11.6789 to the nearest hundredth without drawing a number line, draw an arrow under the hundredths place to identify the round-off place.

11.6789
$\quad\;\uparrow$

We must choose between 11.67 and 11.68. Because the digit to the right of the round-off position is 8, the number is more than halfway to 11.68. So we choose the larger number.

$11.6789 \approx 11.68$

CAUTION

Do not replace the dropped digits with zeros if the round-off place is to the right of the decimal point. $11.6789 \approx 11.6800$ indicates a round-off position of ten-thousandths.

To round a decimal number to a given place value

1. Draw an arrow under the given place value. (After enough practice, you will be able to round mentally and will not need the arrow.)
2. If the digit to the right of the arrow is 5, 6, 7, 8, or 9, add 1 to the digit above the arrow. That is, round to the larger number.
3. If the digit to the right of the arrow is 0, 1, 2, 3, or 4, keep the digit above the arrow. That is, round to the smaller number.
4. Write whatever zeros are necessary after the arrow so that the number above the arrow has the same place value as the original. See Example H.

This method is sometimes called the "four-five rule." Although this rounding procedure is the one most commonly used, it is not the only way to round. Many government agencies round by *truncation*—that is, by dropping the digit after the decimal point. Thus, $\$87.52 \approx \$87$. It is common for retail stores to round up for any amounts smaller than one cent. Thus, $\$3.553 \approx \$3.56$. There is also a rule for rounding numbers in science, which is sometimes referred to as the "even/odd rule." You might learn to use a different round-off rule, depending on what kind of work you are doing.

EXAMPLES G–J

DIRECTIONS: Round as indicated.

STRATEGY: Draw an arrow under the given place value. Examine the digit to the right of the arrow to determine whether to round up or down.

G. Round 0.8239 to the nearest hundredth.

$0.8239 \approx 0.82$ (arrow under the 2) The digit to the right of the round-off place is 3, so round down.

H. Round 7843.9 to the nearest thousand.

$7843.9 \approx 8000$ (arrow under the 7) Three zeros must be written after the 8 to keep it in the thousands place.

I. Round 389.6 to the nearest unit.

$389.6 \approx 390$ The number to the right of the round-off place is 6, so we round up by adding 389 + 1 = 390.

J. Round 52.7635 and 7.9958 to the nearest unit, the nearest tenth, the nearest hundredth, and the nearest thousandth.

	Unit		Tenth		Hundredth		Thousandth
52.7635 ≈	53	≈	52.8	≈	52.76	≈	52.764
7.9958 ≈	8	≈	8.0	≈	8.00	≈	7.996

CAUTION

The zeros following the decimal in 8.0 and 8.00 are necessary to show that the original was rounded to the nearest tenth and hundredth, respectively.

WARM-UPS G–J

G. Round 0.41678 to the nearest thousandth.

H. Round 883.89 to the nearest ten.

I. Round 4519.8 to the nearest unit.

J. Round 34.7835 to the nearest unit, the nearest tenth, the nearest hundredth, and the nearest thousandth.

Answers to Warm-Ups

G. 0.417 **H.** 880 **I.** 4520
J. 35; 34.8; 34.78; 34.784

EXERCISES 4.1

OBJECTIVE 1 Write word names from place value names and place value names from word names.

A.

Write the word name.

1. 0.17 **2.** 0.67 **3.** 0.267 **4.** 0.712

5. 4.0005 **6.** 6.104 **7.** 6.47 **8.** 18.07

Write the place value name.

9. Sixteen hundredths **10.** Forty-five hundredths

11. One hundred thirteen thousandths **12.** Seven hundred twelve thousandths

13. Two and nineteen thousandths **14.** One and six hundredths

15. Thirty-two ten-thousandths **16.** Nine ten-thousandths

B.

Write the word name.

17. 0.703 **18.** 7.03 **19.** 70.03 **20.** 7.003

21. 18.0205 **22.** 45.0051 **23.** 60.009 **24.** 600.090

Write the place value name.

25. Twelve thousandths **26.** Twelve thousand

27. Eight hundred and forty-two thousandths **28.** Three hundred and fifteen thousandths

29. Five hundred five and five thousandths **30.** Five and five hundred five thousandths

31. Twenty-three and twenty-three hundredths **32.** Two hundred eight and two hundred eight thousandths

 Name ______ Class ______ Date ______

OBJECTIVE 2 Round a given decimal.

A.

Round to the nearest unit, tenth, and hundredth.

	Unit	Tenth	Hundredth
33. 23.777			
34. 51.666			
35. 639.875			
36. 344.333			
37. 0.8156			
38. 0.92937			

Round to the nearest cent.

39. \$33.5374 **40.** \$84.4167 **41.** \$398.6135 **42.** \$795.6485

B.

Round to the nearest ten, hundredth, and thousandth.

	Ten	Hundredth	Thousandth
43. 27.6647			
44. 42.1689			
45. 245.2454			
46. 118.0752			
47. 0.7728			
48. 0.5979			

Round to the nearest dollar.

49. \$22.63 **50.** \$49.49 **51.** \$1129.38 **52.** \$3178.48

C.

Exercises 53–56. The graph shows the precipitation for four weeks in a southern city.

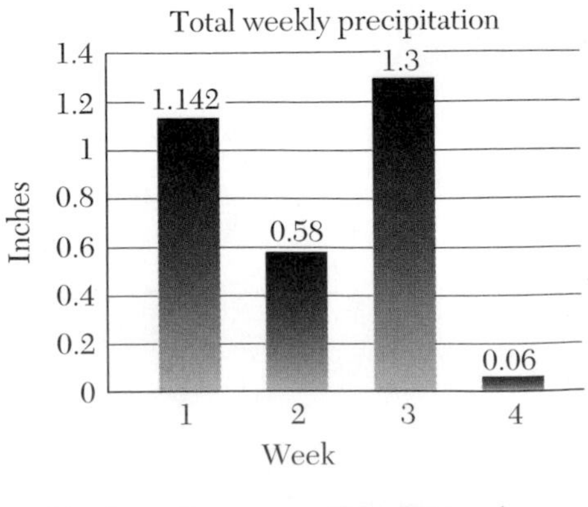

53. Write the word name for the number of inches of precipitation during the third week.

54. Write the word name for the number of inches of precipitation during the first week.

55. Round the amount of precipitation in week 1 to the nearest tenth of an inch.

56. Round the amount of precipitation in week 4 to the nearest tenth of an inch.

57. Dan Ngo buys a deep fryer that has a marked price of \$64.79. What word name does he write on the check?

58. Fari Alhadet buys a truckload of organic fertilizer for her yard. The price of the load is \$106.75. What word name does she write on the check?

Exercises 59–62. Use the graph of the number line.

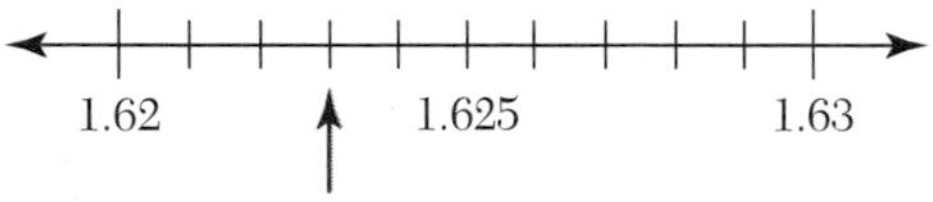

59. What is the position of the arrow to the nearest hundredth?

60. What is the position of the arrow to the nearest thousandth?

61. What is the position of the arrow to the nearest tenth?

62. What is the position of the arrow to the nearest unit?

Write the word name.

63. 567.9023

64. 6055.6055

65. The computer at Grant's savings company shows that his account, including the interest he has earned, has a value of \$1617.37099921. Round the value of the account to the nearest cent.

66. In doing her homework, Catherine finds that her calculator shows the answer to a division exercise as 34.78250012. If she is to round the answer to the nearest thousandth, what answer does she report?

Write the place value name.

67. Two hundred thirteen and one thousand, one hundred one ten-thousandths

68. Eleven thousand five and one thousand fifteen hundred-thousandths

69. In 2001, the Sultanate of Oman had a population density of about 30.895 people per square mile. Round the population density to the nearest whole person per square mile.

70. One text lists the mean (average) distance from Earth to the sun as 92,960,000 miles. To what place does this number appear to have been rounded?

71. According to the 2000 Census figures, the city of San Jose, California, had the third largest number of people per household in the country, with an average household size of 3.20. To what place does this number appear to have been rounded?

Round to the indicated place value.

	Hundred	Hundredth	Ten-thousandth
72. 3742.80596			
73. 982.678456			
74. 5790.350024			
75. 78,042.38875			

Exercises 76–79 relate to the chapter application.

76. In January 1906, a Stanley car with a steam engine set a 1-mile speed record by going 127.659 miles per hour. Round this rate to the nearest tenth of a mile per hour.

77. In March 1927, a Sunbeam set a 1-mile speed record by going 203.790 mph. What place value was this rate rounded to?

78. In October 1970, a Blue Flame set a 1-mile speed record by going 622.407 mph. Explain why it is incorrect to round the rate to 622.5 mph.

79. On October 15, 1997, Andy Green recorded the first supersonic land 1-mile record in a Thrust SSC with a speed of 763.035 mph. Round this speed to the nearest whole mile per hour and to the nearest tenth of a mile per hour.

STATE YOUR UNDERSTANDING

80. Explain the difference between an exact decimal value and an approximate decimal value. Give an example of each.

81. Explain in words the meaning of the value of the 4s in the numerals 43.29 and 18.64. Include some comment on how and why the values of the digit 4 are alike and how and why they are different.

82. Consider the decimal represented by abc.defg. Explain how to round this number to the nearest hundredth.

CHALLENGE

83. Round 8.28282828 to the nearest thousandth. Is the rounded value less than or greater than the original value? Write an inequality to illustrate your answer.

84. Round 7.7777777 to the nearest unit. Is the rounded value less than or greater than the original value? Write an inequality to illustrate your answer.

85. Write the place value name for two hundred fifteen ten-millionths.

86. Write the word name for 40,715.300519.

GROUP WORK

87. Discuss with the members of the group cases where you think rounding by "truncating" is the best way to round. Rounding by truncating means to drop all digit values to the right of the rounding position. For example, $2.77 \approx 2$, $\$19.33 \approx \$19$, $347.99 \approx 347$. Can each of the groups think of a situation where such rounding is actually used?

MAINTAIN YOUR SKILLS

Find the missing numerator.

88. $\dfrac{6}{7} = \dfrac{?}{28}$

89. $\dfrac{5}{12} = \dfrac{?}{132}$

90. $\dfrac{42}{125} = \dfrac{?}{10{,}000}$

91. $\dfrac{17}{40} = \dfrac{?}{1000}$

True or false.

92. $4567 < 4559$

93. $30{,}951 > 30{,}899$

List the numbers from smallest to largest.

94. 367, 401, 393, 363, 390

95. 4117, 4123, 4118, 4121, 4110

96. During a recent candy sale, Mia sold 14 boxes of candy, each containing 24 candy bars. Chia sold 11 bags of candy, each containing 30 candy bars. Who sold the most candy bars?

97. Jose answers 21 of 25 problems on a test correctly. If the next test has 100 problems, how many must he get correct to have the same score?

4.2 Changing Decimals to Fractions; Listing in Order

HOW & WHY

OBJECTIVES

1. Change a decimal to a fraction.
2. List a set of decimals from smallest to largest.

OBJECTIVE 1 Change a decimal to a fraction.

The word name of a decimal is also the word name of a fraction.

Consider 0.615.

READ: six hundred fifteen thousandths

WRITE: $\frac{615}{1000}$

So, $0.615 = \frac{615}{1000} = \frac{123}{200}$

Some other examples are shown in the Table 4.3.

TABLE 4.3 DECIMALS TO FRACTIONS

Place Value Name	Word Name	Fraction
0.43	Forty-three hundredths	$\frac{43}{100}$
0.7	Seven tenths	$\frac{7}{10}$
0.175	One hundred seventy-five thousandths	$\frac{175}{1000} = \frac{7}{40}$
0.16	Sixteen hundredths	$\frac{16}{100} = \frac{4}{25}$

To change a decimal to a fraction

1. Read the decimal word name.
2. Write the fraction that has the same value.
3. Simplify.

Notice that because of place value, the number of decimal places in a decimal tells us the number of zeros in the denominator of the fraction. This fact can be used as another way to write the fraction or to check that the fraction is correct:

$$\underbrace{4.58}_{\text{two decimal places}} = \underbrace{4\frac{58}{100}}_{\text{two zeros}} = 4\frac{29}{50} \quad \text{or} \quad 4.58 = \frac{458}{100} = 4\frac{29}{50}$$

WARM-UPS A–C

A. Change 0.63 to a fraction.

B. Write 0.375 as a fraction.

C. Write as a mixed number: 12.65

EXAMPLES A–C

DIRECTIONS: Change the decimal to a fraction or mixed number.

STRATEGY: Say the word name to yourself and write the fraction or mixed number that is equivalent. Simplify.

A. Change 0.77 to a fraction.

Seventy-seven hundredths — Say the word name.

$\frac{77}{100}$ — Write as a fraction.

B. Write 0.525 as a fraction.

Five hundred twenty-five thousandths — Say the word name.

$\frac{525}{1000} = \frac{21}{40}$ — Write as a fraction and simplify.

C. Write as a mixed number: 85.35

Eighty-five and thirty-five hundredths — Say the word name.

$85\frac{35}{100} = 85\frac{7}{20}$ — Write as a mixed number and simplify.

HOW & WHY

OBJECTIVE 2 List a set of decimals from smallest to largest.

Fractions can be listed in order when they have a common denominator by ordering the numerators. This idea can be extended to decimals when they have the same number of decimal places. For instance, $0.36 = \frac{36}{100}$ and $0.47 = \frac{47}{100}$ have a common denominator when written in fractional form. So 0.36 is less than 0.47 or $0.36 < 0.47$.

We can see that $1.7 < 3.1$ because 1.7 is to the left of 3.1 on the number line. See Figure 4.3.

The decimals 0.4 and 0.23 have a common denominator when a zero is placed after the 4. Thus

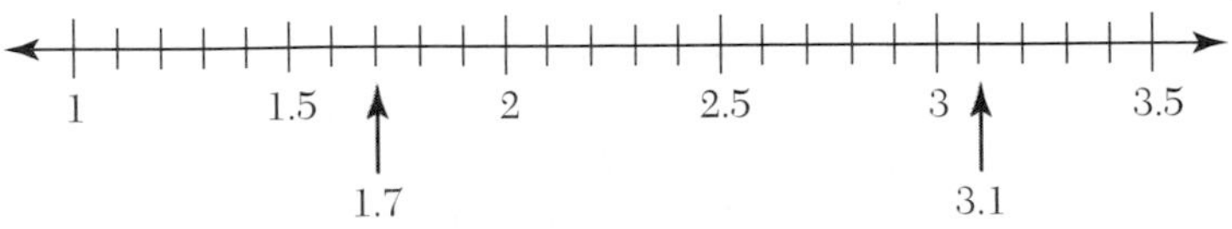

FIGURE 4.3

$$0.4 = \frac{4}{10} = \frac{4}{10} \cdot \frac{10}{10} = \frac{40}{100} = 0.40$$

so that

$$0.4 = \frac{40}{100} \quad \text{and} \quad 0.23 = \frac{23}{100}$$

Then, because $\frac{40}{100} > \frac{23}{100}$, we conclude that $0.4 > 0.23$.

Answers to Warm-Ups

A. $\frac{63}{100}$ **B.** $\frac{3}{8}$ **C.** $12\frac{13}{20}$

Many forms for decimal numbers are equivalent. For example,

$$6.3 = 6.30 = 6.300 = 6.3000 = 6.30000$$
$$0.85 = 0.850 = 0.8500 = 0.85000 = 0.850000$$
$$45.982 = 45.9820 = 45.98200 = 45.982000 = 45.9820000$$

The zeros to the right of the decimal point following the last nonzero digit do not change the value of the decimal. Usually these extra zeros are not written, but they are useful when operating with decimals.

To list a set of decimals from smallest to largest

1. Make sure that all numbers have the same number of decimal places to the right of the decimal point by writing zeros to the right of the last digit when necessary.
2. Write the numbers in order as if they were whole numbers.
3. Remove the extra zeros.

EXAMPLES D–F

DIRECTIONS: Is the statement true or false?

STRATEGY: Write each numeral with the same number of decimal places. Compare the values without regard to the decimal point.

D. True or false: $0.55 > 0.52$?

$55 > 52$ is true. Compare the numbers without regard to the decimal point.

So, $0.55 > 0.52$ is true.

E. True or false: $0.986 < 0.98$?

$0.986 < 0.980$ Write with the same number of decimal places.
$986 < 980$ is false. Without regard to the decimal point, 986 is larger.

So, $0.986 < 0.98$ is false.

F. True or false: $32.008 > 32.09$?

$32.008 > 32.090$ Write with the same number of decimal places.
$32008 > 32090$ is false.

So, $32.008 > 32.09$ is false.

WARM-UPS D–F

D. True or false: $0.23 < 0.22$?

E. True or false: $0.63 < 0.6299$?

F. True or false: $66.7 > 66.683$?

Answers to Warm-Ups
* **D.** false **E.** false **F.** true

EXAMPLES G–I

DIRECTIONS: List the decimals from smallest to largest.

STRATEGY: Write zeros on the right so that all numbers have the same number of decimal places. Compare the numbers as if they were whole numbers and then remove the extra zeros.

G. List 0.62, 0.637, 0.6139, and 0.621 from smallest to largest.

0.6200
0.6370
0.6139
0.6210

First, write all numbers with the same number of decimal places by inserting zeros on the right.

0.6139, 0.6200, 0.6210. 0.6370 Second, write the numbers in order as if they were whole numbers.

0.6139, 0.62, 0.621, 0.637 Third, remove the extra zeros.

WARM-UPS G–I

G. List 0.75, 0.692, 0.748, and 0.732 from smallest to largest.

Answers to Warm-Ups
G. 0.692, 0.732, 0.748, 0.75

H. List 3.03, 3.0033, 3.0333, and 3.0303 from smallest to largest.

I. Roberto and Rachel measure the same coin. Roberto measures 0.916 and Rachel measures 0.9153. Whose measure is larger?

H. List 8.537, 8.631, 8.5334, and 8.538 from smallest to largest.

8.537 = 8.5370		Step 1
8.631 = 8.6310		
8.5334 = 8.5334		
8.538 = 8.5380		
8.5334, 8.5370, 8.5380, 8.6310		Step 2
8.5334, 8.537, 8.538, 8.631		Step 3

I. Using a micrometer to measure the diameter of a foreign coin, Mike measures 0.6735 and Mildred measures 0.672. Whose measure is larger?

0.6735, 0.6720 — Write with the same number of decimal places and compare as whole numbers.

* Mike's measure is larger because $6735 > 6720$.

Answers to Warm-Ups

H. 3.0033, 3.03, 3.0303, 3.0333

I. Roberto's measure is larger.

EXERCISES 4.2

OBJECTIVE 1 Change a decimal to a fraction.

A. *Change each decimal to a fraction and simplify, if possible.*

1. 0.53 **2.** 0.73 **3.** 0.75 **4.** 0.8

5. Two hundred thirteen thousandths **6.** Seven hundred three thousandths

7. 0.82 **8.** 0.64 **9.** 0.72 **10.** 0.45

B. *Change the decimal to a fraction or mixed number and simplify.*

11. 5.63 **12.** 23.61 **13.** 0.875 **14.** 0.475

15. 5.36 **16.** 32.32 **17.** 7.492 **18.** 3.650

19. Eight hundred thousandths **20.** Twenty-five hundred-thousandths

OBJECTIVE 2 List a set of decimals from smallest to largest.

A.

List the set of decimals from smallest to largest.

21. 0.6, 0.7, 0.1 **22.** 0.07, 0.03, 0.025

23. 0.15, 0.61, 0.17 **24.** 0.14, 0.18, 0.09

25. 4.16, 4.161, 4.159 **26.** 7.18, 7.183, 7.179

Are the following statements true or false?

27. $0.29 < 0.2$ **28.** $0.82 < 0.90$

29. $9.54 > 9.45$ **30.** $6.89 > 6.98$

B.

List the set of decimals from smallest to largest.

31. 0.0533, 0.053003, 0.053, 0.052, 0.053016 **32.** 7.008, 0.708, 0.7009, 0.70109

33. 0.888, 0.88799, 0.8881, 0.88579 **34.** 8.36, 8.2975, 8.3599. 8.3401

35. 32.006, 32.06, 32.059, 32.095 **36.** 92.2708, 92.0278, 92.0728, 92.8207

Are the following statements true or false?

37. $3.1231 < 3.1213$ **38.** $4.1243 > 4.124$ **39.** $47.3406 < 47.4046$ **40.** $77.3045 > 77.3405$

C.

41. The probability that a flipped coin will come up heads three times in a row is 0.125. Write this as a reduced fraction.

42. The probability that a flipped coin will come up heads twice and tails once out of three flips is 0.375. Write this as a reduced fraction.

43. The Davis Meat Company bids 98.375¢ per pound to provide meat to the Beef and Bottle Restaurant. Circle K Meats puts in a bid of 98.35¢, and J & K Meats makes a bid of 98.3801¢. Which is the best bid for the restaurant?

44. Charles loses 2.165 pounds during the week. Karla loses 2.203 pounds and Mitchell loses 2.295 pounds during the same week. Who loses the most weight this week?

Exercises 45–47. The graph shows the weekly precipitation for one month in a small city in Wisconsin.

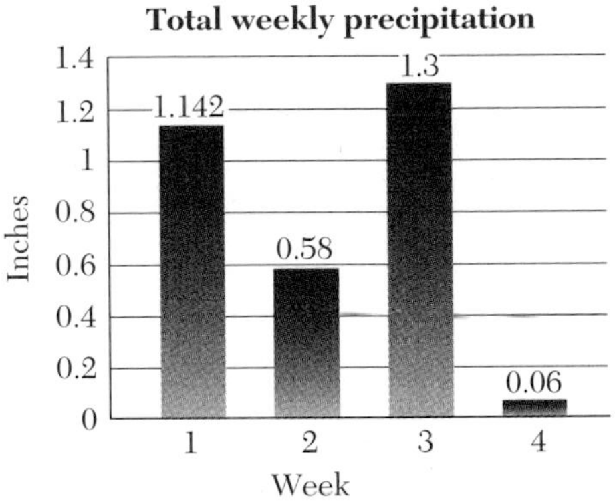

45. Write a reduced fraction to show the number of inches of precipitation in the fourth week.

46. Write a mixed number to show the number of inches of precipitation in the first week.

47. Which week has the least precipitation? The most precipitation?

Change the decimal to a fraction or mixed number and simplify.

48. 0.1705

49. 0.5425

50. 403.304

51. 65.0075

52. Gerry may choose a 0.055 raise in pay or a $\frac{1}{20}$ increase. Which value will yield more money? Compare in fraction form.

53. A chemistry class requires 0.667 mg of soap for each student. Hoa has 0.67 mg of soap. Does she need more or less soap?

List the decimals from smallest to largest.

54. 0.00829, 0.0083001, 0.0082, 0.0083, 0.0083015

55. 5.0009, 5.001, 5.00088. 5.00091, 5.00101

56. 49.457, 49.449, 49.501, 49.576, 49.491, 49.5011

57. 37.96, 38.01, 37.95, 37.59, 38.15, 38.25, 37.90, 37.79

58. Maria is using a pattern from Europe to make her wedding dress. The pattern calls for 9.3125 yards of material. Change the measure to a mixed number and simplify.

59. For a bridesmaid's bow, Maria may choose 0.825 yard or $\frac{6}{7}$ yard for the same price. Which should she choose to get the most fabric? Compare in fraction form.

60. One synodic day on Jupiter (midday to midday) is about 9.925933 hours, whereas one sidereal day (measured by apparent star movements) is about 9.925 hours. Which is longer?

61. The population density of Cape Verde is about 250.839 people per square mile, and the population density of Austria is about 250.875 people per square mile. Which country has fewer people per square mile?

62. Betty Crocker cake mixes, when prepared as directed, have the following decimal fractions of calories per slice from fat: Apple Cinnamon, 0.36; Butter Pecan, 0.4; Butter Recipe/Chocolate, 0.43; Chocolate Chip, 0.42; Spice, 0.38; Golden Vanilla, 0.45. If each slice contains 280 calories, which cake has the most calories from fat? Fewest calories from fat?

63. Hash brown potatoes have the following numbers of fat grams per serving: frozen plain, 7.95 g; frozen with butter sauce, 8.9 g; and homemade with vegetable oil, 10.85 g. Write the fat grams as mixed numbers reduced to lowest terms. Which serving of hash browns has the least amount of fat?

Exercises 64–67 relate to the chapter application.

64. At one point in the 1997–1998 NBA season, the listed teams had won the following decimal fraction of their games: Seattle, 0.789; L.A. Lakers, 0.757; San Antonio, 0.684; Utah, 0.667; Miami, 0.667; New York, 0.556; Indiana, 0.694; Chicago, 0.692. Rank these teams from best record to worst record.

65. At one point in the 1997–1998 NBA season, Tim Hardaway had hit 0.835 of his free throws. Express his record as a fraction. What fraction of his free throws did he miss?

66. The table displays the top ten professional bowling averages for the central region in 2000.

Top Ten PBA Averages, Central Region, 2000

Bowler	Number of Games	Average
Paul Ashby	56	207.85
Pat Cumbrowski	54	208.01
Mike Eaton	42	207.88
Ted Hannah	69	219.04
Jim Johnson	56	210.03
Bob Learn	46	220.26
Chris Matthews	32	208.87
Sean Quinn	46	215.97
Jeff Wellis	33	208.37
Jeff Zaffio	42	209.21

Which bowler had the highest average for the year?

67. Sort the table in Exercise 66 so the averages are displayed from highest to lowest.

STATE YOUR UNDERSTANDING

68. Explain how the number line can be a good visual aid for determining which of two numerals has the larger value.

CHALLENGE

69. Change 0.44, 0.404, and 0.04044 to fractions and reduce.

70. Determine whether each statement is true or false.

a. $7.44 < 7\frac{7}{18}$

b. $8.6 > 8\frac{5}{9}$

c. $3\frac{2}{7} < 3.285$

d. $9\frac{3}{11} > 9.271$

GROUP WORK

71. Find a pattern, a vehicle manual, and/or a parts list whose measurements are given in decimal form. Change the measurements to fraction form.

72. Have each member of the group write one fraction and one decimal each with values between 3 and 4. Then, as a group, list all the fractions and decimal values from smallest to largest.

MAINTAIN YOUR SKILLS

Add or subtract.

73. $567 + 8123 + 75 + 4911$

74. $94{,}703 + 6002 + 34{,}645 + 709$

75. $15{,}869 - 12{,}971$

76. $17{,}001 - 14{,}643$

77. $\frac{1}{2} + \frac{3}{4} + \frac{1}{8}$

78. $\frac{1}{3} + \frac{7}{12} + \frac{5}{6}$

79. $\frac{31}{32} - \frac{7}{8}$

80. $\frac{19}{20} - \frac{5}{12}$

81. Pedro did a traffic count on the Sunset Highway. He had the following counts: 6–7 A.M., 123; 7–8 A.M., 231; 8–9 A.M., 546; 9–10 A.M., 472; and 10–11 A.M., 205. How many vehicles did he count during the 5 hours?

82. Joanna has $893 in her bank account. She writes checks for $45, $93, $132, $15, and $321. What balance does she now have in her account?

4.3 Adding and Subtracting Decimals

OBJECTIVES

1. Add decimals.
2. Subtract decimals.
3. Estimate the sum or difference of decimals.

HOW & WHY

OBJECTIVE 1 Add decimals.

What is the sum of 11.3 + 16.4? We use the expanded form of the decimal to explain addition.

$$\begin{array}{rl} 11.3 = & 1 \text{ ten} + 1 \text{ ones} + 3 \text{ tenths} \\ +16.4 = & 1 \text{ ten} + 6 \text{ ones} + 4 \text{ tenths} \\ \hline & 2 \text{ tens} + 7 \text{ ones} + 7 \text{ tenths} = 27.7 \end{array}$$

We use the same principle for adding decimals that we use for whole numbers. That is, we add like units. The vertical form gives us a natural grouping of the tens, ones, and tenths. By inserting zeros so all the numbers have the same number of decimal places, we write the addition 5.8 + 17.6 + 8.52 as 5.80 + 17.60 + 8.52.

$$\begin{array}{r} 5.80 \\ 17.60 \\ +\ 8.52 \\ \hline 31.92 \end{array}$$

To add decimals

1. Write in columns with the decimal points aligned. Insert extra zeros to help align the place values.
2. Add the decimals as if they were whole numbers.
3. Align the decimal point in the sum with those above.

EXAMPLES A–C

DIRECTIONS: Add.

STRATEGY: Write each numeral with the same number of decimal places, align the decimal points, and add.

A. Add: 6.1 + 37.82 + 54 + 0.04

$$\begin{array}{r} 6.10 \\ 37.82 \\ 54.00 \\ +\ 0.04 \\ \hline 97.96 \end{array}$$

Write each numeral with two decimal places. The extra zeros help line up the place values.

CALCULATOR EXAMPLE

B. Add: 4.7435 + 0.0448 + 7.59 + 21.863

STRATEGY: The extra zeros do not need to be inserted. The calculator will automatically align the like place values when adding.

The sum is 34.2413.

C. What is the total cost of an automobile tire if the retail price is $84.95, the federal excise tax is $3.13, the state sales tax is $4.25, and the local sales tax is $0.68?

WARM-UPS A–C

A. Add:
7.8 + 57.14 + 33 + 0.03

B. Add: 5.8117 + 0.0668 + 6.18 + 53.993

C. What is the total cost of a pair of emerald earrings if the retail price is $146.98, the federal tax is $5.74, the state sales tax is $8.82, and the city sales tax is $1.15?

STRATEGY: Add the retail price and the taxes.

$$\begin{array}{r} \$84.95 \\ 3.13 \\ 4.25 \\ +\ \ 0.68 \\ \hline \$93.01 \end{array}$$

The tire costs $93.01.

HOW & WHY

OBJECTIVE 2 Subtract decimals.

What is the difference 9.47 − 5.24? To find the difference, we write the numbers in column form, aligning the decimal points. Now subtract as if they are whole numbers.

$$\begin{array}{r} 9.47 \\ -5.24 \\ \hline 4.23 \end{array}$$

The decimal point in the difference is aligned with those above.

When necessary, we can regroup, or borrow, as with whole numbers. What is the difference 7.835 − 3.918?

$$\begin{array}{r} 7.835 \\ -3.918 \\ \hline \end{array}$$

We need to borrow one from the hundredths column (1 hundredth = 10 thousandths), and we need to borrow one from the ones column (1 one = 10 tenths).

$$\begin{array}{r} \scriptstyle 6\ \ 18\ 2\ 15 \\ \not{7}.\not{8}\not{3}\not{5} \\ -3.9\ 1\ 8 \\ \hline 3.9\ 1\ 7 \end{array}$$

So the difference is 3.917.

Sometimes it is necessary to write zeros on the right so the numbers have the same number of decimal places. See Example E.

To subtract decimals

1. Write the decimals in columns, with the decimal points aligned. Insert extra zeros to align the place values.
2. Subtract the decimals as if they were whole numbers.
3. Align the decimal point in the difference with those above.

Answers to Warm-Ups

A. 97.97 **B.** 66.0515
C. The earrings cost $162.69.

WARM-UPS D–H

D. Subtract: 11.367 − 9.279

Answers to Warm-Ups

D. 2.088

EXAMPLES D–H

DIRECTIONS: Subtract.

STRATEGY: Write each numeral with the same number of decimal places, align the decimal points, and subtract.

D. Subtract: 6.793 − 0.596

$$\begin{array}{r} 6.793 \\ -0.596 \\ \hline \end{array}$$

Line up the decimal points so the place values are aligned.

$$\begin{array}{r} \scriptstyle 8\ 13 \\ 6.7\not{9}\not{3} \\ -0.5\ 9\ 6 \\ \hline \end{array}$$

Borrow 1 hundredth from the 9 in the hundredths place. (1 hundredth = 10 thousandths)

```
   6 18
    8 13
  6.7 9 3      Borrow 1 tenth from the 7 in the tenths place.
 −0.5 9 6      (1 tenth = 10 hundredths)
  6.1 9 7
```

CHECK:

```
  0.596     Check by adding.
 +6.197
  6.793
```

The difference is 6.197.

E. Subtract 4.37 from 8.

```
  8.00    We write 8 as 8.00 so that both numerals will have the same number
 −4.37    of decimal places.

  7 10
  8.00    We need to borrow to subtract in the hundredths place. Because there
 −4.37    is a 0 in the tenths place, we start by borrowing 1 from the ones place.
          (1 one = 10 tenths)
   9 10
  7 10    Now borrow 1 tenth to add to the hundredths place.
  8.00    (1 tenth = 10 hundredths)
 −4.37
  3.63    Subtract.
```

CHECK:

```
  4.37
 +3.63
  8.00
```

The difference is 3.63.

F. Find the difference of 7.193 and 4.845. Round to the nearest tenth.

```
  6 11 8 13
  7. 1 9 3
 −4. 8 4 5
  2. 3 4 8    The check is left for the student.
```

The difference is 2.3 to the nearest tenth.

CAUTION

Do not round before subtracting. Note the difference if we do:

$7.2 - 4.8 = 2.4$

CALCULATOR EXAMPLE

G. Subtract: 637.9673 − 456.929

STRATEGY: The calculator automatically lines up the decimal points.

The difference is 181.0383.

H. Marta purchases a small radio for $37.74. She gives the clerk two $20 bills to pay for the radio. How much change does she get?

STRATEGY: Because two $20 bills are worth $40, subtract the cost of the radio from $40.

```
  $40.00
 − 37.74
  $ 2.26
```

Marta gets $2.26 in change.

E. Subtract 6.756 from 9.

F. Find the difference of 8.493 and 3.736. Round to the nearest tenth.

G. Subtract: 832.7445 − 578.399

H. Mickey buys a video for $18.69. She gives the clerk a $20 bill to pay for the cassette. How much change does she get?

Answers to Warm-Ups

E. 2.244 **F.** 4.8 **G.** 254.3455
H. Mickey gets $1.31 in change.

Clerks sometimes make change by counting backward—that is, by adding to $37.74 the amount necessary to equal $40.

$37.74 + a penny = $37.75
$37.75 + a quarter = $38.00
$38.00 + 1 dollar = $39.00
$39.00 + 1 dollar = $40.00

So the change is $0.01 + $0.25 + $1 + $1 = $2.26.

HOW & WHY

OBJECTIVE 3 Estimate the sum or difference of decimals.

The sum or difference of a group of decimals can be estimated by rounding each to a specified place value in the group and then adding or subtracting the rounded values. For instance,

0.756	0.8	Round each number to the nearest tenth.
0.092	0.1	
0.0072	0.0	
+0.0205	+0.0	
	0.9	

The estimate of the sum is 0.9. One use of the estimate is to see if the sum of the group of numbers is correct. If the calculated sum is not close to the estimated sum, 0.9, you should check the addition by readding. In this case, the calculated sum, 0.8757, is close to the estimate.

To estimate the sum or difference of a group of decimals

1. Round each number in the group to the same place value.
2. Add or subtract the rounded numbers.

WARM-UP 1

1. Estimate the difference by rounding to the nearest thousandth:
0.00472 − 0.00193

EXAMPLE 1

DIRECTION: Estimate the difference by rounding to the specified place value.

STRATEGY: Round each number to the hundredths place. Then subtract and compare.

1. 0.8556 − 0.459 (hundredth)

$$\begin{array}{r} 0.86 \\ -0.46 \\ \hline 0.40 \end{array}$$ Round to the nearest hundredth.

Now subtract and compare.

$$\begin{array}{r} \scriptstyle 7\,14 \quad \\ \scriptstyle 4\,15 \\ 0.\not{8}\,\not{5}\,\not{5}\,6 \\ -0.4\,5\,9\,0 \\ \hline 0.3\,9\,6\,6 \end{array}$$ Insert a zero so we have the same number of decimal places.

The difference is 0.3966 and is close to the estimate, 0.40.

Answer to Warm-Up
1. 0.003; 0.00279

EXERCISES 4.3

OBJECTIVE 1 Add decimals.

Add

A.

1. 0.6 + 0.3 **2.** 0.7 + 0.4 **3.** 2.5 + 1.1 **4.** 6.7 + 2.1

5. 1.6 + 2.3 + 5.1 **6.** 4.1 + 1.5 + 2.3 **7.** 23.3 + 4.13 **8.** 18.9 + 5.67

9. To add 4.5, 6.78, 9.342, and 23, first rewrite each with ______ decimal places.

10. The sum of 6.7, 8.93, 5.4321, and 45.72 has ______ decimal places.

B.

11. $\begin{array}{r} 10.6 \\ +12.732 \\ \hline \end{array}$ **12.** $\begin{array}{r} 15.9 \\ +\ 7.83 \\ \hline \end{array}$ **13.** $\begin{array}{r} 8.28 \\ 0.28 \\ 12.3 \\ +\ 2.54 \\ \hline \end{array}$ **14.** $\begin{array}{r} 9.06 \\ 0.82 \\ 11.5 \\ +\ 4.35 \\ \hline \end{array}$

15. 0.517 + 0.359 + 1.821

16. 3.146 + 3.461 + 0.614

17. 0.0017 + 1.007 + 7 + 1.071

18. 1.0304 + 1.4003 + 1.34 + 0.403

19. 48.006 + 46.008 + 4.86 + 4.608

20. 67.004 + 6.74 + 4.067 + 46.77

21. $\begin{array}{r} 9.3 \\ 18.415 \\ +62.3894 \\ \hline \end{array}$ **22.** $\begin{array}{r} 15.07 \\ 189.981 \\ +6904.4063 \\ \hline \end{array}$ **23.** $\begin{array}{r} 43.524 \\ 12.8 \\ +774.943 \\ \hline \end{array}$ **24.** $\begin{array}{r} 314.143 \\ 712.217 \\ +333.444 \\ \hline \end{array}$

25. Find the sum of 11.08, 8.1, 0.346, and 9.07.

26. Find the sum of 2.6933, 5.895, 0.99, and 5.328.

OBJECTIVE 2 Subtract decimals.

Subtract.

A.

27. 0.9 − 0.6 **28.** 2.7 − 2.2 **29.** 7.8 − 5.7 **30.** 0.84 − 0.53

31. $\begin{array}{r} 8.31 \\ -3.21 \\ \hline \end{array}$ **32.** $\begin{array}{r} 17.48 \\ -\ 6.23 \\ \hline \end{array}$ **33.** $\begin{array}{r} 20.14 \\ -13.66 \\ \hline \end{array}$ **34.** $\begin{array}{r} 21.18 \\ -17.39 \\ \hline \end{array}$

35. Subtract 11.14 from 32.01.

36. Find the difference of 23.465 and 9.9.

B.

37. $\begin{array}{r} 0.723 \\ -0.457 \\ \hline \end{array}$ **38.** $\begin{array}{r} 9.308 \\ -5.639 \\ \hline \end{array}$ **39.** $\begin{array}{r} 2.712 \\ -1.148 \\ \hline \end{array}$ **40.** $\begin{array}{r} 7.303 \\ -0.178 \\ \hline \end{array}$

41. 12.145 − 8.146

42. 6.842 −3.458

43. 134.98 − 67.936

44. 405.4057 − 316.316

45. Subtract 14.15 from 21.09.

46. Subtract 4.717 from 9.3.

47. Find the difference of 7.539 and 7.322.

48. Find the difference of 93.101 and 87.456.

OBJECTIVE 3 Estimate the sum or difference of decimals.

A. *Estimate the sum or difference by rounding to the largest nonzero place value.*

49. 0.0635 + 0.0296 + 0.0555

50. 0.00843 + 0.00297 + 0.00671

51. 0.831 − 0.462

52. 0.067 − 0.049

53. 3.895 + 2.045 + 0.34 + 0.045

54. 0.45 + 0.0021 + 0.0043 + 0.1001

55. 0.075 − 0.0023

56. 0.00675 − 0.000984

57.
$$\begin{array}{r} 0.636 \\ 0.419 \\ 0.53821 \\ +0.8217 \\ \hline \end{array}$$

58.
$$\begin{array}{r} 0.0631 \\ 0.091 \\ 0.03581 \\ +0.04003 \\ \hline \end{array}$$

B. *Estimate the sum or difference by rounding to the largest nonzero place value. Add or subtract.*

59.
$$\begin{array}{r} 0.0764 \\ -0.03621 \\ \hline \end{array}$$

60.
$$\begin{array}{r} 0.0056982 \\ -0.003781 \\ \hline \end{array}$$

61. 0.0781 + 0.00932 + 0.07639 + 0.00759

62. 7.006 + 0.9341 + 0.03952 + 4.0444

63. 41.8341 − 34.6152

64. 7.9342 − 2.78932

65. 0.067 + 0.456 + 0.0964 + 0.5321 + 0.112

66. 4.005 + 0.875 + 3.96 + 7.832 + 4.009

67. 0.0643 + 0.8143 + 0.513 − (0.4083 + 0.7114)

68. 7.619 + 3.048 − (1.699 + 2.539 + 4.87)

C.

69. On a vacation trip, Manuel stopped for gas four times. The first time he bought 9.2 gallons. At the second station, he bought 11.9 gallons, and at the third, he bought 15.4 gallons. At the last stop, he bought 12.6 gallons. How much gas did he buy on the trip?

70. Heather wrote five checks in the amounts of $45.78, $23.90, $129.55, $7.75, and $85. She has $295.67 in her checking account. Does she have enough money to cover the five checks?

71. Find the sum of 63.812, 218.9, 101.333, 79.42, and 139.883. Round the sum to the nearest tenth.

72. Find the sum of 641.85, 312.963, 18.4936, 29.0049, and 6.1945. Round the sum to the nearest hundredth.

Exercises 73–75. The table shows the annual annuity sales in billions of dollars.

Annuity Sales (in billions of dollars)

1985	\$4.5	1990	\$12
1986	\$8.1	1991	\$17.3
1987	\$9.3	1992	\$28.5
1988	\$7.2	1993	\$46.6
1989	\$9.8	1994	\$50.4

73. Find the total sales for the 10-year period.

74. How many more dollars were invested in 1992 than in 1989?

75. How many more dollars were invested in the 1990s than in the 1980s?

76. Doris makes a gross salary (before deductions) of \$2796 per month. She has the following monthly deductions: federal income tax, \$254.87; state income tax, \$152.32; Social Security, \$155.40; Medicare, \$35.61; retirement contribution, \$82.45; union dues, \$35; and health insurance, \$134.45. Find her actual take-home (net) pay.

77. Jack goes shopping with \$75 in cash. He pays \$17.99 for a T-shirt, \$6.49 for a CD, and \$30.50 for a sweater. On the way home, he buys \$19.58 worth of gas. How much money does he have left?

78. In 1996, the average interest on a 30-year home mortgage dropped from 8.17% to 7.56% in 1 week. What was the drop in interest rate?

79. What is the total cost of a cart of groceries that contains bread for \$2.18, bananas for \$2.37, cheese for \$4.97, cereal for \$4.28, coffee for \$5.16, and meat for \$11.89?

Exercises 80–81. The table shows the lengths of railway tunnels in various countries.

World's Longest Railway Tunnels

Tunnel	Length (km)	Country
Seikan	53.91	Japan
English Channel tunnel	49.95	UK–France
Dai-shimizu	22.53	Japan

80. How much longer is the longest tunnel than the second longest tunnel?

81. What is the total length of the Japanese tunnels?

Exercises 82–84. The table shows projections for the number of families without children under 18.

Projected Number of Families Without Children Under 18

	Year			
	1995	2000	2005	2010
Families without children under 18 (in millions)	35.8	38.6	42.0	45.7

SOURCE: U.S. Census Bureau.

82. What is the projected change in the number of families in the United States without children under 18 between 1995 and 2010?

83. Which 5-year period is projected for the largest change?

84. What could explain the increase indicated in the table?

85. How high from ground level is the top of the smokestack shown here? Round to the nearest foot.

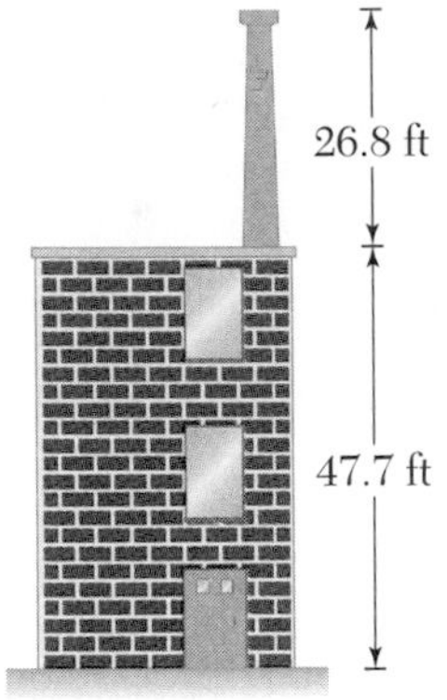

86. Find the length of the piston skirt (A) shown in the figure if the other dimensions are as follows: B = 0.3125 in., C = 0.250 in., D = 0.3125 in., E = 0.250 in., F = 0.3125 in., G = 0.375 in., H = 0.3125.

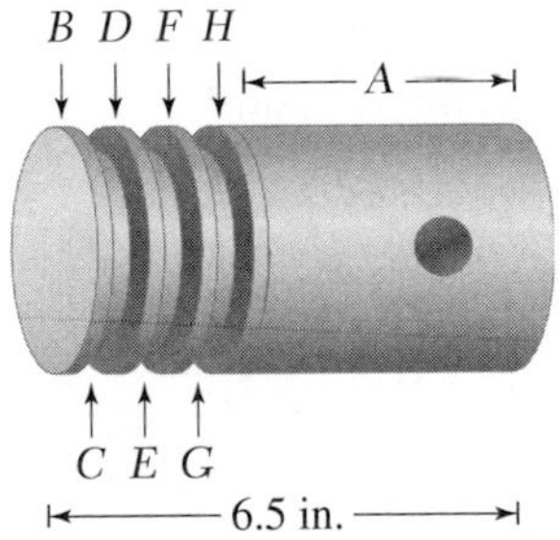

87. What is the center-to-center distance, A, between the holes in the diagram?

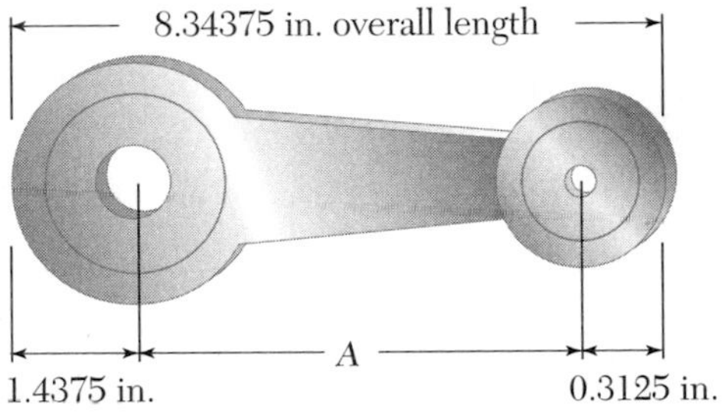

88. Find the total length of the connecting bar shown here.

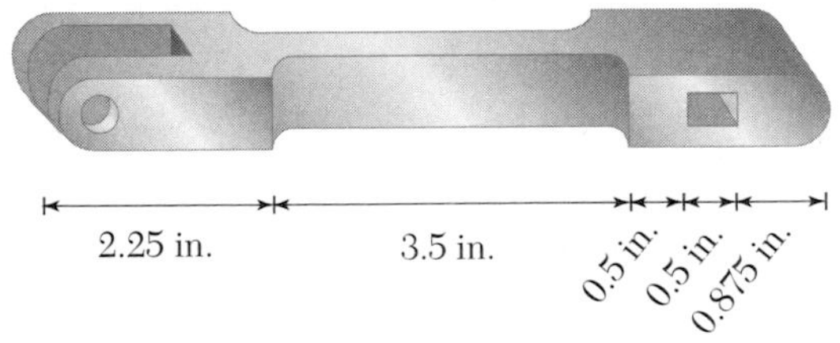

Exercises 89–93 relate to the chapter application.

89. Muthoni runs a race in 12.16 seconds, whereas Sera runs the same race in 11.382 seconds. How much faster is Sera?

90. A skier posts a race time of 1.257 minutes. A second skier posts a time of 1.32 minutes. A third skier completes the race in 1.2378 minutes. Find the difference between the fastest and the slowest times.

91. A college men's 4-×-100-m relay track team has runners with individual times of 9.35 sec, 9.91 sec, 10.04 sec, and 9.65 sec. What is the time for the relay?

92. A high school girl's swim team has a 200-yd freestyle relay, in which swimmers have times of 21.79 sec, 22.64 sec, 22.38 sec, and 23.13 sec. What is the time for the relay?

93. A high school girl's track coach knows that the rival school's team in the 4-×-100-m relay has a time of 52.78 sec. If the coach knows that her top three sprinters have times of 12.83, 13.22, and 13.56 sec, how fast does the fourth sprinter need to be in order to beat the rival school's relay team?

STATE YOUR UNDERSTANDING

94. Explain the procedure for adding 2.005, 8.2, 0.00004, and 3.

95. Explain the similarities between subtracting decimals and subtracting fractions.

96. Copy the table and fill it in.

Operation on Decimals	Procedure	Example
Addition		
Subtraction		

CHALLENGE

97. How many 8.75s must be added to have a sum that is greater than 200?

98. Find the missing number in the sequence: 0.4, 0.8, 1.3, ___, 2.6, 3.4, 4.3, 5.3.

99. Find the missing number in the sequence: 0.2, 0.19, 0.188, ___, 0.18766, 0.187655.

100. Which number in the following group is 11.1 less than 989.989: 999,999, 989.999, 988.889, 979.889, or 978.889?

101. Write the difference between $6\frac{9}{16}$ and 4.99 in decimal form.

102. Round the sum of 9.8989, 8.9898, 7.987, and 6.866 to the nearest tenth.

GROUP WORK

103. As a group, review the multiplication of fractions. Have each member make up a pair of fractions whose denominators are in the list: 10, 100, 1000, and 10,000. Find the product of each pair and change it to decimal form. In group discussion, make up a rule for multiplying decimals.

MAINTAIN YOUR SKILLS

Multiply.

104. 45(143)

105. 904(731)

106. $\begin{array}{r} 8367 \\ \times\ 507 \\ \hline \end{array}$

107. (56)(34)(14)

108. $\frac{3}{5} \cdot \frac{7}{10}$

109. $\frac{36}{75} \cdot \frac{15}{16} \cdot \frac{40}{27}$

110. $4\frac{2}{5} \cdot 2\frac{4}{5}$

111. $\left(3\frac{1}{2}\right)\left(6\frac{3}{4}\right)$

112. A nursery plants one seedling per square foot of ground. How many seedlings can be planted in a rectangular plot of ground that measures 235 ft by 347 ft?

113. The Candy Basket puts 45 pieces of chocolate peanut clusters in its Ultra Box. How many pieces of chocolate peanut clusters are needed to fill an order for 225 Ultra Boxes?

Getting Ready for Algebra

HOW & WHY

OBJECTIVE

Solve equations that involve addition and subtraction of decimals.

We solve equations that involve addition and subtraction of decimals in the same way as we do equations with whole numbers and fractions.

To solve an equation using addition or subtraction

1. Add the same number to both sides of the equation to isolate the variable.
2. Or subtract the same number from both sides of the equation to isolate the variable.

EXAMPLES A–E

DIRECTIONS: Solve.

STRATEGY: Isolate the variable by adding or subtracting the same number to or from both sides.

A. $7.8 = x + 5.6$

$7.8 = x + 5.6$

$7.8 - 5.6 = x + 5.6 - 5.6$ Eliminate the addition by subtracting 5.6 from both sides of the equation. Because subtraction is the inverse of addition, the variable will be isolated.

$2.2 = x$ Simplify.

CHECK: $7.8 = 2.2 + 5.6$ Substitute 2.2 for x in the original equation and simplify.

$7.8 = 7.8$ True

The solution is $x = 2.2$.

B. $z - 14.9 = 32.7$

$$\begin{aligned} z - 14.9 &= 32.7 \\ +14.9 &= +14.9 \\ \hline z &= 47.6 \end{aligned}$$

Eliminate the subtraction by adding 14.9 to both sides of the equation. Because subtraction is the inverse of addition, the variable will be isolated.

CHECK: $47.6 - 14.9 = 32.7$ Substitute 47.6 for z in the original equation and simplify.

$32.7 = 32.7$ True

The solution is 47.6.

C. $b + 17.325 = 34.6$

$b + 17.325 = 34.6$

$b + 17.325 - 17.325 = 34.6 - 17.325$ Subtract 17.325 from both sides and simplify.

$b = 17.275$

CHECK: $17.275 + 17.325 = 34.6$ Substitute 17.275 for b in the original equation.

$34.6 = 34.6$ True

The solution is $b = 17.275$.

D. $y - 6.233 = 8.005$

$$\begin{aligned} y - 6.233 &= 8.005 \\ +6.233 &= +6.233 \\ \hline y &= 14.238 \end{aligned}$$

Add 6.233 to both sides and simplify.

WARM-UPS A–E

A. $11.7 = p + 4.2$

B. $t - 13.6 = 29.5$

C. $c + 56.785 = 62$

D. $w - 33.17 = 12.455$

Answers to Warm-Ups

A. $p = 7.5$ **B.** $t = 43.1$
C. $c = 5.215$ **D.** $w = 45.625$

CHECK: $14.238 - 6.233 = 8.005$ Substitute 14.238 for y in the original equation.

$8.005 = 8.005$ True

The solution is $y = 14.238$.

E. The price of a graphing calculator decreased by \$19.30 over the past year. What was the price a year ago if the calculator now sells for \$81.95?

First write the English version of the equation:

Cost last year − decrease in cost = cost this year

Let x represent the cost last year.

$x - 19.30 = 81.95$	Translate to algebra.
$x - 19.30 + 19.30 = 81.95 + 19.30$	Add 19.30 to both sides.
$x = 101.25$	Simplify.

Because $101.25 - 19.30 = 81.95$, the cost of the calculator last year was \$101.25.

E. A farmer practicing "sustainable" farming reduced his soil erosion by 1.58 tons in 1 year. If he lost 3.94 tons of topsoil this year to erosion, how many tons did he lose last year?

Answers to Warm-Ups

E. The farmer lost 5.52 tons of topsoil to erosion.

EXERCISES

Solve.

1. $16.3 = x + 5.2$
2. $6.904 = x + 3.5$
3. $y - 0.64 = 13.19$
4. $w - 0.08 = 0.713$
5. $t + 0.03 = 0.514$
6. $x + 14.7 = 28.43$
7. $x - 7.3 = 5.21$
8. $y - 9.3 = 0.42$
9. $7.33 = w + 0.13$
10. $14 = x + 7.6$
11. $t - 8.37 = 0.08$
12. $w - 0.03 = 0.451$
13. $5.78 = a + 1.94$
14. $55.9 = w - 11.8$
15. $6.6 = x - 9.57$
16. $7 = 5.9 + x$
17. $a + 82.3 = 100$
18. $b + 45.76 = 93$
19. $s - 2.5 = 4.773$
20. $r - 6.7 = 5.217$
21. $c + 432.8 = 1029.16$
22. $d - 316.72 = 606.5$

23. The price of an energy-efficient hot water heater decreased by $52.75 over the past 2 years. What was the price 2 years ago if the heater now sells for $374.98?

24. In one state, the use of household biodegradable cleaners increased by 2444.67 lb per month because of state laws banning phosphates. How many pounds of these cleaners were used before the new laws if the average use now is 5780.5 lb?

25. The selling price of a personal computer is $1033.95. If the cost is $875.29, what is the markup?

26. The selling price of a new tire is $78.95. If the markup on the tire is $22.84, what is the cost (to the store) of the tire?

27. A shopper needs to buy a bus pass and some groceries. The shopper has $43 with which to make both purchases. If the bus pass costs $19, write and solve an equation that represents the shopper's situation. How much can the shopper spend on groceries?

28. In a math class, the final grade is determined by adding the test scores and the homework scores. If a student has a homework score of 18 and it takes a total of 90 to receive a grade of A, what total test score must the student have to receive a grade of A? Write an equation and solve it to determine the answer.

4.4 Multiplying Decimals

OBJECTIVES

1. Multiply decimals.
2. Estimate the product of decimals.

HOW & WHY

OBJECTIVE 1 Multiply decimals.

The "multiplication table" for decimals is the same as for whole numbers. In fact, decimals are multiplied the same way as whole numbers with one difference: the location of the decimal point in the product. To discover the rule for the location of the decimal point, we use what we already know about multiplication of fractions. First, change the decimal form to fractional form to find the product. Next, change the product back to decimal form and observe the number of decimal places in the product. Consider the examples in Table 4.4. We see that the product in decimal form has the same number of decimal places as the total number of places in the decimal factors.

TABLE 4.4 MULTIPLICATION OF DECIMALS

Decimal Form	Fractional Form	Product of Fractions	Product as a Decimal	Number of Decimal Places in Product
0.3×0.8	$\frac{3}{10} \times \frac{8}{10}$	$\frac{24}{100}$	0.24	two
11.2×0.07	$\frac{112}{10} \times \frac{7}{100}$	$\frac{784}{1000}$	0.784	three
0.02×0.13	$\frac{2}{100} \times \frac{13}{100}$	$\frac{26}{10000}$	0.0026	four

The shortcut is to multiply the numbers and insert the decimal point. If necessary, insert zeros so there are enough decimal places. The product of 0.1×0.4 has two decimal places because tenths multiplied by tenths yields hundredths.

$$0.1 \times 0.4 = 0.04 \quad \text{because} \quad \frac{1}{10} \times \frac{4}{10} = \frac{4}{100}$$

To multiply decimals

1. Multiply the numbers as if they were whole numbers.
2. Locate the decimal point by counting the number of decimal places (to the right of the decimal point) in both factors. The total of these two counts is the number of decimal places the product must have.
3. If necessary, insert zeros to the left of the numeral so there are enough decimal places (see Example D).

When multiplying decimals, it is not necessary to align the decimal points in the decimals being multiplied.

WARM-UPS A–F

A. Multiply: (7)(0.3)

B. Find the product of 0.6 and 0.53.

C. Find the product of (22.9) and (0.51).

D. Multiply 0.07 times 0.067.

E. Find the product: 67.91(43.058)

F. If 12 strips, each 6.45 centimeters wide, are to be cut from a piece of sheet metal, what is the narrowest piece of sheet metal that can be used?

Answers to Warm-Ups

A. 2.1 **B.** 0.318 **C.** 11.679
D. 0.00469 **E.** 2924.06878
F. The sheet metal must be 77.4 centimeters wide.

EXAMPLES A–F

DIRECTIONS: Multiply.

STRATEGY: First multiply the numbers, ignoring the decimal points. Place the decimal point in the product by counting the number of decimal places in the two factors. Insert zeros if necessary to produce the number of required places.

A. Multiply: (0.8)(12)

(0.8)(12) = 9.6 — Multiply 8 and 12. The total number of decimal places in both factors is one (1), so there is one decimal place in the product.

So, (0.8)(12) = 9.6.

B. Find the product of 0.7 and 0.34.

(0.7)(0.34) = 0.238 — There are three decimal places in the product because the total number of places in the factors is three.

So the product of 0.7 and 0.34 is 0.238.

C. Find the product of (7.92) and (5.2).

```
   7.92
 ×  5.2
 ------
   1584
  3960
 ------
 41.184
```

Multiply the numbers as if they were whole numbers. There are three decimal places in the product.

So the product of 7.92 and 5.2 is 41.184.

D. Multiply 4.3 times 0.0007.

```
      4.3
 × 0.0007
 --------
  0.00301
```

Because 4.3 has one decimal place and 0.0007 has four decimal places, the product must have five decimal places. We must insert two zeros at the left so there are enough places in the answer.

So 4.3 times 0.0007 is 0.00301.

CALCULATOR EXAMPLE

E. Find the product: (42.73)(39.481)

STRATEGY: The calculator will automatically place the decimal point in the correct position.

The product is 1687.02313.

F. If exactly eight strips of metal, each 3.875 in. wide, are to be cut from a piece of sheet metal, what is the smallest (in width) piece of sheet metal that can be used?

STRATEGY: To find the width of the piece of sheet metal, we multiply the width of one of the strips by the number of strips needed.

```
   3.875
 ×     8
 -------
  31.000
```

The extra zeros can be dropped.

The pieces must be 31 in. wide.

HOW & WHY

OBJECTIVE 2 Estimate the product of decimals.

The product of two decimals can be estimated by rounding each factor and then multiplying the rounded factors. For instance,

$$\begin{array}{r} 0.0673 \\ \times\ \ 0.79 \\ \hline \end{array} \qquad \begin{array}{r} 0.07 \\ \times\ \ 0.8 \\ \hline 0.056 \end{array}$$

Round to the nearest hundredth.
Round to the nearest tenth.
Multiply.

The estimate of the product is 0.056. One use of the estimate is to see if the product is correct. If the calculated product is not close to 0.056, you should check the multiplication. In this case, the actual product is 0.053167, which is close to the estimate.

To estimate the product of two decimals

1. Round each number to a specified place value.
2. Multiply the rounded numbers.

EXAMPLE G

DIRECTIONS: Estimate the product. Then find the product and compare.

STRATEGY: Round each number to the specified place value. Multiply the rounded numbers. Multiply the original numbers.

G. 0.316 and 9.107 Round both numbers to the nearest tenth.

$$\begin{array}{r} 0.3 \\ \times\ 9.1 \\ \hline 3 \\ 270 \\ \hline 2.73 \end{array}$$

Round to the nearest tenth.
Round to the nearest tenth.

$$\begin{array}{r} 0.316 \\ \times 9.107 \\ \hline 2\ 212 \\ 31\ 60 \\ 2\ 844 \\ \hline 2.877812 \end{array}$$

The product is 2.877812 and is close to the estimated product of 2.73.

WARM-UP G

G. Estimate the product; then find the product and compare. 3.871 and 0.0051 Round 3.871 to the nearest tenth and 0.0051 to the nearest thousandth.

Answer to Warm-Up

G. 0.0195; 0.0197421

EXERCISES 4.4

OBJECTIVE 1 Multiply decimals.

Multiply.

A.

1. 0.5×9 **2.** 0.8×7 **3.** 1.5×6 **4.** 2.1×8

5. 3×0.09 **6.** 0.08×3 **7.** 0.5×0.4 **8.** 0.4×0.2

9. 0.07×0.6 **10.** 0.9×0.006 **11.** $0.16 \times (0.4)$ **12.** $1.2 \times (0.05)$

13. The number of decimal places in the product of 3.511 and 6.2 is ______.

14. In the product $0.34 \times ? = 0.408$, the number of decimal places in the missing factor is ________.

B.

15. 8.94×0.006 **16.** 2.17×0.004 **17.** 1.45×4.6 **18.** 4.23×3.2

19. 6.97×0.36 **20.** 27.7×0.13 **21.** 0.346×7.8 **22.** 0.073×9.6

23. Find the product of 7.65 and 2.98.

24. Find the product of 5.87 and 3.44.

25. Multiply: 4.6(0.8)(0.13)

26. Multiply: 6.3(1.1)(0.15)

OBJECTIVE 2 Estimate the product of decimals.

A. *Estimate the product. Round each number to its largest nonzero place value.*

27. 32(0.845) **28.** 680(0.00231) **29.** 0.069(0.0537)

30. 0.00531(0.0892) **31.** (15.69)(0.0792) **32.** (0.00623)(47.83)

33. 0.0875×0.021 **34.** 0.2753×0.9631

B. *Estimate the product and then multiply.*

35. 42.7×0.53 **36.** 38.5×0.21 **37.** 0.356×0.067 **38.** 0.832×0.041

39. 0.0416×4.02 **40.** 0.00831×6.73 **41.** 0.825×0.0054 **42.** 0.575×0.00378

C.

Exercises 43–47. The table shows the amount of gas purchased by Grant and the price he paid per gallon for five fill-ups.

Gasoline Purchases

Number of Gallons	Price per Gallon
20.7	\$1.679
20.4	\$1.709
19.3	\$1.749
18.9	\$1.699
18.4	\$1.829

43. What is the total number of gallons of gas that Grant purchased?

44. To the nearest cent, how much did he pay for the second fill-up?

45. To the nearest cent, how much did he pay for the fifth fill-up?

46. To the nearest cent, what is the total amount he paid for the five fill-ups?

47. At which price per gallon did he pay the least for his fill-up?

Estimate the product by rounding each number to its largest place value. Then multiply.

48. (469.5)(7.12)

49. (89.76)(4.61)

50. (313.17)(8.73)

51. (10.5)(760.02)

Multiply.

52. (6.95)(4.71)(32.14) Round to the nearest hundredth.

53. (1.26)(171.9)(13.021) Round to the nearest thousandth.

54. (4.3)(6.12)(0.33)(11.9) Round to the nearest tenth.

55. (8.6)(0.86)(8.06)(80.6) Round to the nearest hundredth.

56. Joe earns \$11.85 per hour. How much does he earn if he works 36.5 hours in 1 week? Round to the nearest cent.

57. If upholstery fabric costs \$41.85 per yard, how much will Joanne pay for 16.25 yards? Round to the nearest cent.

Exercises 58–61. The table shows the cost of renting a car from a local agency.

Cost of Car Rentals

Type of Car	Cost per Day	Price per Mile Driven
Compact	\$19.95	\$0.25
Midsize	\$28.72	\$0.32
Luxury	\$43.85	\$0.42

58. What does it cost to rent a compact car for 4 days if it is driven 324 miles?

59. What does it cost to rent a midsize car for 3 days if it is driven 312 miles?

60. What does it cost to rent a luxury car for 6 days if it is driven 453 miles?

61. Which costs less, renting a midsize car for 3 days or a compact car for 5 days if both are driven 345 miles? How much less does it cost?

62. Tiffany can choose any of the following ways to finance her new car. Which method is the least expensive in the long run?

$650 down and $305.54 per month for 5 years
$350 down and $343.57 per month for 54 months
$400 down and $386.42 per month for 4 years

63. A new freezer-refrigerator is advertised at three different stores as follows:

Store 1: $75 down and $85.95 for 18 months
Store 2: $125 down and $63.25 per month for 24 months
Store 3: $300 down and $109.55 per month for 12 months

Which store is selling the freezer-refrigerator for the least total cost?

Exercises 64–66. The table shows calories expended for some physical activities.

Calorie Expenditure for Selected Physical Activities

Activity	Step Aerobics	Running (7 min/mi)	Cycling (10 mph)	Walking (4.5 mph)
Calories per pound of body weight per minute	0.070	0.102	0.050	0.045

64. Vanessa weighs 132 lb and does 50 min of step aerobics three times per week. How many calories does she burn per week?

65. Steve weighs 163 lb and runs 20 min five times per week at a 7 min/mi pace. How many calories does he burn per week?

66. Sephra weighs 150 lb and likes to walk daily at 4.5 mph. Her friend Dana weighs 125 lb and prefers to cycle daily at 10 mph. If both women exercise the same amount of time, who burns more calories?

67. An order of 43 bars of steel is delivered to a machine shop. Each bar is 17.325 ft long. Find the total linear feet of steel in the order.

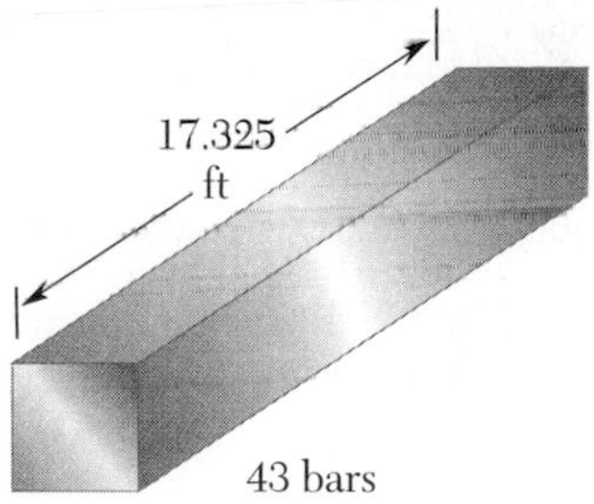

68. From a table in a machinist's handbook, it is determined that hexagon steel bars 1.325 in. across weigh 4.3 lb per running foot. Using this constant, find the weight of a 1.325-in. hexagon steel bar that is 18.875 ft long.

69. In 1970, the per capita consumption of red meat was 132 pounds. In 1980, the consumption was 126.4 pounds. In 1990, the amount consumed was 112.3 pounds per person. Compute the total weight of red meat consumed by a family of four using the rates for each of these years. Discuss the reasons for the change in consumption.

70. The fat content in a 3 oz serving of common meats and fish is as follows: beef rib, 7.4 g; beef top round, 3.4 g; beef top sirloin, 4.8 g; dark meat chicken without skin, 8.3 g; white meat chicken without skin, 3.8 g; pink salmon, 3.8 g; and Atlantic cod, 0.7 g. Which contains the most grams of fat: 3 servings of beef ribs, 6 servings of beef top round, 4 servings of beef top sirloin, 2 servings of dark meat chicken, 6 servings of white meat chicken, 5 servings of pink salmon, or 25 servings of Atlantic cod?

71. Older models of toilets use 5.5 gallons of water per flush. Models made in the 1970s use 3.5 gallons per flush. The new low-flow models use 1.55 gallons per flush. Assume each person flushes the toilet an average of five times per day. Determine the amount of water used in a town with a population of 41,782 in 1 day for each type of toilet. How much water is saved by using the low-flow model as opposed to the pre-1970s model?

72. The annual property tax bills arrive in early November in some states. The McNamara house is assessed for \$117,750. The annual tax rate is \$0.199 per hundred dollars of assessment. Find what the family owes in taxes.

73. Find the property tax on the Gregory estate, which is assessed at \$1,665,700. The tax rate in the area is \$1.135 per hundred dollars of assessment. Round to the nearest dollar.

Exercises 74–76 relate to the chapter application. In Olympic diving, seven judges rate each dive using a whole or half number between 0 and 10. The high and low scores are thrown out and the remaining scores are added together. (If a high or low score occurs more than once, only one is thrown out.) The sum is then multiplied by 0.6 and then by the difficulty factor of the dive to obtain the total points awarded.

74. A diver does a reverse $1\frac{1}{2}$ somersault with $2\frac{1}{2}$ twists, a dive with a difficulty factor of 2.9. She receives scores of 6.0, 6.5, 6.5, 7.0, 6.0, 7.5, and 7.0. What are the total points awarded for the dive?

75. Another diver also does a reverse $1\frac{1}{2}$ somersault with $2\frac{1}{2}$ twists. This diver receives scores of 7.5, 6.5, 7.5, 8.0, 8.0, 7.5, and 8.0. What are the total points awarded for the dive?

76. A cut-through reverse $1\frac{1}{2}$ somersault has a difficulty factor of 2.6. What is the highest number of points possible with this dive?

STATE YOUR UNDERSTANDING

77. Explain how to determine the number of decimal places needed in the product of two decimals.

78. Suppose you use a calculator to multiply (0.006)(3.2)(68) and get 13.056. Explain, using placement of the decimal point in a product, how you can tell that at least one of the numbers was entered incorrectly. How can such errors occur? How can you estimate the answers before using the calculator so you can avoid such errors?

CHALLENGE

79. What is the smallest whole number you can multiply 0.66 by to get a product that is greater than 55?

80. What is the largest whole number you can multiply 0.78 by to get a product that is less than 58.9?

81. Find the missing number in the following sequence: 1.8, 0.36. 0.108, 0.0432, ________.

82. Find the missing number in the following sequence: 3.1, 0.31, ______, 0.0000031, 0.0000000031.

GROUP WORK

83. Visit a grocery store or use newspaper ads to "purchase" the items listed. Have each member of your group use a different store or chain. Which members of your group "spent" the most? Least? Which group "spent" the most? Least?

Three 12-packs of Diet Pepsi
Eight gallons of 2% milk
Five pounds of hamburger
Four cans of the store-brand creamed corn
Five 4-roll packs of toilet paper
Seven pound of butter
72 hamburger rolls
12 large boxes of Cheerios

MAINTAIN YOUR SKILLS

Multiply or divide as indicated.

84. 445(100)

85. 75(10,000)

86. 78,300 ÷ 100

87. 15,000,000 ÷ 100,000

88. 12(1,000,000)

89. 803×10^3

90. $67{,}000 \div 10^2$

91. $380{,}000 \div 10^4$

92. 23×10^7

93. $4{,}210{,}000{,}000 \div 10^5$

4.5 Multiplying and Dividing by Powers of 10; Scientific Notation

VOCABULARY

Recall that a **power of 10** is the value obtained when 10 is written with an exponent.

Scientific notation is a special way to write numbers as a product using a number between 1 and 10 and a power of 10.

OBJECTIVES

1. Multiply or divide a number by a power of 10.
2. Write a number in scientific notation or change a number in scientific notation to its place value name.

HOW & WHY

OBJECTIVE 1 Multiply or divide a number by a power of 10.

The shortcut used in Section 1.5 for multiplying and dividing by a power of 10 works in a similar way with decimals. Consider the following products:

$$\begin{array}{r} 0.6 \\ \times\ 10 \\ \hline 0 \\ 6\,0 \\ \hline 6.0 \end{array} = 6 \qquad \begin{array}{r} 0.34 \\ \times\ \ 10 \\ \hline 0 \\ 3\,40 \\ \hline 3.40 \end{array} = 3.4 \qquad \begin{array}{r} 7.21 \\ \times\ \ \ 10 \\ \hline 0 \\ 72\,10 \\ \hline 72.10 \end{array} = 72.1$$

Note in each case that multiplying a decimal by 10 has the effect of moving the decimal point one place to the right.

Because $100 = 10 \cdot 10$, multiplying by 100 is the same as multiplying by 10 two times in succession. So, multiplying by 100 has the effect of moving the decimal point two places to the right. For instance

$$(0.71)(100) = (0.71)(10 \cdot 10) = (0.71 \cdot 10) \cdot 10 = 7.1 \cdot 10 = 71$$

Because $1000 = 10 \cdot 10 \cdot 10$, the decimal point will move three places to the right when multiplying by 1000. Because $10{,}000 = 10 \cdot 10 \cdot 10 \cdot 10$, the decimal point will move four places to the right when multiplying by 10,000, and so on in the same pattern.

$$(0.004523)(10{,}000) = 45.23$$

Zeros may have to be placed on the right in order to move the correct number of decimal places:

$$(6.3)(1000) = 6.300 = 6300$$

In this problem, two zeros are placed on the right.

Because multiplying a decimal by 10 has the effect of moving the decimal point one place to the right, dividing a number by 10 must move the decimal point one place to the left. Again, we are using the fact that multiplication and division are inverse operations. Division by 100 will move the decimal point two places to the left, and so on. Thus,

$$591.5 \div 100 = 591.5 = 5.915$$

$$0.852 \div 1000 = 0.000852$$

Three zeros are placed on the left so that the decimal point may be moved three places to the left.

To multiply a number by a power of 10

Move the decimal point to the right. The number of places to move is shown by the number of zeros in the power of 10.

To divide a number by a power of 10

Move the decimal point to the left. The number of places to move is shown by the number of zeros in the power of 10.

WARM-UPS A–G

A. Multiply: 78.33(10)

B. Multiply: 0.51(100)

C. Find the product of 49.4 and 10^4.

D. Divide: $772 \div 10$

E. Divide: $786.7 \div 100$

F. Find the quotient: $67.395 \div 10^4$

Answers to Warm-Ups

A. 783.3 **B.** 51 **C.** 494,000 **D.** 77.2 **E.** 7.867 **F.** 0.0067395

EXAMPLES A–G

DIRECTIONS: Multiply or divide as indicated.

STRATEGY: To multiply by a power of 10, move the decimal point to the right, inserting zeros as needed. To divide by a power of 10, move the decimal point to the left. The exponent of 10 specifies the number of places to move the decimal point.

A. Multiply: 7.357(10)

$7.357(10) = 73.57$ Multiplying by 10 moves the decimal point one place to the right.

So, $7.357(10) = 73.57$.

B. Multiply: 0.497(100)

$0.497(100) = 49.7$ Multiplying by 100 moves the decimal point two places to the right.

So, $0.497(100) = 49.7$.

C. Find the product of 13.7 and 10^3.

$13.7(10^3) = 13{,}700$ Multiplying by 10^3 moves the decimal point three places to the right. Two zeros must be inserted on the right to make the move.

So the product of 13.7 and 10^3 is 13,700.

D. Divide: $39.6 \div 10$

$39.6 \div 10 = 3.96$ Dividing by 10 moves the decimal point one place to the left.

So, $39.6 \div 10 = 3.96$.

E. Divide: $452.3 \div 100$

$452.3 \div 100 = 4.523$ Dividing by 100 moves the decimal point two places to the left.

So, $452.3 \div 100 = 4.523$.

F. Find the quotient: $366.2 \div 10^4$

$366.2 \div 10^4 = 0.03662$ Move the decimal point four places to the left. A zero is inserted on the left so we can make the move.

So, $366.2 \div 10^4 = 0.03662$.

G. A stack of sheet metal contains 1000 sheets. The stack is 93.2 inches high. How thick is each sheet of metal?

$93.2 \div 1000 = 0.0932$ — To find the thickness of each sheet, divide the height by the number of sheets.

Each sheet is 0.0932 inch thick.

G. One thousand sheets of clear plastic is 0.5 inch thick. How thick is each sheet? (This is the thickness of some household plastic wrap.)

HOW & WHY

OBJECTIVE 2 Write a number in scientific notation or change a number in scientific notation to its place value name.

Scientific notation is widely used in science, technology, and industry to write large and small numbers. Every scientific calculator has a key for entering numbers in scientific notation. This notation makes it possible for a calculator or computer to deal with much larger or smaller numbers than those that take up 8, 9 , or 10 spaces on the display.

For example, see Table 4.5.

TABLE 4.5 SCIENTIFIC NOTATION

Word Name	Place Value (Numeral Form)	Scientific Notation	Calculator or Computer Display
One million	1,000,000	1×10^6	1.06 or 1 E 6
Five billion	5,000,000,000	5×10^9	5.09 or 5 E 9
One trillion, three billion	1,003,000,000,000	1.003×10^{12}	1.003 12 or 1.003 E 12

Small numbers are shown by writing the power of 10 using a negative exponent. (You will learn more about this when you take a course in algebra.) For now, remember that multiplying by a negative power of 10 is the same as dividing by a power of 10, which means you will be moving the decimal point to the left. See Table 4.6.

TABLE 4.6 SCIENTIFIC NOTATION

Word Name	Place Value (Numeral Form)	Scientific Notation	Calculator or Computer Display
Eight thousandths	0.008	8×10^{-3}	8. −03 or 8 E −3
Seven ten-millionths	0.0000007	7×10^{-7}	7. −07 or 7 E −7
Fourteen hundred-billionths	0.00000000014	1.4×10^{-10}	1.4 −10 or 1.4 E −10

The shortcut for multiplying by a power of 10 is to move the decimal to the right, and the shortcut for dividing by a power of 10 is to move the decimal point to the left.

To write a number in scientific notation

1. Move the decimal point right or left so that only one digit remains to the left of the decimal point. The result will be a number between 1 and 10. If the choice is 1 or 10 itself, use 1.
2. Multiply the decimal found in step 1 by a power of 10. The exponent of 10 to use is one that will make the new product equal to the original number.
   - **a.** If you had to move the decimal to the left, multiply by the same number of 10s as the number of places moved.
   - **b.** If you had to move the decimal to the right, divide (by writing a negative exponent) by the same number of 10s as the number of places moved.

Scientific Notation

A number in scientific notation is written as the product of two numbers. The first number is between 1 and 10 (including 1 but not 10) and the second number is a power of 10. You must use an × to show the multiplication.

Answers to Warm-Ups

G. Each sheet of plastic is 0.0005 inch thick.

To change from scientific notation to place value name

1. If the exponent of 10 is positive, multiply by as many 10s (move the decimal point to the right as many places) as the exponent shows.
2. If the exponent of 10 is negative, divide by as many 10s (move the decimal point to the left as many places) as the exponent shows.

For numbers larger than 1:

Place value name:	15,000	7,300,000	18,500,000,000	
Numbers between 1 and 10:	1.5	7.3	1.85	Move the decimal (which is after the units place) to the left until the number is between 1 and 10 (one digit to the left of the decimal).
Scientific notation:	1.5×10^4	7.3×10^6	1.85×10^{10}	Multiply each by a power of 10 that shows how many places left the decimal moved, or how many places you would have to move to the right to recover the original number.

For numbers smaller than 1:

Place value name:	0.000074	0.00000009	0.0000000000267	
Numbers between 1 and 10:	7.4	9.	2.67	Move the decimal to the right until the number is between 1 and 10.
Scientific notation:	7.4×10^{-5}	9×10^{-8}	2.67×10^{-11}	Divide each by a power of 10 that shows how many places right the decimal moved. Show this division by a negative power of 10.

It is important to note that scientific notation is not rounding. The scientific notation has exactly the same value as the original name.

WARM-UPS H–J

H. 65,000,000

I. 0.00000543

Answers to Warm-Ups

H. 6.5×10^7 **I.** 5.43×10^{-6}

EXAMPLES H–J

DIRECTIONS: Write in scientic notation.

STRATEGY: Move the decimal point so there is one digit to the left. Multiply or divide this number by the appropriate power of 10 so the value is the same as the original number.

H. 943,000,000

9.43 is between 1 and 10. — Move the decimal eight places to the left.

$9.43 \times 100{,}000{,}000$ is 943,000,000. — Moving the decimal left is equivalent to dividing by 10 for each place.

$943{,}000{,}000 = 9.43 \times 10^8$ — To keep the values the same, we multiply by 10 eight times.

I. 0.00000087

8.7 is between 1 and 10. — Move the decimal seven places to the right.

$8.7 \div 10{,}000{,}000$ is 0.00000087. — Moving the decimal right is equivalent to multiplying by 10 for each place.

$0.00000087 = 8.7 \times 10^{-7}$ — To keep the values the same we divide by 10 seven times.

J. Approximately 12,000,000 people in the United States have type 2 diabetes. Write this number in scientific notation.

1.2 is between 1 and 10.

$1.2 \times 10{,}000{,}000 = 12{,}000{,}000$

$12{,}000{,}000 = 1.2 \times 10^7$

In scientific notation the number of people with type 2 diabetes is 1.2×10^7.

J. The age of a 22-year-old student is approximately 694,000,000 seconds. Write this number in scientific notation.

Answers to Warm-Ups

J. The age of a 22-year-old student is approximately 6.94×10^8 seconds.

EXAMPLES K–L

DIRECTIONS: Write the place value name.

STRATEGY: If the exponent is positive, move the decimal point to the right as many places as shown in the exponent. If the exponent is negative, move the decimal point to the left as many places as shown by the exponent.

K. Write the place value name for 6.7×10^{-6}.

$6.7 \times 10^{-6} = 0.0000067$ — The exponent is negative, so move the decimal point six places to the left. That is, divide by 10 six times.

L. Write the place value name for 4.033×10^9.

$4.033 \times 10^9 = 4{,}033{,}000{,}000$ — The exponent is positive, so move the decimal point nine places to the right. That is, multiply by 10 nine times.

WARM-UPS K–L

K. Write the place value name for 3.45×10^{-8}.

L. Write the place value name for 9.1×10^{10}.

Answers to Warm-Ups

K. 0.0000000345 **L.** 91,000,000,000

Name ______________________ Class ______________ Date __________

EXERCISES 4.5

OBJECTIVE 1 Multiply or divide a number by a power of 10.

Multiply or divide.

A.

1. $3.67 \div 10$

2. $47.13 \div 10$

3. $(8.21)(100)$

4. $63.7(100)$

5. $(0.62833)(1000)$

6. $(34.6211)(1000)$

7. $\dfrac{135.9}{100}$

8. $\dfrac{898.3}{1000}$

9. $\dfrac{7913}{100}$

10. $\dfrac{4582}{1000}$

11. 0.87×10^4

12. 4.3×10^5

13. To multiply 4.56 by 10^4, move the decimal point four places to the ________.

14. To divide 4.56 by 10^5, move the decimal point five places to the ________.

B.

15. $(12.563)(1000)$

16. $21.63(100)$

17. $3.95 \div 10$

18. $187.3 \div 1000$

19. $36.9(1000)$

20. $0.6783(10)$

21. $\dfrac{9077.5}{10{,}000}$

22. $\dfrac{6351.42}{100{,}000}$

23. $2.93\ (100{,}000)$

24. $7.16(1{,}000{,}000)$

25. $1367.94 \div 100$

26. $78.94 \div 1000$

27. $581.9 \div 100{,}000$

28. $14.631 \div 1{,}000{,}000$

OBJECTIVE 2 Write a number in scientific notation or change a number in scientific notation to its place value name.

A.

Write in scientific notation.

29. 420,000

30. 7400

31. 0.00035

32. 0.0000521

33. 4195.3

34. 82710.3

Write in place value form.

35. 7×10^4

36. 9×10^6

37. 8×10^{-3}

38. 6×10^{-6}

39. 8.35×10^4

40. 9.02×10^5

B.

Write in scientific notation.

41. 82,000

42. 6,930,000

43. 0.0000072

44. 0.000317

45. 0.0000000000821

46. 0.00000002977

47. 975.002

48. 3496.701

Write in place value notation.

49. 2.912×10^{-5}

50. 7.052×10^{-4}

51. 7.11×10^{9}

52. 8.032×10^{8}

53. 3.11×10^{-8}

54. 5.6×10^{-9}

55. 5.6723×10^{2}

56. 7.892111×10^{5}

C.

57. Ken's Shoe Store buys 100 pairs of shoes that cost $27.68 per pair. What is the total cost of the shoes?

58. If Mae's Shoe Store buys 100 pairs of shoes for a total cost of $4355, what is the cost of each pair of shoes?

59. Ms. James buys 100 acres of land at a cost of $1100 per acre. What is the total cost of her land?

60. If 1000 bricks weigh 6100 lb, how much does each brick weigh?

61. The total land area of Earth is approximately 52,000,000 square miles. What is the total area written in scientific notation?

62. A local computer store offers a small computer with 1152K (1,152,000) bytes of memory. Write the number of bytes in scientific notation.

63. The length of a red light ray is 0.000000072 cm. Write this length in scientific notation.

64. The time it takes light to travel 1 kilometer is approximately 0.0000033 second. Write this time in scientific notation.

65. The speed of light is approximately 1.116×10^{7} miles per minute. Write this speed in place value notation.

66. Earth is approximately 1.5×10^{8} kilometers from the sun. Write this distance in place value form.

67. The shortest wavelength of visible light is approximately 4×10^{-5} cm. Write this length in place value form.

68. A sheet of paper is approximately 1.3×10^{-3} inch thick. Write the thickness in place value form.

69. A family in the Northeast used 3.276×10^{8} BTUs of energy during 1989. A family in the Midwest used 3.312×10^{8} BTUs in the same year. A family in the South used 3.933×10^{8} BTUs, and a family in the West used 1.935×10^{8} BTUs. Write the total energy usage for the four families in place value form.

70. In 1990, the per capita consumption of fish was 15.5 pounds. In the same year, the per capita consumption of poultry was 63.6 pounds and of red meat was 112.3 pounds. Write the total amount in each category consumed by 100,000 people in scientific notation.

71. The population of Cabot Cove was approximately 10,000 in 1995. During the year, the community consumed a total of 276,000 gallons of milk. What was the per capita consumption of milk in Cabot Cove in 1995?

72. In 1980, $24,744,000,000 was spent on air pollution abatement. Ten years later, $26,326,000,000 was spent. In scientific notation, how much more money was spent in 1990 than in 1980? What is the average amount of increase per year during the period?

Exercises 73–75 relate to the chapter application. In baseball, a hitter's batting average is calculated by dividing the number of hits by the number of times at bat. Mathematically, this number is always between zero and one.

73. In 1999 and 2000, the American League batting champion was Nomar Garciaparra of the Boston Red Sox. Garciaparra's 2000 batting average was 0.372, but fans would say his average was "three hundred seventy-two." Mathematically, what are the fans doing to the actual number?

74. Explain why the highest possible batting average is 1.0.

75. The major league player with the highest season batting average in the last century was Rogers Hornsby of St. Louis. In 1924 he batted 424. Change this to the mathematically calculated number of his batting average.

STATE YOUR UNDERSTANDING

76. Find a pair of numbers whose product is larger than ten trillion. Explain how scientific notation makes it possible to multiply these factors on a calculator. Why is it not possible without scientific notation?

CHALLENGE

77. A parsec is a unit of measure used to determine distance between stars. One parsec is approximately 206,265 times the average distance of Earth from the sun. If the average distance from Earth to the sun is approximately 93,000,000 miles, find the approximate length of 1 parsec. Write the length in scientific notation. Round the number in scientific notation to the nearest hundredth.

78. Light will travel approximately 5,866,000,000,000 miles in 1 year. Approximately how far will light travel in 8 years? Write the distance in scientific notation. Round the number in scientific notation to the nearest thousandth.

Simplify.

79. $\dfrac{(3.25 \times 10^{-3})(2.4 \times 10^{3})}{(4.8 \times 10^{-4})(2.5 \times 10^{-3})}$

80. $\dfrac{(3.25 \times 10^{-7})(2.4 \times 10^{6})}{(4.8 \times 10^{4})(2.5 \times 10^{-3})}$

GROUP WORK

81. Find the 1990 population for the ten largest cities and the five smallest cities in your state. Round these numbers to the nearest thousand. Find the total number of pounds of fruit, at the rate of 92.3 pounds per person, and the total number of pounds of vegetables, at the rate of 11.2 pounds per person, consumed in each of these fifteen cities.

MAINTAIN YOUR SKILLS

Divide.

82. $38\overline{)18{,}050}$

83. $76\overline{)8208}$

84. $103\overline{)21{,}527}$

85. $\dfrac{312}{11}$

86. $\dfrac{3467}{23}$

87. Find the quotient of 297,168 and 123.

88. Find the quotient of 231,876 and 203.

89. Find the quotient of 2,529,948 and 525. Round to the nearest ten.

90. Find the perimeter of a rectangular field that is 567 ft long and 288 ft wide.

91. The area of a rectangle is 648 in^2. If the length of the rectangle is 3 ft, find the width.

4.6 Dividing Decimals

HOW & WHY

OBJECTIVES

1. Divide decimals.
2. Estimate the quotient of decimals.

OBJECTIVE 1 Divide decimals.

Division of decimals is the same as division of whole numbers, with one difference. The difference is the location of the decimal point in the quotient.

As with multiplication, we examine the fractional form of division to discover the method of placing the decimal point in the quotient. First, change the decimal form to fractional form to find the quotient. Next, change the quotient to decimal form. Consider the information in Table 4.7.

TABLE 4.7 DIVISION BY A WHOLE NUMBER

Decimal Form	Fractional Form	Division: Fractional Form	Division: Decimal Form
$3\overline{)0.36}$	$\frac{36}{100} \div 3$	$\frac{36}{100} \div 3 = \frac{\overset{12}{\cancel{36}}}{100} \cdot \frac{1}{\underset{1}{\cancel{3}}} = \frac{12}{100}$	$\begin{array}{r} 0.12 \\ 3\overline{)0.36} \end{array}$
$8\overline{)0.72}$	$\frac{72}{100} \div 8$	$\frac{72}{100} \div 8 = \frac{\overset{9}{\cancel{72}}}{100} \cdot \frac{1}{\underset{1}{\cancel{8}}} = \frac{9}{100}$	$\begin{array}{r} 0.09 \\ 8\overline{)0.72} \end{array}$
$5\overline{)0.3}$	$\frac{3}{10} \div 5$	$\frac{3}{10} \div 5 = \frac{3}{10} \cdot \frac{1}{5} = \frac{3}{50} = \frac{6}{100}$	$\begin{array}{r} 0.06 \\ 5\overline{)0.3} \end{array}$

We can see from Table 4.7 that the decimal point for the quotient of a decimal and a whole number is written directly above the decimal point in the dividend. It may be necessary to insert zeros to do the division. See Example B.

When a decimal is divided by 7, the division process may not have a remainder of zero at any step:

$$\begin{array}{r} 0.97 \\ 7\overline{)6.85} \\ \underline{6\,3} \\ 55 \\ \underline{49} \\ 6 \end{array}$$

At this step, we can write zeros to the right of the digit 5 because 6.85 = 6.850 = 6.8500 = 6.85000 = 6.850000.

$$\begin{array}{l} 0.97857 \\ 7\overline{)6.85000} \\ \underline{6\,3} \\ 55 \\ \underline{49} \\ 60 \\ \underline{56} \\ 40 \\ \underline{35} \\ 50 \\ \underline{49} \\ 1 \end{array}$$

It appears that we might go on inserting zeros and continuing endlessly. This is indeed what happens. Such decimals are called "nonterminating, repeating decimals." For example, the quotient of this division is sometimes written

$0.97857142857142\ldots$ or $0.97\overline{857142}$

The bar written above the sequence of digits 857142 indicates that these digits are repeated endlessly.

In practical applications, we stop the division process one place value beyond the accuracy required by the situation and then round. Therefore

```
  0.97                    0.9785
7)6.85                  7)6.8500
  6 3                     6 3
   55                      55
   49                      49
    6   Stop.               60
                            56
                             40
                             35
                              5   Stop.
```

$6.85 \div 7 \approx 1.0$ rounded to the nearest tenth.

$6.85 \div 7 \approx 0.979$ rounded to the nearest thousandth.

Now let's examine division when the divisor is also a decimal. We use what we already know about division with a whole-number divisor. See Table 4.8.

TABLE 4.8 DIVISION BY A DECIMAL

Decimal Form	Conversion to a Whole-Number Divisor	Decimal Form of the Division
$0.3\overline{)0.36}$	$\frac{0.36}{0.3}\cdot\frac{10}{10}=\frac{3.6}{3}=1.2$	$0.3\overline{)0.3\,6}$ = 1.2
$0.4\overline{)1.52}$	$\frac{1.52}{0.4}\cdot\frac{10}{10}=\frac{15.2}{4}=3.8$	$0.4\overline{)1.5\,2}$ = 3.8
$0.08\overline{)0.72}$	$\frac{0.72}{0.08}\cdot\frac{100}{100}=\frac{72}{8}=9$	$0.08\overline{)0.72}$ = 9.
$0.25\overline{)0.3}$	$\frac{0.3}{0.25}\cdot\frac{100}{100}=\frac{30}{25}=1.2$	$0.25\overline{)0.30\,0}$ = 1.2
$0.006\overline{)4.8}$	$\frac{4.8}{0.006}\cdot\frac{1000}{1000}=\frac{4800}{6}=800$	$0.006\overline{)4.800}$ = 800.

We see from Table 4.8 that we move the decimal point in both the divisor and the dividend the number of places to make the divisor a whole number. Then divide as before.

To divide two numbers

1. If the divisor is not a whole number, move the decimal point in both the divisor and dividend to the right the number of places necessary to make the divisor a whole number.
2. Place the decimal point in the quotient above the decimal point in the dividend.
3. Divide as if both numbers were whole numbers.
4. Round to the given place value. (If no round-off place is given, divide until the remainder is zero or round as appropriate in the problem. For instance, in problems with money, round to the nearest cent.)

EXAMPLES A–G

DIRECTIONS: Divide. Round as indicated.

STRATEGY: If the divisor is not a whole number, move the decimal point in both the divisor and the dividend to the right the number of places necessary to make the divisor a whole number. The decimal point in the quotient is found by writing it directly above the decimal (as moved) in the dividend.

A. Divide: $16\overline{)33.008}$

$$\begin{array}{r}
2.063 \\
16\overline{)33.008} \\
\underline{32} \\
1\,0 \\
\underline{0\,0} \\
100 \\
\underline{96} \\
48 \\
\underline{48} \\
0
\end{array}$$

The numerals in the answer are lined up in columns that have the same place value as those in the dividend.

CHECK:

$$\begin{array}{r}
2.063 \\
\times \quad 16 \\
\hline
12\,378 \\
\underline{20\,63} \\
33.008
\end{array}$$

So the quotient is 2.063.

CAUTION

Write the decimal point for the quotient directly above the decimal point in the dividend.

B. Find the quotient of 7.41 and 6.

STRATEGY: Recall that the quotient of a and b is written $a \div b$ or $b\overline{)a}$.

WARM-UPS A–G

A. Divide: $12\overline{)61.236}$

B. Find the quotient of 42.39 and 18.

$$
\begin{array}{r}
1.23 \\
6\overline{)7.41} \\
\underline{6} \\
1\,4 \\
\underline{1\,2} \\
21 \\
\underline{18} \\
3
\end{array}
$$

Here the remainder is not zero, so the division is not complete. We write a zero on the right (7.410) without changing the value of the dividend and continue dividing.

$$
\begin{array}{r}
1.235 \\
6\overline{)7.410} \\
\underline{6} \\
1\,4 \\
\underline{1\,2} \\
21 \\
\underline{18} \\
30 \\
\underline{30} \\
0
\end{array}
$$

Both the quotient (1.235) and the rewritten dividend (7.410) have three decimal places. Check by multiplying 6×1.235:

$$
\begin{array}{r}
1.235 \\
\underline{\times \quad 6} \\
7.410
\end{array}
$$

The quotient is 1.235.

C. Divide 489.9 by 34 and round the quotient to the nearest hundredth.

C. Divide 673.7 by 21 and round the quotient to the nearest hundredth.

$$
\begin{array}{r}
32.080 \\
21\overline{)673.700} \\
\underline{63} \\
43 \\
\underline{42} \\
17 \\
\underline{00} \\
170 \\
\underline{168} \\
20 \\
\underline{00} \\
20
\end{array}
$$

It is necessary to place two zeros on the right in order to round to the hundredths place because the division must be carried out one place past the place to which you wish to round.

The quotient is approximately 32.08.

D. Divide $5.87 \div 0.7$ and round to the nearest hundredth.

D. Divide $6.75 \div 0.7$ and round to the nearest hundredth.

$$0.7\overline{)6.75}$$

First move both decimal points one place to the right so the divisor is the whole number 7. The same result is obtained by multiplying both divisor and dividend by 10:

$$
\begin{array}{r}
9.6 \\
7\overline{)67.5} \\
\underline{63} \\
4\,5 \\
\underline{4\,2} \\
3
\end{array}
$$

$$\frac{6.75}{0.7} \times \frac{10}{10} = \frac{67.5}{7}$$

$$
\begin{array}{r}
9.642 \\
7\overline{)67.500} \\
\underline{63} \\
4\,5 \\
\underline{4\,2} \\
30 \\
\underline{28} \\
20 \\
\underline{14} \\
6
\end{array}
$$

The number of zeros you place on the right depends on either the directions for rounding or your own choice of the number of places. Here we find the approximate quotient to the nearest hundredth.

The quotient is approximately 9.64.

Answers to Warm-Ups

A. 5.103 **B.** 2.355 **C.** 14.41
D. 8.39

E. Divide 0.57395 by 0.067 and round to the nearest thousandth.

0.067)0.57395

```
      8.5664
67)573.9500
   536
    37 9
    33 5
     4 45
     4 02
       430
       402
        280
        268
         12
```

Move both decimals three places to the right. It is necessary to insert two zeros on the right in order to round to the the thousandths place.

The quotient is approximately 8.566.

CALCULATOR EXAMPLE

F. Find the quotient of 87.6587 and 5.834 and round to the nearest thousandth.

$87.6587 \div 5.834 \approx 15.025488$

The quotient is 15.025 to the nearest thousandth.

G. What is the cost per ounce of a 19.25-oz box of cereal that costs $3.29? This is called the "unit price" and is used for comparing prices. Many stores are required to show this price for the food they sell.

STRATEGY: To find the unit price (cost per ounce), we divide the cost by the number of ounces. Round to the nearest tenth of a cent.

```
         0.1709
19.25)3.29 0000
      1 92 5
      1 36 50
      1 34 75
         1 850
         0 000
         1 8500
         1 7325
           1175
```

The cereal costs approximately $0.171, or 17.1¢ per ounce.

E. Divide 0.85697 by 0.083 and round to the nearest thousandth.

F. Find the quotient of 603.6033 and 9.704 and round to the nearest thousandth.

G. What is the unit price of corn chips if a 8.5-oz bag costs $2.34? Round to the nearest tenth of a cent.

HOW & WHY

OBJECTIVE 2 Estimate the quotient of decimals.

The quotient of two decimals can be estimated by rounding each number to a specified place value and then dividing the rounded numbers. For instance,

0.27)0.006345 0.3)0.006 Round the divisor to the nearest tenth and the dividend to the nearest thousandth.

```
  0.02
3)0.06
   6
   0
```

Multiply by 10 so the divisor is a whole number. Divide.

Answers to Warm-Ups

E. 10.325 **F.** 62.201

G. The unit price of corn chips is $0.275, or 27.5¢.

The estimate of the quotient is 0.02. One use of the estimate is to see if the quotient is correct. If the calculated quotient is not close to 0.02, you should check the division. In this case, the actual quotient is 0.0235, which is close to the estimate.

The estimated quotient will not always come out even. When this happens, divide to find the first two nonzero digits in the quotient. If necessary, write zeros on the right so that the digits have the correct place value. For instance,

$7.41\overline{)38.63574}$ $\quad 7\overline{)39}$ Round each number to the nearest unit.

$$\begin{array}{r} 5.5 \\ 7\overline{)39.0} \\ \underline{35} \\ 4\,0 \\ \underline{3\,5} \end{array}$$

The estimated quotient is 5.5. The exact quotient is 5.214.

To estimate the quotient of decimals

1. Round the divisor to one digit and the dividend to two digits.
2. Divide the rounded numbers.
3. If the first partial quotient has a remainder, continue the division so the quotient has two digits. Write zeros on the right, if necessary, for the digits to have the correct place value.

WARM-UP H

H. Divide 0.0039 by 0.723.

EXAMPLE H

DIRECTIONS: Estimate the quotient and find the quotient to the nearest ten-thousandth.

STRATEGY: Round the divisor to one digit and the dividend to two digits and then divide the rounded numbers. If the first partial quotient has a remainder, continue the division so the quotient has two digits.

H. Divide 0.05682 by 0.908.

$0.908\overline{)0.05682}$ $\quad 0.9\overline{)0.057}$ Round.

$$\begin{array}{r} 0.063 \\ 9\overline{)0.570} \\ \underline{54} \\ 30 \\ \underline{27} \end{array}$$ Divide.

$$\begin{array}{r} 0.06257 \\ 908\overline{)56.82} \\ \underline{54\,48} \\ 2\,340 \\ \underline{1\,816} \\ 5240 \\ \underline{4540} \\ 7000 \\ \underline{6356} \\ 644 \end{array}$$

The estimated quotient is 0.063 and the quotient is 0.0626 to the nearest ten-thousandth.

Answer to Warm-Up

H. 0.0055; 0.0054

*

EXERCISES 4.6

OBJECTIVE 1 Divide decimals.

A. *Divide.*

1. $9\overline{)6.3}$ **2.** $7\overline{)5.6}$ **3.** $3\overline{)19.8}$ **4.** $5\overline{)21.5}$

5. $0.1\overline{)18.31}$ **6.** $0.01\overline{)3.79}$ **7.** $242.4 \div 0.12$ **8.** $446.6 \div 0.11$

9. To divide 27.8 by 0.6, we first multiply both the dividend and the divisor by 10 so we are dividing by a ______________.

10. To divide 0.4763 by 0.287, we first multiply both the dividend and the divisor by ________.

B.

Divide.

11. $70\overline{)938}$ **12.** $70\overline{)172.9}$ **13.** $16.64 \div 32$ **14.** $3.936 \div (32)$

Divide and round to the nearest tenth.

15. $9\overline{)4.78}$ **16.** $8\overline{)0.499}$ **17.** $1.7\overline{)12.863}$ **18.** $7.8\overline{)49.381}$

Divide and round to the nearest thousandth.

19. $2.7\overline{)41.01}$ **20.** $37.4\overline{)75.126}$

OBJECTIVE 2 Estimate the quotient of decimals.

A. *Estimate the quotient.*

21. $6\overline{)0.5934}$ **22.** $8\overline{)0.0693}$ **23.** $0.2\overline{)2.67}$ **24.** $0.3\overline{)5.823}$

25. $0.793 \div 0.413$ **26.** $0.6341 \div 0.0285$ **27.** $34 \div 0.0756$ **28.** $76 \div 0.08659$

B. *Estimate the quotient. Divide and round to the nearest thousandth.*

29. $73\overline{)7105.36}$ **30.** $23\overline{)103.916}$

31. $4.16\overline{)0.06849}$ **32.** $3.46\overline{)0.5699}$

33. $0.13 \div 0.009$ **34.** $0.39 \div 0.0087$

C.

35. Find the quotient of 12.13 and 0.15, and round to the nearest hundredth.

36. Find the quotient of 0.596 and 86, and round to the nearest hundredth.

Exercises 37–43. The table shows some prices from a grocery store.

Grocery Prices

Item	Quantity	Price	Item	Quantity	Price
Oranges	3 lb	$4.42	Potatoes	15 lb	$1.16
Strawberries	6 pt	$4.77	Rib steak	3.24 lb	$9.62
Syrup	36 oz	$2.77	Ham	4.6 lb	$16.05

37. Find the unit price (price per pound) of oranges. Round to the nearest tenth of a cent.

38. Find the unit price (price per ounce) of syrup. Round to the nearest tenth of a cent.

39. Find the unit price of rib steak. Round to the nearest tenth of a cent.

40. Find the unit price of potatoes. Round to the nearest tenth of a cent.

41. Using the unit price, find the cost of 7 lb of ham. Round to the nearest cent.

42. Using the unit price, find the cost of 11 pt of strawberries. Round to the nearest cent.

43. Using unit pricing, find the total cost of 6 lb of rib steak, 2 pt of strawberries, 4 lb of potatoes, and 5 lb of oranges.

44. One hundred twelve alumni of Boston University donated $67,635 to the university. To the nearest cent, what was the average donation?

Estimate the quotient and divide. Round the actual quotient to the nearest hundredth.

45. $78.619 \div 0.411$

46. $110.61 \div 0.3482$

47. $0.3491 \div 0.0562$

48. $0.67321 \div 0.03599$

49. June drove 542.6 miles on 18.3 gallons of gas. What is her mileage (miles per gallon)? Round to the nearest mile.

50. A 65-gallon drum of cleaning solvent in an auto repair shop is being used at the rate of 1.94 gallons per day. At this rate, how many days will the drum last? Round to the nearest day.

51. The Williams Construction Company uses cable that weighs 3.2 pounds per foot. A partly filled spool of the cable is weighed. The cable itself weighs 923 pounds after taking off for the weight of the spool. To the nearest foot, how many feet of cable are on the spool?

52. A plumber connects four buildings' sewers to the public sewer line. The total bill for the job is $6355.48. What is the average cost for each connection?

53. A carpenter works 214.5 hours in one month. How many hours did she work each day if the month contained 22 workdays? (Assume that she worked the same number of hours each day.)

54. The 1-ft I beam, shown here, weighs 32.7 lb. What is the length of a beam weighing 630.6 lb? Find the length to the nearest tenth of a foot.

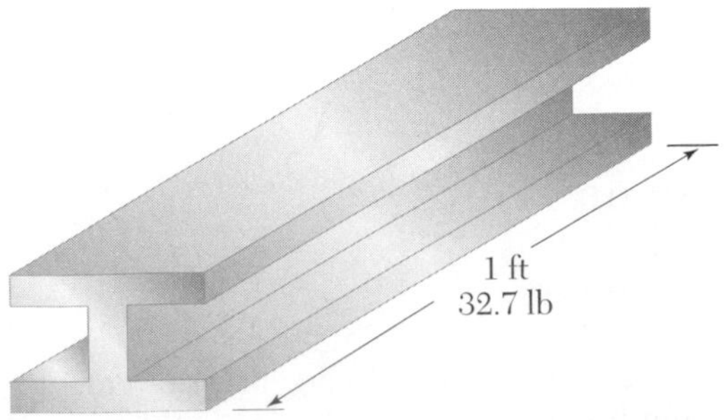

55. A contractor estimates the labor cost of pouring 75 cubic yards of concrete to be $4050. How much does he allow for each cubic yard of concrete?

56. Allowing 0.125 in. of waste for each cut, how many bushings, which are 1.35 in. in length, can be cut from a 10-in. length of bronze? What is the length of the piece that is left?

Exercises 57–59. The table shows population facts in Scandinavia.

Population and Area in Scandinavia

Country	Area (in square miles)	Population
Denmark	16,639	5,336,394
Finland	130,100	5,167,486
Norway	125,200	4,481,162
Sweden	173,732	8,873,052

57. Which country has the smallest area and which has the smallest population?

58. Population density is the number of people per square mile. Calculate the population density for each country, rounded to the nearest hundredth. Add another column to the table with this information.

59. What do you conclude about how crowded the countries are?

Exercises 60–64 relate to the chapter application. In baseball, a pitcher's earned run average (ERA) is calculated by dividing the number of earned runs by the quotient of the number of innings pitched and 9. The lower a pitcher's ERA, the better.

60. Suppose a pitcher allowed 26 earned runs in 80 innings of play. Calculate his ERA and round to the nearest hundredth.

61. A pitcher allows 15 earned runs in 100 innings. Calculate his ERA, rounding to the nearest hundredth.

62. A runner's stolen base average is the quotient of the number of bases stolen and the number of attempts. As with the batting averages, this number is usually rounded to the nearest thousandth. Calculate the stolen base average of a runner who stole 18 bases in 29 attempts.

63. A good stolen base average is 0.700 or higher. Express this as a fraction and say in words what the fraction represents.

64. The combined height of the NBA's 348 players at the start of the 2000–2001 season was 2294.67 feet or about twice the height of the Empire State Building. Find the average height of an NBA player. Round to the nearest tenth.

STATE YOUR UNDERSTANDING

65. Describe a procedure for determining the placement of the decimal in a quotient. Include an explanation for the justification of the procedure.

66. Explain how to find the quotient of $4.1448 \div 0.0012$.

67. Copy the table and fill it in.

Division Procedure

Operation on Decimals	Procedure	Example
Division		

CHALLENGE

68. What will be the value of \$3000 invested at 6% interest, compounded quarterly, at the end of 1 year? ("Compounded quarterly" means that the interest earned for the quarter, the annual interest divided by 4, is added to the principal and then earns interest for the next quarter.) How much more is earned by compounding quarterly instead of annually?

GROUP WORK

69. Determine the distance each member of your group travels to school each day. Find the average distance to the nearest hundredth of a mile for your group. Compare these results with those of the class. Find the average distance for the entire class. Recalculate the class average after throwing out the longest and the shortest distances. Are the averages different? Why?

70. In diving, a back $3\frac{1}{2}$ somersault, tuck position, has a difficulty factor of 3.3. What do the 2 judges need to score this dive in order for it to be worth the same number of points as a perfect cut-through reverse $1\frac{1}{2}$ somersault, which has a difficulty factor of 2.6? (See Exercises 74–76 in Section 4.4.)

MAINTAIN YOUR SKILLS

Simplify.

71. $\frac{95}{114}$

72. $\frac{168}{216}$

Write as an improper fraction.

73. $4\frac{8}{11}$

74. $18\frac{5}{7}$

Write as a mixed number.

75. $\frac{215}{12}$

76. $\frac{459}{25}$

Find the missing numerator.

77. $\frac{17}{25} = \frac{?}{100}$

78. $\frac{9}{40} = \frac{?}{1000}$

Write as a fraction or mixed number and simplify.

79. $24 \div 40$

80. $135 \div 30$

Getting Ready for Algebra

HOW & WHY

OBJECTIVE

Solve equations that involve multiplication and division of decimals.

We solve equations that involve multiplication and division of decimals in the same way as equations with whole numbers and fractions.

To solve an equation using multiplication or division

1. Divide both sides of the equation by the same number to isolate the variable.
2. Or multiply both sides of the equation by the same number to isolate the variable.

EXAMPLES A–E

DIRECTIONS: Solve.

STRATEGY: Isolate the variable by multiplying or dividing both sides by the same number.

A. $1.7x = 86.7$

$$\frac{1.7x}{1.7} = \frac{86.7}{1.7}$$

Because x is multiplied by 1.7, we divide both sides by 1.7. The division is usually written in fractional form. Because division is the inverse of multiplication, the variable is isolated.

$$x = 51$$

Simplify.

CHECK: $1.7(51) = 86.7$ Substitute 51 for x in the original equation and simplify.

$86.7 = 86.7$ True

The solution is $x = 51$.

B. $9.8 = \frac{a}{11.6}$

$$11.6(9.8) = 11.6\left(\frac{a}{11.6}\right)$$

$$113.68 = a$$

Because a is divided by 11.6, we multiply both sides by 11.6. Because multiplication is the inverse of division, the variable is isolated.

CHECK: $9.8 = \frac{113.68}{11.6}$ Substitute 113.68 for a in the original equation and simplify.

$9.8 = 9.8$ True

The solution is $a = 113.68$.

C. $34.6y = 186.84$

$$\frac{34.6y}{34.6} = \frac{186.84}{34.6}$$

$$y = 5.4$$

Divide both sides by 34.6 to eliminate the multiplication and simplify.

CHECK: $34.6(5.4) = 186.84$ Substitute 5.4 for y in the original equation and simplify.

$186.84 = 186.84$ True

The solution is $y = 5.4$.

WARM-UPS A–E

A. $1.2t = 96$

B. $13.5 = \frac{r}{9.7}$

C. $0.18a = 1.6632$

Answers to Warm-Ups

A. $t = 80$ **B.** $130.95 = r$ **C.** $a = 9.24$

D. $\dfrac{x}{0.481} = 7.2$

D. $\dfrac{c}{0.508} = 3.1$

$$0.508\left(\frac{c}{0.508}\right) = 0.508(3.1)$$ Multiply both sides by 0.508 and simplify.

$$c = 1.5748$$

CHECK: $\dfrac{1.5748}{0.508} = 3.1$ Substitute 1.5748 for c in the original equation and simplify.

$3.1 = 3.1$ True

The solution is $c = 3.1$.

E. Use the formula in Example E to find the number of calories per serving if there is a total of 3253.6 calories in 28 servings.

E. The total number of calories T is given by the formula $T = sC$, where s represents the number of servings and C represents the number of calories per serving. Find the number of calories per serving if 7.5 servings contain 739.5 calories.

First substitute the known values into the formula.

$T = sC$

$739.5 = 7.5C$ Substitute $T = 739.5$ and $s = 7.5$.

$\dfrac{739.5}{7.5} = \dfrac{7.5C}{7.5}$ Divide both sides by 7.5.

$98.6 = C$

* Because $7.5(8.6) = 739.5$, the number of calories per serving is 98.6.

Answers to Warm-Ups

D. $x = 3.4632$

E. There are 116.2 calories per serving.

EXERCISES

Solve.

1. $2.7x = 18.9$
2. $2.3x = 0.782$
3. $0.04y = 12.34$
4. $0.06w = 0.942$
5. $0.9476 = 4.12t$
6. $302.77 = 13.7x$
7. $3.3m = 0.198$
8. $0.008p = 12$
9. $0.016q = 9$
10. $11 = .025w$
11. $9 = 0.32h$
12. $2.6x = 35.88$
13. $\frac{y}{9.5} = 0.28$
14. $0.07 = \frac{b}{0.73}$
15. $0.312 = \frac{c}{0.65}$
16. $\frac{w}{0.12} = 1.35$
17. $0.0325 = \frac{x}{32}$
18. $0.17 = \frac{t}{8.23}$
19. $\frac{s}{0.07} = 0.345$
20. $\frac{y}{16.75} = 2.06$
21. $\frac{z}{21.02} = 4.08$
22. $\frac{c}{10.7} = 2.055$
23. The total number of calories T is given by the formula $T = sC$, where s represents the number of servings and C represents the number of calories per serving. Find the number of servings if the total number of calories is 3617.9 and there are 157.3 calories per serving.
24. Use the formula in Exercise 23 to find the number of servings if the total number of calories is 10,628.4 and there are 312.6 calories per serving.
25. Ohm's law is given by the formula $E = IR$, where E is the voltage (number of volts), I is the current (number of amperes), and R is the resistance (number of ohms). What is the current in a circuit if the resistance is 18 ohms and the voltage is 135 volts?
26. Use the formula in Exercise 25 to find the current in a circuit if the resistance is 16 ohms and the voltage is 145 volts.
27. Find the length of a rectangle that has a width of 13.6 ft and an area of 250.24 ft^2.
28. Find the width of a rectangular plot of ground that has an area of 3751.44 m^2 and a length of 127.6 m.
29. Each student in a certain instructor's math classes hands in 20 homework assignments. During the term, the instructor has graded a total of 3500 homework assignments. How many students does this instructor have in all her classes? Write and solve an equation to determine the answer.
30. Twenty-four plastic soda bottles were recycled and made into one shirt. At this rate, how many shirts can be made from 910 soda bottles. Write and solve an equation to determine the answer.

4.7 Changing Fractions to Decimals

HOW & WHY

OBJECTIVE Change fractions to decimals.

Every decimal can be written as a whole number times the place value of the last digit on the right:

$$0.87 = 87 \times \frac{1}{100} = \frac{87}{100}$$

The fraction has a power of 10 for the denominator. Any fraction that has only prime factors of 2 and 5 in the denominator can be written as a decimal by building the denominator to a power of 10.

$$\frac{4}{5} = \frac{4}{5} \cdot \frac{2}{2} = \frac{8}{10} = 0.8$$
$$\frac{9}{20} = \frac{9}{20} \cdot \frac{5}{5} = \frac{45}{100} = 0.45$$

Every fraction can be thought of as a division problem $\left(\frac{4}{5} = 4 \div 5\right)$. Therefore, a second method for changing fractions to decimals is division. As you discovered in the previous section, many division problems with decimals do not have a zero remainder at any point. If the denominator of the fraction has prime factors other than 2 or 5, the quotient could be a nonterminating decimal. The fraction $\frac{1}{6}$ is an example:

$$\frac{1}{6} = 0.166666666.\ .\ .\ . \quad \text{or} \quad 0.1\overline{6}$$

The bar over the 6 indicates that the decimal repeats the number 6 forever. In the exercises for this section, round the division to the indicated decimal place or use the repeat bar as directed.

OBJECTIVE

Change fractions to decimals.

CAUTION

Be careful to use an equal sign (=) when your conversion is exact and an approximately equal sign (≈) when you have rounded.

To change a fraction to a decimal

Divide the numerator by the denominator.

To change a mixed number to a decimal

Change the fractional part to a decimal and add to the whole-number part.

WARM-UPS A–G

A. Change $\frac{17}{20}$ to a decimal.

B. Change $8\frac{7}{25}$ to a decimal.

C. Change $\frac{11}{16}$ to a decimal.

Answers to Warm-Ups

A. 0.85 **B.** 8.28 **C.** 0.6875

EXAMPLES A–G

DIRECTIONS: Change the fraction or mixed number to a decimal.

STRATEGY: Divide the numerator by the denominator. Round as indicated. If the number is a mixed number, add the decimal to the whole number.

A. Change $\frac{17}{40}$ to a decimal.

$$\begin{array}{r} 0.425 \\ 40\overline{)17.000} \\ \underline{16\,0} \\ 1\,00 \\ \underline{80} \\ 200 \\ \underline{200} \\ 0 \end{array}$$

Divide the numerator, 17, by the denominator, 40.

Therefore, $\frac{17}{40} = 0.425$.

B. Change $11\frac{43}{50}$ to a decimal.

$$\begin{array}{r} 0.86 \\ 50\overline{)43.00} \\ \underline{40\,0} \\ 3\,00 \\ \underline{3\,00} \\ 0 \end{array}$$

Divide the numerator, 43, by the denominator, 50.

$11\frac{43}{50} = 11.86$ Add the decimal to the whole number.

or

$\frac{43}{50} = \frac{43}{50}\cdot\frac{2}{2} = \frac{86}{100} = 0.86$ A fraction with a denominator that has only 2s or 5s for prime factors can be changed to a decimal that terminates by building.

So, $11\frac{43}{50} = 11 + 0.86 = 11.86$.

C. Change $\frac{15}{32}$ to a decimal.

$$\begin{array}{r} 0.46875 \\ 32\overline{)15.00000} \\ \underline{12\,8} \\ 2\,20 \\ \underline{1\,92} \\ 280 \\ \underline{256} \\ 240 \\ \underline{224} \\ 160 \\ \underline{160} \\ 0 \end{array}$$

This fraction can be changed by building to a denominator of 100,000, but the factor is not easily recognized unless we use a calculator.

$\frac{15}{32} = \frac{15}{32}\cdot\frac{3125}{3125} = \frac{46{,}875}{100{,}000} = 0.46875$

So, $\frac{15}{32} = 0.46875$.

> **CAUTION**
>
> Most fractions cannot be changed to terminating decimals because the denominators contain factors other than 2 and 5. In these cases, we round to the indicated place value or use a repeat bar.

D. Change $\frac{17}{21}$ to a decimal rounded to the nearest hundredth.

$$\begin{array}{r} 0.809 \\ 21\overline{)17.000} \\ \underline{16\,8} \\ 20 \\ \underline{00} \\ 200 \\ \underline{189} \\ 11 \end{array}$$

Divide 17 by 21. Carry out the division to three decimal places and round to the nearest hundredth.

So, $\frac{17}{21} \approx 0.81$.

E. Change $\frac{5}{9}$ to a decimal. Do not round.

$$\begin{array}{r} .5555 \\ 9\overline{)5.0000} \\ \underline{4\,5} \\ 50 \\ \underline{45} \\ 50 \\ \underline{45} \\ 50 \\ \underline{45} \\ 5 \end{array}$$

We see that the division will not have a zero remainder, so we use the repeat bar to show the quotient.

So $\frac{5}{9} = 0.\overline{5}$.

CALCULATOR EXAMPLE

F. Change $\frac{254}{385}$ to a decimal rounded to the nearest ten-thousandth.

$254 \div 385 \approx 0.6597402$

So $\frac{254}{385} \approx 0.6597$ to the nearest ten-thousandth.

D. Change $\frac{6}{7}$ to a decimal rounded to the nearest hundredth.

E. Change $\frac{3}{11}$ to a decimal. Do not round.

F. Change $\frac{571}{692}$ to a decimal rounded to the nearest thousandth.

Answers to Warm-Ups

D. 0.86 **E.** $0.\overline{27}$ **F.** 0.825

G. Change the measurements on the given pattern to the nearest tenth for use with a ruler marked in tenths.

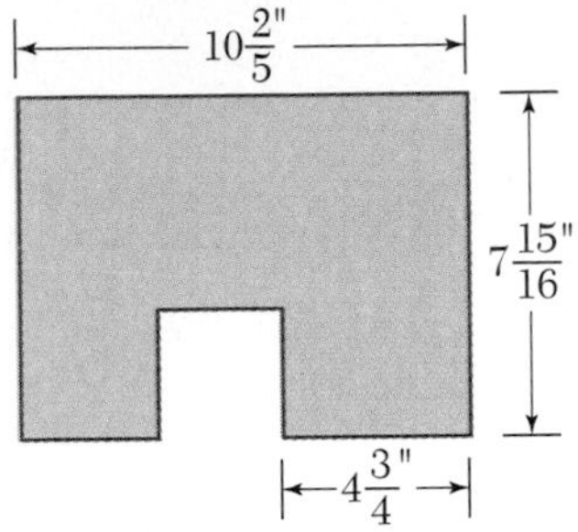

G. Jan needs to make a pattern of the shape as shown. Her ruler is marked in tenths. Change all the measurements to tenths so she can make an accurate pattern.

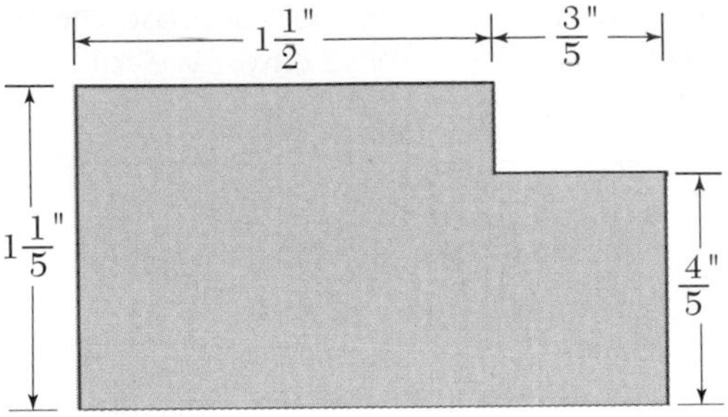

So that Jan can use her ruler for more accurate measure, each fraction is changed to a decimal rounded to the nearest tenth.

$$1\frac{1}{2} = 1\frac{5}{10} = 1.5$$

$$\frac{3}{5} = \frac{6}{10} = 0.6$$

$$1\frac{1}{5} = 1\frac{2}{10} = 1.2$$

$$\frac{4}{5} = \frac{8}{10} = 0.8$$

Each fraction and mixed number can be changed by either building each to a denominator of 10 as shown or by dividing the numerator by the denominator. The measurements on the drawing can be labeled:

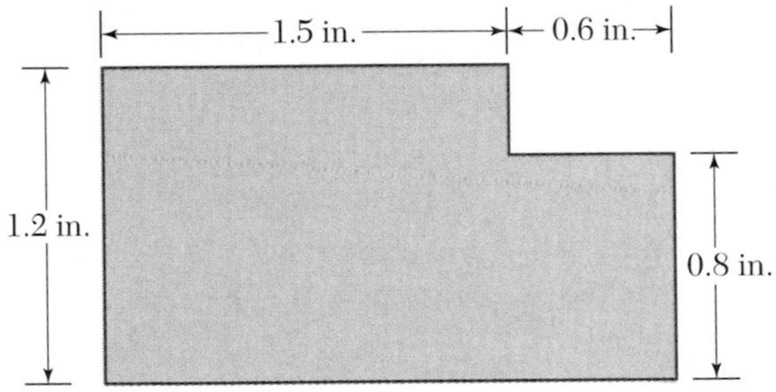

Answers to Warm-Ups

G. The decimal measures are $10\frac{2}{5}$ in. = 10.4 in.; $7\frac{15}{16}$ in. ≈ 7.9 in.; and $4\frac{3}{4}$ in. ≈ 4.8 in.

EXERCISES 4.7

OBJECTIVE Change fractions to decimals.

A. *Change the fraction or mixed number to a decimal.*

1. $\frac{3}{4}$ **2.** $\frac{7}{10}$ **3.** $\frac{3}{8}$ **4.** $\frac{5}{8}$

5. $\frac{11}{16}$ **6.** $\frac{7}{32}$ **7.** $3\frac{17}{20}$ **8.** $6\frac{13}{20}$

9. $11\frac{61}{125}$ **10.** $12\frac{9}{50}$

B.

Change to a decimal rounded to the indicated place value.

	Tenth	Hundredth
11. $\frac{5}{7}$		
12. $\frac{7}{9}$		
13. $\frac{9}{11}$		
14. $\frac{5}{6}$		
15. $\frac{7}{13}$		
16. $\frac{3}{14}$		
17. $\frac{8}{15}$		
18. $\frac{7}{19}$		
19. $5\frac{13}{18}$		
20. $11\frac{9}{17}$		

Change each of the following fractions to decimals. Use the repeat bar.

21. $\frac{8}{11}$ **22.** $\frac{5}{22}$ **23.** $\frac{5}{12}$ **24.** $\frac{8}{15}$

C.

Change each of the following fractions to decimals to the nearest indicated place value.

	Hundredth	Thousandth
25. $\frac{17}{53}$		
26. $\frac{15}{33}$		
27. $\frac{29}{65}$		
28. $\frac{78}{93}$		

Change to a decimal. Use the repeat bar.

29. $\frac{5}{13}$ **30.** $\frac{7}{33}$ **31.** $\frac{6}{7}$ **32.** $\frac{23}{26}$

33. A piece of blank metal stock is $1\frac{5}{8}$ in. in diameter. A micrometer measures in decimal units. If the stock is measured with the micrometer, what will the reading be?

34. A wrist pin is $\frac{15}{16}$ in. in diameter. What would the micrometer reading be?

35. Convert the measurements in the figure to decimals.

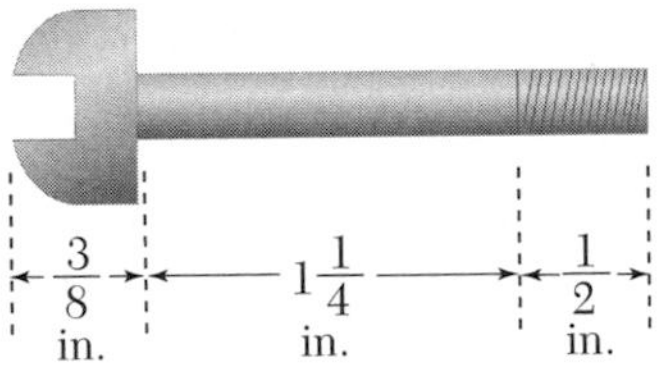

36. Stephen needs $\frac{27}{40}$ inch of rope. What is the decimal equivalent?

Change to a decimal. Round as indicated.

	Hundredths	Thousandths	Ten-thousandths
37. $\frac{42}{47}$			
38. $37\frac{18}{53}$			
39. $6\frac{42}{165}$			
40. $\frac{777}{1435}$			

41. A remnant of material $1\frac{1}{4}$ yards long costs \$6.08. Find the cost per yard of the fabric using fractions. Recalculate the same cost using decimals. Which is easier? Why?

42. A telephone lobbyist works $37\frac{3}{4}$ hours during one week. If he is paid \$45.40 per hour, compute his gross wages for the week. Did you use decimals or fractions to do the calculation? Why?

43. In a pole vault meet conducted by telephone, the highest jump in Iowa is $23\frac{7}{16}$ ft. The highest jump in Texas is 23.439 ft. Which state has the winning jump?

44. Ronald is writing a paper for his philosophy class. He is to word-process the paper, and the instructor has specified that all margins should be $1\frac{1}{4}$ inches. The software requires that the margins be specified in decimal form, rounded to the nearest tenth of an inch. What number does Ronald specify for the margins?

45. On June 1 in Portland, the sun rose at 5:25 A.M. and set at 8:52 P.M. Express the number of daylight hours as a decimal.

Exercises 46–47 relate to the chapter application.

46. In the 2000 Summer Olympics, Virgilijis Alekna of Lithuania won the discus throw with a toss of 227 ft 4 in. Convert this distance to a mixed number of feet; then convert your distance to decimal form.

47. In the 2000 Summer Olympics, Naoko Takahashi of Japan won the women's marathon with a time of 2 hr, 23 min, 14 sec. Convert this time to a mixed number of hours, and then convert the time to decimal form.

STATE YOUR UNDERSTANDING

48. Write a short paragraph on the uses of decimals and of fractions. Include examples of places where fractions are more useful and examples where decimals are more fitting.

CHALLENGE

49. Which is larger, 0.0036 or $\frac{2}{625}$?

50. Which is larger, 2.5×10^{-4} or $\frac{3}{2000}$?

51. First, decide whether the fraction $\frac{1.23}{80}$ is more or less than 0.1. Then change the fraction to a decimal. Were you correct in your estimate?

52. First, decide whether the fraction $\frac{62}{0.125}$ is more or less than 100. Then change the fraction to a decimal. Were you correct in your estimate?

GROUP WORK

53. Select ten stocks from the NYSE and assume you have 1000 shares of each. From today's paper, calculate the current value of your holdings. Using the reported changes, calculate the value of your holdings yesterday. How much money did you make or lose?

MAINTAIN YOUR SKILLS

Perform the indicated operations.

54. $6 \cdot 12 \div 4 + 2 - 4$

55. $(8 - 5) \cdot 5 - 4 + 6 \div 2$

56. $6^2 \cdot 4 - 3 \cdot 7 + 15$

57. $(7 - 4)^2 - 9 \div 3 + 7$

Find the average.

58. 456, 892, 703, 1021

59. 13, 607, 431, 904, 75

60. $\frac{2}{3}, \frac{5}{8}$, and $\frac{2}{9}$.

61. $\frac{5}{12}, \frac{7}{8}, \frac{1}{6}, \frac{4}{3}$

62. Mr. Lewis buys 350 books for $60 at an auction. He sells two-fifths of them for $25, and he sells 25 books at $1.50 each, 45 books at $1 each, and he gives away the rest. How many books does he give away? What is his total profit if his handling cost is $15?

63. John C. Scott Realty sold six houses last week at the following prices: $145,780, $234,700, $195,435, $389,500, $275,000, $305,677. What was the average sale price of the houses?

4.8 Order of Operations; Average, Median, and Mode

HOW & WHY

OBJECTIVES

1. Do any combination of operations with decimals.
2. Find the average, median, or mode of a set of decimals.

OBJECTIVE 1 Do any combination of operations with decimals.

The order of operations for decimals is the same as that for whole numbers and fractions.

Order of Operations

To simplify an expression with more than one operation, follow these steps

1. Parentheses: Do the operations within grouping symbols first (parentheses, fraction bar, and so on) in the order given in steps 2, 3, and 4.
2. Exponents: Do the operations indicated by exponents.
3. Multiply and divide: Do only multiplication and division as they appear from left to right.
4. Add and subtract: Do addition and subtraction as they appear from left to right.

EXAMPLE A–E

DIRECTIONS: Perform the indicated operations.

STRATEGY: Use the same order of operations as for whole numbers and fractions.

A. Simplify: $0.87 - 0.32(0.35)$

$0.87 - 0.32(0.35) = 0.87 - 0.112$ Multiplication is done first.
$= 0.758$ Subtract.

So $0.87 - 0.32(0.35) = 0.758$.

B. Simplify: $0.66 \div 0.22(4.05)$

$0.66 \div 0.22(4.05) = 3(4.05)$ Division is done first because it occurs first.
$= 12.15$ Multiply.

So $0.66 \div 0.22(4.05) = 12.15$.

C. Simplify: $(3.4)^2 - (2.6)^2$

$(3.4)^2 - (2.6)^2 = 11.56 - 6.76$ Exponents are done first.
$= 4.8$ Subtract.

So $(3.4)^2 - (2.6)^2 = 4.8$.

CALCULATOR EXAMPLE

D. Simplify: $8.736 \div 2.8 + (4.57)(5.9) + 12.67$

STRATEGY: All but the least expensive calculators have algebraic logic. The operations can be entered in the same order as they appear in the exercise.

So $8.736 \div 2.8 + (4.57)(5.9) + 12.67 = 42.753$.

WARM-UPS A–E

A. Simplify: $0.93 - 0.45(0.62)$

B. Simplify: $0.43 \div 0.5(2.55)$

C. Simplify: $(5.4)^2 - (1.6)^2$

D. Simplify: $102.92 \div 8.3 + (0.67)(34.7) - 21.46$

Answers to Warm-Ups

A. 0.651 **B.** 2.193
C. 26.6 **D.** 14.189

E. Nuyen buys the following tickets for an upcoming Pops concerts at the local symphony: 3 tickets at \$45.75 each; 2 tickets at \$39.50 each; 4 tickets at \$42.85 each; 3 tickets at \$48.50 each; and 5 tickets at \$40.45 each. For buying more than 10 tickets, Nuyen gets \$3.00 off each ticket purchased. What was Nuyen's total cost for the tickets?

E. Ellen buys the following items at the grocery store: 3 cans of soup at \$1.23 each; 2 cans of peas at \$0.89 each; 1 carton of orange juice at 2 for \$5.00; 3 cans of tuna fish at \$2.79 each; and 1 jar of peanut butter at \$3.95. Ellen had a coupon for \$2.00 off when you purchase 3 cans of tuna. What did Ellen pay for the groceries?

STRATEGY: Find the sum of the cost of each item And then subtract the coupon savings. To find the cost of each type of food, multiply the unit price by the number of items.

$3(\$1.23) + 2(\$0.89) + 1(\$5.00 \div 2) + 3(\$2.79) + 1(\$3.95) - \$2.00$	To find the unit price of the orange juice, we must divide the price for two by 2.
$\$3.69 + \$1.78 + 1(\$2.50) + \$8.37 + \$3.95 - \$2.00$	Multiply and divide.
$\$3.69 + \$1.78 + \$2.50 + \$8.37 + \$3.95 - \$2.00$	Multiply.
$\$18.29$	Add and subtract.

Ellen spent \$18.29 for the groceries.

HOW & WHY

OBJECTIVE 2 Find the average, median, or mode of a set of decimals.

The method for finding the average, median, or mode of a set of decimals is the same as that for whole numbers and fractions.

To find the average of a set of numbers

1. Add the numbers.
2. Divide the sum by the number of numbers in the set.

To find the median of a set of numbers

1. List the numbers in order from smallest to largest.
2. If there are an odd number of numbers in the set, the median is the middle number.
3. If there are an even number of numbers in the set, the median is the average (mean) of the two middle numbers.

To find the mode of a set of numbers

1. Find the number or numbers that occur most often.
2. If all the numbers occur the same number of times, there is no mode.

Answers to Warm-Ups

E. Nuyen pays \$684.40 for the tickets.

WARM-UPS F–H

EXAMPLES F–H

DIRECTIONS: Find the average, median, or mode.

STRATEGY: Use the same procedures as for whole numbers and fractions.

F. Find the average of 0.57, 0.38, 2.91, and 0.72.

$0.57 + 0.38 + 2.91 + 0.72 = 4.58$ First add the numbers.

$4.58 \div 4 = 1.145$ Second, divide by 4, the number of numbers.

So the average of 0.57, 0.38, 2.91, and 0.72 is 1.145.

G. Pedro's grocery bills for the past 5 weeks were:

Week 1: $155.72
Week 2: $172.25
Week 3: $134.62
Week 4: $210.40
Week 5: $187.91

What is Pedro's average and median cost of groceries per week for the 5 weeks?

Average:

Add the weekly totals and divide by 5, the number of weeks.

$$\begin{array}{r} 155.72 \\ 172.25 \\ 134.62 \\ 210.40 \\ +187.91 \\ \hline 860.90 \end{array}$$

$860.90 \div 5 = 172.18$

Median:

134,62, 155.72, 172.25, 187.91, 210.40 List the numbers from smallest to largest.

172.25 The median is the middle number.

Pedro's average weekly cost of groceries is $172.18, and the median cost is $172.25.

H. During an 8-month period of time, the highest recorded price per share of Oracle in each month was: $21.95, $19.72, $18.05, $14.55, $16.64, $18.05, $14.85, $13.12. Find the average, median, and mode for the highest price per month.

Average:

$21.95 + 19.72 + 18.05 + 14.55 + 16.64 + 18.05 + 14.85 + 13.12 = 136.93$ Add the numbers.

$136.93 \div 8 \approx 17.12$ Divide by the number of numbers and round to the nearest cent.

Median:

13.12, 14.55, 14.85, 16.64, 18.05, 18.05, 19.72, 21.95 List the numbers from smallest to largest.

$(16.64 + 18.05) \div 2 \approx 17.35$ Find the average of the two middle numbers and round to the nearest cent.

Mode:

18.05 The number that appears most often

The average highest price for a share of Oracle is about $17.12, the median price is about $17.35, and the mode price is $18.05.

F. Find the average of 9.6, 0.27, 9.3, 2.45, and 0.07.

G. Mary's weekly car expenses, including parking, for the past 6 weeks were:

Week 1: $37.95
Week 2: $43.15
Week 3: $28.65
Week 4: $59.14
Week 5: $61.72
Week 6: $50.73

What is Mary's average and median weekly expense for the 6 weeks?

H. Over a 6-month period of time, Jose was able to save the following amount each month: $123.25, $110.60, $145.85, $123.25, $150.10, $162.34. Find the average, median, and mode for his savings per month.

Answers to Warm-Ups

F. 4.338

G. Mary's average weekly car expense is $46.89, and the median expense is $46.94.

H. Jose's average savings are about $135.90, the median savings are $134.55, and mode of the savings is $123.25.

EXERCISES 4.8

OBJECTIVE 1 Do any combination of operations with decimals.

Perform the indicated operations.

A.

1. $0.8 + 0.6 - 0.4$
2. $0.9 - 0.5 + 0.7$
3. $0.28 \div 4 - 0.02$
4. $0.56 \div 4 + 0.13$
5. $2.4 - 3(0.7)$
6. $3.6 + 3(0.2)$
7. $3(1.3) + 4(3.1)$
8. $5(1.3) - 0.4(10)$
9. $0.16 + (0.2)^2$
10. $0.52 - (0.4)^2$

B.

11. $6.27 - 4.83 + 1.55 - 2.91$
12. $0.37 + 2.91 - 3.07 + 0.77$
13. $6.3 \div 3.5(8.1)$
14. $39.1 \div 6.8(0.34)$
15. $(15.6)(2.5) \div (0.3)$
16. $(7.5)(3.42) \div 0.15$
17. $(6.1)^2 - 4.3(3.7)$
18. $(4.9)^2 + 8.3 \div 0.25$
19. $(6.7)(1.4)^3 \div 0.7$
20. $(3.1)^3 - (0.8)^2 + 4.5$

OBJECTIVE 2 Find the average, median, or mode of a set of decimals.

A. *Find the average.*

21. 4.5, 6.5
22. 3.6, 5.2
23. 6.7, 9.5
24. 11.7, 16.3
25. 4.5, 3.3, 1.2
26. 1.2, 3.6, 1.8
27. 8.1, 8.5, 8.6
28. 3.9, 5.2, 6.5
29. 3.1, 2.1, 5.1, 2.1
30. 4.2, 5.2, 6.2, 9.2

B. *Find the average and median.*

31. 7.8, 9.08, 3.9, 5.7
32. 4.87, 6.93, 4.1, 9.6
33. 16.4, 8.72, 11.5, 3.23
34. 3.4, 8.2, 5.43, 7.2
35. 9.3, 12.4, 8.6, 27.4, 11.8
36. 61.8, 41.9, 50.7, 11.9, 32.2
37. 0.481, 0.327, 0.793, 0.889
38. 0.321, 0.843, 0.553, 0.213
39. 0.4523, 0.5421, 0.8674, 0.9234, 0.4535
40. 2.67, 11.326, 17.53, 1.095, 22.344

C.

41. Elmer goes shopping and buys 3 cans of cream style corn at 89¢ per can, 4 cans of tomato soup at 99¢ per can, 2 bags of corn chips at \$2.49 per bag, and 6 candy bars at 75¢ each. How much did Elmer spend?

42. Christie buys school supplies for her children. She buys 6 pads of paper at \$1.10 each, 5 pens at 89¢ each, 4 erasers at 39¢ each, and 4 boxes of crayons at \$1.49 each. How much does she spend?

43. The common stock of Oracle Corporation closed at \$16.75, \$17.375, \$17.75, \$15.875, and \$16.325 during 1 week in 2001. What was the average closing price of the stock?

44. A consumer watch group priced a can of a certain brand of soup at six different grocery stores. The group found the following prices: \$2.19, \$2.25, \$1.95, \$1.89, \$2.05, and \$2.40. What are the average and median selling prices of a can of the soup? Round to the nearest cent.

Perform the indicated operations.

45. $(7.3)(6.5) - (4.8)(5.4) + (5.6)^2$

46. $14.7 + 2.49(3.1) - 6.8(1.33) + 34$

47. $10.075 - [(3.4)(2.56) + 1.01]$

48. $6.3 - [(1.3)^2 - 0.95]$

49. $7.6(2.77 + 5.98 - 4.35) - 3.25(1.71)$

50. $9.3(8.71 - 6.91 + 0.78) - 4.5(3.19)$

51. The Adams family had the following natural gas bills for last year:

January	\$145.75	July	\$12.75
February	132.83	August	8.93
March	84.87	September	14.62
April	69.85	October	29.91
May	55.43	November	34.72
June	25.67	December	83.45

The gas company will allow them to make equal payments this year equal to the monthly average of last year. How much will the payment be? Round to the nearest cent.

52. Crescent City police reported the following number of automobile accidents for a week:

Monday	32	Tuesday	37
Wednesday	27	Thursday	41
Friday	18	Saturday	20
Sunday	8		

To the nearest tenth, what was the average number of accidents reported per day? What was the median and mode of the number of accidents per day?

53. Tim Raines, the fullback for the East All-Stars, gained the following yards in six carries: 5.6 yd, 3.5 yd, 7.8 yd, 12 yd, 10.5 yd, and 1 yd. What was the average gain per carry? Round to the nearest tenth of a yard.

54. The price per gallon of the same grade of gasoline at eight different service stations is 1.785, 1.699, 1.809, 1.785, 1.615, 1.795, 1.625, and 1.878. What are the average, median, and mode for the price per gallon at the eight stations? Round to the nearest thousandth.

Exercises 55–59. The table gives the high and low temperatures on 1 day for cities in Canada.

Temperatures in Major Canadian Cities, 5/31/01

City	High (°C)	Low (°C)
Calgary	19	10
Edmonton	22	2
Montreal	17	6
Ottawa	18	5
Quebec	18	3
Toronto	19	4
Vancouver	18	11
Winnipeg	23	9

55. What was the average high temperature for the cities, to the nearest tenth?

56. What was the average low temperature for the cities, to the nearest tenth?

57. What was the average range of temperatures for the cities, to the nearest tenth? Do any of the cities have the average range?

58. What was the mode for the high temperatures?

59. What was the median of the low temperatures?

60. The wholesale cost of shampoo is \$1.11 per bottle, whereas the wholesale cost of conditioner is \$0.89. The Fancy Hair Beauty Salon sells the shampoo for \$8.49 a bottle and the conditioner for \$8.19 a bottle. What is the net income on the sale of a case, 24 bottles, of each product?

Exercises 61–62 refer to the chapter application. The table gives a summary of the 2000 World Series between the New York Yankees and the New York Mets.

2000 World Series

Game	Team	Runs	Hits
1	Mets	3	10
1	Yankees	4	12
2	Mets	5	7
2	Yankees	6	12
3	Mets	4	9
3	Yankees	2	8
4	Mets	2	6
4	Yankees	3	8
5	Mets	2	8
5	Yankees	4	7

61. How many runs did the Yankees average over the five games? How many runs did the Mets average?

62. How many hits did the Yankees average over the five games? How many hits did the Mets average?

STATE YOUR UNDERSTANDING

63. Explain the difference between evaluating $0.3(5.1)^2 + 8.3 \div 5$ and $[0.3(5.1)^2 + 8.3] \div 5$. How do the brackets alter the order of the operations?

CHALLENGE

Insert grouping symbols to make each statement true.

64. $2 \cdot 8.1 \div 5 - 1 = 4.05$

65. $3.62 \div 0.02 + 72.3 \cdot 0.2 = 0.25$

66. $3.62 \div 0.02 + 8.6 \cdot 0.51 = 96.696$

67. $1.4^2 - 0.8^2 = 1.3456$

68. The average of 4.56, 8.23, 16.5, and a missing number is 8.2975. Find the missing number.

GROUP WORK

69. Go to the library and use the daily newspaper to get data to calculate the following:

a. average daily high temperature
b. average daily low temperature
c. average daily rainfall
d. average daily minutes of daylight.

Use data from your town over the last 7 days.

MAINTAIN YOUR SKILLS

Change to a decimal.

70. $\frac{15}{16}$ **71.** $\frac{25}{32}$ **72.** $\frac{17}{80}$ **73.** $\frac{47}{25}$

Change to a fraction or mixed number and simplify.

74. 0.92 **75.** 0.728 **76.** 1.045 **77.** 4.64

78. The sale price of a hand-held calculator is \$22.75. If the sale price is marked down \$7.28 from the original price, what was the original price?

79. The price of a Tomaya VCR is \$348.95. The store is going to put it on sale at a discount of \$59.50. What price should the clerk put on the VCR for the sale?

Getting Ready for Algebra

HOW & WHY

OBJECTIVE

Solve equations that require more than one operation.

We solve equations that require more than one operation in the same way as we do with equations with whole numbers and fractions.

To solve an equation that requires more than one operation

1. Eliminate the addition or subtraction by performing the inverse operation.
2. Eliminate the multiplication by dividing both sides by the same number—that is, perform the inverse operation.

EXAMPLES A–C

DIRECTIONS: Solve.

STRATEGY: Isolate the variable by performing the inverse operations.

A. $2.6x + 4.8 = 25.6$

$$2.6x + 4.8 - 4.8 = 25.6 - 4.8$$
$$2.6x = 20.8$$

Eliminate the addition by subtracting 4.8 from both sides.

$$\frac{2.6x}{2.6} = \frac{20.8}{2.6}$$
$$x = 8$$

Eliminate the multiplication by dividing both sides by 2.6.

CHECK: $2.6(8) + 4.8 = 25.6$
$20.8 + 4.8 = 25.6$
$25.6 = 25.6$

The solution is $x = 8$.

B. $8.3 = 1.25x + 4.65$

$$8.3 = 1.25x + 4.65$$
$$-4.65 = \qquad - 4.65$$
$$3.65 = 1.25x$$

Subtract 4.65 from both sides.

$$\frac{3.65}{1.25} = \frac{125x}{1.25}$$
$$2.92 = x$$

Divide both sides by 1.25.

CHECK: $8.3 = 1.25(2.92) + 4.65$
$8.3 = 3.65 + 4.65$
$8.3 = 8.3$

The solution is $x = 2.92$.

WARM-UPS A–C

A. $0.07y - 3.8 = 0.4$

B. $5.72 = 3.25t + 5.33$

Answers to Warm-Ups
A. $y = 60$ **B.** $0.12 = t$

C. Use the formula in Example C to find the Celsius temperature that corresponds to 122.9°F.

C. The formula relating temperature measured in degrees Fahrenheit and degrees Celsius is $F = 1.8C + 32$. Find the Celsius temperature that corresponds to 58.19°F.

First, substitute the known values into the formula.

$$F = 1.8C + 32$$
$$58.19 = 1.8C + 32 \quad \text{Substitute } F = 58.19.$$
$$58.19 - 32 = 1.8C + 32 - 32 \quad \text{Subtract 32 from both sides.}$$
$$26.19 = 1.8C$$
$$\frac{26.19}{1.8} = \frac{1.8C}{1.8} \quad \text{Divide both sides by 1.8.}$$
$$14.55 = C$$

Because $1.8(14.55) + 32 = 58.19$, the temperature is 14.55°C.

Answers to Warm-Ups

C. The temperature is 50.5°C.

EXERCISES

Solve.

1. $2.5x - 7.6 = 12.8$
2. $0.25x - 7.3 = 0.95$
3. $1.8x + 6.7 = 12.1$
4. $15w + 0.006 = 49.506$
5. $4.115 = 2.15t + 3.9$
6. $10.175 = 1.25y + 9.3$
7. $0.03x - 18.7 = 3.53$
8. $0.08r - 5.62 = 72.3$
9. $7x + 0.06 = 2.3$
10. $13x + 14.66 = 15.7$
11. $3.65m - 122.2 = 108.115$
12. $22.5t - 657 = 231.75$
13. $5000 = 125y + 2055$
14. $3700 = 48w + 1228$
15. $60p - 253 = 9.5$
16. $17.8 = 0.66y + 7.9$
17. $8.551 = 4.42 + 0.17x$
18. $14 = 0.25w - 8.6$
19. $45 = 1.75h - 1.9$
20. $4000 = 96y + 1772.8$
21. $1375 = 80c + 873$
22. $7632 = 90t - 234$
23. The formula relating temperatures measured in degrees Fahrenheit and degrees Celsius is $F = 1.8C + 32$. Find the Celsius temperature that corresponds to 248°F.
24. Use the formula in Exercise 23 to find the Celsius temperature that corresponds to 45.5°F.
25. The formula for the balance of a loan D is $D + NP = B$, where P represents the monthly payment, N represents the number of payments made, and B represents the amount of money borrowed. Find the number of the monthly payment Gina has made if she borrowed \$1764, has a remaining balance of \$661.50, and pays \$73.50 per month.
26. Use the formula in Exercise 25 to find the number of payments made by Morales if he borrowed \$8442, has a balance of \$3048.50, and makes a monthly payment of \$234.50.
27. Catherine is an auto mechanic. She charges \$36 per hour for her labor. The cost of parts needed is in addition to her labor charge. How many hours of labor result from a repair job in which the total bill (including \$137.50 for parts) is \$749.50? Write and solve an equation to determine the answer.
28. A car rental agency charges \$28 per day plus \$0.27 per mile to rent a car. Determine how many miles were driven by a customer after a 3-day rental that cost \$390.45. Write and solve an equation to determine the answer.

Name ______ Class ______ Date ______

CHAPTER 4

KEY CONCEPTS

SECTION 4.1 DECIMALS: READING, WRITING, AND ROUNDING

Definitions and Concepts	Examples
Decimal numbers are another way of writing fractions and mixed numbers.	1.3 One and three tenths 2.78 Two and seventy-eight hundredths 5.964 Five and nine hundred sixty-four thousandths
To round a decimal to a given place value: • Mark the given place value. • If the digit on the right is 5 or more, add 1 to the marked place and drop all digits on the right. • If the digit on the right is 4 or less, drop all digits on the right. • Write zeros on the right if necessary so that the marked digit still has the same place value.	Round 4.792 to the nearest tenth. ↑ 4.792 ≈ 4.8 Round 4.792 to the nearest hundredth. ↑ 4.792 ≈ 4.79 Round 563.79 to the nearest ten. ↑ 563.78 ≈ 560

SECTION 4.2 CHANGING DECIMALS TO FRACTIONS; LISTING IN ORDER

Definitions and Concepts	Examples
To change a decimal to a fraction: • Read the decimal word name.	0.45 is read "forty-five hundredths"
• Write the fraction that has the same name.	$0.45 = \frac{45}{100}$
• Simplify.	$= \frac{9}{20}$
To list decimals in order: • Insert zeros on the right so all the decimals have the same number of decimal places. • Write the numbers in order as if they were whole numbers. • Remove the extra zeros.	List 1.46, 1.3, 1.427 in order from smallest to largest. 1.46 = 1.460 1.3 = 1.300 1.427 = 1.427 $1.300 < 1.427 < 1.460$ So, $1.3 < 1.427 < 1.46$.

SECTION 4.3 ADDING AND SUBTRACTING DECIMALS

Definitions and Concepts	Examples
To add or subtract decimals: • Write in columns with the decimal points aligned. Insert zeros on the right if necessary. • Add or subtract. • Align the decimal point in the answer with those above.	2.67 + 10.9 8.5 − 3.64 $\begin{array}{r} 2.67 \\ +10.90 \\ \hline 13.57 \end{array}$ $\begin{array}{r} 8.50 \\ -3.64 \\ \hline 4.86 \end{array}$

SECTION 4.4 MULTIPLYING DECIMALS

Definitions and Concepts	Examples	
To multiply decimals: • Multiply the numbers as if they were whole numbers. • Count the number of decimal places in each factor. The total of the decimal places is the number of decimal places in the product. Insert zeros on the left if necessary.	4.2×0.12 $$\begin{array}{r} 4.2 \\ \times\ 0.12 \\ \hline 84 \\ 42 \\ \hline 0.504 \end{array}$$ (Three decimal places needed)	0.03×0.007 $$\begin{array}{r} 0.03 \\ \times\ 0.007 \\ \hline 0.00021 \end{array}$$ (Five decimal places needed)

SECTION 4.5 MULTIPLYING AND DIVIDING BY POWERS OF 10; SCIENTIFIC NOTATION

Definitions and Concepts	Examples
To multiply by a power of 10: • Move the decimal point to the right the same number of places as there are zeros in the power of 10.	$3.45(10{,}000) = 34{,}500$ (Move four places right.)
To divide by a power of 10: • Move the decimal point to the left the same number of places as there are zeros in the power of 10.	$3.45 \div 1000 = 0.00345$ (Move three places left.)
Scientific notation is a special way to write numbers as a product of a number between 1 and 10 and a power of 10.	$34{,}500 = 3.45 \times 10^4$ $0.00345 = 3.45 \times 10^{-3}$

SECTION 4.6 DIVIDING DECIMALS

Definitions and Concepts	Examples
To divide decimals: • If the divisor is not a whole number, move the decimal point in both the divisor and the dividend to the right as many places as necessary to make the divisor a whole number. • Place the decimal point in the quotient above the decimal point in the dividend. • Divide as if both numbers were whole numbers. • Round as appropriate.	$0.04\overline{)5.3} = 004\overline{)530.0}$ (Move two places right.) Quotient: 132.5 $$\begin{array}{r} 4 \\ \hline 13 \\ 12 \\ \hline 10 \\ 8 \\ \hline 20 \\ 20 \\ \hline 0 \end{array}$$

SECTION 4.7 CHANGING FRACTIONS TO DECIMALS

Definitions and Concepts	Examples
To change a fraction to a decimal, divide the numerator by the denominator. Round as appropriate.	Change $\frac{5}{8}$ to a decimal. $$\begin{array}{r} 0.625 \\ 8\overline{)5.000} \\ 4\,8 \\ \hline 20 \\ 16 \\ \hline 40 \\ 40 \\ \hline \end{array}$$

SECTION 4.8 ORDER OF OPERATIONS; AVERAGE, MEDIAN, AND MODE

Definitions and Concepts	Examples
The order of operations for decimals is the same as that for whole numbers: • Parentheses • Exponents • Multiplication / Division • Addition / Subtraction	$14.8 - 0.2(8.3 + 4.76) = 14.8 - 0.2(13.06)$ $= 14.8 - 2.612$ $= 12.188$
Finding the average of a set of decimals is the same as for whole numbers: • Add the numbers. • Divide by the number of numbers.	Find the average of 5.8, 6.12, and 7.394. $5.8 + 6.12 + 7.394 = 19.314$ $19.314 \div 3 = 6.438$ The average is 6.438.
Finding the median of a set of decimals is the same as for whole numbers: • List the numbers in order from smallest to largest. • If there are an odd number of numbers in the set, the median is the middle number. • If there are an even number of numbers in the set, the median is the average of the middle two.	Find the median of 5.8, 6.12, 7.394, 9.6, and 7.01. 5.8, 6.12, 7.01, 7.394, 9.6 The median is 7.01.
Finding the mode of a set of decimals is the same as for whole numbers: • Find the number or numbers that occur most often. • If all the numbers occur the same number of times, there is no mode.	Find the mode of 5.8, 6.12. 7.03, 8.2, and 6.12. The mode is 6.12.

CHAPTER 4

REVIEW EXERCISES

Section 4.1

Write the word name.

1. 5.18

2. 0.762

3. 10.032

4. 0.0000012

Write the place value name.

5. Seventeen and five hundredths

6. One hundred nine ten-thousandths

7. Four hundred and four hundredths

8. One hundred twenty-five and forty-five thousandths

Exercises 9–11. Round the numbers to the nearest tenth, hundredth, and thousandth.

	Tenth	Hundredth	Thousandth
9. 34.7648			
10. 4.8298			
11. 0.467215			

12. The display on Mary's calculator shows 67.94672199 as the result of a division exercise. If she is to round the answer to the nearest thousandth, what answer does she report?

Section 4.2

Change the decimal to a fraction or mixed number and simplify.

13. 0.64

14. 3.015

15. 0.00256

16. 0.505

List the set of decimals from smallest to largest.

17. 0.95, 0.89, 1.01

18. 0.08, 0.083. 0.0799

19. 5.007, 5.012, 0.515, 5.101

20. 34.023, 34.103, 34.0204, 34.0239

Is the statement true or false?

21. $6.1774 < 6.1780$

22. $67.0123 > 67.0219$

Section 4.3

Add.

23.
$$\begin{array}{r} 8.356 \\ 0.45 \\ 13.082 \\ +\ 7.4 \\ \hline \end{array}$$

24.
$$\begin{array}{r} 12.0678 \\ 7.012 \\ 56.0921 \\ +\ 0.0045 \\ \hline \end{array}$$

Subtract.

25.
$$\begin{array}{r} 16.0754 \\ -\ 9.5623 \\ \hline \end{array}$$

26.
$$\begin{array}{r} 45.096 \\ -23.68845 \\ \hline \end{array}$$

Estimate the sum or difference by rounding to the nearest ten. Add or subtract and round the sum or difference to the nearest tenth.

27. 23.005 + 6.073 + 0.34 + 12.7 + 123.087

28. 607.706 − 453.837

Find the perimeter of the following figures.

29.

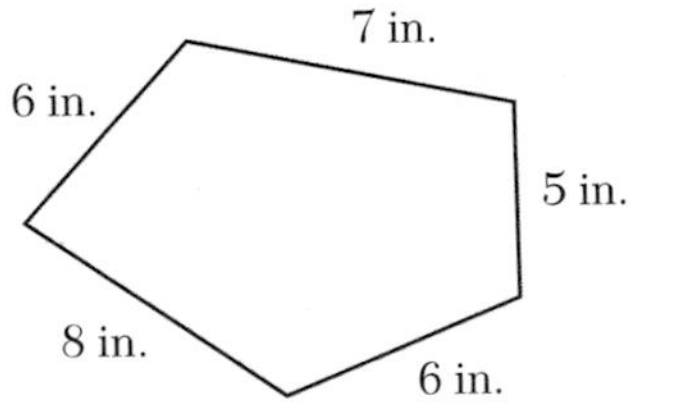

30.

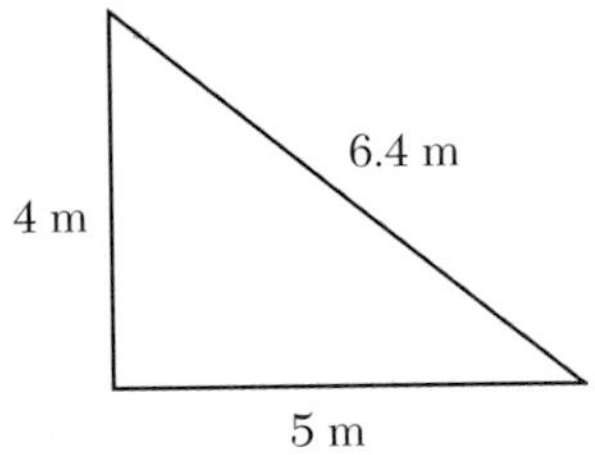

31. Hilda makes a gross salary (before deductions) of $4365 per month. She has the following monthly deductions: federal income tax, $873; state income tax, $392.85; Social Security, $240.07; Medicare, $55.44; retirement contribution, $218.25; union dues, $35; and health insurance, $210.15. Find her actual take-home (net) pay.

32. During 1 week in 2001, the price of a share of Oracle went from $19.67 to $15.93. What was the loss for the week per share?

Section 4.4

Multiply.

33.
$$\begin{array}{r} 7.06 \\ \times\ 2.3 \\ \hline \end{array}$$

34.
$$\begin{array}{r} 11.24 \\ \times\ 3.5 \\ \hline \end{array}$$

35.
$$\begin{array}{r} 0.00238 \\ \times\ 5.17 \\ \hline \end{array}$$

36.
$$\begin{array}{r} 12.057 \\ \times\ 8.08 \\ \hline \end{array}$$

Estimate the product by rounding each factor to its largest nonzero place value and multiply.

37. 0.074(2.004)

38. (0.0056)(23.7)

39. Multiply: (0.03)(4.12)(0.015). Round to the nearest ten-thousandth.

40. Find the area of the rectangle.

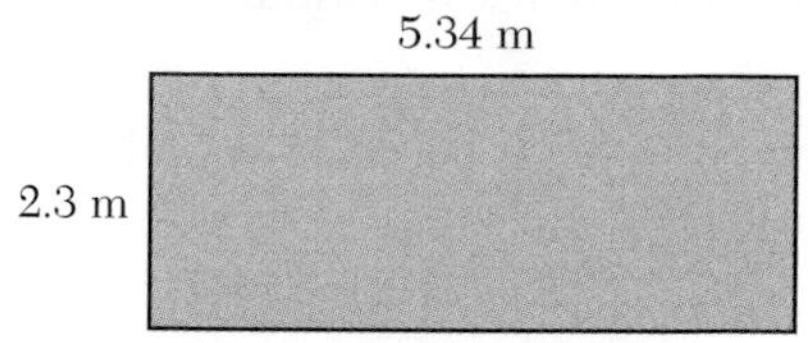

41. Pauline selects an upholstery fabric that costs $52.35 per meter. How much will Pauline pay for 23.75 m? Round to the nearest cent.

42. Agnelo can choose any of the following ways to finance his new car. Which method is the least expensive in the long run?

$850 down and $401.64 per month for 5 years
$475 down and $443.10 per month for 54 months
$600 down and $495.30 per month for 4 years

Section 4.5

Multiply or divide.

43. $76.93 \div 10^4$

44. 5.008×10^6

45. 0.7321(100,000)

46. $9.503 \div 100$

Write in scientific notation.

47. 0.0078

48. 34.67

49. 0.0000042

50. 4700.9

Write the place value name.

51. 7×10^7

52. 9.75×10^{-4}

53. 8.08×10^{-2}

54. 3.505×10^3

55. Sole Support Shoes buys 100 pairs of shoes for $9735. What is the average price of the shoes?

56. During the bear market of 2001, the stock market at one time was down $50 billion. Write this loss in scientific notation.

Section 4.6

Divide.

57. $0.3\overline{)0.0111}$

58. $90\overline{)32.4}$

59. $56.7 \div 0.32$

60. $0.14\overline{)0.00504}$

61. $0.456\overline{)0.38304}$

62. $2.4299 \div 1.034$

Estimate the quotient. Divide and round to the nearest hundredth.

63. $2.8\overline{)212.789}$

64. $0.068\overline{)0.01956}$

65. One hundred thirty-seven alumni of Notre Dame University donated \$345,986 to the university. To the nearest cent, what was the average donation?

66. Carol drove 467.8 miles on 13.9 gallons of gas. What is her mileage (miles per gallon)? Round to the nearest mile per gallon.

Section 4.7

Change the fraction or mixed number to a decimal.

67. $\frac{9}{16}$ 68. $\frac{3}{20}$ 69. $17\frac{37}{125}$

Change to a decimal rounded to the indicated place value.

70. $\frac{8}{37}$ (tenth) 71. $\frac{57}{93}$ (hundredth) 72. $\frac{42}{61}$ (thousandth)

Change to a decimal. Use the repeat bar.

73. $\frac{9}{13}$ 74. $\frac{7}{48}$

75. The value of a share of Microsoft is $63\frac{7}{32}$. What is the value in decimal form? Round to the nearest hundredth.

76. In a shot put meet conducted by telephone, the longest put in Georgia was $60\frac{11}{66}$ ft. The longest put in Idaho was 60.799 ft. Which state had the winning put?

Section 4.8

Perform the indicated operations.

77. $0.45 + 4.09 - 2.83 + 1.04$ 78. $13.8 \div 0.12 \times 4.03$

79. $(5.6)^2 - (3.4)(1.43)$ 80. $(5.5)(2.4)^3 \div 9.9$

81. $(7.2)(6.02) - (3.7)(0.05) + 8.29$ 82. $6.2(3.45 - 2.07 + 0.98) - 3.1(1.45)$

Find the average and median.

83. 4.56, 11.93, 13.4, 1.58, 8.09 84. 56.78, 45.32, 81.3, 90.04

85. 0.5672, 0.6086, 0.3447, 0.5555 86. 67.9, 54.3, 72.9, 83.8, 45.7

87. David goes shopping and buys 5 cans of baked beans at \$1.29 per can, 7 cans of tomato soup at 89¢ per can, 3 bags of baked corn chips at \$3.29 per bag, and 5 bars of soap at 79¢ per bar. How much did David spend?

88. The Metropolis Police Department reported the following number of robberies for the week:

Monday	12	Tuesday	21	Wednesday	5	Thursday	18
Friday	46	Saturday	67	Sunday	17		

To the nearest tenth, what is the average number of robberies reported per day?

CHAPTER 4

TRUE / FALSE CONCEPT REVIEW

ANSWERS

Check your understanding of the language of basic mathematics. Tell whether each of the following statements is true (always true) or false (not always true). For each statement you judge to be false, revise it to make a statement that is true.

1. The word name for 0.709 is seven hundred and nine thousandths. 1. __________

2. .348 and 0.348 name the same number. 2. __________

3. To write 0.85 in expanded form, we write $\frac{85}{100}$. 3. __________

4. Because 0.265 is read "two hundred sixty-five thousandths," we write $\frac{265}{1000}$ and reduce to change the decimal to a fraction. 4. __________

5. True or false: $0.732687 > 0.74$? 5. __________

6. Because $4.6 > 3.9$ is true, 3.9 is to the left of 4.6 on the number line. 6. __________

7. To list a group of decimals in order, we need to write or think of all the numbers as having the same number of decimal places. 7. __________

8. Decimals are either exact or approximate. 8. __________

9. To round 356.7488 to the nearest tenth, we write 356.8 because the 4 in the hundredths place rounds up to 5 because it is followed by an 8. 9. __________

10. The sum of 0.6 and 0.73 is 1.33. 10. __________

11. $9.7 - 0.2 = 7.7$ 11. __________

12. The final answer of a multiplication problem will always contain the same number of decimal places as the total number of places in the two numbers being multiplied. 12. __________

13. To multiply a number by a positive power of 10, move the decimal point in the number to the right the same number of places as the number of zeros in the power of 10. 13. __________

14. To divide a number by a positive power of 10 that is written in exponent form, move the decimal the same number of places to the right as the exponent indicates. 14. __________

15. To change 3.57×10^{-5} to place value form, move the decimal five places to the right. 15. __________

16. To divide a number by a decimal, first change the decimal to a whole number by moving the decimal point to the right. 16. ______

17. All fractions can be changed to exact terminating decimals. 17. ______

18. The decimal 0.7849 rounds to 0.79 to the nearest hundredth. 18. ______

19. The order of operations for decimals is the same as for whole numbers. 19. ______

20. To find the average of a group of decimals, find their sum and divide by the number of decimals in the group. 20. ______

Name ______________________ Class ______________________ Date ____________

CHAPTER 4

TEST

ANSWERS

1. Divide. Round the answer to the nearest thousandth: $0.94\overline{)3.7881}$. 1. ____________

2. List the following decimals from the smallest to the largest: 0.638, 0.642, 0.6389, 0.6299, 0.6307 2. ____________

3. Write the word name for 56.072. 3. ____________

4. Multiply: 7.34(5.8) 4. ____________

5. Write as a decimal: $\frac{11}{125}$ 5. ____________

6. Round to the nearest hundredth: 62.996 6. ____________

7. Subtract: $87 - 14.837$ 7. ____________

8. Change to a mixed number with the fraction part simplified: 12.525 8. ____________

9. Write in scientific notation: 0.00000341 9. ____________

10. Write as an approximate decimal to the nearest hundredth: $\frac{14}{19}$ 10. ____________

11. Round to the nearest hundred: 34,987.495 11. ____________

12. Perform the indicated operations: $7.23 \div (0.24 \times 2.5) + 9.34$ 12. ____________

13. Subtract: $\begin{array}{r} 35.426 \\ -28.519 \\ \hline \end{array}$ 13. ____________

14. Change to place value notation: 7.45×10^{-4} 14. ____________

15. Write the place value name for four hundred thirteen and eighty-five thousandths. 15. ____________

16. Multiply: 0.0000523(10,000) 16. ____________

17. Write in scientific notation: 63,920 17. ____________

18. Add: $7.98 + 0.467 + 14.92 + 8.3$ 18. ____________

19. Multiply: 87.4(0.00072) 19. ____________

20. Divide: $55\overline{)0.4114}$

20. ____________

21. For each of the four Sundays of February, the offering at the Chapel on the Hill was \$3267.75, \$3695.50, \$4010.25, and \$4231.50. What was the average Sunday offering?

21. ____________

22. Add:

$$\begin{array}{r} 567.84 \\ 45.862 \\ 4009.62 \\ 8.3 \\ 23.67 \\ +\ 298.361 \\ \hline \end{array}$$

22. ____________

23. During a canned vegetable sale, Grant buys 23 cans of various vegetables. If the sale price is 4 cans for \$2.68, how much does Grant pay for the canned vegetables?

23. ____________

24. At the beginning of the 1995–1996 NBA season, Michael Jordan was the all-time leader in points scored per game, with 32.2. How many games had he played if he scored a total of 21,998 points (to the nearest game)?

24. ____________

25. In baseball, the slugging percentage is calculated by dividing the number of total bases (a double is worth two bases) by the number of times at bat and then multiplying by1000. What is the slugging percentage of a player who has 231 bases in 302 times at bat? Round to the nearest whole number.

25. ____________

26. Harold and Jerry go on diets. Initially, Harold weighed 267.8 lb and Jerry weighed 209.4 lb. After 1 month on the diet, Harold weighed 254.63 lb and Jerry weighed 196.2 lb. Who lost the most weight and by how much?

26. ____________

CHAPTER 4

GROUP PROJECT (2–3 weeks)

OPTIONAL

The NFL keeps many statistics regarding its teams and players. Because quarterbacks play an important part in the overall team effort, much time and attention has been given to keeping statistics on quarterbacks. But all these statistics do not necessarily make it easy to decide which quarterback is the best. Consider the following statistics from the 2000 season.

Players with 10 Games or More in 2000 Season

Player	Passes Attempted	Passes Completed	Yards Gained	Touchdown Passes	Interceptions
Brian Griese, Denver	336	216	2688	19	4
Kurt Warner, St. Louis	347	235	3429	21	18
Daunte Culpepper, Minnesota	474	297	3937	33	16
Jeff Garcia, San Francisco	561	355	4278	31	10
Peyton Manning, Indiana	571	357	4413	33	15

1. Which quarterback deserved to be rated as the top quarterback of the year? How did you make your choice?

The NFL has developed a rating system for quarterbacks that combines all the statistics in the table and gives each quarterback a single numeric "grade," so they can easily be compared. Although the exact calculations used by the NFL are complicated, Randolph Taylor of Las Positas College in California has developed the following formula, which closely approximates the NFL ratings:

Let A = the number of passes attempted
C = the number of passes completed
Y = the number of yards gained passing
T = the number of touchdown passes
I = the number of interceptions

$$\text{Rating} = \frac{5}{6}\left(\frac{C}{A}\cdot 100\right) + \frac{25}{6}\left(\frac{Y}{A}\right) + \frac{10}{3}\left(\frac{T}{A}\cdot 100\right) - \frac{25}{6}\left(\frac{I}{A}\cdot 100\right) + \frac{25}{12}$$

2. Use the rating formula to calculate ratings for the quarterbacks in the table. Use your calculator and do not round except at the end, rounding to the nearest hundredth.

3. Explain why everything in the formula is added except $\frac{25}{6}\left(\frac{I}{A}\cdot 100\right)$.

4. According to your calculations, who was the best quarterback for the 2000 season? How does this compare to your answer to question 1?

5. In the 2000 season, Kevin Thompson of Cleveland played in one game, made one attempt at a pass, and completed it for 8 yards. He had no touchdowns or interceptions. Calculate his rating and comment on how he compares with the other quarterbacks in the table.

6. What are the drawbacks to using the rating formula as the only measure of a quarterback's performance?

7. (Optional) Have your group compile a list of the five all-time best quarterbacks. Find statistics for each of the quarterbacks on your list (use almanacs or the web) and compute their ratings. Comment on your results.

GOOD ADVICE FOR STUDYING

Learning to Learn Math

Learning mathematics is a building process. For example, if you have not mastered fractions, rational expressions are difficult to learn because they require an understanding of the rules for fractions. Therefore, if you are having difficulty with the current topic, you may not have mastered a previous skill that you need. It will be necessary to go back and learn/relearn this skill before you can continue.

Learning math also means learning not just skills, but how and where and when to **apply** the skills. For example, if it takes 16 gallons of gas to travel 320 miles, how many miles to the gallon are you getting? What skill would you use to solve this problem? (Answer: dividing) Reading the application problems and thinking of situations where you have used or could use these concepts helps integrate the concept into your experience.

Learning mathematics is learning something basic to daily life and to virtually every field of science and business. The examples in the book state the problem and a **strategy** for solving the problem. This strategy applies to several related problems. Learning mathematics is learning strategies to solve related problems.

The more you begin to appreciate mathematics as relevant to your life, the more you will see mathematics as worthy of your time, and the more committed you will feel to studying mathematics. Here is an activity that may give you fresh opportunities to see how much mathematics relates to your life. First, create a simple "web" or "map" with "math" at the center and spokes out from this center naming areas in life where math comes up—areas where math is useful. Capture as many areas as you can in a 5-minute period.

Second, turn the page over and construct a second "web," but this time choose one of your "math areas" for the center and create spokes that capture subtopics or subheadings in this particular math area. Take another 5 minutes for this second web. Next, create a problem that you believe to be solvable, from your own experience, and that might be enticing for someone else to solve.

If you are working with a partner, trade problems and see if you can solve each others' problems. Talk about how you might approach the problems and whether they are stated clearly and believably. Here are some criteria for a good problem:

- Enough data and information to solve the problem
- Clear statement of what you need to find
- Not too many questions included
- Appropriate reading level and clearly written
- Appears solvable and not too scary
- Makes the reader care about wanting to solve the problem

Going through this activity may help you become more aware of mathematics in your daily life and give you a greater understanding of problems and problem solving.

5

RATIO AND PROPORTION

SECTIONS

APPLICATION

One of the ways we relate to our surroundings is by judging the size of various objects relative to our bodies. Have you ever gone up a flight of stairs in which the stairs were taller than normal? You are acutely aware that something is not right because your body is expecting a certain amount of space between the stairs. We tend to trip on the first few stairs because we do not lift our feet high enough. But then our body makes an adjustment, and we successfully negotiate the rest of the stairs but with more effort than is usually required. Have you ever observed a toddler climbing stairs? Try to imagine how difficult it would be to climb stairs that were as large compared to our bodies as normal stairs are compared to a toddler's body. This concept of relative size is referred to, both mathematically and in the real world, as "proportionality."

Complete the following table.

	Adult	Toddler
Height of a stair in inches	8	8
Body height in inches		

Write a fraction using the number in the Adult column. Use the stair measurement as the numerator and the body height measurement as the denominator. State what it means in the context of this situation. Do the same for the numbers in the Toddler column. Which fraction is larger? Explain what this means in context.

Many movies stem from the premise that someone or something is proportionally larger or smaller than normal. Early films include the King Kong movies. More recent examples include *Honey, I Shrunk the Kids* and its sequels. Such movies are special challenges to the set designers, who must ensure that all the surroundings are proportionally correct so the premise is visually believable. In this chapter, we investigate some of the proportions used in the movie *Honey, I Shrunk the Kids*.

5.1 Ratio and Rate

OBJECTIVES

1. Write a fraction that shows a ratio comparison of two like measurements.
2. Write a fraction that shows a rate comparison of two unlike measurements.
3. Write a unit rate.

VOCABULARY

A **ratio** is a comparison of two measurements by division.

Like measurements have the same unit of measure.

Unlike measurements have different units of measure.

A **rate** is a comparison of two unlike measurements by division.

A **unit rate** is a rate with a denominator of one unit.

HOW & WHY

OBJECTIVE 1 Write a fraction that shows a ratio comparison of two like measurements.

Two numbers can be compared by subtraction or division. If we compare 30 and 10 by subtraction, $30 - 10 = 20$, we can say that 30 is 20 more than 10. If we compare 30 and 10 by division, $30 \div 10 = 3$, we can say that 30 is 3 times larger than 10.

The indicated division, $30 \div 10$, is called a *ratio*. These are common ways to write the ratio comparing 30 and 10:

$30:10 \qquad 30 \div 10 \qquad 30 \text{ to } 10 \qquad \frac{30}{10}$

Because we are comparing 30 to 10, 30 is written first or placed in the numerator of the fraction.

Here we write ratios as fractions. Because a ratio is a fraction, it can often be simplified. The ratio $\frac{12}{16}$ is simplified to $\frac{3}{4}$. If the ratio contains two like measurements, it can be simplified in the same way as a fraction.

$\frac{\$7}{\$70} = \frac{7}{10}$ The units, \$ or dollars, are dropped because they are the same.

$\frac{15 \text{ miles}}{35 \text{ miles}} = \frac{3}{7}$ The units, miles, are dropped and the fraction is simplified.

WARM-UPS A–C

A. Write the ratio of 70 to 112.

EXAMPLES A–C

DIRECTIONS: Write a ratio in simplified form.

STRATEGY: Write the ratio as a simplified fraction.

A. Write the ratio of 88 to 110.

$\frac{88}{110} = \frac{4}{5}$ Write 88 in the numerator and simplify.

The ratio of 88 to 110 is $\frac{4}{5}$.

Answers to Warm-Ups

A. $\frac{5}{8}$

B. Write the ratio of the length of a room to its width if the room is 30 ft by 24 ft.

$\frac{30 \text{ ft}}{24 \text{ ft}} = \frac{30}{24} = \frac{5}{4}$ Write 30 ft in the numerator, drop the common units, and simplify.

The ratio of the length of the room to its width is $\frac{5}{4}$.

C. Write the ratio of 5 dimes to 5 quarters. Compare in cents.

$\frac{5 \text{ dimes}}{5 \text{ quarters}} = \frac{50 \text{ cents}}{125 \text{ cents}} = \frac{50}{125} = \frac{2}{5}$

The ratio of 5 dimes to 5 quarters is $\frac{2}{5}$.

B. Write the ratio of the length of a meeting room to its width if the meeting room is 45 ft by 35 ft.

C. Write the ratio of 42 feet to 35 yards. Compare in feet.

HOW & WHY

OBJECTIVE 2 Write a fraction that shows a rate comparison of two unlike measurements.

Fractions are used to compare unlike measurements as well as like measurements. Such a comparison is called a *rate*. The rate of $\frac{27 \text{ children}}{10 \text{ families}}$ compares the unlike measurements "27 children" and "10 families." A familiar application of a rate occurs in the computation of gas mileage. For example, if a car travels 192 miles on 8 gallons of gas, we compare miles to gallons by writing $\frac{192 \text{ miles}}{8 \text{ gallons}}$. This rate can be simplified but the units are *not dropped* because they are unlike.

$$\frac{192 \text{ miles}}{8 \text{ gallons}} = \frac{96 \text{ miles}}{4 \text{ gallons}} = \frac{24 \text{ miles}}{1 \text{ gallon}} = 24 \text{ miles per gallon} = 24 \text{ mpg}$$

CAUTION

When measurement units are different, they are not dropped.

Answers to Warm-Ups

B. $\frac{9}{7}$ C. $\frac{2}{5}$

EXAMPLES D–F

DIRECTIONS: Write a rate in simplified form.

STRATEGY: Write the simplified fraction and keep the unlike units.

D. Write the rate of 12 chairs to 11 people.

$\frac{12 \text{ chairs}}{11 \text{ people}}$ The units must be kept because they are different.

E. Write the rate of 14 cars to 10 homes.

$\frac{14 \text{ cars}}{10 \text{ homes}} = \frac{7 \text{ cars}}{5 \text{ homes}}$

The rate is $\frac{7 \text{ cars}}{5 \text{ homes}}$.

WARM-UPS D–F

D. Write the rate of 11 people to 7 tables.

E. Write the rate of 24 TV sets to 10 homes.

Answers to Warm-Ups

D. $\frac{11 \text{ people}}{7 \text{ tables}}$

E. The rate is $\frac{12 \text{ TV sets}}{5 \text{ homes}}$

F. The following spring the urban committee repeated the tree program. This time they sold 70 oak trees and 440 birch trees.

1. What is the rate of oak trees to birch trees sold?
2. What is the rate of birch trees to the total number of trees sold?

F. An urban environmental committee urged the local citizens to plant deciduous trees around their homes as a means of conserving energy. They determined that 825 oak trees and 675 birch trees had been sold.

1. What is the rate of the number of oak trees to the number of birch trees sold?
2. What is the rate of the number of oak trees to the total number of trees sold?

1. STRATEGY: Write the first measurement, 825 oak trees, in the numerator and the second measurement, 675 birch trees, in the denominator.

$$\frac{825 \text{ oak trees}}{675 \text{ birch trees}} = \frac{11 \text{ oak trees}}{9 \text{ birch trees}} \qquad \text{Simplify.}$$

The rate is $\frac{11 \text{ oak trees}}{9 \text{ birch trees}}$—that is, 11 oak trees were sold for every 9 birch trees sold.

2. STRATEGY: Write the first measurement, 825 oak trees, in the numerator and the second measurement, total number of trees, in the denominator.

$$\frac{825 \text{ oak trees}}{1500 \text{ total trees}} = \frac{11 \text{ oak trees}}{20 \text{ total trees}} \qquad \text{Simplify.}$$

The rate is $\frac{11 \text{ oak trees}}{20 \text{ total trees}}$—that is, 11 out of every 20 trees sold were oak trees.

HOW & WHY

OBJECTIVE 3 Write a unit rate.

When a rate is simplified so that the denominator is a 1-unit measurement, then we have a *unit rate*.

For example,

$$\frac{204 \text{ miles}}{8 \text{ gallons}} = \frac{51 \text{ miles}}{2 \text{ gallons}} = \frac{25.5 \text{ miles}}{1 \text{ gallon}} \qquad \text{Read this as "25.5 miles per gallon."}$$

Simplifying rates to unit rates can lead to statements such as "There are 2.6 children per family in the state," because

$$\frac{26 \text{ children}}{10 \text{ families}} = \frac{2.6 \text{ children}}{1 \text{ family}}$$

The unit rate is a comparison, not a fact, because no family has 2.6 children.

To write a unit rate given a rate

1. Do the indicated division.
2. Keep the unlike units.

Answers to Warm-Ups

F. 1. The rate is $\frac{7 \text{ oak trees}}{44 \text{ birch trees}}$.

2. The rate is $\frac{44 \text{ birch trees}}{51 \text{ total trees}}$.

EXAMPLES G–I

DIRECTIONS: Write as a unit rate.

STRATEGY: Do the indicated division so the denominator is a 1-unit measurement.

G. Write the unit rate for $\frac{\$2.04}{3 \text{ cans of peas}}$.

STRATEGY: Do the indicated division so the denominator is 1 unit.

$$\frac{\$2.04}{3 \text{ cans of peas}} = \frac{\$0.68}{1 \text{ can of peas}}.$$

The unit rate is 68¢ per can.

H. Write the unit rate for $\frac{802.5 \text{ miles}}{12.5 \text{ hours}}$.

$$\frac{802.5 \text{ miles}}{12.5 \text{ hours}} = \frac{64.2 \text{ miles}}{1 \text{ hour}}$$ Divide numerator and denominator by 12.5.

The unit rate is 64.2 miles per hour.

CALCULATOR EXAMPLE

I. The population density of a region is a unit rate. The rate is the number of people per 1 square mile of area.

1. Find the population density of the city of Cedar Crest in Granite County if the population is 4100 and the area of the city is 52 square miles. Round to the nearest tenth.
2. Find the population density of Granite County if the population is 13,650 and the area of the county is 1600 square miles. Round to the nearest tenth.

1. STRATEGY: Write the rate and divide the numerator by the denominator using your calculator.

$$\text{Density} = \frac{4100 \text{ people}}{52 \text{ square miles}}$$

$$\approx \frac{78.8461538 \text{ people}}{1 \text{ square mile}}$$ Divide.

$$\approx \frac{78.8 \text{ people}}{1 \text{ square mile}}$$ Round to the nearest tenth.

The density is 78.8 people per square mile, to the nearest tenth.

2. STRATEGY: Write the rate and divide the numerator by the denominator using your calculator.

$$\text{Density} = \frac{13{,}650 \text{ people}}{1600 \text{ square miles}}$$

$$\approx \frac{8.53125 \text{ people}}{1 \text{ square mile}}$$ Divide.

$$\approx \frac{8.5 \text{ people}}{1 \text{ square mile}}$$ Round to the nearest tenth.

The density is 8.5 people per square mile, to the nearest tenth.

WARM-UPS G–I

G. Write the unit rate for $\frac{528 \text{ lb}}{12 \text{ in.}^2}$

H. Write the unit rate for $\frac{485 \text{ mi}}{25 \text{ gal}}$.

I. 1. Find the approximate population density of the city of Los Angeles in 2000 if the population is estimated at 3,600,000 and the area is 467 square miles. Round to the nearest whole number.

2. What was the approximate population density of the county of Los Angeles in 2000 if the population is estimated at 9,900,000 and the area is 1110 square miles? Round to the nearest whole number.

Answers to Warm-Ups

G. The unit rate is 44 lb per in.^{2}.
H. The unit rate is 19.4 mpg.
* **I.** 1. The population density was 7709 people per mi^2.
2. The population density was 8919 people per mi^2.

Name ______________________ Class ______________ Date __________

EXERCISES 5.1

OBJECTIVE 1 Write a fraction that shows a ratio comparison of two like measurements.

Write as a ratio in simplified form.

A.

1. 8 to 56

2. 6 to 60

3. 14 ft to 42 ft

4. 24 tsp to 18 tsp

5. 40 cents to 45 cents

6. 30 dimes to 54 dimes

B.

7. 1 dime to 4 nickels (compare in cents)

8. 3 quarters to 5 dimes (compare in cents)

9. 3 ft to 40 in. (compare in inches)

10. 2 yd to 8 ft (compare in feet)

11. 200 min to 3 hrs (compare in minutes)

12. 20 days to 50 wks (compare in days)

OBJECTIVE 2 Write a fraction that shows a rate comparison of two unlike measurements.

Write as a rate and simplify.

A.

13. 13 cars to 6 families

14. 18 children to 6 families

15. 110 miles in 2 hours

16. 264 km in 3 hr

17. 63 miles to 3 gallons

18. 100 km to 4 gal

19. 253 trees in 22 rows

20. $280 in 16 hr

B.

21. 18 trees to 63 ft

22. 91 TVs to 52 houses

23. 38 books to 95 students

24. 750 people for 3000 tickets

25. 774 students to 36 rooms

26. 89 satellite dishes to 267 houses

27. 345 pies to 46 sales

28. $17.68 per 34 lb of apples

 Name ____________ Class ____________ Date ____________

OBJECTIVE 3 Write a unit rate.

A. *Write a rate and simplify to a unit rate.*

29. 152 students to 4 classes

30. 630 km to 6 hr

31. 36 ft to 9 sec

32. 75 m to 3 min

33. 132 yd to 2 billboards

34. 60 yd to 15 posts

35. 5000 miles to 4 quarts

36. $15.40 per 35 dozen eggs

B. *Write a unit rate. Round to the nearest tenth.*

37. 36 children to 14 families

38. 52 cars to 25 families

39. 1000 ft to 12 sec

40. 1000 yd to 15 min

41. 12,095 lb to 45 mi^2

42. 5486 kg to 290 cm^2

43. 2225 gal per 3 hr

44. 4872 plants in 78 rows

C.

45. A Jackson and Perkins catalog advertised miniature roses for $10.95 each or a special deal of three roses for $29.95.
a. If Carol orders the 3-rose special, what is the price per rose? Round to the nearest cent.
b. How much savings is this compared to buying three separate roses?

46. A Jackson and Perkins catalog advertised mixed color foxgloves at either 6 for $24.95 or 12 for $39.95.
a. What is the price per plant if you buy six? Round to the nearest cent.
b. What is the price per plant if you buy 12?
c. How much would Ted save if he bought the 12-foxglove package as compared to buying two 6-foxglove packages?

47. Safeway sells Nestle's chocolate chips in a 12-oz bag for $1.98 and a super size 60-oz bag for $6.99.
a. What is the price per ounce for each sized bag?
b. How much can Jerry save by buying the large bag instead of buying 60 oz in small bags?

48. The Publix has Coke products on sale for two 12 packs of 12-oz cans for $5. They also have the 2-liter bottles of Coke products on sale for $1.09.
a. What is the price per ounce for the 12 packs of cans? Round to the nearest thousandth of a cent.
b. Two liters contain approximately 67.6 oz. What is the price per ounce for the 2-liter bottles? Round to the nearest thousandth of a cent.
c. What is the price of a 2-liter bottle that has the same cost per ounce as the 12 pack of cans?

49. The parking lot in the lower level of the Senter Building has 18 spaces for compact cars and 24 spaces for larger cars.

a. What is the ratio of compact spaces to larger spaces?

b. What is the ratio of compact spaces to the total number of spaces?

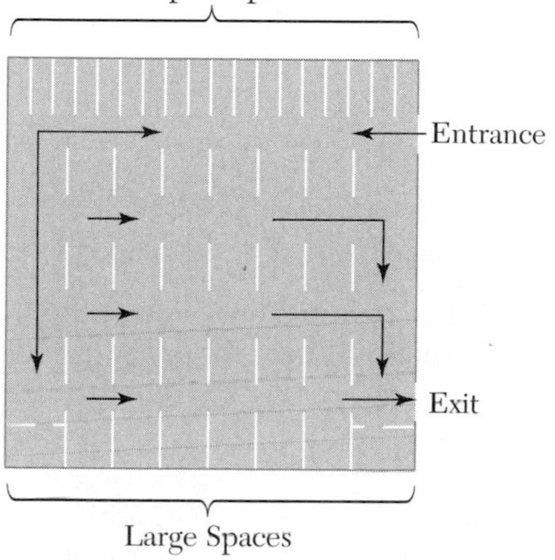

50. The Reliable Auto Repair Service building has eight stalls for repairing automobiles and four stalls for repairing small trucks.

a. What is the ratio of the number of stalls for small trucks to the number of stalls for automobiles?

b. What is the ratio of the number of stalls for small trucks to the total?

51. A coat is regularly priced at \$99.99, but during a sale its price is \$66.66. What is the ratio of the sale price to the regular price?

52. In Exercise 51, what is the ratio of the discount to the regular price?

53. What is the population density of Dryton City if there are 22,450 people and the area is 230 square miles? Simplify to a unit comparison, rounded to the nearest tenth.

54. What is the population density of Struvaria if 950,000 people live there and the area is 18,000 square miles? Simplify to a unit comparison, rounded to the nearest tenth.

55. What was the population density of your city in 2000?

56. What was the population density of your state in 2000?

57. In the United States, four people use an average of 250 gallons of water per day. One hundred gallons are used to flush toilets, 80 gallons are used in baths or showers, 35 gallons for doing laundry, 15 gallons for washing dishes, 12 gallons for cooking and drinking, and 8 gallons are used in the bathroom sink.

a. Write the ratio of laundry use to toilet use.

b. Write the ratio of bath or shower use to dishwashing use.

58. Use Exercise 57.

a. Write the ratio of cooking and drinking use to dishwashing use.

b. Write the ratio of laundry use per person.

59. Drinking water is considered to be polluted when a pollution index of 0.05 mg of lead per liter is reached. At that rate, how many milligrams of lead are enough to pollute 25 L of drinking water?

60. Available figures indicate that 3 of every 20 rivers in the United States showed an increase in water pollution in a recent 10-year period. Determine how many rivers are in your state. At the same rate, determine how many of those rivers had an increased pollution level during the same period.

Exercises 61–63. It is often difficult to compare the prices of food items, frequently because of the packaging. Is a 14-oz can of pears for $0.89 a better buy than a 16-oz can of pears for $1.00? To help consumers compare, unit pricing *is often posted. Mathematically, we write the information as a rate and simplify to a 1-unit comparison.*

61. Write a ratio for a 14-oz can of pears that sells for $0.89 and simplify it to a unit price (price per 1 oz of pears). Do the same with the 16-oz can of pears for $1.00. Which is the better buy?

62. Which is the best buy: a 15-oz box of Cheerios for $2.49, a 20-oz box for $3.29, or a 2-lb, 3-oz box for $5.39?

63. Which is the better buy: 5 lb of granulated sugar on sale for $4.95 or 25 lb of sugar for $24.90?

Exercises 64–65. Some food items have the same unit price regardless of the quantity purchased. Other food items have a decreasing unit price as the size of the container increases. In order to determine which category a food falls into, find the unit price for each item.

64. Is the unit price of hamburger the same if 2.35 lb costs $4.44 and 3.52 lb costs $6.65?

65. Is the unit price of frozen orange juice the same if a 12-oz can costs $1.09 and a 16-oz can costs $1.30?

66. List five items that usually have the same unit price regardless of quantity purchased and five that do not. What circumstances could cause an item to change categories?

67. A Quantum Professional PR600C fishing reel can retrieve 105 in. of fishing line in 5 turns of the handle. Write this as a ratio and then calculate the retrieve rate (measured in inches per turn).

68. A Quantum Professional PR600CX fishing reel can retrieve 127 in. of fishing line in 5 turns of the handle. Write this as a ratio and then calculate the retrieve rate.

69. Full-time equivalency (FTE) is a method by which colleges calculate enrollment. The loads of all students are added together and then divided into theoretical full-time students. Three River Community College requires its full-time students to take 15 credits per term. One term there were 645 students enrolled, taking a total of 4020 credits. Find the average number of credits per student and the FTE for the term.

Exercises 70–71 relate to the chapter application.

70. In the movie *Honey, I Shrunk the Kids*, two teenagers and their little brothers are accidentally reduced to $\frac{1}{4}$ in. tall. Assume that the average teenager in this country is 5′6″. Write a ratio that describes this relationship.

71. The kids encounter an ant, and one of them rides on the ant's back as though riding on a horse. Estimate the length of a horse and the length of an ant. Write a ratio between these two lengths, which should have the same value as the ratio in Exercise 70. Reduce each of your ratios to decimal form to decide if they are the same or at least reasonably close. Is it consistent mathematically that the shrunken kids could ride an ant as though they were riding a horse? Explain.

STATE YOUR UNDERSTANDING

72. Write a short paragraph explaining why ratios are useful ways to compare measurements.

73. Explain the difference among a ratio, a rate, and a unit rate. Give an example of each.

CHALLENGE

74. Give an example of a rate that is not a ratio. Give an example of a ratio that is not a rate.

75. The ratio of noses to persons is $\frac{1}{1}$ or 1 to 1. Find three examples of 2-to-1 ratios and three examples of 3-to-1 ratios.

76. Each gram of fat contains 9 calories, Chicken sandwiches at various fast-food places contain the following total calories and grams of fat.

	Calories	Grams of Fat
a. RB's Light Roast Chicken Sandwich	276	7
b. KB's Boiler Chicken Sandwich	267	8
c. Hard B's Chicken Filet	370	13
d. LJS's Baked Chicken Sandwich	130	4
e. The Major's Chicken Sandwich	482	27
f. Mickey's Chicken	415	19
g. Tampico's Soft Chicken Taco	213	10
h. Winston's Grilled Chicken Sandwich	290	7

Find the ratio of fat calories to total calories for each sandwich.

GROUP WORK

77. Have each member of your group select a country other than the United States. Each member is to use the library, the Internet, or other resource to find the population and area of the country selected. Calculate the population density for each country and compare your findings. Which country has the greatest population density. The least?

78. The golden ratio of 1.618 to 1 has been determined by artists to be very pleasing aesthetically. The ratio has been discovered to occur in nature in many places, including the human body. In particular, the ratio applies to successive segments of the fingers.

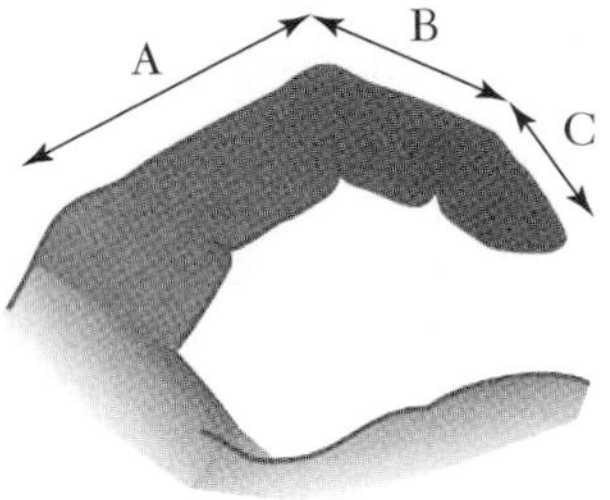

Measure as accurately as possible at least three fingers of everyone in the group. Calculate the ratio of successive segments, and fill in the table.

Name	Finger	A	B	C	$\frac{A}{B}$	$\frac{B}{C}$

Whose fingers come closest to the golden ratio? Can you find other body measures that have this ratio?

MAINTAIN YOUR SKILLS

Simplify.

79. $\frac{3}{8} \cdot \frac{4}{21}$

80. $\frac{5}{12} \div \frac{15}{28}$

81. 6.5(0.03)

82. $12.85 \div 2.5$

83. $\frac{7}{8}(0.6)$

84. $1\frac{3}{8} + 4\frac{3}{10}$

85. $9 - 4\frac{7}{15}$

86. $31.2 \div 1000$

Which is larger?

87. $\frac{3}{4}$ or $\frac{94}{125}$

88. $\frac{7}{10}$ or $\frac{29}{40}$

5.2 Solving Proportions

VOCABULARY

A **proportion** is a statement that two ratios are equal.

In a proportion, **cross multiplication** means multiplying the numerator of each ratio times the denominator of the other.

Cross products are the products obtained from cross multiplication.

Solving a proportion means finding a missing number, usually represented as a letter or **variable**, that will make the proportion true.

OBJECTIVES

1. Determine whether a proportion is true or false.
2. Solve a proportion.

HOW & WHY

OBJECTIVE 1 Determine whether a proportion is true or false.

The statement $\frac{21}{12} = \frac{28}{16}$ is a proportion. A proportion states that two rates or ratios are equal. To check whether the proportion is true or false, we use cross multiplication.

The proportion $\frac{21}{12} = \frac{28}{16}$ is true if the cross products are equal.

$$\frac{21}{12} \overset{?}{=} \frac{28}{16}$$

$21(16) \overset{?}{=} 12(28)$ Find the cross products.

$336 = 336$ The cross products are equal.

The proportion is true.

The cross-multiplication test is actually a shortcut for converting both fractions to equivalent fractions with common denominators and checking that the numerators match. Let us examine the same proportion using the formal method.

$$\frac{21}{12} \overset{?}{=} \frac{28}{16}$$

$\frac{16}{16} \cdot \frac{21}{12} \overset{?}{=} \frac{28}{16} \cdot \frac{12}{12}$ A common denominator is $16 \cdot 12 = 192$.

$\frac{336}{192} = \frac{336}{192}$ Multiply. The numerators are the same.

The proportion is true.

The cross products in the shortcut (cross multiplication) are the numerators in the formal method. Therefore, checking the cross products is a valid procedure for determining the truth of a proportion.

To check whether a proportion is true or false

1. Check that the ratios or rates have the same unit(s).
2. Cross multiply.
3. If the cross products are equal, the proportion is true.

WARM-UPS A–C

A. Is $\frac{7}{9} = \frac{56}{72}$ true or false?

B. Is $\frac{2.2}{2.5} = \frac{2}{5}$ true or false?

C. Is $\frac{1.20 \text{ dollars}}{2 \text{ quarters}} = \frac{72 \text{ nickels}}{15 \text{ dimes}}$ true or false?

Answers to Warm-Ups

A. true **B.** false **C.** true

EXAMPLES A–C

DIRECTIONS: Determine whether a proportion is true or false.

STRATEGY: Check the cross products. If they are equal, the proportion is true.

A. Is $\frac{6}{5} = \frac{78}{65}$ true or false?

$$\frac{6}{5} = \frac{78}{65}$$
$$6(65) = 5(78)$$
$$390 = 390 \qquad \text{True}$$

The proportion is true.

B. Is $\frac{2.1}{7.1} = \frac{2}{7}$ true or false?

$$\frac{2.1}{7.1} = \frac{2}{7} \qquad \text{Find the cross products.}$$
$$2.1(7) = 7.1(2)$$
$$14.7 = 14.2 \qquad \text{False}$$

The proportion is false.

C. Is $\frac{1 \text{ dollar}}{3 \text{ quarters}} = \frac{8 \text{ dimes}}{12 \text{ nickels}}$ true or false?

STRATEGY: The units in the rates are not the same. We change all units to cents and simplify.

$$\frac{1 \text{ dollar}}{3 \text{ quarters}} = \frac{8 \text{ dimes}}{12 \text{ nickels}}$$
$$\frac{100 \text{ cents}}{75 \text{ cents}} = \frac{80 \text{ cents}}{60 \text{ cents}}$$
$$\frac{100}{75} = \frac{80}{60} \qquad \text{Like units may be dropped.}$$
$$100(60) = 75(80)$$
$$6000 = 6000 \qquad \text{True}$$

The proportion is true.

HOW & WHY

OBJECTIVE 2 Solve a proportion.

Proportions are used to solve many problems in science, technology, and business. There are four numbers or measures in a proportion. If three of the numbers are known, we can find the missing number. For example,

$$\frac{x}{6} = \frac{14}{21}$$
$$21x = 6(14) \qquad \text{Cross multiply.}$$
$$21x = 84$$

Every multiplication fact can be written as a related division fact. The product divided by one factor gives the other factor, so $21x = 84$ can be written as $x = 84 \div 21$.

$$x = 84 \div 21 \qquad \text{Rewrite as division.}$$
$$x = 4$$

CHECK: $\frac{4}{6} = \frac{14}{21}$ Substitute 4 for x in the original proportion.

$4(21) = 6(14)$ Cross multiply (or observe that both fractions simplify to $\frac{2}{3}$).

$84 = 84$

The missing number is 4.

To solve a proportion

1. Cross multiply.
2. Do the related division problem to find the missing number.

EXAMPLES D–G

DIRECTIONS: Solve the proportion.

STRATEGY: Cross multiply, then write the related division and simplify.

D. Solve: $\frac{4}{15} = \frac{8}{x}$

$4x = 15(8)$	Cross multiply.
$4x = 120$	Simplify.
$x = 120 \div 4$	Rewrite as division.
$x = 30$	Simplify.

The missing number is 30.

E. Solve: $\frac{0.8}{z} = \frac{0.4}{1.4}$

$0.8(1.4) = 0.4z$	Cross multiply.
$1.12 = 0.4z$	Simplify.
$1.12 \div 0.4 = z$	Rewrite as division.
$z = 2.8$	Simplify.

The missing number is 2.8.

F. Solve: $\frac{\frac{3}{4}}{\frac{5}{8}} = \frac{\frac{1}{2}}{w}$

$\frac{3}{4}w = \frac{5}{8}\left(\frac{1}{2}\right)$	Cross multiply.
$\frac{3}{4}w = \frac{5}{16}$	Simplify.
$w = \frac{5}{16} \div \frac{3}{4}$	Rewrite as division.
$w = \frac{5}{16} \cdot \frac{4}{3}$	Invert the divisor.
$w = \frac{5}{12}$	Simplify.

The missing number is $\frac{5}{12}$.

WARM-UPS D–G

D. Solve: $\frac{5}{8} = \frac{10}{y}$

E. Solve: $\frac{0.6}{c} = \frac{0.5}{0.75}$

F. Solve: $\frac{\frac{3}{4}}{1\frac{2}{3}} = \frac{\frac{1}{2}}{x}$

Answers to Warm-Ups

D. $y = 16$ **E.** $c = 0.9$

F. $x = \frac{10}{9}$ or $1\frac{1}{9}$

G. Solve $\frac{8}{v} = \frac{2.82}{7.31}$ and round to the nearest hundredth.

CALCULATOR EXAMPLE

G. Solve $\frac{2}{t} = \frac{5.8}{6.52}$ and round to the nearest hundredth.

$2(6.52) = 5.8t$	Cross multiply.
$2(6.52) \div 5.8 = t$	Rewrite as division.
$2.24827568 \approx t$	Simplify using a calculator.
$2.25 \approx t$	Round.

* The missing number is 2.25 to the nearest hundredth.

Answers to Warm-Ups

G. $v \approx 20.74$

EXERCISES 5.2

OBJECTIVE 1 Determine whether a proportion is true or false.

True or false?

A.

1. $\frac{15}{3} = \frac{55}{11}$
2. $\frac{3}{4} = \frac{18}{24}$
3. $\frac{3}{2} = \frac{9}{4}$
4. $\frac{3}{4} = \frac{9}{16}$
5. $\frac{2}{3} = \frac{10}{15}$
6. $\frac{6}{8} = \frac{9}{12}$

B.

7. $\frac{16}{14} = \frac{24}{21}$
8. $\frac{16}{12} = \frac{12}{9}$
9. $\frac{35}{22} = \frac{30}{20}$
10. $\frac{24}{32} = \frac{36}{38}$
11. $\frac{27}{45} = \frac{30}{60}$
12. $\frac{36}{45} = \frac{20}{25}$
13. $\frac{13}{4} = \frac{9.75}{3}$
14. $\frac{25}{6} = \frac{12.5}{3}$

OBJECTIVE 2 Solve a proportion.

Solve.

A.

15. $\frac{1}{3} = \frac{a}{24}$
16. $\frac{1}{4} = \frac{b}{28}$
17. $\frac{2}{6} = \frac{c}{18}$
18. $\frac{2}{9} = \frac{x}{18}$
19. $\frac{3}{y} = \frac{4}{8}$
20. $\frac{8}{z} = \frac{4}{10}$
21. $\frac{14}{28} = \frac{5}{c}$
22. $\frac{8}{12} = \frac{6}{d}$
23. $\frac{2}{3} = \frac{12}{x}$
24. $\frac{3}{6} = \frac{8}{y}$
25. $\frac{p}{8} = \frac{2}{32}$
26. $\frac{q}{4} = \frac{2}{24}$

B.

27. $\frac{x}{7} = \frac{3}{2}$
28. $\frac{y}{5} = \frac{3}{4}$
29. $\frac{2}{z} = \frac{5}{11}$
30. $\frac{12}{x} = \frac{16}{3}$
31. $\frac{14}{21} = \frac{w}{14}$
32. $\frac{9}{12} = \frac{y}{15}$
33. $\frac{15}{16} = \frac{12}{a}$
34. $\frac{28}{7} = \frac{50}{b}$
35. $\frac{0.1}{c} = \frac{0.2}{1.2}$
36. $\frac{0.5}{z} = \frac{0.2}{0.6}$
37. $\frac{\frac{3}{5}}{b} = \frac{8}{5}$
38. $\frac{\frac{2}{3}}{c} = \frac{\frac{8}{9}}{1\frac{7}{9}}$

39. $\dfrac{0.8}{3.5} = \dfrac{0.08}{x}$

40. $\dfrac{1.4}{2.5} = \dfrac{3.5}{y}$

41. $\dfrac{7}{4.2} = \dfrac{w}{0.03}$

42. $\dfrac{0.8}{0.3} = \dfrac{b}{2.4}$

43. $\dfrac{y}{3} = \dfrac{9}{\frac{1}{8}}$

44. $\dfrac{s}{40} = \dfrac{\frac{3}{4}}{5}$

45. $\dfrac{t}{24} = \dfrac{3\frac{1}{2}}{10\frac{1}{2}}$

46. $\dfrac{w}{4\frac{1}{4}} = \dfrac{3\frac{1}{3}}{2\frac{1}{2}}$

Solve. Round to the nearest tenth.

47. $\dfrac{3}{11} = \dfrac{w}{5}$

48. $\dfrac{3}{11} = \dfrac{x}{15}$

49. $\dfrac{9}{35} = \dfrac{14}{y}$

50. $\dfrac{9}{35} = \dfrac{24}{z}$

Solve. Round to the nearest hundredth.

51. $\dfrac{1.5}{5.5} = \dfrac{a}{0.8}$

52. $\dfrac{1.5}{5.5} = \dfrac{b}{2.8}$

53. $\dfrac{\frac{3}{7}}{c} = \dfrac{9}{20}$

54. $\dfrac{\frac{3}{7}}{d} = \dfrac{9}{32}$

C.

Fill in the boxes to make statement true. Explain your answer.

55. If $\dfrac{\square}{120} = \dfrac{x}{12}$, then $x = 1$.

56. If $\dfrac{\square}{25} = \dfrac{y}{20}$, then $y = 4$.

57. Find the error(s) in the statement: If $\dfrac{2}{5} = \dfrac{x}{19}$, then $2x = 5(19)$. Correct the statement. Explain how you would avoid this error.

58. Find the error(s) in the statement: If $\dfrac{3}{w} = \dfrac{7}{9}$, then $3w = 7(9)$. Correct the statement. Explain how you would avoid this error.

59. Current recommendations from the American Heart Association are that the ratio of total cholesterol to HDL ("good" cholesterol) for women be no more than 4.5 to 1. Fran has a total cholesterol level of 214. The proportion $\dfrac{214}{H} = \dfrac{4.5}{1}$ gives the minimum allowable level, H, of HDL for Fran to stay within the guidelines. What is this level, rounded to the nearest whole number?

60. The American Heart Association's recommendation for men is a ratio of total cholesterol to HDL of no more than 4.0 to 1. Jim's total cholesterol level is 168, and the proportion $\dfrac{168}{H} = \dfrac{4.0}{1}$ gives his minimum allowable HDL level, H. What is Jim's minimum allowable HDL level?

61. In a recent year, the average pupil–teacher ratio in public elementary schools was 19 to 1. Westridge Elementary has 650 students. The proportion $\frac{650}{t} = \frac{19}{1}$ gives the expected number of teachers, t, in keeping with the national average. How many teachers would you expect Westridge to have, rounded to the nearest whole teacher?

62. The highest marriage rate this century in the United States occurred in 1946, when the rate was 118 marriages per 1000 unmarried women per year. In a city with 8450 unmarried women in 1946, the proportion $\frac{m}{8450} = \frac{118}{1000}$ gives m, the number of expected marriages. How many marriages were expected in the city in 1946?

63. For every ten people in the United States, it is estimated that seven suffer from some form of migraine headache.

a. If three times as many women as men suffer from migraines, find the number of men that suffer migraines in a group where 306 woman are affected. To determine the number of men, solve the proportion $\frac{1}{3} = \frac{M}{306}$, where M represents the number of men in the group.

b. Find the number of migraine sufferers in a group of 350 people. To determine the number of sufferers, solve the proportion $\frac{7}{10} = \frac{N}{350}$, where N represents the number of migraine sufferers in the group.

c. Some researchers are of the opinion that headaches in 8 of every 40 migraine sufferers are related to diet. At this rate, determine the number of migraine headaches that may be related to diet in a group of 350 sufferers. To find the number, solve the proportion $\frac{8}{40} = \frac{D}{350}$, where D represents the number of headaches related to diet.

Exercises 64–66 relate to the chapter application.

64. Early in the movie, the shrunken kids encounter a butterfly. One of them comments that the butterfly has a 42-ft wingspan. This measure, of course, is not physically true. It is the shrunken kid's perception relative to his new size. To check the accuracy of this statement, we complete the table below and use it to set up a proportion.

	Shrunken Kid	Butterfly
Actual Size	0.25 in.	3 in.
Perceived Size	5.5 ft	42 ft

From the table, we set up the proportion $\frac{0.25}{5.5} = \frac{3}{42}$.

Is this proportion true or false?

65. Perhaps the proportion in Exercise 62 is false because of the assumption that a butterfly has a 3-in. wingspan. Experiment with butterflies of other sizes until you find a size that makes the proportion true. Do you think this part of the movie is realistic mathematically?

66. Later in the film, the kids find a Lego and crawl into the holes in the back of it to sleep. The diameter of the holes appears to be about 4 ft, relative to the shrunken kids. Measure the diameter of the hole of a real Lego to complete the following table.

	Shrunken Kid	Lego (hole diameter)
Actual Size	0.25 in.	
Perceived Size	5.5 ft	4 ft

Set up a proportion from the table, as in Exercise 64. Is your proportion true or false? How realistic is the set design at this point?

 Name ______________________ Class ______________ Date __________

STATE YOUR UNDERSTANDING

67. Explain how to solve $\frac{3.5}{\frac{1}{4}} = \frac{7}{y}$.

68. Look up the word *proportion* in the dictionary and write two definitions that differ from the mathematical definition used in this section. Write three sentences using the word *proportion*, that illustrate each of the meanings.

CHALLENGE

Solve.

69. $\frac{9 + 3}{9 + 6} = \frac{8}{a}$

70. $\frac{5(9) - 2(5)}{8(6) - 3(2)} = \frac{8}{a}$

Solve. Round to the nearest thousandth.

71. $\frac{7}{w} = \frac{18.92}{23.81}$

72. $\frac{7}{t} = \frac{18.81}{23.92}$

GROUP WORK

73. Five ounces of decaffeinated coffee contain approximately 3 mg of caffeine, whereas 5 oz of regular coffee contain an average of 120 mg of caffeine. Five ounces of tea brewed for 1 min contain an average of 21 mg of caffeine. Twelve ounces of regular cola contain an average of 54 mg of caffeine. Six ounces of cocoa contain an average of 11 mg of caffeine. Twelve ounces of iced tea contain an average of 72 mg of caffeine. Determine the total amount of caffeine each member of your group consumed yesterday. Make a chart to illustrate this information. Combine this information with that of other groups in your class to make a class amount. Make a class chart to illustrate the totals. Determine the average amount of caffeine consumed by each member of the group and then by each member of the class. Compare these averages by making ratios. Discuss the similarities and the differences.

MAINTAIN YOUR SKILLS

74. Find the difference of 620.3 and 499.9781.

75. Find the average of 1.8, 0.006, 17, and 8.5.

76. Find the average of 6.45, 7.13, and 5.11.

77. Multiply 4.835 by 10,000.

78. Divide 4.835 by 10,000.

79. Multiply 0.875 by 29.

80. Multiply 12.75 by 8.09.

81. Divide 0.70035 by 0.35.

82. Divide 3.6 by 0.72.

83. If gasoline is $1.769 per gallon, how much does Quan pay for 12.8 gallons? Round your answer to the nearest cent.

5.3 Applications of Proportions

HOW & WHY

OBJECTIVE Solve word problems using proportions.

OBJECTIVE

Solve word problems using proportions.

If the ratio of two quantities is constant, the ratio can be used to find the missing part of a second ratio. For instance, if 2 lb of bananas cost \$0.48, what will 12 lb of bananas cost? Table 5.1 organizes this information.

TABLE 5.1

	Case I	Case II
Pounds of Bananas	2	12
Cost in Dollars	0.48	

In Table 5.1, the cost in Case II is missing. Call the missing value y, as in Table 5.2.

TABLE 5.2

	Case I	Case II
Pounds of Bananas	2	12
Cost in Dollars	0.48	y

Write the proportion using the ratios shown in Table 5.2.

$$\frac{2 \text{ lb of bananas}}{\$0.48} = \frac{12 \text{ lb of bananas}}{\$y}$$

The units are the same on each side of the equation, so we can drop them.

$$\frac{2}{0.48} = \frac{12}{y}$$

$2y = 0.48(12)$ Cross multiply.

$2y = 5.76$

$y = 5.76 \div 2$ Rewrite as division.

$y = 2.88$ Simplify.

So 12 lb of bananas will cost \$2.88.

Using a table forces the units of a proportion to match. Therefore, we usually do not write the units in the proportion itself. We *always* use the units in the answer.

To solve word problems involving proportions

1. Write the two ratios and form the proportion. A table with three columns and two rows will help organize the data. The proportion will be shown in the boxes of the table.
2. Solve the proportion.
3. Write the solution including the appropriate units.

WARM-UPS A–D

A. A pro shop advertises golf balls at 6 for \$7.25. At this rate, what will 4 dozen balls cost?

B. A house has a property tax of \$1260 and is valued at \$45,000. At the same rate, what will be the property tax on a house valued at \$65,000?

C. On a road map of Josephine County, $\frac{1}{4}$ in. represents 30 mi. How many miles are represented by $2\frac{2}{5}$ in.?

Answers to Warm-Ups

A. The cost of 4 dozen golf balls is \$58.
B. The property tax will be \$1820.
C. On the map, $2\frac{2}{5}$ in. represents 288 mi.

EXAMPLES A–D

DIRECTIONS: Solve the following problems using proportions.

STRATEGY: Make a table with three columns and two rows. Label the last two columns "Case I" and "Case II." Label the rows with the units in the problem. Fill in the table with the quantities given and assign a variable to the unknown quantity. Write the proportion shown in the table and solve it. Write the solution, including units of measure.

A. If 3 cans of tuna fish sell for \$3.58, what is the cost of 48 cans of tuna fish?

Make a table.

	Case I	Case II
Cans of Tuna Fish	3	48
Cost in Dollars	3.58	C

$$\frac{3}{3.58} = \frac{48}{C}$$ Write the proportion.

$$3C = (3.58)(48)$$ Cross multiply.

$$3C = 171.84$$

$$C = 57.28$$ Divide.

The cost of 48 cans of tuna fish is \$57.28.

B. Mary Alice pays \$2250 property tax on her house, which is valued at \$123,000. At the same rate, what is the property tax on a house valued at \$164,000?

Make a table.

	Case I	Case II
Tax	\$2250	T
Value	\$123,000	\$164,000

$$\frac{2250}{123{,}000} = \frac{T}{164{,}000}$$ Write the proportion.

$$2250(164{,}000) = 123{,}000T$$ Cross multiply.

$$369{,}000{,}000 = 123{,}000T$$

$$3000 = T$$ Divide.

The tax on the \$164,000 house is \$3000.

C. On a road map of Alaska, $\frac{1}{4}$ in. represents 75 miles. How many miles are represented by $1\frac{1}{3}$ in.?

Make a table.

	Case I	Case I
Inches	$\frac{1}{4}$	$1\frac{1}{3}$
Miles	75	M

$$\frac{\frac{1}{4}}{75} = \frac{1\frac{1}{3}}{M}$$ Write the proportion.

$$\frac{1}{4}M = 75\left(1\frac{1}{3}\right)$$ Cross multiply.

$$\frac{1}{4}M = 75\left(\frac{4}{3}\right)$$ Change to an improper fraction.

$$\frac{1}{4}M = 100$$ Simplify.

$$M = 400$$ Divide.

On the map, $1\frac{1}{3}$ inches represents 400 miles.

D. The city fire code requires a school to have at least 50 ft^2 of floor space for every 3 students who are in the class. What is the minimum number of square feet needed for 30 students?

	Case I	Case II
Number of Students	3	30
Square Feet of Space	50	S

$$\frac{3}{50} = \frac{30}{S}$$ Write the proportion.

$$3S = 50(30)$$ Cross multiply.

$$3S = 1500$$

$$S = 500$$

The room must have at least 500 ft^2 for 30 students.

D. In another city, the fire code requires a school to have at least 86 ft^2 for every 5 students. What is the minimum area needed for 30 students?

Answers to Warm-Ups

D. The minimum area needed is 516 ft^2.

EXERCISES 5.3

OBJECTIVE Solve word problems using proportions.

A.

Exercises 1–6. A photograph that measures 6 in. wide and 4 in. high is to be enlarged so the width will be 15 in. What will be the height of the enlargement?

	Case I	Case II
Width (in.)	(a)	(b)
Height (in.)	(c)	(d)

1. What goes into box (a)?

2. What goes into box (b)?

3. What goes into box (c)?

4. What goes into box (d)?

5. What is the proportion for the problem?

6. What is the height of the enlargement?

Exercises 7–12. If a fir tree is 35 ft tall and casts a shadow of 15 ft, how tall is a tree that casts a shadow of 24 ft?

	First Tree	Second Tree
Height (ft)	(1)	(2)
Shadow	(3)	(4)

7. What goes into box (1)?

8. What goes into box (2)?

9. What goes into box (3)?

10. What goes into box (4)?

11. What is the proportion for the problem?

12. How tall is the second tree?

Exercises 13–18. Jean and Jim are building a fence around their yard. From past experience they know that they are able to build 48 ft in 8 hr. If they work at the same rate, how many hours will it take them to complete the job if the perimeter of the yard is 288 ft?

	Case I	Case II
Time (hr)	(5)	(6)
Length of fence (ft)	(7)	(8)

13. What goes into box (5)?

14. What goes into box (6)?

15. What goes into box (7)?

16. What goes into box (8)?

17. What is the proportion for the problem?

18. How many hours will it take to build the fence?

B.

Exercises 19–23. The Centerburg Junior High School expects a fall enrollment of 910 students. The district assigns teachers at the rate of 3 teachers for every 65 students. The district currently has 38 teachers assigned to the school. How many additional teachers does the district need to assign to the school?

	Case I	Case II
Teachers	3	(e)
Students	65	(f)

19. What goes into box (e)?

20. What goes into box (f)?

21. What is the proportion for the problem?

22. How many teachers will be needed at the school next year?

23. How many additional teachers will need to be assigned?

Exercises 24–26. The average restaurant in Universeville produces 30 lb of garbage in $1\frac{1}{2}$ days. How many pounds of garbage does the average restaurant produce in 2 weeks? Let x represent the missing number of pounds.

	Case I	Case II
Days		
Garbage (lb)		

24. What goes into each of the four boxes?

25. What is the proportion for the problem?

26. How many pounds of garbage does the average restaurant have at the end of 2 weeks?

C.

27. Idamae is knitting a sweater. The knitting gauge is eight rows to the inch. How many rows must she knit to complete $12\frac{1}{2}$ in. of the sweater?

28. In 1950, 19 of every 25 residences in the United States had flush toilets. In a town of 14,700 residences, how many would be expected to have flush toilets?

29. In 1997, there were 776 cars, buses, and trucks per 1000 people in the United States. In that year, Los Angeles had a population of 3,598,000. How many cars, buses, and trucks would you expect Los Angeles to have had?

30. In 1900 in the United States, there were 40 deaths from diphtheria for every 100,000 people. How many diphtheria deaths would be expected in a town of 60,000 people? (Zero cases of diphtheria were reported in 1995.)

31. In 1950 in the United States, families spent \$3 of every \$10 of family income on food. What would you expect a family to spend on food in 1950 if their income was \$30,000?

32. By 1998, families only spent \$3 of every \$20 on food. What would you expect a family to spend on food in 1998 if their income was \$45,000?

33. If 30 lb of fertilizer will cover 1500 ft^2 of lawn, how many square feet will 50 lb of fertilizer cover?

Exercises 34–36. The Logan Community College basketball team won 12 of its first 15 games. At this rate, how many games will they win if they play a 30-game schedule?

	Case I	Case II
Games Won		
Games Played		

34. What goes into each of the four boxes?

35. What is the proportion for the problem?

36. How many games should they win with a 30-game schedule?

37. Hazel sells cars at the Quality Used Car Co. If she sells a car for $1600, she earns $80. If she sells a car for $2550, how much does she earn?

38. If gasoline sells for $1.759 per gallon, how many gallons can be purchased for $35.18?

39. In Jeena's Vegetable Market, onions are priced at 2 lb for $0.62. If Mike buys 5 lb, what does he pay?

40. Twenty-five pounds of tomatoes cost $22.70 at the farmer's market. At this rate, what is the cost of 10 lb?

41. A new car is driven 451 miles in 8.2 hr. At the same rate, how long will it take to drive 935 mi?

42. Celia earns a salary of $900 per month, from which she saves $45 each month. Her salary is increased to $990 per month. If she keeps the same rate of savings, how much will she save?

43. Ginger and George have a room in their house that needs a new carpet. It will take 33 yd^2 of carpet to cover the floor. Hickson's Carpet Emporium will install 33 yd^2 of carpet for $526.35. Ginger and George decide to have a second room of their house carpeted. This room requires 22 yd^2 of carpet. At the same rate, how much will it cost to have the second room carpeted?

44. A 16-oz can of pears costs $0.98 and a 29-oz can costs $1.69. Is the price per ounce the same in both cases? If not, what should the price of the 29-oz can be to make the price per ounce equivalent?

45. Jim's doctor gives instructions to Ida, a nurse, to prepare a hypodermic containing 8 mg of a drug. The drug is in a solution that contains 20 mg in 1 cm^3 (1 cc). How many cubic centimeters should Ida use for the injection?

46. If a 24-ft beam of structural steel contracts 0.0036 in. for each drop of 5 degrees in temperature, then, at same rate, how much will a 50-ft beam of structural steel contract for each drop of 5 degrees in temperature?

47. A package of gumdrops weighing 1.5 oz sells for 45¢. At the same rate, what is the cost of 1 lb (16 oz) of gumdrops?

48. The ratio of girls to boys taking a math class is 5 to 4. How many girls are in a class of 81 students? *Hint*: Complete the table.

	Case I	Case II
Number of Girls		
Number of Students	9	81

49. Betty prepares a mixture of nuts that has cashews and peanuts in a ratio of 3 to 7. How many pounds of each will she need to make 40 lb of the mixture?

50. A local health food store is making a cereal mix that has nuts to cereal in a ratio of 2 to 7. If the store makes 126 oz of the mix, how many ounces of nuts will be needed?

51. Debra is making green paint by using 3 quarts of blue paint for every 4 quarts of yellow paint. How much blue paint will she need to make 98 quarts of green paint?

52. A concrete mix contains 3 bags of cement, 2 bags of sand, and 3 bags of gravel. How many bags of cement are necessary for 68 bags of the concrete mix?

53. Mario makes meatballs for his famous spaghetti sauce by using 10 lb of ground round to 3 lb of spice additives. How many pounds of ground round will he need for 91 lb of meatballs?

54. If \$1 is worth 210 drachma (Greek currency) and a used refrigerator costs \$247, what is the cost of the refrigerator in drachmas?

55. If \$1 is worth £0.65 (British pound) and a computer costs \$1300, what is the cost of the computer in pounds?

56. If \$1 is worth 1455 lire (Italian currency) and a pair of shoes costs 69,840 lire, what is the cost in dollars?

57. Auto batteries are sometimes priced proportionally to the number of years they are expected to last. If a \$35.85 battery is expected to last 36 months, what is the comparable price of a 60-month battery?

58. In 1970, only 7.1 lb of every 100 lb of waste was recovered. In 1980, this rose to 9.7 lb. By 1990, the amount was 13.1 lb. In 2000, the amount was up to 15.2 lb. Determine the amount of waste recovered from 56,000,000 lb of waste in each of these years.

59. The quantity of ozone contained in 1 m^3 of air may not exceed 235 mg or the air is judged to be polluted. What is the maximum quantity of ozone that can be contained in 12 m^3 of air before the air is judged to be polluted?

60. A 5.5-oz can of Alpo cat food is priced at three cans for \$1.00. A 13-oz can is \$0.59. The store manager wants to put the smaller cans on sale so they are the same unit price as the larger cans. What price should the smaller cans be marked?

61. A large box of brownie mix that makes four batches of brownies costs \$4.79 at a warehouse outlet. A box of brownie mix that makes one batch costs \$1.29 in a grocery store. By how much should the grocery store reduce each box so that its prices are competitive with those of the warehouse outlet?

Exercises 62–64 relate to the chapter application.

62. You are the set designer for the movie *Honey, I Shrunk the Kids*. If the backyard were actually 60 ft long, how long would it be in the perception of the shrunken kids? *Hint*: Complete the table and be careful with the units.

	Shrunken Kid	Backyard
Actual Size	0.25 in.	
Perceived Size	5.5 ft	

63. One of the kids is nearly eaten when he falls into a bowl of Cheerios. How big does the set designer make a Cheerio if it is to be mathematically accurate?

64. Along with the kids, a sofa was also shrunk. If the sofa was 6 ft long, how small did it become?

STATE YOUR UNDERSTANDING

65. What is a proportion? Write three examples of situations that are proportional.

66. Look on the label of any food package to find the number of calories in one serving. Use this information to create a problem that can be solved by a proportion. Write the solution of your problem in the same way as the examples in this section are written.

67. From a consumer's viewpoint, explain why it is not always an advantage for costs of goods and services to be proportional.

CHALLENGE

68. In 1982, there were approximately 25 California condors. This low population was the result of hunting, habitat loss, and poisoning. The U.S. Fish and Wildlife Service instituted a program that resulted in 73 condors in 1992. If this increase continues proportionally, how many condors do you think there will be in 2017?

69. The tachometer of a sports car shows the engine speed to be 2800 revolutions per minute. The transmission ratio (engine speed to drive shaft speed) for the car is 2.5 to 1. Find the drive shaft speed.

70. Two families rented a mountain cabin for 19 days at a cost of \$1905. The Santini family stayed for 8 days and the Nguyen family stayed for 11 days. How much did it cost each family? Round the rents to the nearest dollar.

GROUP WORK

71. A $13\frac{1}{2}$-oz bag of Cheetos costs \$2.69 and a 24-oz bag costs \$4.09. You have a coupon for \$0.50 that the store will double. Divide the group into two teams. One team will formulate an argument that using the coupon on the smaller bag results in a better value. The other team will formulate an argument that using the coupon on the larger bag is better. Present your arguments to the whole group and select the one that is most convincing. Share your results with the rest of the class.

72. List all the types of recycling done by you and your group members. Determine how many people participate in each type of recycling. Determine ratios for each kind of recycling. Find the population of your city or county. Using your class ratio, determine how many people in your area are recycling each type of material. Make a chart to illustrate your findings. Contact your local recycling center to see how your ratios compare to their estimates. Explain the similarities and differences.

MAINTAIN YOUR SKILLS

73. Round 37.4145 to the nearest hundredth and to the nearest thousandth.

74. Round 73.4195 to the nearest hundredth and to the nearest thousandth.

75. Compare the decimals 0.00872 and 0.011. Write the result as an inequality.

76. Compare the decimals 0.06 and 0.15. Write the result as an inequality.

77. What is the total cost of 11.9 gallons of gasoline that costs \$1.599 per gallon? Round to the nearest cent.

78. A barrel of liquid weighs 429.5 lb. If the barrel weighs 22.5 lb and the liquid weighs 7.41 lb per gallon, how many gallons of liquid are in the barrel to the nearest gallon?

Change each decimal to a simplified fraction.

79. 0.865

80. 0.01125

Change each fraction to a decimal rounded to the nearest thousandth.

81. $\frac{123}{220}$

82. $\frac{37}{350}$

CHAPTER 5

KEY CONCEPTS

SECTION 5.1 RATIO AND RATE

Definitions and Concepts	Examples
A ratio is a comparison of two measurements by division.	The ratio of the length of a room to its width is $\frac{12\text{ ft}}{9\text{ ft}} = \frac{12}{9} = \frac{4}{3}$.
A rate is a comparison of two unlike measurements by division.	The rate of a biker who rides 21 mi in 2 hr is $\frac{21\text{ mi}}{2\text{ hr}}$.
A unit rate is a rate with a denominator of one unit.	The unit rate of a biker who rides 21 mi in 2 hr is $\frac{21\text{ mi}}{2\text{ hr}} = \frac{10.5\text{ mi}}{1\text{ hr}} = 10.5$ mph.

SECTION 5.2 SOLVING PROPORTIONS

Definitions and Concepts	Examples
A proportion is a statement that two ratios are equal.	$\frac{6}{12} = \frac{1}{2}$ is a proportion.
To solve a proportion: • Cross multiply. • Do the related division problem to find the missing number.	Solve $\frac{3}{x} = \frac{15}{43}$. $3(43) = 15x$ $129 = 15x$ $129 \div 15 = x$ $8.6 = x$

SECTION 5.3 APPLICATIONS OF PROPORTIONS

Definitions and Concepts

To solve word problems involving proportion:
- Make a table to organize the information.
- Write a proportion from the table.
- Solve the proportion.
- Write the solution, including appropriate units.

Examples

If 3 cans of cat food sell for \$3.69, how much will 8 cans cost?

	Case I	Case II
Cans	3	8
Cost	\$3.69	C

$$\frac{3}{3.69} = \frac{8}{C}$$
$$3C = 29.52$$
$$C = 29.52 \div 3$$
$$C = 9.84$$

The 8 cans of cat food will cost \$9.84.

CHAPTER 5

REVIEW EXERCISES

Section 5.1

Write as a ratio in simplified form.

1. 7 to 35

2. 9 to 54

3. 12 m to 10 m

4. 12 km to 9 km

5. 1 dollar to 25 nickels (compare in nickels)

6. 660 ft to 1 mi (compare in feet)

7. 16 in. to 2 ft (compare in inches)

8. 3 ft to 3 yd (compare in feet)

Write a rate and simplify.

9. 9 people to 10 chairs

10. 23 miles to 3 hikes

11. 3 videotapes to 21 DVDs

12. 132 kilometers to 4 runs

13. 88 lb to 33 ft

14. 36 buttons to 24 bows

15. 765 people to 27 committees

16. 8780 households to 6 cable companies

Write a rate and simplify to a unit rate.

17. 50 mi to 2 hr

18. 60 mi to 4 min

19. 90¢ per 10 lb of potatoes

20. $1.17 per 3 lb of broccoli

Write a unit rate. Round to the nearest tenth.

21. 825 miles per 22 gallons

22. 13,266 km per 220 gal

23. 225 gal per 14 min

24. 850 L per 14 min

25. One section of the country has 3500 TV sets per 1000 households. Another section has 500 TV sets per 150 households. Are the rates of the TV sets to the number of households the same in both parts of the country?

26. In Pineberg, there are 5000 automobiles per 3750 households. In Firville, there are 6400 automobiles per 4800 households. Are the rates of the number of automobiles to the number of households the same?

Section 5.2

True or false?

27. $\frac{6}{3} = \frac{72}{36}$

28. $\frac{2}{3} = \frac{26}{39}$

29. $\frac{18}{14} = \frac{12}{10}$

30. $\frac{16}{25} = \frac{10}{15}$ **31.** $\frac{2.8125}{3} = \frac{15}{16}$ **32.** $\frac{9.375}{3} = \frac{25}{8}$

Solve.

33. $\frac{1}{2} = \frac{r}{18}$ **34.** $\frac{1}{3} = \frac{s}{18}$ **35.** $\frac{2}{t} = \frac{5}{10}$

36. $\frac{8}{v} = \frac{2}{5}$ **37.** $\frac{f}{7} = \frac{2}{28}$ **38.** $\frac{g}{2} = \frac{2}{12}$

39. $\frac{16}{24} = \frac{r}{16}$ **40.** $\frac{s}{10} = \frac{15}{16}$ **41.** $\frac{8}{12} = \frac{t}{8}$

42. $\frac{7}{5} = \frac{w}{7}$

Solve. Round to the nearest tenth.

43. $\frac{5}{6} = \frac{a}{5}$ **44.** $\frac{5}{6} = \frac{6}{b}$ **45.** $\frac{16}{5} = \frac{c}{12}$ **46.** $\frac{16}{5} = \frac{12}{d}$

47. A box of Tide that is sufficient for 18 loads of laundry costs \$3.65. What is the most that a store brand of detergent can cost if the box is sufficient for 25 loads and is more economical to use than Tide? To find the cost, solve the proportion $\frac{\$3.65}{18} = \frac{c}{25}$, where c represents the cost of the store brand.

48. Available figures show that it takes the use of 18,000,000 gasoline-powered mowers to produce the same amount of air pollution as 3,000,000 new cars. Determine the number of gasoline-powered lawn mowers that will produce the same amount of air pollution as 50,000 new cars. To find the number of lawn mowers, solve the proportion $\frac{18{,}000{,}000}{3{,}000{,}000} = \frac{L}{50{,}000}$, where L represents the number of lawn mowers.

Section 5.3

49. For every 2 hr a week that Merle is in class, she plans to spend 5 hr a week doing her homework. If she is in class 15 hr each week, how many hours will she plan to be studying each week?

50. If 16 lb of fertilizer will cover 1500 ft^2 of lawn, how much fertilizer is needed to cover 2500 ft^2?

51. Juan must do 25 hr of work to pay for the tuition for three college credits. If Juan intends to sign up for 15 credit hours in the fall, how many hours will he need to work to pay for his tuition?

52. In Exercise 51, if Juan works 40 hr per week, how many weeks will he need to work to pay for his tuition? (Any part of a week counts as a full week.)

53. Larry sells men's clothing at the University Men's Shop. For \$120 in clothing sales, Larry makes \$15. How much does he make on a sale of \$350 worth of clothing?

54. Dissolving 1.5 lb of salt in 1 gal of water makes a brine solution. At this rate, how many gallons of water are needed to made a brine solution with 12 lb of salt?

CHAPTER 5

TRUE / FALSE CONCEPT REVIEW

ANSWERS

Check your understanding of the language of basic mathematics. Tell whether each of the following statements is true (always true) or false (not always true). For each statement you judge to be false, revise it to make a statement that is true.

1. A fraction can be regarded as a ratio. **1.** ______

2. A ratio is a comparison of two fractions **2.** ______

3. $\frac{18\text{ miles}}{1\text{ gallon}} = \frac{54\text{ miles}}{2\text{ hr}}$ **3.** ______

4. To solve a proportion, we must know the values of three of the four numbers. **4.** ______

5. If $\frac{8}{5} = \frac{t}{2}$, then $t = 3\frac{1}{5}$ **5.** ______

6. In a proportion, two ratios are equal. **6.** ______

7. Three feet and 1 yard are like measures. **7.** ______

8. Ratios always compare like units. **8.** ______

9. To determine whether a proportion is true or false, the ratios must have the same units. **9.** ______

10. If a fir tree that is 18 ft tall casts a shadow of 17 ft, how tall is a tree that casts a shadow of 25 ft? The following table can be used to solve this problem. **10.** ______

	First Tree	Second Tree
Height	17	18
Shadow	x	25

CHAPTER 5

TEST

ANSWERS

1. Write a ratio to compare 12 yards to 15 yards. 1. ________

2. On a test, Ken answered 20 of 32 questions correctly. At the same rate, how many would he answer correctly if there were 72 questions on a test? 2. ________

3. Solve the proportion: $\frac{4.8}{12} = \frac{0.36}{w}$ 3. ________

4. Is the following proportion true or false: $\frac{16}{35} = \frac{24}{51}$? 4. ________

5. Solve the proportion: $\frac{13}{36} = \frac{y}{18}$ 5. ________

6. Is the following proportion true or false: $\frac{9 \text{ in.}}{2 \text{ ft}} = \frac{6 \text{ in.}}{16 \text{ in.}}$? 6. ________

7. If Mary is paid $49.14 for 7 hr of work, how much should she be paid for 12 hr of work? 7. ________

8. Write a ratio to compare 8 hr to 3 days (compare in hours). 8. ________

9. There is a canned food sale at the supermarket. A case of 24 cans of peas is priced at $19.68. At the same rate, what is the price of 10 cans? 9. ________

10. If 40 lb of beef contains 7 lb of bones, how many pounds of bones may be expected in 100 lb of beef? 10. ________

11. Solve the proportion: $\frac{0.4}{0.5} = \frac{0.5}{x}$ 11. ________

12. A charter fishing boat has been catching an average of 3 salmon for every 4 people they take fishing. At that rate, how many fish will they catch if over a period of time they take a total of 32 people fishing? 12. ________

13. On a trip home, Jennie used 12.5 gal of gas. The trip odometer on her car registered 295 mi for the trip. She is planning a trip to see a friend who lives 236 mi away. How much gas will Jennie need for the trip? 13. ________

14. Solve the proportion: $\frac{a}{8} = \frac{4.24}{6.4}$ 14. ________

15. Is the following a rate: $\frac{130 \text{ mi}}{2 \text{ hr}}$? 15. ________

16. If a 20-ft tree casts a 15-ft shadow, how long a shadow is cast by a 14-ft tree? 16. ________

17. What is the population density of a town that is 150 square miles and has 5580 people? Reduce to a 1-square-mile comparison. 17. ________

18. Solve the proportion and round your answer to the nearest hundredth: $\frac{4.78}{y} = \frac{32.5}{11.2}$ **18.** ______

19. A landscape firm has a job that it takes a crew of three $4\frac{1}{2}$ hr to complete. How many of these jobs could the crew of three complete in 117 hr? **19.** ______

20. The ratio of males to females in a literature class is 3 to 5. How many females are in a class of 48 students? **20.** ______

CHAPTER 5

GROUP PROJECT

OPTIONAL

The human body is the source of many common proportions. Artists have long studied the human figure in order to portray it accurately. Your group will be investigating how each member compares to the standard and how various artists have used the standards.

Most adult bodies can be divided into eight equal portions. The first section is from the top of the head to the chin. Next is from the chin to the bottom of the sternum. The third section is from the sternum to the navel, and the fourth is from the navel to the bottom of the torso. The bottom of the torso to the bottom of the knee is two sections long, and the bottom of the knee to the bottom of the foot is the last two sections. (Actually, these last two sections are a little short. Most people agree that the body is actually closer to 7.5 sections, but because this is hard to judge proportionally, we use eight sections and leave the bottom one short.)

Complete the following table for each group member.

Section	Length (in cm)	Ratio of Section to Head (actual)	Ratio of Section to Head (expected)
Head			
Chin to Sternum			
Sternum to Navel			
Navel to Torso Bottom			
Torso Bottom to Knee			
Knee to Foot			

Explain how your group arrived at the values in the last column. Which member of the group comes closest to the standards? Did you find any differences between the males and females in your group? Either draw a body using the standard proportions, or get a copy of a figure from a painting and analyze how close the artist came to the standards.

A slightly different method of dividing the upper torso is to start at the bottom of the torso and divide into thirds at the waist and the shoulders. In this method, there is a pronounced difference between males and females. In females, the middle third between the waist and shoulders is actually shorter than the other two. In males, the bottom third from waist to bottom of the torso is shorter than the others. For each member of your group, fill out the following table.

Section	Length (in cm)	Ratio of Section to Entire Upper Torso (actual)	Ratio of Section to Entire Upper Torso (expected)
Head to Shoulders			
Shoulders to Waist			
Waist to Bottom of Torso			

Explain how your group arrived at the values in the last column (these will depend on gender). Which member of your group comes closest to the standards? Either draw a body using the standard proportions, or get a copy of a figure from a painting and analyze how close the artist came to the standards.

Children have different body proportions than adults, and these proportions change with the age of the child. Measure three children who are the same age. Use their head measurement as one unit, and compute the ratio of head to entire body. How close are the three children's ratios to each other? Before the Renaissance, artists usually depicted children as miniature adults. This means that the proportions fit those in the first table rather than those you just discovered. Find a painting from before the Renaissance that contains a child. Calculate the child's proportions and comment on them. Be sure to reference the painting you use.

CHAPTERS 1–5

CUMULATIVE REVIEW

Add or subtract.

1. $356 + 781 + 910 + 57$

2. $6700 - 3456$

3.
$$\begin{array}{r} 9623 \\ -6489 \\ \hline \end{array}$$

4.
$$\begin{array}{r} 78 \\ 310 \\ 295 \\ +\ \ 9 \\ \hline \end{array}$$

Multiply or divide.

5.
$$\begin{array}{r} 304 \\ \times\ 37 \\ \hline \end{array}$$

6. $1071 \div 17$

7. $145(61)$

8. $23\overline{)2944}$

9. Find the value of 2^7.

10. Multiply: 34×10^4

11. Simplify: $34 + 18 \times 2 \div 3 + 3$

12. Find the average, median, and mode of 35, 56, 72, 35, 18, 22, 91.

Exercises 13–14. The graph shows the distribution of students taking algebra at a local high school.

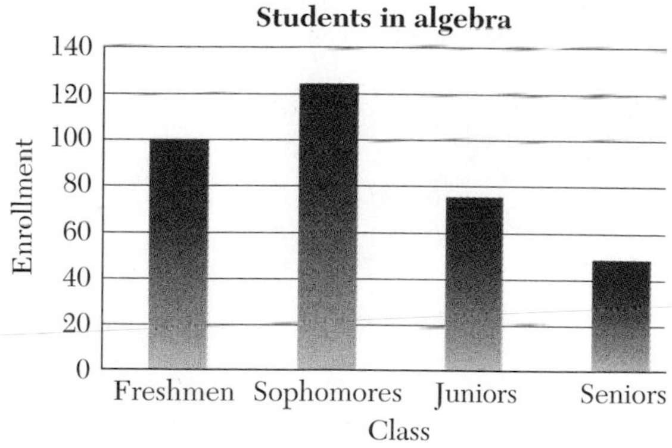

13. Which class had the most students taking algebra?

14. How many seniors were taking algebra?

15. Draw a line graph to display the traffic count at the intersection of Elm and 3rd Avenues, as given in the table.

Traffic Count at Elm and 3rd

Monday	Tuesday	Wednesday	Thursday	Friday
425	500	375	450	550

16. Is 1245 divisible by 2, 3, or 5?

17. Is 3060 divisible by 6, 9, or 10?

18. List the first 5 multiples of 21.

19. Is 678 a multiple of 24?

20. Write 170 as a product of two factors in all possible ways.

21. List all the factors of 190.

22. Find the greatest common factor (GCF) of 60 and 75.

23. Is 303 prime or composite?

24. Write the prime factorization of 820.

25. Find the least common multiple (LCM) and the greatest common factor (GCF) of 116, 84, and 36.

26. Change to a mixed number: $\frac{23}{5}$

27. Change to an improper fraction: $11\frac{4}{9}$

Simplify completely.

28. $\frac{124}{144}$

29. $\frac{640}{760}$

Multiply or divide. Simplify completely.

30. $\frac{15}{28} \cdot \frac{21}{45}$

31. $\frac{56}{80} \cdot \frac{48}{49} \cdot \frac{35}{40}$

32. $\frac{18}{21} \div \frac{45}{77}$

33. $\frac{33}{34} \div \frac{88}{51}$

34. $4\frac{1}{5} \div 3\frac{2}{3}$

35. $9\frac{1}{7} \cdot 4\frac{1}{5}$

36. List the fractions from smallest to largest: $\frac{8}{17}, \frac{11}{21}, \frac{7}{15}, \frac{3}{5}$

Add and subtract. Simplify completely.

37. $\frac{14}{15} + \frac{7}{20} + \frac{1}{6}$

38. $\frac{16}{45} - \frac{8}{75}$

39. $\frac{13}{40} + \frac{11}{72}$

40. $\frac{34}{42} - \frac{17}{35}$

41. $2\frac{2}{3} + 3\frac{3}{5} + 1\frac{3}{4}$

42. $7\frac{3}{5} - 5\frac{5}{7}$

Perform the indicated operations. Simplify completely.

43. $\frac{3}{7} \cdot \frac{5}{6} + \frac{1}{7} \div \frac{8}{9}$

44. $\frac{2}{3}\left(\frac{1}{4} - \frac{1}{5}\right) + \frac{3}{4}\left(\frac{4}{5} - \frac{2}{5}\right)^2$

45. Find the average of $\frac{11}{12}, \frac{9}{10}, \frac{8}{15}, \frac{19}{20}$, and $\frac{4}{5}$.

46. An article is priced to sell for $96 at the Aquarium Gift Store. It is sale priced at $\frac{1}{3}$ off. What is its sale price?

47. Jamie answers $\frac{4}{5}$ of the problems correctly on her Chapter I test. If there are 40 problems on the Chapter III test, how many must she get correct to answer the same fractional amount?

48. At the end of one year, John has grown $\frac{5}{12}$ in. During the same time, his sister has grown $\frac{3}{8}$ in. Who has grown the most? By how much?

49. A machinist needs a bar that is $1\frac{5}{8}$ in. thick. If he cuts it from a bar that is $1\frac{27}{32}$ in. thick, how much must he cut off?

Write the word name.

50. 16.034

51. 0.00512

Write the place value name.

52. Two hundred forty-seven and six hundredths

53. Four thousand and four ten thousandths

Change each decimal to a fraction and simplify if possible.

54. 0.44

55. 0.035

List the decimals from smallest to largest.

56. 0.34, 0.087, 0.29, 0.125, 0.031

57. 0.042, 0.62, 0.2, 0.091, 0.009

Add and subtract.

58.
$$\begin{array}{r} 456.023 \\ 43.9 \\ 103.045 \\ +\ \ 9.32 \\ \hline \end{array}$$

59.
$$\begin{array}{r} 75.342 \\ -9.677 \\ \hline \end{array}$$

60. 0.6734 − 0.519

61. 13.45 + 8.002 + 104.003 + 0.67 + 1.345

62. Pete wrote checks in the amounts of $34.95, $16.75, $234.50, $132.60, and $45.45. Pete has $461.98 in his checking account. Does he have enough money to cover the checks?

63. Jose goes shopping with \$124 in cash. He pays \$24.95 for a polo shirt, \$23.95 for a DVD, \$11.55 for lunch, and \$44.95 for a pair of tennis shoes. On the way home, he buys \$16.50 worth of gas. How much money does he have left?

Multiply or divide.

64. $\begin{array}{r} 23.45 \\ \times\ 12.6 \\ \hline \end{array}$

65. $\begin{array}{r} 0.0346 \\ \times\ \ 0.71 \\ \hline \end{array}$

66. $4.5\overline{)23.58}$

67. $54.321 \div 0.17$ (Round to the nearest tenth.)

68. $56.341(1000)$

69. $56.341 \div 1000$

Write in scientific notation.

70. 0.00381

71. 360.9

72. The home of Joe Tyler is assessed at \$345,000. The property tax rate in the area is \$1.315 per hundred dollars of assessment. Find the property tax that Joe pays, rounded to the nearest dollar.

73. A 55-gallon drum of cleaning solvent at Five Star Cleaners is being used at the rate of 0.73 gallon per day. At this rate, how many days will the solvent last? Round to the nearest day?

74. A rental car costs \$21.95 a day plus \$0.29 per mile driven. What is the cost of renting a car for 1 day if it is driven 369 miles?

Change the fraction or mixed number to a decimal.

75. $8\frac{17}{20}$

76. $\frac{12}{83}$ (Round to the nearest thousandth.)

Perform the indicated operations.

77. $0.4(0.6 - 0.34) + 5.5 - (3.2 + 1.18)$

78. $0.01(4.3)^2 + 7.05 \div 0.15 + 8.2$

Find the average, median, and mode of the group of numbers.

79. 13.45, 12.11, 9.05, 12.11, 16.7, 18.9

80. 0.678, 0.032, 0.56, 0.93, 0.92

81. Andy buys school supplies for the start of the new semester. He buys 3 notebooks at \$3.25 each, 7 pens at \$1.19 each, 4 packs of notepaper at \$0.98 each, and a calculator at \$74.80. How much does he spend?

82. The camera system at the intersection of Hall and Cedar Hills Boulevard caught the following number of drivers going through a red light during 1 week: Monday, 56; Tuesday, 84; Wednesday, 102; Thursday, 92; Friday, 123; Saturday, 178; Sunday, 64. What was the average number of tickets issued per day? Round to the nearest tenth.

83. Write as a ratio in simplified form: 8 in. to 2 ft.

84. Write as a rate and simplify: 99 lb to 44 ft.

85. Write as a unit rate, rounded to the nearest tenth: 244 mi to 13.4 gal.

True or false?

86. $\frac{3}{15} = \frac{8}{40}$

87. $\frac{9}{16} = \frac{12}{25}$

Solve.

88. $\frac{4}{24} = \frac{7}{a}$

89. $\frac{15}{4} = \frac{x}{12}$

Solve. Round to the nearest tenth.

90. $\frac{5}{11} = \frac{b}{13}$

91. $\frac{29}{80} = \frac{54}{c}$

Exercises 92–97. If a flagpole 45 ft tall casts a shadow of 30 ft, how tall is a flagpole that casts a shadow of 12 ft?

	First Pole	Second Pole
Height (ft)	(1)	(2)
Shadow (ft)	(3)	(4)

92. What goes into box (1)?

93. What goes into box (2)?

94. What goes into box (3)?

95. What goes in box (4)?

96. What is the proportion for the problem?

97. How tall is the second flagpole?

98. Joanne saves \$76 a month from her take-home pay of \$1275. If her take-home pay increases to \$1550 per month, how much will she save at the same rate? Round to the nearest dollar.

99. Greg sells men's clothing at the Emporium. If he sells \$150 worth of clothes, he earns \$22.50. At the same rate, how much does he make if he sells \$610 worth of clothes?

100. Dan buys \$2000 worth of stock on Wednesday and sells it on Friday for \$2800. If Dan had invested \$5000 in the same stock on Wednesday, what would it have been worth on Friday?

Preparing for Tests

Testing usually causes the most anxiety for students. By studying more effectively, you can eliminate many of the causes of anxiety. But there are also other ways to prepare that will help relieve your fears.

If you are math anxious, you actually may study too much out of fear of failure and not allow enough time for resting and nurturing yourself. Every day, allow yourself some time to focus on your concerns, feelings, problems, or anything that might distract you when you try to study. Then, when these thoughts distract you, say to yourself, "I will not think about this now. I will later at __ o'clock. Now I have to focus on math." If problems become unmanageable, make an appointment with a college counselor.

Nurturing is any activity that will help you recharge your energy. Choose an activity that makes you feel good, such as going for a walk, daydreaming, reading a favorite book, doing yard work, taking a bubble bath, or playing basketball.

Other ways to keep your body functioning effectively under stress are diet and exercise. Exercise is one of the most beneficial means of relieving stress. Try to eat healthy foods and drink plenty of water. Avoid caffeine, nicotine, drugs, alcohol, and "junk food."

Plan to have all your assignments finished two days before the test, if possible. The day before the test should be completely dedicated to reviewing and practicing for the test.

Many students can do the problems but cannot understand the instructions and vocabulary, so they do not know where to begin. Review any concepts that you have missed or any that you were unsure of or "guessed at."

When you feel comfortable with all of the concepts, you are ready to take the practice test at the end of the chapter. You should simulate the actual testing situation as much as possible. Have at hand all the tools you will use on the real test: sharpened pencils, eraser, and calculator (if your instructor allows). Give yourself the same amount of time as you'll be given on the actual test. Plan a time for your practice test when you can be sure there will be no interruptions. Work each problem slowly and carefully. Remember, if you make a mistake by rushing through a problem and have to do it over, it will take more time than doing the problem carefully in the first place.

After taking the test, go back and study topics referenced with the answers you missed or that you feel you do not understand. You should now know if you are ready for the test. If you have been studying effectively and did well on the practice test, you should be ready for the real test. You are prepared!

6

PERCENT

APPLICATION

The price we pay for everyday items such as food and clothing is theoretically simple. The manufacturer of the item sets the price based on how much it costs to produce and adds a small profit. The manufacturer then sells the item to a retail store, which in turn marks it up and sells it to you, the consumer. But as you know, it is rarely as simple as that. The price you actually pay for an item also depends on the time of year, the availability of raw materials, the amount of competition among manufacturers of comparable items, the economic circumstances of the retailer, the geographic location of the retailer, and many other factors.

SECTIONS

GROUP ACTIVITY

Select a common item whose price is affected by the following factors:

1. Time of year
2. Economic circumstances of the retailer
3. Competition of comparable products

Discuss how the factor varies and how the price of the item is affected. For each factor, make a plausible bar graph that shows the change in price as the factor varies. (You may estimate specific price levels.)

6.1 The Meaning of Percent

OBJECTIVE

Write a percent to express a comparison of two numbers.

VOCABULARY

When ratios are used to compare numbers, the denominator is called the **base unit**. In comparing 90 to 100 $\left(\text{as the ratio } \frac{90}{100}\right)$, 100 is the base unit. The **percent comparison**, or just the **percent**, is a ratio with a base unit of 100. The percent $\frac{90}{100} = (90)\left(\frac{1}{100}\right)$ is usually written 90%. The symbol % is read "percent," and $\% = \frac{1}{100} = 0.01$.

HOW & WHY

OBJECTIVE Write a percent to express a comparison of two numbers.

The word *percent* means "by the hundred." It is from the Roman word *percentum.* In Rome, taxes were collected by the hundred. For example, if you had 100 cattle, the tax collector might take 14 of them to pay your taxes. Hence, 14 per 100, or 14 percent, would be the tax rate.

Look at Figure 6.1 to see an illustration of the concept of "by the hundred." The base unit is 100, and 24 of the 100 parts are shaded. The ratio of shaded parts to total parts is $\frac{24}{100} = 24\left(\frac{1}{100}\right) = 24\%$. We say that 24% of the unit is shaded.

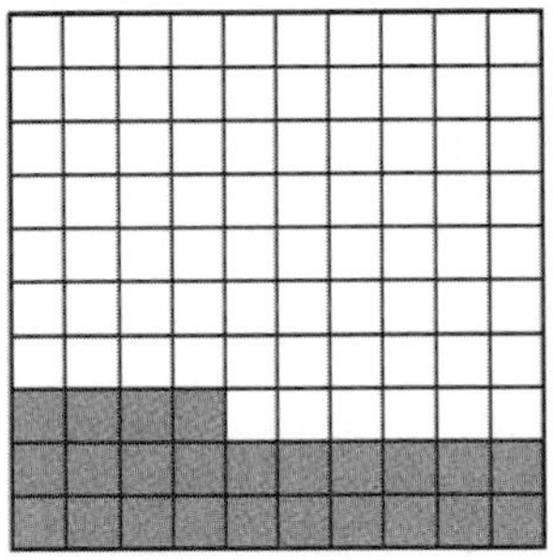

FIGURE 6.1

Figure 6.1 also illustrates that if the numerator is smaller than the denominator, then not all of the base unit is shaded, and hence the comparison is less than 100%. If the numerator equals the denominator, the entire unit is shaded and the comparison is 100%. If the numerator is larger than the denominator, more than one entire unit is shaded, and the comparison is more than 100%.

The ratio of two numbers can be used to find the percent when the base unit is not 100. Compare 7 to 20. The ratio is $\frac{7}{20}$. Now find the equivalent ratio with a denominator of 100.

$$\frac{7}{20} = \frac{35}{100} = 35 \cdot \frac{1}{100} = 35\%.$$

If the equivalent ratio with a denominator of 100 cannot be found easily, solve as a proportion. See Example F.

To find the percent comparison of two numbers

1. Write the ratio of the first number to the base number.
2. Find the equivalent ratio with denominator 100.
3. $\dfrac{\text{numerator}}{100} = \text{numerator} \cdot \dfrac{1}{100} = \text{numerator } \%$

EXAMPLES A–C

DIRECTIONS: Write the percent of each region that is shaded.

STRATEGY: (1) Count the number of parts in each unit. (2) Count the number of parts that are shaded. (3) Write the ratio of these as a fraction, and build the fraction to a denominator of 100. (4) Write the percent using the numerator in step 3.

A. What percent of the unit is shaded?

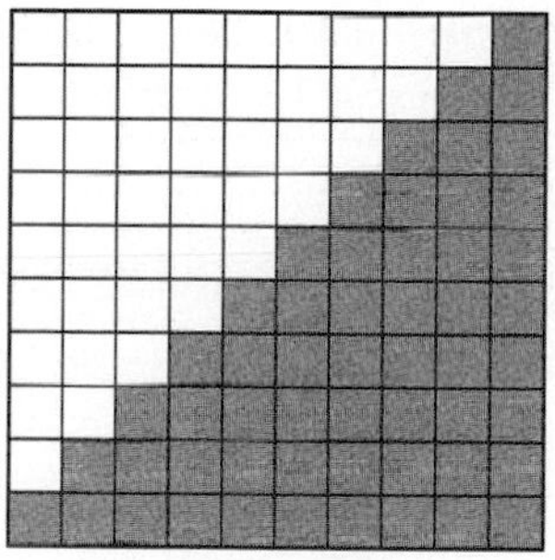

1. 100 parts in the region
2. 55 parts are shaded.
3. $\dfrac{55}{100}$
4. 55%

So 55% of the region is shaded.

B. What the percent of the region is shaded?

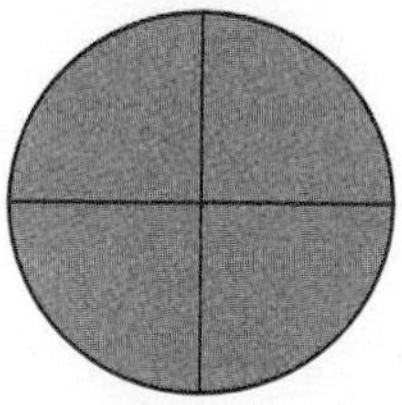

1. 4 parts in the region
2. 4 parts are shaded.
3. $\dfrac{4}{4} = \dfrac{100}{100}$
4. 100%

So 100% of the region is shaded.

WARM-UPS A–C

A. What percent of the unit is shaded?

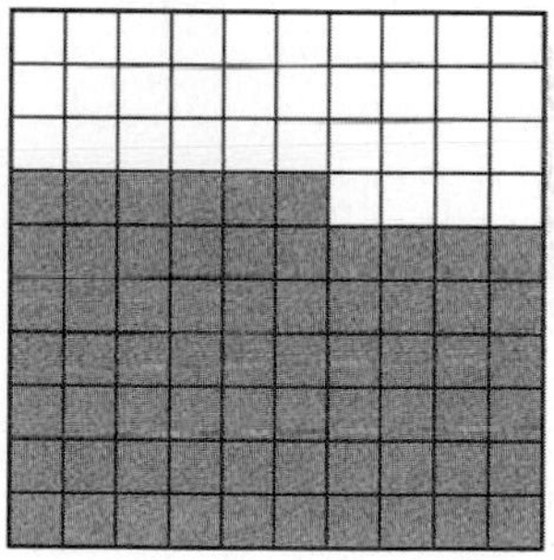

B. What the percent of the region is shaded?

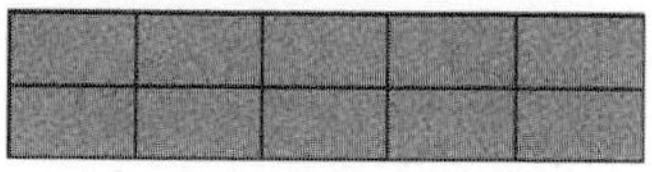

Answers to Warm-Ups

A. 66% **B.** 100%

C. What percent of the region is shaded?

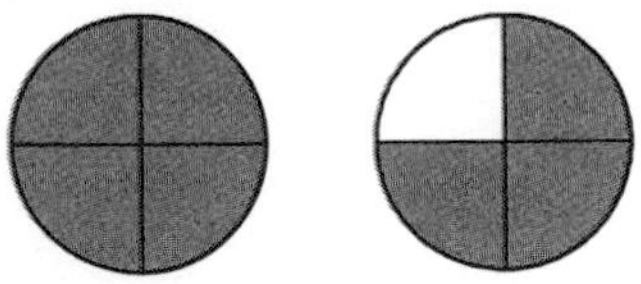

One unit One unit

Answers to Warm-Ups

C. 175%

C. What percent of the region is shaded?

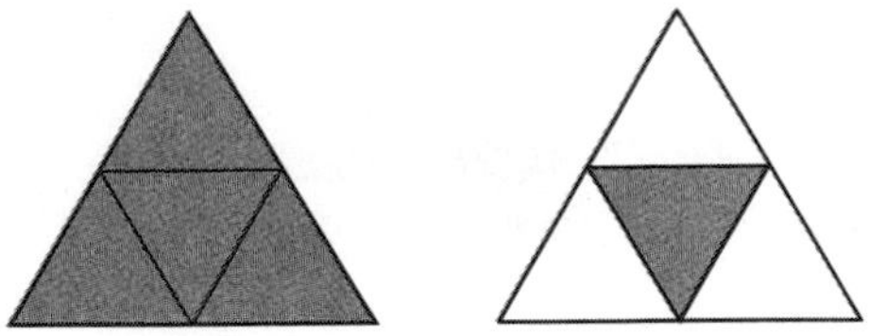

1. 4 parts in each unit
2. 5 parts are shaded.
3. $\frac{5}{4} = \frac{125}{100}$ Write as a fraction with a denominator of 100.
4. 125% Write as a percent.

So 125% of a unit in the region is shaded.

WARM-UPS D–H

D. At the last soccer match of the season, of the first 100 tickets sold, 81 were student tickets. What percent are student tickets?

E. Write the ratio of 65 to 50 as a percent.

F. Write the ratio of 10 to 12 as a percent.

Answers to Warm-Ups

D. The percent of student tickets is 81%.

E. 130% **F.** $83\frac{1}{3}\%$

EXAMPLES D–H

DIRECTIONS: Write the percent for the comparison.

STRATEGY: Write the comparison in fraction form. Build the fraction to hundredths or solve a proportion and write the percent using the numerator.

D. At a football game, 32 women are among the first 100 fans to enter. What percent of the first 100 fans are women?

$\frac{32}{100} = 32 \cdot \frac{1}{100} = 32\%$ The comparison of women to first 100 fans is 32 to 100. Write the fraction and change to a percent.

So 32% of the first 100 fans are women.

E. Write the ratio of 33 to 25 as a percent.

$\frac{33}{25} = \frac{132}{100}$ Write the ratio and build this fraction to one with a denominator of 100.

$= 132 \cdot \frac{1}{100}$

$= 132\%$ Change to a percent.

So 33 is 132% of 25.

F. Write the ratio of 15 to 21 as a percent.

$\frac{15}{21} = \frac{R}{100}$ Because we cannot build the fraction to one with a denominator of 100 using whole numbers, we write a proportion to find the percent.

$15(100) = 21R$ Cross multiply.

$\frac{1500}{21} = R$

$71\frac{3}{7} = R$

So, $\frac{15}{21} = \frac{71\frac{3}{7}}{100}$

$= 71\frac{3}{7} \cdot \frac{1}{100}$

$= 71\frac{3}{7}\%$

So 15 is $71\frac{3}{7}\%$ of 21.

CALCULATOR EXAMPLE

G. Compare 63 to 560 as a percent.

$$\frac{63}{560} = \frac{R}{100}$$ Write as a proportion.

$560R = 63(100)$ Solve.

$R = 63(100) \div 560$ Evaluate using a calculator.

$R = 11.25$

So 63 is 11.25% of 560.

H. After a strong antismoking campaign in one state, 384 restaurants out of 400 banned smoking. What percent of the restaurants banned smoking?

$$\frac{384}{400} = \frac{96}{100}$$ Write the ratio comparison and simplify.

$$= 96 \cdot \frac{1}{100}$$

$$= 96\%$$

So, 96% of the restaurants banned smoking during the campaign.

G. Compare 180 to 480 as a percent.

H. Of 300 fish caught in Olive Lake, 87 were tagged. What percent of the fish were tagged?

Answers to Warm-Ups

G. 37.5%

H. The percent of fish that were tagged was 29%.

Name ______________________ Class ______________________ Date ______________

EXERCISES 6.1

OBJECTIVE Write a percent to express a comparison of two numbers.

A.

What percent of each of the following regions is shaded?

1.

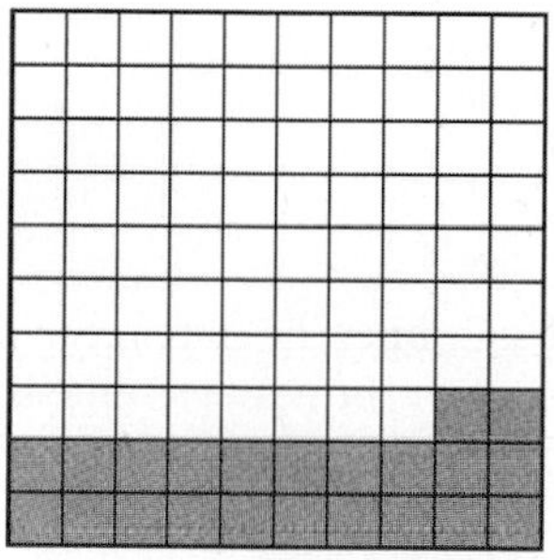

2.

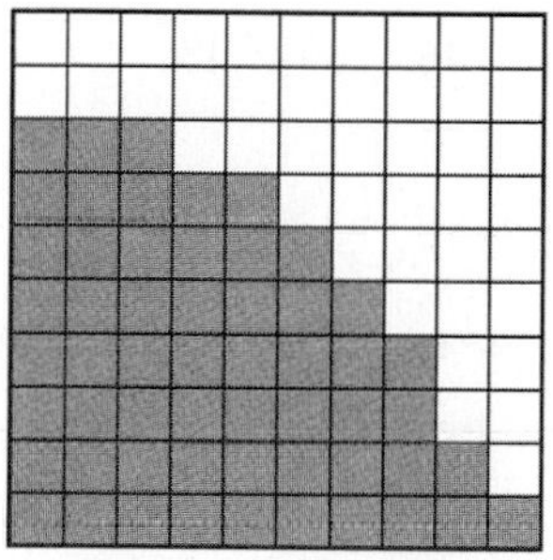

3.

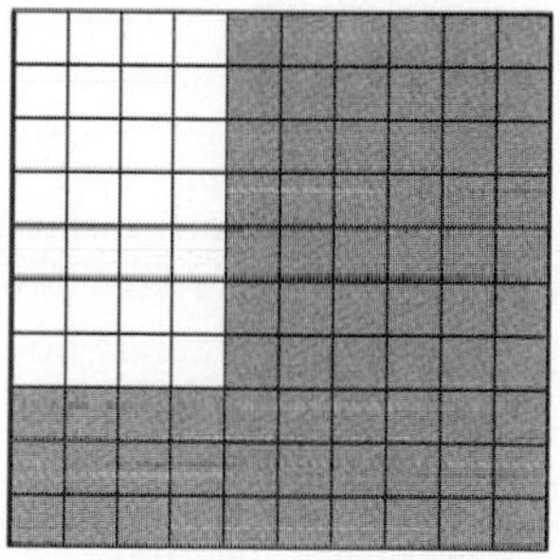

4.

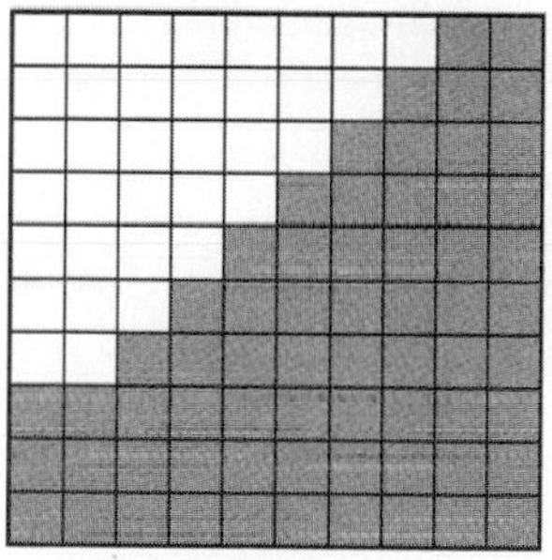

5.

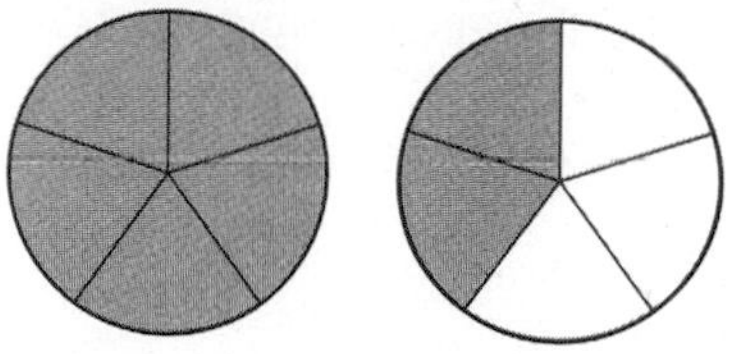

6.

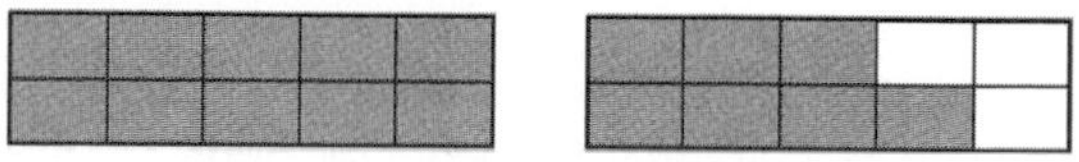

Write an exact percent for these comparisons; use fractions when necessary.

7. 39 out of 100 **8.** 79 per 100 **9.** 7 to 100 **10.** 23 to 100

11. 33 per 50 **12.** 17 per 50 **13.** 11 out of 25 **14.** 21 to 25

15. 9 per 20 **16.** 17 per 20

B. *Write an exact percent for these comparisons; use fractions when necessary.*

17. 17 to 10 **18.** 450 to 120 **19.** 163 out of 163 **20.** 57 to 57

21. 24 to 16 **22.** 36 to 16 **23.** 85 to 200 **24.** 69 to 200

25. 12 per 15 **26.** 73 per 500 **27.** 75 per 80 **28.** 93 per 80

29. 35 to 42 **30.** 23 to 30 **31.** 76 to 114 **32.** 11 to 15

C.

33. The fact that 12% of all people are blonde indicates that ___?___ out of 100 people are blonde.

34. In a recent election there was a 65% turnout of registered voters. This indicates that ___?___ out of 100 registered voters turned out to vote.

35. In a recent mail-in election, out of every 100 eligible voters, 82 cast their ballots. What percent of the eligible voters exercised their right to vote?

36. Of the people who use Shiny toothpaste, 37 out of 100 report fewer cavities. Out of every 100 people who report, what percent does not report fewer cavities?

Write an exact percent for these comparisons; use fractions when necessary.

37. 105 to 336

38. 204 to 480

39. 115 to 25

40. 123 to 15

41. 319 to 600

42. 46 to 900

43. If a telephone tax is 8 cents per dollar, what percent is this?

44. For every $100 spent on gasoline, the state receives $7.50 tax. What percent of the price of gasoline is the state tax?

45. A bank pays $5.65 interest per year for every $100 in savings. What is the annual interest rate?

46. James has $500 in his savings account. Of that amount, $35 is interest that was paid to him. What percent of the total amount is the interest?

47. Last year Mr. and Mrs. Johanson were informed that the property tax rate on their home was $2.75 per $100 of the house's assessed value. What percent is the tax rate?

48. Beginning in the early 1970s, women in the armed forces were treated the same as men with respect to training, pay, and rank. As a result, the number of women in the armed forces nearly tripled over the levels of the late 1960s. In the year 2000, about 7 out of every 50 officers were women. Express this as a percent.

49. In 1968, the number of juveniles (under age 18) arrested for property crimes was at the highest level for the century. Juveniles accounted for 11 out of every 20 arrests for property crimes in that year. What percent of property crime arrests were juveniles in 1968?

50. In 1968, what percent of property crime arrests were adults? (See Exercise 49.)

51. In the year 2000, 14 out of every 25 people earning a bachelor's degree in the United States were female. What percent of the people earning bachelor's degrees were women in 2000?

52. In the year 2000, what percent of the people earning bachelor's degrees were men? (See Exercise 51.)

Exercises 53–56 are related to the chapter application.

53. Carol spends $67 on a new outfit. If she has $100, what percent of her money does she spend on the outfit?

54. A graphing calculator originally priced at $100 is on sale for $89. What is the percent of discount? (Discount is the difference between the original price and the sales price.)

55. Mickie bought a TV and makes monthly payments to pay for it. Last year, she paid a total of of $900. Of the total that she paid, $135 was interest. What percent of the total was interest?

56. Pablo buys a suit that was originally priced at $100. He buys it for 35% off the original price. What does he pay for the suit?

STATE YOUR UNDERSTANDING

57. What is a percent? How is it related to fractions and decimals?

58. Explain the difference in meaning of the symbols 25% and 125%. In your explanation, use diagrams to illustrate the meanings. Contrast similarities and differences in the diagrams.

CHALLENGE

59. Write the ratio of 101 to 500 as a fraction and as a percent.

60. Write the ratio of 503 to 800 as a fraction and as a percent.

GROUP WORK

61. Have the members of your group use the resource center to find some background about the percent symbol (%). Divide the task so that one member looks in a large dictionary, some look in different encyclopedias, and others look in other mathematics books. Together, make a short report to the rest of the class on your findings.

MAINTAIN YOUR SKILLS

Multiply.

62. 5.67(100)

63. 23.985×100

64. 0.00578(1000)

65. 176.3×1000

66. 1.006(10)

67. 14.6×100

68. $\frac{15}{17} \cdot \frac{17}{15}$

69. $\frac{3}{11} \cdot \frac{11}{3}$

70. Bill goes to the store with $15. He uses his calculator to keep track of the money he is spending. He decides that he can make the following purchases. Is he correct?

Article	Cost
2 loaves of bread	$0.99 each
5 cans of soup	$0.89 each
1 box of crackers	$1.29
2 lb hamburger	$1.59 per lb
6 cans of root beer	6 cans for $2.79

71. Ms. Henderson earns $14.85 per hour and works the following hours during 1 month. How much are her monthly earnings?

Week	Hours
1	35
2	30.25
3	25.5
4	36.75
5	6.5

6.2 Changing Decimals to Percents

HOW & WHY

OBJECTIVE Write a given decimal as a percent.

OBJECTIVE
Write a given decimal as a percent.

In multiplication, where one factor is $\frac{1}{100}$, the indicated multiplication can be read as a percent. That is, $75\left(\frac{1}{100}\right) = 75\%$, $0.8\left(\frac{1}{100}\right) = 0.8\%$, and $\frac{3}{4}\left(\frac{1}{100}\right) = \frac{3}{4}\%$.

To write a number as a percent, multiply by $100 \cdot \frac{1}{100}$, a name for 1. This is shown in Table 6.1.

TABLE 6.1 CHANGE A DECIMAL TO A PERCENT

Number	Multiply by 1 $100\left(\frac{1}{100}\right) = 1$	Multiply by 100	Percent
0.74	$0.74(100)\left(\frac{1}{100}\right)$	$74.\left(\frac{1}{100}\right)$	74%
0.6	$0.6(100)\left(\frac{1}{100}\right)$	$60.\left(\frac{1}{100}\right)$	60%
4	$4(100)\left(\frac{1}{100}\right)$	$400.\left(\frac{1}{100}\right)$	400%

In each case the decimal point is moved two places to the right and the percent symbol (%) is inserted.

To change a decimal to a percent

1. Move the decimal point two places to the right. (Write zero on the right if necessary.)
2. Write the percent symbol (%) on the right.

EXAMPLES A–G

DIRECTIONS: Change the decimal to a percent.

STRATEGY: Move the decimal point two places to the right and write the percent sign on the right.

A. Write 0.53 as a percent.

0.53 = 53% — Move the decimal point two places to the right. Write the percent symbol on the right.

So, 0.53 = 53%.

B. Change 0.04 to a percent.

0.04 = 004% = 4% — Because the zeros are to the left of 4, we can drop them.

So, 0.04 = 4%.

WARM-UPS A–G

A. Write 0.37 as a percent.

B. Change 0.08 to a percent.

Answers to Warm-Ups
A. 37% **B.** 8%

C. Write 0.612 as a percent.

C. Write 0.458 as a percent.

0.458 = 45.8%

So, 0.458 = 45.8%.

D. Change 0.0023 to a percent.

D. Change 0.0011 to a percent.

0.0011 = 000.11% = 0.11% This is eleven hundredths of 1 percent.

So, 0.0011 = 0.11%.

E. Write 7 as a percent.

E. Write 14 as a percent.

14 = 14.00 = 1400% Insert two zeros on the right so we can move two decimal places. Fourteen hundred percent is 14 times 100%.

So, 14 = 1400%.

F. Change $0.29\overline{6}$ to a percent.

F. Change $0.84\overline{3}$ to a percent.

$0.84\overline{3} = 84.\overline{3}\% = 84\frac{1}{3}\%$ The repeating decimal $0.\overline{3} = \frac{1}{3}$.

So, $0.84\overline{3} = 84.\overline{3}\% = 84\frac{1}{3}\%$.

G. The tax code lists the tax rate on a zone 3 lot at 0.029. What is the the tax rate expressed as a percent?

G. The tax rate on a building lot is given as 0.036. What is the tax rate expressed as a percent?

0.036 = 003.6% = 3.6%

So the tax rate expressed as a percent is 3.6%.

Answers to Warm-Ups

C. 61.2% D. 0.23% E. 700%

F. $29.\overline{6}\%$ or $29\frac{2}{3}\%$

G. The tax rate is 2.9%.

EXERCISES 6.2

OBJECTIVE Write a given decimal as a percent.

Write each decimal as a percent.

A.

1. 0.67 **2.** 0.53 **3.** 6.51 **4.** 8.64

5. 0.05 **6.** 0.03 **7.** 1.39 **8.** 6.98

9. 17 **10.** 21 **11.** 0.0035 **12.** 0.0017

13. 0.732 **14.** 0.376 **15.** 0.45 **16.** 0.77

17. 6 **18.** 11 **19.** 0.416 **20.** 0.712

B.

21. 0.0731 **22.** 0.0716 **23.** 80 **24.** 45

25. 17.81 **26.** 4.311 **27.** 0.00034 **28.** 0.00236

29. 7.1 **30.** 2.39 **31.** 0.0345 **32.** 0.0213

33. 0.00047 **34.** 0.0031 **35.** 0.2054 **36.** 0.3608

37. $0.8116\overline{6}$ **38.** $0.043\overline{3}$ **39.** 0.1005 **40.** 0.30607

C.

41. If the tax rate on a person's income is 0.27, what is the rate expressed as a percent?

42. The completion rate in a certain math class is 0.67. What is the rate as a percent?

43. A merchant sold 0.46 of her sale items in the first day of the sale. What percent of the sale items was sold?

44. The sales tax in an eastern state is 0.068. Express this as a percent.

45. If 0.178 of the contestants in a race withdraw, what is the percent of withdrawals?

46. Mary Ellen measured the July rainfall. She found it was 0.067 of the year's total. Express this as a percent.

47. During the East Coast blizzard of 2001, the price of a snow blower increased by a factor of 1.37. Express this as a percent.

48. The price of the new 2002 Accord is 1.13 times the price of a 2000 model. Express this as a percent.

49. A swim team wins 0.715 of its meets. Write this as a percent.

50. A WNBA basketball player makes 0.798 of her free throws. Express this as a percent.

51. In July 2001, the night that Barry Bonds hit two home runs and tied Mickey Mantle for 9th on the career home runs list, his batting average was 0.300. Express this as a percent and state its meaning.

52. In the middle of the 2001 season, Jeremy Giambi of the Oakland Athletics had a batting average of 0.318 when he was a designated hitter and a batting average of 0.327 when he was batting as the first baseman. Express these as percents.

53. The Bureau of Labor Statistics expects that in the decade from 1996 to 2006 there will be about 81,000 new physical therapy jobs. The total number of physical therapy jobs will be 1.704 times the number of jobs in 1996. Express this as a percent.

54. A nautical mile is about 1.15 times the length of a statute (land) mile. Express this as a percent.

55. Social Security paid by employees is found by multiplying gross wages by 0.062. The Medicare payment is found by multiplying gross wages by 0.0145. Express the sum of these amounts as a percent.

56. At one point in 2001, the interest rate on new homes dropped to 0.0615. Express the interest rate as a percent.

Exercises 57–58 relate to the chapter application.

57. The sale price of a can of beans is 0.89 of what it was before the sale. Express this as a percent. What "percent off" will the store advertise?

58. Mary spends 0.325 of her monthly income on groceries. What percent of her monthly income is spent on groceries?

STATE YOUR UNDERSTANDING

59. Explain how the decimal form and the percent form of a number are related. Give an example of each form.

CHALLENGE

Write as percents.

60. 0.0004507

61. 18,000

Write as percents without using repeating decimals.

62. 0.024 and $0.02\overline{4}$

63. 0.425 and $0.42\overline{5}$

GROUP WORK

64. Baseball batting averages are written as decimals. A batter with an average of 238 has hit an average of 238 times out of 1000 times at bat (0.238). Find the batting averages of the top five players in the American and National Leagues for the last 5 years. Express these averages as percents.

Name ______________________ Class ______________________ Date ____________

MAINTAIN YOUR SKILLS

Divide.

65. $12.45 \div 100$

66. $0.0672 \div 1000$

67. $1743 \div 10^4$

68. $0.9003 \div 10^2$

69. $\dfrac{34.6}{100}$

70. $\dfrac{31}{1000}$

71. Round to the nearest thousandth: 3.87264

72. Round to the nearest ten-thousand: 345,891.62479

Perform the indicated operations.

73. $(0.37)(0.4) + 2.5 - (0.04)(0.02)$

74. $[2.4^2 + 7.3(0.2)](0.3)^2 - 0.041$

6.3 Changing Percents to Decimals

HOW & WHY

OBJECTIVE Write a given percent as a decimal.

The percent symbol indicates multiplication by $\frac{1}{100}$, so

$$23\% = 23 \cdot \frac{1}{100} = \frac{23}{100} = 23 \div 100$$

As we learned in Section 4.5, dividing a number by 100 is done by moving the decimal point two places to the left.

$$23\% = 23 \div 100 = 0.23$$

OBJECTIVE

Write a given percent as a decimal.

To change a percent to a decimal

1. Move the decimal point two places to the left. (Write zeros on the left if necessary.)
2. Drop the percent symbol (%).

EXAMPLES A–F

DIRECTIONS: Change a percent to a decimal.

STRATEGY: Move the decimal point two places to the left and drop the percent symbol.

A. Change 28.7% to a decimal.

28.7% = 0.287 Move the decimal point two places left. Drop the percent symbol.

So, 28.7% = 0.287.

B. Write 92% as a decimal.

92% = 0.92

So, 92% = 0.92.

C. Change 561% to a decimal.

561% = 5.61 A value over 100% becomes a mixed number or whole number.

So, 561% = 5.61.

D. Write $43\frac{7}{10}\%$ as a decimal.

$43\frac{7}{10}\% = 43.7\%$ Change the fraction to a decimal.

$= 0.437$ Change the percent to a decimal.

So, $43\frac{7}{10}\% = 0.437$.

WARM-UPS A–F

A. Change 48.3% to a decimal.

B. Write 38% as a decimal.

C. Change 833% to a decimal.

D. Write $72\frac{2}{5}\%$ as a decimal.

Answers to Warm-Ups

A. 0.483 **B.** 0.38 **C.** 8.33 **D.** 0.724

E. Change $19\frac{4}{15}\%$ to a decimal. Round to the nearest thousandth.

E. Change $16\frac{4}{11}\%$ to a decimal. Round to the nearest thousandth.

$16\frac{4}{11}\% = 16.\overline{36}\%$ By division, $\frac{4}{11} = 0.\overline{36}$.

$= 0.16\overline{36}$ Change to a decimal.

≈ 0.164 Round to the nearest thousandth.

So, $16\frac{4}{11}\% \approx 0.164$.

F. When ordering cement, a contractor orders 3.1% more than is needed to allow for waste. What decimal will she enter into the computer to calculate the extra amount to be added to the order?

F. When ordering fresh vegetables, a grocer orders 9.3% more than is needed to allow for spoilage. What decimal is entered into the computer to calculate the amount of extra vegetables to be added to the order?

$9.3\% = 0.093$ Change the percent to a decimal.

So the grocer will enter 0.093 in the computer.

Answers to Warm-Ups

E. 0.193

F. The contractor will enter 0.031 in the computer.

EXERCISES 6.3

OBJECTIVE Write a given percent as a decimal.

Write each of the following as a decimal.

A.

1. 19% **2.** 81% **3.** 53% **4.** 23%

5. 73% **6.** 57% **7.** 6.41% **8.** 8.12%

9. 503% **10.** 222% **11.** 216.5% **12.** 156.7%

13. 0.012% **14.** 0.034% **15.** 213% **16.** 529%

17. 0.19% **18.** 0.672% **19.** 2340% **20.** 1511%

B.

21. 0.037% **22.** 0.0582% **23.** 100% **24.** 300%

25. 212% **26.** 691% **27.** $\frac{1}{2}\%$ **28.** $\frac{1}{4}\%$

29. $\frac{5}{8}\%$ **30.** $\frac{4}{5}\%$ **31.** $37\frac{3}{4}\%$ **32.** $19\frac{3}{5}\%$

33. 0.00014% **34.** 0.0000709% **35.** $\frac{7}{8}\%$ **36.** $\frac{5}{16}\%$

37. 183.554% **38.** 491.07% **39.** 475.5% **40.** 213.9%

C.

41. Employees just settled their new contract and got a 8.25% raise over the next 2 years. Express this as a decimal.

42. When bidding for a job, an estimator adds 7.85% to the bid to cover unexpected expenses. What decimal part is this?

43. Interest rates are expressed as percents. The Credit Union charges 9.75% interest on new auto loans. What decimal will the Credit Union use to compute the interest?

44. What decimal is used to compute the interest on a mortgage that has an interest rate of 7.65%?

45. Unemployment is down 0.35%. Express this as a decimal.

46. The cost of living rose 0.48% during June. Express this as a decimal.

Change to a decimal rounded to the nearest thousandth.

47. $\frac{6}{11}\%$

48. $\frac{7}{9}\%$

49. $62\frac{4}{15}\%$

50. $57\frac{5}{13}\%$

51. $\frac{23}{6}\%$

52. $\frac{15}{8}\%$

53. In industrialized countries, 60% of the river pollution is due to agricultural runoff. Change this to a decimal.

54. Recycling aluminum cans consumes 95% less energy than smelting new stocks of metal. Change this to a decimal.

55. In 1993, Americans had 4% lower blood cholesterol than in 1981. Change this to a decimal.

56. One serving of Quaker Toasted Oatmeal contains 8% of the daily recommended intake of sodium and 14% of the recommended intake of dietary fiber. Express these as decimals.

57. Find today's interest rates for home mortgages for 15 and 30 years. Express these as decimals.

58. Find the interest being paid in savings accounts at three different financial institutions in your area. Express these as decimals.

59. One yard is about 91.4% of a meter. Express this as a decimal.

60. The Moscow subway system has the largest number of riders of any subway system in the world. It has 115.3% of the riders of the Tokyo subway system, the next largest system. Express this as a decimal.

61. The Petronas Towers in Kuala Lumpur, Malaysia, are the tallest buildings in the world. They are about 102% of the height of the Sears Tower in Chicago, the next tallest building. Express this as a decimal.

62. Pluto is the smallest planet in our solar system, with a diameter that is about 27.4% of the diameter of Earth. Express this as a decimal.

63. During the 2001 season, Roger Clemens of the New York Yankees, won 87% of the games in which he was involved in the decision. Express this as a decimal.

64. During the 2001 major league baseball season, the Seattle Mariners won 71.6% of the games they played. Express this as a decimal.

Exercises 65–66 relate to the chapter application.

65. The May Company advertises a sale at 45% off originally marked prices. What decimal will Carol use to calculate the savings on an item originally priced at $78.95?

66. Nordstrom advertises a sale of an additional 15% off sale prices. What decimal will Houng use to calculate the savings on a sale item priced at $55?

STATE YOUR UNDERSTANDING

67. When changing a percent to a decimal, how can you tell when the decimal will be greater than 1?

CHALLENGE

68. Change $11\frac{4}{17}\%$ to a decimal rounded to the nearest tenth and the nearest thousandth.

69. Change $56\frac{11}{12}\%$ to a decimal rounded to the nearest tenth and the nearest thousandth.

GROUP WORK

70. Research the major causes of the greenhouse effect. Find out which substances cause the greenhouse effect and the percent contributed by each. Write these percents in decimal form. Discuss ways to reduce the greenhouse effect in class and write group reports on your findings.

MAINTAIN YOUR SKILLS

Change to a decimal.

71. $\frac{7}{8}$

72. $\frac{9}{24}$

73. $\frac{19}{16}$

74. $\frac{24}{75}$

75. $\frac{27}{15}$

76. $\frac{117}{65}$

Change to a fraction.

77. 0.715

78. 0.1025

79. In 1 week, Greg earns $245. His deductions (income tax, Social Security, and so on) total $38.45. What is his take-home pay?

80. The cost of gasoline is reduced from $0.695 per liter to $0.629 per liter. How much money is saved on an automobile trip that requires 340 liters?

6.4 Changing Fractions to Percents

HOW & WHY

OBJECTIVE Change a fraction or mixed number to a percent.

OBJECTIVE
Change a fraction or mixed number to a percent.

We already know how to change fractions to decimals and decimals to percent. We now combine the two ideas to change fractions to a percent.

To change a fraction or mixed number to a percent

1. Change to a decimal. The decimal is rounded or carried out as directed.
2. Change the decimal to percent.

Unless directed to round, the division is completed or else the quotient is written as a repeating decimal.

EXAMPLES A–H

DIRECTIONS: Change the fraction or mixed number to a percent.

STRATEGY: Change the number to a decimal and then to a percent.

A. Change $\frac{4}{5}$ to a percent.

$\frac{4}{5} = 0.8$ Divide 4 by 5 to change the fraction to a decimal.

$= 80\%$ Change the decimal to a percent.

So, $\frac{4}{5} = 80\%$.

B. Write $\frac{11}{16}$ as a percent.

$\frac{11}{16} = 0.6875$ Change to a decimal.

$= 68.75\%$

So, $\frac{11}{16} = 68.75\%$.

C. Change $\frac{3}{11}$ to a percent.

$\frac{3}{11} = 0.27\overline{27}$ Write $\frac{3}{11}$ as a repeating decimal.

$= 27.\overline{27}\%$ Change to a percent.

$= 27\frac{3}{11}\%$ The repeating decimal $0.\overline{27} = \frac{3}{11}$.

So, $\frac{3}{11} = 27.\overline{27}\%$ or $27\frac{3}{11}\%$.

WARM-UPS A–H

A. Change $\frac{3}{10}$ to a percent.

B. Write $\frac{3}{8}$ as a percent.

C. Change $\frac{5}{9}$ to a percent.

Answers to Warm-Ups

A. 30% **B.** 37.5%

C. $55.\overline{5}\%$ or $55\frac{5}{9}\%$

D. Write $1\frac{12}{25}$ as a percent.

D. Write $2\frac{7}{20}$ as a percent.

$$2\frac{7}{20} = 2.35$$
$$= 235\%$$

So, $2\frac{7}{20} = 235\%$.

E. Change $\frac{2}{17}$ to a percent. Round to the nearest tenth of a percent.

E. Change $\frac{5}{13}$ to a percent. Round to the nearest tenth of a percent.

CAUTION

One tenth of a percent is a thousandth; that is, $\frac{1}{10}$ of $\frac{1}{100}$ is $\frac{1}{1000} = 0.001$.

STRATEGY: To write the percent rounded to the nearest tenth of a percent, we need to change the fraction to a decimal rounded to the nearest thousandth. (That is, we need three decimal places.)

$$\frac{5}{13} \approx 0.385$$ Write as a decimal rounded to the nearest thousandth.
$$\approx 38.5\%$$

So, $\frac{5}{13} \approx 38.5\%$.

F. Write $\frac{79}{230}$ as a decimal rounded to the nearest tenth of a percent.

F. Write $\frac{247}{382}$ as a percent rounded to the nearest tenth of a percent.

$$\frac{247}{382} \approx 0.647$$ Write as a decimal rounded to the nearest thousandth.
$$\approx 64.7\%$$

So, $\frac{247}{382} \approx 64.7\%$.

CALCULATOR EXAMPLE

G. Change $4\frac{19}{40}$ to a percent.

G. Change $7\frac{73}{80}$ to a percent.

STRATEGY: Change the mixed number to a fraction; then use the calculator to change to a percent.

$$7\frac{73}{80} = \frac{633}{80}$$ Change to an improper fraction.
$$= 7.9125$$
$$= 791.25\%$$

So, $7\frac{73}{80} = 791.25\%$.

Answers to Warm-Ups

D. 148% **E.** 11.8%
F. 34.3% **G.** 447.5%

H. A motor that needs repair is turning only $\frac{7}{16}$ the number of revolutions per minute that is normal. What percent of the normal rate is this?

$$\frac{7}{16} = 0.4375$$
$$= 43.75\%$$

The motor is turning at 43.75% of its normal rate.

H. An 8-cylinder motor has only seven of its cylinders firing. What percent of the cylinders are firing?

Answers to Warm-Ups

H. The percent of cylinders firing is 87.5%.

EXERCISES 6.4

OBJECTIVE Change a fraction or mixed number to a percent.

A. *Change each fraction to a percent.*

1. $\frac{77}{100}$ **2.** $\frac{27}{100}$ **3.** $\frac{37}{50}$ **4.** $\frac{7}{10}$

5. $\frac{13}{25}$ **6.** $\frac{17}{25}$ **7.** $\frac{1}{2}$ **8.** $\frac{3}{5}$

9. $\frac{11}{25}$ **10.** $\frac{7}{50}$ **11.** $\frac{3}{20}$ **12.** $\frac{3}{25}$

13. $\frac{13}{10}$ **14.** $\frac{27}{20}$ **15.** $\frac{11}{8}$ **16.** $\frac{19}{16}$

17. $\frac{43}{1000}$ **18.** $\frac{91}{1000}$ **19.** $\frac{13}{20}$ **20.** $\frac{11}{20}$

B.

Change each fraction or mixed number to a percent.

21. $4\frac{3}{5}$ **22.** $6\frac{1}{4}$ **23.** $\frac{2}{3}$

24. $\frac{1}{6}$ **25.** $\frac{23}{6}$ **26.** $\frac{14}{3}$

27. $2\frac{5}{6}$ **28.** $4\frac{7}{12}$ **29.** $\frac{49}{400}$

30. $\frac{67}{200}$

Change each fraction or mixed number to a percent. Round to the nearest tenth of a percent.

31. $\frac{12}{17}$ **32.** $\frac{21}{36}$ **33.** $\frac{5}{9}$

34. $\frac{7}{9}$ **35.** $\frac{11}{13}$ **36.** $\frac{9}{23}$

37. $11\frac{9}{23}$ **38.** $24\frac{13}{15}$ **39.** $\frac{2}{21}$

40. $\frac{2}{23}$

C.

41. Kobe Bryant made 19 out of 20 free throw attempts in one game. What percent of the free throws did he make?

42. Maureen gets 21 problems correct on a 25-problem test. What percent is correct?

43. Four-sevenths of the eligible voters turned out for a recent Hillsboro city election. What percent of the voters turned out, to the nearest tenth of a percent?

44. In a supermarket, two eggs out of 11 dozen are lost because of cracks. What percent of the eggs must be discarded, to the nearest tenth of a percent?

45. The local soccer club won 13 out of their first 17 games. What percent of the games did they win, rounded to the nearest tenth of a percent?

46. Joe Lucky won 15 out of 23 hands at the blackjack table. What percent of the hands did he win, rounded to the nearest tenth of a percent?

Change each of the following fractions or mixed numbers to a percent rounded to the nearest hundredth of a percent.

47. $\frac{63}{400}$

48. $\frac{341}{8000}$

49. $\frac{143}{5000}$

50. $\frac{73}{1300}$

51. $6\frac{19}{33}$

52. $13\frac{12}{17}$

53. A vitamin C tablet is listed as fulfilling $\frac{1}{8}$ the recommended daily consumption of vitamin C. Miquel takes 13 of these tablets per day to ward off a cold. What percent of the average recommended consumption is he taking?

54. In July of the 2001 season, the Seattle Mariners had won 68 games and lost 26. Write a fraction that gives the number of games won compared to total games played, and then convert this to a percent.

55. In the year 2000, the number of post offices was about 28,000, whereas in 1900, the number of post offices was about 77,000. Write a fraction that represents the current number of post offices as a portion of the number in 1900, and change this fraction to a percent.

56. According to the Bureau of Labor Statistics, the number of desktop publishing jobs is expected to go from 30,000 in 1996 to 53,000 in 2006. Write a fraction that represents the expected number of desktop publishing jobs in 2006 as a portion of the jobs in 1996, and change this fraction to a percent.

57. The island of St. Croix, one of the Virgin Islands, is 84 mi^2. The total area of the Virgin Islands is 140 mi^2. Write a fraction that represents the area of St. Croix as a portion of the area of the Virgin Islands, and change this fraction to a percent.

58. A chicken sandwich contains 7 grams of fat. Each gram of fat has 9 calories. If the entire sandwich contains 290 calories, what percent of the calories comes from the fat content? Round to the nearest percent.

59. In 1995, one area of California had 211 smoggy days. What was the percent of smoggy days? In 1998, there were only 167 smoggy days. What was the percent that year? Compare these percents and discuss the possible reasons for this decline. Round the percents to the nearest tenth of a percent.

Exercises 60–61 relate to the chapter application.

60. Consumer reports indicate that the cost of food is $1\frac{1}{12}$ what it was 1 year ago. Express this as a percent. Round to the nearest tenth of a percent.

61. A department store advertises one-third off the regular price on Monday and an additional one-seventh off the original price on Tuesday. What percent is taken off the original price if the item is purchased on Tuesday? Round to the nearest whole percent.

STATE YOUR UNDERSTANDING

62. Explain why not all fractions can be changed to a whole number percent. What is special about those fractions that can?

CHALLENGE

63. Change $2\frac{4}{13}\%$ to the nearest tenth of a percent.

64. Change $\frac{6}{13}\%$ to the nearest hundredth of a percent.

GROUP WORK

65. Keep a record of everything you eat for one entire day. Use exact amounts as much as possible. With a calorie and fat counter, compute the percent of fat in each item. Then find the percent of fat you consumed that day. The latest recommendations suggest that the fat content not exceed 30% per day. How did you do? Which foods have the highest and which the lowest fat content? Was this a typical day for you? Continue this exercise for 1 week and compare your daily percent of fat with others in your group. Compute a weekly average individually and as a group.

MAINTAIN YOUR SKILLS

Change to a fraction or mixed number.

66. 0.84

67. 0.132

68. 4.065

69. 16.48

70. 0.0045

71. 0.1055

Multiply.

72. (19)(0.0115)

73. (8.003)(0.87)

Divide.

74. $0.38\overline{)5.738}$

75. Divide 62 by 480 and round to the nearest thousandth.

6.5 Changing Percents to Fractions

HOW & WHY

OBJECTIVE Change percents to fractions or mixed numbers.

The expression 65% is equal to $65 \times \frac{1}{100}$. This leads to a very efficient method for changing a percent to a fraction. See Example A.

OBJECTIVE

Change percents to fractions or mixed numbers.

To change a percent to a fraction or a mixed number

1. Replace the percent symbol (%) with the fraction $\left(\frac{1}{100}\right)$.
2. If necessary, rewrite the other factor as a fraction.
3. Multiply and simplify.

EXAMPLES A–E

DIRECTIONS: Change the percent to a fraction or mixed number.

STRATEGY: Change the percent symbol to the fraction $\frac{1}{100}$ and multiply.

A. Change 65% to a fraction.

$65\% = 65 \cdot \frac{1}{100}$ Replace the percent symbol (%) with $\frac{1}{100}$.

$= \frac{65}{100}$ Multiply.

$= \frac{13}{20}$ Simplify.

CAUTION

You need to multiply by $\frac{1}{100}$, not just write it down.

So, $65\% = \frac{13}{20}$.

B. Change 436% to a mixed number.

$436\% = 436 \cdot \frac{1}{100}$ $\% = \frac{1}{100}$

$= \frac{436}{100}$ Multiply.

$= \frac{109}{25}$ Simplify.

$= 4\frac{9}{25}$ Write as a mixed number.

So, $463\% = 4\frac{9}{25}$.

WARM-UPS A–E

A. Change 85% to a fraction.

B. Change 515% to a mixed number.

Answers to Warm-Ups

A. $\frac{17}{20}$ **B.** $5\frac{3}{20}$

C. Change 19.5% to a fraction.

C. Change 17.5% to a fraction.

$$17.5\% = 17.5 \cdot \frac{1}{100}$$

$$= 17\frac{1}{2} \cdot \frac{1}{100}$$ Change the decimal to a mixed number.

$$= \frac{35}{2} \cdot \frac{1}{100}$$ Change the mixed number to a fraction.

$$= \frac{35}{200} = \frac{7}{40}$$ Multiply and simplify.

So, $17.5\% = \frac{7}{40}$.

D. Change $5\frac{5}{6}\%$ to a fraction.

D. Change $15\frac{1}{3}\%$ to a fraction.

$$15\frac{1}{3}\% = 15\frac{1}{3} \cdot \frac{1}{100}$$

$$= \frac{46}{3} \cdot \frac{1}{100}$$

$$= \frac{46}{300} = \frac{23}{150}$$

So, $15\frac{1}{3}\% = \frac{23}{150}$.

E. Greg scores 88% on a math test. What fraction of the questions does he get incorrect?

E. A biological study shows that spraying a forest for gypsy moths is 92% successful. What fraction of the moths survives the spraying?

STRATEGY: Subtract the 92% from 100% to find the percent of the moths that survived. Then change the percent that survive to a fraction.

$$100\% - 92\% = 8\%$$

$$8\% = 8 \cdot \frac{1}{100}$$

$$= \frac{8}{100} = \frac{2}{25}$$

So, $\frac{2}{25}$ or 2 out of 25, gypsy moths survived the spraying.

Answers to Warm-Ups

C. $\frac{39}{200}$ **D.** $\frac{7}{120}$

E. Greg gets $\frac{3}{25}$ of the questions incorrect.

EXERCISES 6.5

OBJECTIVE Change percents to fractions or mixed numbers.

Change each of the following percents to fractions or mixed numbers.

A.

1. 12% **2.** 20% **3.** 45% **4.** 85%

5. 130% **6.** 180% **7.** 500% **8.** 900%

9. 80% **10.** 30% **11.** 72% **12.** 56%

13. 75% **14.** 90% **15.** 35% **16.** 65%

17. 150% **18.** 225% **19.** 67% **20.** 41%

B.

21. 35.5% **22.** 44.4% **23.** 6.8% **24.** 3.5%

25. 30.5% **26.** 40.8% **27.** $\frac{1}{4}\%$ **28.** $\frac{5}{7}\%$

29. $7\frac{1}{2}\%$ **30.** $6\frac{1}{4}\%$ **31.** $\frac{8}{9}\%$ **32.** $\frac{5}{6}\%$

33. $52\frac{1}{2}\%$ **34.** $28\frac{3}{4}\%$ **35.** $49\frac{4}{9}\%$ **36.** $33\frac{4}{7}\%$

37. 3.75% **38.** 1.25% **39.** $221\frac{2}{3}\%$ **40.** $173\frac{1}{3}\%$

C.

41. George and Ethel paid 24% of their annual income in taxes last year. What fractional part of their income went to taxes?

42. One summer Judy earned 45% of her college expenses working at a local food processing plant. What fractional part of her expenses did she earn that summer?

43. The Kilroys spend 32% of their monthly income on their mortgage payment. What fractional part goes toward the mortgage?

44. Jorge spends 26% of his salary on his car each month. What fractional part does he spend on his car?

45. Grant Hill made 84% of his shots during an NBA game. What fractional part of his shots did Grant make?

46. The offensive team for the Kansas City Chiefs was on the field 56% of the time during a game with the Dallas Cowboys. For what fractional part of the game was the offense on the field?

Write each of the following as a fraction.

47. $15\frac{1}{5}\%$

48. $26\frac{1}{4}\%$

49. $4\frac{4}{9}\%$

50. $1\frac{1}{9}\%$

51. 0.725%

52. 0.475%

53. The enrollment at City Community College this year is 108% of last year's enrollment. What fraction of last year's enrollment does this represent?

54. The city budget is 124% over last year's budget. What fraction does this represent?

55. A census determines that $37\frac{1}{2}\%$ of the residents of a city are age 40 or over and that 45% are age 25 or under. What fraction of the residents is between the ages of 25 and 40?

56. Brooks and Foster form a partnership. If Brooks's investment is $62\frac{1}{4}\%$ of the total, what fraction of the total is Foster's share?

57. The spraying for the Mediterranean fruit fly is found to be 88% successful. What fraction of these flies is eliminated?

58. The salmon run in an Oregon stream has dropped to 42% of what it was 10 years ago. What fractional part of the run was lost during the 10 years?

59. The literacy rate in Vietnam is 94%. Convert this to a fraction and explain its meaning.

60. The population of South Africa is 75.2% black. Convert this to a fraction and explain its meaning.

61. In 1998, the country with the highest percent of Internet usage was Iceland, with a rate of 32%. Convert this to a fraction and explain its meaning.

62. According to the National Park Service, the Blue Ridge Parkway in Virginia and North Carolina was the most visited national park in 1998. In fact, the Parkway accounted for 22% of all visitors to the ten most popular national parks. Convert this to a fraction and explain its meaning.

Exercises 63–64 relate to the chapter application.

63. Juanita can save 28% on canned food if she buys in quantities of 12 or more cans of each type of food. If she buys in quantity, what fraction of the original price is she paying?

64. Houng buys light bulbs for 15% off, plus he sends in the coupon for a 12% manufacturer's rebate. What fraction of the original price does he pay for the bulbs after he receives the rebate?

STATE YOUR UNDERSTANDING

65. Name two circumstances that can be described by either a percent or by a fraction. Compare the advantages or disadvantages of using percents or fractions.

66. Explain how to change between the fraction and percent forms of a number. Give examples of each.

CHALLENGE

67. Change 0.00025% to a fraction.

68. Change 0.0005% to a fraction.

69. Change 150.005% to a mixed number.

70. Change 180.05% to a mixed number.

GROUP WORK

71. Have each member of your group read the ads for the local department stores in the weekend paper. Record the "% off" in as many ads as you can find. Convert the percents to fractions. In which form is it easier to estimate the savings because of the sale?

MAINTAIN YOUR SKILLS

Change to a decimal.

72. $\frac{33}{40}$

73. $\frac{443}{640}$

Change to a fraction.

74. 0.365

75. 0.208

Change to a percent.

76. 0.567

77. 5.007

Change to a decimal.

78. 8.13%

79. 112.8%

80. Write a ratio to compare $39,000,000 to 75,000,000 people.

81. Write a ratio to compare 14,765,000 gallons of water to 100,000 people.

6.6 Fractions, Decimals, Percents: A Review

HOW & WHY

OBJECTIVE Given a decimal, fraction, or percent, rewrite in a related form.

Decimals, fractions, and percents can each be expressed in terms of the others. We can:

Write a percent as a decimal and as a fraction
Write a fraction as a percent and as a decimal
Write a decimal as a percent and as a fraction

For example:

$$40\% = 40 \cdot \frac{1}{100} = \frac{40}{100} = \frac{2}{5} \quad \text{and} \quad 40\% = 0.4$$

$$\frac{5}{8} = 5 \div 8 = 0.625 \quad \text{and} \quad \frac{5}{8} = 0.625 = 62.5\%$$

$$0.95 = 95\% \quad \text{and} \quad 0.95 = \frac{95}{100} = \frac{19}{20}$$

Table 6.2 shows some common fractions and their particular decimal equivalents. Some of the decimals are repeating decimals. Remember that a repeating decimal is shown by the bar over the digits that repeat. These fractions occur often in applications of percents. You should memorize them so you can recall the patterns when they appear.

OBJECTIVE

Given a decimal, fraction, or percent, rewrite in a related form.

TABLE 6.2 COMMON FRACTIONS AND DECIMALS

$\frac{1}{2} = 0.5$					
$\frac{1}{3} = 0.\overline{3}$	$\frac{2}{3} = 0.\overline{6}$				
$\frac{1}{4} = 0.25$		$\frac{3}{4} = 0.75$			
$\frac{1}{5} = 0.2$	$\frac{2}{5} = 0.4$	$\frac{3}{5} = 0.6$	$\frac{4}{5} = 0.8$		
$\frac{1}{6} = 0.1\overline{6}$				$\frac{5}{6} = 0.8\overline{3}$	
$\frac{1}{8} = 0.125$		$\frac{3}{8} = 0.375$		$\frac{5}{8} = 0.625$	$\frac{7}{8} = 0.875$

For example, $2.\overline{6} = 2\frac{2}{3}$, $8.1\overline{6} = 8\frac{1}{6}$, and $17.125 = 17\frac{1}{8}$.

WARM-UPS A–B

A.

Fraction	Decimal	Percent
$\frac{2}{3}$		
		27%
	0.72	
$\frac{73}{100}$		
		160%
	1.3	

B. In Example B, write the percent that is recycled as a decimal and as a fraction.

EXAMPLES A–B

DIRECTIONS: Fill in the empty spaces with the related percent, decimal, or fraction.

STRATEGY: Use the procedures of the previous sections.

A.

Fraction	Decimal	Percent
		30%
$\frac{7}{8}$		
	0.62	

Fraction	Decimal	Percent
$\frac{3}{10}$	0.3	30%
$\frac{7}{8}$	0.875	87.5% or $87\frac{1}{2}$%
$\frac{31}{50}$	0.62	62%

$30\% = 0.30 = \frac{30}{100} = \frac{3}{10}$

$\frac{7}{8} = 0.875 = 87.5\% = 87\frac{1}{2}\%$

$0.62 = 62\% = \frac{62}{100} = \frac{31}{50}$

B. The average American uses about 200 pounds of plastic a year. Approximately 60% of this is used for packaging and about 5% of it is recycled. Write the percent used for packaging as a decimal and a fraction.

$$60\% = 0.60 = \frac{60}{100} = \frac{3}{5}$$

* So, 0.6, or $\frac{3}{5}$, of the plastic is used for packaging.

Answer to Warm-Ups

A.

Fraction	Decimal	Percent
$\frac{2}{3}$	$0.\overline{6}$	$66\frac{2}{3}$%
$\frac{27}{100}$	0.27	27%
$\frac{18}{25}$	0.72	72%
$\frac{73}{100}$	0.73	73%
$1\frac{3}{5}$	1.6	160%
$1\frac{3}{10}$	1.3	130%

B. Of the 200 pounds of plastic, 0.05, or $\frac{1}{20}$, is recycled.

EXERCISES 6.6

OBJECTIVE Given a decimal, fraction, or percent, rewrite in a related form.

Fill in the empty spaces with the related percent, decimal, or fraction.

	Fraction	Decimal	Percent		Fraction	Decimal	Percent
1.	$\frac{1}{10}$			**2.**			30%
3.		0.75		**4.**	$\frac{9}{10}$		
5.			145%	**6.**	$\frac{3}{8}$		
7.		0.001		**8.**		1	
9.	$2\frac{1}{4}$			**10.**		0.8	
11.			$5\frac{1}{2}\%$	**12.**		0.875	
13.			$\frac{1}{2}\%$	**14.**		0.6	
15.			$62\frac{1}{2}\%$	**16.**			50%
17.		0.86		**18.**	$\frac{5}{6}$		
19.		0.08		**20.**	$\frac{2}{3}$		
21.			25%	**22.**		0.2	
23.			40%	**24.**			$33\frac{1}{3}\%$
25.		0.125		**26.**	$\frac{7}{10}$		

27. Louis goes to buy new tires for his truck. He finds them on sale for $\frac{1}{4}$ off. What percent is this?

28. Michael went on a diet. He now weighs $66\frac{2}{3}\%$ of his original weight. What fraction is this?

29. During the month of August, Super Value Grocery has a special on sweet corn: Buy any number of ears and get $\frac{3}{8}$ more for one cent. During the same time period, Hank's Super Market also runs a special on sweet corn: Buy any number of ears and get 35% more for one cent. Which store offers the best deal?

30. Three multivitamins contain the following amounts of the RDA (recommended daily allowance) of calcium: Brand A, 15%; Brand B, 0.149; and Brand C, $\frac{1}{7}$. Which brand contains the most calcium?

31. Teresa is negotiating a business deal. Vendor A has offered a deal that increases the price by 6%, whereas her boss has authorized up to 0.0575 more. Vendor B has offered Teresa a deal that increases the price by $\frac{1}{18}$. Who has offered the best deal? Does either or both meet the boss's authorized amount?

32. During the first full month of play in the local high school basketball season, the top three shooters had the following statistics: Shari made 78.3% of her shots, Bev made 0.805 of her shots, and Teri made $\frac{58}{73}$ of her shots. Who was the most accurate shooter?

33. During one baseball season, Evan got a hit 29% of the times he was at bat, Nick got a hit $\frac{7}{25}$ of the times he was at bat, and Edward's batting average was 0.287. Which man had the best batting average?

34. During a promotional sale, Rite-Aid advertises Coppertone sunscreen for $\frac{1}{3}$ off the suggested retail price, whereas Walgreen's advertises it for 30% off. Which is the better deal?

35. In Peru, the literacy rate is 89%. In Jamaica, the literacy rate is 851 per 1000. Which country has the higher rate?

36. In Germany, the annual death rate is about 10.5 per 1000 people. In Denmark, the annual death rate is about 1.1%. Which country has the lower annual death rate?

Exercises 37–40 relate to the chapter application.

37. George buys a new VCR at a 40%-off sale. What fraction is this?

38. A local department store is having its red tag sale. All merchandise will now be 20% off the the original price. What decimal is this?

39. Melinda is researching the best place to buy a computer. On the same-priced computer, Family Computers offers $\frac{1}{8}$ off, The Computer Store will give a 12% discount, and Machines Etc. will allow a 0.13 discount. Where does Melinda get the best deal?

40. Randy is trading in his above-the-ground swimming pool for a larger model. Prices are the same for Model PS+ and Model PT. PS+ holds 11% more water than his old pool, whereas PT holds $\frac{1}{9}$ more water. Which should he buy to get the most additional water for his new pool?

STATE YOUR UNDERSTANDING

41. Write a short paragraph with examples that illustrate when to use fractions, when to use decimals, and when to use percents to show comparisons.

CHALLENGE

42. Fill in the table. Be exact.

Fraction	Decimal	Percent
$\frac{14}{3}$		
	1.1875	
	0.063125	6.3125%

GROUP WORK

43. Have each member of your group make up a table like the one in Example A. Exchange your table with the other group members and fill in the blanks. Check your answers with the rest of the group.

MAINTAIN YOUR SKILLS

Solve the following proportions.

44. $\frac{25}{30} = \frac{x}{45}$

45. $\frac{45}{81} = \frac{23}{y}$

46. $\frac{x}{72} = \frac{40}{9}$

47. $\frac{x}{8.3} = \frac{117}{260}$

48. $\frac{\frac{1}{2}}{100} = \frac{A}{21}$

49. $\frac{85}{100} = \frac{170}{B}$

50. $\frac{R}{100} = \frac{9}{34}$; round to the nearest tenth.

51. $\frac{23}{100} = \frac{A}{41.3}$; round to the nearest tenth.

52. Sean drives 536 miles and uses 10.6 gallons of gasoline. At that rate, how many gallons will he need to drive 1150 miles? Round to the nearest tenth of a gallon.

53. The Bacons' house is worth $124,500 and is insured so that the Bacons will be paid four-fifths of the value of any damage. One-third of the value of the house is destroyed by fire. How much insurance money should they collect?

6.7 Solving Percent Problems

VOCABULARY

In the statement "R% of B is A,"

R is the **rate** of percent
B is the **base** unit and follows the word "of"
A is the **amount** that is compared to B

To solve a percent problem means to do one of the following:

1. Find A, given R and B.
2. Find B, given R and A.
3. Find R, given A and B.

OBJECTIVES

1. Solve percent problems using the formula.
2. Solve percent problems using a proportion.

HOW & WHY

OBJECTIVE 1 Solve percent problems using the formula.

We show two methods for solving percent problems. We refer to these as

The percent formula, $R \times B = A$ (see Examples A–E)
The proportion method (see Examples F–H)

In each method, we must identify the rate of percent (R), the base (B), and the amount (A). To help determine these, keep in mind that

R, the rate of percent, includes the percent symbol (%)
B, the base, follows the words *of* or *percent of*
A, the amount, sometimes called the *percentage*, is the amount compared to B and follows the word *is*.

The method you choose to solve percent problems should depend on

1. the method your instructor recommends
2. your major field of study
3. how you use percent in your day-to-day activities

What percent of B is A? The word *of* in this context and in other places in mathematics indicates multiplication. The word *is* describes the relationship "is equal to" or "=." Thus we write:

$$\begin{array}{l} R \text{ of } B \text{ is } A \\ \quad\ \downarrow \quad \downarrow \\ R \times B = A \end{array}$$

When solving percent problems identify the rate (%) first, the base (of) next, and the amount (is) last. For example, what percent of 30 is 6?

R of B is A. The rate R is unknown, the base B (B follows the word "of") is 30, and the amount A (A follows the word "is") is 6.

$R \times B = A$
$R(30) = 6$ Substitute 30 for B and 6 for A.
$R = 6 \div 30$ Divide.
$= 0.2$
$= 20\%$ Change to percent.

So, 6 is 20% of 30.

All percent problems can be solved using $R \times B = A$. However, two other forms can speed up the process.

$A = R \times B$
$R = A \div B$
$B = A \div R$

The triangle in Figure 6.2 is a useful device to help you select the correct form of the formula to use.

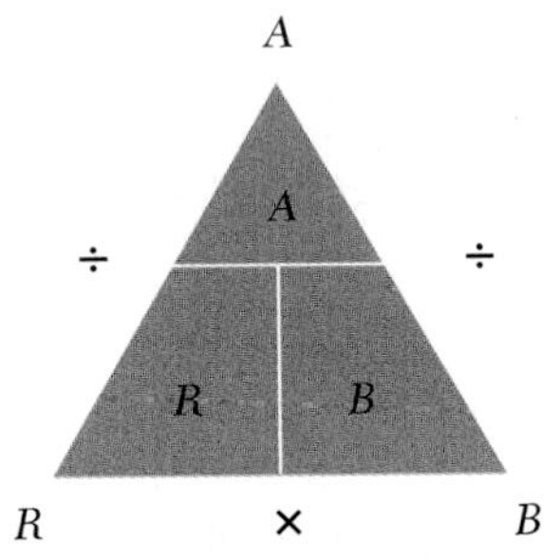

FIGURE 6.2

When the unknown value is covered, the positions of the uncovered (known) values help us remember how to see what operation to use:

When A is covered, we see $R \times B$, reading from left to right.
When B is covered, we see $A \div R$, reading from top to bottom.
When R is covered, we see $A \div B$, reading from top to bottom.

For example, 34% of what number is 53.04? The rate (R) is 34%, B is unknown (B follows the word *of*), and A (A follows the word *is*) is 53.04. Fill in the triangle and cover B. See Figure 6.3.

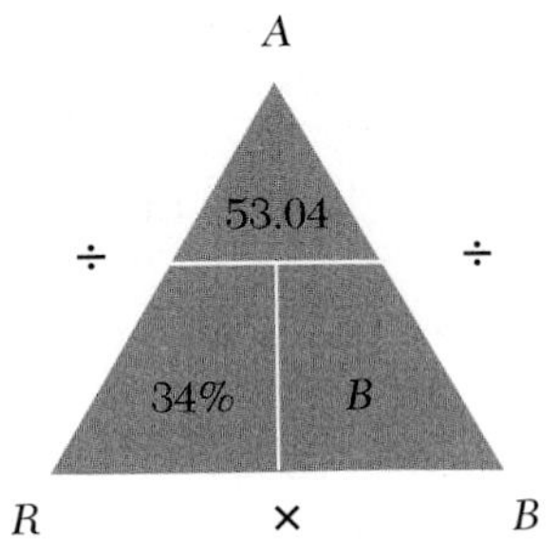

FIGURE 6.3

Reading from the top, we see that A is divided by R. Therefore,

$B = A \div R$
$= 53.04 \div 0.34$ Substitute 53.04 for A and 0.34 for R.
$= 156$

So, 34% of 156 is 53.04.

CAUTION

Remember to use the decimal form of the rate, R, when solving percent problems.

EXAMPLES A–E

DIRECTIONS: Solve the percent problems using the formula.

STRATEGY: Identify R, B, and A. Select a formula. Substitute the known values and find the unknown value.

A. 48% of what number is 27?

48% of B is 27. The % symbol follows 48, so R = 48%. The base B (following the word "of") is unknown. A follows the word "is," so A = 27.

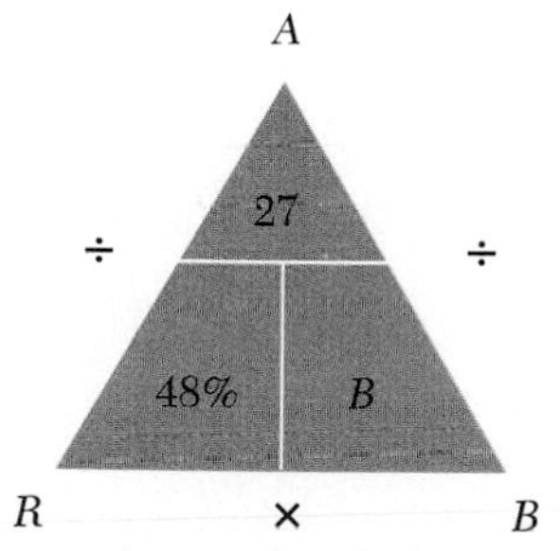

$B = A \div R$ Cover the unknown, B, in the triangle and read "A ÷ R."
$B = 27 \div 0.48$ Use the decimal form of the rate, 48% = 0.48.
$B = 56.25$

So, 48% of 56.25 is 27.

B. 16 is what percent of 64?

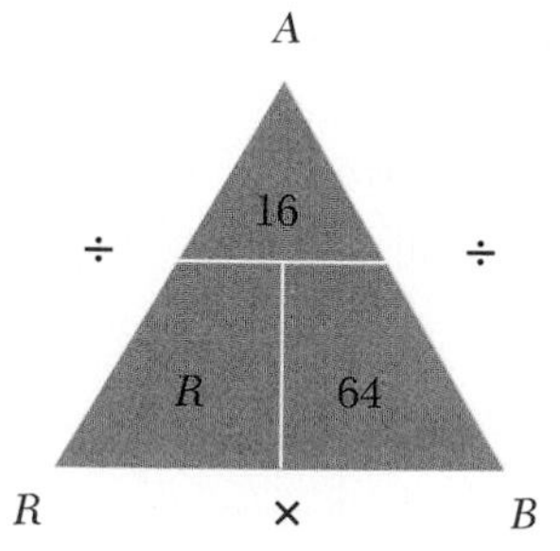

The rate R is unknown. B follows "of," so B = 64, and A = 16.

$R = A \div B$ Cover the unknown, R, in the triangle and read "A ÷ B."
$R = 16 \div 64$
$R = 0.25 = 25\%$ Divide. Change the decimal to a percent.

So, 16 is 25% of 64.

WARM-UPS A–E

A. 56% of what number is 35?

B. 13 is what percent of 65?

Answers to Warm-Ups
A. 62.5 **B.** 20%

C. What is $66\frac{2}{3}\%$ of 87?

C. What is $33\frac{1}{3}\%$ of 843?

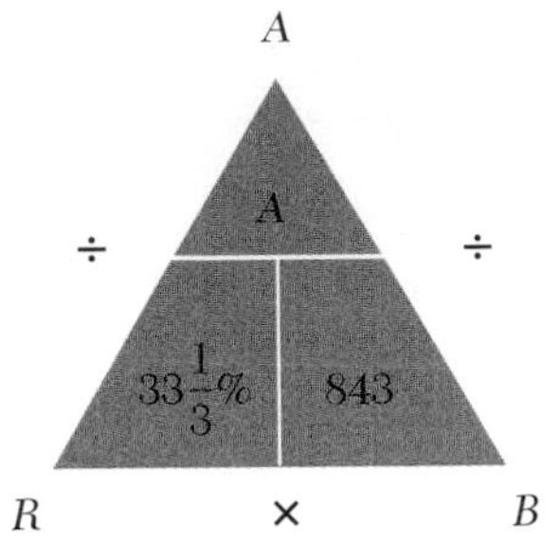

The % symbol follows $33\frac{1}{3}$, so $R = 33\frac{1}{3}\%$. The base B (following the word "of") is 843. A is unknown.

$A = R \times B$ — Cover A in the triangle and read "R × B."

$A = \frac{1}{3}(843)$ — Change the percent to a fraction.

$$33\frac{1}{3}\% = 33\frac{1}{3}\left(\frac{1}{100}\right) = \frac{100}{3}\left(\frac{1}{100}\right) = \frac{1}{3}$$

$A = 281$ — Multiply and simplify.

So, 281 is $33\frac{1}{3}\%$ of 843.

D. 43% of what number is 709.5?

D. 55% of what number is 83.6?

STRATEGY: We use the formula $R \times B = A$.

$R \times B = A$	Formula
$0.55B = 83.6$	$R = 55\%$ and $A = 83.6$ B is unknown.
$B = 83.6 \div 0.55$	Rewrite as division.
$= 152$	

So, 55% of 152 is 83.6.

E. 12.3 is what percent of 294? Round to the nearest tenth of a percent.

E. 45.7 is what percent of 111? Round to the nearest tenth of a percent.

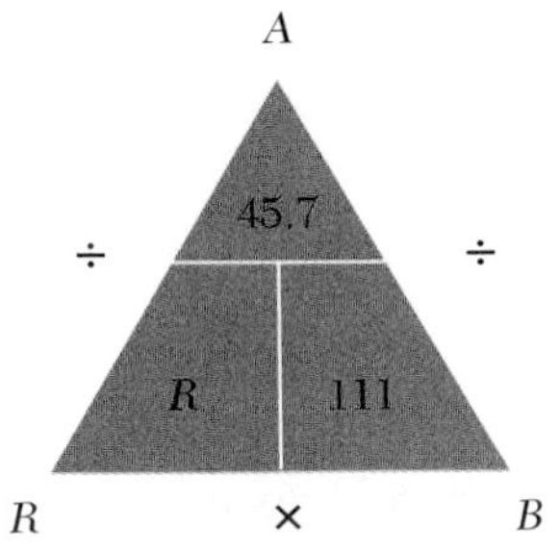

$A = 45.7$ and $B = 111$. R is unknown.

$R = A \div B$	In the percent triangle, cover R and read "A ÷ B."
$R = 45.7 \div 111$	
$R = 0.4117117\ldots$	
$R \approx 0.412$	To round to the nearest tenth of a percent, we round the decimal to the nearest thousandth.
$R \approx 41.2\%$	

So, 45.7 is 41.2% of 111.

Answers to Warm-Ups
C. 58 **D.** 1650 **E.** 4.2%

HOW & WHY

OBJECTIVE 2 Solve percent problems using a proportion.

Because R is a comparison of A to B—and we saw earlier that a comparison like this can be written as a ratio—we can write the percent ratio equal to the ratio of A and B. In writing the percent as a ratio, we let $R = X\%$. We can now write the proportion: $\frac{X}{100} = \frac{A}{B}$.

When any one of the values of R, A, and B is unknown, it can be found by solving the proportion.

For example, what percent of 105 is 33.6?

$A = 33.6$, $B = 105$, and $R = X\% = ?$

$$\frac{X}{100} = \frac{A}{B}$$
$$\frac{X}{100} = \frac{33.6}{105}$$
$$105X = 100(33.6) \quad \text{Cross multiply.}$$
$$105X = 3360$$
$$X = 3360 \div 105$$
$$X = 32$$
$$R = X\% = 32\%$$

So, 32% of 105 is 33.6.

EXAMPLES F–H

DIRECTIONS: Solve the percent problem using a proportion.

STRATEGY: Write the proportion, $R = \frac{X}{100} = \frac{A}{B}$, fill in the known values, and solve.

F. 45% of 96 is what number?

$$\frac{X}{100} = \frac{A}{B} \quad \text{Proportion for solving percent problems}$$
$$\frac{45}{100} = \frac{A}{96} \quad R = X\% = 45\%, \text{ so } X = 45, B = 96, \text{ and A is unknown.}$$
$$45(96) = 100A \quad \text{Cross multiply.}$$
$$4320 = 100A$$
$$4320 \div 100 = A$$
$$43.2 = A$$

So, 45% of 96 is 43.2.

G. 132% of __?__ is 134.64

$$\frac{X}{100} = \frac{A}{B} \quad \text{Proportion for solving percent problems}$$
$$\frac{132}{100} = \frac{134.64}{B} \quad R = X\% = 132\%, \text{ so } X = 132, \text{ B is unknown, and } A = 134.64.$$
$$132(B) = 100(134.64) \quad \text{Cross multiply.}$$
$$132(B) = 13464$$
$$B = 13464 \div 132$$
$$B = 102$$

So, 132% of 102 is 134.64.

WARM-UPS F–H

F. 63% of 90 is what number?

G. 145% of __?__ is 304.5

Answers to Warm-Ups

F. 56.7 **G.** 210

H. 49 is what percent of 135, rounded to the nearest tenth of a percent?

H. 81 is what percent of 215? Round to the nearest tenth of a percent.

$$\frac{X}{100} = \frac{A}{B}$$

$$\frac{X}{100} = \frac{81}{215}$$ $R = X\%$ is unknown, $A = 81$, and $B = 215$.

$215(X) = 100(81)$ Cross multiply.

$215(X) = 8100$

$X = 8100 \div 215$

$X \approx 37.67$ Carry out the division to two decimal places.

$X \approx 37.7$ Round to the nearest tenth.

$R \approx 37.7\%$ $R = X\%$

So, 81 is 37.7% of 215, to the nearest tenth of a percent.

Answers to Warm-Ups

H. 36.3%

EXERCISES 6.7

OBJECTIVES
1. Solve percent problems using the formula.
2. Solve percent problems using a proportion.

Solve.

A.

1. 17 is 50% of ___?___.

2. 17 is 20% of ___?___.

3. What is 70% of 60?

4. What is 40% of 90?

5. 6 is ___?___% of 3.

6. 8 is ___?___% of 2.

7. ___?___% of 90 is 45.

8. ___?___% of 80 is 20.

9. 80% of ___?___ is 32.

10. 30% of ___?___ is 18.

11. 70% of 63 is ___?___.

12. ___?___ is 60% of 86

13. 72 is ___?___% of 90.

14. ___?___% of 45 is 9.

15. 47% of ___?___ is 47.

16. 47 is ___?___% of 47.

17. $\frac{1}{3}$% of 600 is ___?___.

18. $\frac{1}{7}$% of 1400 is ___?___.

19. 125% of 80 is ___?___.

20. 175% of 60 is ___?___.

B.

21. 12.5% of 60 is ___?___.

22. 45.5% of 80 is ___?___.

23. 0.504 is ___?___% of 42.

24. 0.234 is ___?___% of 78.

25. 72 is 48% of ___?___.

26. 81 is 18% of ___?___.

27. 47.5% of 66 is ___?___.

28. 33.4% of 52 is ___?___.

29. 82% of ___?___ is 12.71.

30. 63% of ___?___ is 59.85.

31. 172% of ___?___ is 559.

32. 132% of ___?___ is 285.12.

33. 96 is ___?___% of 125.

34. 135 is ___?___% of 160.

35. 5.38% of 220 is ___?___.

36. 4.95% of 610 is ___?___.

37. 2.05 is ___?___% of 3.28.

38. 6.09 is ___?___% of 17.4.

39. $11\frac{1}{9}$% of 1845 is ___?___.

40. $16\frac{2}{3}$% of 3522 is ___?___.

C.

41. What percent of 87 is 13? Round to the nearest tenth of one percent.

42. What percent of 421 is 147? Round to the nearest tenth of one percent.

43. Ninety-six is 23.8% of what number? Round to the nearest hundredth.

44. Forty-one is 35.2 % of what number? Round to the nearest hundredth.

45. Thirty-two and seven-tenths percent of 695 is what number?

46. Seventy-three and twelve-hundredths percent of 35 is what number?

47. Forty-seven is what percent of 29? Round to the nearest tenth of one percent.

48. One hundred sixty-nine is what percent of 54? Round to the nearest tenth of one percent.

Solve.

49. $6\frac{3}{4}\%$ of $12\frac{2}{3}$ is ____?____. (as a fraction)

50. ____?____ is $9\frac{7}{8}\%$ of $61\frac{3}{5}$. (as a decimal)

51. $6\frac{4}{15}\%$ of 850 is ____?____. (as a mixed number)

52. $17\frac{1}{9}\%$ of ____?____ is 54.95. (as a decimal rounded to the nearest hundredth)

53. ____?____% of 29.07 is 16.15. (round to the nearest tenth of one percent)

54. ____?____% of $12\frac{4}{15}$ is $7\frac{2}{3}$. (as a mixed number)

STATE YOUR UNDERSTANDING

55. Explain the inaccuracies in this statement: "Starbuck industries charges 70¢ for a part that costs them 30¢ to make. They're making 40% profit."

56. Explain how to use the *RAB* triangle to solve percent problems.

CHALLENGE

57. $\frac{1}{2}\%$ of $45\frac{1}{3}$ is what fraction?

58. $\frac{3}{7}\%$ of $21\frac{1}{9}$ is what fraction?

GROUP WORK

59. Divide up the task of computing these percents: 45% of 37, 37% of 45, 18% of 80, 80% of 18, 130% of 22, 22% of 130, 0.6% of 5.5, 5.5% of 0.6. Compare your answers and together write up a statement about the answers.

60. Divide up the task of computing these percents: 30% of the number that is 80% of 250, 80% of the number that is 30% of 250, 60% of the number that is 20% of 340, 20% of the number that is 60% of 230, 150% of the number that is 200% of 40, 200% of the number that is 150% of 40. Compare your answers and together write up a statement about the answers.

MAINTAIN YOUR SKILLS

Solve the proportions.

61. $\frac{18}{24} = \frac{x}{60}$

62. $\frac{6.2}{2.5} = \frac{93}{y}$

63. $\frac{a}{\frac{5}{8}} = \frac{1\frac{1}{2}}{3\frac{3}{4}}$

64. $\frac{1\frac{1}{2}}{t} = \frac{5\frac{5}{8}}{1\frac{2}{3}}$

65. $\frac{1.4}{0.21} = \frac{w}{3.03}$

66. $\frac{2.6}{0.07} = \frac{3.51}{t}$

Exercises 67–70. On a certain map, $1\frac{3}{4}$ in. represents 70 mi.

67. How many miles are represented by $2\frac{7}{8}$ in.?

68. How many miles are represented by $3\frac{3}{16}$ in.?

69. How many inches are needed to represent 820 mi?

70. How many inches are needed to represent 22 mi?

6.8 Applications of Percent

VOCABULARY

When a value B is increased by an amount A $(B + A)$, the rate of percent R is called the **percent of increase**.

When a value B is decreased by an amount A $(B - A)$, the rate of percent R is called the **percent of decrease**.

OBJECTIVES

1. Solve percent word problems.
2. Read data from a circle graph or construct a circle graph from data.

HOW & WHY

OBJECTIVE 1 Solve percent word problems.

When a word problem is translated to the simpler form, "What percent of what is what?" the unknown value can be found by using one of the methods of Section 6.7. Both methods are used in the examples.

EXAMPLES A–E

DIRECTIONS: Solve the percent word problem.

STRATEGY: Write the problem in the form, "What percent of what is what?" Fill in the known values and find the unknown value.

A. Rod buys a racing bicycle with a 15% 1-year loan. The interest payment is \$84. How much is his loan?

STRATEGY: Use the percent formula.

15% of the loan is \$84. The \$84 interest is 15% of the loan, so $R = 15\% = 0.15$, and $A = \$84$. Substitute these values into the triangle and solve for B.

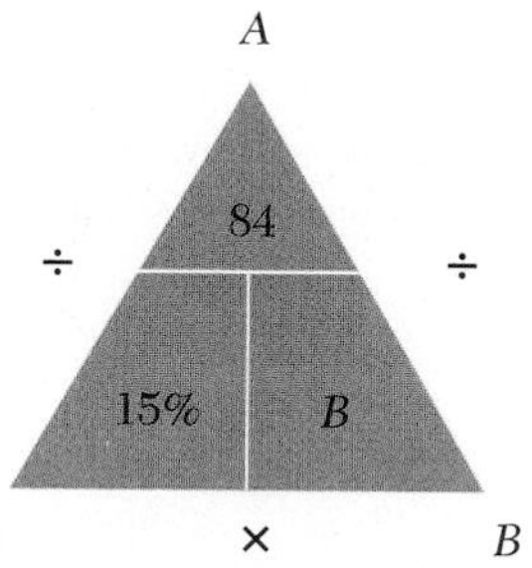

$B = A \div R$
$B = 84 \div 0.15$
$B = 560$

Rod's loan is for \$560.

B. This year the population of Deschutes County is 162% of its population 10 years ago. The population 10 years ago was 142,000. What is the population this year?

STRATEGY: Use the proportion equation.

162% of 142,000 is the current population.

WARM-UPS A–E

A. Jamie buys a motorcycle with an 18% 1-year loan. If the interest payment is \$216, how much is the loan?

B. The cost of a certain model of Ford is 135% of what it was 5 years ago. If the cost of the automobile 5 years ago was \$20,400, what is the cost today?

Answers to Warm-Ups

A. Jamie's loan is \$1200. **B.** The cost of the Ford today is \$27,540.

$$\frac{162}{100} = \frac{A}{142,000}$$

$R = X\% = 162\%$, so $X = 162$, and $B = 142,000$. Substitute these values into the proportion, $\frac{X}{100} = \frac{A}{B}$, and solve.

$$142,000(162) = 100A$$
$$23,004,000 = 100A$$
$$23,004,000 \div 100 = A$$
$$230,040 = A$$

Cross multiply.

The population this year is 230,040.

C. The bakery also has 260 packages of day-old buns to sell. If the price was originally $1.89 per package and the bakery sells them for $1.05 per package, what percent discount, based on the original price, should the bakery advertise?

C. The Davidson Bakery has 620 loaves of day-old bread it wants to sell. If the price was originally $2.49 a loaf and the bakery sells it for $1.69 a loaf, what percent discount, based on the original price, should the bakery advertise?

STRATEGY: Use the percent formula.

$2.49 − $1.69 = $0.80 — First determine the discount amount.

What % of $2.49 is $0.80? — $A = 0.80$ and $B = 2.49$ because we are asking the percent of discount based on the original price. Fill in the triangle and solve for R.

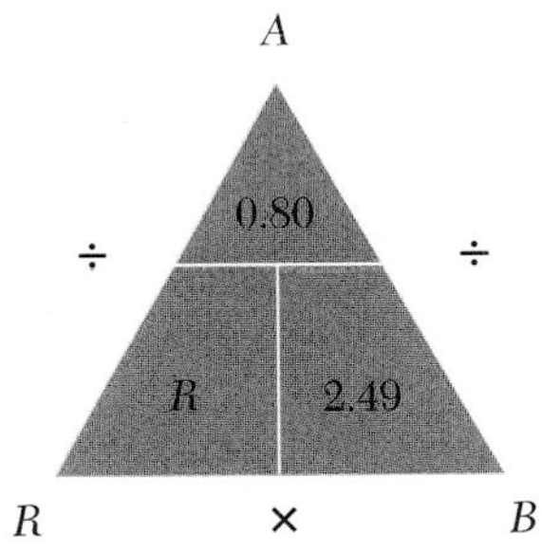

$R = A \div B$
$R = 0.80 \div 2.49$
$R \approx 0.321$ — Round to the nearest thousandth.
$R \approx 32.1\%$ — Change the decimal to percent.

The discount is about 32.1%, so the bakery will probably advertise "over 32% off."

D. A list of grades in a math class revealed that 9 students received As, 14 received Bs, 27 received Cs, and 7 received Ds. What percent of the students received a grade of C?

D. A student newspaper polls a group of students. Five of them say they walk to school, 11 say they ride the bus, 15 ride in car pools, and 4 drive their own cars. What percent of the group rides in car pools?

STRATEGY: Use the proportion to solve.

$5 + 11 + 15 + 4 = 35$ — First find the number in the group.

What % of 35 is 15? — There are 35 in the group, so $B = 35$; 15 ride in car pools, so $A = 15$

$$\frac{X}{100} = \frac{15}{35}$$
$35(X) = 1500$ — Cross multiply.
$X = 1500 \div 35$
$X \approx 42.9$ — Round to the nearest tenth.
$R \approx 42.9\%$ — $R = X\%$

Approximately 43% of the students ride in car pools.

Answers to Warm-Ups

C. The bakery will advertise a discount of 44.4%, or over 44%.

D. The percent of students receiving a C grade is approximately 47%.

E. In a statistical study of 545 people, 215 said they preferred eating whole wheat bread. What percent of the people surveyed preferred eating whole wheat bread? Round to the nearest whole percent.

STRATEGY: Use the percent formula to solve.

What percent of 545 is 215? B = 545 and A = 215

$R \times B = A$ Percent formula.
$R(545) = 215$ Substitute.
$R = 215 \div 545$
$R \approx 0.39$ Round to the nearest hundredth.
$R \approx 39\%$

So approximately 39% of the people surveyed preferred eating whole wheat bread. *

E. In a similar study of 615 people, 185 said they jog for exercise. What percent of those surveyed jog? Round to the nearest whole percent.

Answers to Warm-Ups

E. Of the 615 people, 30% jog.

EXAMPLES F–G

DIRECTIONS: Find the percent of increase or decrease.

STRATEGY: Use one of the two methods to solve for R.

F. Find the percent of increase from 380 to 551.

STRATEGY: Use the percent formula.

$551 - 380 = 171$ The difference, 171, is the amount of increase. The percent of increase is calculated from the amount of increase, 171, based on the original amount, 380.

What percent of 380 is 171? B = 380 and A = 171

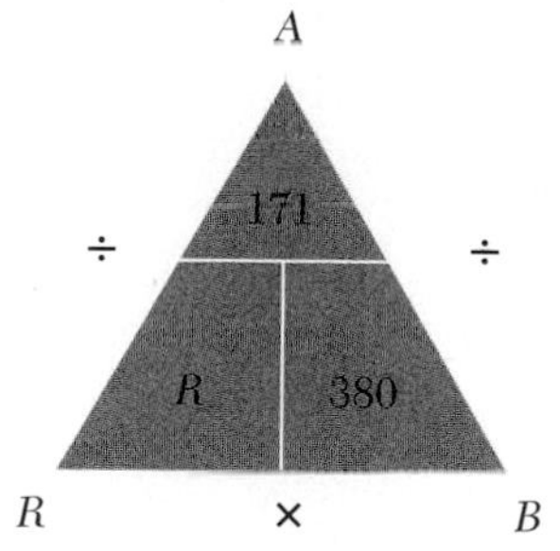

$R = A \div B$
$R = 171 \div 380$
$R = 0.45$
$R = 45\%$

There is a 45% increase from 380 to 551.

G. Find the percent of decrease from 784 to 509.6.

STRATEGY: Use the percent formula.

$784 - 509.6 = 274.4$ The difference, 274.4, is the amount of decrease from 784 to 509.6. The percent of decrease is the amount, 274.4, based on 784.

What percent of 784 is 274.4? B = 784 and A = 274.4

WARM-UPS F–G

F. Find the percent increase from 165 to 231.

G. Find the percent decrease from 550 to 132.

Answers to Warm-Ups

F. The percent of increase is 40%.
G. The percent of decrease is 76%.

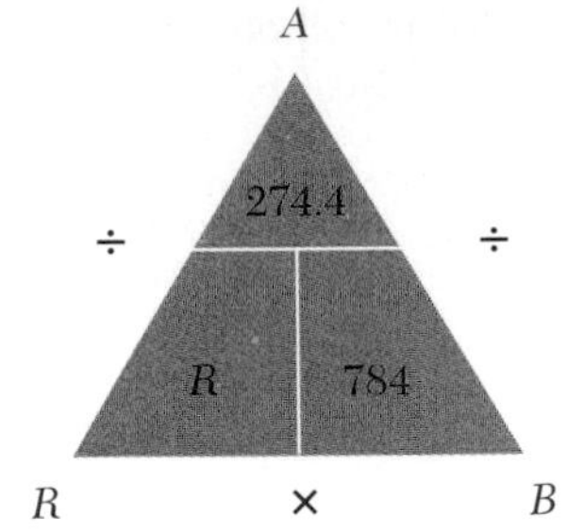

$R = A \div B$
$R = 274.4 \div 784$
$R = 0.35$
$R = 35\%$

There is a 35% decrease from 784 to 509.6.

HOW & WHY

OBJECTIVE 2 Read data from a circle graph or construct a circle graph from data.

A circle graph, or pie chart, is used to show how a whole unit is divided into parts. The area of the circle represents the entire unit, and each subdivision is represented by a sector. Percents are often used as the unit of measure of the subdivision. Consider the pie chart in Figure 6.4.

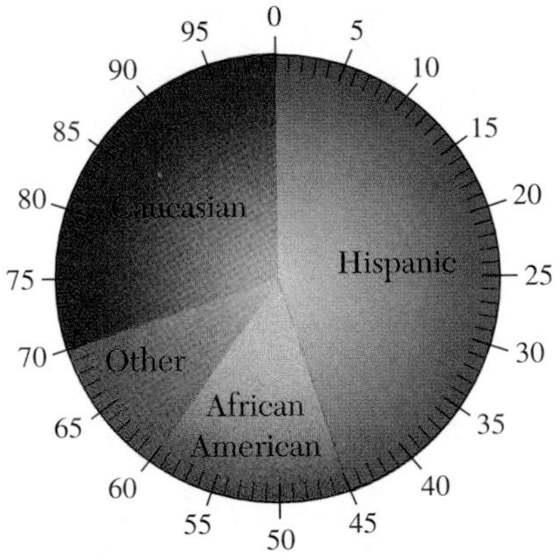

Percent of population of El Centro by ethnic group

FIGURE 6.4

From the circle graph, we can conclude:

1. The largest ethnic group in El Centro is Hispanic.
2. The Caucasian population is twice the African-American population.
3. The African-American and Hispanic populations are 60% of the total.

If the population of El Centro is 125,000, we can also compute the approximate number in each group. For instance, the number of Hispanics is found by:

$R \times B = A \qquad R = 45\% = 0.45,\ B = 125{,}000$
$(0.45)(125{,}000) = A$
$56{,}250 = A$

There are approximately 56,250 Hispanics in El Centro.

To construct a circle graph, determine what fractional part or percent each subdivision is, compared to the total. Then draw a circle and divide it accordingly. We can draw a pie chart of the data in Table 6.3.

TABLE 6.3 POPULATION BY AGE GROUP

	Age Groups		
	0–21	22–50	Over 50
Population	14,560	29,120	14,560

Begin by adding two rows and a column to Table 6.3 to create Table 6.4.

TABLE 6.4 POPULATION BY AGE GROUP

	Age Groups			
	0–21	22–50	Over 50	Total
Population	14,560	29,120	14,560	58,240
Fractional part	$\frac{1}{4}$	$\frac{1}{2}$	$\frac{1}{4}$	1
Percent	25%	50%	25%	100%

The third row is computed by writing each age group as a fraction of the total population and reducing. For example, the 0–21 age group is

$$\frac{14{,}560}{58{,}240} = \frac{1456}{5824} = \frac{364}{1456} = \frac{1}{4}, \text{ or } 25\%$$

Now draw the circle graph and label it. See Figure 6.5.

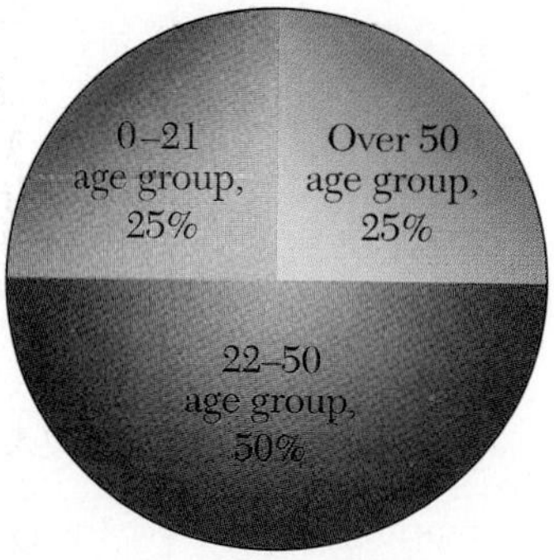

Group distribution by age

FIGURE 6.5

Sometimes circle graphs are drawn using 1 degree as the unit of measure for the sectors. This is left for a future course.

EXAMPLE H

DIRECTIONS: Answer the questions associated with the graph.

STRATEGY: Examine the graph to determine the size of the related sector.

WARM-UP H

H. The sales of items at Grocery Mart are displayed in the circle graph.

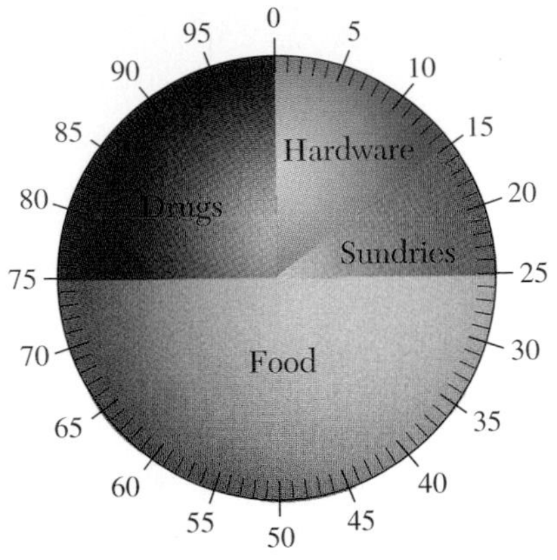

Items sold

i. What is the area of highest sales?
ii. What percent of total sales is from sundries and drugs?
iii. What percent of total sales is from food and hardware?

H. The sources of City Community College's revenue are displayed in the circle graph.

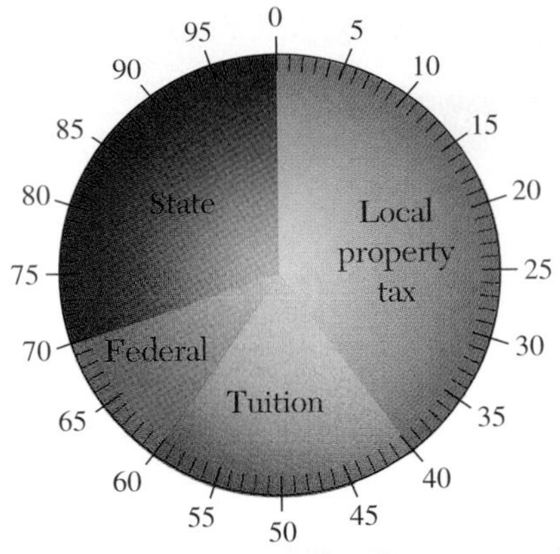

Revenue sources

i. What percent of the revenue is from the federal government?
ii. What percent of the revenue is from tuition and property taxes?
iii. What percent of the revenue is from federal and state government?

i. 10% Read directly from the graph.
ii. 60% Add the percents for tuition and property taxes.
iii. 40% Add the percents for state and federal sources.

Answers to Warm-Up

H. **i.** The highest sales are in food.
ii. Sundries and drugs account for 35% of sales.
iii. Food and hardware account for 65% of sales.

WARM-UP I

I. Make a circle graph to illustrate that in a survey of 20 people, 7 like football best, 9 like basketball best, and 4 like baseball best.

EXAMPLE I

DIRECTIONS: Make a circle graph that illustrates the information.

STRATEGY: Use the information to calculate the percents. Divide the circle accordingly and label.

I. Make a circle graph to illustrate that out of 25 students in Frau Heinker's German class, 4 are seniors, 5 are juniors, 14 are sophomores, and 2 are freshmen.

Seniors: $\frac{4}{25} = 16\%$ Compute the percents.

Juniors: $\frac{5}{25} = 20\%$

Sophomores: $\frac{14}{25} = 56\%$

Freshmen: $\frac{2}{25} = 8\%$

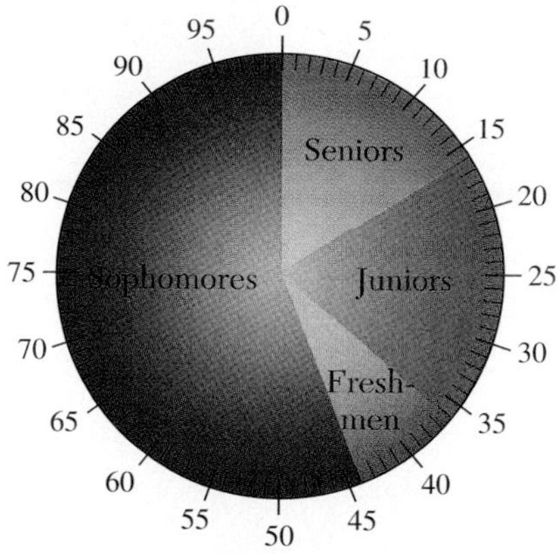

Distribution of students in Frau Heinker's German class

Construct and label the graph.

Answer to Warm-Up

1.

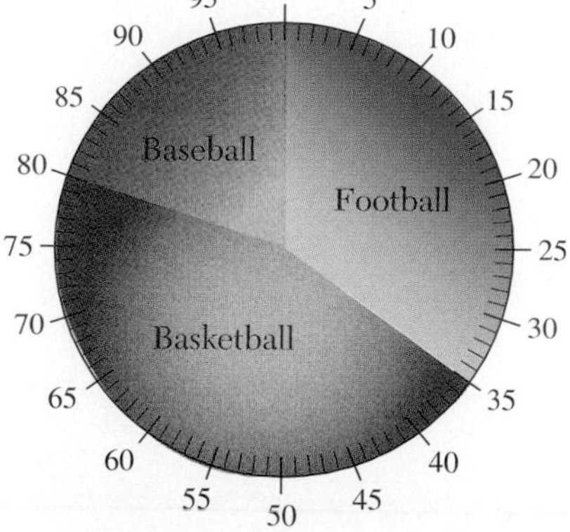

Sport preference

EXERCISES 6.8

OBJECTIVE 1 Solve percent word problems.

1. Sarah buys a television set that costs \$395.95. The sales tax is 4.5% of the cost. How much does Sarah pay in sales taxes?

2. Dan bought a used motorcycle for \$1158. He made a down payment of 20%. How much cash did he pay as a down payment?

3. Last year Joan had 14% of her salary withheld for taxes. If the total amount withheld was \$2456.30, what was Joan's yearly salary?

4. The manager of a fruit stand lost $16\frac{2}{3}\%$ of his bananas to spoilage and sold the rest. He discarded four boxes of bananas in 2 weeks. How many boxes did he have in stock?

5. Frank sells magazine subscriptions. He keeps 18% of the cost of each subscription. How many dollars' worth of subscriptions must he sell to earn \$243.90?

6. Maria sells candy from door to door. She keeps 15% of all her sales. How many dollars' worth of sales did Maria have if she earned \$315?

7. Debbi's base rate pay is \$8.78 per hour. She receives time and a half for all hours over 40 that she works in 1 week. What were her total earnings if last week she worked a total of 52 hours?

8. Truong's base rate of pay is \$7.88 per hour. He receives time and a half for all hours over 40 that he works in 1 week. What were his earnings if last week he worked a total of 48 hours?

9. A states sales tax is 6.5%. Rich paid \$615.57 for a new TV, which included the sales tax. What was the price of the TV?

10. Margo bought a new SUV for \$32,875.91, which includes a 5.5% sales tax. What was the price of the SUV?

11. Ilga bought a new van for \$23,670. She paid 16% down. How much does she still owe?

12. Janet paid \$455 as a down payment on her new appliances. The appliances cost \$3500. What percent did she pay as a down payment? How much does she still owe?

13. John got 28 problems correct on a 30-problem test. What was his percent score to the nearest whole-number percent?

14. To pass a test to qualify for a job interview, Whitney must score at least 70%. If there are 45 questions on the test, how many must she get correct to score 70%?

15. Vera's house is valued at \$115,000 and rents for \$10,750 per year. What percent of the value of the house is the annual income from rent (to the nearest tenth of a percent)?

16. Eddie and his family went to a restaurant for dinner. The dinner check was \$54.45. He left the waiter a tip of \$8. What percent of the check was the tip, to the nearest whole-number percent?

17. Adams High School's basketball team finished the season with a record of 17 wins and 8 losses. What percent of the games played were won?

18. The town of Verboort has a population of 17,850, of which 48% is male. Of the men, 32% are 40 years or older. How many men are there in Verboort who are younger than 40?

19. If you use a 45¢ coupon to buy a \$3.19 box of ice cream sandwiches, what percent off is this, to the nearest whole-number percent?

20. A store advertises \$20 off every coat in stock. What percent savings is this on a coat that is regularly \$84.65, to the nearest tenth of 1%.

21. In preparing a mixture of concrete, Susan uses 300 pound of gravel, 100 pounds of cement, and 200 pounds of sand. What percent of the mixture is gravel?

22. St. Joseph's Hospital has 8 three-bed wards, 20 four-bed wards, 12 two-bed wards, and 10 private rooms. What percent of the capacity of St. Joseph's Hospital is in private rooms (to the nearest tenth of a percent)?

23. It is claimed that in 15,000 hours, or 6 years, a gasoline engine will be down 256 hours for routine maintenance, whereas a diesel engine will be down only 104 hours. What is the difference in the percent of down times, to the nearest whole percent?

24. During a 6-year period, the cost of maintaining a diesel engine averages \$2650 and the cost of maintaining a gasoline engine averages \$4600. What is the percent of saving of the diesel compared with the gasoline engine? Find to the nearest whole percent.

25. Good driving habits can increase mileage and save on gas. If good driving causes a car's mileage to go from 27.5 mpg to 29.6 mpg, what is the percent of increased mileage? Round to the nearest whole-number percent.

26. According to the Bureau of Labor Statistics, there were 90,000 physical therapists in the United States in 1992. Physical therapy is one of the fastest-growing occupations, with the Census Bureau predicting an 88% increase over 1992 levels by the year 2005. How many physical therapists are predicted by the year 2005? Round to the nearest thousand.

27. The number of human services workers is growing at an even faster rate than that of physical therapists. The Census Bureau is predicting a 135% increase over 1992 levels by the year 2005. How many human services workers are predicted in the year 2005 if there were 189,000 in the United States in 1992? Round to the nearest thousand.

28. For customers who use a bank's credit card, there is a $1\frac{3}{4}\%$ finance charge on monthly accounts that have a balance of \$400 or less. Merle's finance charge for August was \$6.09. What was the amount of her account for August?

29. A fast-food hamburger has 342 calories from fat. This is approximately 54.3% of the total number of calories in the hamburger. How many calories are in one hamburger? Round to the nearest calorie.

30. In 1980, 64% of all workers drove to work alone. In 1990, 73% drove alone. Find the population of your community and compute the number of single-car drivers for both of these years.

31. Find the percent of the students in your class who drove alone to class today. Compare this percent with the percents in Exercise 30. How does the class compare to the national average for both years?

32. The label in the figure shows the nutrition facts for one serving of Toasted Oatmeal. Use the information on the label to determine the recommended daily intake of (a) total fat, (b) sodium, (c) potassium, and (d) dietary fiber. Use the percentages for cereal alone. Round to the nearest whole number.

Nutrition Facts
Serving Size 1 cup (49g)
Servings per Container about 9

Amount Per Serving	Cereal alone	with 1/2 cup Vitamin A&D Fortified Skim Milk
Calories	190	230
Calories from Fat	25	25
		% Daily Values**
Total Fat 2.5g*	4%	4%
Saturated Fat 0.5g	3%	3%
Polyunsaturated Fat 0.5g		
Monounsaturated Fat 1g		
Cholesterol 0mg	0%	0%
Sodium 220mg	9%	12%
Potassium 180mg	5%	11%
Total Carbohydrate 39g		13%
		15%
Dietary Fiber 3g	14%	14%
Soluble Fiber 1g		
Insoluble Fiber 2g		
Sugars 11g		
Other Carbohydrates 24g		
Protein 5g		

33. The table shows the calories per serving of the item, along with the number of fat grams per serving.

Calories and Fat Grams per Serving

Item	Calories per Serving	Fat Grams per Serving
Light Mayonnaise	50	4.5
Cocktail Peanuts	170	14.0
Wheat Thin Crackers	120	4.0
Cream Sandwich Cookies	110	2.5

Each fat gram is equivalent to 10 calories. Find the percent of calories from fat for each item. Round to the nearest whole percent.

OBJECTIVE 2 Read data from a circle graph or construct a circle graph from data.

For Exercises 34–35, the figure shows the ethnic distribution of children enrolled in Head Start.

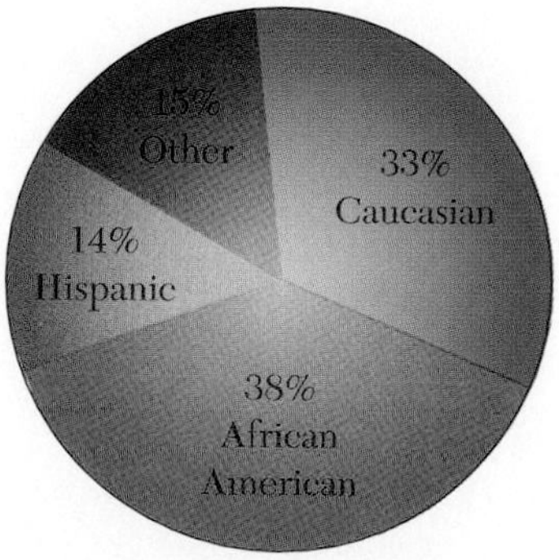

Ethnic distribution

34. Which group has the smallest number of children in Head Start?

35. What percent of children in Head Start are non-Caucasian?

For Exercises 36–39, the figure shows how dollars are spent in a particular industry.

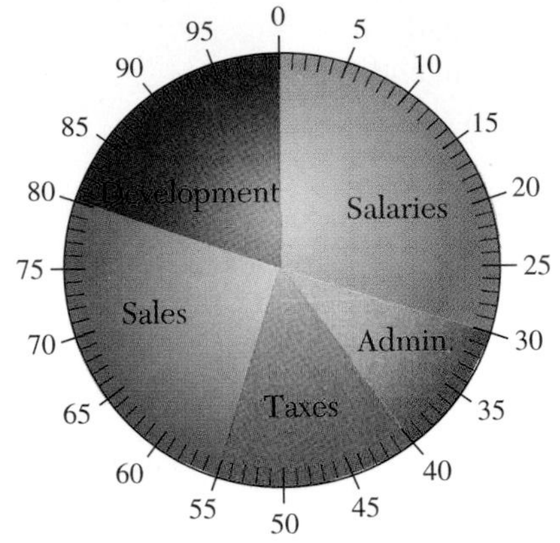

Distribution of expenditures

36. What item has the greatest expenditure of funds?

37. Which is more costly, development or taxes?

38. If the total expenditures of the industry were \$3,200,000, how much was spent on development?

39. Using the expenditures in Exercise 38, determine how much was spent on salaries.

40. In a family of three children, there are eight possibilities of boy and girl combinations. One possibility is that they are all girls. Another possibility is that they are all boys. There are three ways for the family to have two girls and a boy. There are also three ways for the family to have two boys and a girl. Use the following circle to make a circle graph illustrating this information.

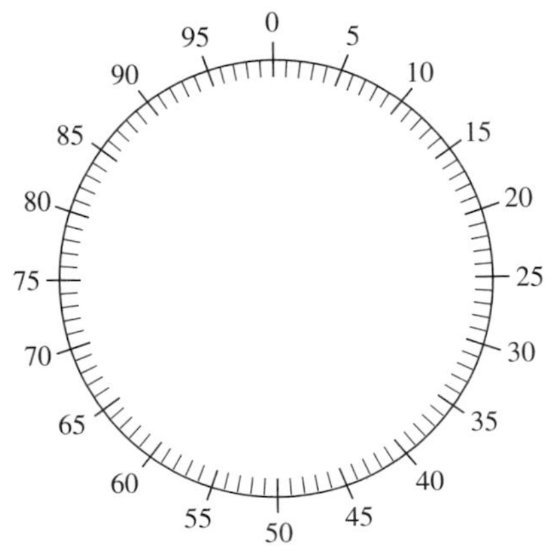

41. According to one state official, 48% of all households have no firearms, 12% of all households store firearms unloaded and have ammunition locked up, 10% of all households keep firearms loaded and unlocked, and 30% store firearms another way. Use the following circle to make a circle graph illustrating this information.

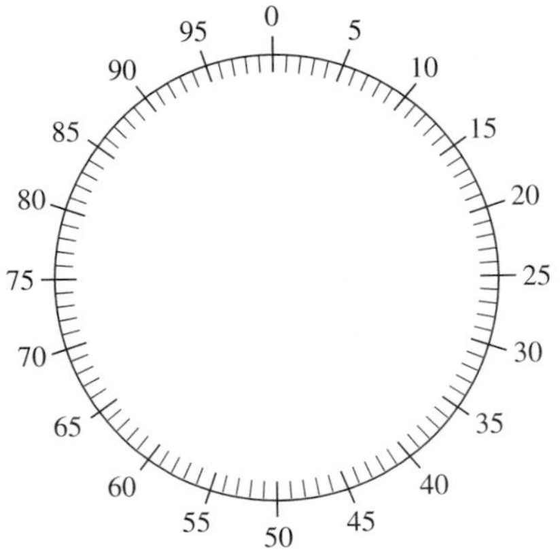

42. The major causes of death worldwide are listed below. Use the given circle to make a circle graph illustrating this information.

Infectious and parasitic diseases	32%
Heart, circulatory diseases, and stroke	19%
Unknown causes	16%
Cancer	12%
Accidents and violence	8%
Infant death	6%
Chronic lung diseases	6%
Other causes	less than 1%

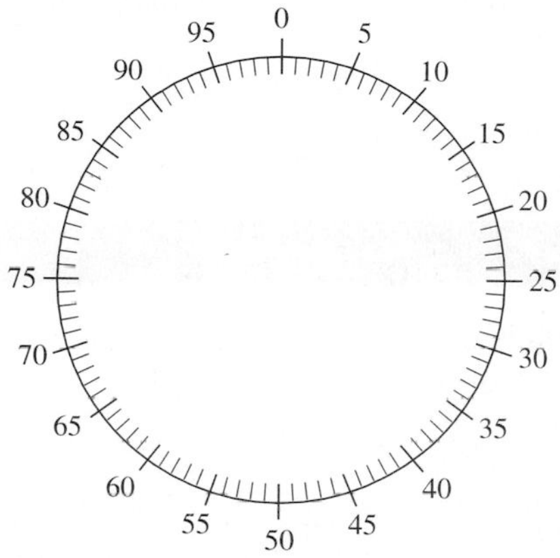

43. In 1900, there were about 10.3 million foreign-born residents in the United States, which represented about 13.6% of the total population. What was the U.S. population in 1900?

44. By 1999, the number of foreign-born residents in the United States increased to about 25.8 million. What percent increase is this from 1900? (See Exercise 43.)

45. In 1999, foreign-born residents were about 9.5% of the total population. What was the total population in the United States in 1999? (See Exercise 44.)

46. In 1900, tuition at Harvard was \$3000 per year. In 1999, Harvard's tuition had risen to \$22,054. What percent increase is this?

47. According to 1990 census figures, the city of Cleveland, Ohio, had a population of about 505,600, 50.5% of which was minority. Approximate the number of minority residents of Cleveland in 1990.

STATE YOUR UNDERSTANDING

48. Percent comparisons go back to the Middle Ages. In your opinion, why are these kinds of comparisons still being used today?

49. When an amount B experiences a percent increase, what is true of R? What is true of A?

50. Explain how you know the following statement is false without checking the math. "The population of Allensville has declined 12% over the last 10 years. In 1993, it was 500 and in 2003 it is 600."

CHALLENGE

51. Carol's baby weighed $7\frac{3}{4}$ lb when he was born. On his first birthday, he weighed $24\frac{3}{8}$ lb. What was the percent of increase during the year? Round to the nearest whole percent.

52. Jose purchases a car for \$12,500. He makes a down payment of \$1500. His payments are \$265 per month for 48 months. What percent of the purchase cost does Jose pay for the car, including the interest? Round to the nearest tenth of a percent.

53. Matilda buys the following items at Safeway: a can of peas, \$0.89; Wheaties, \$2.46; butter, \$1.78; Ivory soap, \$1.15; broom, \$4.97; steak, \$6.78; chicken, \$3.21; milk, \$1.89; eggs, \$1.78; bread, \$1.56; peanut butter, \$2.35; stamps, \$6.80; potatoes, \$1.98; lettuce, \$2.07; orange juice, \$2.89. What percent of the cost was in nonfood items? What percent of the cost was in meat products? Round to the nearest tenth of a percent.

GROUP WORK

54. Researchers are testing the effectiveness of a new drug. There were 50,000 people in the test, half receiving drug A (the old drug) and half receiving drug B (the new one). Of those receiving drug A, 18,400 had relief from their symptoms. Of those receiving drug B, 18,900 had relief from their symptoms. How would you describe the results of the test if you wanted to make it appear that drug B was much better than drug A? How would you describe the results of the test if you wanted to make it appear that drug B was only marginally better than drug A?

MAINTAIN YOUR SKILLS

55. Sally and Rita are partners. How much does each receive of the income if they are to share \$12,600 in a ratio of 6 to 4 (Sally 6, Rita 4)?

56. A 9.6-m board is to be cut into two pieces in a ratio of 6 to 2. What is the length of each board after the cut?

57. Joe buys a new car for \$2100 down and \$321 per month for 48 months. What is the total amount paid for the car?

58. Pia has a savings account balance of \$3892. She deposits \$74 per month for a year. What is her account balance, not including interest earned?

59. An engine with a displacement of 400 cubic inches develops 260 horsepower. How much horsepower is developed by an engine with a displacement of 175 cubic inches?

60. Harry pays federal income taxes of \$11,450 on an income of \$57,250. In the same tax bracket, what would be the tax on an income of \$48,000?

61. Peter attended 18 of the 20 G.E.D. classes held last month. What percent of the classes did he attend?

62. A family spends \$120 for food out of a budget of \$500. What percent goes for food?

63. There are 20 problems on an algebra test. What is the percent score for 17 problems correct?

64. There are 23 questions on a test for volleyball rules. What is the percent score for 18 correct answers, to the nearest percent?

6.9 Business Applications of Percent

VOCABULARY

The **cost** of an article is the amount paid for the item by the business. The **markup** is the amount added to the cost of an article by the store, so the store can pay its expenses and make a profit. The **selling price** of an article is the cost plus the markup. When a store has a sale, the **amount of discount** is the amount subtracted from the selling price (regular price). **Net income** is the amount left after expenses have been paid. **Commission** is the money salespeople earn based on the dollar value of the goods sold. **Depreciation** is the name given to the decrease in value caused by age or use.

OBJECTIVE

Solve business-related percent problems.

HOW & WHY

OBJECTIVE Solve business-related percent problems.

Businesses use percent in a variety of ways. Among these are percent of markup, percent of discount, percent of profit, interest rates, taxes, salary increases, and commissions.

EXAMPLES A–F

DIRECTIONS: Solve business-related percent problems.

STRATEGY: Write the problem in the simpler word form, "What percent of what is what?" Fill in the known values and find the unknown value.

A. The cost of a toaster is \$34.50. The markup is 44% of the cost. What is the selling price of the toaster?

STRATEGY: *Markup* is the amount added to the cost of an article by the store, so the store can pay its expenses and make a profit. Let M represent the markup. Use the percent formula.

Simpler Word Form

What % of the cost is the markup?

44% of \$34.50 is M. $R = 44\% = 0.44$, $B = \$34.50$, and $A = M$.

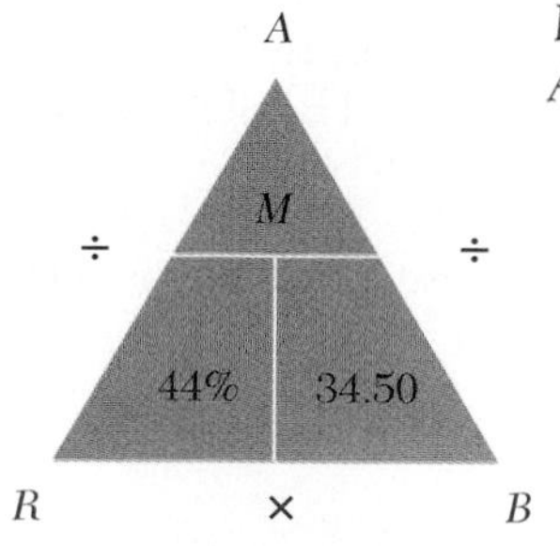

$M = R \times B$
$M = 0.44(34.50)$
$M = 15.18$

So the markup is \$15.18.

S.P. = cost + markup
= 34.50 + \$15.18
= \$49.68

To find the selling price, add the cost and the markup. Let S.P. represent the selling price.

The store will sell the toaster for \$49.68.

WARM-UPS A–F

A. The cost of an electric blanket is \$38.40. The markup is 55% of the cost. What is the selling price of the electric blanket?

Answers to Warm-Ups

A. The electric blanket sells for \$59.52.

B. The regular selling price of a portable TV is \$62.95. What is the sale price of the TV if it is discounted 33%? Round to the nearest cent.

STRATEGY: The *amount of discount* is the amount subtracted from the regular price. The *percent of discount* is 33%. Use the percent formula.

Simpler Word Form

33% of the selling price is the discount.

$$
\begin{aligned}
R \times B &= A & \quad & R = 33\% = 0.33 \\
& & & B = \$62.95 \\
0.33(62.95) &= A \\
20.77 &\approx A & & \text{Round to the nearest cent (hundredth).}
\end{aligned}
$$

The discount is \$20.77.

STRATEGY: To find the sale price, subtract the discount from the selling price.

$$
\begin{aligned}
\text{Sale price} &= \$62.95 - \$20.77 \\
&= \$42.18
\end{aligned}
$$

The sale price is \$42.18.

B. The selling price of a clothes dryer is \$415.65. What is the sale price of the dryer if the percent of discount is 25%? Round to the nearest cent.

C. A business pays 0.68% of its net income as a transportation tax to the local transit agency. What is the transit tax on a business with a net income of \$10,775,000?

STRATEGY: Net income is the amount after expenses have been deducted. Use the proportion.

Simpler Word Form

0.68% of the net income is the transit tax.

$$
\begin{aligned}
\frac{X}{100} &= \frac{A}{B} & \quad & \text{Percent proportion} \\
\frac{0.68}{100} &= \frac{A}{10{,}775{,}000} & & R = X\% = 0.68\%, \text{ so } X = 0.68, \\
& & & \text{and } B = 10{,}775{,}000. \\
0.68(10{,}775{,}000) &= 100(A) & & \text{Cross multiply.} \\
7{,}327{,}000 &= 100(A) \\
7{,}327{,}000 \div 100 &= A \\
73{,}270 &= A
\end{aligned}
$$

The business pays \$73,270 in transit taxes.

C. A business withholds 6.5% of each employee's gross salary for state taxes. How much is withheld from an employee earning \$65,800?

D. A phone company buys a four-wheel-drive SUV to use while inspecting phone lines. After 1 year, the SUV depreciated to 85% of its original value. The SUV is now valued at \$33,133. What did it cost originally?

STRATEGY: *Depreciation* is the name given to the decrease in value caused by age or use. Use the percent formula.

Simpler Word Form

85% of the original cost is \$33,133.

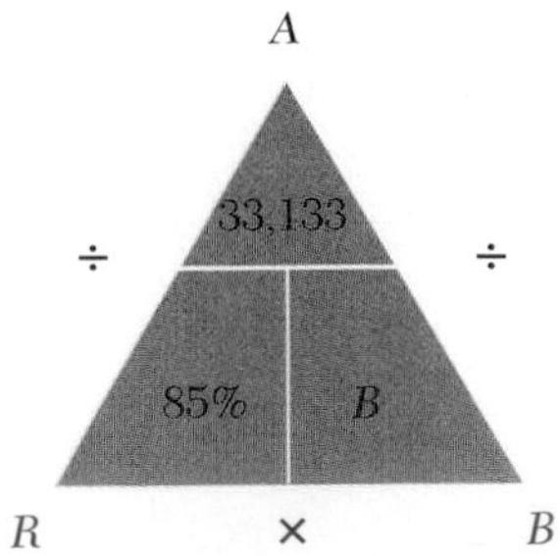

$R = 85\%$
$A = \$33{,}133$

D. A CEO's Lexus has depreciated to 65% of its original cost. If the value of the Lexus is now \$31,785, what did it cost originally?

Answers to Warm-Ups

B. The sale price of the clothes dryer is \$311.74.
C. The state taxes withheld are \$4277.
D. The Lexus originally cost \$48,900.

$B = A \div R$
$B = 33{,}133 \div 0.85$ Change 85% to a decimal, 0.85.
$B = 38{,}980$

The original cost of the SUV was $38,980.

E. Jean's rate of pay is $15.85 per hour. She gets time and one-half for each hour over 40 hours worked in 1 week. What are her earnings if she worked 52.5 hours in 1 week?

STRATEGY: Time and one-half means that she will earn 1.5 times or 150% of her regular hourly pay. First find her overtime pay rate, and then calculate her earnings at each rate and add. To find her overtime rate use the percent formula.

Simpler Word Form

150% of her pay rate is the overtime pay rate.

$R \times B = A$ $R = 150\% = 1.5$
$B = \$15.85$

$1.50(15.85) = A$
$23.775 = A$

Jean's overtime pay rate is $23.775.

$\text{Earnings} = 40(\$15.85) + 12.5(\$23.775)$ Jean worked 40 hours at the regular rate and 12.5 hours at the overtime rate. Round pay to the nearest cent.

$\approx \$634 + \$297.19$
$\approx \$931.19$

Jean earns $931.19 for the week.

F. Mr. Jordan sells men's clothing. He receives a salary of $375 per week plus a commission of 9.75% of his total sales. One week he sells $12,865 worth of clothing. What are his total earnings for the week?

STRATEGY: *Commission* is the money salespeople earn based on goods sold. To find the total earnings add the commission to the base salary. Use the formula.

Simpler Word Form

9.75% of sales is the commission.

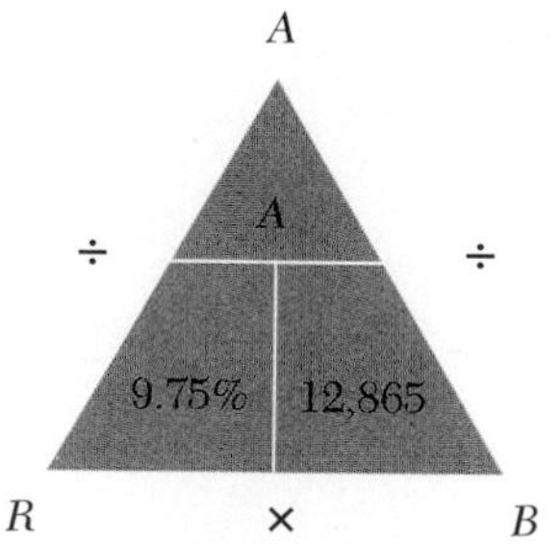

$R = 9.75\% = 0.0975$
$B = \$12{,}865$

$A = R \times B$
$A = 0.0975(12{,}865)$
$A \approx 1254.34$ Round to the nearest hundredth (cent).

The salesman earned a commission of $1254.34.

$\text{Total earnings} = \text{salary} + \text{commission}$
$= \$375 + \$1254.34$
$= \$1629.34$

Mr. Jordan's total earnings for the week are $1629.34.

E. What are Jean's earnings for a week in Example E if her hourly rate is $21.10?

F. What are Mr. Jordan's total earnings for a week in which his sales total $8565, if his salary and commission are the same as in Example F?

Answers to Warm-Ups

E. Jean's earnings for the week are $1239.63.
F. Mr. Jordan's earnings for the week are $1210.09.

EXERCISES 6.9

OBJECTIVE Solve business-related percent problems.

1. An article that costs the store owner $18.60 is marked up $4.65. What is the percent of markup based on the cost?

2. An article that costs the store owner $38.64 is sold for $55.20. What is the percent of markup based on the selling price?

3. A tool that costs a hardware merchant $9.20 is marked up 45% of the cost. What is the selling price?

4. An article is priced to sell at $48.95. If the markup is 28% of the selling price, how much is the markup to the nearest cent?

5. The Hugh TV and Appliance Store regularly sells a television set for $536.95. An advertisement in the paper shows that it is on sale at a discount of 25%. What is the sale price to the nearest cent?

6. A competitor of the store in Exercise 5 has the same TV set on sale. The competitor normally sells the set for $539.95 and has it advertised at a 26% discount. To the nearest cent, what is the sale price of the TV? Which is the better buy and by how much?

7. Maria earns a 9% commission on all of her sales. How much did she earn last week if her total sales were $7285?

8. Martin earns a 12.5% commission on all sales. How much did he earn last week if his total sales were $9300?

9. Matthew receives a weekly salary of $210 plus a commission of 8% on his sales. Last week he earned $625. What were his total sales for the week?

10. Amy receives a weekly salary of $185 plus a commission of 9% on her total sales. How much did she earn last week if her total sales were $3128?

11. The Top Company offered a 6% rebate on the purchase of the company's best model of canopy. If the regular price is $445.60, what is the amount of the rebate to the nearest cent?

12. The Stihl Company is offering a $24 rebate on the purchase of a chain saw that sells for $610.95. What is the percent of the rebate to the nearest whole percent?

13. The taxes on a piece of property are $4281, and the property has an assessed value of $356,750. What is the percent tax rate in the district where the property is located?

14. When preparing the payroll, Jane must deduct 7.51% of the earnings for Social Security. How much will Jane deduct on a monthly salary of $6755?

15. Hazel contributes 12% of her salary to her retirement fund and 1.25% to the United Way. How much does she contribute to both monthly if her monthly salary is $4875?

16. Last year, the president of the Sweet Tooth Candy Company received a bonus of $23,760. If the president receives 3.75% of the profits as a bonus, what were the company's profits last year?

17. The CEO of the Big Boy Bicycle Company is to receive a bonus of 5.25% of the profits. The profits for the last quarter were $134,700. What will be the CEO's bonus for the quarter?

18. A new automobile depreciated 16% during the first year. If its original cost was $28,679, what is its depreciated value at the end of the first year?

19. A new machine was purchased by the Hot-Blast Foundry at a cost of $43,876. At the end of the first year, it had depreciated to a value of $37,733.36. What was the rate of depreciation for the first year?

20. Three people resolve to become business associates. The first associate pays 42% for a share of a golf course, the second associate pays 31%, and the third associate pays for the remaining 27% of the golf course. The profits are to be divided among the three according to each associate's share of the ownership. The profits for last month were $24,800. How much does each associate receive?

21. Four people resolve to become business associates. The first associate pays 32% for a share of a restaurant, the second associate pays 26%, the third associate pays 24%, the fourth associate pays for the remaining 18% of the restaurant. The profits are to be divided among the four according to each associate's share of the ownership. The profits for last month were $19,800. How much does each associate receive?

22. During one week, Ms. James sold a total of $26,725 worth of hardware to the stores in her territory. She receives a 4% commission on sales of $2000 or less, 5% on that portion of her sales from $2001 to $15,000, and 6% on all sales over $15,000. What was her total commission for the week?

23. If Ms. James, in Exercise 22, had sales of $22,455 the next week, how much did she earn in commissions?

24. The retail price (selling price) of a dining room table is $1560. If the markup is 50% of the cost, what is the cost?

25. The retail price (selling price) of a china hutch is $1033.50. If the markup is 30% of the cost, what is the cost?

26. The tax rate in Rural Township is $3.20 per $1000 of assessed value. What is the tax expressed as a percent?

27. Using the tax rate from Exercise 26, what would be the taxes on a house that has an assessed value of $134,750?

28. A labor union renegotiated a contract with the Bobcat Manufacturing Company. The average wage under the new contract will increase 7.82%. If the average wage under the old contract were $12.55 per hour, what will be the average wage under the new contract?

29. A labor union negotiated a contract with the Swift Copier Company. The average wage under the new contract will increase 5.75%. If the average wage under the old contract was $9.35 per hour, what will be the average wage under the new contract?

30. A retailer bought a small color television set for $62.45 and sold it for $110.50. The cost to the retailer increased to $72; the retailer in turn increased the selling price to $129.95. Will the retailer make a larger or smaller percent of markup?

31. A discount house advertises that it sells all appliances at cost plus 10%. If Kathy buys a refrigerator for $979, what is the profit for the store?

32. In 1990, use of office paper resulted in over 11 tons of solid waste. Of this amount, 15% was recycled. How many tons were recycled?

33. In 1900, full-time workers in manufacturing worked an average of 53 hours per week. By the year 2000, the average had decreased to 42 hours per week. What percent decrease is this?

34. In 1996, the Bureau of Labor Statistics estimated that the labor force will increase by about 11% to a level of 149 million by the year 2006. What was the number of workers in 1996?

35. In 1996, the Bureau of Labor Statistics predicted that the fastest-growing occupation would be database administrators/computer support specialists/computer scientists. The number of jobs was expected to go from 212,000 in 1996 to 461,000 in 2006. What percent increase is this?

36. On October 6, 1931, the Dow Jones Industrial Average experienced its greatest percent gain in a single day. The average closed at 99.34 points, which was a 14.87% increase over the previous day. What was the close on the previous day?

37. On April 14, 2000, the Dow Jones Industrial Average experienced its greatest point loss in a single day. The average closed at 10,305.77 points, which was a 5.66% loss from the previous day. What was the close on the previous day?

Exercises 38–45 relate to the chapter application.

38. Corduroy overalls that are regularly $42.75 are on sale for 25% off. What is the sale price of the overalls?

39. A pair of New Balance cross trainers, which regularly sells for $87.99, goes on sale for $74.99. What percent off is this? Round to the nearest whole percent.

40. A bag of Tootsie Rolls is marked "20% more free—14.5 oz for the price of 12 oz." Assuming there has not been a change in price, is the claim accurate? Explain.

41. A department store puts a blazer, which was originally priced at $167.50, on sale for 20% off. At the end of the season, the store has an "Additional 40% off everything that is already reduced" sale. What is the price of the blazer? What percent savings does this represent over the original price?

42. A shoe store advertises "Buy one pair, get 50% off a second pair of lesser or equal value." The mother of twin boys buys a pair of basketball shoes priced at $55.99 and a pair of hiking shoes priced at $42.98. How much did she pay for the two pairs of shoes? What percent savings is this to the nearest tenth of a percent?

43. The Klub House advertises on the radio that all merchandise is on sale at 25% off. When you go in to buy a set of golf clubs that originally sold for $1375, you find that the store is giving an additional 10% discount off the original price. What is the price you will pay for the set of clubs?

44. In Exercise 43, if the salesperson says that the 10% discount can be applied only to the sale price, what is the price of the clubs?

45. A store advertises "30% off all clearance items." A boy's knit shirt is on a clearance rack that is marked 20% off. How much is saved on a knit shirt that was originally priced $25.99?

STATE YOUR UNDERSTANDING

46. Explain why the percent of increase from 500 to 750 (50%) is not the same as the percent of decrease from 750 to 500 $\left(33\frac{1}{3}\%\right)$.

47. Explain why a 10% increase in $100 followed by a 10% decrease does *not* yield $100.

48. Write a few sentences clarifying the following statements: (a) "a price increase of 100% is double the original price," and (b) "an enlargement of a floor area by 200% of the original is triple the original area."

CHALLENGE

49. The markup on furniture at NuMart is 35%, based on the selling price. The cost of a lounge chair is $68. What is the selling price?

50. A retailer buys a shipment of athletic shoes for $42.64 a pair and sells them for $61.95 a pair. On the next shipment, the cost is increased by 8%, and the retailer, in turn, increases the selling price to $66.45. Is the percent of markup more or less after the price increase? By how much, rounded to the nearest whole percent?

51. During a 6-year period, the cost of maintaining a diesel engine averages \$2650 and the cost of maintaining a gasoline engine averages \$4600. What is the percent of savings of the diesel compared with the gasoline engine? Round to the nearest whole percent.

GROUP WORK

52. During 1 month in 1991, the interest rate on a 30-year home mortgage averaged 9.29%. A 15-year mortgage averaged 8.98%. Ten years later, the interest rate on a 30-year mortgage averaged 6.625% and a 15-year mortgage averaged 6.0%. Find the difference in simple interest costs for 1 year on houses that cost \$85,000, \$110,000, \$145,000, and \$266,000. Find the assessed value of your own dwelling and then compute the costs at these different rates. Determine the average interest for the members of your group and for the class. Compare the values with each other, with the other groups, and with the values for 1991 and 2001.

MAINTAIN YOUR SKILLS

Add.

53. 456 + 387 + 1293 + 781

54. 32.67 + 45.098 + 102.5 + 134.76

Subtract.

55. 34,761 − 29,849

56. 134.56 − 98.235

Multiply.

57. (341)(56)

58. (56.72)(0.023)

Divide.

59. 9324 ÷ 36

60. 76.4 ÷ 1.34 Round to the nearest thousandth.

61. Geri works at two jobs each week. Last week she worked 25 hours at the job paying \$7.82 per hour and 23 hours at the job that pays \$10.42 per hour. How much did she earn last week?

62. Mary left her estate to her seven nieces and nephews. The executor of the estate received a 5% commission for handling the estate. How much did each receive if the estate was worth \$1,456,000?

CHAPTER 6

KEY CONCEPTS

SECTION 6.1 THE MEANING OF PERCENT

Definitions and Concepts	Examples
A percent is a ratio with a base unit (the denominator) of 100.	$75\% = \frac{75}{100}$
The symbol % means $\frac{1}{100}$ or 0.01.	If 22 out of 100 people are left-handed, what percent is this?
$100\% = \frac{100}{100} = 1$	$\frac{22}{100} = 22\%$

SECTION 6.2 CHANGING DECIMALS TO PERCENTS

Definitions and Concepts	Examples
To change a decimal to a percent: • Move the decimal point two places to the right (add zeros if necessary). • Write % on the right.	$0.23 = 23\%$ $5.7 = 570\%$

SECTION 6.3 CHANGING PERCENTS TO DECIMALS

Definitions and Concepts	Examples
To change a percent to a decimal: • Move the decimal point two places to the left (add zeros if necessary). • Drop the percent symbol (%).	$67\% = 0.67$ $2.8\% = 0.028$

SECTION 6.4 CHANGING FRACTIONS TO PERCENTS

Definitions and Concepts	Examples
To change a fraction or mixed number to a percent: • Change to a decimal. • Change the decimal to a percent.	$\frac{6}{7} \approx 0.857 \approx 85.7\%$ $3\frac{4}{25} = 3.16 = 316\%$

SECTION 6.5 CHANGING PERCENTS TO FRACTIONS

Definitions and Concepts	Examples
To change a percent to a fraction: • Replace the percent symbol (%) with $\frac{1}{100}$. • If necessary, rewrite the other factor as a fraction. • Multiply and simplify.	$45\% = 45 \cdot \frac{1}{100} = \frac{\overset{9}{\cancel{45}}}{\underset{20}{\cancel{100}}} = \frac{9}{20}$ $6.5\% = 6.5 \cdot \frac{1}{100} = \frac{13}{2} \cdot \frac{1}{100} = \frac{13}{200}$

SECTION 6.6 FRACTIONS, DECIMALS, PERCENTS: A REVIEW

Definitions and Concepts

Every number has three forms: fraction, decimal, and percent.

Examples

Fraction	Decimal	Percent
$\frac{7}{10}$	0.7	70%
$5\frac{3}{8}$	5.375	537.5%

SECTION 6.7 SOLVING PERCENT PROBLEMS

Definitions and Concepts

The percent formula is $R \times B = A$, where R is the rate of percent, B is the base, and A is the amount.

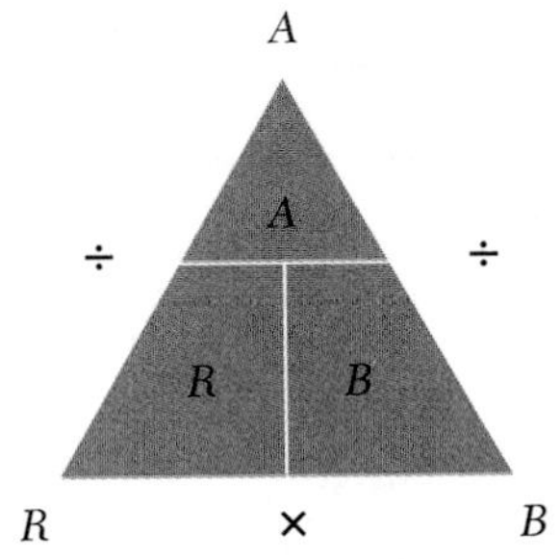

To solve percent problems using a proportion, set up the proportion.

$\frac{X}{100} = \frac{A}{B}$, where $R = X\%$

Examples

6 is 24% of what number?
$A = 6, R = 0.24, B = ?$

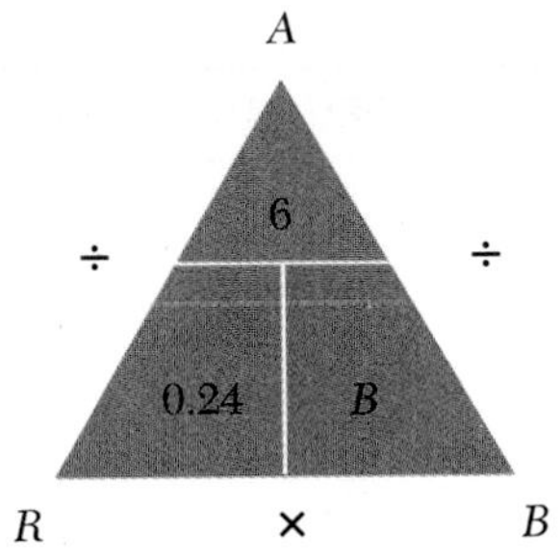

$B = A \div R$
$B = 6 \div 0.24$
$B = 25$

So, 6 is 24% of 25.

8 is what percent of 200?
$A = 8, B = 200, R = X\% = ?$

$$\frac{8}{200} = \frac{X}{100}$$
$$8 \cdot 100 = 200X$$
$$800 \div 200 = X$$
$$4 = X$$

So, 8 is 4% of 200.

SECTION 6.8 APPLICATIONS OF PERCENT

Definitions and Concepts

To solve percent applications:

- Restate the problem as a simple percent statement.
- Identify values for A, R, and B.
- Use the percent formula or a proportion.

Examples

Of 72 students in Physics 231, 32 are women. What percent of the students are women?

Restate: 32 is what percent of 72?
$A = 32, B = 72, R = X\% = ?$

$$\frac{32}{72} = \frac{X}{100}$$
$$32 \cdot 100 = 72X$$
$$3200 \div 72 = X$$
$$X \approx 44.4$$

So, about 44.4% of the students are women.

SECTION 6.9 BUSINESS APPLICATIONS OF PERCENT

Definitions and Concepts	Examples
Cost + markup = selling price	A lamp costs \$15.00 and is marked up 25%. Selling price = 15 + 25% of 15 = 15 + 3.75 = 18.75
Regular price − amount of discount = sale price	A microwave oven priced at \$160 goes on sale for 30% off. Sale price = 160 − 30% of 160 = 160 − 48 = 112
Commission is a percentage of the value of goods sold.	A realtor earns a 7% commission on a house she sells for \$180,000. Commission = 7% of 180,000 = 12,600

Name ______________________ Class ______________________ Date ____________

CHAPTER 6

REVIEW EXERCISES

Section 6.1

What percent of each of the following regions is shaded?

1. 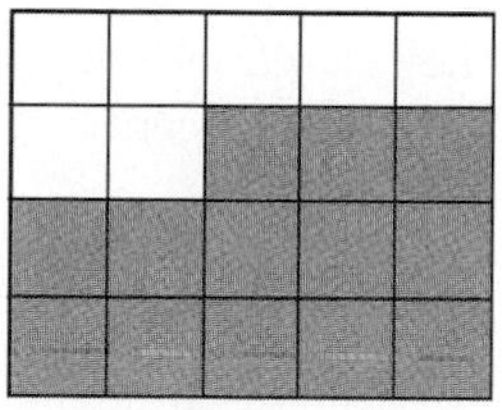

2. 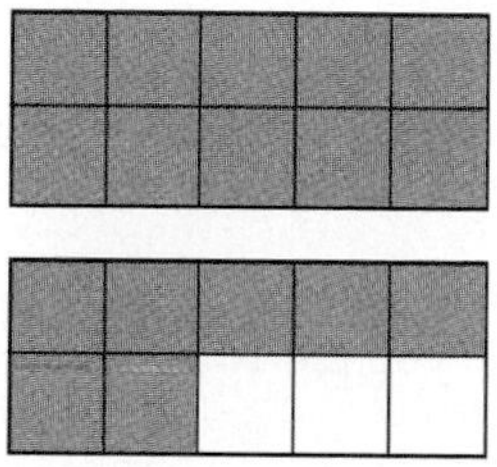

Write an exact percent for these comparisons. Use fractions when necessary.

3. 41 per 50 **4.** 324 per 120 **5.** 44 per 77

6. A Gallup Poll showed that 23 of 100 teenagers had used illegal drugs during the past year. What percent of the teenagers had used illegal drugs?

Section 6.2

Write each decimal as a percent.

7. 0.438 **8.** 0.082 **9.** 0.309

10. 5.671 **11.** 0.00017 **12.** 55

13. The Seattle Mariners won 0.839 of their first 31 games in 2001. Write this as a percent.

14. The sale price on a new stereo is 0.73 of the original price. Express this as a percent. What "percent off" will the store advertise?

Section 6.3

Write each of the following as a decimal.

15. 67% **16.** 461% **17.** 7.47%

18. 0.032% **19.** $\frac{1}{16}\%$ **20.** $81\frac{3}{4}\%$

21. What decimal is used to compute the interest on a credit card balance that has an interest rate of 19.785%?

22. Thirty-year home mortgages are being offered at 6.35%. What decimal will be used to compute the interest?

Section 6.4

Change each fraction or mixed number to a percent.

23. $\frac{91}{100}$ **24.** $\frac{13}{16}$ **25.** $5\frac{3}{64}$

Change each fraction or mixed number to a percent. Round to the nearest tenth of a percent.

26. $\frac{13}{17}$ **27.** $\frac{5}{37}$ **28.** $7\frac{8}{39}$

29. A Little League baseball team won 19 of the 27 games they played . What percent of the games did they win, rounded to the nearest hundredth of a percent?

30. Jose got 37 problems correct on a 42-problem exam. What percent of the problems did Jose get correct? Round to the nearest whole percent.

Section 6.5

Change each of the following percents to fractions or mixed numbers.

31. 48% **32.** 165% **33.** 7.8%

34. $\frac{1}{8}\%$ **35.** 21.5% **36.** 964%

37. The offensive team for the Chicago Bears was on the field 38% of the time during a recent game with the Green Bay Packers. What fractional part of the game was the Bears defense on the field?

38. Spraying for the gypsy moth is found to be 92.4% effective. What fraction of the gypsy moths is eliminated?

Section 6.6

Exercises 39–46. Fill in the table with the related percent, decimal, or fraction.

Fraction	Decimal	Percent
$\frac{9}{25}$	**39.**	**40.**
41.	0.72	**42.**
43.	**44.**	1.5%
$3\frac{11}{40}$	**45.**	**46.**

Section 6.7

Solve.

47. 14% of 235 is ________. **48.** 14 is 35% of ________.

49. 17 is ________% of 80. **50.** 37 is ________% of 125.

51. 6.4% of 820 is ________. **52.** 20.93 is 23% of ________.

53. What percent of 523 is 46? Round to the nearest tenth of a percent.

54. One hundred thirty-four is 141.6% of what number? Round to the nearest hundredth.

Section 6.8

55. Last year Melinda had 23.5% of her salary withheld for taxes. If the total amount withheld was $3156.75, what was Melinda's yearly salary?

56. The population of Arlington grew from 3564 to 5721 over the last 5 years. What was the percent of increase in the population? Round to the nearest tenth of a percent.

57. The work force at Omni Plastics grew by 32% over the last 3 years. If the company had 250 employees 3 years ago, how many employees does it have now?

58. Mrs. Hope's third grade class has the following ethnic distribution: Hispanic, 13; African American, 7; Asian, 5; Caucasian, 11. What percent of her class is African American? Round to the nearest tenth of a percent.

59. During the 2000–2001 season, the Portland Trailblazers lost 17 of their last 25 games. What percent of the games did they lose?

60. To qualify for an interview, Toni had to get a minimum of 72% on a preemployment test. Toni got 45 of 62 question correct. Does Toni qualify for an interview?

Exercises 61–62. The figure shows the grade distribution in an algebra class.

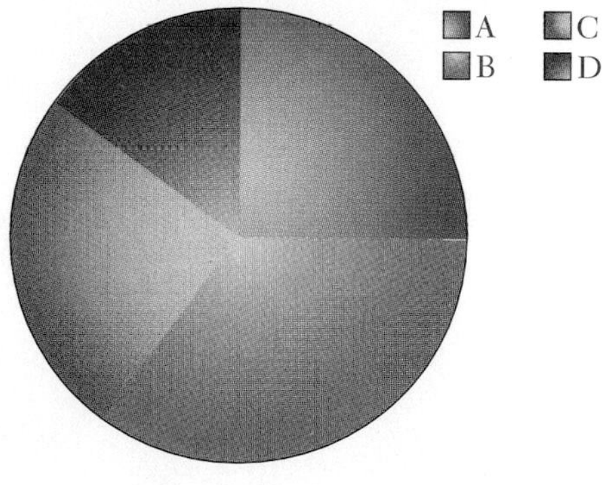

Algebra grades

61. Which grade was received by the most students?

62. Were more A and C grades given than B and D grades?

Section 6.9

63. A golf club bag that costs the store $32.85 is marked up by $26.28. What is the percent of markup based on the cost?

64. A Queen Anne chair is priced to sell at \$563.80. If the markup is 45% of the selling price, how much is the markup?

65. The May Company advertises a sale at 25% off on men's suits. The store then offers a coupon that gives an additional 20% off the sale price. What is the cost to the consumer of a suit that was originally priced at \$675.90?

66. A salesclerk earns a base salary of \$1500 per month plus a commission of 11.5% on all sales over \$9500. What is her salary in a month where her sales totaled \$21,300?

67. The taxes on a downtown high rise is \$38,714. If the property has an assessed value of \$1,345,750, what is the percent tax rate in the district where the high rise is located? Round to the nearest tenth of a percent.

68. A real estate salesperson earns a commission of 3.5% on the sale of a home. What is his commission if the home sold for \$267,780?

69. A pair of New Balance walking shoes, which regularly sell for \$110.95, go on sale for \$79.49. What percent off is this? Round to the nearest whole percent.

70. In 2001, screenwriters agreed on a new contract with the major studios that called for a 37% increase in wages over a 5-year period. What was a screenwriter earning in 2001 if her salary in 2006 will be \$82,675?

CHAPTER 6

TRUE/FALSE CONCEPT REVIEW

ANSWERS

Check your understanding of the language of basic mathematics. Tell whether each of the following statements is true (always true) or false (not always true). For each statement you judge to be false, revise it to make a statement that is true.

1. Percent means per one hundred. — 1. ____________

2. The symbol % is read "percent." — 2. ____________

3. To change a fraction to a percent, move the decimal point in the numerator two places to the left and add the percent symbol. — 3. ____________

4. To change a decimal to a percent, move the decimal point two places to the left. — 4. ____________

5. Percent is a ratio. — 5. ____________

6. In percent, the base unit can be more than 100. — 6. ____________

7. To change a percent to a decimal, drop the percent symbol and move the decimal point two places to the left. — 7. ____________

8. A percent can be equal to a whole number. — 8. ____________

9. To solve a problem written in the form A is R of B, we can use the proportion $\frac{B}{A} = \frac{X}{100}$, where $R = X\%$. — 9. ____________

10. To solve the problem, "If there is a 5% sales tax on a radio selling for \$64.49, how much is the tax?" the simpler word form could be "5% of \$64.40 is what?" — 10. ____________

11. $1.2 = 120\%$ — 11. ____________

12. $7\frac{1}{2} = 7.5\%$ — 12. ____________

13. $0.009\% = 0.9$ — 13. ____________

14. If 0.3% of B is 84, then $B = 280$. — 14. ____________

15. If some percent of 24 is 12, then the percent is 50%. — 15. ____________

16. If $2\frac{4}{5}\%$ of 300 is A, then $A = 8.4$. — 16. ____________

17. Two consecutive decreases of 15% is the same as a decrease of 30%. — 17. ____________

 Name ______________________ Class ______________________ Date ____________

18. If Selma is given a 10% raise on Monday, but her salary is cut 10% on Wednesday, her salary is the same as it was Monday before the raise. 18. ____________

19. It is possible to increase a city's population by 110%. 19. ____________

20. If the price of a stock increases 100% for each of 3 years, the value of $1 of stock is worth $8 at the end of 3 years. 20. ____________

21. A 50% growth in population is the same as 150% of the original population. 21. ____________

22. $\frac{1}{2}\% = 0.5$ 22. ____________

CHAPTER 6

TEST

ANSWERS

1. A computer regularly sells for $1285. During a sale, the dealer discounts the price $312.50. What is the percent of discount? Round to the nearest tenth of a percent. 1. ________

2. Write as a percent: 0.07436 2. ________

3. If 21 of every 100 people in a certain town are Hispanic, what percent of the population is Hispanic? 3. ________

4. What percent of $6\frac{5}{8}$ is $4\frac{3}{4}$? Round to the nearest tenth of a percent. 4. ________

5. Change to a percent: $\frac{11}{16}$ 5. ________

6. One hundred twenty-three percent of what number is 75.03? 6. ________

7. Change to a fraction: $34\frac{2}{7}\%$ 7. ________

8. Write as a percent: 7.91 8. ________

9. What number is 3.9% of 124? 9. ________

10. Change to a percent (to the nearest tenth of a percent): $3\frac{8}{11}$ 10. ________

11. Change to a fraction or mixed number: 272% 11. ________

12. Write as a decimal: 43.71% 12. ________

13. The Siekas family spends 17.3% of their monthly income on food. If their monthly income is $5786.42, how much do they spend on food? 13. ________

Exercises 14–19. Complete the following table:

Fraction	Decimal	Percent
$\frac{13}{16}$	14.	15.
16.	0.624	17.
18.	19.	18.5%

20. 51.6 is ________% of 78.7 (to the nearest tenth of a percent). 20. ________

21. Write as a decimal: 0.619% 21. ________

22. If a tire costs the dealer $42.44 and the markup is 44% of the cost, what is the selling price of the tire?

22. ______

23. Nordstrom offers a sale on Tommy Hilfiger jackets at a discount of 23%. What is the sale price of a jacket that originally sold for $135.95?

23. ______

24. The population of a county in Mississippi grew from 256,821 people to 334,723 people over the last 10 years. What is the percent of increase in the population (to the nearest tenth of a percent)?

24. ______

25. The following graph shows the distribution of grades in an American history class. What percent of the class received a B grade? Round to the nearest tenth of a percent.

26. ______

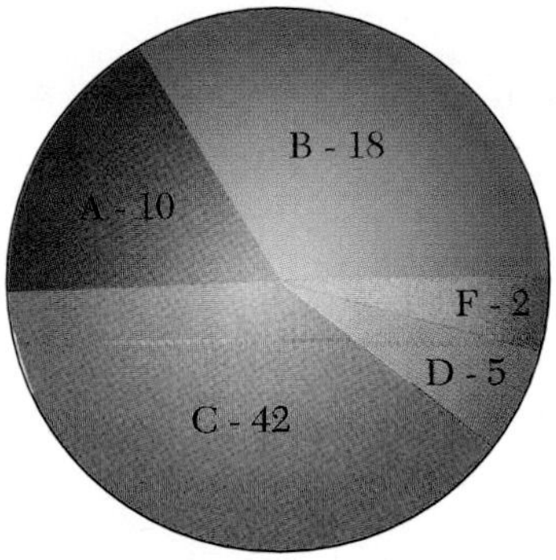

26. Gas prices dropped from summer to winter from $2.199 to $1.899. What is the percent of decrease? Round to the nearest percent.

26. ______

27. What is the selling price of a suit if the markup is 80% of the cost and the cost is $310.40?

27. ______

28. If a hamburger contains 780 calories and 37 g of fat, what percent of the calories is from fat? Assume each gram of fat contains 10 calories. Round to the nearest tenth of a percent.

28. ______

29. In a mathematics class of 80 students, 22.5% withdrew from the course prior to the final exam. How many students took the final exam?

29. ______

30. The price of a radial tire went from $65.85 to $73.85 over a 2-year period. What was the percent of increase in price over the 2 years? Find to the nearest tenth of a percent.

30. ______

CHAPTER 6

GROUP PROJECT (1–2 weeks)

OPTIONAL

Much research has been conducted in the area of human nutrition, and yet there are still many unanswered questions. Most nutritionists agree, however, that the American diet has too much fat and refined carbohydrates (primarily sugar).

You are about to become a nutritional consultant to John and Jane Doe. John is a 32-year-old male, 5′10″, 165 lb, who consumes 2800 calories per day. Jane is a 31-year-old female, 5′5″, 130 lb, who consumes 2000 calories per day. Both exercise regularly and try to be careful about their diet. However, neither of them likes to cook, so they eat out a lot.

Experts disagree about the exact ratio of carbohydrates, protein, and fat in an optimal diet. Some recommend a 40% carbohydrate, 30% protein, 30% fat ratio. Others would shift the amounts to 50%, 25%, and 25%, respectively. Almost no one recommends more than 30% fat.

All foods sold in grocery stores are required by the government to include nutritional labeling. Among other things, the label specifies the total calories per serving and the number of grams of carbohydrates, protein, and fat per serving. Most fast-food chains also provide this information. A gram of carbohydrate and a gram of protein each contain about 4 calories. Assume a gram of fat contains 10 calories.

Your job is to identify exactly what John and Jane will eat for a day so they meet their nutritional requirements entirely with food purchased from fast-food restaurants. Make a second day's plan for each of them with food purchased from a grocery, bearing in mind that the only cooking they do is heating up in a microwave oven.

Include in your final report:

1. a rationale for the ratio of carbohydrates, protein, and fat selected
2. two days of menus for John and Jane
3. a table for each person and each day that shows the total calories of carbohydrates, protein, and fat, and the final percentages

GOOD ADVICE FOR STUDYING

Low-Stress Tests

It's natural to be anxious before an exam. In fact, a little anxiety is actually good: It keeps you alert and on your toes. Obviously, too much stress over tests is not good. Here are some proven tips for taking low-stress tests.

1. Before going to the exam, find a place on campus where you can physically and mentally relax. Don't come into the classroom in a rush.
2. Arrive at the classroom in time to arrange all the tools you will need for the test: sharpened pencils, eraser, plenty of scratch paper, and a water bottle. Try to avoid talking with classmates about the test. Instead, concentrate on deep breathing and relaxation.
3. Before starting the test, on a separate piece of paper, write all the things you may forget while you are busy at work: formulas, rules, definitions, and reminders to yourself. Doing so relieves the load on your short-term memory.
4. Read all of the test problems and mark the easiest ones. Don't skip reading the directions. Note point values so you don't spend too much time on problems that count only a little, at the expense of problems that count a lot.
5. Do the easiest problems first; do the rest of the problems in order of difficulty.
6. Estimate a reasonable answer before you make calculations. When you finish the problem, check to see that your answer agrees with your estimate.
7. If you get stuck on a problem, mark it and come back to it later.
8. When you have finished trying all the problems, go back to the problems you didn't finish and do what you can. Show all steps because you may get partial credit even if you cannot complete a problem.
9. When you are finished, go back over the test to see that all the problems are as complete as possible and that you have indicated your final answer. Use all of the time allowed, unless you are sure there is nothing more that you can do.
10. Turn in your test and be confident that you did the best job you could. Congratulate yourself on a low-stress test!

If you find yourself feeling anxious, try a 3×5 "calming" card. It may include the following: (a) a personal coping statement such as "I have studied hard and prepared well for this test, I will do fine"; (b) a brief description of your peaceful scene; and (c) a reminder to stop, breathe, and relax your tense muscles.

By now, you should be closer to taking control of math instead of allowing math to control you. You are avoiding learned helplessness (believing that other people or influences control your life). Perfectionism, procrastination, fear of failure, and blaming others are also ineffective attitudes that block your power of control. Take responsibility and believe that you have the power within to control your life situation.

7 MEASUREMENT AND GEOMETRY

SECTIONS

7.1 Measuring Length
7.2 Measuring Capacity, Weight, and Temperature
7.3 Perimeter
7.4 Area
7.5 Volume
7.6 Square Roots and the Pythagorean Theorem

APPLICATION

Paul and Barbara have just purchased a row house in the Georgetown section of Washington, D.C. The backyard is rather small and completely fenced. They decide to take out all the grass and put in a brick patio and formal rose garden. The plans for the patio and garden are shown here.

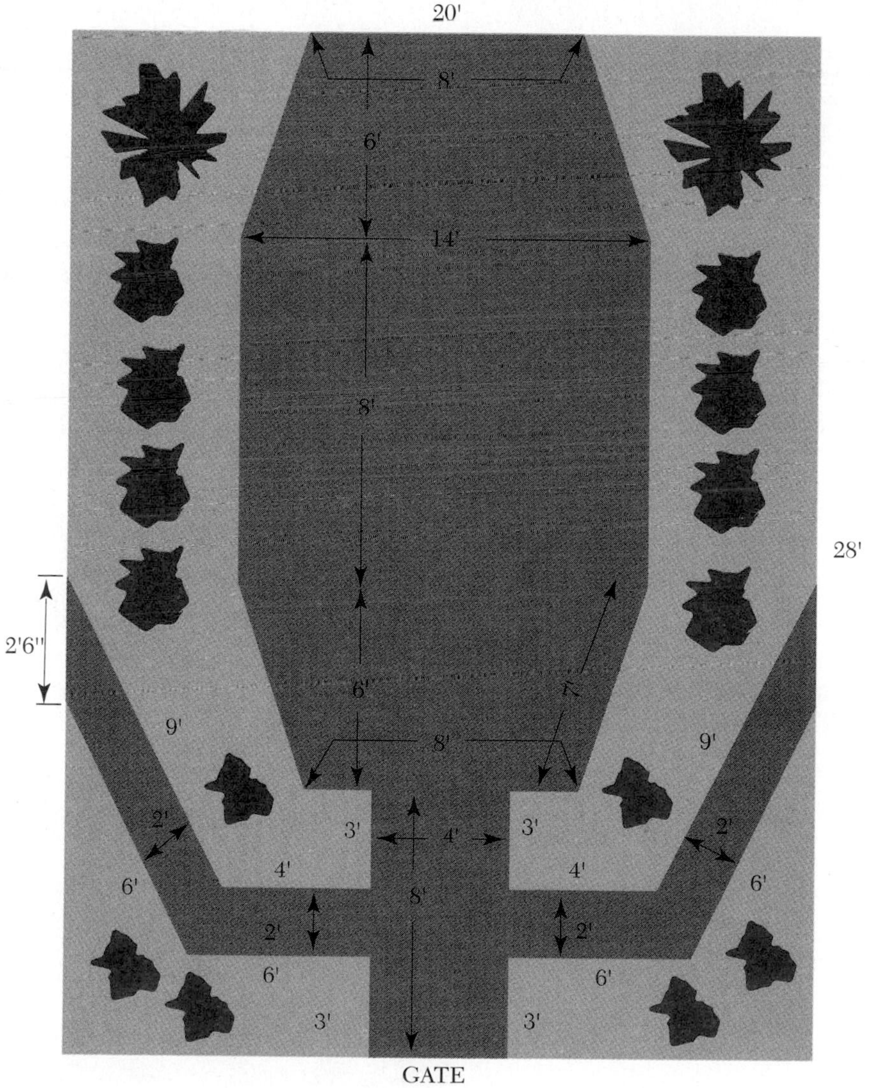

7.1 Measuring Length

OBJECTIVES

1. Recognize and use appropriate units of length from the English and metric measuring systems.
2. Convert units of length.
   a. Unit fractions within the same system.
   b. Between systems.
   c. Moving the decimal point in the metric system.
3. Perform operations on measurements.

VOCABULARY

A **unit of measure** is the name of a fixed quantity that is used as a standard.

A **measurement** is a number together with a unit of measure.

Equivalent measurements are measures of the same amount that use different units.

A **unit fraction** is a fraction whose numerator and denominator are equivalent measurements.

The **English system** is the measurement system commonly used in the United States.

The **metric system** is the measurement system used by most of the world.

HOW & WHY

OBJECTIVE 1 Recognize and use appropriate units of length from the English and metric measuring systems.

One of the main ways of describing an object is to give its measurements. We measure how long an object is, how much it weighs, how much space it occupies, how long it has existed, how hot it is, and so forth. Units of measure are universally defined so we all mean the same thing when we use a measurement. There are two major systems of measurement in use in the United States. One is the English system, so named because we adopted what was used in England at the time. This system is a mixture of units from various countries and cultures, and it is the system most Americans are familiar with. The second is the metric system, which is currently used by almost the entire world.

Measures of length answer questions like "how long?" or "how tall?" or "how deep?" We need units of length to measure small distances, medium distances, and long distances. In the English system, we use inches to measure small distances, feet to measure medium distances, and miles to measure long distances. Other, less common units of length in the English system include yards, rods, fathoms, and light-years.

The metric system was invented by French scientists in 1799. Their goal was to make a system that was easy to learn and would be used worldwide. They based the system for length on the meter and related it to Earth by defining it as 1/10,000,000 of the distance between the North Pole and the equator.

Meter

A meter is currently defined by international treaty in terms of wavelengths of the orange-red radiation of an isotope of krypton (^{86}Kr) in a vacuum.

To make the metric system easy to use, scientists based all conversions on powers of 10 and gave the same suffix to all units of measure for the same characteristic. All units of length in the metric system end in "-meter." Furthermore, multiples of the base unit are indicated by the prefix. So any measure beginning with "kilo-" means 1000 of the base unit. Any measure beginning with "centi-" means 1/100 of the base unit, and any measure beginning with "milli-" means 1/1000 of the base unit.

Table 7.1 lists common units of length in both systems.

TABLE 7.1 COMMON UNITS OF LENGTH

Size	English	Metric
Large	Mile (mi)	Kilometer (km)
Medium	Foot (ft) or yard (yd)	Meter (m)
Small	Inch (in.)	Centimeter (cm)
Tiny		Millimeter (mm)

EXAMPLES A–C

DIRECTIONS: Write both an English unit and a metric unit to measure the following.

STRATEGY: Decide on the size of the object and pick the appropriate units.

A. The distance from Baltimore, Maryland, to Washington, D.C.

This is a long distance, so it is measured in miles or kilometers.

B. The width of a calculator

This is a small distance, so it is measured in inches or centimeters.

C. The width of a dining room

This is a medium distance, so it is measured in feet or meters.

WARM-UPS A–C

A. The distance from San Francisco to Los Angeles

B. The width of a camera

C. The length of a bedroom

HOW & WHY

OBJECTIVE 2 Convert units of length (unit fractions within the same system).

Using Table 7.2, we can convert measurements in each system.

Check out the MSDS Hyper Glossary for unit conversions and a slick converter. http://www.ilpi.com/msds/ref/distanceunits.htm/

TABLE 7.2 EQUIVALENT LENGTH MEASUREMENTS

English	Metric
12 inches (in.) = 1 foot (ft)	1000 millimeters (mm) = 1 meter (m)
3 feet (ft) = 1 yard (yd)	100 centimeters (cm) = 1 meter (m)
5280 feet (ft) = 1 mile (mi)	1000 meters (m) = 1 kilometer (km)

Because 12 in. = 1 ft, they are equivalent measurements. A fraction with these measures in the numerator and denominator is equal to 1 because the numerator and denominator are equivalent.

$$\frac{12\text{ in.}}{1\text{ ft}} = \frac{1\text{ ft}}{12\text{ in.}} = 1$$

To convert a measurement from one unit to another, we multiply by the unit fraction: $\frac{\text{desired unit of measure}}{\text{original unit of measure}}$. Because we are multiplying by 1, the measurement is unchanged but the units are different.

For example, to convert 48 in. to feet, we choose a unit fraction that has feet, the desired units in the numerator, and inches, the original units, in the denominator. In this case, we use $\frac{1\text{ ft}}{12\text{ in.}}$ for the conversion.

$$48\text{ in.} = \frac{48\text{ }\cancel{\text{in.}}}{1} \cdot \frac{1\text{ ft}}{12\text{ }\cancel{\text{in.}}}$$ Multiply by the appropriate unit fraction.

$$= \frac{48\text{ ft}}{12}$$ Multiply.

$$= 4\text{ ft}$$ Simplify.

In some cases, it is necessary to multiply by more than one unit fraction to get the desired results. For example, to convert 5.2 km to centimeters, we use the unit fraction $\frac{1000\text{ m}}{1\text{ km}}$ to convert the kilometers to meters, and then $\frac{100\text{ cm}}{1\text{ m}}$ to convert the meters to centimeters.

Answers to Warm-Ups

A. miles or kilometers **B.** inches or centimeters **C.** feet or meters

$$5.2\text{ km} = \frac{5.2\text{ km}}{1} \cdot \frac{1000\text{ m}}{1\text{ km}} \cdot \frac{100\text{ cm}}{1\text{ m}}$$ Multiply by the appropriate unit fractions.

$$= 520{,}000\text{ cm}$$ Simplify.

To convert units of length

1. Multiply by the unit fraction that has the desired units in the numerator and the original units in the denominator.
2. Simplify.

WARM-UPS D–E

D. Convert 27 cm to meters.

E. Convert 12 mi to inches.

EXAMPLES D–E

DIRECTIONS: Convert units of measure.

STRATEGY: Multiply the given measure by the appropriate unit fractions and simplify.

D. Convert 4 mm to meters.

$$4\text{ mm} = \frac{4\text{ mm}}{1} \cdot \frac{1\text{ m}}{1000\text{ mm}}$$ Multiply by the appropriate unit fraction.

$$= \frac{4\text{ m}}{1000}$$ Simplify.

$$= 0.004\text{ m}$$

So, 4 mm = 0.004 m.

E. Convert 7.8 mi to inches.

$$7.8\text{ mi} = \frac{7.8\text{ mi}}{1} \cdot \frac{5280\text{ ft}}{1\text{ mi}} \cdot \frac{12\text{ in.}}{1\text{ ft}}$$ Convert miles to feet and then feet to inches.

$$= 494208\text{ in.}$$ Simplify.

* So, 7.8 mi = 494,208 in.

OBJECTIVE 2 Convert units of length (between systems).

Sometimes we need to convert lengths from one system to another. Most conversions between systems are approximations, as in Table 7.3. The method of using unit fractions for converting is the same as when converting within the same system.

TABLE 7.3 LENGTH CONVERSIONS BETWEEN ENGLISH AND METRIC SYSTEMS

1 inch ≈ 2.5400 centimeters	1 centimeter ≈ 0.3937 inch
1 foot ≈ 0.3048 meter	1 meter ≈ 3.2808 feet
1 yard ≈ 0.9144 meter	1 meter ≈ 1.0936 yards
1 mile ≈ 1.6093 kilometers	1 kilometer ≈ 0.6214 mile

Because of the rounding in Table 7.3, it is possible to obtain slightly different answers when converting using different unit fractions. For everyday measurements, this is not a problem, as we usually only measure to the nearest tenth or hundredth. However, in a laboratory or industrial setting, where more precision is necessary, you may need to use more accurate conversions than those in Table 7.3.

Answers to Warm-Ups

D. 0.27 m **E.** 760,320 in.

CAUTION

1. Because the conversions in Table 7.3 are all rounded to the nearest ten-thousandth, we cannot expect closer accuracy when using them.
2. Because two different unit fractions can often be used for a single conversion, it is possible that answers will vary slightly.

EXAMPLES F–G

DIRECTIONS: Convert units of measure. Round to the nearest hundredth.

STRATEGY: Multiply the given measure by the appropriate unit fraction(s) and simplify.

F. Convert 2 ft to meters.

$$2 \text{ ft} \approx \frac{2\ \cancel{\text{ft}}}{1}\cdot\frac{0.3048 \text{ m}}{1\ \cancel{\text{ft}}}$$ Multiply by the appropriate unit fraction.

$$\approx 0.6096 \text{ m}$$ Simplify.

So, 2 ft ≈ 0.61m.

ALTERNATE SOLUTION: Use the unit fraction $\frac{1 \text{ m}}{3.2808 \text{ ft}}$ (see Table 7.3).

$$2 \text{ ft} \approx \frac{2\ \cancel{\text{ft}}}{1}\cdot\frac{1 \text{ m}}{3.2808\ \cancel{\text{ft}}}$$ Multiply by the appropriate unit fraction.

$$\approx \frac{2}{3.2808} \text{ m}$$ Simplify.

$$\approx 0.6096 \text{ m}$$ Simplify.

So, 2 ft ≈ 0.61 m.

Note that we get slightly different answers using different unit fractions, but they are the same, rounded to the nearest hundredth.

G. Convert 1.6 km to feet.

STRATEGY: Convert kilometers to miles and then miles to feet.

$$1.6 \text{ km} \approx \frac{1.6\ \cancel{\text{km}}}{1}\cdot\frac{0.6214\ \cancel{\text{mi}}}{1\ \cancel{\text{km}}}\cdot\frac{5280 \text{ ft}}{1\ \cancel{\text{mi}}}$$ Convert kilometers to miles and then miles to feet.

$$\approx 5249.59 \text{ ft}$$ Simplify.

So, 1.6 km ≈ 5249.59 ft.

WARM-UPS F–G

F. Convert 5 m to feet.

G. Convert 6 ft to centimeters.

OBJECTIVE 2 Convert units of length (moving the decimal point in the metric system).

The metric system is based on powers of 10 and it is easy to multiply and divide by powers of 10. Therefore, we can shortcut metric-to-metric conversions by multiplying or dividing by powers of 10.

We have already seen how kilometers, centimeters, and millimeters relate to the base unit, the meter. Other, less frequently used units are also part of the metric system. Consider the following.

Answers to Warm-Ups

F. 16.41 ft **G.** 182.88 cm

Kilometer	Hecto-meter	Deka-meter	Meter	Deci-meter	Centi-meter	Millimeter
km	hm	dam	m	dm	cm	mm
1000 m	100 m	10 m	1 m	0.1 m	0.01 m	0.001 m

Reminder

To multiply a number by a power of 10, move the decimal point to the right. The number of places to move is the number of zeros in the power of 10 or the exponent of 10.

To divide a number by a power of 10, move the decimal point to the left. The number of places to move is the number of zeroes in the power of ten, or the exponent of 10.

To convert within the metric system, simply determine how many places you must move to get from the given unit to the desired unit. For example, to change kilometers to meters, we must move three places to the right.

km	hm	dam	m	dm	cm	mm

To change 4 km to meters, we multiply 4 km by 10^3 or 1000. So, 4 km = 4000 m. To change centimeters to meters, we must move two places left.

Kilometer	Hecto-meter	Dekameter	Meter	Deci-meter	Centi-meter	Millimeter
km	hm	dam	m	dm	cm	mm

To change 85 cm to meters, we must divide by 100 or 10^2. So, 85 cm = 0.85 m.

To convert units within the metric system

Move the decimal point the same number of places and in the same direction as you do to go from the given units to the desired units on the chart.

WARM-UPS H–I

H. Change 69 km to meters.

I. Change 346 mm to centimeters.

EXAMPLES H–I

DIRECTIONS: Convert units as indicated.

STRATEGY: Use the chart for a shortcut.

H. Change 47 m to millimeters.

Meters to millimeters is three places right on the chart. Therefore, we move the decimal three places to the right.

47 m = 47,000 mm

I. Change 13 mm to meters.

Millimeters to meters is three places left on the chart. Therefore we move the decimal three places to the left.

13 mm = 0.013 m

HOW & WHY

OBJECTIVE 3 Perform operations on measurements.

When we write 4 inches, we mean four units that are 1 inch in length. Mathematically, we write this as 4 inches = 4 · (1 inch). This way of interpreting measurements makes it

Answers to Warm-Ups

H. 69,000 m **I.** 34.6 cm

easy to find multiples of measurements. Consider 3 boards, each 5 feet long. The total length of the boards is

$$\begin{aligned} 3 \cdot (5 \text{ feet}) &= 3 \cdot 5 \cdot (1 \text{ foot}) \\ &= 15 \cdot (1 \text{ foot}) \\ &= 15 \text{ feet} \end{aligned}$$

Similarly, a bolt of ribbon has 5 yd on it. If it is cut into ten equal pieces, how long is each piece?

$$\begin{aligned} \frac{5 \text{ yd}}{10} &= \frac{5 \cdot (1 \text{ yd})}{10} \\ &= \frac{5}{10} \cdot (1 \text{ yd}) \\ &= \frac{1}{2} \cdot (1 \text{ yd}) \\ &= \frac{1}{2} \text{ yd} \end{aligned}$$

To multiply or divide a measurement by a number

Multiply or divide the two numbers and write the unit of measure.

EXAMPLES J–K

DIRECTIONS: Solve.

STRATEGY: Describe each situation with a statement involving measurements and simplify.

J. What is the total height of three storage units stacked on one another if each unit is 60 cm tall?

STRATEGY: To find the total height, multiply the height of one unit by 3.

$$\begin{aligned} \text{Total height} &= 3 \cdot (60 \text{ cm}) \\ &= 180 \text{ cm} \end{aligned}$$

The total height is 180 cm.

K. A fence 24 ft long is to be constructed in four sections. How long is each section?

STRATEGY: To find the length of each section, divide the total length by 4.

$$\begin{aligned} \frac{24 \text{ ft}}{4} &= \frac{24 \cdot (1 \text{ ft})}{4} \\ &= 6 \text{ ft} \end{aligned}$$

So each section is 6 ft long.

WARM-UPS J–K

J. What is the total length of six copper pipes that are each 8 ft long?

K. An 8-km race is divided into five legs. How long is each leg?

The expression "You can't add apples and oranges" applies to adding and subtracting measurements. Only measurements with the same units of measure may be added or subtracted.

$$\begin{aligned} 10 \text{ mm} + 6 \text{ mm} &= (10 + 6) \text{ mm} \\ &= 16 \text{ mm} \end{aligned}$$

Answers to Warm-Ups

J. 48 ft **K.** 1.6 km

CAUTION

3 m + 4 cm $\neq$ 7 m

To add or subtract measurements

1. If the units of measure are unlike, convert to like measures.
2. Add or subtract the numbers and write the unit of measure.

WARM-UPS L–M

L. In a bookcase, the height of the bottom shelf is 15 in., the next shelf is 12 in., and the top shelf is 10 in. What is the total height of the bookcase?

M. One wall of a den measures 10 ft 4 in. The door is 2 ft 8 in. wide. What is the remaining width?

EXAMPLES L–M

DIRECTIONS: Solve.

STRATEGY: Describe each situation with a statement involving measurements and simplify.

L. Ben is 1.88 m tall, Nate is 1.79 m tall, and Tony is 1.72 m tall. What is the total height of the three boys?

STRATEGY: To find the total height, add the three heights.

$$\begin{aligned}\text{Total height} &= 1.88\text{ m} + 1.79\text{ m} + 1.72\text{ m}\\ &= (1.88 + 1.79 + 1.72)\text{ m}\\ &= 5.39\text{ m}\end{aligned}$$

So, the total height of the three boys is 5.39 m.

M. If a carpenter cuts a piece of board that is 2 ft 5 in. from a board that is 8 ft 3 in. long, how much board is left? (Disregard the width of the cut.)

STRATEGY: Subtract the length of the cut piece from the length of the board.

$$\begin{array}{r} 8\text{ ft } 3\text{ in.}\\ -2\text{ ft } 5\text{ in.}\\ \hline \text{remaining board}\end{array}$$

$$\begin{array}{r} 7\text{ ft } 1\text{ ft } 3\text{ in.}\\ -2\text{ ft } \quad 5\text{ in.}\\ \hline \end{array}$$ Borrow 1 ft from the 8 ft (1 ft = 12 in.)

$$\begin{array}{r} 7\text{ ft } 15\text{ in.}\\ -2\text{ ft } \ 5\text{ in.}\\ \hline 5\text{ ft } 10\text{ in.}\end{array}$$ 1 ft 3 in. = 15 in.
Subtract.

* There are 5 ft 10 in. of board remaining.

Answers to Warm-Ups

L. 37 in. **M.** 7 ft 8 in.

EXERCISES 7.1

OBJECTIVE 1 Recognize and use appropriate units of length from the English and metric measuring systems.

Give both an English unit and a metric unit to measure the following.

A.

1. The height of a skyscraper

2. The dimensions of a sheet of typing paper

3. The thickness of a fingernail

4. The length of a table

5. The width of a slat on a mini-blind

6. The distance from home to school

7. The height of a tree

8. The cruising altitude of a 747 jet

B.

9. The depth of the ruins of the Titanic

10. The thickness of a wire

11. The length of a shoelace

12. The width of a human hair

13. The height of the ceiling in a room

14. The distance from the moon to Earth

15. The length of a swimming pool

16. The height of Mt. Everest

OBJECTIVE 2 Convert units of length.

A. *Convert as indicated.*

17. 5 ft to inches

18. 27 cm to millimeters

19. 32 km to meters

20. 1 mile to inches

21. 14 yd to feet

22. 312 cm to meters

23. 82 m to kilometers

24. 123 ft to yards

25. 6 km to centimeters

26. 3 mi to feet

B. *Convert as indicated. Round to the nearest hundredth if necessary.*

27. 1000 ft to miles

28. 1000 ft to meters

29. 27 in. to feet

30. 32,000 cm to kilometers

31. 827 in. to centimeters

32. 50 mi to kilometers

33. 120 km to miles

34. 46 cm to inches

35. 46 cm to feet

36. 39 in. to meters

OBJECTIVE 3 Perform operations on measurements.

A. *Do the indicated operations.*

37. 6(13 ft)

38. 13 cm + 25 cm

39. 32 in. ÷ 8

40. 18 m − 13 m

41. 25(8 mm)

42. 36 mi ÷ 4

43. 3 ft + 4 ft + 9 ft

44. 85 cm − 30 cm

45. 100(26 mi)

46. 4 km + 6 km + 2 km

B. *Do the indicated operations.*

47. (5 ft 2 in.) + 27 in.

48. 7 m − 38 cm

49. 14 mi ÷ 8

50. 45 km + 27 m

51. 75(321 mm)

52. 645 cm ÷ 30

53.
  9 ft 3 in.
−2 ft 8 in.

54.
  8 yd 2 ft 7 in.
+3 yd 2 ft 10 in.

55. (544 mi)23

56. 12 yd − 13 in.

C.

57. (7 ft 3 in.) + (2 ft 9 in.) + (9 ft 7 in.)

58. (5 yd 1 ft 4 in.) − (3 yd 2 ft 7 in.)

59. (2 yd 2 ft 1 in.) · 3

60.
  3 yd 2 ft 5 in.
−2 yd 1 ft 9 in.

61. Martha wants to construct four picture frames that are each 14 inches square. How much molding should she buy?

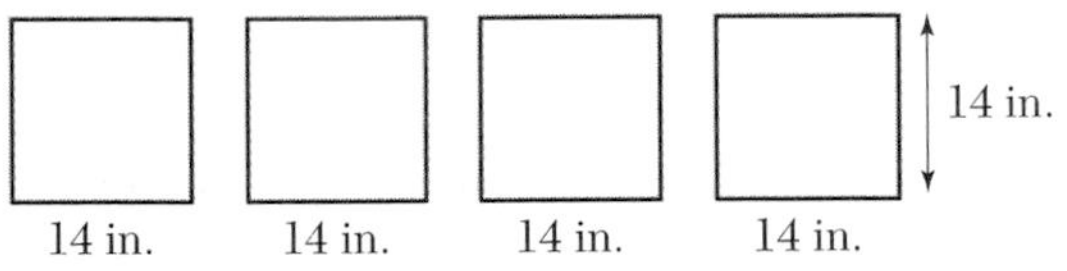

62. Enrique, who lives in Seattle, is going on vacation to visit several friends who live in Atlanta, Houston, and Los Angeles. The distance between Seattle and Atlanta is 2182 miles. The distance from Atlanta to Houston is 689 miles. The distance from Houston to Los Angeles is 1374 miles and the distance from Los Angeles and Seattle is 959 miles. How many miles does Enrique fly on his vacation?

63. Rosa buys 2 m of linen to make napkins. If she can make a napkin from each 30 cm of fabric, how many napkins can she make and how much fabric will be left over?

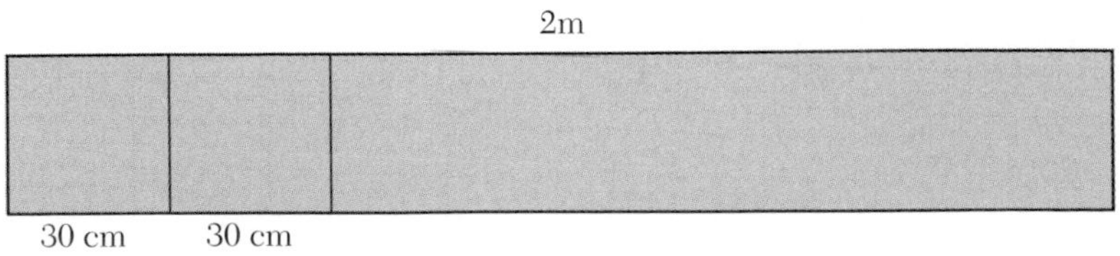

64. Estimate the length of your shoe in both inches and centimeters. Measure your shoe in both units.

65. Estimate the length of your middle finger in both in. and cm. Measure your middle finger in both units.

66. During one round of golf, Rick made birdie putts of 3 ft 8 in., 12 ft 10 in., 20 ft 8 in., and 7 ft 4 in. What was the total length of all the birdie putts?

67. The swimming pool at Tualatin Hills is 50 m long. How many meters of lane dividers should be purchased in order to separate the pool into 9 lanes?

68. A decorator is wallpapering. If each length of wallpaper is 7 ft 4 in., eight lengths are needed to cover a wall, and three walls are to be covered, how much total wallpaper is needed for the project?

Exercises 69–72. The table lists the longest rivers in the world.

River	Length (mi)
Nile (Africa)	4160
Amazon (S. America)	4000
Chang Jiang (Asia)	3964
Ob-Irtysh (Asia)	3362
Huang (Asia)	2903
Congo (Africa)	2900

69. Which, if any, of these figures appear to be estimates? Why?

70. What is the total length of the five longest rivers in the world?

71. The São Francisco River in S. America is the twentieth longest river in the world, with a length of 1988 miles. Write a sentence relating its length to that of the Amazon, using multiplication.

72. Write a sentence relating the lengths of the Nile River and the Congo River using addition or subtraction.

Exercises 73–75 refer to the chapter application.

73. What are the dimensions of Paul and Barbara's backyard?

74. How wide is the patio? How long is it?

75. How wide are the walkways?

STATE YOUR UNDERSTANDING

76. Give two examples of equivalent measures.

77. Explain how to add or subtract measures.

78. Explain how to multiply or divide a measure by a number.

GROUP WORK

79. Measure, as accurately as you can, at least five parts of a typical desk in the classroom. Give measurements in both the English system and the metric system. Compare your results with the other groups. Do you all agree? Give some possible reasons for the variations in measurements.

MAINTAIN YOUR SKILLS

Do the indicated operations.

80. 4.78(10,000)

81. 4.78 ÷ 100,000

82. $\frac{3}{7} + \frac{3}{4}$

83. $\frac{12}{25} \div \frac{28}{15}$

84. $5\frac{1}{3} - 2\frac{5}{6}$

85. $12\frac{9}{10} + 4\frac{5}{8}$

86. $(3.63)^2$

87. (4.8)(5.2)

88. Find the average (mean) of 54, 78, 112, 162.

89. Find the median of 54, 78, 112, 162.

7.2 Measuring Capacity, Weight, and Temperature

VOCABULARY

Units of **capacity** measure liquid quantities.

Units of **weight** measure heaviness.

OBJECTIVES

1. Convert and perform operations on units of capacity.
2. Convert and perform operations on units of weight.
3. Convert units of temperature and time.

HOW & WHY

OBJECTIVE 1 Convert and perform operations on units of capacity.

Measures of capacity answer the question "How much liquid?" We use teaspoons to measure vanilla for a recipe, buy soda in 2-liter bottles, and measure gasoline in gallons or liters. As is true for length, there are units of capacity in both the English and metric systems. The basic unit of capacity in the metric system is the liter (L), which is slightly more than 1 quart. Table 7.4 gives the most commonly used measures of capacity in each system, and Table 7.5 gives some equivalencies between the systems.

TABLE 7.4 MEASURES OF CAPACITY

English	Metric
3 teaspoons (tsp) = 1 tablespoon (Tbsp)	1000 milliliters (mL) = 1 liter (L)
2 cups (c) = 1 pint (pt)	1000 liters (L) = 1 kiloliter (kL)
2 pints (pt) = 1 quart (qt)	
4 quarts (qt) = 1 gallon (gal)	

TABLE 7.5 MEASURES OF CAPACITY

English–Metric	Metric–English
1 teaspoon ≈ 4.929 milliliters	1 milliliter ≈ 0.203 teaspoon
1 quart ≈ 0.946 liter	1 liter ≈ 1.057 quarts
1 gallon ≈ 3.785 liters	1 liter ≈ 0.264 gallon

To convert units of capacity, we use the same technique of unit fractions that we used for units of length. See Section 7.1.

CAUTION

The accuracy of conversions between systems depends on the accuracy of the conversion factors.

To convert units of capacity

1. Multiply by the unit fraction that has the desired units in the numerator and the original units in the denominator.
2. Simplify.

WARM-UPS A–B

A. Convert 33 qt to gallons.

B. Convert 4827 mL to quarts.

EXAMPLES A–B

DIRECTIONS: Convert units of measure. Round to the nearest thousandth.

STRATEGY: Multiply by the unit fraction with the desired units in the numerator and the original units in the denominator. Simplify.

A. Convert 259 mL to liters.

$$259 \text{ mL} = \frac{259 \cancel{\text{mL}}}{1} \cdot \frac{1 \text{ L}}{1000 \cancel{\text{mL}}}$$ Use appropriate unit fraction.

$$= \frac{259 \text{ L}}{1000}$$ Simplify.

$$= 0.259 \text{ L}$$

So, 259 mL = 0.259 L.

ALTERNATE SOLUTION: Because this is a metric-to-metric conversion, we can use a chart as a shortcut.

kL	hL	daL	L	dL	cL	mL

To change from milliliters to liters, we move the decimal point three places left. Therefore, 259 mL = 0.259 L.

B. Convert 27 pt to liters

$$27 \text{ pt} \approx \frac{27 \cancel{\text{pt}}}{1} \cdot \frac{1 \cancel{\text{qt}}}{2 \cancel{\text{pt}}} \cdot \frac{0.946 \text{ L}}{1 \cancel{\text{qt}}}$$ Use appropriate unit fractions.

$$\approx \frac{27(0.946)}{2} \text{ L}$$ Simplify.

$$\approx 12.771 \text{ L}$$

* So, 27 pt ≈ 12.771 L.

Operations with units of capacity are performed using the same procedures as those for units of length. See Section 7.1.

To multiply or divide a measurement by a number

Multiply or divide the two numbers and write the unit of measure.

To add or subtract measurements

1. If the units of measure are unlike, convert to like measures.
2. Add or subtract the numbers and write the unit of measure.

Answers to Warm-Ups

A. 8.25 gal **B.** 5.102 qt

WARM-UPS C–D

EXAMPLES C–D

DIRECTIONS: Solve.

STRATEGY: Describe each situation with a statement involving measurements and simplify.

C. A chemistry instructor has 450 mL of an acid solution that he needs to divide into equal amounts for four lab groups. How much acid does each group receive?

$$\begin{aligned}\text{Acid for one group} &= \frac{450\text{ mL}}{4} \quad \text{Divide.}\\ &= \frac{450}{4}\text{ mL}\\ &= 112.5\text{ mL}\end{aligned}$$

Each group receives 112.5 mL of acid solution.

D. Clayton has 5 gal 3 qt of motor oil to be recycled. His friend Ernie adds another 7 qt of oil. How much oil is being recycled?

$$\begin{aligned}\text{Recycled oil} &= (5\text{ gal }3\text{ qt}) + 7\text{ qt}\\ &= 5\text{ gal} + 3\text{ qt} + 7\text{ qt}\\ &= 5\text{ gal} + 10\text{ qt}\\ &= 5\text{ gal} + 2\text{ gal} + 2\text{ qt}\\ &= 7\text{ gal }2\text{ qt}\end{aligned}$$

The total amount of oil to be recycled is 7 gal 2 qt.

C. What is the total capacity of five vases that each hold 3.5 c of water?

D. Cynthia made 4 pt 1 c of strawberry jam and Bonnie made 7 c of raspberry jam. How much jam did the two women make?

*

HOW & WHY

OBJECTIVE 2 Convert and perform operations on units of weight.

Units of weight answer the question "How heavy is it?" In the English system, we use ounces to measure the weight of baked beans in a can, pounds to measure the weight of a person, and tons to measure the weight of a ship. In the metric system, the basic unit for measuring heaviness is the gram, which weighs about as much as a small paper clip. We use grams to measure the weight of cans of baked beans, kilograms to measure the weight of a person, and metric tons to measure the weight of ships. Very small weights like doses of vitamins are measured in milligrams. Table 7.6 lists the most commonly used measures of weight in each system, and Table 7.7 shows some equivalencies between the systems.

TABLE 7.6 MEASURES OF WEIGHT

English system	Metric system
16 ounces (oz) = 1 pound (lb)	1000 milligrams (mg) = 1 gram (g)
2000 pounds (lb) = 1 ton	1000 grams (g) = 1 kilogram (kg)
	1000 kilograms = 1 metric ton

TABLE 7.7 MEASURES OF WEIGHT BETWEEN SYSTEMS

English–Metric	Metric–English
1 ounce ≈ 28.350 grams	1 gram ≈ 0.035 ounce
1 pound ≈ 453.592 grams	1 kilogram ≈ 2.205 pounds
1 pound ≈ 0.454 kilogram	

(*Note*: The weight of an object is different from its mass. An object has the same mass everywhere in space. Weight depends on gravity, so an object has different weights on Earth and on the moon. On Earth, gravity is approximately uniform and so weight and mass are often used interchangeably. Technically, a gram is a unit of mass.)

Answers to Warm-Ups
C. 17.5 c **D.** 8 pt

WARM-UPS E–F

E. Convert 200 g to ounces.

F. If 140 oz of peanut brittle is divided equally among five sacks, how much goes into each sack?

EXAMPLES E–F

DIRECTIONS: Solve as indicated. Round to the nearest hundredth.

STRATEGY: Follow the strategies given for each example.

E. Convert 73 kg to pounds.

STRATEGY: Multiply by the unit fraction with pounds in the numerator and kilograms in the denominator and simplify.

$$73 \text{ kg} \approx \frac{73 \not{\text{kg}}}{1} \cdot \frac{2.205 \text{ lb}}{1 \not{\text{kg}}}$$ Use appropriate unit fraction.

$$\approx 160.97 \text{ lb}$$ Simplify.

So, 73 kg ≈ 160.97 lb.

F. A can of peaches weighs 2 lb 5 oz and a can of pears weighs 1 lb 14 oz. What is the total weight of the two cans?

STRATEGY: Add the weight of the two cans and simplify.

$$\begin{array}{r} 2 \text{ lb} \quad 5 \text{ oz} \\ +1 \text{ lb} \ 14 \text{ oz} \\ \hline 3 \text{ lb} \ 19 \text{ oz} \end{array} = 3 \text{ lb} + 1 \text{ lb } 3 \text{ oz} = 4 \text{ lb } 3 \text{ oz}$$

The cans weigh 4 lb 3 oz.

Answers to Warm-Ups

E. 7.00 oz **F.** 28 oz

HOW & WHY

OBJECTIVE 3 Convert units of temperature and time.

Temperature is measured in degrees. There are two major scales for measuring temperature. The English system uses the Fahrenheit scale, which sets the freezing point of water at 32°F and the boiling point of water at 212°F. The Fahrenheit scale was developed by a physicist named Gabriel Daniel Fahrenheit in the early 1700s.

The metric system uses the Celsius scale, which sets the freezing point of water at 0°C and the boiling point of water at 100°C. The Celsius scale is named after Swedish astronomer Anders Celsius, who lived in the early 1700s and invented a thermometer using the Celsius scale.

Converting between scales is often done using special conversion formulas.

To convert from Fahrenheit to Celsius

Use the formula: $C = \frac{5}{9} \cdot (F - 32)$

or

Subtract 32 from the Fahrenheit temperature and multiply by $\frac{5}{9}$.

WARM-UP G

EXAMPLE G

DIRECTIONS: Convert as indicated.

STRATEGY: Use the conversion formula.

G. What is 68°F on the Celsius scale?

$$C = \frac{5}{9} \cdot (F - 32) \quad \text{Formula}$$
$$= \frac{5}{9} \cdot (68 - 32) \quad \text{Substitute 68 for } F.$$
$$= \frac{5}{9} \cdot 36 \quad \text{Simplify.}$$
$$= 20$$

So, 68°F = 20°C.

G. What is 50°F on the Celsius scale?

To convert from Celsius to Fahrenheit

Use the formula: $F = \frac{9}{5} \cdot C + 32$

or

Multiply the Celsius temperature by $\frac{9}{5}$ and add 32.

Answer to Warm-Up

G. 10°C

EXAMPLE H

DIRECTIONS: Convert as indicated.

STRATEGY: Use the conversion formula.

H. Convert 50°C to degrees Fahrenheit.

$$F = \frac{9}{5} \cdot C + 32 \quad \text{Formula}$$
$$= \frac{9}{5} \cdot 50 + 32 \quad \text{Substitute 50 for C.}$$
$$= 90 + 32 \quad \text{Simplify.}$$
$$= 122$$

So, 50°C = 122°F.

WARM-UP H

H. Convert 30°C to degrees Fahrenheit.

The same units of time are used in both the English and metric systems. We are all familiar with seconds, minutes, hours, days, weeks, and years. The computer age has contributed some new units of time that are very short, such as milliseconds $\left(\frac{1}{1000} \text{ of a second}\right)$ and nanoseconds $\left(\frac{1}{1{,}000{,}000{,}000} \text{ of a second}\right)$. Table 7.8 gives the commonly used time conversions.

TABLE 7.8 TIME CONVERSIONS

60 seconds (sec) = 1 minute (min)	7 days = 1 week
60 minutes (min) = 1 hour (hr)	365 days ≈ 1 year*
24 hours (hr) = 1 day	

*Technically, a solar year (the number of days it takes Earth to make one complete revolution around the sun) is 365.2422 days. Our calendars account for this by having a leap day once every 4 years.

Answer to Warm-Up

H. 86°F

WARM-UPS I–J

I. The three members of a Marine training team had individual times for an obstacle course of: 5 min 37 sec, 4 min 56 sec, 5 min 48 sec. What was the total time for the team?

J. Bacteria in a lab have been growing for 12 days. How many seconds is this?

EXAMPLES I–J

DIRECTIONS: Solve.

STRATEGY: Follow the strategies given for each example.

I. The four members of an 800-m freestyle relay team had individual times of: 1 min 54 sec, 1 min 59 sec, 2 min 8 sec, 1 min 49 sec. What was the total time for the relay team?

STRATEGY: Add the times together.

$$\begin{array}{r} 1\text{ min } 54\text{ sec} \\ 1\text{ min } 59\text{ sec} \\ 2\text{ min } 8\text{ sec} \\ +1\text{ min } 49\text{ sec} \\ \hline 5\text{ min } 170\text{ sec} \end{array}$$

Add the times.

$$5\text{ min } 170\text{ sec} = 5\text{ min} + 2\text{ min } 50\text{ sec} = 7\text{ min } 50\text{ sec}$$

Convert seconds to minutes.

The relay team's time was 7 min 50 sec.

J. Jerry is 8 years old. How many minutes old is he?

STRATEGY: Use the appropriate unit fractions.

$$8\text{ yr} = \frac{8\text{ }\cancel{\text{yr}}}{1} \cdot \frac{365\text{ }\cancel{\text{day}}}{1\text{ }\cancel{\text{yr}}} \cdot \frac{24\text{ }\cancel{\text{hr}}}{1\text{ }\cancel{\text{day}}} \cdot \frac{60\text{ min}}{1\text{ }\cancel{\text{hr}}} = 4{,}204{,}800\text{ min}$$

Use appropriate unit fraction.

Note: Jerry has lived through two leap years, which are not accounted for in the conversion. Therefore, we must add two days to the total.

$$2\text{ day} = \frac{2\text{ }\cancel{\text{day}}}{1} \cdot \frac{24\text{ }\cancel{\text{hr}}}{1\text{ }\cancel{\text{day}}} \cdot \frac{60\text{ min}}{1\text{ }\cancel{\text{hr}}} = 2880\text{ min}$$

* Jerry has been alive 4,204,800 min + 2880 min = 4,207,680 min.

Answers to Warm-Ups

I. The team's time was 16 min 21 sec.

J. The bacteria have been growing for 1,036,800 sec.

Name ______________ Class ______________ Date ______________

EXERCISES 7.2

OBJECTIVE 1 Convert and perform operations on units of capacity.

Do the indicated operations. Round decimal answers to the nearest hundredth.

A.

1. 15 qt + 20 qt

2. (4 c) · 5

3. (400 mL) ÷ 50

4. 60 L − 14 L

5. (80 gal) ÷ 20

6. 5 kL + 9 kL + 10 kL

7. Convert 6 qt to cups.

8. Convert 4 L to milliliters.

9. Convert 18 qt to gallons.

10. Convert 500 mL to liters.

B.

11. 23 · (18 mL)

12. 380 kL − 175 kL

13. 13 c + 46 c

14. (8498 gal) ÷ 14

15. 35 qt − 27 qt − 8 qt

16. 38 mL + 11 mL + 22 mL + 5 mL

17. Convert 7 gal to cups.

18. Convert 2.6 kL to milliliters.

19. Convert 35 gal to liters.

20. Convert 585 mL to cups.

OBJECTIVE 2 Convert and perform operations on units of weight.

Do the indicated operations. Round decimal answers to the nearest hundredth.

A.

21. 32 g − 12 g

22. 3 · (17 oz)

23. 7 lb + 13 lb

24. (66 mg) ÷ 6

25. (15 kg)(3)

26. 3 oz + 5 oz + 7 oz

27. Convert 8 kg to grams.

28. Convert 3 lb to ounces.

29. Convert 7000 mg to grams.

30. Convert 80 oz to pounds.

B.

31. (2912 lb) ÷ 14

32. 170 kg − 89 kg

33. 129 oz + 4599 oz

34. 16(24 mg)

35. 34 g + 674 g − 58 g

36. (5 lb 13 oz) · 3

37. Convert 36 mg to grams.

38. Convert 14 oz to pounds.

39. Convert 75 g to ounces.

40. Convert 140 lb to kilograms.

 Name ______________________ Class ______________________ Date ____________

OBJECTIVE 3 Convert units of temperature and time.

Convert as indicated. Round to the nearest tenth.

A.

41. 32°F to degrees Celsius

42. 45°C to degrees Fahrenheit

43. 41°F to degrees Celsius

44. 4 min to seconds

45. 3 days to hours

46. 15°C to degrees Fahrenheit

47. 35°C to degrees Fahrenheit

48. 104°F to degrees Celsius

49. 10 weeks to days

50. 5 hr to minutes

B.

51. 100°F to degrees Celsius

52. 16 sec to minutes

53. 48°F to degrees Celsius

54. 8°C to degrees Fahrenheit

55. 31°C to degrees Fahrenheit

56. 350°F to degrees Celsius

57. 2 yr to hours

58. 17°C to degrees Fahrenheit

59. 15 min to days

60. 365 days to weeks

C.

61. (7 lb 8 oz) + (2 lb 13 oz) + (11 lb 1 oz)

62. (6 gal 2 qt) + (3 gal 3 qt) + (7 gal 2 qt)

63. (3 days 20 hr 48 min) + (4 days 18 hr 52 min)

64.
$$\begin{array}{r} 2 \text{ gal } 3 \text{ qt } 1 \text{ pt} \\ +4 \text{ gal } 2 \text{ qt } 1 \text{ pt} \\ \hline \end{array}$$

65.
$$\begin{array}{r} 17 \text{ lb } \ \ 3 \text{ oz} \\ -12 \text{ lb } 14 \text{ oz} \\ \hline \end{array}$$

66. The Corner Grocery sold 20 lb 6 oz of hamburger on Wednesday, 13 lb 8 oz on Thursday, and 21 lb 9 oz on Friday. How much hamburger was sold during the three days?

67. Normal body temperature is considered to be 98.6°F. What is normal body temperature on the Celsius scale?

68. Lewis, who is a lab assistant, has 312 mL of acid that is to be divided equally among 24 students. How many milliliters will each student receive?

69. A doctor prescribes allergy medication of two tablets, 20 mg each, to be taken three times per day for a full week. How many milligrams of medication will the patient get in a week?

70. A bank time–temperature sign is stuck on 28°C. What is the Fahrenheit reading (round to the nearest whole degree)?

71. A 400-m relay team has splits of 48.3 sec, 53.8 sec, 50.2 sec, and 47.9 sec. What is the total time for the team in minutes and seconds?

72. Mikal combines two cans of soup for lunch. The chicken vegetable soup is 1 lb 10 oz and the chicken noodle is 14 oz. How much soup does Mikal have?

73. Denika has a 2-liter bottle of 7-Up that she needs to divide evenly among 10 children at a birthday party. How much 7-Up does each child get?

74. The following table gives the daily high temperature for Ocala, Florida, for 1 week in December. Find the average daily high temperature for the week.

Sunday	Monday	Tuesday	Wednesday	Thursday	Friday	Saturday
62°F	54°F	57°F	64°F	68°F	69°F	67°F

75. If a bag contains 298 g of potato chips, how many grams are contained in 7 bags?

76. An elevator has a maximum capacity of 2500 lb. A singing group of eight men and eight women get on. The average weight of the women is 125 lb, and the average weight of the men is 190 lb. Can they ride safely together?

77. The weight classes for Olympic wrestling (both freestyle and Greco-Roman) are given in the table. Fill in the equivalent pound measures, rounded to the nearest whole pound.

Olympic Wrestling Weight Classes

Kilograms	54	58	63	69	76	85	97	130
Pounds								

78. In the shipping industry, most weights are measured in long tons, which are defined as 2240 lb. The largest tanker in the world is the Jahre Viking, which weighs 662,420 (long) tons fully loaded. How much does the Jahre Viking weigh in pounds?

79. How much does the Jahre Viking weigh in kilograms? (See Exercise 78.) Round to the nearest million.

80. A rule of thumb for a person's daily protein requirement is that the number of grams of protein needed is the same as half the person's weight in pounds. Steve weighs 71 kg. How much protein should he eat per day?

81. One serving of Life cereal has 3 g of protein, and putting $\frac{1}{2}$ c of 1% milk on it adds 4.5 g of protein. If Steve eats only milk and cereal for a day, how many servings does he need in order to get sufficient protein? (See Exercise 80.)

82. Scientists give the average surface temperature of Earth as 59°F. They estimate that the temperature increases about 1°F for every 200-ft drop in depth below the surface. What is the temperature 1 mile below the surface?

STATE YOUR UNDERSTANDING

83. Can you add 4 g to 5 in.? Explain how or explain why you can't do it.

84. If 8 in. + 10 in. = 1 ft 6 in., why isn't it true that 8 oz + 10 oz = 1 lb 6 oz?

GROUP WORK

Use the information in the table to answer Exercises 85–87. Compare your answers with those of another group. Precious metals and gems are measured in troy weight according to the following:

1 pennyweight (dwt) = 24 grains
1 ounce troy (oz t) = 20 pennyweights
1 pound troy (lb t) = 12 ounces troy

85. How many grains are in 1 ounce troy? How many grains are in 1 pound troy?

86. Suppose you have a silver bracelet that you want a jeweler to melt down and combine with the silver of two old rings to create a medallion that weighs 5 oz t 14 dwt. The bracelet weighs 3 oz t 18 dwt and one ring weighs 1 oz t 4 dwt. What does the second ring weigh?

87. Kayla has an ingot of platinum that weighs 2 lb t 8 oz t 15 dwt. She wants to divide it equally among her six grandchildren. How much will each piece weigh?

 Name ____________ Class ____________ Date ____________

MAINTAIN YOUR SKILLS

Do the indicated operations.

88. $54 + 20 + 97$

89. $451 + 88 + 309$

90. $3.7 + 12.99$

91. $34.7 + 6.8 + 0.44$

92. $4\frac{3}{8} + 2\frac{7}{8}$

93. $12\frac{3}{5} + 32\frac{3}{4}$

94. $\frac{4}{9} + \frac{3}{8} + \frac{11}{12}$

95. $5.2 + 148 + \frac{1}{4}$

96. Convert 4 yd 2 ft 11 in. into inches.

97. Convert 4.82 km into meters.

7.3 Perimeter

VOCABULARY

A **polygon** is any closed figure whose sides are line segments.

Polygons are named according to the number of sides they have. Table 7.9 lists some common polygons.

Quadrilaterals are polygons with four sides. Table 7.10 lists the characteristics of common quadrilaterals.

The **perimeter** of a polygon is the distance around the outside of the polygon.

The **circumference** of a circle is the distance around the circle.

The **radius** of a circle is the distance from the center to any point on the circle.

The **diameter** of a circle is twice the radius.

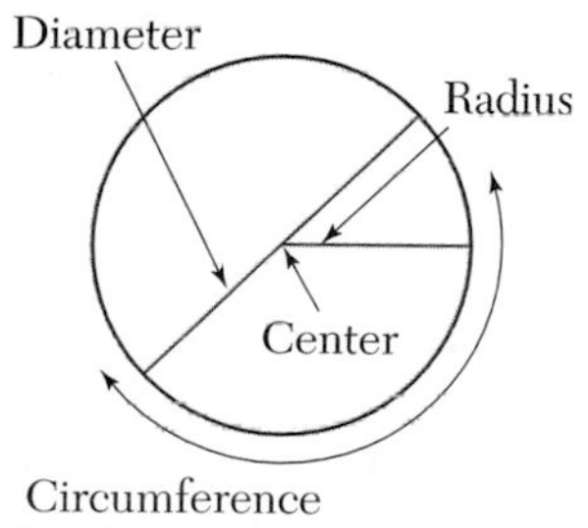

OBJECTIVES

1. Find the perimeter of a polygon.
2. Find the circumference of a circle.

TABLE 7.9 COMMON POLYGONS

Number of sides	Name	Picture
3	Triangle ABC	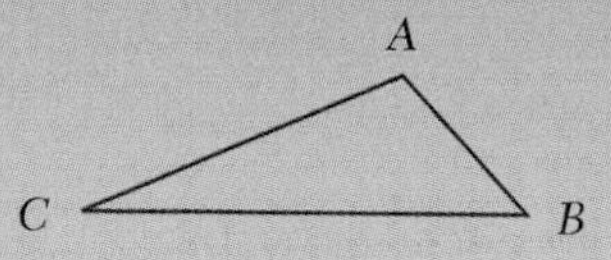
4	Quadrilateral ABCD	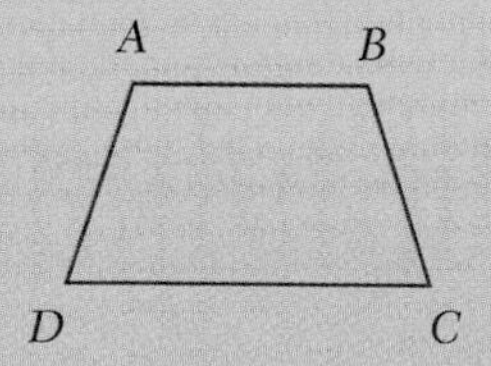
5	Pentagon ABCDE	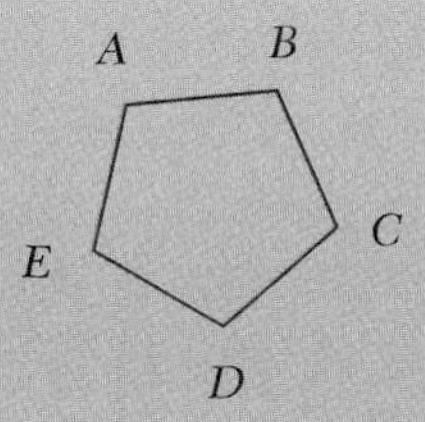
6	Hexagon ABCDEF	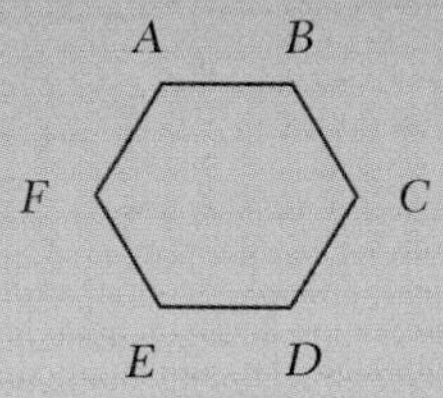
8	Octagon ABCDEFGH	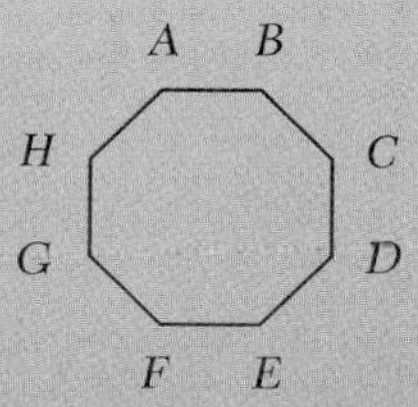

TABLE 7.10 COMMON QUADRILATERALS

Trapezoid	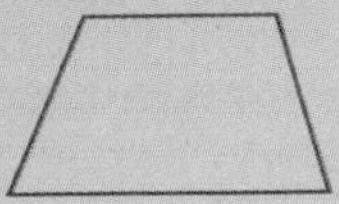	One pair of parallel sides
Parallelogram		Two pairs of equal parallel sides
Rectangle		A parallelogram with four right angles
Square		A rectangle with all sides equal

HOW & WHY

OBJECTIVE 1 Find the perimeter of a polygon.

The perimeter of a figure can be thought of in terms of the distance traveled by walking around the outside of it or by the length of a fence around the figure. The units of measure used for perimeters are length measures such as inches, feet, meters. Perimeter is calculated by adding the length of all the individual sides.

For example, to calculate the perimeter of this figure, we add the lengths of the sides.

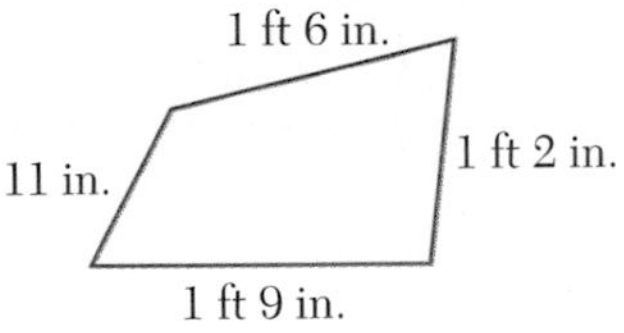

$$\begin{array}{r} 1\text{ ft } \ \ 6\text{ in.} \\ 1\text{ ft } \ \ 2\text{ in.} \\ 1\text{ ft } \ \ 9\text{ in.} \\ +\qquad 11\text{ in.} \\ \hline 3\text{ ft } 28\text{ in.} \end{array}$$

$$\begin{aligned} 3\text{ ft } 28\text{ in.} &= 3\text{ ft} + 2\text{ ft } 4\text{ in.} \qquad 28\text{ in.} = 2\text{ ft } 4\text{ in.} \\ &= 5\text{ ft } 4\text{ in.} \end{aligned}$$

The perimeter is 5 ft 4 in.

To find the perimeter of a polygon

Add the lengths of the sides.

To find the perimeter of a square

Multiply the length of one side by 4.

$P = 4s$

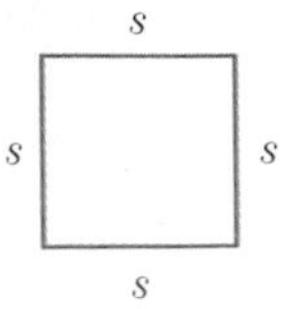

To find the perimeter of a rectangle

Add twice the length and twice the width.

$P = 2\ell + 2w$

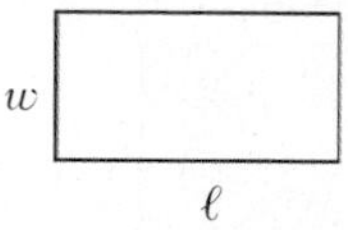

EXAMPLES A–D

DIRECTIONS: Find the perimeters of the given polygons.

STRATEGY: Add the lengths of the sides.

A. Find the perimeter of the triangle.

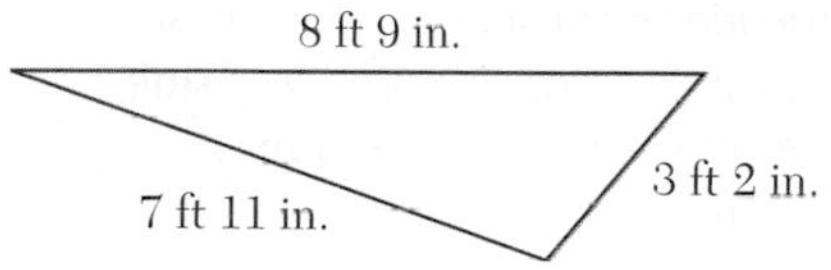

$$\begin{array}{r} 8 \text{ ft} \;\; 9 \text{ in.} \\ 7 \text{ ft} \; 11 \text{ in.} \\ + \; 3 \text{ ft} \;\; 2 \text{ in.} \\ \hline 18 \text{ ft} \; 22 \text{ in.} \end{array} = 19 \text{ ft } 10 \text{ in.}$$

Add the lengths.

22 in. = 1 ft 10 in.

The perimeter is 19 ft 10 in.

B. Find the perimeter of the rectangle.

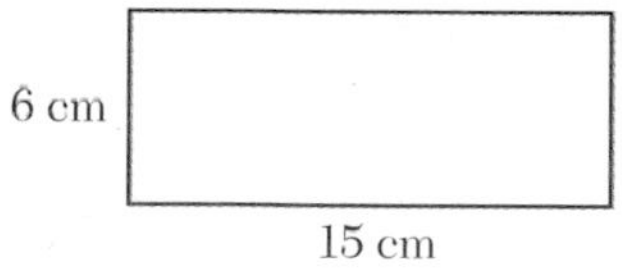

$P = 2\ell + 2w$	Perimeter formula for rectangles
$= 2(15 \text{ cm}) + 2(6 \text{ cm})$	Substitute.
$= 30 \text{ cm} + 12 \text{ cm}$	Multiply.
$= 42 \text{ cm}$	Add.

The perimeter of the rectangle is 42 cm.

WARM-UPS A–D

A. Find the perimeter of the trapezoid.

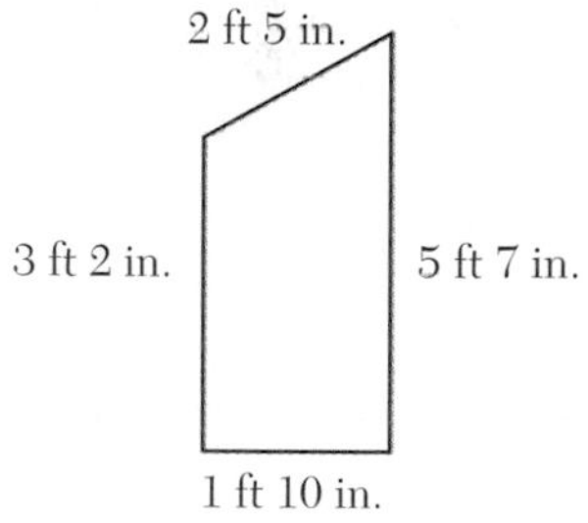

B. Find the perimeter of the square.

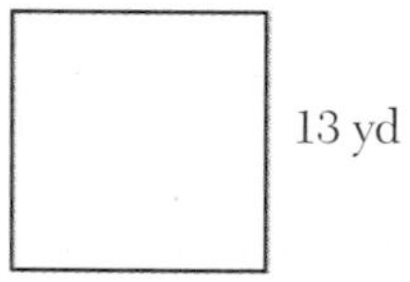

Answers to Warm-Ups

A. 13 ft **B.** 52 yd

C. Find the perimeter of the polygon.

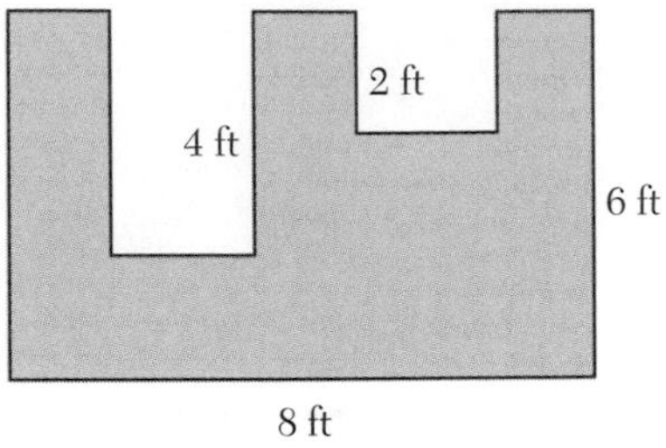

C. Find the perimeter of the polygon.

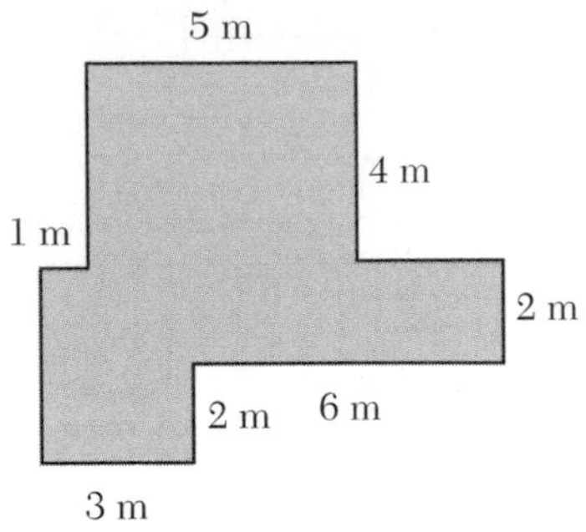

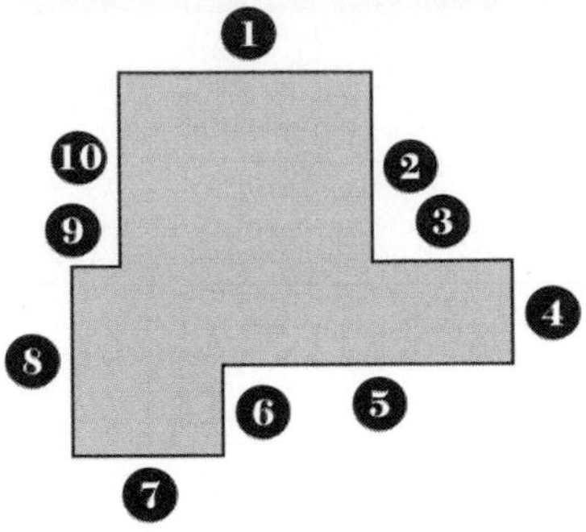

Side	Length
1	5 m
2	4 m
3	?
4	2 m
5	6 m
6	2 m
7	3 m
8	?
9	1 m
10	?
3	3m
8 + 10	8 m

To find the perimeter, number the sides (there are 10) and write down their lengths.

To find the length of side 3, notice that

side 3 = side 5 + side 7 − side 9 − side 1
= 6 m + 3 m − 1 m − 5 m
= 3 m

Sides 8 and 10 are not given, but

side 8 + side 10 = side 2 + side 4 + side 6
= 4 m + 2 m + 2 m
= 8 m

So,

$P = 5\text{ m} + 4\text{ m} + 3\text{ m} + 2\text{ m} + 6\text{ m} + 2\text{ m} + 3\text{ m} + 1\text{ m} + 8\text{ m}$
$= 34\text{ m}$

The perimeter of the polygon is 34 m.

D. How much baseboard lumber is needed for the room pictured?

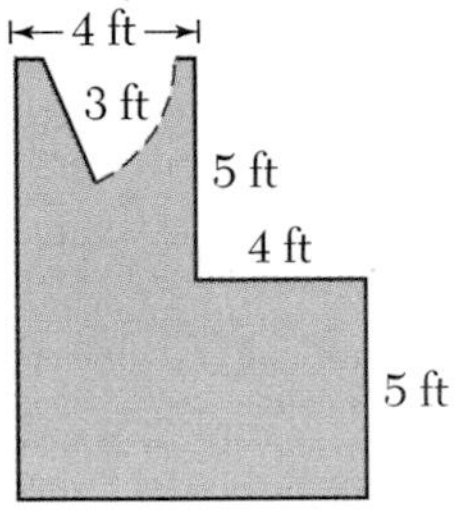

D. A carpenter is replacing the baseboards in a room. The floor of the room is pictured. How many feet of baseboard are needed?

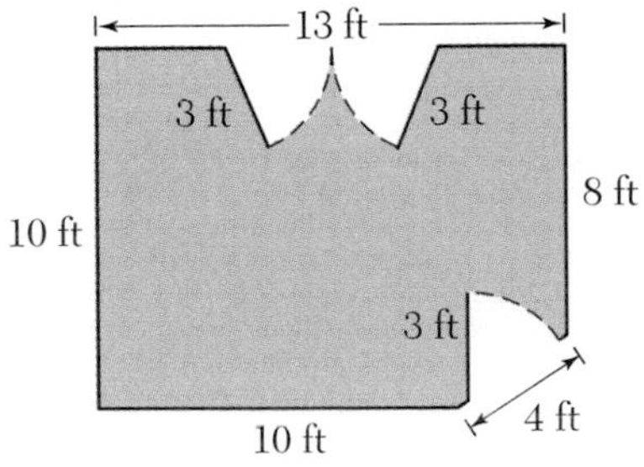

$P = 13\text{ ft} + 10\text{ ft} + 10\text{ ft} + 4\text{ ft} + 8\text{ ft}$ — Find the perimeter of the room, including all the doors.
$P = 45\text{ ft}$
Baseboard $= 45\text{ ft} - (6\text{ ft} + 3\text{ ft})$ — Subtract the combined width of the doors.
$= 45\text{ ft} - 9\text{ ft}$
$= 36\text{ ft}$ — Simplify.

The carpenter needs 36 ft of baseboard.

Answers to Warm-Ups

C. 40 ft **D.** The carpenter needs 33 ft of baseboard.

HOW & WHY

OBJECTIVE 2 Find the circumference of a circle.

Formulas for geometric figures that involve circles contain the number called pi (π). The number π is the quotient of the circumference of (distance around) the circle and its diameter. This number is the same for every circle no matter how large or small. This remarkable fact was discovered over a long period of time, and during that time a large number of approximations have been used. Today, the most commonly used approximations are

$$\pi \approx 3.14 \qquad \text{or} \qquad \pi \approx \frac{22}{7}$$

Calculator note: Scientific and graphing calculators have a $\boxed{\pi}$ key that generates a decimal value for π with 8, 10, or 12 decimal places, depending on the calculator. Using the $\boxed{\pi}$ key instead of one of the approximations increases the accuracy of your calculations.

To find the circumference of a circle

If C is the circumference, d is the diameter, and r is the radius of a circle, then the circumference is the product of π and the diameter or the product of π and twice the radius.

$$C = \pi d \qquad \text{or} \qquad C = 2\pi r$$

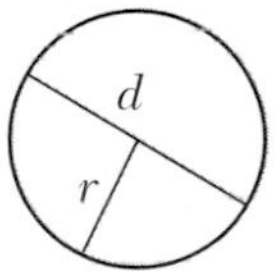

Because the radius, diameter, and circumference of a circle are all lengths, they are measured in units of length.

EXAMPLES E–F

DIRECTIONS: Find the circumference of the circle.

STRATEGY: Use the formula $C = \pi d$ or $C = 2\pi r$ and substitute.

E. Find the circumference of the circle.

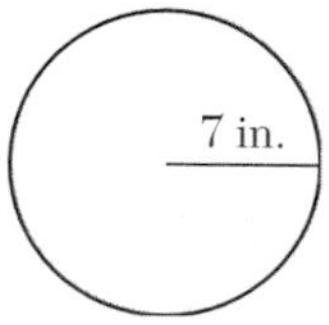

$C = 2\pi r$	Formula for circumference
$C \approx 2(3.14)(7 \text{ in.})$	Substitute. Because we are using an approximation for the value of π, the value of the circumference is also an approximation. We show this using $\approx$.
$C \approx 43.96$	Simplify.

The circumference is about 43.96 in.

WARM-UPS E–F

E. Find the circumference of the circle. Let $\pi \approx 3.14$.

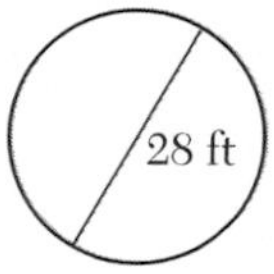

Answers to Warm-Ups

E. 87.92 ft

ALTERNATIVE SOLUTION: Use a calculator and the $\boxed{\pi}$ key.

$C = 2\pi r$	Formula for circumference
$C = 2\pi(7 \text{ in.})$	Substitute. Use the $\boxed{\pi}$ key.
$C \approx 43.9822971503 \text{ in.}$	
$C \approx 43.98 \text{ in.}$	Round.

The circumference is about 43.98 in. Because a more accurate approximation for π was used, we conclude that this is a more accurate approximation for the circumference than the first calculation.

F. How much duct tape is needed to tape around an irrigation pipe that has a radius of $1\frac{1}{2}$ inches? Use the $\boxed{\pi}$ key, and round to the nearest tenth.

F. Chia-Ching wants to install a border of tile around her new in-ground spa. The spa is circular, with a diameter of 4.25 meters. What is the circumference of the spa?

$C = \pi d$	Formula for circumference
$C = \pi(4.25 \text{ m})$	Substitute. Use the $\boxed{\pi}$ key.
$C \approx 13.3517687778 \text{ m}$	
$C \approx 13.35 \text{ m}$	Round.

The circumference of the spa is about 13.35 m.

Answers to Warm-Ups

F. About 9.4 inches of duct tape are needed.

EXERCISES 7.3

OBJECTIVE 1 Find the perimeter of a polygon.

A.

Find the perimeter of the following polygons.

1.

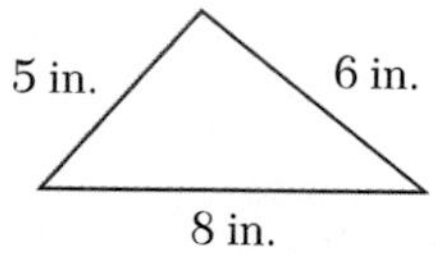

2.

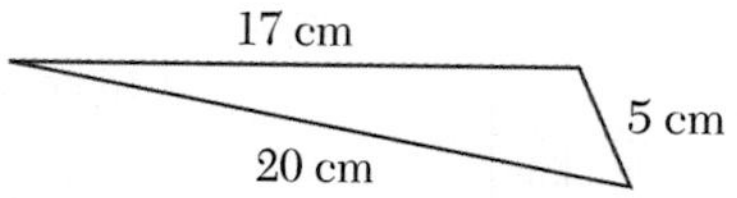

3.

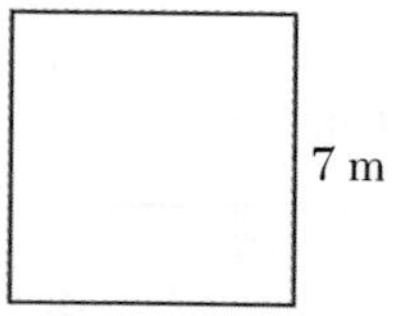

4.

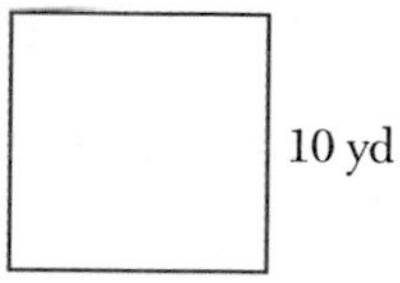

5.

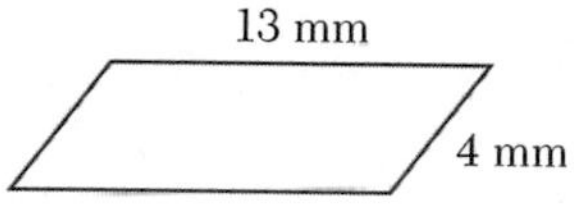

6.

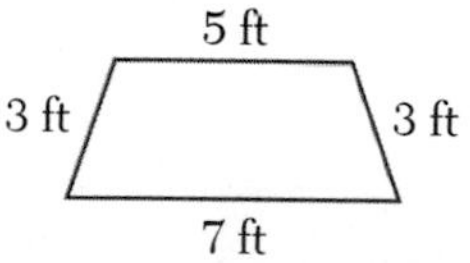

7.

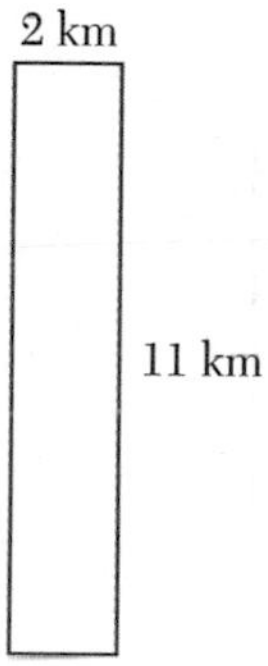

8.

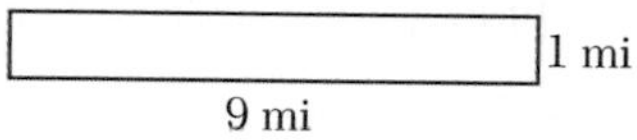

B.

9. Find the perimeter of a triangle with sides 16 mm, 27 mm, and 40 mm.

10. Find the perimeter of a square with sides 230 ft.

11. Find the distance around a rectangular field with length 50 m and width 35 m.

12. Find the distance around a rectangular swimming pool with width 20 yd and length 25 yd.

Find the perimeter of the following polygons.

13.

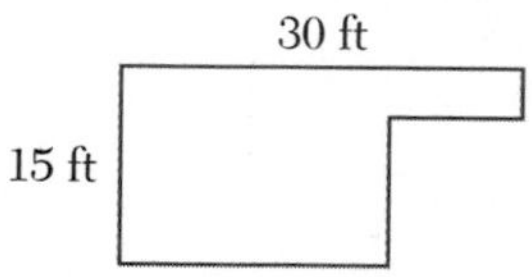

14.

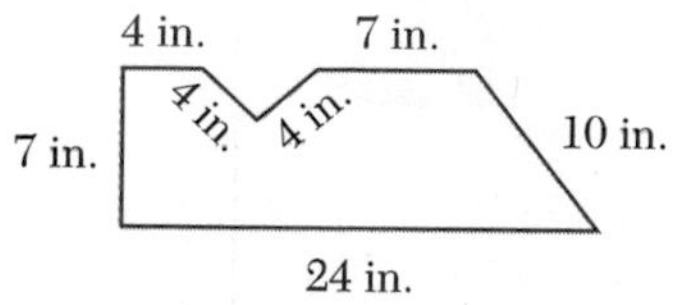

15.

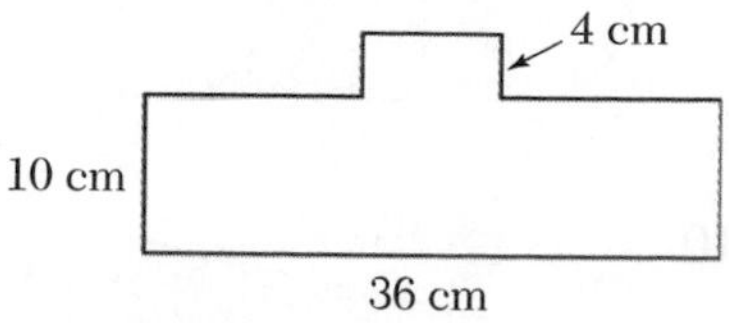

16.

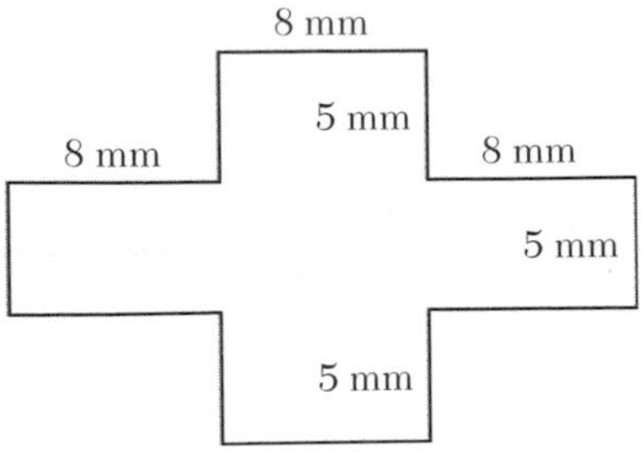

17.

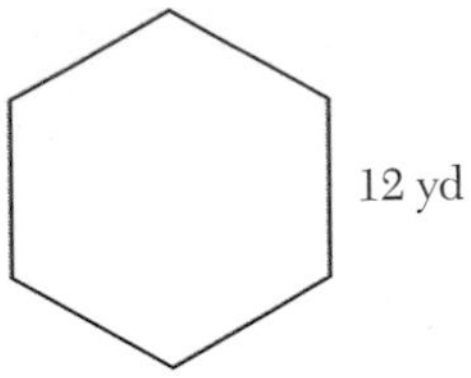

18.

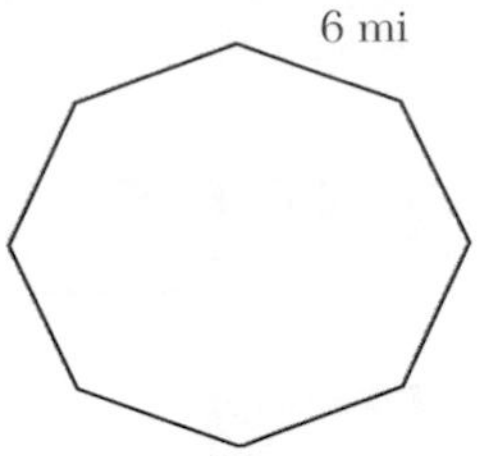

19.

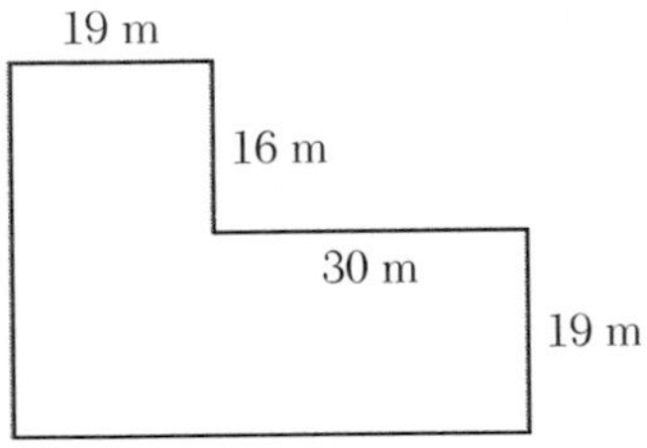

20.

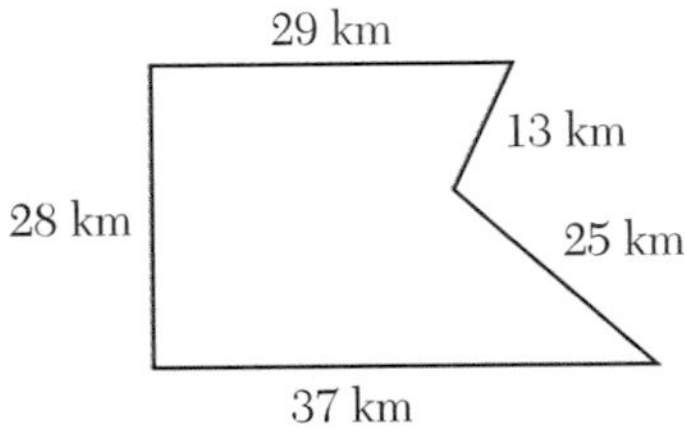

OBJECTIVE 2 Find the circumference of a circle.

A. *Find the circumference of the given circles. Use 3.14 for π.*

21.

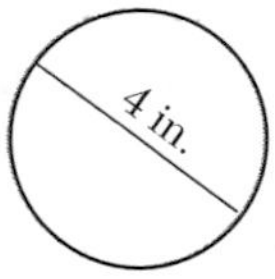

22.

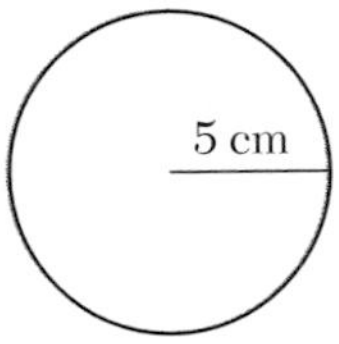

23.

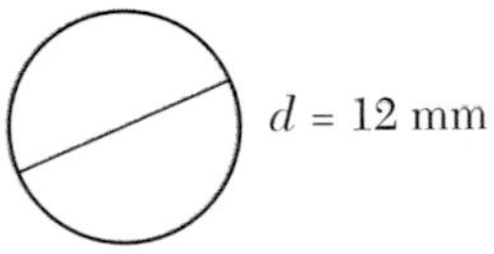

24.

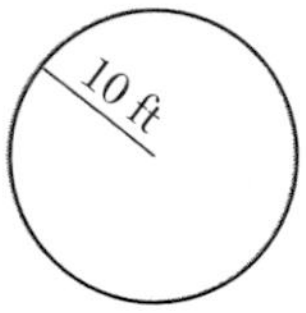

25.

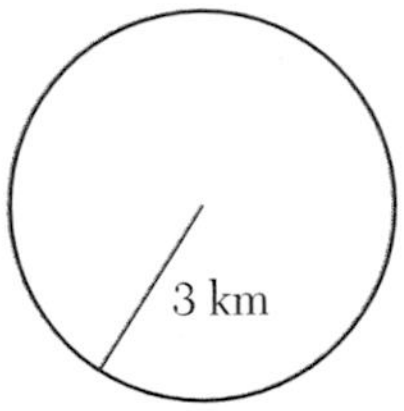

26.

 $d = \frac{1}{2}$ cm

B.

Find the perimeter of each shaded figure. Use the [π] *key, and round to the nearest hundredth.*

27.

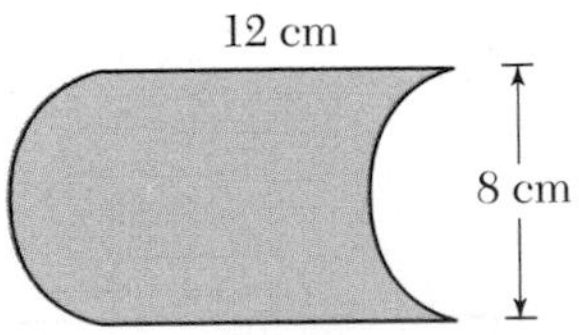

28.

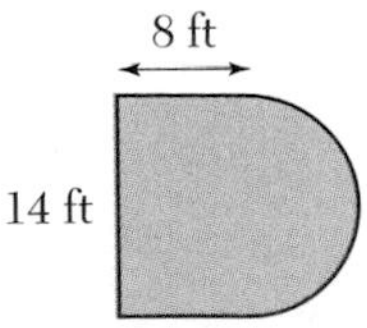

29.

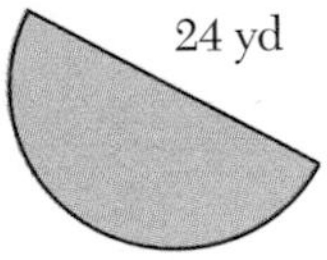

30.

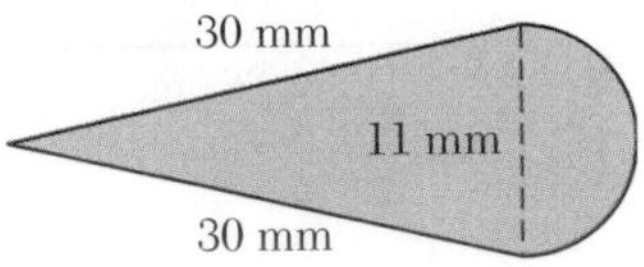

Name ______________________ Class ______________________ Date ______________

Find the sum of the interior and exterior perimeters.

31.

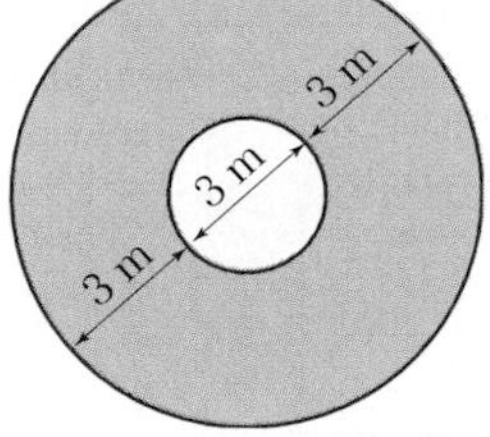

32.

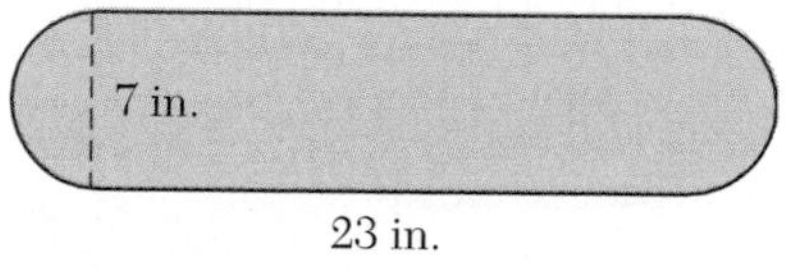

C.

Find the perimeter of the shaded regions. Use the π *key, and round to the nearest hundredth when necessary.*

33.

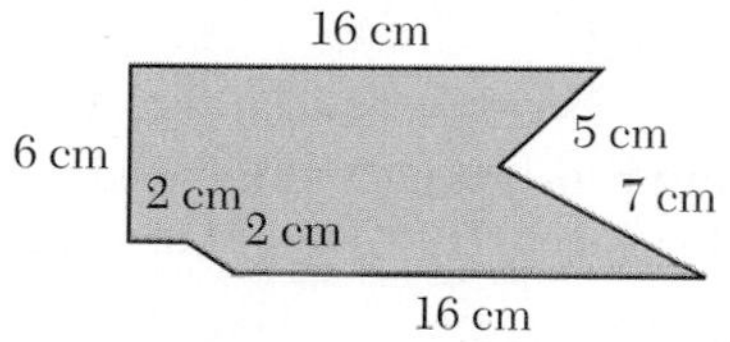

34.

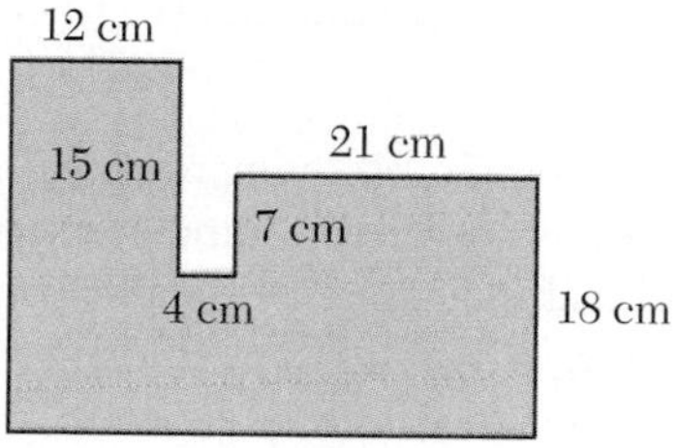

35.

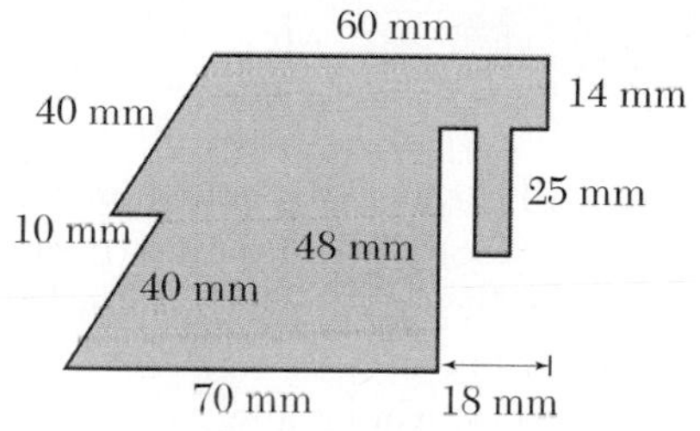

36.

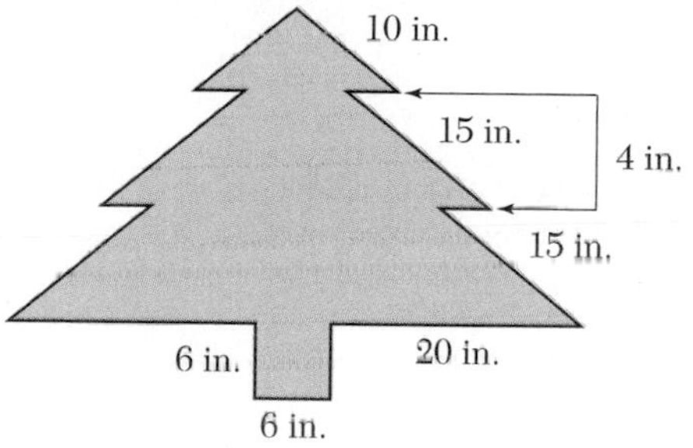

37.

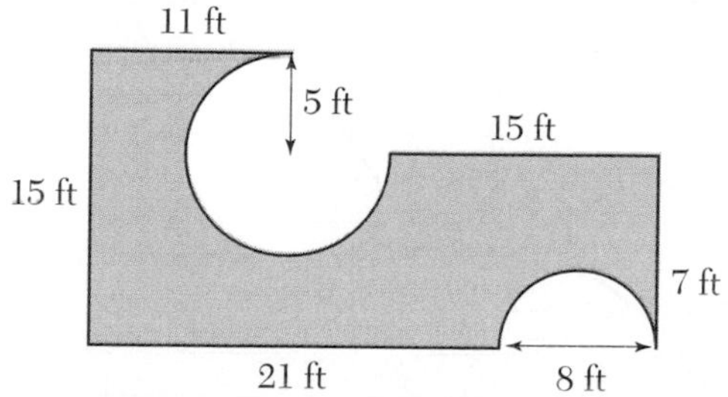

38.

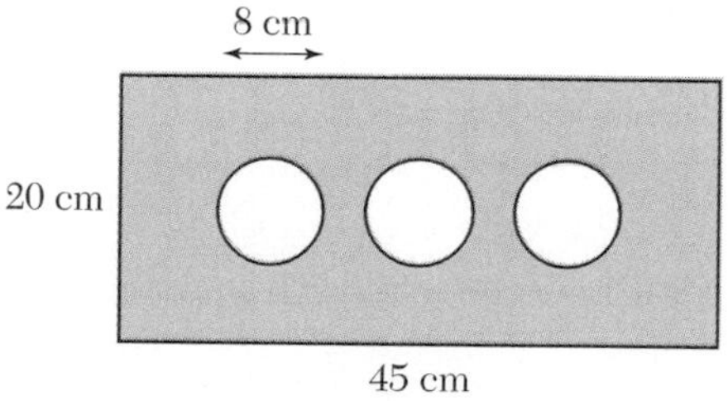

39. How many feet of picture molding are needed to frame five pictures, each measuring 8 in. by 10 in.? The molding is wide enough to require an extra inch added to each dimension to allow for the corners to be mitred.

40. If fencing costs $12 per meter, what will be the cost of fencing a rectangular ranch that is 120 km long and 24 km wide?

41. If Hazel needs 2 minutes to put 1 foot of binding on a rug, how long will it take her to put the binding on a rug that is 15 ft by 12 ft?

42. Annessa is making a three-tiered bridal veil. Each tier is rectangular, with one of the short ends being gathered into the headpiece. The other three sides of each tier are to be trimmed in antique lace. How many yards of lace does Annessa need?

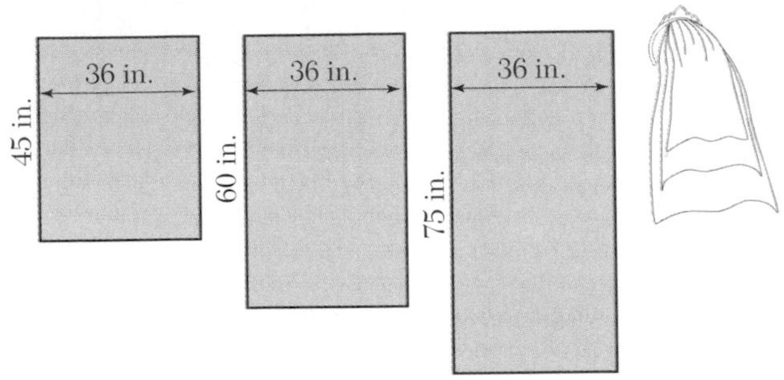

43. Holli has a watercolor picture that is 14 in. by 20 in. She puts it in a mat that is 3 in. wide on all sides. What are the inside dimensions of the frame she needs to buy?

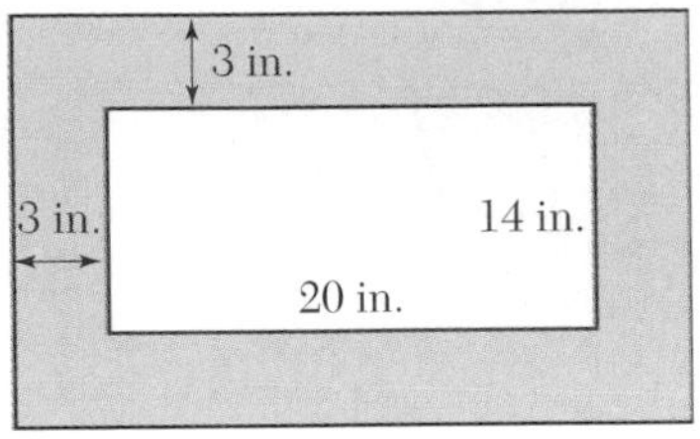

44. Jorge is lining the windows in his living room with Christmas lights. He has one picture window that is 5 ft 8 in. by 4 ft. On each side there is a smaller window that is 2 ft 6 in. by 4 ft. What length of Christmas lights does Jorge need for the three windows?

45. Jenna and Scott just bought a puppy and need to fence their backyard. How much fence should they order?

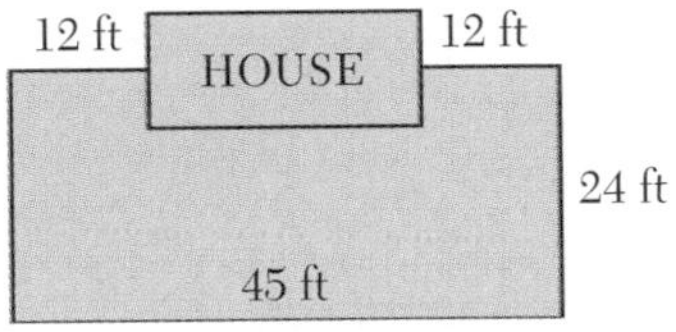

46. As a conditioning exercise, a soccer coach has his team run around the outside of the field three times. If the field measures 60 yd × 100 yd, how far did the team run?

47. A high school football player charted the number of laps he ran around the football field during the first ten days of practice. If the field measures 120 yd by 53 yd, how far did he run during the ten days? Convert your answer to the nearest whole mile.

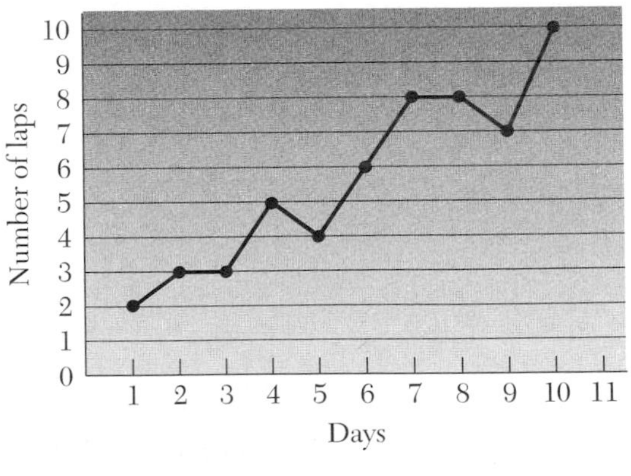

48. A carpenter is putting baseboards in the family room/dining room pictured here. How many feet of baseboard molding are needed?

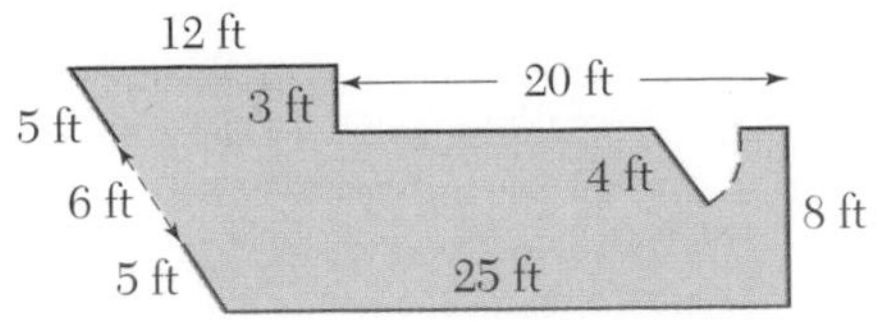

Exercises 49–51 refer to the following. Marylane is constructing raised beds in her backyard for her vegetable garden. Each bed is made from a frame of 2-×-12-inch pressure-treated lumber. Marylane is planning three rectangular frames that measure 4 ft by 8 ft. The lumber comes in 8-ft lengths and costs $12.49 per board.

49. How much lumber is needed for the project?

50. How many boards are needed for the project?

51. What is the total cost of the lumber for the project?

52. Yolande has lace that is 2 inches wide, which she plans to sew around the edge of a circular tablecloth that is 50 inches in diameter. How much lace does she need?

53. Yolande wants to sew a second row of her 2-inch-wide lace around the tablecloth in Exercise 52. How much lace does she need for the second row?

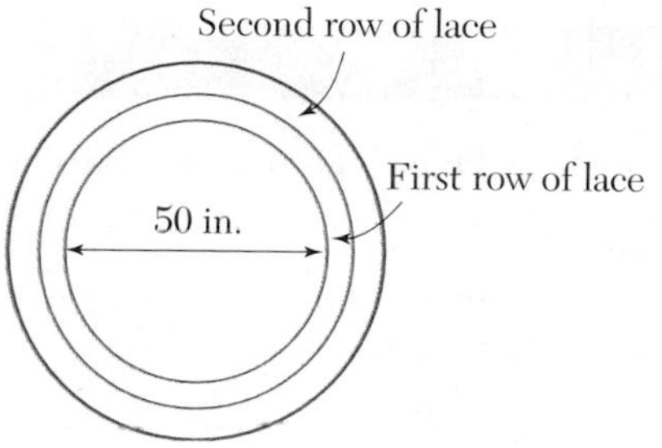

Exercises 54–57 refer to the chapter application.

54. What is the shape of the patio?

55. What is the perimeter of the patio?

56. Estimate the perimeter of the flower bed that wraps around the right side of the patio.

57. Draw a line down the center of the backyard. What do you notice about the two halves? Mathematicians call figures like this symmetrical because a center line divides the figure into "mirror images." Colonial architecture typically makes use of symmetry.

58. Are rectangles symmetrical around a line drawn lengthwise through the center? Are squares symmetrical?

59. Draw a triangle that is symmetrical and another that is not symmetrical.

60. Draw a circle that is symmetrical and another that is not symmetrical.

STATE YOUR UNDERSTANDING

61. Explain what perimeter is and how the perimeter of the following figure is determined.

What possible units (both English and metric) would the perimeter be likely to be measured in if the figure is a national park? If the figure is a room in a house? If the figure is a scrap of paper?

62. Is a rectangle a parallelogram? Why or why not?

63. Explain the difference between a square and a rectangle.

 Name ______________ Class ______________ Date ______________

CHALLENGE

64. A farmer wants to build the goat pens pictured here. Each pen will have a gate 2 ft 6 in. wide in the end. What is the total cost of the pens if the fencing is $3.00 per linear foot and each gate is $15.00?

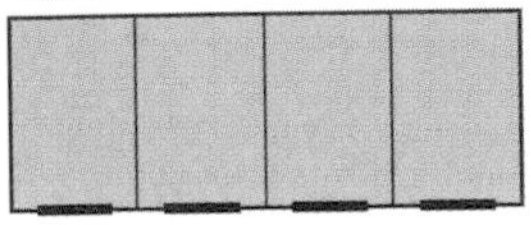

GROUP WORK

65. Federal Express will accept packages according to the formula:

$L_1 + 2L_2 + 2L_3 \leq 160$ in.

in which L_1 is the longest side, L_2 is the next longest side, and L_3 is the shortest side.

Make a table of dimensions of boxes that can be shipped by Federal Express. Try to find the maximum sizes possible.

Longest Side	Next Longest Side	Shortest Side	Total

MAINTAIN YOUR SKILLS

66. Write the place value name for nine hundred thousand, fifty.

67. Write the place value name for nine hundred fifty thousand.

68. Round 32,571,600 to the nearest ten thousand.

69. Find the difference between 733 and 348.

70. Find the product of 733 and 348.

71. Find the quotient of 153,204 and 51.

72. Find the sum of 2 lb 3 oz and 3 lb 14 oz.

73. Find the difference of 6 ft and 7 in.

74. Find the average bowling score for Fred if he bowled games of 167, 182, and 146.

75. What is the total cost for a family of one mother and three children to attend a hockey game if adult tickets are $15, student tickets are $10, and the parking fee is $6?

7.4 Area

VOCABULARY

Area is a measure of surface that is the amount of space inside a two-dimensional figure. It is measured in square units.

The **base** of a geometric figure is a side parallel to the horizon.

The **altitude** or **height** of a geometric figure is the perpendicular distance from the base to the highest point of the figure.

Table 7.11 shows the base and altitude of some common geometric figures.

OBJECTIVES

1. Find the area of common polygons and circles.
2. Convert units of area measure.

TABLE 7.11 BASE AND HEIGHT OF COMMON GEOMETRIC FIGURES

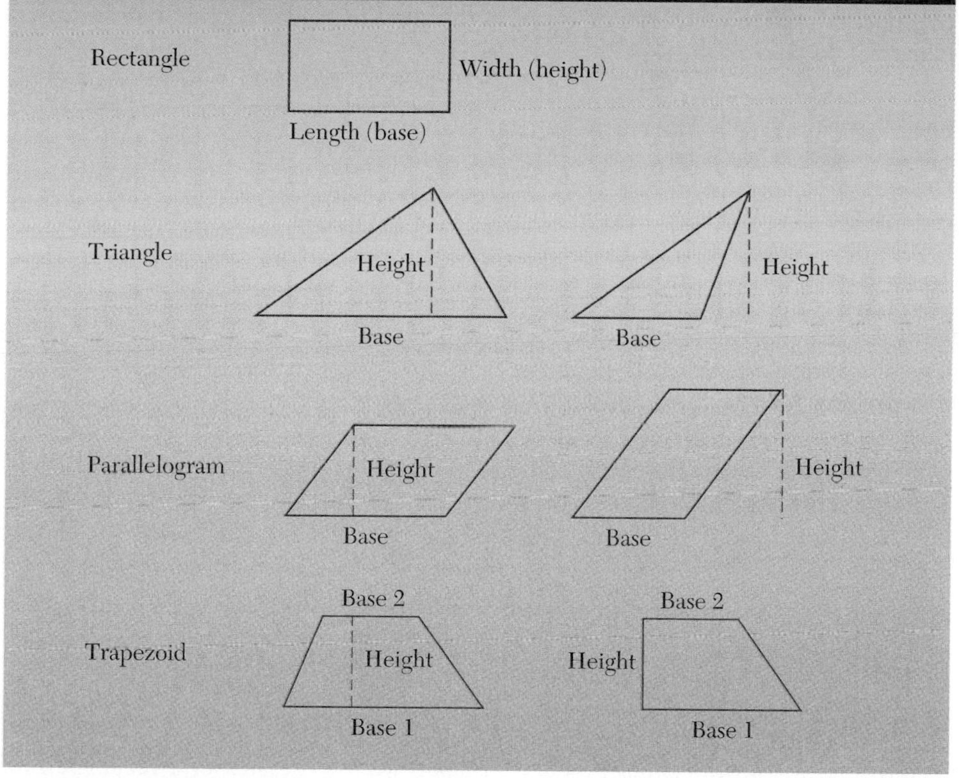

HOW & WHY

OBJECTIVE 1 Find the area of common polygons and circles.

Suppose you wish to tile a rectangular bathroom floor that measures 5 ft by 6 ft. The tiles are 1-ft-by-1-ft squares. How many tiles do you need?

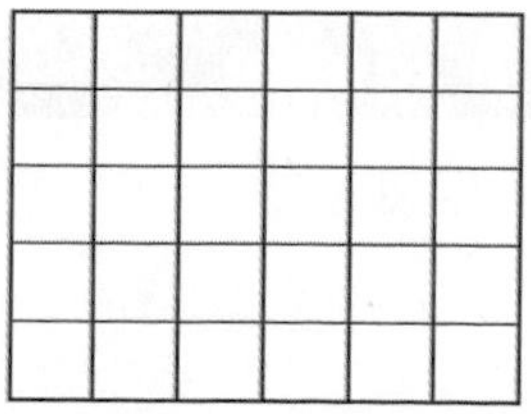

FIGURE 7.1

Using Figure 7.1 as a model of the tiled floor, you can count that 30 tiles are necessary.

Area is a measure of the surface—that is, the amount of space inside a two-dimensional figure. It is measured in square units. Square units are literally squares that measure one unit of length on each side. Figure 7.2 shows two examples.

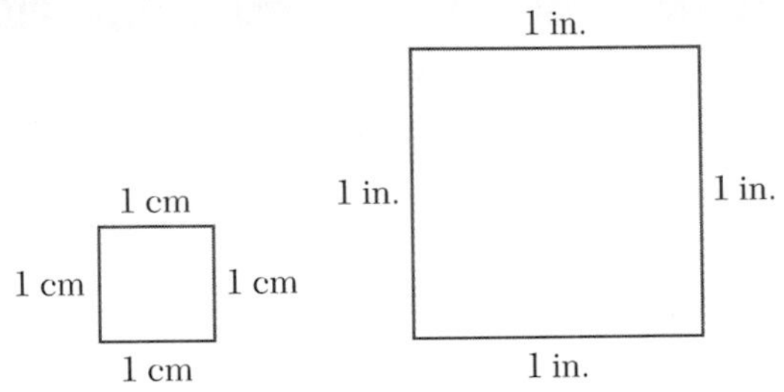

FIGURE 7.2

The measure of the square area on the left is 1 square centimeter, abbreviated cm^2. The measure of the square area on the right is 1 square inch, abbreviated in.^2. Recall that the superscript in the abbreviation is part of the unit's name. It does not indicate an exponential operation.

CAUTION

$10\text{ cm}^2 \neq 100\text{ cm}$

In the bathroom floor example, each tile has an area of 1 ft^2. So, the number of tiles needed is the same as the area of the room, 30 ft^2. We could have arrived at this number by multiplying the length of the room by the width. This is not a coincidence. It works for all rectangles.

To find the area of a rectangle

Multiply the length by the width.

$A = \ell w$

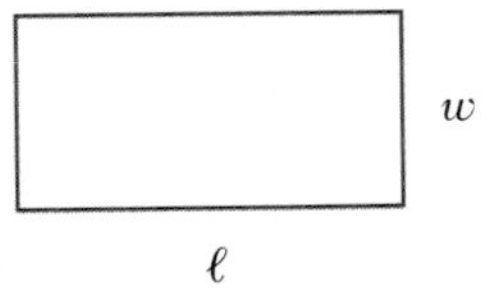

To find the area of other geometric shapes, we use the area of a rectangle as a reference.

Because a square is a special case of a rectangle, with all sides equal, the formula for the area of a square is

$$\begin{aligned} A &= \ell w \\ &= s(s) \\ &= s^2 \end{aligned}$$

To find the area of a square

Square the length of one of the sides.

$A = s^2$

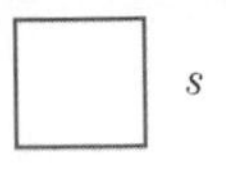

Now let's consider the area of a triangle. We start with a right triangle—that is, a triangle with one 90° angle. See Figure 7.3.

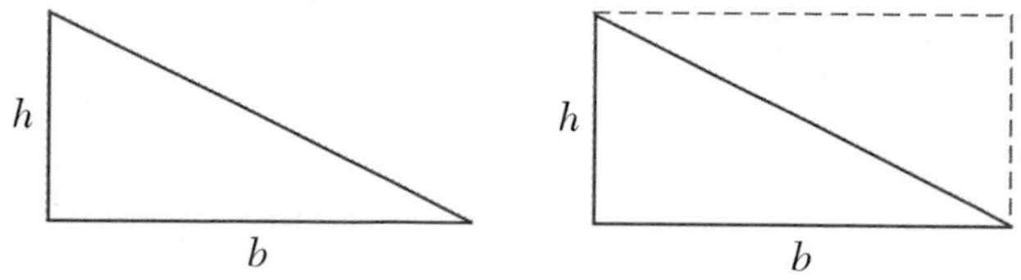

FIGURE 7.3

The triangle on the left has base b and height h. The figure on the right is a rectangle with length b and width h. According to the formula for rectangles, the area is $A = \ell w$ or $A = bh$. However, the rectangle is made up of two triangles, both of which have a base of b and a height of h. Consequently, the area of the rectangle (bh) is exactly twice the area of the triangle. So, we conclude that the area of the triangle is

$$\frac{1}{2}bh \quad \text{or} \quad \frac{bh}{2} \quad \text{or} \quad bh \div 2.$$

Now let's consider a more general triangle. See Figure 7.4.

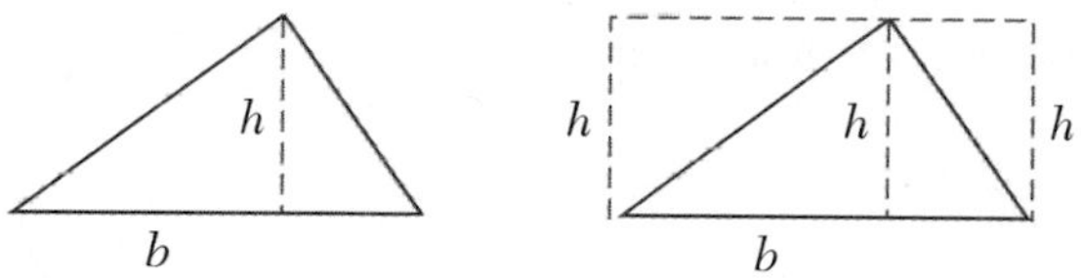

FIGURE 7.4

Again, the area on the left is a triangle with base b and height h. And the figure on the right is a rectangle with length b and height h. Can you see that the rectangle must be exactly twice the area of the original triangle? So again we conclude that the area of the triangle is $\frac{1}{2}bh$ or $\frac{bh}{2}$ or $bh \div 2$.

To find the area of a triangle

Multiply the base times the height and divide by 2.

$$A = \frac{1}{2}bh \quad \text{or} \quad A = \frac{bh}{2} \quad \text{or} \quad A = bh \div 2$$

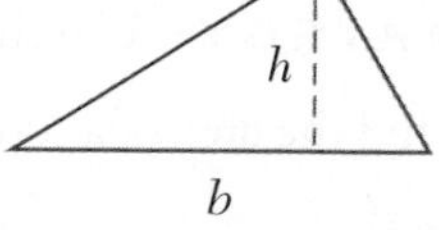

It is possible to use rectangles to find the formulas for the areas of parallelograms and trapezoids. This is left as an exercise (see Exercise 79).

To find the area of a parallelogram

Multiply the base times the height.

$A = bh$

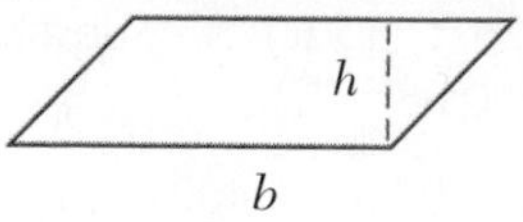

To find the area of a trapezoid

Add the two bases together, multiply by the height; then divide by 2.

$$A = \frac{(b_1 + b_2)h}{2}$$

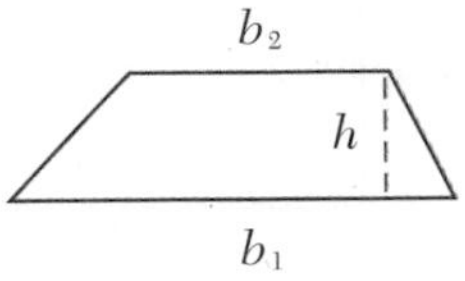

The lengths, widths, bases, sides, and heights must all be measured using the same units before the area formulas may be applied. If the units are different, convert to a common unit before calculating area.

The formula for the area of a circle is less intuitive than the formulas discussed thus far. Nevertheless, the formula for the area of a circle has been known and used for many years. Even though the sides of a circle are not line segments, we still measure the area in square units.

To find the area of a circle

Square the radius and multiply by π.

$$A = \pi r^2$$

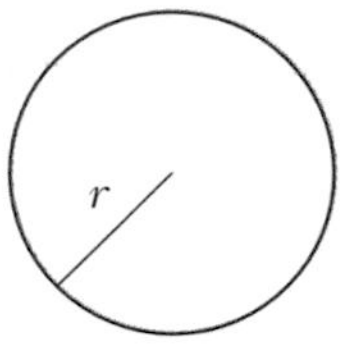

Reminder: You can use 3.14 or $\frac{22}{7}$ to approximate π or use the $\boxed{\pi}$ key on your calculator for a more accurate approximation.

WARM-UPS A–G

A. Find the area of a square that is 12 cm on each side.

B. A decorator found a 15 yd^2 remnant of carpet. Will it be enough to carpet a 5-yd-by-4-yd playroom?

Answers to Warm-Ups

A. The area is 144 cm^2.

B. The remnant will not be enough because the decorator needs 20 yd^2.

EXAMPLES A–G

DIRECTIONS: Find the area.

STRATEGY: Use the area formulas.

A. Find the area of a square that is 7 in. on each side.

$A = s^2$ Formula
$= (7 \text{ in.})^2$ Substitute.
$= 49 \text{ in.}^2$

The area is 49 in.^2.

B. A gallon of deck paint will cover 400 ft^2. A contractor needs to paint a rectangular deck that is 26 ft long and 15 ft wide. Will one gallon of paint be enough?

$A = \ell w$ Formula
$= (26 \text{ ft})(15 \text{ ft})$ Substitute.
$= 390 \text{ ft}^2$

Because $390 \text{ ft}^2 < 400 \text{ ft}^2$, 1 gallon of paint will be enough.

C. Find the area of this triangle.

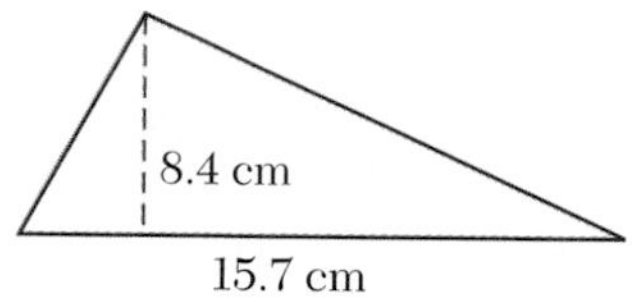

$$
\begin{aligned}
A &= \frac{bh}{2} && \text{Formula}\\
&= \frac{(15.7\text{ cm})(8.4\text{ cm})}{2} && \text{Substitute.}\\
&= \frac{131.88\text{ cm}^2}{2} && \text{Simplify.}\\
&= 65.94\text{ cm}^2
\end{aligned}
$$

The area is 65.94 cm^2.

D. Find the area of the parallelogram with a base of 1 ft and a height of 4 in.

$$
\begin{aligned}
A &= bh && \text{Formula}\\
&= (1\text{ ft})(4\text{ in.}) && \text{Substitute.}\\
&= (12\text{ in.})(4\text{ in.}) && \text{Convert so that units match.}\\
&= 48\text{ in.}^2 && \text{Simplify.}
\end{aligned}
$$

The area of the parallelogram is 48 in.^2.

E. Find the area of the top of a coffee can that has a radius of 102 mm. Let $\pi \approx 3.14$.

$$
\begin{aligned}
A &= \pi r^2 && \text{Formula}\\
&\approx (3.14)(102\text{ mm})^2 && \text{Substitute.}\\
&\approx (3.14)(10404\text{ mm}^2) && \text{Simplify.}\\
&\approx 32{,}668.56\text{ mm}^2
\end{aligned}
$$

The area of the top of the coffee can is about $32{,}668.56\text{ mm}^2$.

F. Find the area of the trapezoid pictured.

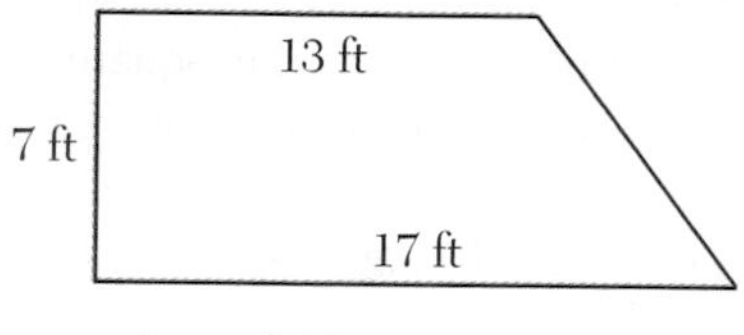

$$
\begin{aligned}
A &= \frac{(b_1 + b_2)h}{2} && \text{Formula}\\
&= \frac{(17\text{ ft} + 13\text{ ft})(7\text{ ft})}{2} && \text{Substitute.}\\
&= \frac{(30\text{ ft})(7\text{ ft})}{2} && \text{Simplify.}\\
&= \frac{210\text{ ft}^2}{2}\\
&= 105\text{ ft}^2
\end{aligned}
$$

The area is 105 ft^2.

C. Find the area of this triangle.

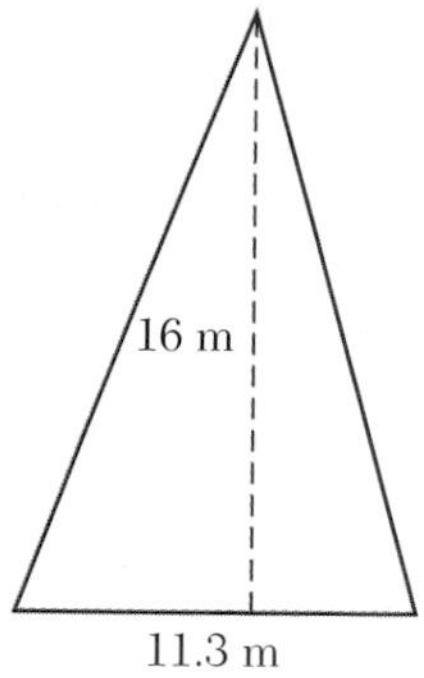

D. Find the area of the parallelogram with a base of 6 yd and a height of 5 ft.

E. What is the area of a quarter, which has a radius of 13 mm? Let $\pi \approx 3.14$.

F. Find the area of a trapezoid with bases of 51 m and 36 m and a height of 11 m.

Answers to Warm-Ups

C. The area is 90.4 m^2.
D. The area is 90 ft^2.
E. The area is about 530.66 mm^2.
F. The area is 478.5 m^2.

G. Find the area of the polygon.

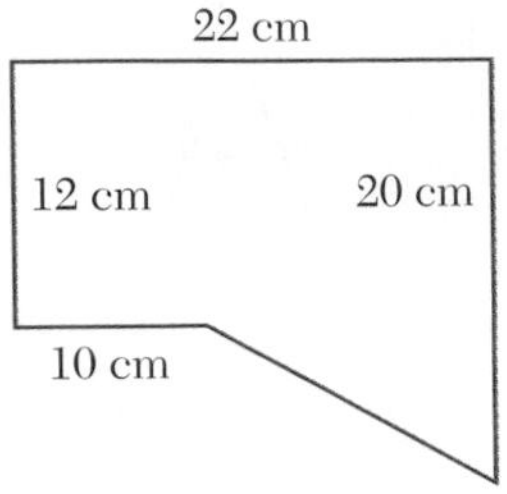

G. Find the area of the polygon.

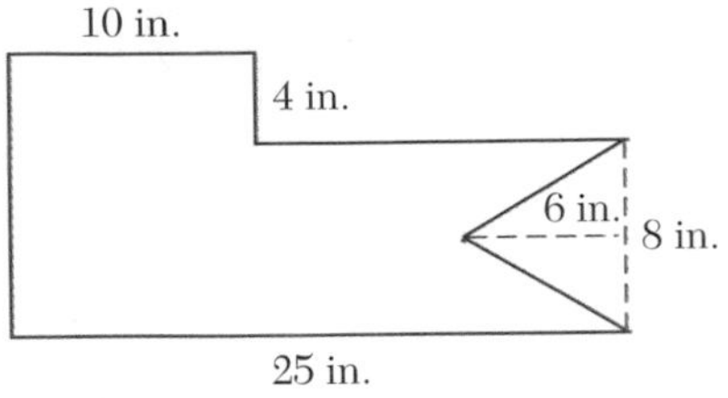

STRATEGY: To find the area of a polygon that is a combination of two or more common figures, first divide it into the common figure components.

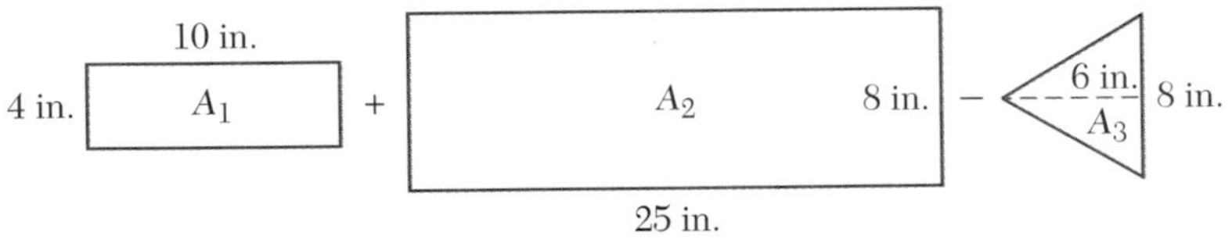

$$\text{Total area} = A_1 + A_2 - A_3$$

Divide into components.

$$A_1 = (10 \text{ in.})(4 \text{ in.}) = 40 \text{ in.}^2$$
$$A_2 = (25 \text{ in.})(8 \text{ in.}) = 200 \text{ in.}^2$$
$$A_3 = \frac{(6 \text{ in.})(8 \text{ in.})}{2} = \frac{48 \text{ in.}^2}{2} = 24 \text{ in.}^2$$

Compute the areas of each component.

$$\text{Total area} = 40 \text{ in.}^2 + 200 \text{ in.}^2 - 24 \text{ in.}^2$$
$$= 216 \text{ in.}^2$$

Combine the areas.

The total area of the figure is 216 in.^{2}.

HOW & WHY

OBJECTIVE 2 Convert units of area measure.

In the United States, we usually measure the dimensions of a room in feet, and so, naturally, we measure its area in square feet. However, carpeting is measured in square yards. Therefore, if we want to buy carpeting for a room, we need to convert units of area measure—in this case, from square feet to square yards. We proceed as before, using unit fractions. The only difference is that we use the appropriate unit fraction twice because we are working with square units.

To change 100 ft^2 to square yards, we write

$$100 \text{ ft}^2 = 100(\cancel{\text{ft}})(\cancel{\text{ft}})\left(\frac{1 \text{ yd}}{3 \cancel{\text{ft}}}\right)\left(\frac{1 \text{ yd}}{3 \cancel{\text{ft}}}\right)$$
$$= \frac{100}{9} \text{ yd}^2$$
$$= 11\frac{1}{9} \text{ yd}^2$$

Answers to Warm-Ups

G. The area is 312 cm^2.

WARM-UPS H–I

EXAMPLES H–I

DIRECTIONS: Convert as indicated.

STRATEGY: Use the appropriate unit fraction as a factor twice.

H. Convert 25 mi^2 to square kilometers. Round to the nearest thousandth.

$$25\text{ mi}^2 \approx 25(\cancel{\text{mi}})(\cancel{\text{mi}})\left(\frac{1.6093\text{ km}}{\cancel{\text{mi}}}\right)\left(\frac{1.6093\text{ km}}{\cancel{\text{mi}}}\right) \quad 1.6093\text{ km} \approx 1\text{ mi}$$

$$\approx 64.746\text{ km}^2 \quad \text{Simplify.}$$

So, 25 $mi^2 \approx 64.746$ km^2.

I. Convert 378 cm^2 to square meters.

$$378\text{ cm}^2 = 378(\cancel{\text{cm}})(\cancel{\text{cm}})\left(\frac{1\text{ m}}{100\ \cancel{\text{cm}}}\right)\left(\frac{1\text{ m}}{100\ \cancel{\text{cm}}}\right) \quad 100\text{ cm} = 1\text{ m}$$

$$= \frac{378}{10{,}000}\text{ m}^2 \quad \text{Simplify.}$$

$$= 0.0378\text{ m}^2$$

So, 378 $cm^2 = 0.0378$ m^2.

H. Convert 5 m^2 to square feet. Round to the nearest hundredth.

I. Convert 3000 m^2 to square kilometers.

*

Answers to Warm-Ups

H. 53.82 ft^2 **I.** 0.003 km^2

EXERCISES 7.4

OBJECTIVE 1 Find the area of common polygons and circles.

A. *Find the area of the following figures. Use* $\frac{22}{7}$ *for* π.

1.

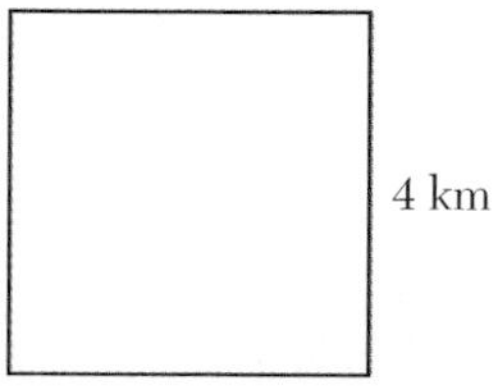

2.

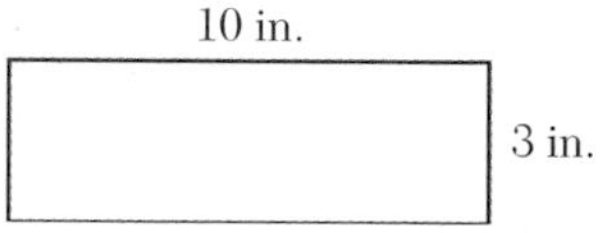

3.

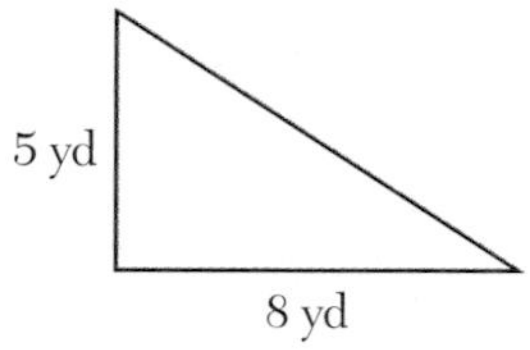

4.

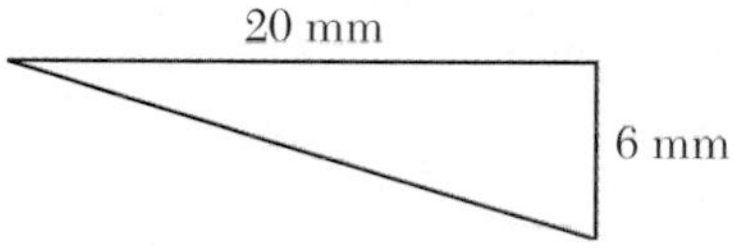

5.

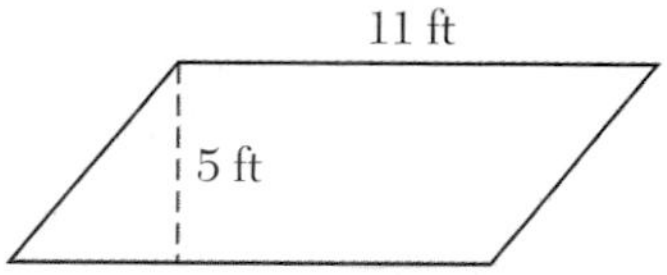

6.

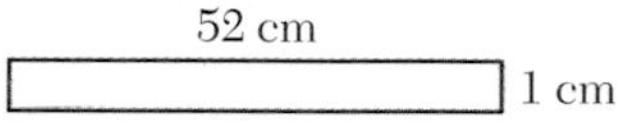

7.

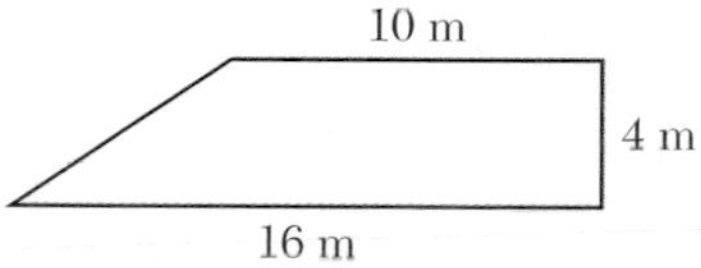

8.

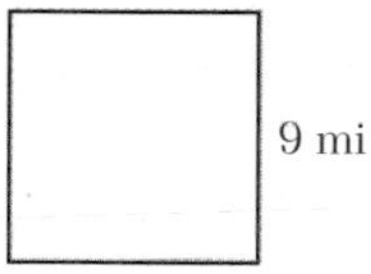

9.

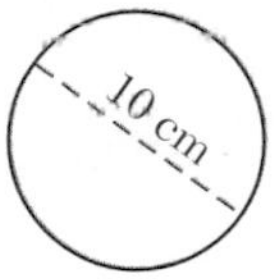

10.

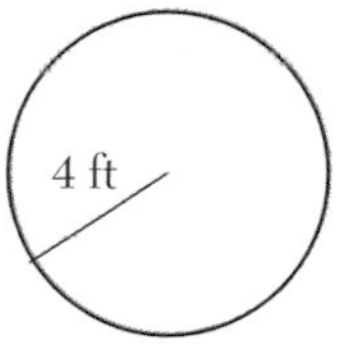

B. *Find the areas of the figures. Use 3.14 for* π.

11. Find the area of a rectangle that has a length of 12 km and a width of 11 km.

12. Find the area of a square with sides of 35 cm.

13. Find the area of a triangle with base 28 yd and height 118 yd.

14. Find the area of a circle with a radius of 6 ft.

15. Find the area of a parallelogram with base 56 in. and height 32 in.

16. Find the area of a circle with a diameter of 30 mm.

17.

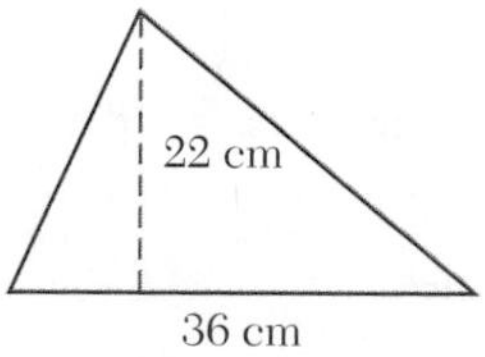

18.

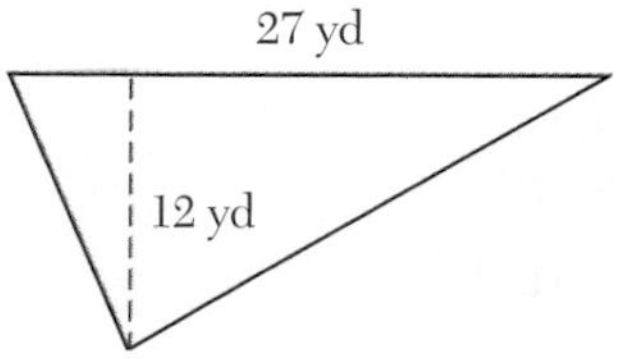

19.

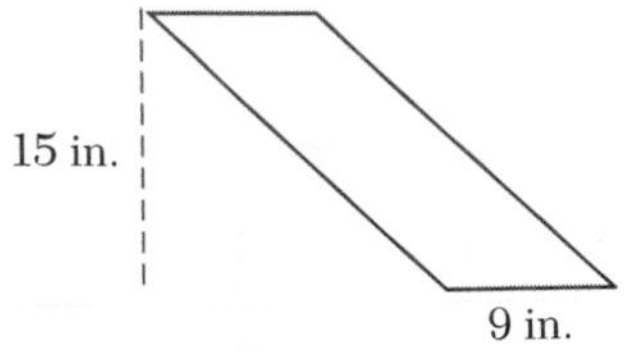

20.

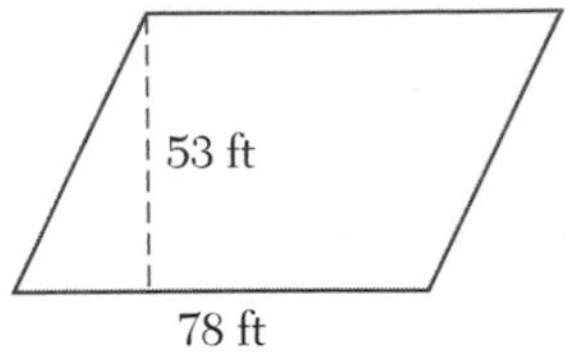

21.

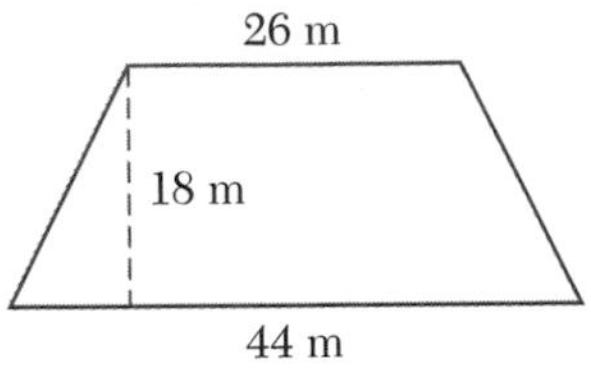

22.

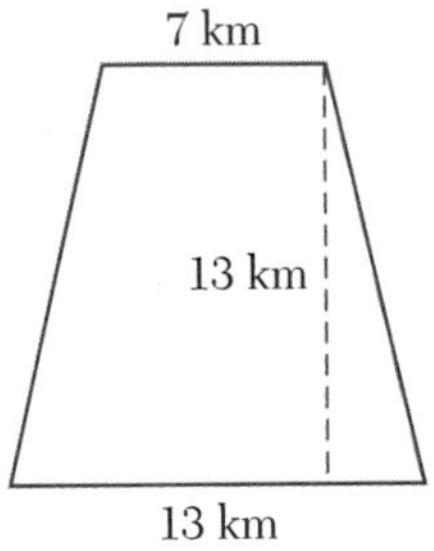

23.

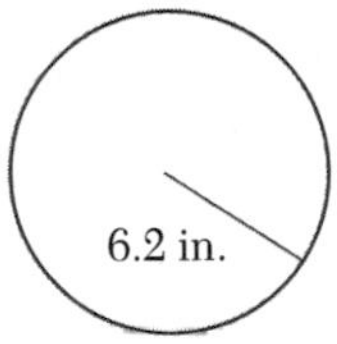

24.

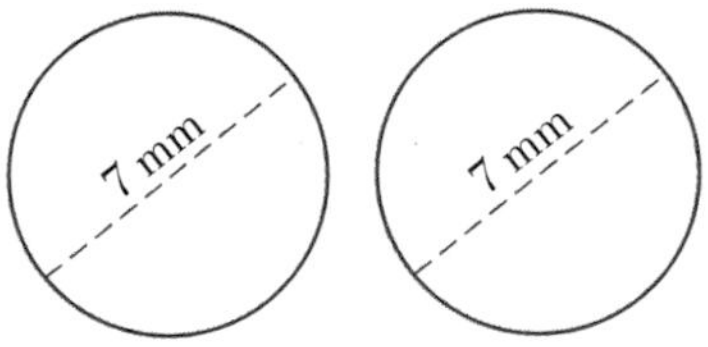

OBJECTIVE 2 Convert units of area measure.

Convert units as indicated. Round to the nearest hundredth.

A.

25. Convert 3 ft^2 to square inches.

26. Convert 5 cm^2 to square millimeters.

27. Convert 4 yd^2 to square feet.

28. Convert 12 m^2 to square centimeters.

29. Convert 1 mi^2 to square feet.

30. Convert 6 km^2 to square meters.

31. Convert 1296 in^2 to square feet.

32. Convert 2,000,000 cm^2 to square meters.

33. Convert 117 ft^2 to square yards.

34. Convert 8700 mm^2 to square centimeters.

B.

35. Convert 4.5 yd^2 to square feet.

36. Convert 62 km^2 to square centimeters.

37. Convert 12,000,000 ft^2 to square miles.

38. Convert 123,456,789 cm^2 to square meters.

39. Convert 250 mm^2 to square inches.

40. Convert 100 $in.^2$ to square centimeters.

41. Convert 3 m^2 to square yards.

42. Convert 80 km^2 to square miles.

43. Convert 15 ft^2 to square centimeters.

44. Convert 100 mi^2 to square kilometers.
Let 1 $mi^2 \approx 2.590$ km^2.

C.

Find the areas of the figures.

45.

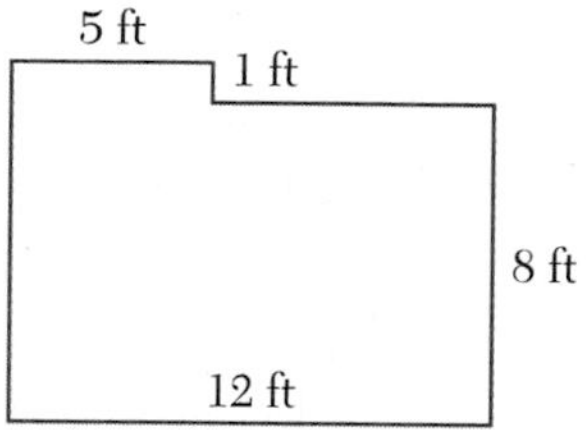

46.

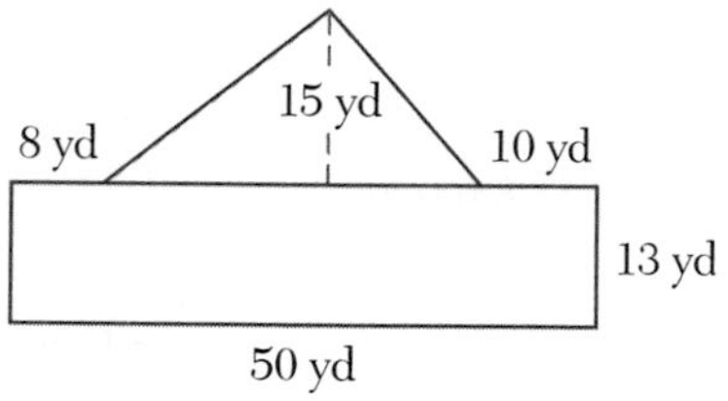

47.

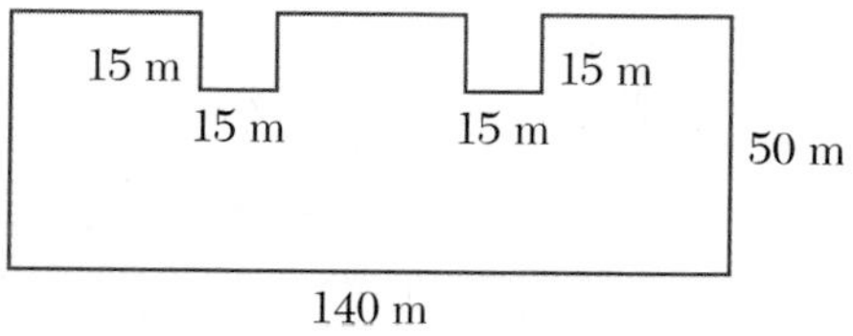

48.

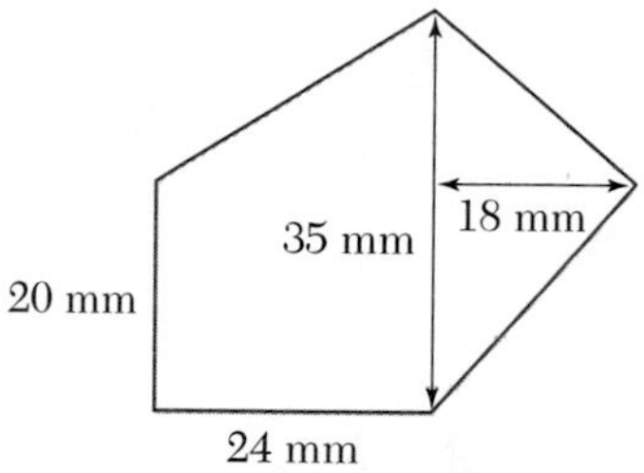

49.

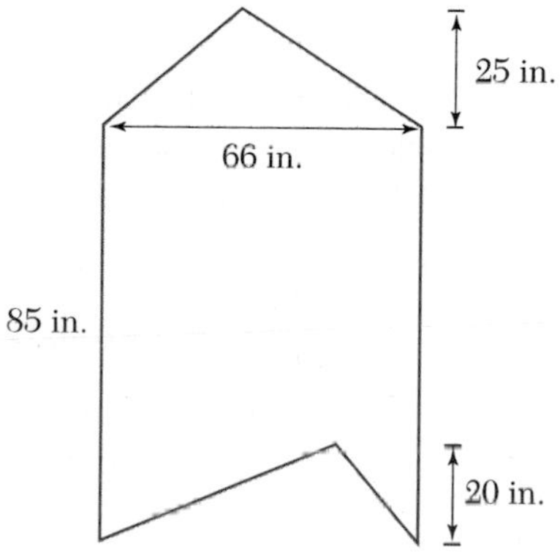

50.

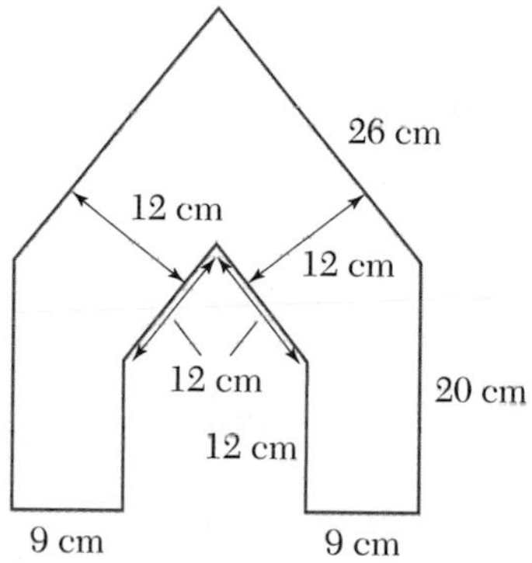

Find the areas of the shaded regions. Use the $\boxed{\pi}$ *key, and round to the nearest hundredth.*

51.

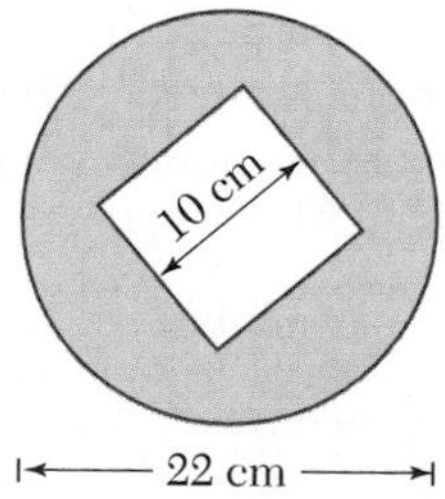

52.

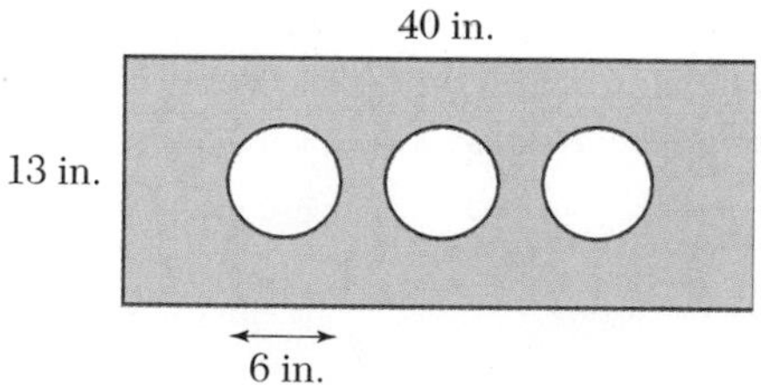

53. The side of Jane's house that has no windows measures 35 ft by 22 ft. If 1 gal of stain will cover 250 ft^2, will 2 gal of stain be enough to stain this side?

54. The south side of Jane's house measures 75 ft by 22 ft and has two windows, each 4 ft by 6 ft. Will 4 gal of the stain be enough for this side?

55. April and Larry are building an in-ground spa. The spa is 4 ft deep and 8 ft in. diameter. They intend to tile the sides and bottom of the spa. How many square feet of tile do they need? *Hint*: The sides of a cylinder form a rectangle whose length is the circumference of the circle. Round to the nearest whole square foot.

56. Ingrid is going to carpet two rooms in her house. The floor in one room measures 30 ft × 24 ft and the floor in the other room measures 22 ft × 18 ft. If the carpet costs \$27.00 per square yard installed, what will it cost Ingrid to have the carpet installed?

57. If 1 oz of weed killer treats 1 m^2 of lawn, how many ounces of weed killer will Debbie need to treat a rectangular lawn that measures 30 m by 12 m?

58. To the nearest acre, how many acres are contained in a rectangular plot of ground if the length is 1850 ft and the width is 682 ft? (43,560 ft^2 = 1 acre)

59. How much glass is needed to replace a set of two sliding glass doors in which each pane measures 3 ft by 6 ft?

60. How many square yards of carpet are needed to cover a rectangular room that is 9 ft by 12 ft?

61. How many square feet of sheathing is needed for the gable end of a house that has a rise of 9 ft and a span of 36 ft? (See the drawing.)

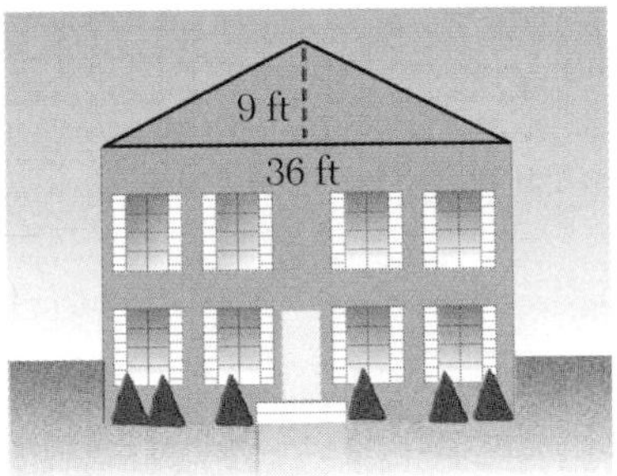

62. A farmer wants to construct a small shed that is 6 ft × 9 ft around the base and 8 ft high. How many gallons of paint will be needed to cover the outside of all four walls of the shed? (Assume 1 gal covers 250 ft^2.)

63. How much padding is needed to make a pad for the hexagonal table pictured?

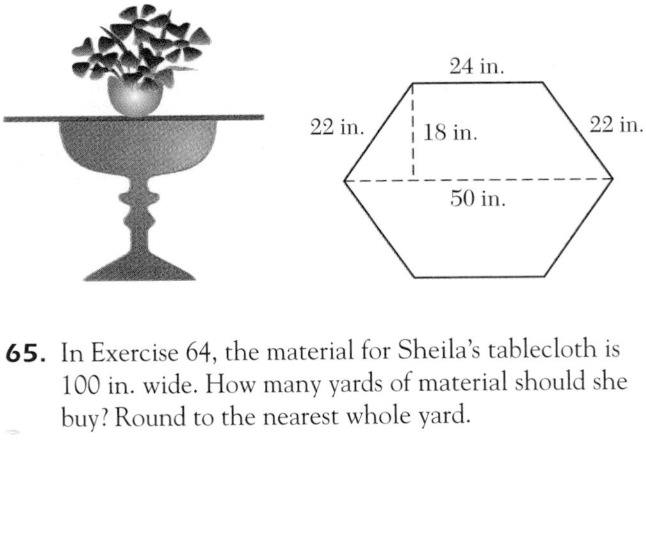

64. Sheila is making a circular tablecloth for an accent table. The top of the table is 20 in. in diameter, and the table is 3 ft tall. Sheila wants the tablecloth to touch the floor all around. Assuming a $\frac{1}{2}$-in. hem, how much material does she need? Use $\pi \approx 3.14$, and round to the nearest square inch.

65. In Exercise 64, the material for Sheila's tablecloth is 100 in. wide. How many yards of material should she buy? Round to the nearest whole yard.

66. A window manufacturer is reviewing the plans of a home to determine the amount of glass needed to fill the order. The number and size of the window panes and sliding glass door panes are listed in the table.

	Dimensions	Number Needed
Windows	3 ft × 3 ft	4
	3 ft × 4 ft	7
	3 ft × 5 ft	2
	4 ft × 4 ft	2
	5 ft × 6 ft	1
Sliding Doors	7 ft × 3 ft	2

How much glass does he need to fill the order?

67. One 2-lb bag of wildflower seed will cover 70 ft^2. How many bags of seed are needed to cover the region shown here? (Do not seed the shaded area.)

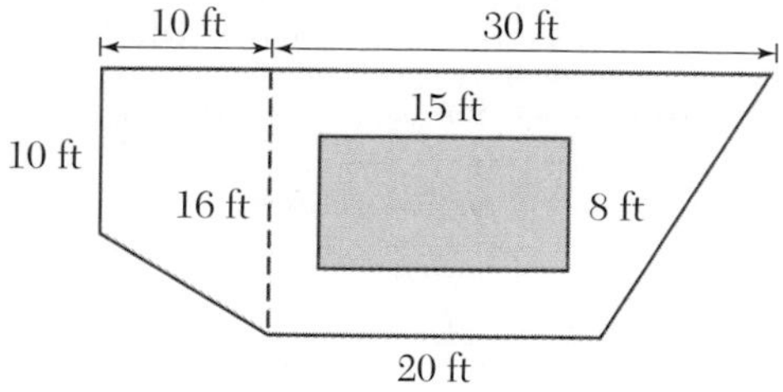

68. How many square yards of carpet are needed to carpet a flight of stairs if each stair is 10 in. by 40 in., the risers are 8 in., and there are 12 stairs and 13 risers in the flight? Round to the nearest whole square yard.

69. How many 8-ft-by-4-ft sheets of paneling are needed to panel two walls of a den that measure 14 ft by 8 ft and 10 ft by 8 ft?

70. Joy is making a king-sized quilt that measures 72 in. by 85 in. She needs to buy fabric for the back of the quilt. How many yards should she purchase if the fabric is 45 in. wide? Round to the nearest whole square yard.

Exercises 71–72 refer to the chapter application.

71. The contractor who was hired to build the brick patio will begin by pouring a concrete slab. Then he will put the bricks on the slab. The estimate of both the number of bricks and the amount of mortar needed is based on the area of the patio and walkways. Subdivide the patio and the walkways into geometric figures; then calculate the total area to be covered in bricks.

72. The number of bricks and amount of mortar needed also depend on the thickness of the mortar between the bricks. The plans specify a joint thickness of $\frac{1}{4}$ inch. According to industry standards, this will require seven bricks per square foot. Find the total number of bricks required for the patio and walkways.

STATE YOUR UNDERSTANDING

73. What kinds of units measure area? Give examples from both systems.

74. Explain how to calculate the area of the figure below. Do not include the shaded portion.

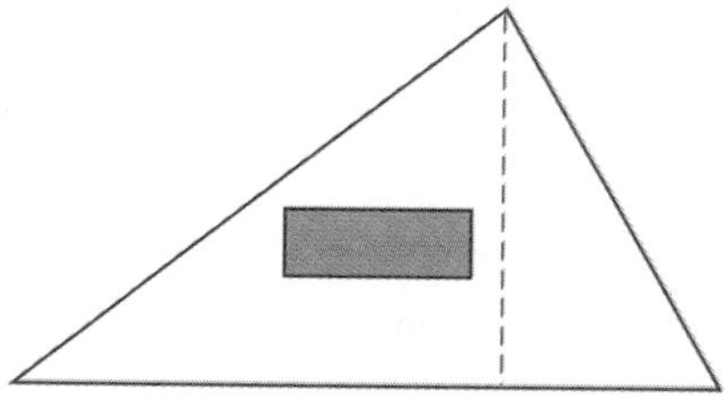

75. Describe how you could approximate the area of a geometric figure using 1-in. squares.

CHALLENGE

76. Joe is going to tile his kitchen floor. Along the outside he will put black squares that are 6 in. on each side. The next (inside) row will be white squares that are 6 in. on each side. The remaining inside rows will alternate between black and white squares that are 1 ft on each side. How many squares of each color will he need for the kitchen floor, which measures 9 ft by 10 ft?

77. A rectangular plot of ground measuring 120 ft by 200 ft is to have a cement walk 5 ft wide placed around the inside of the perimeter. How much of the area of the plot will be used by the walk and how much of the area will remain for the lawn?

78. Nellie and Mike decide that their spa should have a built-in ledge to sit on that is 18 inches wide. The ledge will be 2 feet off the bottom. How many square feet of tile do they need for the entire inside of the spa? See Exercise 55. Round to the nearest whole square foot.

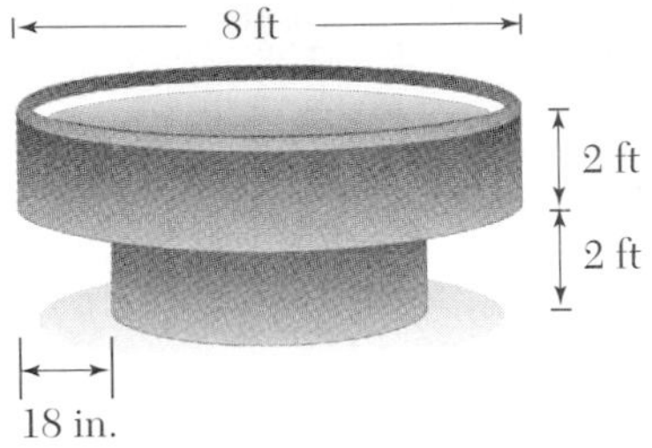

GROUP WORK

79. Use the formula for area of a rectangle to find the area of a parallelogram and a trapezoid. Draw pictures to illustrate your argument.

80. Determine the coverage of one gallon of semi-gloss paint. How much of this paint is needed to paint your classroom, excluding chalkboards, windows, and doors? What would it cost? Compare your results with the other groups in the class. Did all the groups get the same results? Give possible explanations for the differences.

MAINTAIN YOUR SKILLS

81. Multiply 45.79 by 10,000.

82. What percent of 568 is 122? Round to the nearest whole percent.

83. Simplify: $2(46 - 28) + 50 \div 5$

84. A family of six attends a weaving exhibition. Parking was \$4, adult admission to the exhibition \$6, senior admission \$4, and child admission \$3. How much does it cost for two parents, one grandmother, and three children to attend the exhibition?

85. Simplify: $3^2 + 2^3$

86. Solve the proportion: $\frac{x}{45} = \frac{13}{16}$. Round to the nearest hundredth.

87. Find the average and median: 16, 17, 18, 19, 210

88. Find the perimeter of a rectangle that is 4 in. wide and 2 ft long.

89. The table lists calories burned per hour for various activities.

Activity	Calories Burned/Hr
Sitting	50
Slow walking	140
Stacking firewood	360
Hiking	500

Make a bar graph that summarizes this information.

90. Convert 4.5 ft to centimeters. Round to the nearest hundredth.

7.5 Volume

VOCABULARY

A **cube** is a three-dimensional geometric solid that has six sides (called **faces**), each of which is a square.

Volume is the name given to the amount of space that is contained inside a three-dimensional object.

OBJECTIVES

1. Find the volume of common geometric shapes.
2. Convert units of volume.

HOW & WHY

OBJECTIVE 1 Find the volume of common geometric shapes.

Suppose you have a shoebox that measures 12 in. long by 4 in. wide by 5 in. high that you want to use to store toy blocks that are 1 in. by 1 in. by 1 in. How many blocks will fit in the box? See Figure 7.5.

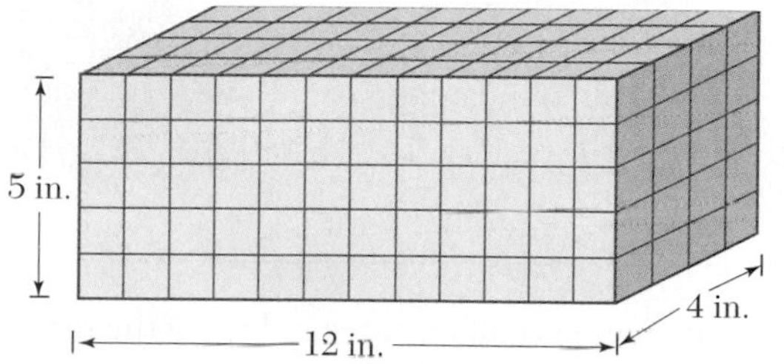

FIGURE 7.5

In each layer there are 12(4) = 48 blocks and there are 5 layers. Therefore, the box holds 12(4)(5) = 240 blocks.

Volume is a measure of the amount of space that is contained in a three-dimensional object. Often, volume is measured in cubic units. These units are literally cubes that measure one unit on each side. For example, Figure 7.6 is a cubic inch (1 in.^{3}) and a cubic centimeter (1 cm^3).

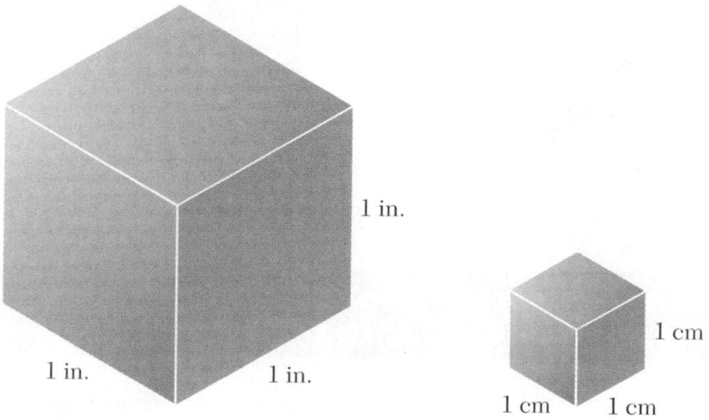

FIGURE 7.6

The shoebox has a volume of 240 in.^{3} because exactly 240 blocks, which have a volume of 1 in.^{3} each, can fit in the box and totally fill it up.

In general, volume can be thought of as the number of cubes that fill up a space. If the space is a rectangular solid, like the shoebox, it is a relatively easy matter to determine the volume by making a layer of cubes that covers the bottom and then deciding how many layers are necessary to fill the box. Note that the number of cubes necessary for the bottom layer is the same as the area of the base of the box, ℓw. The number of layers needed is the same as the height of the box, h. So we come to the following volume formulas.

To find the volume of a rectangular solid

Multiply the length by the width by the height.

$V = \ell wh$

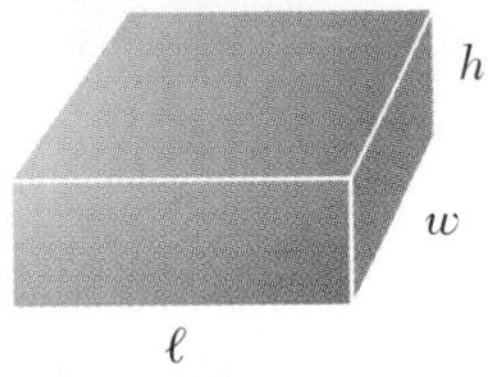

To find the volume of a cube

Cube one of the sides.

$V = s^3$

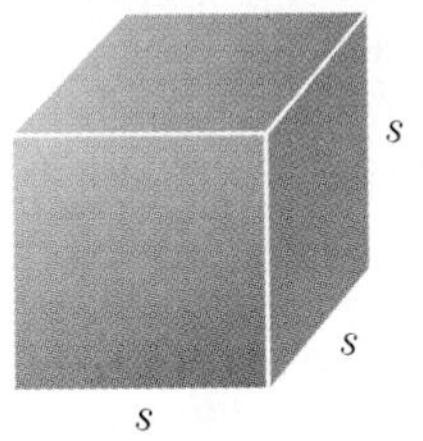

The length, width, and height must all be measured using the same units before the volume formulas may be applied. If the units are different, convert to a common unit before calculating the volume.

The principle used for finding the volume of a box can be extended to any solid with sides that are perpendicular to the base. The area of the base gives the number of cubes necessary to make the bottom layer, and the height gives the number of layers necessary to fill the solid. See Figure 7.7.

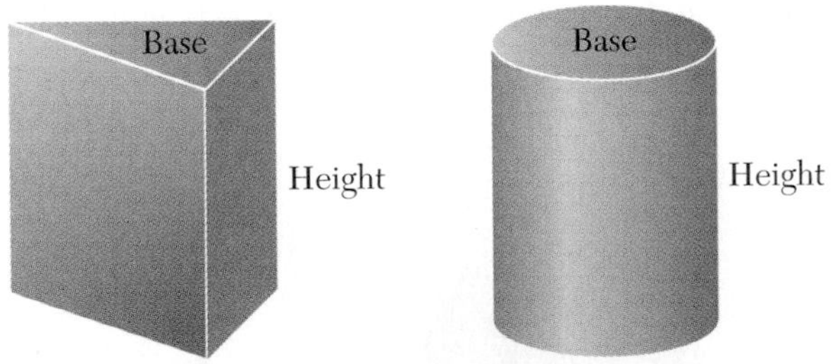

FIGURE 7.7

To find the volume of a solid with sides perpendicular to the base

Multiply the area of the base by the height:

$V = Bh$

where B is the area of the base.

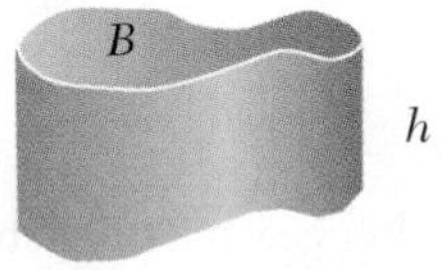

Because cylinders, spheres, and cones contain circles, their volume formulas also contain the number π. Even though these shapes are curved instead of straight, their volumes are measured in cubic units.

Volume formulas

Name	Picture	Formula
Cylinder (right circular cylinder)	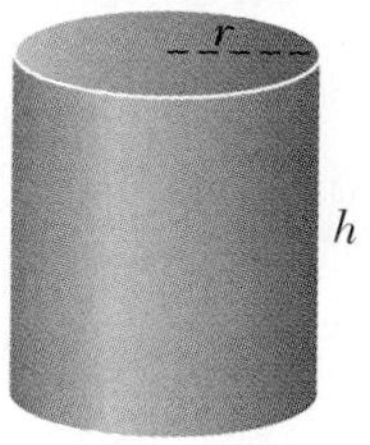	$V = \pi r^2 h$ r = radius of the base h = height of the cylinder
Sphere	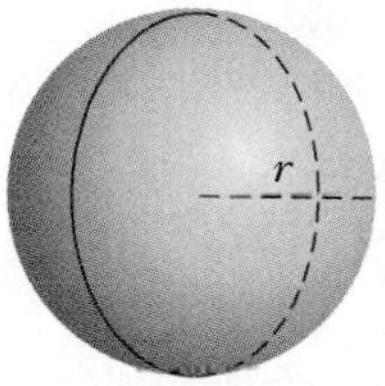	$V = \frac{4}{3}\pi r^3$ r = radius of the sphere
Cone (right circular cone)	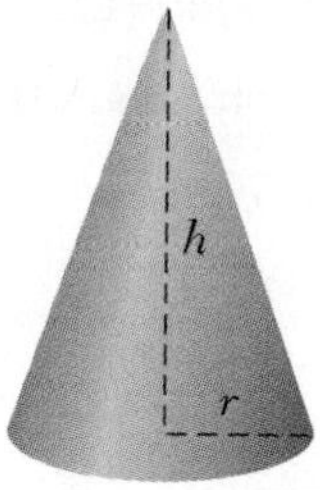	$V = \frac{1}{3}\pi r^2 h$ r = radius of the cone h = height of the cone

EXAMPLES A–D

DIRECTIONS: Find the volume.

STRATEGY: Use the volume formulas.

A. Find the volume of a cube that is 5 mm on each side.

$V = s^3$ Formula
$= (5 \text{ mm})^3$ Substitute 5 for s.
$= 125 \text{ mm}^3$ Simplify.

The volume of the cube is 125 mm^3.

B. How much concrete is needed to pour a step that is 4 ft long, 3 ft wide, and 6 in. deep?

$V = \ell wh$ Formula
$= (4 \text{ ft})(3 \text{ ft})(6 \text{ in.})$ Substitute.
$= (4 \text{ ft})(3 \text{ ft})(0.5 \text{ ft})$ Convert inches to feet.
$= 6 \text{ ft}^3$ Simplify.

The step requires 6 ft^3 of concrete.

ALTERNATE SOLUTION:

$V = \ell wh$ Formula
$= (4 \text{ ft})(3 \text{ ft})(6 \text{ in.})$ Substitute.
$= (48 \text{ in.})(36 \text{ in.})(6 \text{ in.})$ Convert feet to inches.
$= 10{,}368 \text{ in.}^3$

The step requires 10,368 in.^3 of concrete.

WARM-UPS A–D

A. Find the volume of a cube that is 8 ft on each side.

B. How much concrete is needed for a rectangular stepping-stone that is 1 ft long, 8 in. wide, and 3 in. deep?

Answers to Warm-Ups

A. 512 ft^3

B. It will take 288 in.^3 of concrete.

C. Find the volume of a cylinder with a radius of 12 cm and a height of 13 cm. Round to the nearest hundredth. Let $\pi \approx 3.14$.

D. Find the volume of a right circular cone that has a base diameter of 3 ft and a height of 7 ft. Use the π key, and round to the nearest hundredth.

C. What is the volume of a can (cylinder) that has a radius of 2 ft and a height of 6 ft? Let $\pi \approx 3.14$.

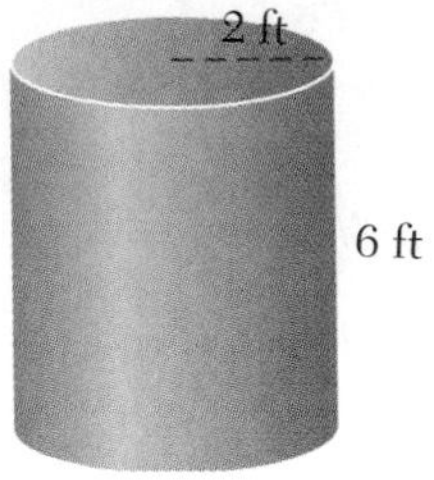

$V = \pi r^2 h$ — Formula

$\approx (3.14)(2\text{ ft})^2(6\text{ ft})$ — Substitute.

$\approx 75.36\text{ ft}^3$ — Simplify.

The volume of the can is about 75.36 ft^3.

ALTERNATE SOLUTION:

$V = \pi r^2 h$ — Formula

$= (\pi)(2\text{ ft})^2(6\text{ ft})$ — Substitute.

$\approx 75.39822\text{ ft}^3$ — Simplify, using the π key on your calculator.

The volume of the can is about 75.40 ft^3, rounded to the nearest hundredth.

D. Find the volume of the right circular cone. Round to the nearest hundredth.

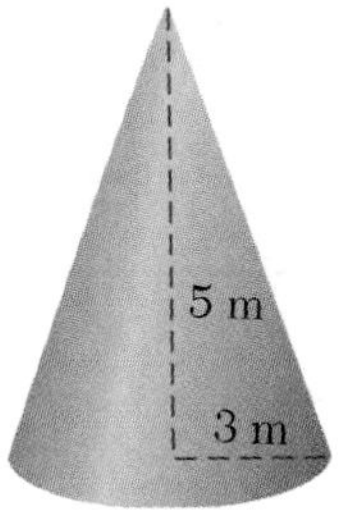

$V = \frac{1}{3}\pi r^2 h$ — Formula

$= \frac{1}{3}\pi(3\text{ m})^2(5\text{ m})$ — Substitute.

$\approx 47.12\text{ m}^3$ — Simplify, using the π key on your calculator. Round to the nearest hundredth.

The volume of the cone is about 47.12 m^3.

HOW & WHY

OBJECTIVE 2 Convert units of volume.

As Example B suggests, it is possible to measure the same volume in both cubic feet and cubic inches. We calculated that the amount of concrete needed for the step is 6 ft^3 or 10,368 in.^3.

To convert units of volume in general, we use unit fractions. However, because we are converting cubic units, we use the unit fraction three times.

$$6\text{ ft}^3 = 6(\text{ft})(\text{ft})(\text{ft})$$
$$= 6\left(\cancel{\text{ft}}\cdot\frac{12\text{ in.}}{1\ \cancel{\text{ft}}}\right)\left(\cancel{\text{ft}}\cdot\frac{12\text{ in.}}{1\ \cancel{\text{ft}}}\right)\left(\cancel{\text{ft}}\cdot\frac{12\text{ in.}}{1\ \cancel{\text{ft}}}\right)$$
$$= 6\cdot 12\cdot 12\cdot 12(\text{in.})(\text{in.})(\text{in.})$$
$$= 10{,}368\text{ in.}^3$$

Answers to Warm-Ups

C. The volume is about 5878.08 cm^3.

D. The volume is about 16.49 ft^3.

When measuring the capacity of a solid to hold liquid, we sometimes use special units. Recall that in the English system, liquid capacity is measured in ounces, quarts, and gallons. In the metric system, milliliters, liters, and kiloliters are used. One cubic centimeter measures the same volume as 1 milliliter. That is, 1 cm^3 = 1 mL. So a container with a volume of 50 cm^3 holds 50 mL of liquid.

EXAMPLES E–H

DIRECTIONS: Convert units of volume.

STRATEGY: Use the appropriate unit fraction three times.

E. Convert 25 yd^3 to cubic feet.

$$25\text{ yd}^3 = 25\,\cancel{\text{yd}^3}\cdot\left(\frac{3\text{ ft}}{\cancel{\text{yd}}}\right)\left(\frac{3\text{ ft}}{\cancel{\text{yd}}}\right)\left(\frac{3\text{ ft}}{\cancel{\text{yd}}}\right)$$

Use the unit fraction $\frac{3\text{ ft}}{1\text{ yd}}$.

$$= 25\cdot 3\cdot 3\cdot 3\text{ ft}^3$$
$$= 675\text{ ft}^3$$

So, 25 yd^3 = 675 ft^3.

F. How many cubic meters in 76 cm^3?

$$76\text{ cm}^3 = 76\,\cancel{\text{cm}^3}\cdot\left(\frac{1\text{ m}}{100\,\cancel{\text{cm}}}\right)\left(\frac{1\text{ m}}{100\,\cancel{\text{cm}}}\right)\left(\frac{1\text{ m}}{100\,\cancel{\text{cm}}}\right)$$
$$= \frac{76}{100\cdot 100\cdot 100}\text{ m}^3$$
$$= 0.000076\text{ m}^3$$

So, 76 cm^3 = 0.000076 m^3.

G. How many milliliters of water does this container hold?

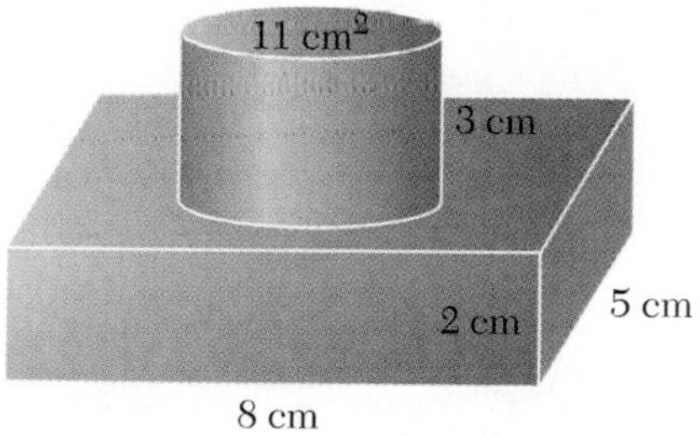

Total volume = $V_1 + V_2$

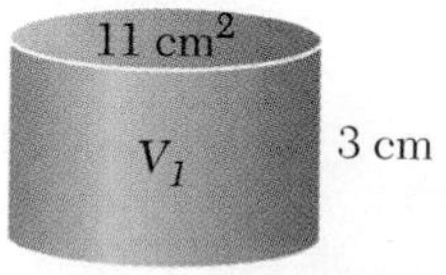

$$V_1 = Bh$$
$$= (11\text{ cm}^2)(3\text{ cm})$$
$$= 33\text{ cm}^3$$

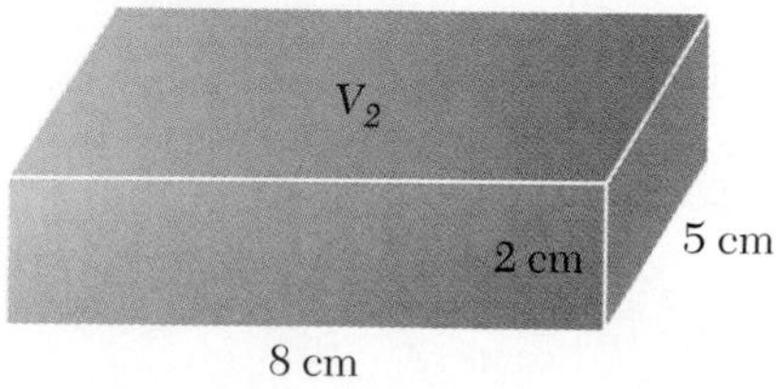

$$V_2 = \ell wh$$
$$= (8\text{ cm})(5\text{ cm})(2\text{ cm})$$
$$= 80\text{ cm}^3$$

$$\text{Total volume} = 33\text{ cm}^3 + 80\text{ cm}^3$$
$$= 113\text{ cm}^3$$
$$= 113\text{ mL}$$

The figure holds 113 mL of water.

WARM-UPS E–H

E. Convert 10 ft^3 to cubic inches.

F. How many cubic kilometers in 450,000 m^3?

G. How many milliliters of water does this container hold?

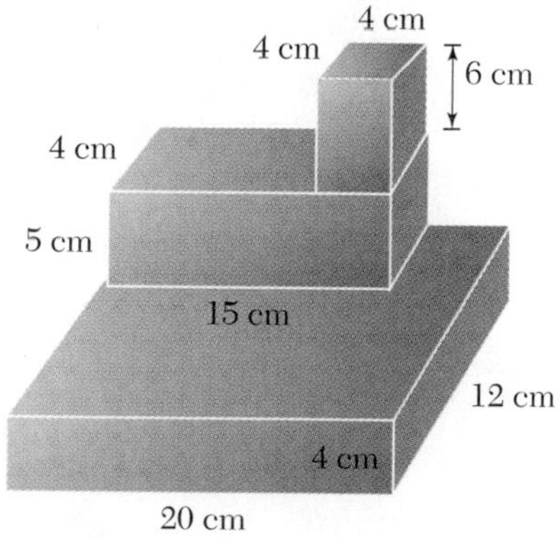

Answers to Warm-Ups

E. 17,280 $in.^3$ **F.** 0.00045 km^3
G. It holds 1356 mL of water.

H. June wants to mulch a flower bed along her house that is 25 ft by 3 ft. She wants the mulch to be 6 in. deep. How many 0.5 yd^3 bags of mulch does she need?

H. Dave has a rose garden that is 6 ft by 15 ft that he wants to cover in mulch 4 in. deep. He can get the mulch in bags that hold 0.5 yd^3 each. How many bags of mulch does Dave need for his garden?

$V = \ell w h$ — Formula

$= (15\text{ ft})(6\text{ ft})(4\text{ in.})$ — Substitute 15 ft for ℓ, 6 ft for w, and 4 in. for h.

$= (15\text{ ft})(6\text{ ft})\left(\frac{1}{3}\text{ ft}\right)$ — Convert inches to feet.

$= 30\text{ ft}^3$ — Simplify.

$= 30(\text{ft})(\text{ft})(\text{ft})\left(\frac{1\text{ yd}}{3\text{ ft}}\right)\left(\frac{1\text{ yd}}{3\text{ ft}}\right)\left(\frac{1\text{ yd}}{3\text{ ft}}\right)$ — Convert cubic feet to cubic yards.

$= \frac{30}{27}\text{ yd}^3$ — Simplify.

$\approx 1.1\text{ yd}^3$

Because $1.1 \div 0.5 = 2.2$, Dave will need three bags of mulch to cover his garden to a depth of 4 in., with some mulch left over. Or, because 1.1 yd^3 is pretty close to 1 yd^3, Dave could choose to buy only two bags of mulch and settle for the mulch being not quite 4 in. deep.

Answers to Warm-Ups

H. June needs 3 bags of mulch.

Name ______________________ Class ______________________ Date ____________

EXERCISES 7.5

OBJECTIVE 1 Find the volume of common geometric shapes.

A.

Find the volume of the figures. Use 3.14 for π.

1.

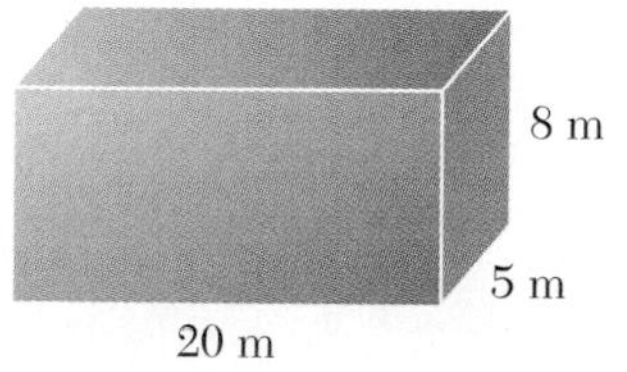

2.

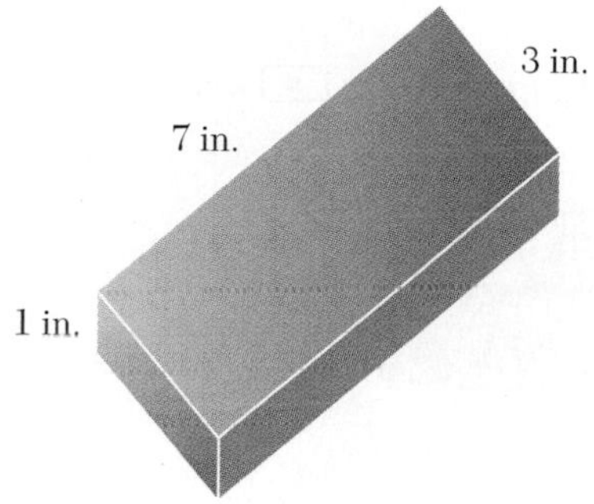

3.

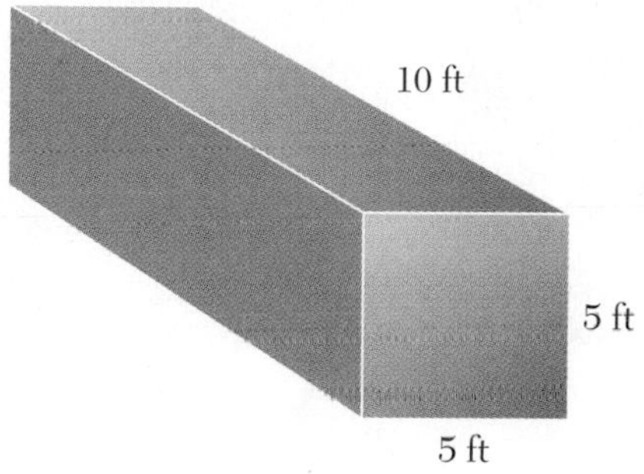

4.

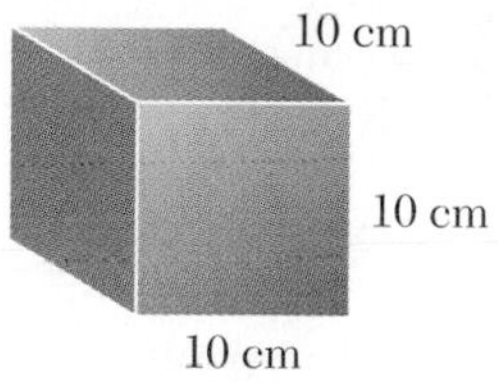

5.

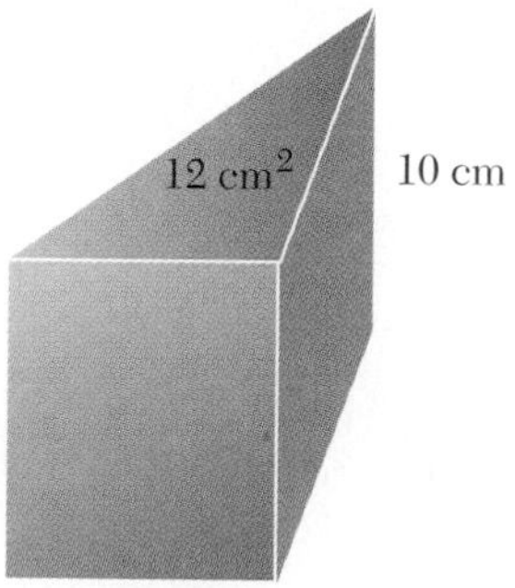

6.

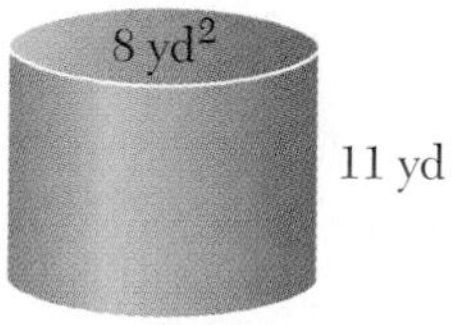

7.

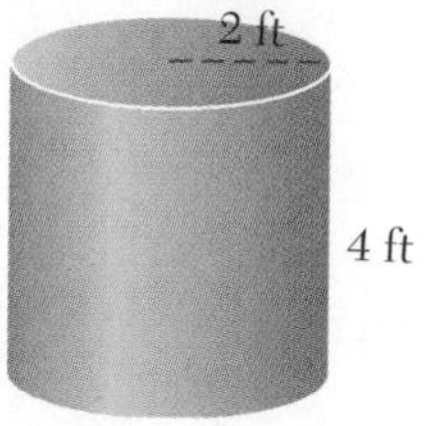

8.

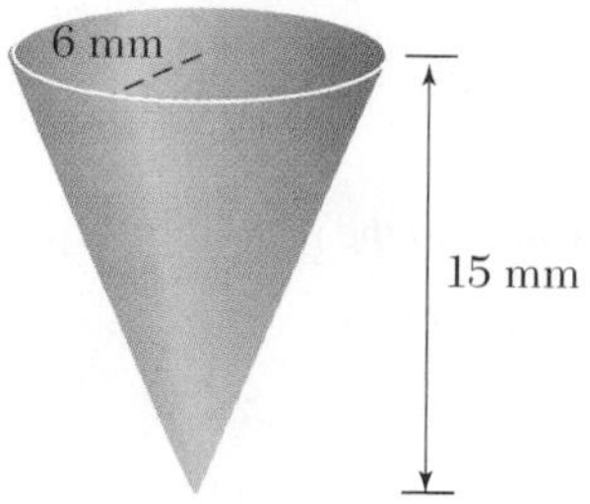

B.

9. How many mL of water will fill up a box that measures 60 cm long, 35 cm wide, and 15 cm high?

10. Find the volume of a cube that measures 24 mi on each side.

11. Find the volume of a garbage can that is 4 ft tall and has a base of 12 ft^2.

12. Find the volume of two identical fuzzy dice tied to the mirror of a 1957 Chevy if one edge measures 5 in.

13. Find the volume of a beach ball that is 18 in. in diameter. Use 3.14 for π.

14. Find the volume of an ice cream cone that is 5 inches tall and has a diameter of 4 inches. Use $\frac{22}{7}$ for π.

Find the volume of the figures. Use the [π] *key, and round to the nearest hundredth.*

15.

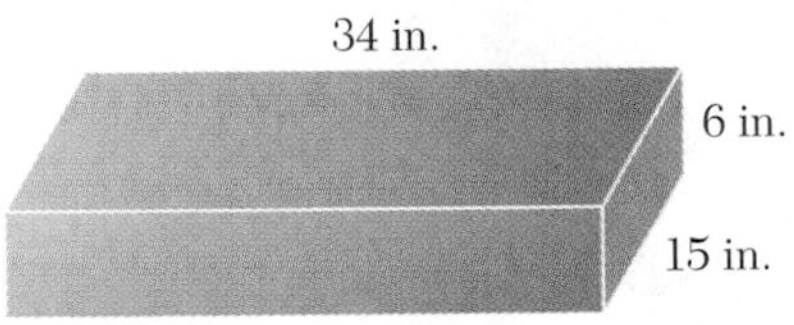

16.

17.

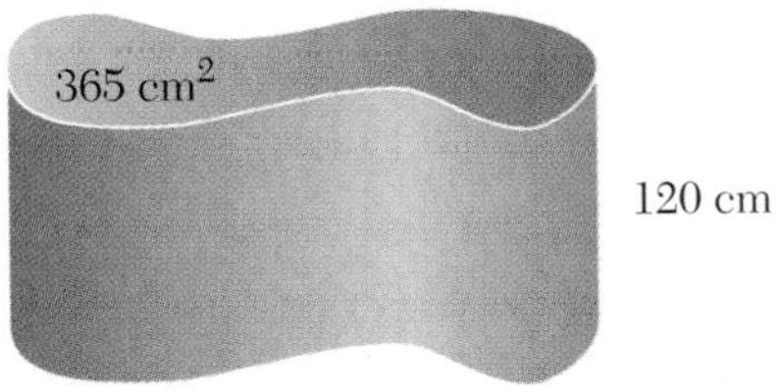

18.

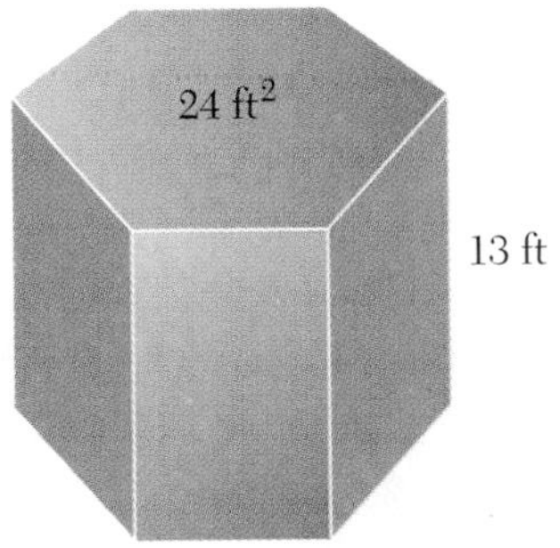

OBJECTIVE 2 Convert units of volume.

A. *Convert units of volume.*

19. Convert 1 ft^3 to cubic inches.

20. Convert 1 m^3 to cubic centimeters.

21. Convert 1 yd^3 to cubic feet.

22. Convert 1 km^3 to cubic meters.

23. Convert 1 yd^3 to cubic inches.

24. Convert 1 km^3 to cubic centimeters.

25. Convert 6912 $in.^3$ to cubic feet.

26. Convert 5000 mm^3 to cubic centimeters.

27. Convert 216 ft^3 to cubic yards.

28. Convert 2,000,000,000 m^3 to cubic kilometers.

B. *Convert. Round decimal values to the nearest hundredth.*

29. Convert 14 ft^3 to cubic inches.

30. Convert 12 m^3 to cubic centimeters.

31. Convert 1 mi^3 to cubic feet.

32. Convert 123,456,789 m^3 to cubic kilometers.

33. Convert 160 in.^{3} to cubic feet.

34. Convert 2 ft^3 to cubic centimeters.

35. Convert 1 mi^3 to cubic kilometers.

36. Convert 350 m^3 to cubic yards.

37. Convert 7800 cm^3 to cubic feet.

38. Convert 25 in.^{3} to cubic millimeters.

C.

39. Find the volume of a can that has a base with area 245 in.^{2} and a height of 2 ft.

40. How many cubic inches of concrete are needed to pour a sidewalk that is 3 ft wide, 4 in. deep, and 54 ft long? Concrete is commonly measured in cubic yards, so convert your answer to cubic yards. *Hint*: Convert to cubic feet first, and then convert that to cubic yards.

Find the volume of the figures.

41.

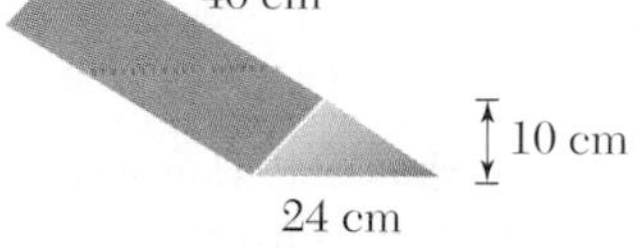

42.

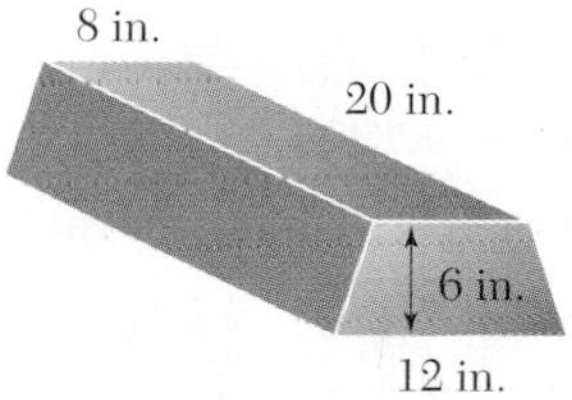

43.

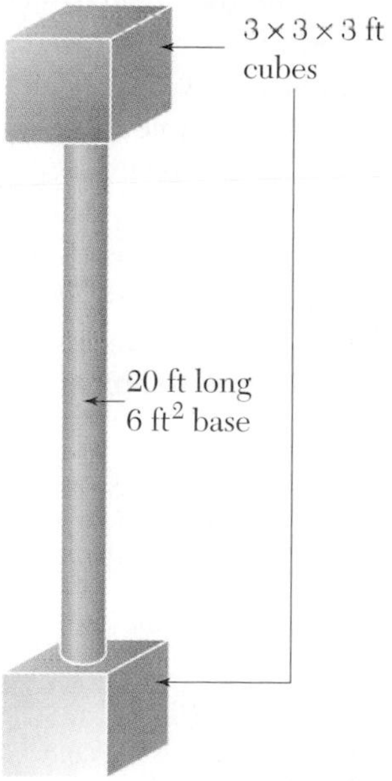

44.

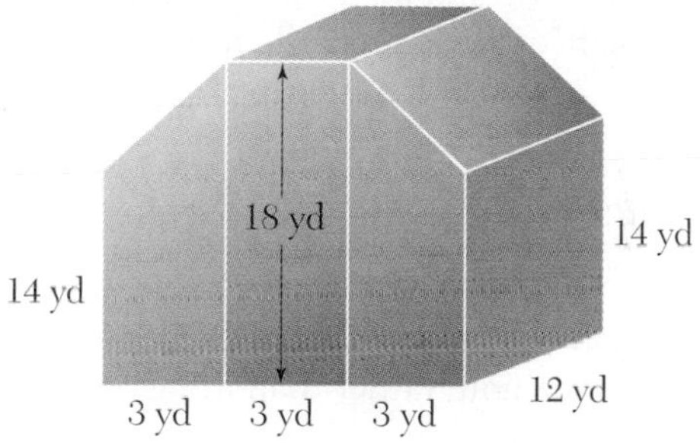

45.

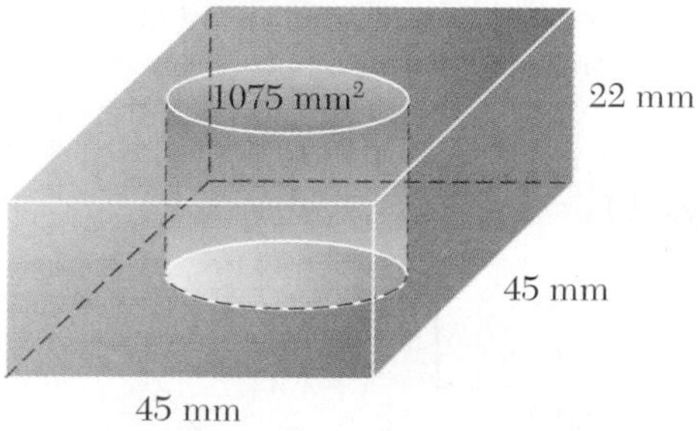

46.

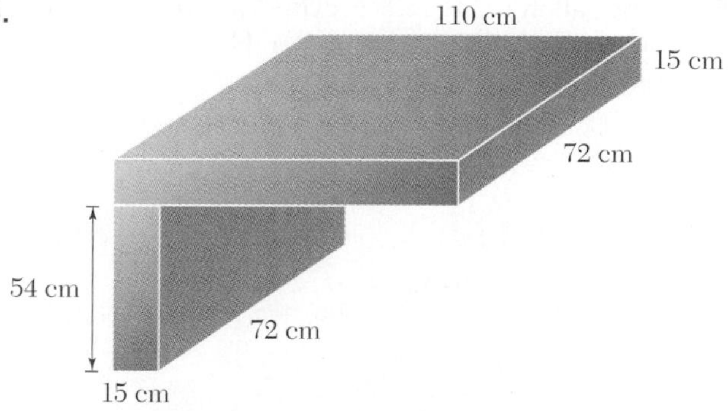

47. A bag of potting soil contains 4000 in.^{3}. How many bags are needed to fill five flower boxes, each of which measures 4 ft long, 8 in. high, and 6 in. deep? Remember that you can only buy whole bags of potting soil.

48. An excavation is being made for a basement. The hole is 24 ft wide, 36 ft long, and 7 ft deep. If the bed of a truck holds 378 ft^3, how many truckloads of dirt will need to be hauled away?

49. A farming corporation is building four new grain silos. The inside dimensions of the silos are given in the table.

	Area of base	Height
Silo A	1800 ft^2	60 ft
Silo B	1200 ft^2	75 ft
Silo C	900 ft^2	80 ft
Silo D	600 ft^2	100 ft

Find the total volume in these silos.

Exercises 50–54 refer to the chapter application.

50. According to industry standards, a joint thickness of $\frac{1}{4}$ inch means the bricklayer will need 9 ft^3 of mortar per 1000 bricks. Find the total amount of mortar needed for the patio and walkways. Round to the nearest whole ft^3. (See Exercise 72 in Section 7.4.)

51. The cement subcontracter orders materials based on the total volume of the slab. The industry standard for patios and walkways is 4–5 in. of thickness. Because the slab will be topped with bricks, the contracter decides on a thickness of 4 in. Find the volume of the slab in cubic inches. Convert this to cubic feet, rounding up to the next whole cubic foot. Convert this to cubic yards. Round up to the next whole cubic yard if necessary. (See Exercise 71 in Section 7.4.)

52. Explain why, in Exercise 51, it is necessary to round up to the next whole unit, rather than using the rounding rule stated in Section 1.1.

53. The cement contractor in Exercise 51 must first build a wood form that completely outlines the slab. How many linear feet of wood are needed to build the form?

54. The landscaper recommends that Barbara and Paul buy topsoil before planting the garden. They must buy enough to be able to spread the topsoil to a depth of 8 in. How many cubic inches of top soil do they need? Because soil is usually sold in cubic yards, round this figure up to the next cubic yard.

55. One gallon is 231 in.^{3}. A cylindrical trash can holds 32 gal. If it is 4 ft tall, what is the area of the base of the trash can?

56. One gallon is 231 in.^{3}. A cylindrical trash can holds 36 gal and is 24 in. in diameter. How tall is the trash can?

57. Norma is buying mushroom compost to mulch her garden. The garden is pictured here. How many cubic yards of compost does she need to mulch the entire garden 4 in. deep? (She cannot buy fractional parts of a cubic yard.)

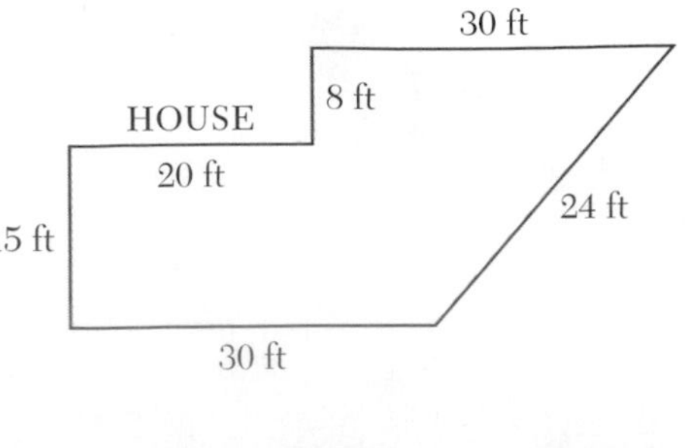

58. One gallon is 231 in.^{3}. How many cubic inches does a 20-gal fish tank hold? Find a possible length, width, and height in inches for a 20-gal fish tank.

59. Lucy is packing water polo balls into a box to ship to a tournament. Each ball has a radius of 6 in. Lucy's box is 3 ft by 2 ft by 2 ft. How many balls can she fit into the box?

60. How much space in Lucy's box is empty, and how much of it is occupied by water polo balls? Use the $\boxed{\pi}$ key, and round to the nearest whole cubic inch. (See Exercise 59.)

STATE YOUR UNDERSTANDING

61. Explain what is meant by volume. Name three occasions in the last week when the volume of an object was relevant.

62. Explain how to find the volume of the following figure.

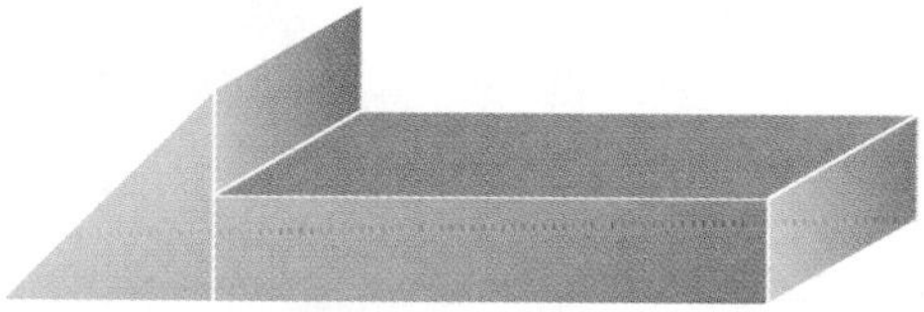

63. Explain why the formula for the volume of a box is a special case of the formula $V = bh$.

CHALLENGE

64. The Bakers are constructing an in-ground pool in their backyard. The pool will be 15 ft wide and 30 ft long. It will be 3 ft deep for 10 ft at one end. It will then drop to a depth of 10 ft at the other end. How many cubic feet of water are needed to fill the pool?

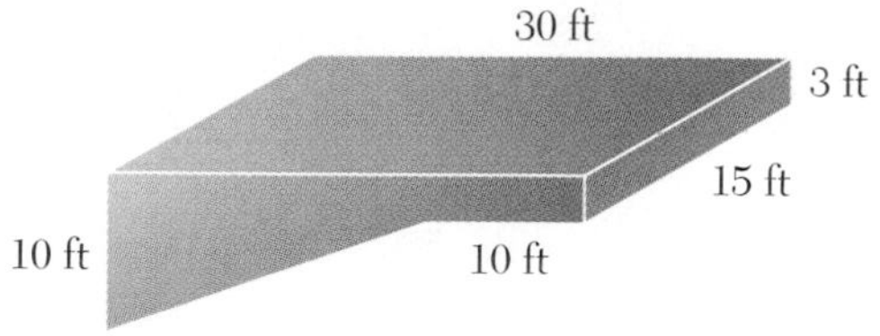

If the pool is to have a capacity of 14 yd^3, how many loads of dirt were dug out and hauled away?

65. Doug is putting a 5-m-by-20-m swimming pool in his backyard. The pool will be 2 m deep. What is the size of the hole for the pool? The hole will be lined with concrete that is 15 cm thick. How much concrete is needed to line the pool?

GROUP WORK

66. The insulating ability of construction materials is measured in R values. Industry standards for exterior walls are currently R-19. An 8-in. thickness of loose fiberglass is necessary to achieve an R-19 value. Calculate the amount of cubic feet of loose figerglass needed to insulate the exterior walls of the mountain cabin pictured. All four side windows measure 2 × 3 ft. Both doors measure 4 × 7 ft. The front window measures 5 × 3 ft.

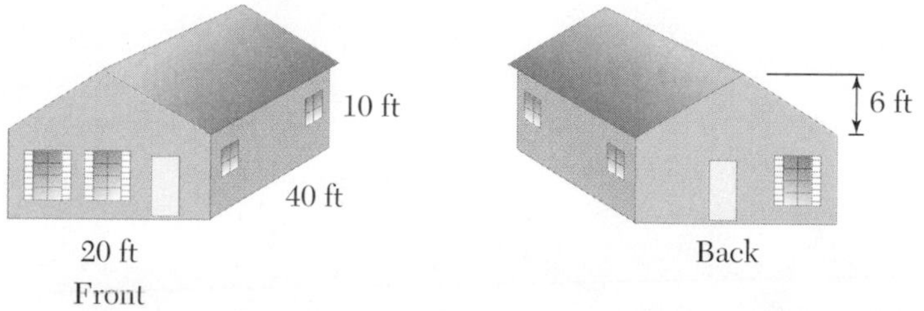

MAINTAIN YOUR SKILLS

Evaluate the following:

67. 8^3

68. $4^2 + 5^2$

69. $2^3 + 3^3 + 4^3$

70. $9^2 - 4^2$

71. $(2^2 + 3^2)^2$

72. $3168 \div 24$

73. $(10^2)(2^2)(3^2)$

74. $5(36 \div 12 + 1 - 2)$

75. Find the area of a rectangle that is 3 in. wide and 2 ft long.

76. A warehouse store sells 5 lb bags of Good and Plenty. Bob buys a bag and stores the candy in 20-oz jars. How many jars does he need?

7.6 Square Roots and the Pythagorean Theorem

VOCABULARY

A **perfect square** is a whole number that is the square of another whole number. For example, $16 = 4^2$, so 16 is a perfect square.

A **square root** of a positive number is a number that is squared to give the original number. The symbol "$\sqrt{\ }$" is used to name a square root of a number.

$\sqrt{16} = 4$ because $4^2 = 4 \cdot 4 = 16$

A **right triangle** is a triangle with one right (90°) angle.

The **hypotenuse** of a right triangle is the side that is across from the right angle.

The **legs** of a right triangle are the sides adjacent to the right angle.

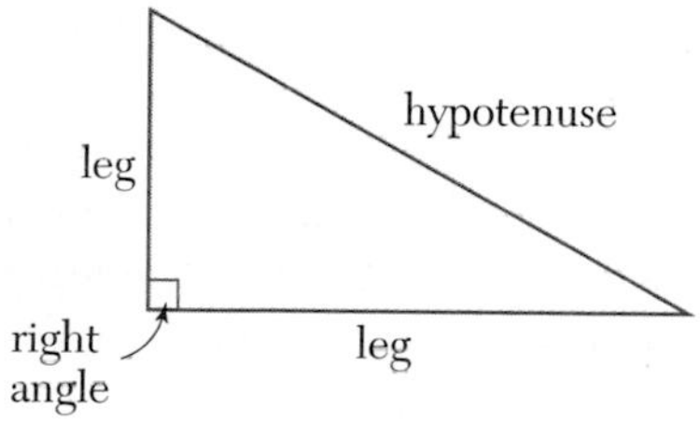

OBJECTIVES

1. Find the square root of a number.
2. Apply the Pythagorean theorem.

HOW & WHY

OBJECTIVE 1 Find the square root of a number.

The relationship between squares and square roots is similar to the relationship between multiples and factors. In Chapter 2, we saw that if one number is a multiple of a second number, then the second number is a factor of the first number. It is also true that if one number is the square of a second number, then the second number is a square root of the first number. For example, the number 81 is the square of 9, so 9 is a square root of 81. We write this symbolically as $9^2 = 81$ so $\sqrt{81} = 9$.

A perfect square is a whole number that is the square of another whole number. It is relatively easy to find the square root of a perfect square or the square root of a fraction whose numerator and denominator are perfect squares. Simply square numbers until you find the correct one.

To find the square root of a perfect square

Use guess and check. Square whole numbers until you find the correct one.

EXAMPLES A–C

DIRECTIONS: Find a square root.

STRATEGY: Use guess and check.

A. $\sqrt{121}$

$\sqrt{121} = 11$ Because $11 \cdot 11 = 121$

WARM-UPS A–C

A. $\sqrt{144}$

Answers to Warm-Ups:

A. 12

B. $\sqrt{729}$

B. $\sqrt{529}$

Trial number	Square	
20	$20 \cdot 20 = 400$	Too small
25	$25 \cdot 25 = 625$	Too large
24	$24 \cdot 24 = 576$	Too large
23	$23 \cdot 23 = 529$	

So, $\sqrt{529} = 23$.

C. $\sqrt{\frac{9}{25}}$

C. $\sqrt{\frac{4}{49}}$

$2 \cdot 2 = 4$ and $7 \cdot 7 = 49$

So, $\sqrt{\frac{4}{49}} = \frac{2}{7}$.

*

When we need to find the square root of a number that is not a perfect square, the method of guess and check is possible but usually too time-consuming to be practical. To further complicate matters, many square roots of nonperfect squares are nonterminating decimals. In these cases, we find an approximation of the desired square root and round to a convenient decimal place.

For instance, consider $\sqrt{5}$.

Guess	So $\sqrt{5}$ is between . . .
$2^2 = 4$ $3^2 = 9$	2 and 3
$2.2^2 = 4.84$ $2.3^2 = 5.29$	2.2 and 2.3
$2.23^2 = 4.9729$ $2.24^2 = 5.0176$	2.23 and 2.24
$2.236^2 = 4.999696$ $2.237^2 = 5.004169$	2.236 and 2.237

At this point, we discontinue the process and conclude that

$\sqrt{5} \approx 2.24$ to the nearest hundredth

Because all but the most basic calculators have a square root key, [√], we usually find the square roots of nonperfect squares by using a calculator. The calculator gives us $\sqrt{5} \approx 2.2360679775$, which is consistent with our previous calculation. (*Note*: Some calculators require that you hit the square root key, [√], before entering the number, and others require that you enter the number first.)

To find a square root of a number

Use the square root key, [√], on a calculator. Round as necessary.

Answers to Warm-Ups:

B. 27 **C.** $\frac{3}{5}$

EXAMPLES D–E

DIRECTIONS: Find a square root. Round to the nearest hundredth if necessary.

STRATEGY: Use the square root key, [√], on a calculator; then round.

D. $\sqrt{95}$

The calculator gives $\sqrt{95} \approx 9.74679434481$

So, $\sqrt{95} \approx 9.75$, to the nearest hundredth.

E. $\sqrt{213.16}$

The calculator gives $\sqrt{213.16} = 14.6$

So, $\sqrt{213.16} = 14.6$.

WARM-UPS D–E

D. $\sqrt{87}$

E. $\sqrt{696.96}$

HOW & WHY

OBJECTIVE 2 Apply the Pythagorean theorem.

A right triangle is a triangle with one right (90°) angle. Figure 7.8 shows the right angle, legs, and hypotenuse of a right triangle.

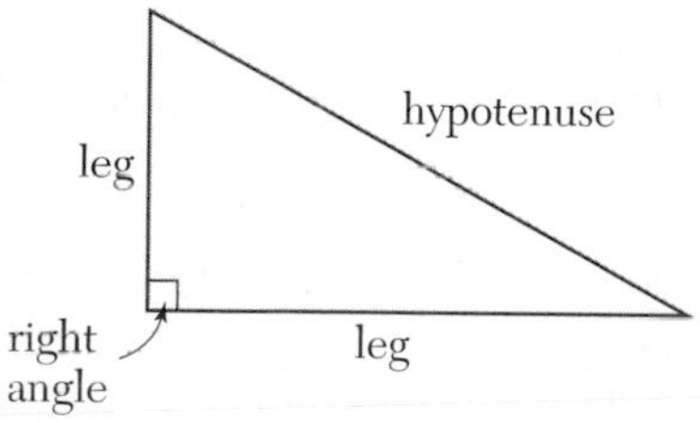

FIGURE 7.8

Thousands of years ago, a Greek named Pythagoras discovered that the lengths of the sides of a right triangle have a special relationship. This relationship is known today as the Pythagorean theorem.

Pythagorean Theorem

In a right triangle, the sum of the squares of the legs is equal to the square of the hypotenuse

or

$(\text{leg}_1)^2 + (\text{leg}_2)^2 = \text{hypotenuse}^2$

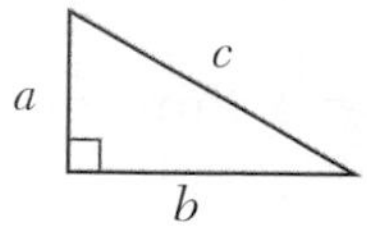

$a^2 + b^2 = c^2$

The Pythagorean theorem enables us to calculate the length of the third side of a right triangle when we know the lengths of the other two sides. Because the Pythagorean theorem involves the squares of the sides, when we want the length of a side, we need to use square roots. The following formulas are used to calculate an unknown side of a right triangle.

To calculate an unknown side of a right triangle

Use the appropriate formula.

$\text{Hypotenuse} = \sqrt{(\text{leg}_1 + (\text{leg}_2)^2}$ or $c = \sqrt{a^2 + b^2}$

$\text{Leg} = \sqrt{(\text{hypotenuse})^2 - (\text{known leg})^2}$ or $a = \sqrt{c^2 - b^2}$

Answers to Warm-Ups:
D. 9.33 **E.** 26.4

WARM-UPS F–H

F. Find the approximate length of the hypotenuse.

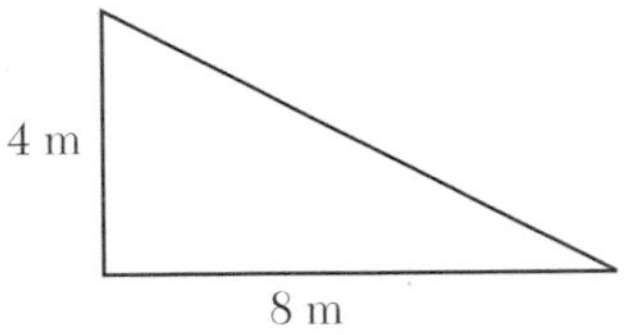

G. Find the length of the unknown leg.

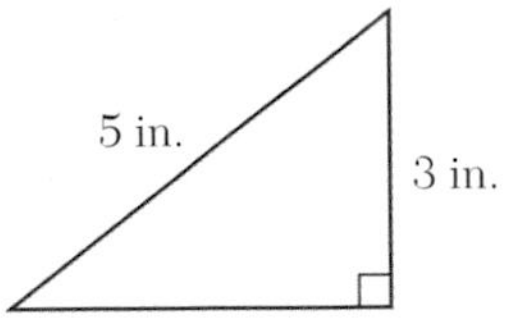

Answers to Warm-Ups:

F. 8.94 m **G.** 4 in.

EXAMPLES F–H

DIRECTIONS: Find the unknown side of the given right triangle.

STRATEGY: Use the formulas. Round decimal values to the nearest hundredth.

F. Find the approximate length of the hypotenuse.

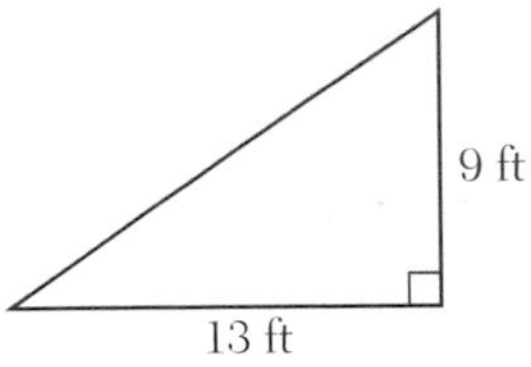

$\text{Hypotenuse} = \sqrt{(\text{leg}_1) + (\text{leg}_2)^2}$	Formula
$= \sqrt{9^2 + 13^2}$	Substitute 9 and 13 for the lengths of leg_1 and leg_2.
$= \sqrt{81 + 169}$	Simplify.
$= \sqrt{250}$	
≈ 15.81	

So, the length of the hypotenuse is about 15.81 ft.

ALTERNATIVE SOLUTION: Use the formula $c = \sqrt{a^2 + b^2}$

$c = \sqrt{a^2 + b^2}$	
$= \sqrt{9^2 + 13^2}$	Substitute 9 for a and 13 for b.
$= \sqrt{81 + 169}$	Simplify.
$= \sqrt{250}$	
≈ 15.81	

The hypotenuse is about 15.81 ft.

G. Find the length of the unknown leg.

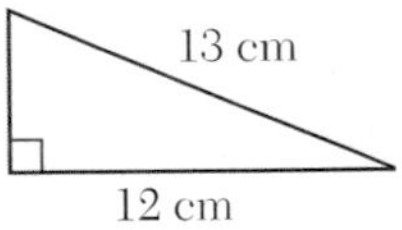

$\text{Leg} = \sqrt{(\text{hypotenuse})^2 - (\text{known leg})^2}$	Formula
$= \sqrt{13^2 - 12^2}$	Substitute 13 for the length of the hypotenuse and 12 for the length of the known leg.
$= \sqrt{169 - 144}$	Simplify.
$= \sqrt{25}$	
$= 5$	

So, the unknown leg is 5 cm.

H. Ted is staking a small tree to keep it growing upright. He wants to attach the rope 3 ft above the ground on the tree and 6 ft from the base of the tree, and he needs one extra foot of rope on each end for tying. How much rope does he need?

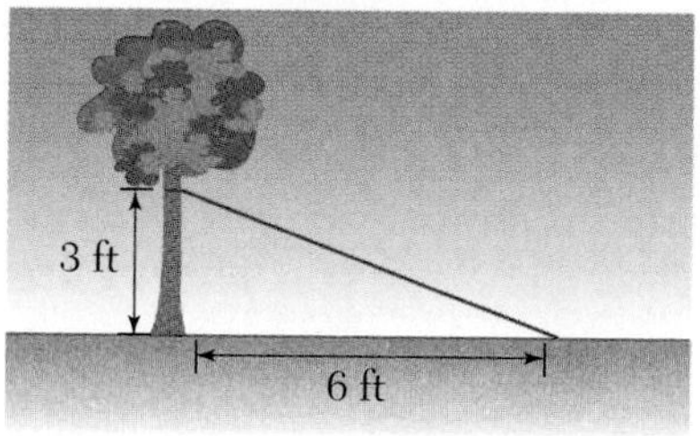

STRATEGY: The tree trunk, the ground, and the rope form a right triangle, with the rope being the hypotenuse of the triangle. We use the formula for finding the hypotenuse.

Hypotenuse	$= \sqrt{(\text{leg}_1) + (\text{leg}_2)^2}$	Formula
	$= \sqrt{3^2 + 6^2}$	Substitute 3 and 6 for the lengths of leg_1 and leg_2.
	$= \sqrt{9 + 36}$	Simplify.
	$= \sqrt{45}$	
	≈ 6.71	

The rope support is about 6.71 ft long, so Ted needs 8.71 ft including the extra 2 ft for the ties.

H. A guy wire for a radio tower is attached 40 ft above the ground and 60 ft from the base of the tower. An additional 6 in. is needed on each end for fastening. How long is the guy wire?

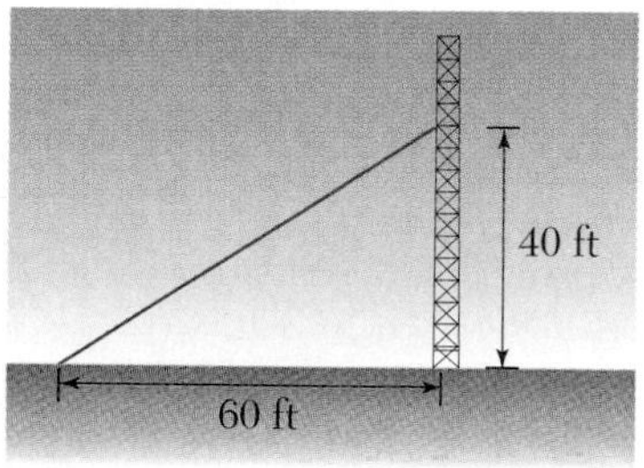

Answers to Warm-Ups:

H. The guy wire is about 73.11 ft.

EXERCISES 7.6

OBJECTIVE 1 Find the square root of a number.

A. *Simplify.*

1. $\sqrt{25}$
2. $\sqrt{49}$
3. $\sqrt{169}$
4. $\sqrt{36}$
5. $\sqrt{225}$
6. $\sqrt{100}$
7. $\sqrt{400}$
8. $\sqrt{900}$
9. $\sqrt{\dfrac{81}{49}}$
10. $\sqrt{\dfrac{25}{64}}$
11. $\sqrt{0.64}$
12. $\sqrt{0.16}$

B. *Simplify. Use the square root key,* $\boxed{\sqrt{\ }}$, *on your calculator and round to the nearest hundredth.*

13. $\sqrt{140}$
14. $\sqrt{999}$
15. $\sqrt{729}$
16. $\sqrt{1681}$
17. $\sqrt{10.67}$
18. $\sqrt{50.41}$
19. $\sqrt{\dfrac{6}{13}}$
20. $\sqrt{\dfrac{625}{2304}}$
21. $\sqrt{11{,}025}$
22. $\sqrt{12{,}000}$

OBJECTIVE 2 Apply the Pythagorean theorem.

Find the unknown side of each right triangle. Round decimal values to the nearest hundredth.

A.

23.
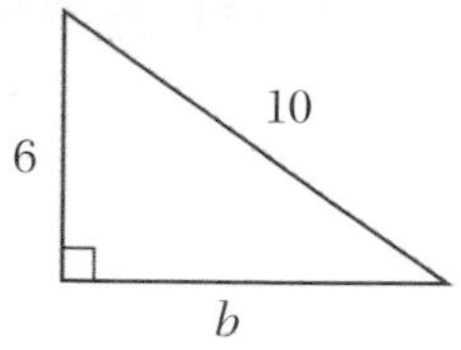

24.
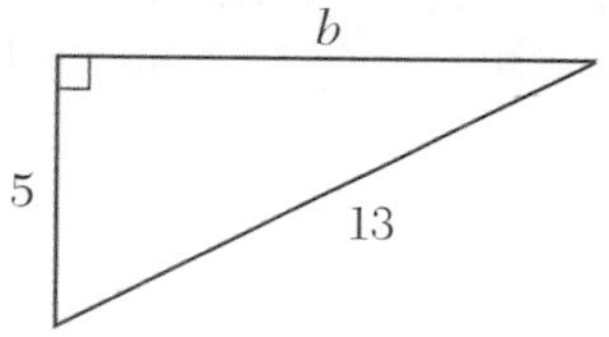

25.
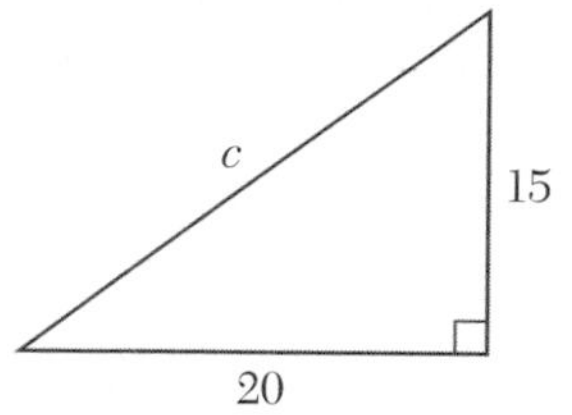

26.
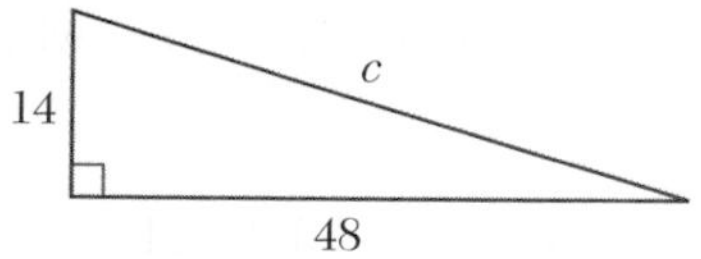

27.
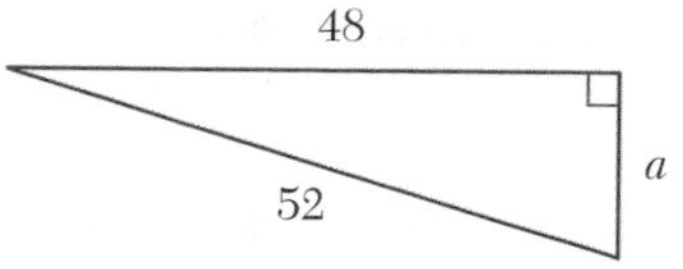

28.
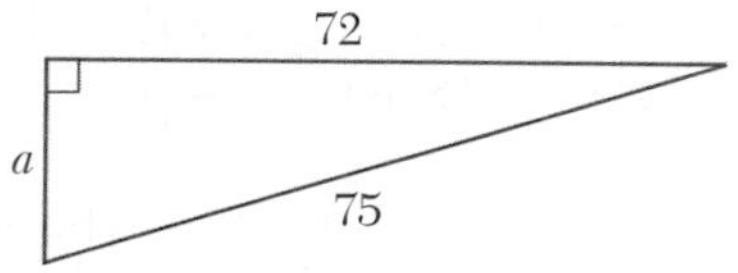

29.
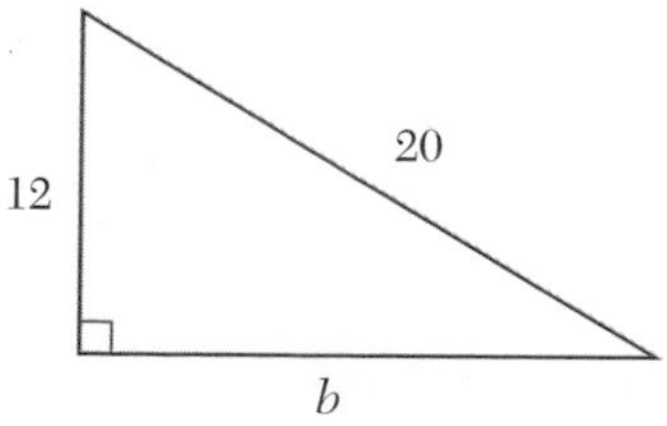

30.
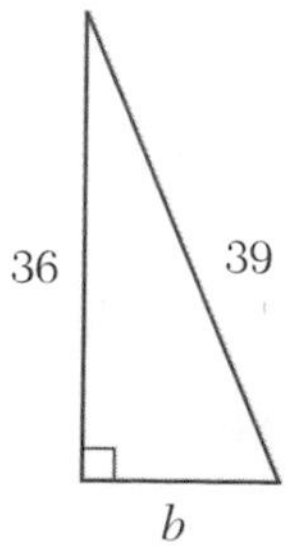

 Name ______________________ Class ______________ Date __________

B.

31.

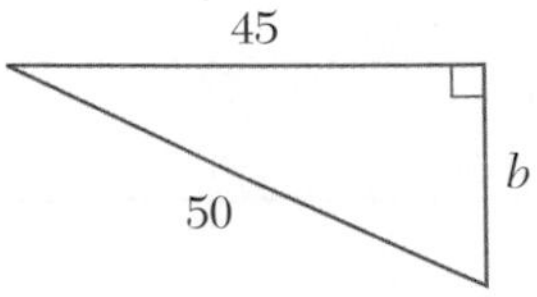

32.

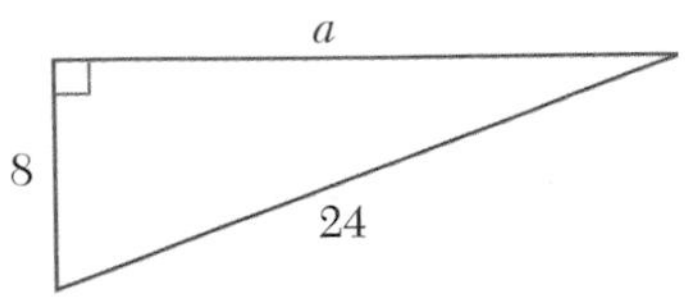

33.

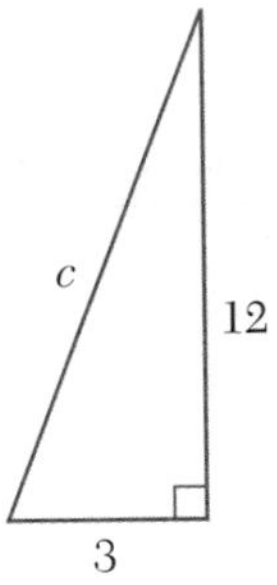

34.

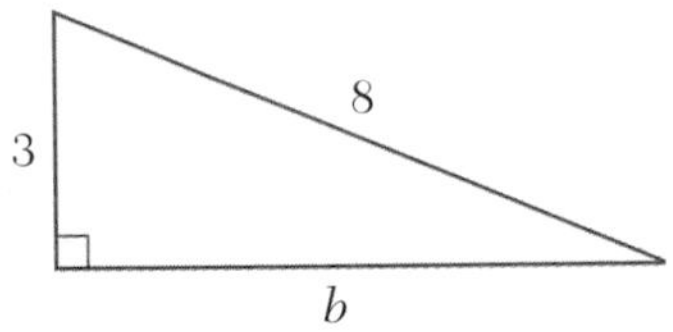

35.

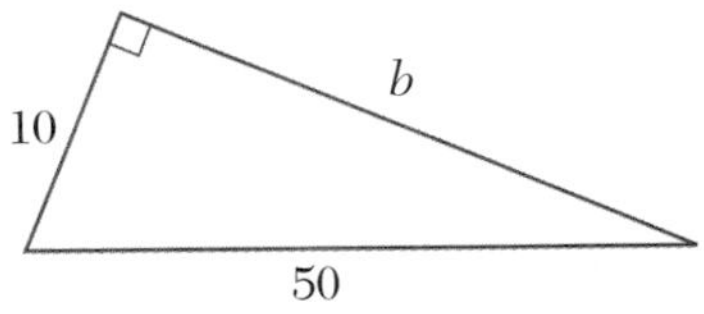

36.

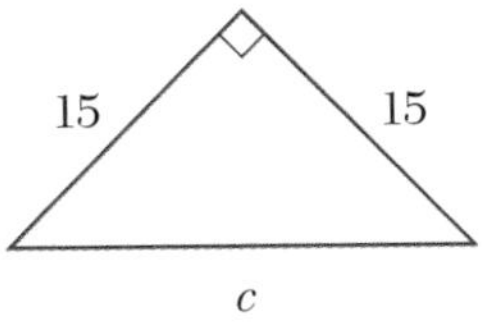

37.

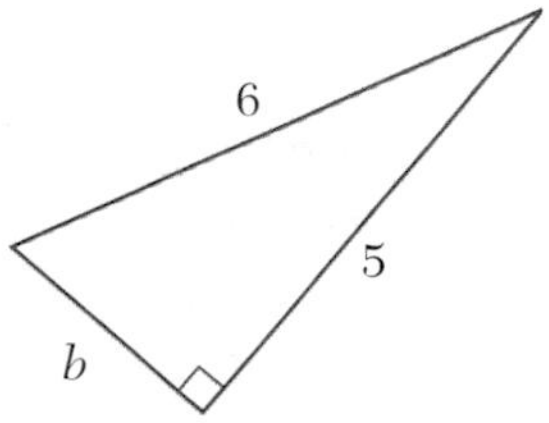

38.

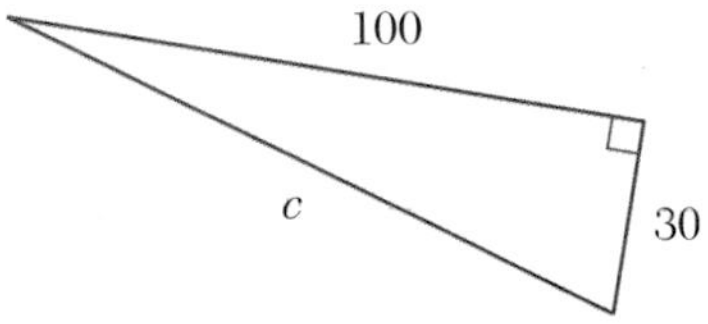

C.

39. Find the hypotenuse of a right triangle that has legs of 28 cm and 96 cm.

40. Find the length of one leg of a right triangle if the other leg is 24 in. and the hypotenuse is 30 in.

41. What is the length of the side of a square whose area is 144 ft^2? (The formula for the length of a side of a square is $S = \sqrt{A}$, where A is the area.)

42. What is the length of the diagonal of a square whose area is 144 ft^2? See Exercise 41.

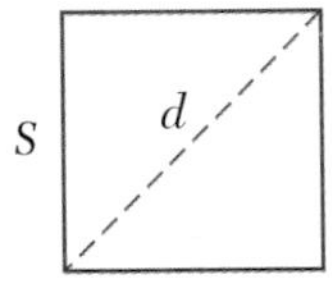

43. What is the length of the side of a square whose area is 289 m^2 ? See Exercise 41.

44. What is the length of the diagonal of a square whose area is 289 m^2?

45. Paul has a 12-ft ladder. If he sets the foot of the ladder 3 ft away from the base of his house, how far up the house is the top of the ladder?

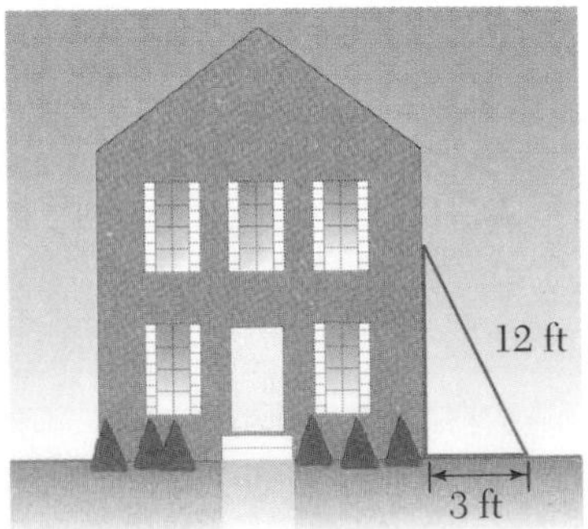

46. Lynzie drives 35 miles due north and then 16 miles due east. How far is she from her starting point?

47. A baseball diamond is actually a square that is 90 ft on a side. How far is it from first base to third base?

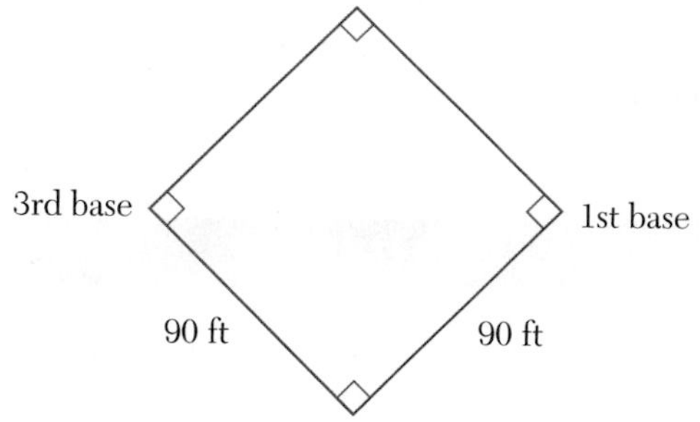

48. How tall is the tree?

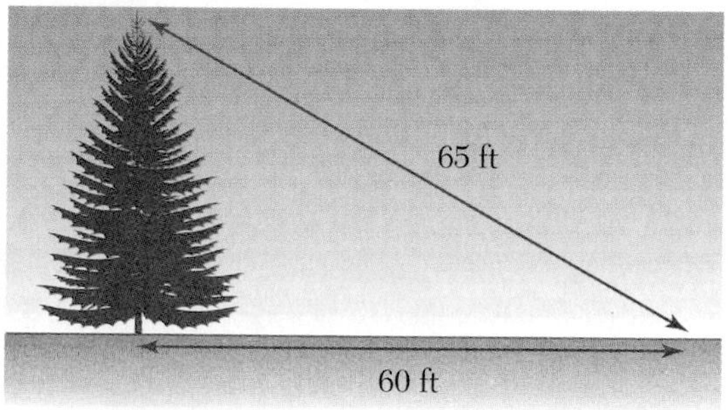

49. Harry wants to install wire supports for the net on his pickleball court. The net is 36 in. high, and he plans to anchor the supports in the ground 4 ft from the base of the net. Find how much wire he needs for two supports, figuring an additional foot of wire for each support to fasten the ends.

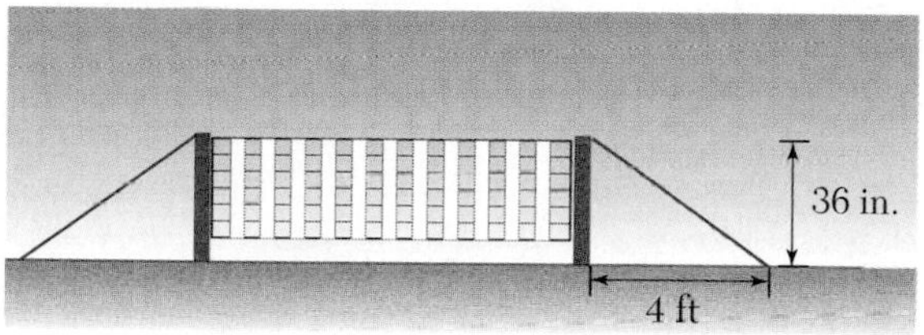

50. The 7th hole on Jim's favorite golf course has a pond between the tee and the green. If Jim can hit the ball 170 yards, should he hit over the pond or go around it?

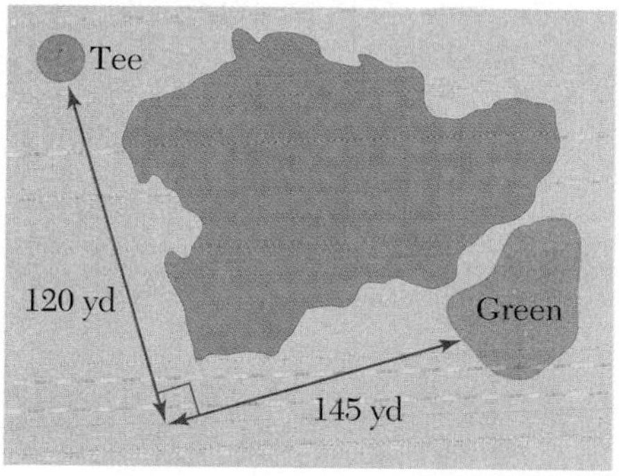

STATE YOUR UNDERSTANDING

51. What is a square root? Explain how to find $\sqrt{289}$ without using a calculator.

52. A carpenter is building a deck on the back of a house. He wants the joists of the deck to be perpendicular to the house, so he attaches a joist to the house, measures 4 ft out on the joist and 3 ft over from the joint on the house. He then wiggles the joist until his two marks are 5 ft apart. At this point the carpenter knows that the joist is perpendicular (makes a right angle) with the house. Explain how he knows.

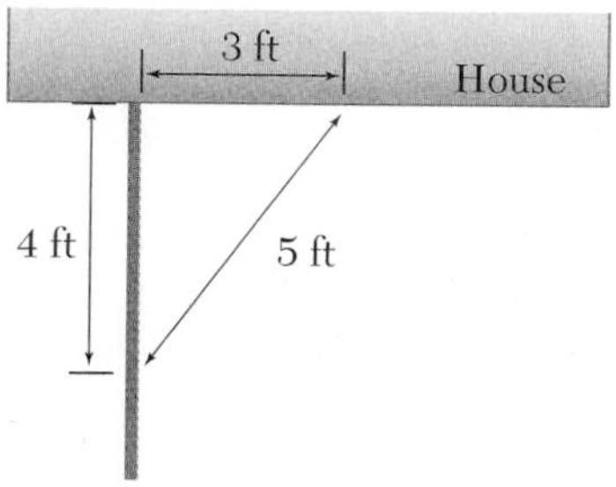

CHALLENGE

53. Sam has a glass rod that is 1 inch in diameter and 35 inches long that he needs to ship across the country. He's having trouble finding a box. His choices are a box that measures 2 feet by 2 feet by 6 inches, or a box that measures 18 inches by 32 inches by 5 inches. Will Sam's rod fit in either of the boxes?

GROUP WORK

54. Trace a circle at least 6 inches in diameter on a piece of paper. Now fold the circle in half (be as exact as you can) and mark the diameter of the fold. Label the ends of the diameter A and B. Pick four points on the circumference of the circle and label them C, D, E, and F. Connect C, D, E, and F to A and B to make four triangles, as in the figure.

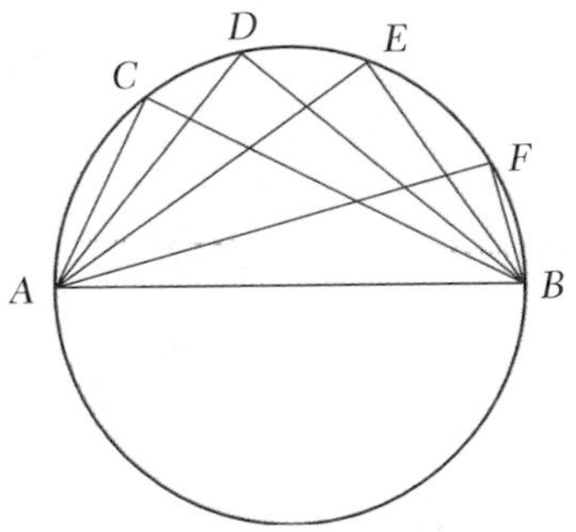

Use a ruler marked in millimeters to measure, as accurately as possible, the sides of all the triangles and record your results in the table.

Triangle	Side 1	Side 2	Side 3	Right Triangle?
Δ ACB				
Δ ADB				
Δ AEB				
Δ AFB				

Geometry tells us that every triangle inscribed in a semicircle is a right triangle. Do your measurements confirm this fact? If not, explain why not.

MAINTAIN YOUR SKILLS

55. What is 15% percent of 45,000?

56. $36.9 + 231 + 0.087$

57. $\left(4\frac{1}{5}\right)\left(3\frac{6}{7}\right)$

58. Find the prime factorization of 80.

59. $23 - 14\frac{3}{8}$

60. $4.65 \div 0.002$

61. At a certain time of day, a 15-ft tree casts a shadow of 8 ft. How tall is another tree if it casts a shadow of 5 ft?

62. Convert 45 ft^2 into square meters. Round to the nearest hundredth.

63. Find the average and median of: 3, 18, 29, 31, 32.

64. Find the volume of a can with diameter 8 in. and height 10 in. Use the [π] key, and round to the nearest hundredth.

CHAPTER 7

KEY CONCEPTS

SECTION 7.1 MEASURING LENGTH

Definitions and Concepts

English	Metric
inch	centimeter
foot	meter
mile	kilometer

To convert units, multiply by the appropriate unit fraction.

Examples

Convert 60 in. to feet.

$$\begin{aligned} 60 \text{ in.} &= \frac{60 \cancel{\text{in.}}}{1} \cdot \frac{1 \text{ ft}}{12 \cancel{\text{in.}}} \\ &= 5 \text{ ft} \end{aligned}$$

To convert within the metric system, use the chart and move the decimal point the same number of places.

Convert 25 cm to meters.

km	hm	dam	m	dm	cm	mm

Move the decimal two places left.

25 cm = 0.25 m

Only like measures can be added or subtracted.

14 cm + 50 cm = 64 cm

8 ft − 3 ft = 5 ft

SECTION 7.2 MEASURING CAPACITY, WEIGHT, AND TEMPERATURE

Definitions and Concepts

	English	Metric
Capacity	quart	liter
Weight	ounce pound	gram kilogram
Temperature	degrees Fahrenheit	degrees Celsius

To convert units, multiply by the appropriate unit fraction.

Examples

Convert 5 qt to liters.

$$\begin{aligned} 5 \text{ qt} &\approx \frac{5 \cancel{\text{qt}}}{1} \cdot \frac{0.946 \text{ L}}{1 \cancel{\text{qt}}} \\ &\approx 4.73 \text{ L} \end{aligned}$$

To convert temperatures, use the formulas

$$C = \frac{5}{9} \cdot (F - 32)$$

$$F = \frac{9}{5} \cdot C + 32$$

Convert 41°F to degrees Celsius.

$$\begin{aligned} C &= \frac{5}{9} \cdot (41 - 32) \\ &= \frac{5}{9} \cdot 9 \\ &= 5 \end{aligned}$$

So, 41°F is 5°C.

SECTION 7.3 PERIMETER

Definitions and Concepts

Perimeter is measured in linear units.

To find the perimeter of a polygon, add the measures of all its sides.

Perimeter of a square:

$P = 4s$

Perimeter of a rectangle:

$P = 2\ell + 2w$

The circumference of a circle is the distance around the circle.

The radius of a circle is the distance from the center to any point on the circle.

The diameter of a circle is twice the radius.

Examples

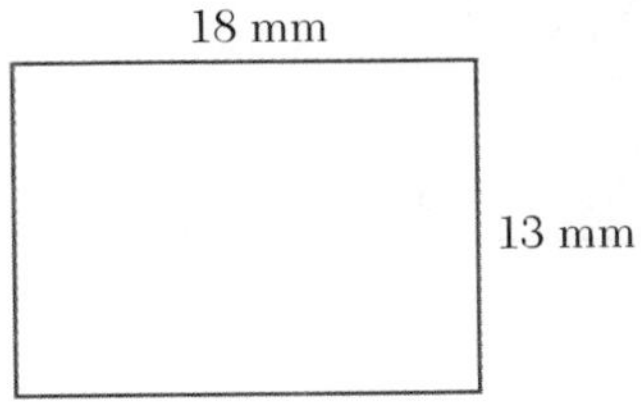

$$\begin{aligned} P &= 2\ell + 2w \\ &= 2(18) + 2(13) \\ &= 36 + 26 \\ &= 62 \end{aligned}$$

The perimeter is 62 mm.

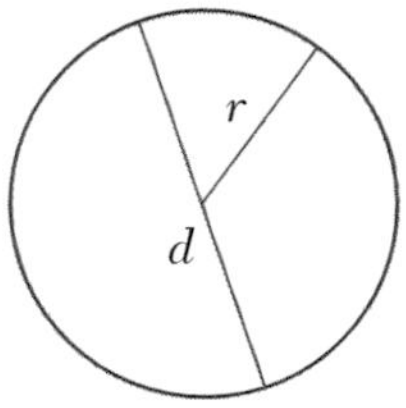

$C = 2\pi r$
$C = \pi d$

SECTION 7.4 AREA

Definitions and Concepts

Area is measured in square units.

Area of a square:

$A = s^2$

Area of a rectangle:

$A = \ell w$

Area of a triangle:

$A = \dfrac{bh}{2}$

Area of a circle:

$A = \pi r^2$

Examples

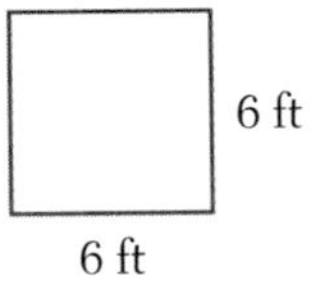

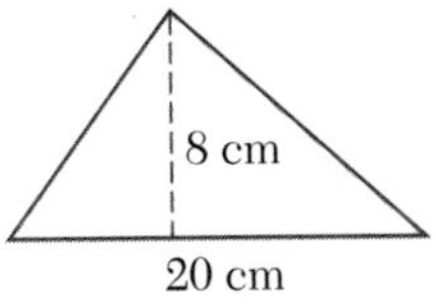

$$\begin{aligned} A &= s^2 \\ &= 6^2 \\ &= 36 \end{aligned}$$

$$\begin{aligned} A &= \frac{bh}{w} \\ &= \frac{20 \cdot 8}{2} \\ &= 80 \end{aligned}$$

The area is 36 ft^2. The area is 80 cm^2.

SECTION 7.5 VOLUME

Definitions and Concepts

Volume is measured in cubic units.

Volume of a cube:

$V = s^3$

Volume of a rectangular solid:

$V = \ell wh$

Volume of a solid with perpendicular sides:

$V = Bh$, where B is the area of the base

Volume of a cylinder:

$V = \pi r^2 h$

Examples

$$\begin{aligned} V &= \ell wh \\ &= (10)(6)(3) \\ &= 180 \end{aligned}$$

The volume is 180 in.^{3}.

SECTION 7.6 SQUARE ROOTS AND THE PYTHAGOREAN THEOREM

Definitions and Concepts

A square root of a number is the number that is squared to give the original number.

To find the square root of a number, use the [√] key on a calculator and round as necessary.

Pythagorean theorem: In a right triangle,

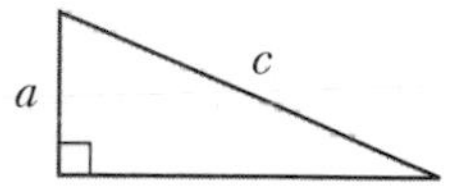

$$a^2 + b^2 = c^2$$

To find a missing side of a right triangle, use the formulas.

$$c = \sqrt{a^2 + b^2} \quad \text{or} \quad a = \sqrt{c^2 - b^2}$$

Examples

$\sqrt{49} = 7$ because $7 \cdot 7 = 49$

$\sqrt{436} \approx 20.88$

Find c.

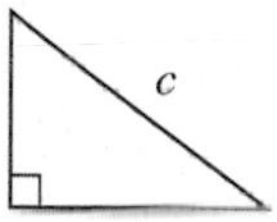

$$\begin{aligned} c &= \sqrt{3^2 + 4^2} \\ &= \sqrt{9 + 16} \\ &= \sqrt{25} \\ &= 5 \end{aligned}$$

Name ______________________ Class ______________________ Date ____________

CHAPTER 7

REVIEW EXERCISES

Section 7.1

Do the indicated operations.

1. $12 \cdot 5$ cm **2.** 300 m ÷ 15 **3.** 4 yd − 22 in.

4. 5 ft + 19 ft + 7 ft **5.** (6 ft 2 in.) − (4 ft 10 in.)

Convert as indicated. Round decimal values to the nearest hundredth.

6. 13 cm to meters **7.** 2 mi to feet

8. 4 km to miles **9.** 15 in. to millimeters

10. Ted is building two plant supports out of copper tubing. Each support requires three vertical tubes that are 8 ft long and five horizontal tubes that are 2 ft long. How much tubing does he need for the two supports?

Section 7.2

Do the indicated operations.

11. 18 gal · 3 **12.** (345 hr) ÷ 15 **13.** 7 oz + 14 oz

14. 35 L − 15 L **15.** (6 hr 42 min) + (3 hr 38 min)

Convert as indicated. Round to the nearest hundredth when necessary.

16. 4 g to milligrams **17.** 24 oz to pounds

18. 3 L to quarts **19.** 59°F to degrees Celsius

20. Gail estimates that she has 45 min to work in her yard for each of the next five days. She is building a retaining wall that should take her 4 hr. How much more time will she need to finish the wall?

Section 7.3

Find the perimeter of the following polygons.

21.

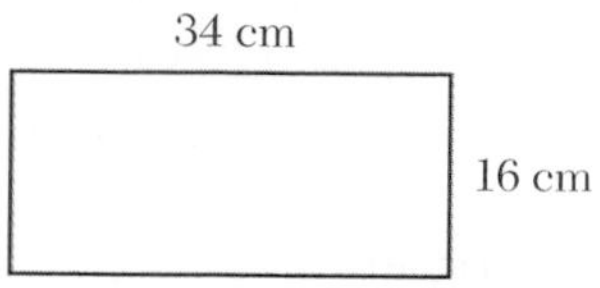

22.

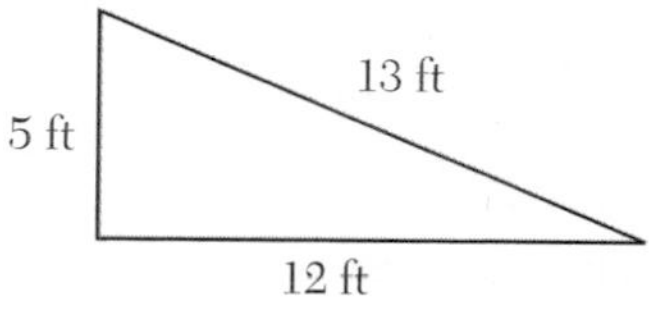

23.

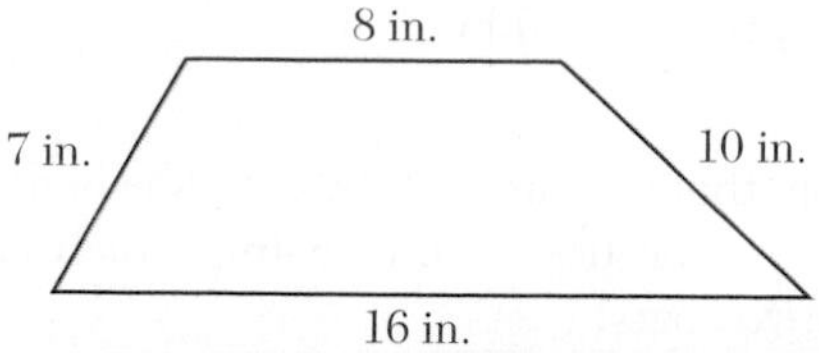

24.

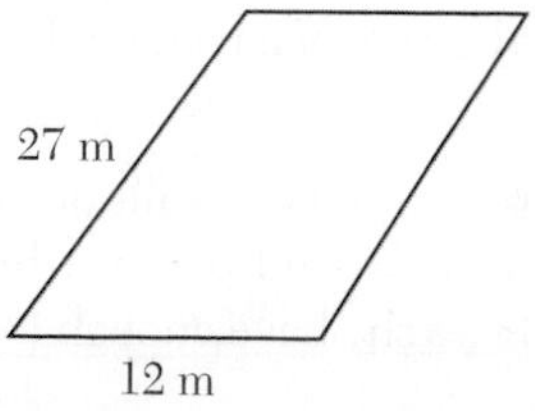

25.

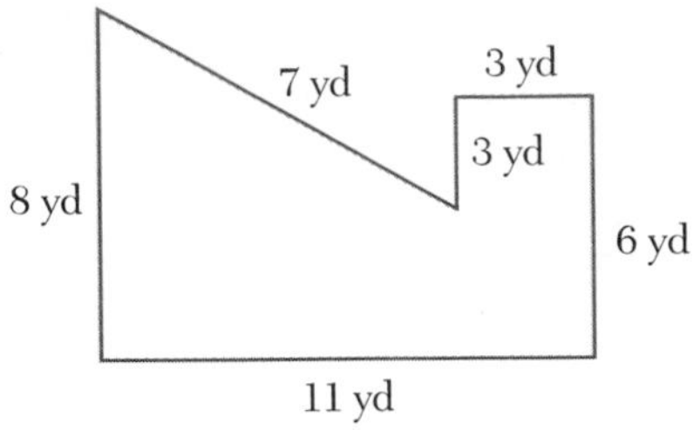

26.

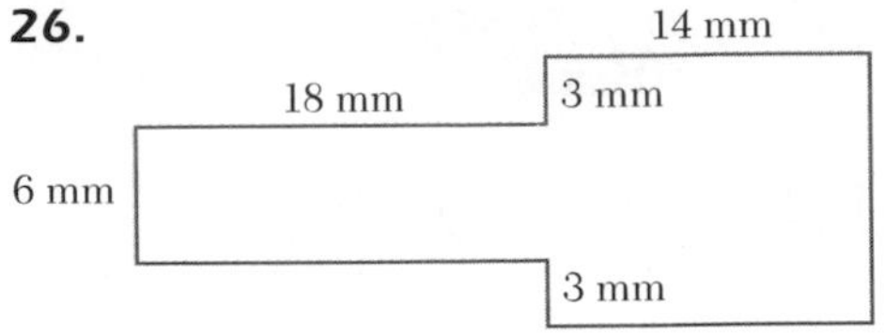

27. Find the perimeter of a square plot of land that is 65 m on each side.

28. Find the distance around a circular hot tub that measures 8 ft in diameter. Let $\pi \approx 3.14$.

29. How much fencing is needed for a rectangular backyard that is 55 ft by 32 ft?

30. Larry is buying new baseboards for his den, which measures 9 ft by 12 ft. How much does he need if the door is 4 ft wide?

Section 7.4

Find the area of the following regions. Use the [π] *key, and round to the nearest hundredth.*

31.

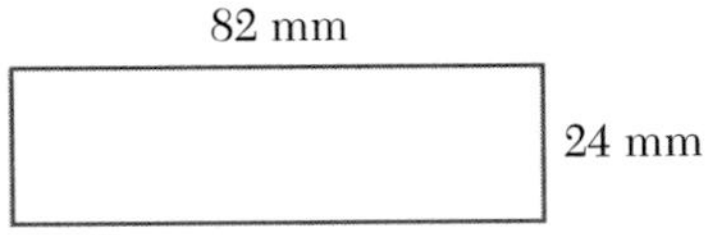

32.

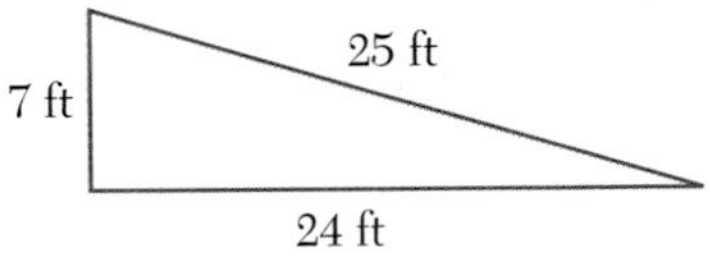

33.

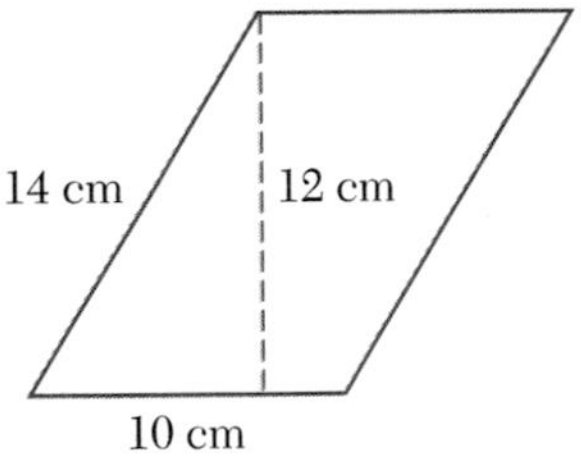

34.

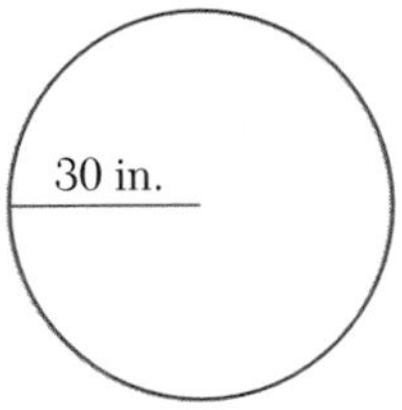

35.

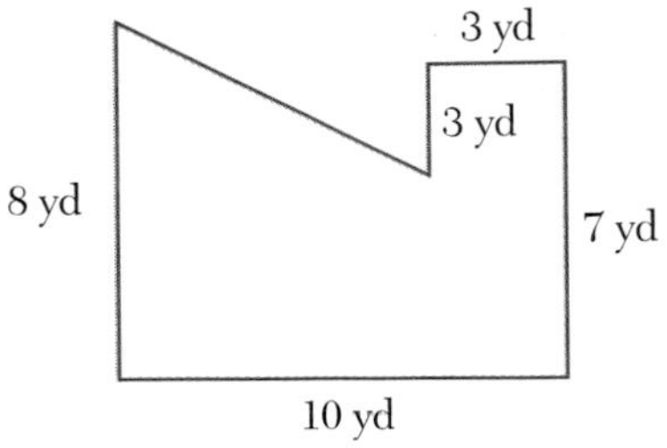

36.

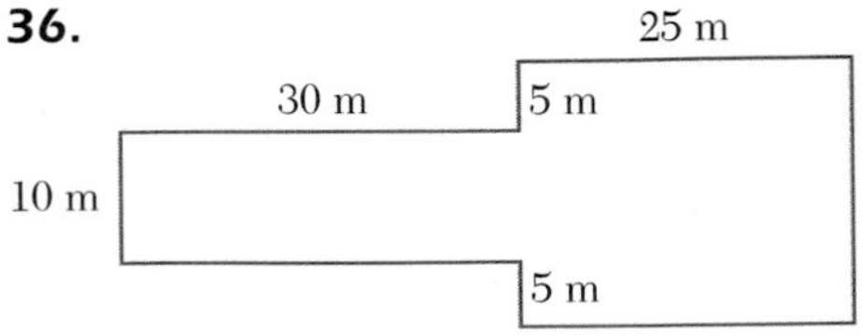

37. What is the area of a square that measures 500 mm on a side?

38. What is the area of a rectangular driveway that is 35 ft by 22 ft?

39. Felicity is painting the walls of a bathroom that measures 6 ft by 8 ft. The bathroom has 8-ft ceilings and one 3-ft-by-6-ft door. She has 1 gal of paint, which covers 350 ft^2. Does she have enough paint for two coats?

Name ______________________ Class ______________________ Date ____________

40. A living room window is a rectangle surmounted by a semicircle. See the figure. How much glass is in the window? Let $\pi \approx 3.14$.

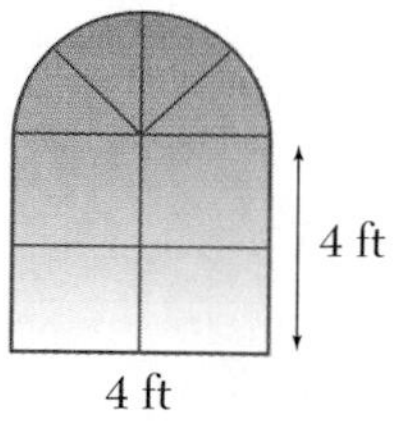

Section 7.5

Find the volume of the following figures. Use the [π] *key, and round to the nearest hundredth.*

41.

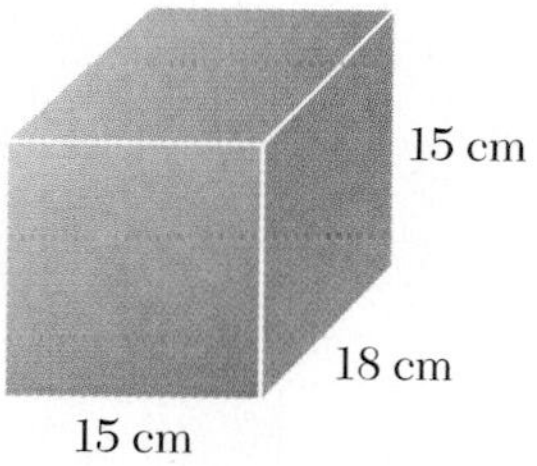

42.

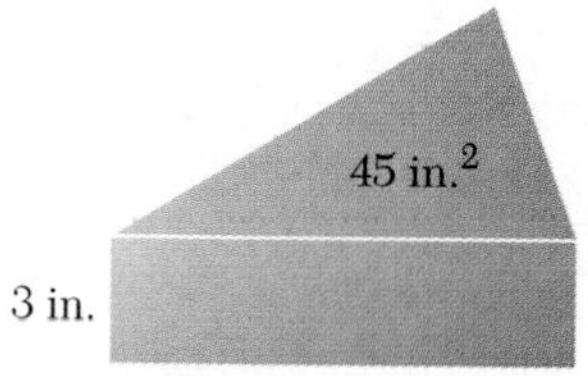

43.

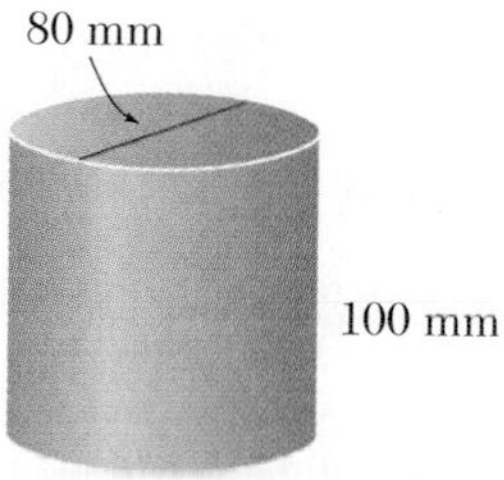

44.

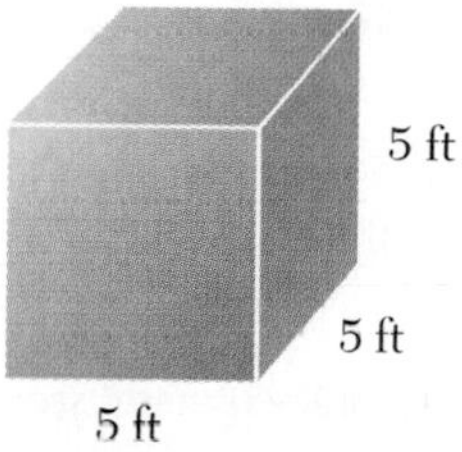

45.

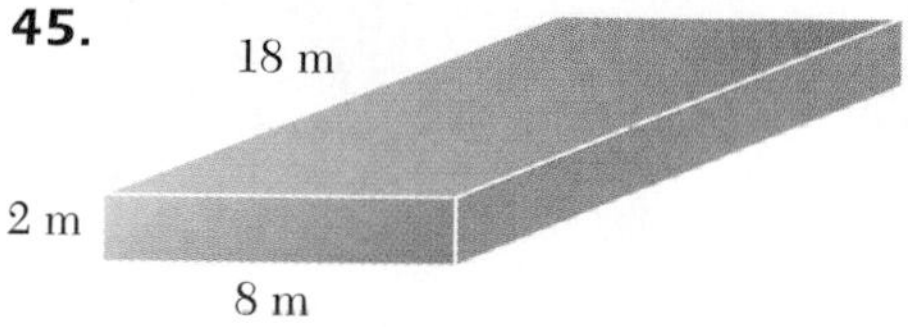

46.

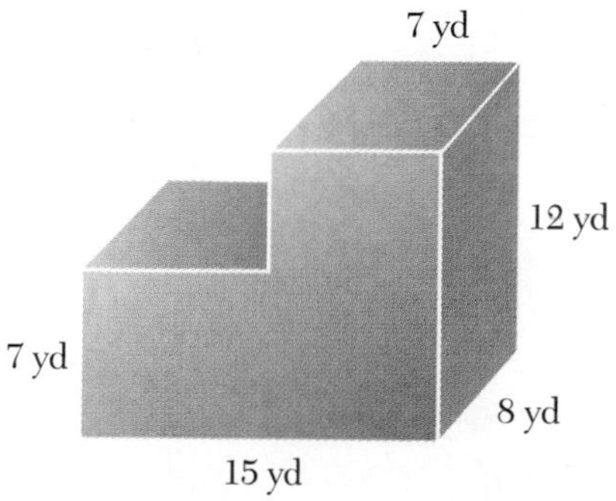

47. What is the volume of a traffic cone that is 28 in. tall and has a base with a radius of 10 in.? Let $\pi \approx 3.14$, and round to the nearest hundredth.

48. What is the volume of a kitchen that measures 14 ft long by 9 ft wide by 8 ft high?

49. How many cubic inches are there in a cubic yard?

50. Will a cubic yard of barkdust be sufficient to cover a rose garden that is 20 ft by 6 ft to a depth of 3 in.?

Section 7.6

Find the square roots, rounding decimal values to the nearest hundredth.

51. $\sqrt{289}$

52. $\sqrt{40}$

53. $\sqrt{\dfrac{49}{81}}$

54. $\sqrt{28.09}$

Find the missing side. Round decimal values to the nearest hundredth.

55.

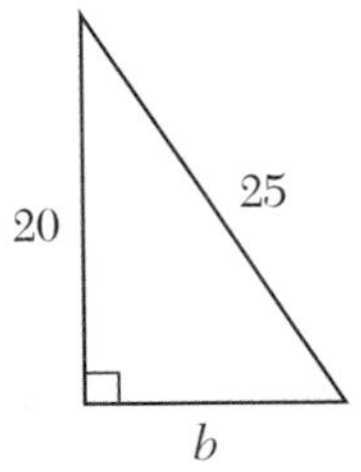

56.

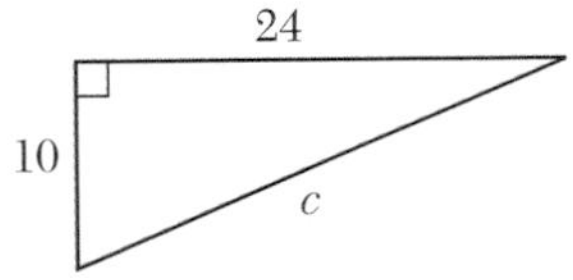

57.

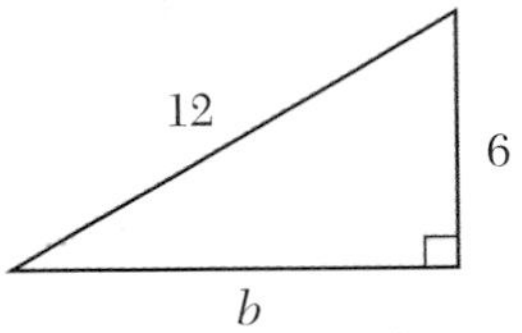

58.

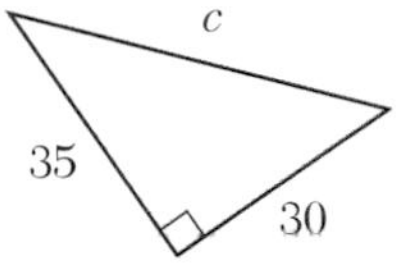

59. A 10-ft ladder is leaning against a wall. If the base of the ladder is 3 ft away from the base of the wall, how high is the top of the ladder?

60. Mark wants to brace a metal storage unit by adding metal strips on the back on both diagonals. If the storage unit is 3 ft wide and 6 ft high, how long are the diagonals and how much metal stripping does he need?

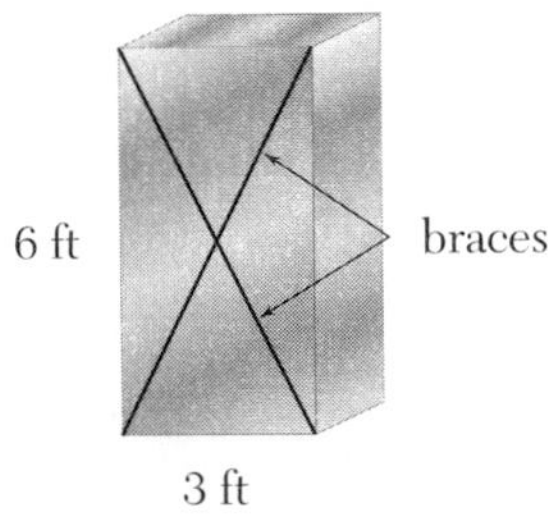

CHAPTER 7

TRUE/FALSE CONCEPT REVIEW

ANSWERS

Check your understanding of the language of algebra and geometry. Tell whether each of the following statements is true (always true) or false (not always true). For each statement you judge to be false, revise it to make a true statement.

1. English measurements are the most commonly used in the world. 1. ______
2. Equivalent measures have different units of measurement. 2. ______
3. A liter is a measure of weight. 3. ______
4. The perimeter of a square can be found in inches, feet, centimeters, or meters. 4. ______
5. Area is the measure of the inside of a solid, such as a box or a can. 5. ______
6. The volume of a square is $V = s^3$. 6. ______
7. The area of a circle is $A = \pi r^2$. 7. ______
8. It is possible to find equivalent measures without remeasuring the original object. 8. ______
9. The metric system uses the base-10 place value system. 9. ______
10. In a right triangle, the longest side is the leg. 10. ______
11. $\sqrt{4} = 16$ 11. ______
12. Weight can be measured in pounds, grams, or kilograms. 12. ______
13. A trapezoid has three sides. 13. ______
14. One milliliter is equivalent to 1 cubic centimeter. 14. ______
15. Measurements can be added or subtracted only when they are expressed with the same unit of measure. 15. ______
16. The distance around a geometric figure is called the perimeter. 16. ______
17. Volume is always measured in cubic units. 17. ______
18. The Pythagorean theorem states that the sum of the lengths of the legs of a right triangle is equal to the length of the hypotenuse. 18. ______
19. $1 \text{ ft}^2 = 12 \text{ in.}^2$ 19. ______
20. The prefix kilo- means 100. 20. ______

TRUE/FALSE CONCEPT REVIEW

CHAPTER 7

TEST

ANSWERS

1. 7 m + 454 mm = ? mm **1.** ____________

2. Find the perimeter of a square that is 34 cm on a side. **2.** ____________

3. Find the volume of a drawer that is 4 in. high, 18 in. wide, and 24 in. deep. **3.** ____________

4. How much vinyl flooring is needed to cover the room pictured? **4.** ____________

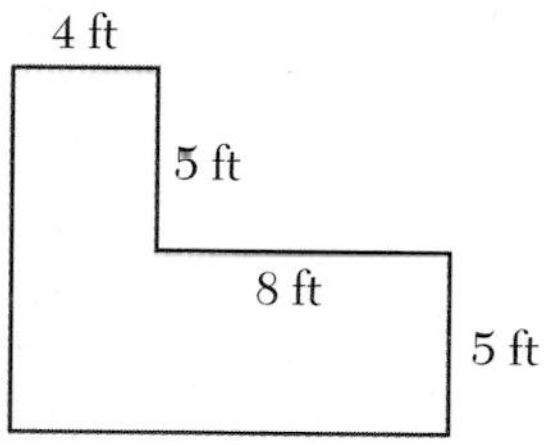

5. Find $\sqrt{200}$ to the nearest thousandth. **5.** ____________

6. Find the perimeter of the figure. **6.** ____________

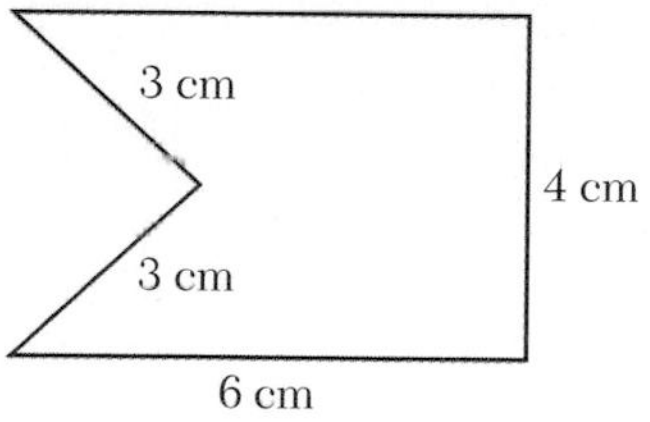

7. Find the area of the triangle. **7.** ____________

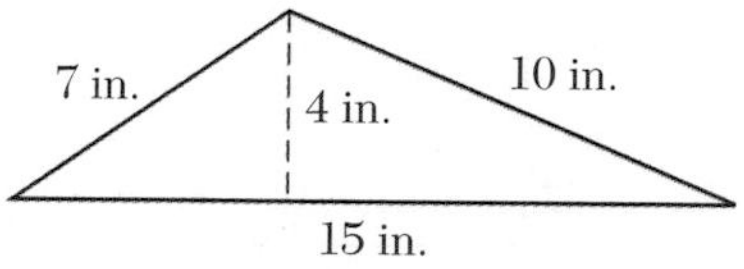

8. Subtract:

$$\begin{array}{r} 5 \text{ gal } 2 \text{ qt} \\ -3 \text{ gal } 2 \text{ qt } 1 \text{ pt} \\ \hline \end{array}$$

8. ____________

9. Name two units of measure in the English system for weight. Name two units of measure in the metric system for weight. **9.** ____________

10. Change 2 ft^2 to square inches. **10.** ____________

11. How much molding is needed to trim a picture window 4 ft wide and 5 ft tall and two side windows each measuring 2 ft wide by 5 ft tall? Assume that the corners will be mitered and that the miters require 2 in. extra on each end of each piece of molding. **11.** ____________

12. Find the volume of the figure. **12.** ______________

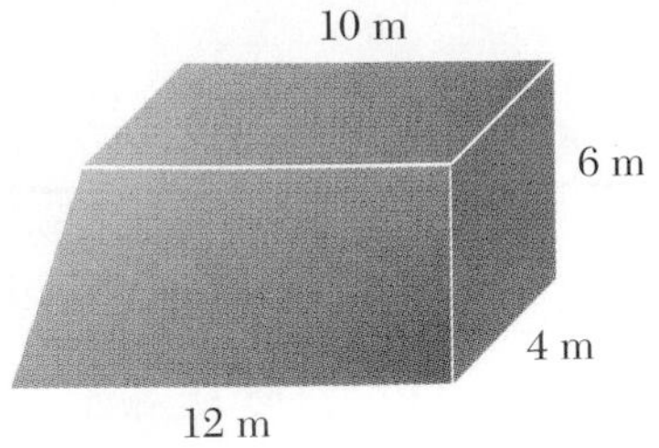

13. Find the area of the parallelogram. **13.** ______________

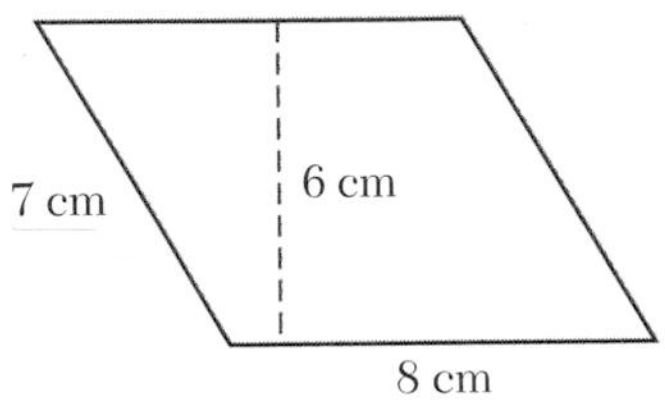

14. Find the volume of a hot water tank that has a circular base with area of 4 ft^2 and a height of 5 ft. **14.** ______________

15. The Golden Silver Company has a bar of silver weighing 684 g. If it is melted down to form six bars of equal weight, what will each bar weigh? **15.** ______________

16. The cost of heavy-duty steel fencing is $2 per linear foot. How much will it cost to build the dog runs pictured? **16.** ______________

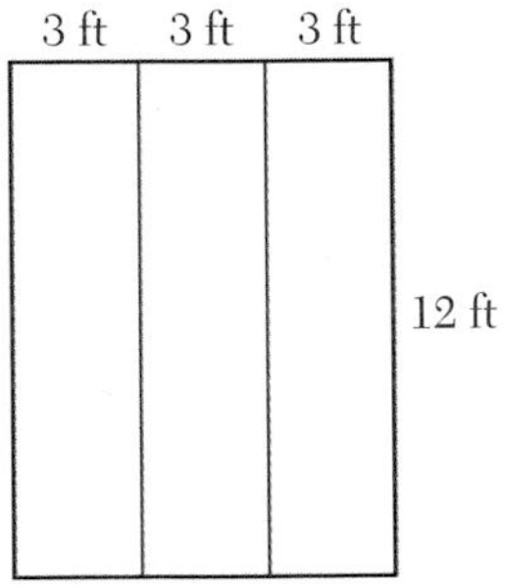

17. Find the missing side. **17.** ______________

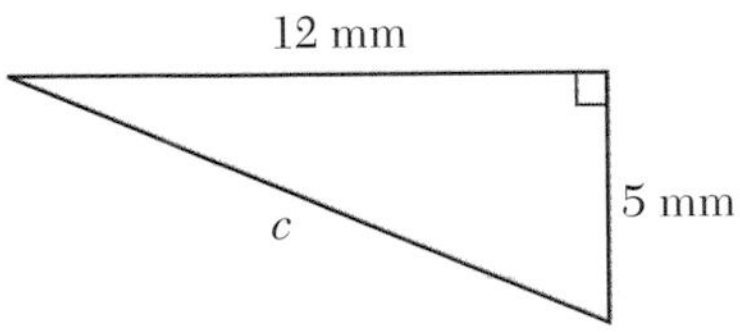

18. Find the area of the figure. Round to the nearest hundredth. Let $\pi \approx 3.14$. **18.** ______________

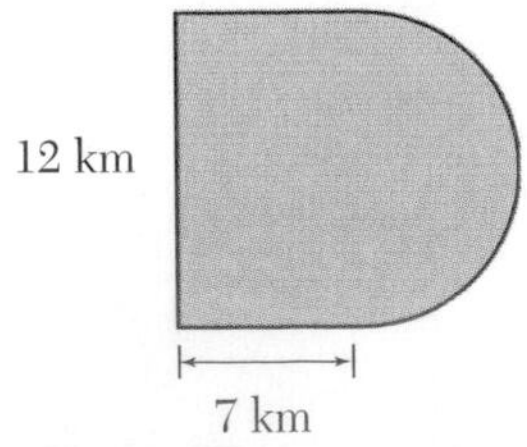

19. A football coach is conditioning his team by having them run around the edge of the field three times each hour. A football field measures 100 yd long by 60 yd wide. How far does each player run in a 2-hr practice? **19.** ____________

20. Both area and volume describe interior space. Explain how they are different. **20.** ____________

21. Jared is lifting weights in preparation for ski season. One week he bench-presses 180 lb on Monday, 185 lb on Tuesday, 195 lb on Wednesday, 205 lb on Thursday, and 200 lb on Friday. What is his average bench-press weight for the week? **21.** ____________

22. Li is buying lace, with which she plans to edge a circular tablecloth. The tablecloth is 80 in. in diameter. How much lace does she need? If the lace is available in whole-yard lengths only, how much must she buy? Let $\pi \approx 3.14$. **22.** ____________

23. Khallil has a 40-lb bag of dog food that is approximately 52 in. long, 16 in. wide, and 5 in. deep. He wants to transfer it to a plastic storage box that is 24 in. long, 18 in. wide, and 12 in. high. Will all the dog food fit into the box? **23.** ____________

24. From the same corner, Clayton rides his bike 10 km due north while Sam walks 3 km due west. How far apart are they? Round to the nearest hundredth. **24.** ____________

25. Neela has a gift box that is 10 in. by 15 in. by 2 in. What minimum amount of wrapping paper does she need to wrap the gift? **25.** ____________

CHAPTER 7

GROUP PROJECT (2–3 weeks) OPTIONAL

You are working for a kitchen design firm that has been hired to design a kitchen for the 10 ft by 12 ft room pictured here.

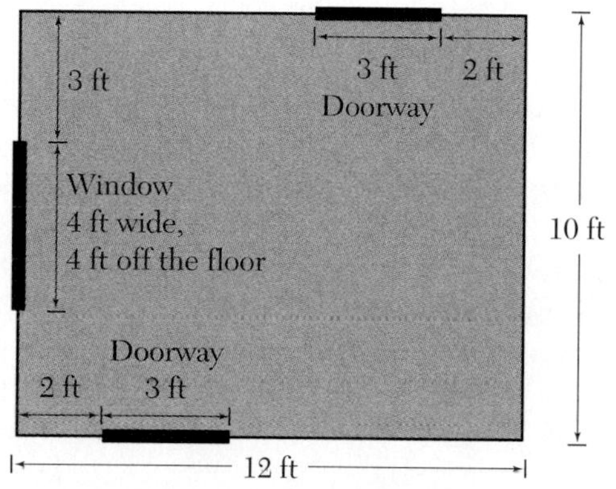

The following table lists appliances and dimensions. Some of the appliances are required and others are optional. All of the dimensions are in inches.

Appliance	High	Wide	Deep	Required
Refrigerator	68	30 or 33	30	Yes
Range/Oven	30	30	26	Yes
Sink	12	36	22	Yes
Dishwasher	30	24	24	Yes
Trash Compactor	30	15	24	No
Built-in Microwave	24	24	24	No

The base cabinets are all 30 in. high and 24 in. deep. The widths can be any multiple of 3 from 12 in. to 36 in. Corner units are 36 in. along the wall in each direction. The base cabinets (and the range, dishwasher, and compactor) will all be installed on 4-in. bases that are 20 in. deep.

The wall (upper) cabinets are all 30 in. high and 12 in. deep. Here, too, the widths can be any multiple of 3 from 12 in. to 36 in. Corner units are 24 in. along the wall in each direction.

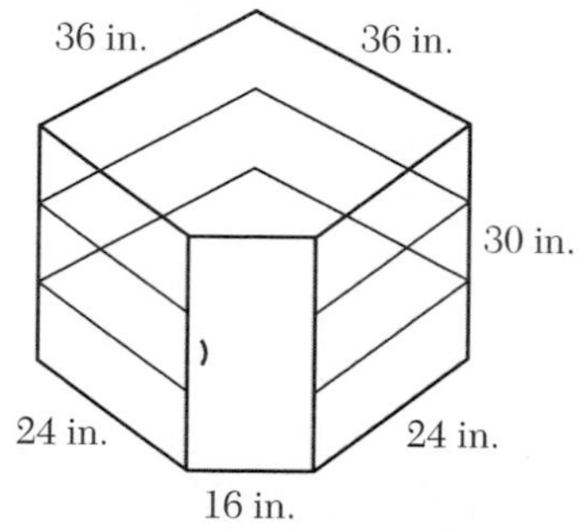

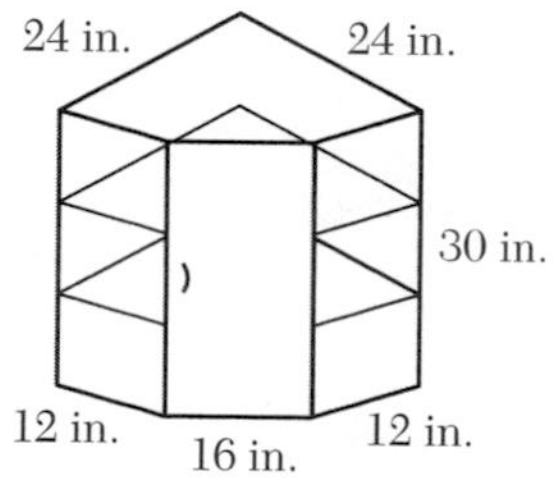

1. The first step is to place the cabinets and appliances. Your client has specified that there must be at least 80 ft^3 of cabinet space. Make a scale drawing of the kitchen and indicate the placement of the cabinets and appliances. Show calculations to justify that your plan satisfies the 80 ft^3 requirement.

2. Countertops measure 25 in. deep with a 4-in. backsplash. The countertops can either be tile or Formica. If the counters are Formica, there will be a 2-in. facing of Formica. If the counters are tile, the facing will be wood that matches the cabinets. (See the figure.) Calculate the amount of tile and wood needed for the counters and the amount of Formica needed for the counters.

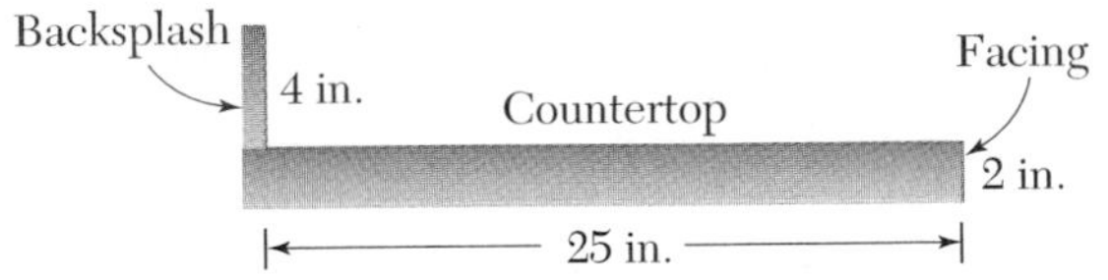

3. The bases under the base cabinets will be covered with a rubber kickplate that is 4 in. high and comes in 8-ft lengths. Calculate the total length of kickplate material needed and the number of lengths of kickplate material necessary to complete the kitchen.
4. Take your plan to a store that sells kitchen appliances, cabinets, and counters. Your goal is to get the best-quality materials for the least amount of money. Prepare at least two cost estimates for the client. Do not include labor in your estimates, but do include the cost of the appliances. Prepare a rationale for each estimate, explaining the choices you made. Which plan will you recommend to the client and why?

Name ______________________ Class ______________ Date __________

CHAPTERS 1–7

CUMULATIVE REVIEW

1. Write the word name for 6035.

2. Round to the nearest hundred: 67,849

3. List from smallest to largest:
381, 390, 375, 384, 309

4. Add:
$$\begin{array}{r} 236 \\ 1511 \\ 46 \\ 2301 \\ +6290 \\ \hline \end{array}$$

5. Subtract: 794 − 358

6. Multiply:
$$\begin{array}{r} 340 \\ \times 212 \\ \hline \end{array}$$

7. Multiply: (47)(15)(11)

8. Multiply. 831(1000)

9. Divide: 74,000 ÷ 100

10. Divide: $854\overline{)53,802}$

11. Find the average, median, and mode of the set of numbers: 22, 46, 31, 37, 46, 29, 34

12. Perform the indicated operations. $6 \cdot 4 + 2 - 5(7 - 4)$

Exercises 13–14. The graph gives the average annual salary for five professions.

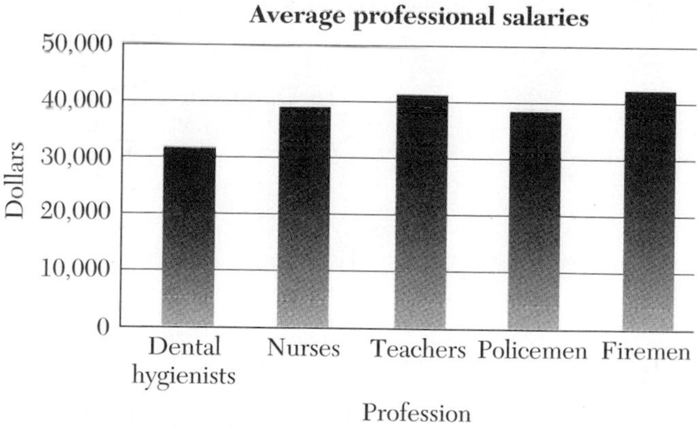

13. Which profession has the lowest average salary?

14. What is the difference in salary between dental hygienists and nurses?

15. During the United Way campaign, Sylvia collected the following pledge amounts: \$150, \$210, \$400, \$125, \$50, \$75, and \$145. What is the total amount of the pledges Sylvia collected?

16. On vacation, Jim drove 573 miles on the first day and 397 miles on the second day. How many more miles did he drive on the first day than on the second day?

17. Millie worked 14 hours at a job paying \$12 per hour and 23 hours at a job that pays \$17 per hour. How much did Millie earn at the two jobs?

18. The 13 starting linemen for the Westside University weighed in as follows: 3 at 210 lb, 2 at 234 lb, 4 at 256 lb, 2 at 285 lb, 1 at 310 lb, 1 at 326 lb. What is the average weight of the starting linemen?

19. List all the factors of 162.

20. What is the greatest common factor (GCF) of 78, 195, and 234?

21. Write the first five multiples of 23.

22. Find the least common multiple (LCM) of 36, 80, and 84.

23. Is 3492 divisible by 2, 3, or 5?

24. Is 2011 prime or composite?

25. Write the prime factorization of 690.

26. Is $\frac{39}{29}$ an improper or proper fraction?

27. Simplify: $\frac{90}{210}$

28. Multiply: $\frac{8}{9} \cdot \frac{12}{15} \cdot \frac{3}{16}$

29. Multiply: $4\frac{1}{2} \cdot 5\frac{2}{3}$

30. Divide: $\frac{9}{14} \div \frac{18}{35}$

31. Divide: $11\frac{1}{5} \div 12\frac{3}{5}$

32. List the numbers from smallest to largest: $\frac{5}{12}, \frac{4}{9}, \frac{7}{15}, \frac{2}{5}, \frac{13}{30}$

33. Add: $\frac{3}{70} + \frac{5}{14} + \frac{7}{10} + \frac{3}{5}$

34. Add: $\begin{array}{r} 23\frac{11}{15} \\ +34\frac{17}{20} \\ \hline \end{array}$

35. Subtract: $\frac{11}{28} - \frac{6}{35}$

36. Subtract: $\begin{array}{r} 72 \\ -53\frac{9}{13} \\ \hline \end{array}$

37. Perform the indicated operations: $\left(\frac{2}{3}\right)\left(\frac{3}{8}\right) + \left(2\frac{1}{4} - 1\frac{7}{8}\right) \div \frac{1}{4}$

38. On January 1, 2002, Huong measured $57\frac{2}{3}$ inches tall. On January 1, 2003, he measured $59\frac{3}{8}$ inches tall. How much did Huong grow during the year?

39. In a class of 235 students, three-fifths of them passed the state competency exam in mathematics. How many students passed the competency exam?

40. Mr. Doty has $2\frac{5}{6}$ quarts of milk to divide among the 34 children in his day care program. How much milk does each child get?

41. Write the word name for 10.047.

42. Round 47.1695 to the nearest ten, tenth, and hundredth.

43. List the decimals from smallest to largest: 1.35, 1.046. 1.4005, 1.307, 1.3491

44. Add: 3.78 + 11.9 + 0.036 + 15.17 + 110

45. Subtract: $589.63 - 493.85$

46. Multiply: 1.695×10^5

47. Write in scientific notation: 0.0000345

48. Multiply: $(16.45)(5.32)$

49. Divide: $436.4525 \div 32.45$

50. Divide and round to the nearest hundredth: $13.6\overline{)22.7}$

51. Write as a decimal: $\frac{11}{16}$

52. Find the average and median of 6.37, 8.91, 7.42, and 5.7.

53. Perform the indicated operations: $(6.2)^2 - 8(1.16) - 3(5.94 - 3.66)$

54. Marilyn needs to buy four textbooks for this semester. She goes to the bookstore to find out how much they cost:

Text	Cost
Algebra	\$39.75
Chemistry	\$48.50
Psychology	\$36.00
American History	\$42.75

What is the total cost of the books that she needs? What is the average cost of the books?

55. Bill is making a tabletop. The two ends of the piece of wood he is using need to be trimmed and sanded. The piece of wood is 53.25 inches long. He requires a length of 48.5 inches. The saw removes $\frac{3}{16}$ inch during each cut, and he wants to allow $\frac{1}{16}$ inch for finishing sanding on each end. If he cuts an equal amount off of each end, how much should he cut from each end?

56. Mr. Hardgrave bought one thousand shares of Microsoft at \$61.34 a share and sold them for \$79.23 a share. If his broker fees totaled \$234.50 for both trades, what was Mr. Hardgrave's profit?

57. In Ms. Adam's algebra class, there are 22 females and 18 males. Write a ratio that represents the female population in the class.

58. Solve: $\frac{3}{8} = \frac{B}{10}$

59. Solve: $\frac{4}{9} = \frac{48}{A}$

60. Safeway advertises bananas at 3 lb for \$1.35. At this rate, what will 14 lb of bananas cost?

61. A flagpole that is 36 ft tall casts a shadow of 14.6 ft. What is the length of a shadow cast by a tree that is 45 ft tall?

62. The property tax on a home valued at \$215,000 is \$1850. At the same tax rate, what will be the tax on a home valued at \$385,000? Round to the nearest dollar.

Change to a decimal.

63. 5.8%

64. 312%

Change to a percent.

65. 0.635

66. 0.0041

Change to a percent.

67. $\frac{11}{32}$

68. $\frac{8}{23}$ (Round to the nearest tenth of a percent.)

Change to a fraction.

69. 68.2%

70. 1.5%

71. What is 11.3% of 63.

72. What percent of 28 is 6.3?

73. 20.37 is 42% of what number?

74. Of the 523 employees of Compugo, 89.5% contributed to the disaster relief fund of 2001. How many employees contributed to the fund? Find to the nearest employee.

75. Helga earns \$150 per week plus an 8% commission on total sales. Last week Helga had sales of \$4590. How much did Helga earn last week?

76. During the fall semester, 356 of the 720 freshmen students at Sunset High School enrolled in algebra classes. What percent of the freshman class enrolled in algebra classes? Round to the nearest tenth of a percent.

Add.

77.
$$\begin{array}{r} 3 \text{ yd } 2 \text{ ft } 10 \text{ in.} \\ +4 \text{ yd } 2 \text{ ft } \ \ 8 \text{ in.} \\ \hline \end{array}$$

78.
$$\begin{array}{r} 3 \text{ qt } 1 \text{ pt } 1 \text{ c} \\ +1 \text{ qt } 1 \text{ pt } 1 \text{ c} \\ \hline \end{array}$$

79. How many mL are there in 0.65 L?

Convert the units of measure.

80. 96 min = ? hr

81. 4.56 mi = ? ft

82. 34.29 cm = ? in.

83. 3.5 lb = ? kg (Round to the nearest hundredth.)

84. 74.1 m^2 = ? ft^2 (Round to the nearest tenth.)

85. The temperature in Michigan set a new high for the day of 109.6°F. What is the new record high for the day measured in Celsius? Round to the nearest tenth.

Find the perimeter of the following figures.

86.

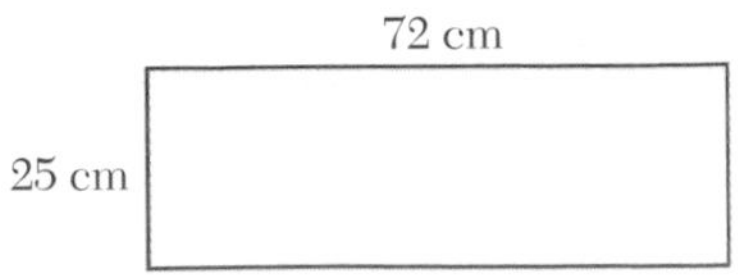

87.

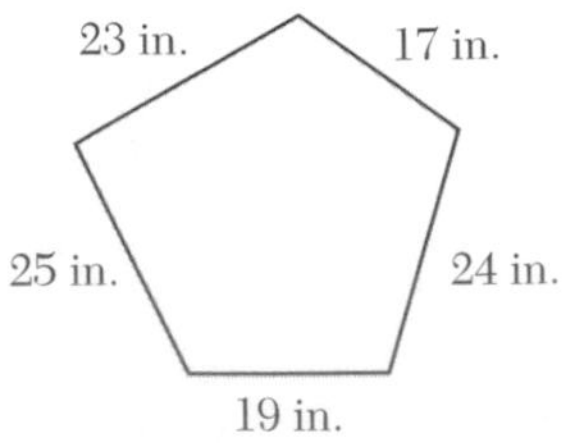

88. Find the circumference of a circle that has a diameter of 31.5 cm. Let $\pi \approx 3.14$.

89. Find the area of a triangle that has a base of 5 ft and a height of 7 ft.

90. Find the area of a circle that has a diameter of 24 cm. Let $\pi \approx 3.14$.

91. Find the area of the polygon.

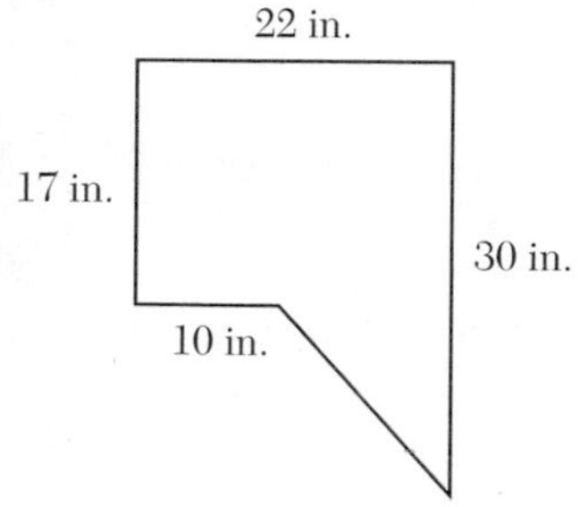

92. Find the volume of a cube that is 5 m on a side. The volume is 125 m^3.

93. Find the volume of a cylinder with a circular base with radius of 6 cm and a height of 9 cm. Let $\pi \approx 3.14$.

94. How many cubic inches of concrete are needed to pour a sidewalk that is 4 ft wide, 4 in. deep, and 65 ft long? Concrete is commonly measured in cubic yards, so convert your answer to cubic yards. Round to the nearest tenth of a cubic yard.

95. Trudy is carpeting three rooms in her house. The rooms measure 18 ft by 15 ft, 22 ft by 20 ft, and 21 ft by 25 ft. She picks out a carpet that sells for $45 a square yard. Assuming that she must buy a whole number of square yards, what is the cost of the carpeting?

Find the square root. If necessary, round to the nearest hundredth.

96. $\sqrt{12{,}321}$ **97.** $\sqrt{36.95}$ **98.** $\sqrt{0.09632}$

99. Find the length of the hypotenuse of a right triangle with legs measuring 36 ft and 27 ft.

100. Find the missing side of the following right triangle. Round to the nearest tenth of a centimeter.

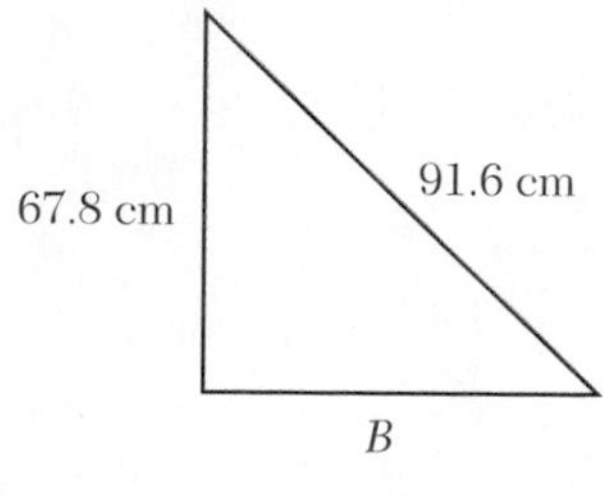

Evaluating Your Performance

Pick up your graded test as soon as possible, while the test is still fresh in your mind. No matter what your score, the time you spend reviewing errors you may have made can help improve future test scores. Don't skip the important step of "evaluating your performance." Begin by categorizing your errors:

1. **Careless errors**. These occur when you know how to do the problem correctly but don't. Careless errors happen when you read the directions incorrectly, make computational errors, or forget to do the problem.
2. **Concept errors**. These occur because you don't grasp the concept fully or accurately. Even if you did the problem again, you would probably make the same error.
3. **Study errors**. These occur when you do not spend enough time studying the material most pertinent to the test.
4. **Application errors**. These usually occur when you are not sure what concept to use.

If you made a careless error, ask yourself if you followed all the suggestions for better test taking. If you didn't, vow to do so on the next test. List what you will do differently next time so that you can minimize these careless errors in the future.

Concept errors need more time to correct. You must review what you didn't understand or you will repeat your mistakes. It is important to grasp the concepts because you will use them later. You may need to seek help from your instructor or a tutor for these kinds of errors.

Avoid study errors by asking the teacher before the test what concepts are most important. Also, pay careful attention to the section objectives because these clarify what you are expected to know.

You can avoid application errors by doing as many of the word problems as you can. Reading the strategies for word problems can help you think about how to start a problem. It also helps to mix up the problems between sections and to do even-numbered problems, for which you do not have answers. It is especially important to try to estimate a reasonable answer before you start calculations.

As a final step in evaluating your performance, take time to think about what you said to yourself during the test. Was this self-talk positive or negative? If the talk was helpful, remember it. Use it again. If not, here are a few suggestions. Be certain that you are keeping a detailed record of your reactions to anxiety. Separate your thoughts from your feelings. Use cue words as signals to begin building coping statements. Your coping statements must challenge your negative belief patterns and they must be believable, not merely pep talk such as "I can do well, I won't worry about it." Make the statements brief and state them in the present tense.

Learn from your mistakes so you can use what you have learned on the next test. If you are satisfied with the results of the test overall, congratulate yourself. You have earned it.

8 ALGEBRA PREVIEW: SIGNED NUMBERS

APPLICATION

The National Geographic Society was founded in 1888 "for the increase and diffusion of geographic knowledge." The Society has supported more than 5000 explorations and research projects with the intent of "adding to knowledge of earth, sea, and sky." Among the many facts catalogued about our planet, the Society keeps records of the elevations of various geographic features. Table 8.1 lists the highest and lowest points on each continent.

When measuring, one has to know where to begin, or the zero point. Notice that when measuring elevations, the zero point is chosen to be sea level (Figure 8.1). All elevations compare the high or low point to sea level.

SECTIONS

TABLE 8.1 CONTINENTAL ELEVATIONS

Continent	Highest Point	Feet Above Sea Level	Lowest Point	Feet Below Sea Level
Africa	Kilimanjaro, Tanzania	19,340	Lake Assal	512
Antarctica	Vinson Massif	16,864	Bentley, Sub-glacial Trench	8327
Asia	Mount Everest, Nepal-Tibet	29,028	Dead Sea, Israel-Jordan	1312
Australia	Mount Kosciusko, New South Wales	7310	Lake Eyre, South Australia	52
Europe	Mount El'brus, Russia	18,510	Caspian Sea, Russia-Azerbaijan	92
North America	Mount McKinley, Alaska	20,320	Death Valley, California	282
South America	Mount Aconcagua, Argentina	22,834	Valdes Peninsula, Argentina	131

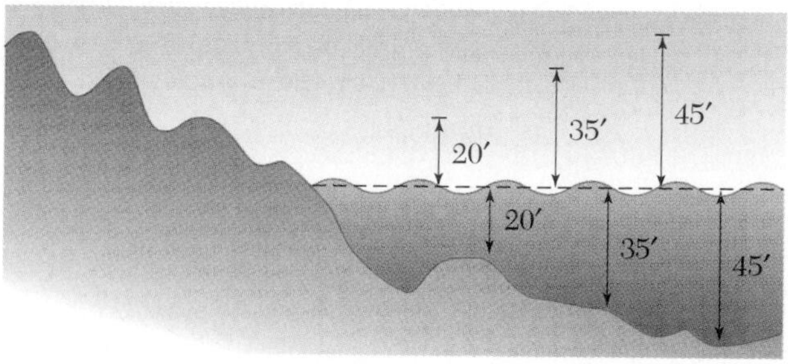

FIGURE 8.1

Mathematically, we represent quantities under the zero point as negative numbers.

What location has the highest continental altitude on Earth? What location has the lowest continental altitude? Is there a location on Earth with a lower altitude? Explain.

8.1 Opposites and Absolute Value

OBJECTIVES

1. Find the opposite of a signed number.
2. Find the absolute value of a signed number.

VOCABULARY

Positive numbers are the numbers of arithmetic and are greater than zero.

Negative numbers are numbers less than zero. Zero is neither positive nor negative. Positive numbers, zero, and negative numbers are called **signed numbers.**

The **opposite**, or **additive inverse**, of a signed number is the number on the number line that is the same distance from zero but on the opposite side. Zero is its own opposite. The opposite of 5 is written -5. This can be read "the opposite of 5" or "negative 5" because they both name the same number.

The **absolute value** of a signed number is the number of units between the number and zero. The expression $|6|$ is read "the absolute value of 6."

HOW & WHY

OBJECTIVE 1 Find the opposite of a signed number.

Exercises such as

$6 - 8$ $\quad$ $9 - 24$ $\quad$ $11 - 12$ $\quad$ and $\quad$ $4 - 134$

do not have answers in the numbers of arithmetic. The answer to each is a signed number. Signed numbers (which include both numbers to the right of zero and to the left of zero) are used to represent quantities with opposite characteristics. For instance,

right and left
up and down
above zero and below zero
gain and loss

A few signed numbers are shown on the number line in Figure 8.2.

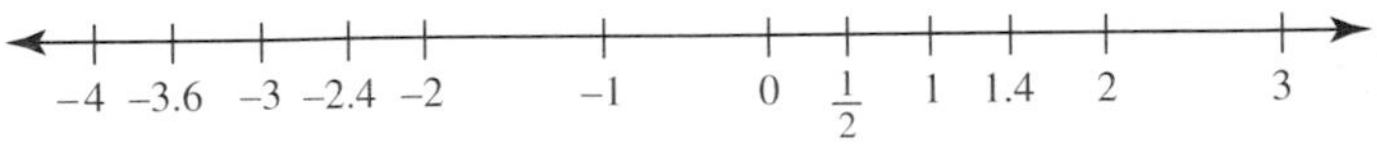

FIGURE 8.2

The negative numbers are to the left of zero. The negative numbers have a dash, or negative sign, in front of them. The numbers to the right of zero are called positive (and may be written with a plus sign). Zero is neither positive nor negative. Here are some signed numbers.

7	Seven, or positive seven
-3	Negative three
-0.12	Negative twelve hundredths
0	Zero is neither positive nor negative.
$+\frac{1}{2}$	One-half, or positive one-half

Positive and negative numbers are used many ways in the physical world—here are some examples:

Positive	Negative
Temperatures above zero (83°)	Temperatures blow zero (−11°)
Feet above sea level (6000 ft)	Feet below sea level (−150 ft)
Profit ($94)	Loss (−$51)
Right (15)	Left (−12)

In any situation where quantities can be measured in opposite directions, positive and negative numbers can be used to show direction.

The dash in front of a number is read in two different ways:

-23 The opposite of 23
-23 Negative 23

The opposite of a number is the number on the number line that is the same distance from zero but on the opposite side. To find the opposite of a number, we refer to a number line. See Figure 8.3.

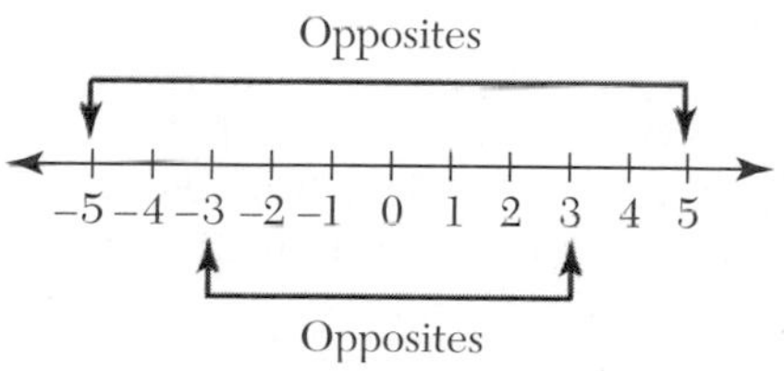

FIGURE 8.3

$-(5) = -5$ The opposite of positive 5 is negative 5.
$-(-3) = 3$ The opposite of negative 3 is positive 3.
$-0 = 0$ The opposite of 0 is 0.

To find the opposite of a positive number

1. Locate a positive number on the number line.
2. Count the number of units it is from zero.
3. Count this many units to the left of zero. Where you stop is the opposite of the positive number.

To find the opposite of a negative number

1. Locate a negative number on the number line.
2. Count the number of units it is from zero.
3. Count this many units to the right of zero. Where you stop is the opposite of the negative number.

The opposite of a positive number is negative and the opposite of a negative number is positive. The opposite of 0 is 0. The expression -5 is sometimes referred to as the "negative of" 5, and 5 is sometimes referred to as the "negative of" -5.

WARM-UPS A–E

A. Find the opposite of 26.

B. Find the opposite of -19.

C. Find the opposite of $-(-13)$.

Answers to Warm-Ups

A. -26 **B.** 19 **C.** -13

EXAMPLES A–E

DIRECTIONS: Find the opposite.

STRATEGY: The opposite of a number is found on the number line the same number of units from zero but on the opposite side of zero.

A. Find the opposite of 14.

STRATEGY: With no sign in front, the number is positive; 14 is read "fourteen" or "positive fourteen."

$-(14) = ?$ Because 14 is 14 units to the right of zero, the opposite of 14 is 14 units to the left of zero.

$-(14) = -14$

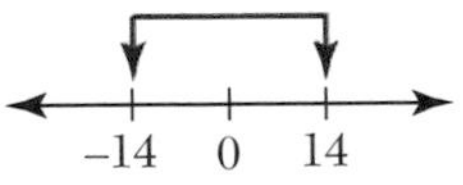

The opposite of 14 is -14.

B. Find the opposite of -28.

STRATEGY: The number -28 can be thought of in two ways: as a negative number that is 28 units to the left of zero or as a number that is 28 units on the opposite side of zero from 28; -28 is read "negative 28" or "the opposite of 28."

$-(-28) = ?$ The opposite of negative 28 is written $-(-28)$ and

$-(-28) = 28$ is found 28 units on the opposite side of zero from -28.

The opposite of -28 is 28.

C. Find the opposite of $-(-11)$.

> **CAUTION**
>
> The dash in front of the parentheses is **not** read "negative." Instead, it is read "the opposite of." The dash directly in front of 11 is read "negative" or "the opposite of."

STRATEGY: First find $-(-11)$. Then find the opposite of that value.

$-(-11)$ is "the opposite of negative 11"

$-(-11) = ?$

$-(-11) = 11$

$-(-11)$ can also be read "the opposite of the opposite of 11." The opposite of -11 is 11 units to the right of zero, which is 11.

So the opposite of $(-11) = 11$.

Continuing, we find the opposite of $-(-11)$, which is written $-[-(-11)]$.

$$-[-(-11)] = -[11]$$
$$= -11$$

The opposite of $-(-11) = -11$.

CALCULATOR EXAMPLE

D. Find the opposite of 19, −2.3, and 4.7.

STRATEGY: The [+/−] key on the calculator will give the opposite of the number.

The opposites are −19, 2.3, and −4.7.

E. If 10% of Americans purchased products with no plastic packaging just 10% of the time, approximately 144,000 lb of plastic would be eliminated (taken out of or decreased) from our landfills.

1. Write this decrease as a signed number.
2. Write the opposite of eliminating (decreasing) 144,000 lb of plastic from our landfills as a signed number.

1. Decreases are often represented by negative numbers. Therefore, a decrease of 144,000 lb is −144,000 lb.
2. The opposite of a decrease is an increase, so −(−144,000 lb) is 144,000 lb.

D. Find the opposite of 13, −7.8, and 7.1.

E. The energy used to produce a pound of rubber is 15,700 BTU. Recycled rubber requires 4100 BTU less.

1. Express this decrease as a signed number.
2. Express the opposite of a decrease of 4100 BTU as a signed number.

HOW & WHY

OBJECTIVE 2 Find the absolute value of a signed number.

The absolute value of a signed number is the number of units between the number and zero on the number line. Absolute value is defined as the number of units only; direction is not involved. Therefore, the absolute value is never negative.

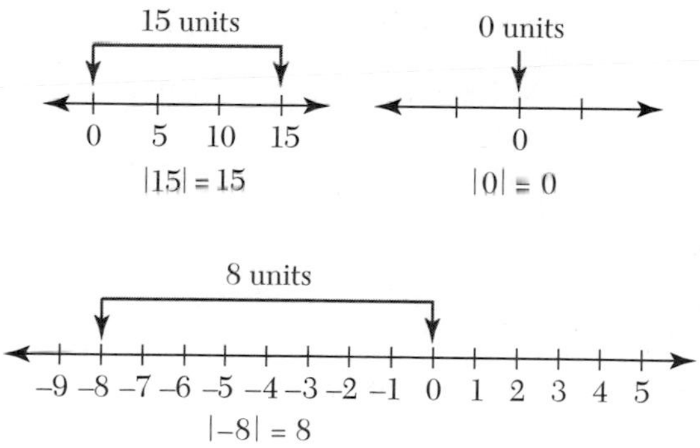

FIGURE 8.4

From Figure 8.4, we can see that the absolute value of a number and the absolute value of the number's opposite are equal. For example,

$|45| = |-45|$ Because both equal 45

The absolute value of a positive number and zero is the number itself. The absolute value of a negative number is its opposite.

To find the absolute value of a signed number

The value is

1. zero if the number is zero
2. the number itself if the number is positive
3. the opposite of the number if the number is negative

Answers to Warm-Ups

D. −13, 7.8, −7.1 **E.** As a signed number, the decrease is −4100 BTU. As a signed number, the opposite of a decrease of 4100 BTU is 4100 BTU.

WARM-UPS F–I

F. Find the absolute value of 29.

G. Find the absolute value of −90.

H. Find the absolute value of −1.

I. Find the absolute value of $-\frac{3}{7}$.

EXAMPLES F–I

DIRECTIONS: Find the absolute value of the number.

STRATEGY: If the number is positive or 0, write the number. If the number is negative, write its opposite.

F. Find the absolute value of 17.

$|17| = 17$ The absolute value of a positive number is the number itself.

G. Find the absolute value of −41.

$$|-41| = -(-41) = 41$$

The absolute value of a negative is its opposite.

H. Find the absolute value of 0.

$|0| = 0$ The absolute value of zero is zero.

I. Find the absolute value of $-\frac{5}{9}$.

$$\left|-\frac{5}{9}\right| = -\left(-\frac{5}{9}\right) = \frac{5}{9}$$

The absolute value of a negative number is its opposite.

Answers to Warm-Ups

F. 29 **G.** 90 **H.** 1 **I.** $\frac{3}{7}$

Name ________________ Class ________________ Date ________________

EXERCISES 8.1

OBJECTIVE 1 Find the opposite of a signed number.

Find the opposite of the signed number.

A.

1. -4 **2.** -8 **3.** 5 **4.** 13

5. -2.3 **6.** -3.3 **7.** $\frac{3}{5}$ **8.** $\frac{3}{4}$

9. $-\frac{6}{7}$ **10.** $-\frac{3}{8}$

11. The opposite of ________ is 61. **12.** The opposite of -23 is ________.

B.

13. -42 **14.** -57 **15.** -3.78 **16.** -6.27

17. $\frac{17}{5}$ **18.** $\frac{23}{7}$ **19.** 0.55 **20.** 0.732

21. 113.8 **22.** 243.7 **23.** -0.0123 **24.** -0.78

OBJECTIVE 2 Find the absolute value of a signed number.

Find the absolute value of the signed number.

A.

25. $|-4|$ **26.** $|-11|$ **27.** $|24|$ **28.** $|61|$

29. $|-3.17|$ **30.** $|-4.6|$ **31.** $\left|\frac{5}{11}\right|$ **32.** $\left|\frac{7}{13}\right|$

33. $\left|-\frac{3}{11}\right|$ **34.** $\left|-\frac{7}{12}\right|$

35. The absolute value of ________ is 32. **36.** The absolute value of ________ is 17.

B.

37. $|0.0065|$ **38.** $|0.0021|$ **39.** $|-355|$ **40.** $|-922|$

41. $\left|-\frac{11}{3}\right|$ **42.** $\left|-\frac{17}{6}\right|$ **43.** $\left|-\frac{33}{7}\right|$ **44.** $\left|-\frac{18}{5}\right|$

45. $|0|$ **46.** $|-100|$ **47.** $|-0.341|$ **48.** $|-4.96|$

C. *Find the value of the following:*

49. Opposite of $\left|-\frac{5}{17}\right|$

50. Opposite of $\left|-\frac{9}{8}\right|$

51. Opposite of $|78|$

52. Opposite of $|-91|$

53. At the New York Stock Exchange, positive and negative numbers are used to record changes in stock prices on the board. What is the opposite of a gain of 1.54?

54. On the NASDAQ, a stock is shown to have taken a loss of three-eighths point (−0.375). What is the opposite of this loss?

55. On a thermometer, temperatures above zero are listed as positive and those below zero as negative. What is the opposite of a reading of 14°C?

56. On a thermometer such as the one in Exercise 55, what is the opposite of a reading of 46°C?

57. The modern calendar counts the years after the birth of Christ as positive numbers (A.D. 2001 or +2001). Years before Christ are listed using negative numbers (2011 B.C. or −2011). What is the opposite of 1915 B.C. or −1915.

58. The empty-weight center of gravity of an airplane is determined. A generator is installed at a moment of −278. At what moment could a weight be placed so that the center of gravity remains the same? (Moment is the product of a quantity such as weight and its distance from a fixed point. In this application, the moments must be opposites to keep the same center of gravity.)

59. A cyclist travels up a mountain 1415 ft and then turns around and travels down the mountain 1143 ft. Represent each trip as a signed number.

60. An energy audit indicates that the Gates family could reduce their average electric bill by $17.65 per month by doing some minor repairs, insulating their attic and crawl space, and caulking around the windows and other cracks in the siding.

a. Express this savings as a signed number.

b. Express the opposite of the savings as a signed number.

61. If 80 miles north is represented by +80, how would you represent 80 miles south?

Exercises 62–64 refer to the chapter application.

62. Rewrite the continental altitudes in the application using signed numbers.

63. The U.S. Department of Defense has extensive maps of the ocean floors because they are vital information for the country's submarine fleet. The table lists the deepest part and the average depth of the world's major oceans.

Ocean	Deepest Part	(ft)	Average Depth (ft)
Pacific	Mariana Trench	35,840	12,925
Atlantic	Puerto Rico Trench	28,232	11,730
Indian	Java Trench	23,376	12,598
Arctic	Eurasia Basin	17,881	3407
Mediterranean	Ionian Basin	16,896	4926

Rewrite the table using signed numbers.

64. Is the highest point on Earth farther away from sea level than the deepest point in the ocean? Explain. What mathematical concept allows you to answer this question? See exercise 63.

Simplify.

65. $-(-24)$

66. $-(81)$

67. $-[-(-14)]$

68. $-[-(33)]$

69. The Buffalo Bills are playing a football game against the Seattle Seahawks. On the first play, the Seahawks lose 8 yd. Represent this as a signed number. What is the opposite of a loss of 8 yd? Represent this as a signed number.

70. The Dallas Cowboys and the New York Giants are having an exhibition game in London, England. The Cowboy offensive team runs a gain of 6 yd, a loss of 8 yd, a gain of 21 yd, and a loss of 15 yd. Represent these yardages as signed numbers.

71. The Golden family is on a vacation in the southwestern United States. Consider north and east as positive directions and south and west as negative directions. On one day they drive north 137 miles, then east 212 miles. The next day they drive west 98 miles, then 38 miles south. Represent each of these distances as signed numbers.

STATE YOUR UNDERSTANDING

72. Is zero the only number that is its own opposite? Justify your answer.

73. Is there a set of numbers for which the absolute value of each number is the number itself? If yes, identify that set and tell why this is true.

74. Explain -4. Draw it on a number line. On the number line, use the concepts of opposites and absolute value as they relate to -4. Give an instance in the world when -4 is useful.

CHALLENGE

Simplify.

75. $|16 - 10| - |14 - 9| + 6$

76. $8 - |12 - 8| - |10 - 8| + 2$

77. If n is a negative number, what kind a number is $-n$?

78. For what numbers is $|-n| = n$ always true?

GROUP WORK

79. Develop a rule (procedure) for adding a positive and a negative integer. *Hint*: You might want to visualize that walking forward represents a positive number and walking backward represents a negative integer. See if you can find answers to: $6 + (-4) = ?$, $7 + (-10) = ?$, and $-6 + 11 = ?$

MAINTAIN YOUR SKILLS

Add.

80. 7 + 11 + 32 + 9

81. 5 + 12 + 21 + 14

Subtract.

82. 17 − 12

83. 45 − 23

Add.

84.
$$\begin{array}{r} 6 \text{ m } 250 \text{ cm} \\ +7 \text{ m } 460 \text{ cm} \\ \hline \end{array} = ?\text{ m}$$

85.
$$\begin{array}{r} 5\text{L } 78 \text{ mL} \\ +2\text{L } 95 \text{ mL} \\ \hline \end{array} = ?\text{ mL}$$

Subtract.

86.
$$\begin{array}{r} 3 \text{ yd } 2 \text{ ft } 8 \text{ in.} \\ -1 \text{ yd } 2 \text{ ft } 11 \text{ in.} \\ \hline \end{array} = ?\text{ in.}$$

87.
$$\begin{array}{r} 7 \text{ gal } 1 \text{ qt } 1 \text{ pt} \\ -4 \text{ gal } 3 \text{ qt } 1 \text{ pt} \\ \hline \end{array} = ?\text{ pt}$$

88. A 12-m coil of wire is to be divided into eight equal parts. How many meters are in each part?

89. Some drug doses are measured in grains. For example, a common aspirin tablet contains 5 grains. If there are 15.43 g in a grain, how many grams do two aspirin tablets contain?

8.2 Adding Signed Numbers

HOW & WHY

OBJECTIVE Add signed numbers.

Positive and negative numbers are used to show opposite quantities:

+482 lb may show 482 lb loaded.
−577 lb may show 577 lb unloaded.
+27 dollars may show 27 dollars earned.
−19 dollars may show 19 dollars spent.

We can use this idea to find the sum of signed numbers. We already know how to add two positive numbers, so we concentrate on adding positive and negative numbers.

Think of 27 dollars earned (positive) and 19 dollars spent (negative). The result is 8 dollars left in your pocket (positive). So

$27 + (-19) = 8$

To get this sum, we subtract the absolute value of −19 (19) from the absolute value of 27 (27). The sum is positive.

Think of 23 dollars spent (negative) and 15 dollars earned (positive). The result is that you still owe 8 dollars (negative). So

$-23 + 15 = -8$

To get this sum, subtract the absolute value of 15 (15) from the absolute value of −23 (23). The sum is negative.

Think of 5 dollars spent (negative) and another 2 dollars spent (negative). The result is 7 spent (negative). So

$-5 + (-2) = -7$

To get this sum, we add 5 to 2 (the absolute value of each). The sum is negative.

The results of the examples lead us to the procedure for adding signed numbers.

To add signed numbers

1. If the signs are alike, add their absolute values and use the common sign.
2. If the signs are not alike, subtract the smaller absolute value from the larger absolute value. The sum will have the sign of the number with the larger absolute value.

As a result of this definition, the sum of a number and its opposite is zero. Thus, $5 + (-5) = 0$, $8 + (-8) = 0$, $-12 + 12 = 0$, and so on.

OBJECTIVE

Add signed numbers.

EXAMPLES A–J

DIRECTIONS: Add.

STRATEGY: If the signs are alike, add the absolute values and use the common sign. If the signs are unlike, subtract their absolute values and use the sign of the number with larger absolute value.

WARM-UPS A–J

A. Add: $56 + (-30)$

A. Add: $45 + (-23)$

STRATEGY: Because the signs are unlike, subtract their absolute values.

$$\begin{aligned}|45| - |-23| &= 45 - 23\\ &= 22\end{aligned}$$

Because the positive number has the larger absolute value, the sum is positive.

So, $45 + (-23) = 22$.

B. Find the sum: $-63 + 44$

B. Find the sum: $-72 + 31$

$$\begin{aligned}|-72| - |31| &= 72 - 31\\ &= 41\end{aligned}$$

$-72 + 31 = -41$ The sum is negative because -72 has the larger absolute value.

C. Add: $42 + (-81)$

C. Add: $37 + (-53)$

$$\begin{aligned}|-53| - |37| &= 53 - 37\\ &= 16\end{aligned}$$

$37 + (-53) = -16$ The sum is negative because -53 has the larger absolute value.

D. Add: $-5.6 + (-2.1)$

D. Add: $-0.12 + (-2.54)$

$$\begin{aligned}|-0.12| + |-2.54| &= 0.12 + 2.54\\ &= 2.66\end{aligned}$$

$-0.12 + (-2.54) = -2.66$ The numbers are negative; therefore the sum is negative.

E. Find the sum: $\frac{2}{3} + \left(-\frac{5}{6}\right)$

E. Find the sum: $\frac{5}{8} + \left(-\frac{3}{4}\right)$

$$\begin{aligned}\left|-\frac{3}{4}\right| - \left|\frac{5}{8}\right| &= \frac{6}{8} - \frac{5}{8}\\ &= \frac{1}{8}\end{aligned}$$

$\frac{5}{8} + \left(-\frac{3}{4}\right) = -\frac{1}{8}$ Because the negative number has the larger absolute value, the sum is negative.

F. Find the sum of -40, 62, -85, and 57.

F. Find the sum of -16, 41, -33, and -14.

STRATEGY: Where there are more than two numbers to add, it may be easier to add the numbers with the same sign first.

$$\begin{array}{r}-16\\ -33\\ \underline{-14}\\ -63\end{array}$$

Add the negative numbers. Add their sum to 41. Because the signs are different, find the difference in their absolute values.

$$\begin{aligned}|-63| - |41| &= 63 - 41\\ &= 22\end{aligned}$$

$-16 + 41 + (-33) + (-14) = -22$ The sum is negative because -63 has the larger absolute value.

Answers to Warm-Ups

A. 26 **B.** -19 **C.** -39 **D.** -7.7 **E.** $-\frac{1}{6}$ **F.** -6

G. Add: $-0.4 + 0.7 + (-5.4) + 1.3$

$$\begin{array}{r} -0.4 \\ -5.4 \\ \hline -5.8 \end{array} \qquad \begin{array}{r} 0.7 \\ 1.3 \\ \hline 2.0 \end{array}$$

The signs are different so subtract their absolute values.

$$\begin{aligned} |-5.8| - |2.0| &= 5.8 - 2.0 \\ &= 3.8 \end{aligned}$$

$-0.4 + 0.7 + (-5.4) + 1.3 = -3.8$ Because -5.8 has the larger absolute value, the sum is negative.

H. Add: $-\frac{5}{6} + \frac{4}{9} + \left(-\frac{5}{18}\right) + \left(-\frac{7}{18}\right)$

STRATEGY: First add the negative numbers.

$$\begin{aligned} -\frac{5}{6} + \left(-\frac{5}{18}\right) + \left(-\frac{7}{18}\right) &= -\frac{15}{18} + \left(-\frac{5}{18}\right) + \left(-\frac{7}{18}\right) \\ &= -\frac{27}{18} \end{aligned}$$

Add the sum of the negative numbers and the positive number. To do this, find the difference of their absolute values.

$$\begin{aligned} \frac{27}{18} - \frac{4}{9} &= \frac{27}{18} - \frac{8}{18} \\ &= \frac{19}{18} \end{aligned}$$

$-\frac{5}{6} + \frac{4}{9} + \left(-\frac{5}{18}\right) + \left(-\frac{7}{18}\right) = -\frac{19}{18}$ The sum is negative because the total of the negative numbers has the larger absolute value.

CALCULATOR EXAMPLE

I. Find the sum: $-89 + 77 + (-104) + (-93)$

Use the $\boxed{+/-}$ on the calculator to indicate negative numbers.

The sum is -209.

J. John owns stock that is traded on the NASDAQ. On Monday, the stock gains \$2.04; on Tuesday, it loses \$3.15; on Wednesday, it loses \$1.96; on Thursday, it gains \$2.21; and on Friday, it gains \$1.55. What is the net price change in the price of the stock for the week?

STRATEGY: To find the net change in the price of the stock, write the daily changes as signed numbers and find the sum of these numbers.

Monday	gains \$2.04	2.04
Tuesday	loses \$3.15	−3.15
Wednesday	loses \$1.96	−1.96
Thursday	gains \$2.21	2.21
Friday	gains \$1.55	1.55

$$\begin{aligned} 2.04 + (-3.15) + (-1.96) + 2.21 + 1.55 &= 5.8 + (-5.11) \\ &= 0.69 \end{aligned}$$

Add the positive numbers and add the negative numbers.

The stock gains \$0.69 during the week.

G. Add: $0.42 + (-5.21) + (-0.94) + 3.65$

H. $-\frac{4}{5} + \frac{1}{2} + \left(-\frac{3}{4}\right) + \left(-\frac{2}{5}\right)$

I. Find the sum: $-88 + 97 + (-117) + (-65)$

J. A second stock that John owns has the following changes for a week: Monday, gains \$1.34; Tuesday, gains \$0.84; Wednesday, loses \$4.24; Thursday, loses \$2.21; Friday, gains \$0.36. What is the net change in the price of the stock for the week?

Answers to Warm-Ups

G. -2.08 **H.** $-\frac{29}{20}$ **I.** -173

J. The stock lost \$3.91 (-3.91) for the week.

EXERCISES 8.2

OBJECTIVE Add signed numbers.

Add.

A.

1. $-7 + 9$ **2.** $-7 + 3$ **3.** $8 + (-6)$ **4.** $10 + (-6)$

5. $-9 + (-6)$ **6.** $-7 + (-8)$ **7.** $-11 + (-3)$ **8.** $-15 + (-9)$

9. $-11 + 11$ **10.** $16 + (-16)$ **11.** $0 + (-12)$ **12.** $-32 + 0$

13. $-23 + (-23)$ **14.** $-13 + (-13)$ **15.** $3 + (-17)$ **16.** $12 + (-15)$

17. $-19 + (-7)$ **18.** $-11 + (-22)$ **19.** $24 + (-17)$ **20.** $-18 + 31$

B.

21. $-5 + (-7) + 6$ **22.** $-5 + (-8) + (-10)$ **23.** $-65 + (-43)$

24. $-56 + (-44)$ **25.** $-98 + 98$ **26.** $108 + (-108)$

27. $-45 + 72$ **28.** $53 + (-38)$ **29.** $-36 + 43 + (-17)$

30. $-37 + 21 + (-9)$ **31.** $62 + (-56) + (-13)$ **32.** $-34 + (-18) + 29$

33. $-4.6 + (-3.7)$ **34.** $-9.7 + (-5.2)$ **35.** $10.6 + (-7.8)$

36. $-13.6 + 8.4$ **37.** $-\frac{5}{6} + \frac{1}{4}$ **38.** $\frac{3}{8} + \left(-\frac{1}{2}\right)$

39. $-\frac{7}{10} + \left(-\frac{7}{15}\right)$ **40.** $-\frac{5}{12} + -\frac{2}{9}$

C.

Simplify.

41. $135 + (-256)$ **42.** $233 + (-332)$

43. $-81 + (-32) + (-76)$ **44.** $-75 + (-82) + (-71)$

45. $-31 + 28 + (-63) + 36$ **46.** $-44 + 37 + (-59) + 45$

47. $49 + (-67) + 27 + 72$ **48.** $81 + (-72) + 33 + 49$

49. $356 + (-762) + (-892) + 541$ **50.** $-923 + 672 + (-823) + (-247)$

51. Find the sum of 542, −481, and −175. **52.** Find the sum of 293, −122, and −211.

Exercises 53–56. The table gives the temperatures recorded by Viking II for a 5-day period at one spot on the surface of Mars.

	Day 1	Day 2	Day 3	Day 4	Day 5
5:00 A.M.	−92°C	−88°C	−115°C	−103°C	−74°C
9:00 A.M.	−57°C	−49°C	−86°C	−93°C	−64°C
1:00 P.M.	−52°C	−33°C	−46°C	−48°C	−10°C
6:00 P.M.	−45°C	−90°C	−102°C	−36°C	−42°C
11:00 P.M.	−107°C	−105°C	−105°C	−98°C	−90°C

53. What is the sum of the temperatures recorded at 9:00 A.M.?

54. What is the sum of the temperatures recorded on day 1?

55. What is the sum of the temperatures recorded at 11:00 P.M.?

56. What is the sum of the temperatures recorded on day 4?

57. An airplane is being reloaded; 963 lb of baggage and mail are removed (−963 lb) and 855 lb of baggage and mail are loaded on (+855 lb). What net change in weight should the cargo master report?

58. At another stop, the plane in Exercise 57 unloads 2341 lb of baggage and mail and takes on 2567 lb. What net change should the cargo master report?

59. The change in altitude of a plane in flight is measured every 10 min. The figures between 3:00 P.M. and 4:00 P.M. are as follows:

3:00	30,000 ft initially	(+30,000)
3:10	increase of 220 ft	(+220 ft)
3:20	decrease of 200 ft	(−200 ft)
3:30	increase of 55 ft	(+55 ft)
3:40	decrease of 110 ft	(−110 ft)
3:50	decrease of 55 ft	(−55 ft)
4:00	decrease of 40 ft	(−40 ft)

What is the altitude of the plane at 4 P.M.? (*Hint*: Find the sum of the initial altitude and the six measured changes between 3 and 4 P.M.)

60. What is the final altitude of the airplane in Exercise 59 if it is initially flying at 23,000 ft with the following changes in altitude?

3:00	23,000 ft initially	(+23,000)
3:10	increase of 315 ft	(+315 ft)
3:20	decrease of 825 ft	(−825 ft)
3:30	increase of 75 ft	(+75 ft)
3:40	decrease of 250 ft	(−250 ft)
3:50	decrease of 85 ft	(−85 ft)
4:00	decrease of 70 ft	(−70 ft)

61. The Pacific Northwest Book Depository handles most textbooks for the local schools. On September 1, the inventory is 34,945 volumes. During the month, the company makes the following transactions (positive numbers represent volumes received, negative numbers represent shipments): 3456, −2024, −3854, 612, −2765. What is the inventory at the end of the month?

62. The Pacific Northwest Book Depository has 18,615 volumes on November 1. During the month, the depository has the following transactions: −4675, 912, −6764, 1612, −950. What is the inventory at the end of the month?

63. The Buffalo Bills made the following consecutive plays during a recent Monday night football game: 8-yd loss, 10-yd gain, and a 7-yd gain. A first down requires a gain of 10 yd. Did they get a first down?

64. The Seattle Seahawks have these consecutive plays one Sunday: 12-yd loss, 19-yd gain, and a 4-yd gain. Do they get a first down?

65. Nordstrom stock has the following changes in one week: up 0.61, down 0.45, down 1.13, up 2.61, and up 0.12.

a. What is the net change for the week?

b. If the stock starts at 72.83 at the beginning of the week, what is the closing price?

66. On a January morning in a small town in upstate New York, the lowest temperature is recorded as 17 degrees below zero. During the following week the daily lowest temperature readings are up 3 degrees, up 7 degrees, down 5 degrees, up 1 degree, no change, and down 3 degrees. What is the low temperature reading for the last day?

67. A new company has the following weekly balances after the first month of business: a loss of \$2376, a gain of \$5230, a loss of \$1055, and a gain of \$3278. What is their net gain or loss for the month?

68. Marie decided to play the state lottery for one month. This meant that she played every Wednesday and Saturday for a total of nine times. This is her record: lost \$10, won \$45, lost \$15, lost \$12, won \$65, lost \$12, lost \$16, won \$10, lost \$6. What is the net result of her playing?

STATE YOUR UNDERSTANDING

69. When adding a positive and a negative number, explain how to determine whether the sum is positive or negative.

CHALLENGE

70. What is the sum of 46 and (-52) increased by 23?

71. What is the sum of (-73) and (-32) added to (-19)?

72. What number added to 47 equals 28?

73. What number added to (-75) equals (-43)?

74. What number added to (-123) equals (-163)?

GROUP WORK

75. Develop a rule for subtracting two signed numbers. (*Hint:* Subtraction is the inverse of addition—that is, it undoes addition. So, if adding $+6$ can be thought of as walking forward 6 paces, then subtracting $+6$ may be thought of as walking backward 6 paces.) See if you can find answers to

$4 - (+6) = ?$ $\quad -5 - (+6) = ?$

and $\quad 8 - (-4) = ?$

MAINTAIN YOUR SKILLS

Subtract:

76. $145 - 67$

77. $356 - 89$

78. $\begin{array}{r} 178 \\ -\ 95 \\ \hline \end{array}$

79. $\begin{array}{r} 897 \\ -698 \\ \hline \end{array}$

80. Find the perimeter of a rectangle that is 42 m long and 29 m wide.

81. Find the circumference of a circle with a radius of 17 cm. The formula is $C = \pi d$, where d is the diameter of the circle. Let $\pi \approx 3.14$.

82. Find the perimeter of a square that is 31 in. on each side.

83. Find the perimeter of a triangle with sides 65 cm, 78 cm, and 52 cm.

84. Bananas are put on sale at 3 lb for \$1.62. What is the cost of 1 pound?

85. The cost of pouring a 3-ft-wide cement sidewalk is estimated to be \$21 per square foot. If a walk is to be placed around a rectangular plot of ground that is 42 ft along the width and 50 ft along the length, what is the cost of pouring the walk?

8.3 Subtracting Signed Numbers

HOW & WHY

OBJECTIVE Subtract signed numbers.

OBJECTIVE
Subtract signed numbers.

The expression $21 - 18 = ?$, in terms of addition, asks $18 + ? = 21$. We know $18 + 3 = 21$, so $21 - 18 = 3$. The expression $-15 - 8 = ?$, in terms of addition, asks $8 + ? = -15$. We know $8 + (-23) = -15$, so $-15 - 8 = -23$. The expression $-25 - (-28) = ?$, in terms of addition, asks $-28 + ? = -25$. We know that $-28 + 3 = -25$, so $-25 - (-28) = 3$.

Compare

$21 - 18$	and	$21 + (-18)$	Both equal 3.
$-15 - 8$	and	$-15 + (-8)$	Both equal -23.
$-25 - (-28)$	and	$-25 + 28$	Both equal 3.

Every subtraction problem can be worked by asking what number added to the subtrahend will yield the minuend. However, when we look at the second column, we see an addition problem that gives the same answer as the original subtraction problem. In each case the opposite (additive inverse) of the number being subtracted is added. Let's look at three more examples.

Answer Obtained by Adding to the Subtrahend	Answer Obtained by Adding the Opposite	
$19 - 7 = 12$	$19 + (-7) = 12$	-7 is the opposite of 7, the number to be subtracted.
$-32 - 45 = -77$	$-32 + (-45) = -77$	-45 is the opposite of 45, the number to be subtracted.
$-21 - (-12) = -9$	$-21 + 12 = -9$	12 is the opposite of -12, the number to be subtracted.

This leads us to the rule for subtracting signed numbers.

To subtract signed numbers

1. Rewrite as an addition problem by adding the opposite of the number to be subtracted.
2. Find the sum.

EXAMPLES A–J

DIRECTIONS: Subtract.

STRATEGY: Add the opposite of the number to be subtracted.

A. Subtract: $42 - 28$

STRATEGY: Rewrite as an addition problem by adding -28, which is the opposite of 28.

$$42 - 28 = 42 + (-28)$$
$$= 14$$

Add. The signs are different, so subtract their absolute values and use the sign of the number with the larger absolute value, which is 42.

WARM-UPS A–J

A. Subtract: $75 - 42$

Answers to Warm-Ups

A. 33

Because both numbers are positive, we can also do the subtraction in the usual manner: $42 - 28 = 14$.

B. Subtract: $-56 - 70$

B. Subtract: $-38 - 61$

$$\begin{aligned} -38 - 61 &= -38 + (-61) \\ &= -99 \end{aligned}$$

Add. Both numbers are negative, so add their absolute values and keep the common sign.

C. Find the difference of -34 and -40.

C. Find the difference of -54 and -12.

$$\begin{aligned} -54 - (-12) &= -54 + 12 \\ &= -42 \end{aligned}$$

Add.

D. Subtract: $32 - 51$

D. Subtract: $63 - 91$

$$\begin{aligned} 63 - 91 &= 63 + (-91) \\ &= -28 \end{aligned}$$

E. Find the difference: $\frac{1}{4} - \left(-\frac{3}{5}\right)$

E. Find the difference: $\frac{5}{8} - \left(-\frac{2}{3}\right)$

$$\begin{aligned} \frac{5}{8} - \left(-\frac{2}{3}\right) &= \frac{5}{8} + \frac{2}{3} \\ &= \frac{15}{24} + \frac{16}{24} \\ &= \frac{31}{24} \end{aligned}$$

Write each fraction with a common denominator and add.

F. Subtract: $45 - (-32) - 24$

F. Subtract: $23 - (-33) - 19$

STRATEGY: Change both subtractions to add the opposite.

$$\begin{aligned} 23 - (-33) - 19 &= 23 + 33 + (-19) \\ &= 56 + (-19) \\ &= 37 \end{aligned}$$

Add 23 and 33.
Add.

G. Subtract: $-0.74 - 1.16 - (-0.93) - (-2.46)$

G. Subtract: $-0.34 - (-1.04) - (-3.02) - 0.81$

STRATEGY: Change all subtractions to add the opposite and add.

$$\begin{aligned} -0.34 - (-1.04) - (-3.02) - 0.81 &= -0.34 + 1.04 + 3.02 + (-0.81) \\ &= 2.91 \end{aligned}$$

H. Subtract: $-\frac{3}{5} - \left(-\frac{3}{8}\right) - \left(-\frac{7}{10}\right) - \frac{7}{20}$

H. Subtract: $-\frac{3}{4} - \frac{7}{8} - \left(-\frac{1}{2}\right) - \left(-\frac{1}{8}\right)$

$$\begin{aligned} -\frac{3}{4} - \frac{7}{8} - \left(-\frac{1}{2}\right) - \left(-\frac{1}{8}\right) &= -\frac{3}{4} + \left(-\frac{7}{8}\right) + \frac{1}{2} + \frac{1}{8} \\ &= -\frac{6}{8} + \left(-\frac{7}{8}\right) + \frac{4}{8} + \frac{1}{8} \\ &= -\frac{8}{8} = -1 \end{aligned}$$

CALCULATOR EXAMPLE

I. Subtract: $-483.34 - (-312.43)$

I. Subtract: $-784.63 - (-532.78)$

STRATEGY: The calculator does not require you to change the subtraction to add the opposite.

The difference is -251.85.

Answers to Warm-Ups

B. -126 **C.** 6 **D.** -19 **E.** $\frac{17}{20}$ **F.** 53 **G.** 1.49 **H.** $\frac{1}{8}$ **I.** -170.91

J. The highest point in North America is Mount McKinley, a peak in central Alaska, which is approximately 20,320 ft above sea level. The lowest point in North America is Death Valley, a deep basin in southeastern California, which is approximately 282 ft below sea level. What is the difference in height between Mount McKinley and Death Valley? (Above sea level is positive and below sea level is negative.)

STRATEGY: To find the difference in height, write each height as a signed number and subtract the lower height from the higher height.

Mount McKinley	20,320 above	20,320
Death Valley	282 below	-282

$20{,}320 - (-282) = 20{,}320 + 282$ Change subtraction to add the opposite and add.

$= 20{,}602$

The difference in height is approximately 20,602 ft.

J. One night last winter the temperature dropped from 24°F to -11°F. What was the difference between the high and the low temperature?

Answers to Warm-Ups

J. The difference in temperatures 35°F.

EXERCISES 8.3

OBJECTIVE Subtract signed numbers.

Subtract.

A.

1. $7 - 3$ **2.** $9 - 5$ **3.** $-7 - 5$ **4.** $-8 - 7$

5. $-8 - (-3)$ **6.** $-7 - (-2)$ **7.** $11 - 8$ **8.** $13 - 9$

9. $-12 - 8$ **10.** $-13 - 6$ **11.** $-23 - (-14)$ **12.** $-17 - (-8)$

13. $19 - (-13)$ **14.** $26 - (-9)$ **15.** $-23 - 11$ **16.** $-24 - 12$

17. $-15 - 13$ **18.** $-33 - 20$ **19.** $-13 - (-14)$ **20.** $-34 - (-35)$

21. $-16 - (-16)$ **22.** $-23 - (-23)$ **23.** $-9 - 9$ **24.** $-31 - 31$

25. $-40 - 40$ **26.** $-14 - 14$

B.

27. $72 - (-46)$ **28.** $54 - (-61)$ **29.** $-57 - 62$

30. $-82 - 91$ **31.** $-48 - (-59)$ **32.** $-66 - (-81)$

33. $-91 - 91$ **34.** $-43 - 43$ **35.** $102 - (-102)$

36. $78 - (-78)$ **37.** $-69 - (-69)$ **38.** $-83 - (-83)$

39. $134 - (-10)$ **40.** $164 - (-20)$ **41.** $132 - (-41)$

42. $173 - (45)$ **43.** $-6.74 - 3.24$ **44.** $-13.34 - 9.81$

45. $-4.65 - (-3.21)$ **46.** $-7.54 - (-8.12)$ **47.** $-23.43 - 32.71$

48. $-18.63 - (-13.74)$

49. Find the difference between 43 and -73.

50. Find the difference between -88 and -97.

51. Subtract 328 from -359.

52. Subtract 145 from -251.

53. *Viking II* records high and low temperatures of -24°C and -109°C for 1 day on the surface of Mars. What is the change in temperature for that day?

54. The surface temperature of one of Jupiter's satellites is measured for 1 week. The highest temperature recorded is -83°C and the lowest is -145°C. What is the difference in the extreme temperatures for the week?

55. At the beginning of the month, Joe's bank account had a balance of \$782.45. At the end of the month, the account was overdrawn by \$13.87 ($-$\$13.87). If there were no deposits during the month, what was the total amount of checks Joe wrote? (*Hint*: Subtract the ending balance from the original balance.)

56. At the beginning of the month Jack's bank account had a balance of \$512.91. At the end of the month the balance was $-$\$67.11. If there were no deposits, find the amount of checks Jack wrote. (Refer to Exercise 55.)

 Name ______________________ Class ______________ Date ____________

Exercises 57–59 refer to the chapter application.

57. The range of a set of numbers is defined as the difference between the largest and the smallest numbers in the set. Calculate the range of altitude for each continent. Which continent has the smallest range and what does this mean in physical terms?

58. What is the difference between the lowest point in the Mediterranean and the lowest point in the Atlantic? See Exercise 63 in Section 8.1.

59. Some people consider Mauna Kea, Hawaii, to be the tallest mountain in the world. It rises 33,476 feet from the ocean floor but is only 13,796 feet above sea level. What is the depth of the ocean floor at this location?

Exercises 60–63 refer to the following table, which shows temperature recordings by a Martian probe for a 5-day period at one point on the surface of Mars.

Temperatures on the Surface of Mars

	Day 1	Day 2	Day 3	Day 4	Day 5
5:00 A.M.	−92°C	−88°C	−115°C	−103°C	−74°C
9:00 A.M.	−57°C	−49°C	−86°C	−93°C	−64°C
1:00 P.M.	−52°C	−33°C	−46°C	−48°C	−10°C
6:00 P.M.	−45°C	−90°C	−102°C	−36°C	−42°C
11:00 P.M.	−107°C	−105°C	−105°C	−98°C	−90°C

60. What is the difference between the high and low temperatures recorded on day 3?

61. What is the difference between the temperatures recorded at 11:00 P.M. on day 3 and day 5?

62. What is the difference between the temperatures recorded at 5 A.M. on day 2 and 6:00 P.M. on day 4?

63. What is the difference between the highest and lowest temperatures recorded during the 5 days?

64. Al's bank account had a balance of $318. He writes a check for $412.75. What is his account balance now?

65. Thomas started with $210.34 in his account. He writes a check for $216.75. What is his account balance now?

66. Carol started school owing her mother $18; by school's end she had borrowed $123 more from her mother. How does her account with her mother stand now?

67. At the beginning of the month, Janna's bank account had a balance of $467.82. At the end of the month, the account was overdrawn by $9.32. If there were no deposits during the month, what was the total amount of checks Janna wrote?

68. What is the difference between the highest point and the lowest point in California?

Highest point: Mount Whitney is 14,494 ft above sea level (+14,494).

Lowest point: Death Valley is 282 ft below sea level (−282).

69. On the first sale of the day Gina makes a profit of $145.67. However, on the next sale, Gina loses $310.12 because of another employee's misquote on the price of an item. After these two sales, what is the status of Gina's sales?

70. The New England Patriots started on their 40-yd line. After three plays they were on their 12-yd line. Did they gain or lose yards? Represent their gain or loss as a signed number.

71. During 1970, the net import of coal was -1.93 quadrillion BTU. In 1980, the net import of coal was -2.39 quadrillion BTU. In 1990, the net import of coal was -2.70 quadrillion BTU. Find the difference in coal imports for each decade.

STATE YOUR UNDERSTANDING

72. Explain the difference between adding and subtracting signed numbers.

73. How would you explain to a 10-year-old how to subtract -8 from 12?

74. Explain why the order in which two numbers are subtracted is important, but the order in which they are added is not.

75. Explain the difference between the problems $5 + (-8)$ and $5 - 8$.

CHALLENGE

76. $-13 - (-54) - |-21|$

77. $-9.46 - [-(-3.22)]$

78. $-76 - (-37) - |-(-55)|$

79. $57 - |-67 - (-51)| - 82$

80. $|17.5 - 13.5| - |21.7 - (-19.6)|$

GROUP WORK

81. Determine the normal daily mean (average) temperature for your city for each month of the year. Find the difference from month to month. Chart this result.

MAINTAIN YOUR SKILLS

Multiply.

82. 16(7)

83. 42(13)

84. $\begin{array}{r} 156 \\ \times\ 37 \\ \hline \end{array}$

85. $\begin{array}{r} 4782 \\ \times\ 362 \\ \hline \end{array}$

86. Find the area of a circle that has a radius of 14 in. (Let $\pi \approx 3.14$.)

87. Find the area of a square that is 16.7 m on each side.

88. Find the area of a rectangle that is 13.6 cm long and 9.4 cm wide.

89. Find the area of a triangle that has a base of 14.4 m and a height of 7.8 m.

90. How many square yards of wall-to-wall carpeting are needed to carpet a rectangular floor that measures 22.5 ft by 30.5 ft?

91. How many square tiles, which are 9 in. on a side, are needed to cover a floor that is 21 ft by 24 ft?

8.4 Multiplying Signed Numbers

HOW & WHY

OBJECTIVES

1. Multiply a positive number and a negative number.
2. Multiply two negative numbers.

OBJECTIVE 1 Multiply a positive number and a negative number.

Consider the following multiplications:

$3(4) = 12$
$3(3) = 9$
$3(2) = 6$
$3(1) = 3$
$3(0) = 0$
$3(-1) = ?$
$3(-2) = ?$

Each product is 3 smaller than the one before it. Continuing this pattern,

$3(-1) = -3$ $\quad$ $3(1) = 3$ and the product is negative.

and

$3(-2) = -6$ $\quad$ $3(2) = 6$ and the product is negative.

The pattern indicates that the product of a positive and a negative number is negative; that is, the opposite of the product of their absolute values.

To find the product of a positive and a negative number

1. Find the product of the absolute values.
2. Make this product negative.

This is sometimes stated, "The product of two unlike signs is negative."

The commutative property of multiplication dictates that no matter the order in which the positive and negative numbers appear, their product is always negative. Thus,

$3(-4) = -12$ $\quad$ and $\quad$ $-4(3) = -12$

EXAMPLES A–F

DIRECTIONS: Multiply.

STRATEGY: Multiply the absolute values and write the opposite of that product.

A. Find the product: $-7(6)$

$-7(6) = -42$ $\quad$ The product of a positive and a negative number is negative.

B. Find the product: $18(-9)$

$18(-9) = -162$ $\quad$ The product of two factors with unlike signs is negative.

C. Find the product: $5(-3.3)$

$5(-3.3) = -16.5$ $\quad$ The product of unlike signs is negative.

WARM-UPS A–F

A. Find the product: $-12(5)$

B. Find the product: $25(-7)$

C. Find the product: $6(-1.9)$

Answer to Warm-Ups

A. -60 **B.** -175 **C.** -11.4

D. Multiply: $\left(-\frac{5}{9}\right)\left(\frac{3}{5}\right)$

E. Multiply: $9(-4)(3)$

F. James wants to lose weight. His goal is to lose an average of 4.5 lb (−4.5 lb) per week. At this rate, how much will he lose in 8 weeks? Express this weight loss as a signed number.

D. Multiply: $\left(-\frac{4}{5}\right)\left(\frac{3}{4}\right)$

$\left(-\frac{\not{4}}{5}\right)\left(\frac{3}{\not{4}}\right) = -\frac{3}{5}$ Simplify. Find the product of the numerators and the denominators. The product is negative.

E. Multiply: $8(-7)(4)$

$8(-7)(4) = -56(4)$ Multiply the first two factors.
$= -224$ Multiply again.

F. In order to attract business, the Family grocery store ran a "loss leader" sale last week. The store sold eggs at a loss of \$0.23 (−\$0.23) per dozen. If 238 dozen eggs were sold last weekend, what was the total loss from the sale of eggs? Express this loss as a signed number.

$238(-0.23) = -54.74$ To find the total loss, multiply the loss per dozen by the number of dozens sold.

Therefore, the loss, written as a signed number, is −\$54.74.

HOW & WHY

OBJECTIVE 2 Multiply two negative numbers.

We use the product of a positive and a negative number to develop a pattern for multiplying two negative numbers.

$-3(4) = -12$
$-3(3) = -9$
$-3(2) = -6$
$-3(1) = -3$
$-3(0) = 0$
$-3(-1) = ?$
$-3(-2) = ?$

Each product is three larger than the one before it. Continuing this pattern,

$-3(-1) = 3$ $3(1) = 3$ and the product is positive.
$-3(-2) = 6$ $3(2) = 6$ and the product is positive.

In each case the product is positive.

To multiply two negative numbers

1. Find the product of the absolute values.
2. Make this product positive.

The product of two like signs is positive. When multiplying more than two signed numbers: If there are an even number of negative factors, the product is positive.

Answer to Warm-Ups

D. $-\frac{1}{3}$ E. −108

F. James's weight loss will be −36 lb.

WARM-UPS G–M

EXAMPLES G–M

DIRECTIONS: Multiply.

STRATEGY: Multiply the absolute values; make the product positive.

G. Multiply: $-11(-7)$

$-11(-7) = 77$ — The product of two negative numbers is positive.

H. Multiply: $-5.2(-0.32)$

$-5.2(-0.32) = 1.664$ — The product of two numbers with like signs is positive.

I. Find the product: $-98(-7)$

$-98(-7) = 686$

J. Find the product of $-\frac{2}{7}$ and $-\frac{3}{8}$.

$$\left(-\frac{\overset{1}{\cancel{2}}}{7}\right)\left(-\frac{3}{\underset{4}{\cancel{8}}}\right) = \frac{3}{28}$$ Negative times negative is positive.

So the product is $\frac{3}{28}$.

K. Multiply: $-16(-3)(-5)$

$-16(-3)(-5) = 48(-5)$ — Multiply the first two factors.

$= -240$ — Multiply again.

L. Find the product of -3, 12, 3, -1, and -5.

STRATEGY: There are an odd number of negative factors; therefore the product is negative.

$(-3)(12)(3)(-1)(-5) = -540$

The product is -540.

CALCULATOR EXAMPLE

M. Multiply: $-82(-9.6)(-12.9)$

The product is $-10{,}154.88$.

G. Multiply: $-12(-8)$

H. Multiply: $-3.6(-2.7)$

I. Find the product: $-111(-6)$

J. Find the product of $-\frac{7}{12}$ and $-\frac{8}{25}$.

K. Multiply: $-13(-3)(-4)$

L. Find the product of -2, 11, 4, -2, and -3.

M. Multiply: $-63(-4.9)(-13.5)$

Answers to Warm-Ups

G. 96 **H.** 9.72 **I.** 666 **J.** $\frac{14}{75}$ **K.** -156 **L.** -528 **M.** -4167.45

EXERCISES 8.4

OBJECTIVE 1 Multiply a positive number and a negative number.

Multiply.

A.

1. $-2(4)$
2. $4(-5)$
3. $-5(2)$
4. $(-6)(8)$
5. $10(-8)$
6. $-13(5)$
7. $-11(7)$
8. $12(-3)$
9. The product of -5 and _____ is -55.
10. The product of 8 and _____ is -72.

B.

11. $-9(34)$
12. $11(-23)$
13. $-17(15)$
14. $23(-18)$
15. $2.5(-3.6)$
16. $-3.4(2.7)$
17. $-0.35(1000)$
18. $2.57(-10000)$
19. $-\frac{2}{3}\cdot\frac{3}{8}$
20. $\left(\frac{3}{8}\right)\left(-\frac{4}{5}\right)$

OBJECTIVE 2 Multiply two negative numbers.

Multiply.

A.

21. $(-1)(-3)$
22. $(-2)(-4)$
23. $-7(-4)$
24. $-6(-5)$
25. $-11(-9)$
26. $(-4)(-3)$
27. $-12(-5)$
28. $-6(-16)$
29. The product of -6 and _____ is 66.
30. The product of -13 and _____ is 39.

B.

31. $-14(-15)$
32. $-23(-17)$
33. $-1.2(-4.5)$
34. $-0.9(-0.72)$
35. $(-5.5)(-4.4)$
36. $(-6.3)(-2.3)$
37. $\left(-\frac{3}{14}\right)\left(-\frac{7}{9}\right)$
38. $\left(-\frac{8}{15}\right)\left(-\frac{5}{24}\right)$
39. $(-0.35)(-4.7)$
40. $(-7.2)(-2.1)$

C.

Multiply.

41. $(-56)(45)$
42. $(16)(-32)$
43. $(15)(31)$
44. $(-23)(71)$
45. $(-1.4)(-5.1)$
46. $(-2.4)(6.1)$
47. $\left(-\frac{9}{16}\right)\left(\frac{8}{15}\right)$
48. $\left(-\frac{8}{21}\right)\left(-\frac{7}{16}\right)$
49. $(-4.01)(3.5)$

50. $(-6.7)(-0.45)$

51. $-3.19(-1.7)(0.1)$

52. $-1.3(4.6)(-0.2)$

53. $-2(4)(-1)(0)(-5)$

54. $(-4)(-7)(3)(-8)(0)$

55. $-0.07(0.3)(-10)(100)$

56. $(0.3)(-0.05)(-10)(-10)$

57. $-2(-5)(-6)(-4)(-1)$

58. $9(-2)(-3)(-5)(-4)$

59. $\left(-\frac{2}{3}\right)\left(-\frac{3}{4}\right)\left(-\frac{4}{5}\right)\left(-\frac{5}{6}\right)$

60. $\left(-\frac{5}{12}\right)\left(\frac{7}{8}\right)\left(\frac{3}{14}\right)\left(-\frac{8}{15}\right)$

61. The formula for converting a temperature measurement from Fahrenheit to Celsius is $C = \frac{5}{9}(F - 32)$. What Celsius measure is equal to 5°F?

62. Use the formula in Exercise 61 to find the Celsius measure that is equal to 86°F.

63. While on a diet for 8 consecutive weeks, Ms. Riles averages a weight loss of 2.6 lb each week. If each loss is represented by -2.6 lb, what is her total loss for the 8 weeks, expressed as a signed number?

64. Mr. Riles goes on a diet for 8 consecutive weeks. He averages a loss of 3.2 lb per week. If each loss is represented by -3.2 lb, what is his total loss, expressed as a signed number?

65. The Dow Jones Industrial Average sustains 12 straight days of a 1.74 decline. What is the total decline during the 12-day period, expressed as a signed number?

66. The Dow Jones Industrial Average sustains 7 straight days of a 2.33 decline. What is the total decline during the 7-day period, expressed as a signed number?

Simplify.

67. $(15 - 8)(5 - 12)$

68. $(17 - 20)(5 - 9)$

69. $(15 - 21)(13 - 6)$

70. $(25 - 36)(5 - 9)$

71. $(-12 + 30)(-4 + 10)$

72. $(11 - 18)(-13 + 5)$

73. Safeway Inc. offers as a loss leader 10 lb of sugar at a loss of 17¢ per bag (−17¢). If 386 bags are sold during the sale, what is the total loss, expressed as a signed number?

74. Albertson's offers a loss leader of coffee at a loss of 23¢ per 3-lb tin (−23¢). If the store sells 412 tins of coffee, find the total loss, expressed as a signed number.

75. Winn Dixie's loss leader is a soft drink on which the store loses 6¢ per six-pack. The store sells 523 of these six-packs. What is Winn Dixie's total loss, expressed as a signed number?

76. Kroger's loss leader is soap powder on which the store loses 28¢ per carton. The store sells 264 cartons. What is Kroger's total loss, expressed as a signed number?

77. Safeway's loss leader is 1 dozen eggs on which the store loses 31¢ per dozen. The store sells 809 dozen eggs that week. Express Safeway's total loss as a signed number.

78. A scientist is studying movement, within its web, of a certain spider. Any movement up is considered to be positive, whereas any movement down is negative. Determine the net movement of a spider that goes up 2 cm five times and down 3 cm twice.

79. A certain junk bond trader purchased 670 shares of stock at 9.34. When she sold her shares, the stock sold for 6.45. What did she pay for the stock? How much money did she receive when she sold this stock? How much did she lose or gain? Represent the loss or gain with a signed number.

80. A company bought 450 items at \$1.23 each. They tried to sell them for \$2.35 and sold only 42 of them. They lowered the price to \$2.10 and sold 105 more. The price was lowered a second time to \$1.35 and 85 were sold. Finally they advertised a close-out price of \$0.95 and sold the remaining items. Determine the net profit or loss for each price. Did they make a profit or lose money on this item overall?

Name ______________________ Class ______________________ Date ______________

Exercises 81–82 refer to the chapter application.

81. Which continent has a low point that is approximately ten times the low point of South America?

82. Which continent has a high point that is approximately twice the absolute value of the lowest point?

STATE YOUR UNDERSTANDING

83. Explain the difference between -3^2 and $(-3)^2$.

84. Explain the procedure for multiplying two signed numbers.

CHALLENGE

Simplify.

85. $|-(-5)|(-9-[-(-5)])$

86. $|-(-8)|(-8-[-(-9)])$

87. Find the product of -8 and the opposite of 7.

88. Find the product of the opposite of 12 and the absolute value of -9.

GROUP WORK

89. Throughout the years, mathematicians have used a variety of examples to show students that the product of two negative numbers is positive. Talk to science and math instructors, and record their favorite explanation. Discuss your findings in class.

MAINTAIN YOUR SKILLS

Divide.

90. $66 \div 11$

91. $816 \div 12$

92. $\frac{1005}{15}$

93. $54\overline{)14{,}472}$

94. $34\overline{)4876}$, round to the nearest hundredth

95. What is the equivalent piecework wage (dollars per piece) if the hourly wage is \$15.86 and the average number of articles completed in 1 hr is 6.1?

96. If carpeting costs \$34.75 per square yard installed, what is the cost of wall-to-wall carpeting needed to cover the floor in a rectangular room that is 24 ft wide by 27 ft long?

97. How far does the tip of the hour hand of a clock travel in 6 hr if the length of the hand is 3 in.? Let $\pi \approx 3.14$.

98. How many square feet of sheet metal are needed to make a box without a top that has measurements of 5 ft 6 in. by 4 ft 6 in. by 9 in.?

99. A mini-storage complex has one unit that is 40 ft by 80 ft and rents for \$1800 per year. What is the cost of a square foot of storage for a year?

8.5 Dividing Signed Numbers

HOW & WHY

OBJECTIVES

1. Divide a positive number and a negative number.
2. Divide two negative numbers.

OBJECTIVE 1 Divide a positive number and a negative number.

To divide two signed numbers, we find the number that when multiplied times the divisor equals the dividend. The expression $-9 \div 3 = ?$ asks $3(?) = -9$; we know $3(-3) = -9$, so $-9 \div 3 = -3$. The expression $24 \div (-6) = ?$ asks $-6(?) = 24$; we know $-6(-4) = 24$, so $24 \div (-6) = -4$.

When dividing unlike signs, we see that the quotient is negative. We use these examples to state how to divide a negative number and a positive number.

To divide a positive and a negative number

1. Find the quotient of the absolute values.
2. Make the quotient negative.

EXAMPLES A–D

DIRECTIONS: Divide.

STRATEGY: Divide the absolute values and make the quotient negative.

A. Divide: $32 \div (-8)$

$32 \div (-8) = -4$ The quotient of two numbers with unlike signs is negative.

B. Divide: $(-4.8) \div 3.2$

$(-4.8) \div 3.2 = -1.5$ When dividing unlike signs, the quotient is negative.

C. Divide: $\left(\frac{4}{35}\right) \div \left(-\frac{4}{5}\right)$

$$\left(\frac{4}{35}\right) \div \left(-\frac{4}{5}\right) = \left(\frac{\overset{1}{\cancel{4}}}{\underset{7}{\cancel{35}}}\right) \cdot \left(-\frac{\overset{1}{\cancel{5}}}{\underset{1}{\cancel{4}}}\right)$$ Multiply by the reciprocal.

$$= -\frac{1}{7}$$ The product is negative.

D. Over a period of 18 weeks, Mr. Rich loses a total of \$4230 (−\$4230) in his stock market account. What is his average loss per week, expressed as a signed number?

STRATEGY: To find the average loss per week, divide the total loss by the number of weeks.

$-4230 \div 18 = -235$

Mr. Rich has an average loss of \$235 (−\$235) per week.

WARM-UPS A–D

A. Divide: $72 \div (-4)$

B. Divide: $(-4.9) \div 1.4$

C. Divide: $\left(\frac{3}{8}\right) \div \left(-\frac{3}{4}\right)$

D. Ms. Rich loses \$6225 in her stock market account over a period of 15 weeks. What is her average loss per week, expressed as a signed number?

HOW & WHY

OBJECTIVE 2 Divide two negative numbers.

To determine how to divide two negative numbers, we again use the relationship to multiplication.

Answers to Warm-Ups

A. −18 **B.** −3.5 **C.** $-\frac{1}{2}$

D. Ms. Rich lost \$415 (−\$415) per week.

The expression $-21 \div (-7) = ?$ asks $(-7)(?) = -21$; we know that $-7(3) = -21$, so $-21 \div (-7) = 3$. The expression $-30 \div (-6) = ?$ asks $(-6)(?) = -30$; we know that $-6(5) = -30$, so $-30 \div (-6) = 5$. We see that in each case, when dividing two negative numbers, the quotient is positive. These examples lead us to the following rule.

To divide two negative numbers

1. Find the quotient of the absolute values.
2. Leave the answer positive.

EXAMPLES E–G

DIRECTIONS: Divide.

STRATEGY: Find the quotient of the absolute values.

E. Find the quotient: $-44 \div (-11)$

$-44 \div (-11) = 4$ The quotient of two negative numbers is positive.

CALCULATOR EXAMPLE

F. Find the quotient: $-7.74 \div (-3.6)$

$-7.74 \div (-3.6) = 2.15$

G. Divide: $\left(-\frac{11}{9}\right)$ by $\left(-\frac{22}{27}\right)$

$$\left(-\frac{11}{9}\right) \div \left(-\frac{22}{27}\right) = \left(-\frac{\overset{1}{\cancel{11}}}{\underset{1}{\cancel{9}}}\right)\left(-\frac{\overset{3}{\cancel{27}}}{\underset{2}{\cancel{22}}}\right)$$ Invert and multiply. Divide out like factors.

$$= \frac{3}{2}$$

WARM-UPS E–G

E. Find the quotient: $-48 \div (-6)$

F. Find the quotient: $-22.75 \div (-2.6)$

G. Divide: $\left(-\frac{9}{30}\right)$ by $\left(-\frac{18}{25}\right)$

Answers to Warm-Ups

E. 8 **F.** 8.75 **G.** $\frac{5}{12}$

EXERCISES 8.5

OBJECTIVE 1 Divide a positive number and a negative number.

Divide.

A.

1. $-10 \div 5$
2. $10 \div (-2)$
3. $-16 \div 4$
4. $15 \div (-3)$
5. $18 \div (-6)$
6. $-18 \div 3$
7. $24 \div (-3)$
8. $-33 \div 11$
9. The quotient of -48 and ____ is -6.
10. The quotient of 70 and ____ is -14.

B.

11. $72 \div (-12)$
12. $84 \div (-12)$
13. $6.06 \div (-3)$
14. $3.05 \div (-5)$
15. $-210 \div 6$
16. $-315 \div 9$
17. $\left(-\frac{6}{7}\right) \div \frac{2}{7}$
18. $\left(-\frac{4}{3}\right) \div \frac{8}{3}$
19. $0.75 \div (-0.625)$
20. $0.125 \div (-0.625)$

OBJECTIVE 2 Divide two negative numbers.

Divide.

A.

21. $-10 \div (-5)$
22. $-10 \div (-2)$
23. $-12 \div (-4)$
24. $-14 \div (-2)$
25. $-28 \div (-4)$
26. $-32 \div (-4)$
27. $-54 \div (-9)$
28. $-63 \div (-7)$
29. The quotient of -105 and ____ is 21.
30. The quotient of -75 and ____ is 15.

B.

31. $-98 \div (-14)$
32. $-88 \div (-11)$
33. $-96 \div (-12)$
34. $-210 \div (-10)$
35. $-12.12 \div (-3)$
36. $-18.16 \div (-4)$
37. $\left(-\frac{3}{8}\right) \div \left(-\frac{3}{4}\right)$
38. $\left(-\frac{1}{2}\right) \div \left(-\frac{5}{8}\right)$
39. $-0.65 \div (-0.13)$
40. $-0.056 \div (-0.4)$

C.

Divide.

41. $-540 \div 12$
42. $-1071 \div 17$
43. $-3364 \div (-29)$
44. $-4872 \div (-48)$
45. $3.735 \div (-0.83)$
46. $-2.352 \div (-0.42)$
47. $0 \div (-35)$
48. $-85 \div 0$
49. $-0.26 \div 100$
50. $-0.56 \div (-100)$
51. $\dfrac{-16{,}272}{36}$
52. $\dfrac{-34{,}083}{-63}$

53. Find the quotient of -384 and -24.

54. Find the quotient of -357 and 21.

55. The membership of the Burlap Baggers Investment Club takes a loss of \$753.90 ($-$753.90) on the sale of stock. If there are six coequal members in the club, what is each member's share of the loss, expressed as a signed number?

56. The temperature in Nome, Alaska, drops from 10° above zero (+10°) to 22° below zero (−22°) in an 8-hour period. What is the average drop in temperature per hour, expressed as a signed number?

57. Mr. Harkness loses a total of 115 lb in 25 weeks. Express the average weekly loss as a signed number.

58. Ms. Harkness loses a total of 65 lb in 25 weeks. Express the average weekly loss as a signed number.

59. A certain stock loses 45.84 points in 12 days. Express the average daily loss as a signed number.

60. A certain stock loses 31.59 points in 9 days. Express the average daily loss as a signed number.

61. Determine the population of Los Angeles in 1980, 1990, and 2000. Determine the population of your city in 1980, 1990, and 2000. Find the average yearly loss or gain for each decade and also for the 20-year period for each city (written as a signed number). List the possible reasons for these changes.

62. Central Electronics lost \$967,140 during one 20-month period. Determine the average monthly loss (written as a signed number). If there are 30 stockholders in this company, determine the total loss per stockholder (written as a signed number).

Exercises 63–64 refer to the chapter application.

63. Which continent has a high point that is approximately one-fourth the height of Mt. Everest?

64. Which continent has a low point that is approximately one-fourth the low point of Africa?

STATE YOUR UNDERSTANDING

65. The sign rules for multiplication and division of signed numbers may be summarized as follows:

If the numbers have the same sign, the answer is positive.
If the numbers have different signs, the answer is negative.

Explain why the rules for division are the same as the rules for multiplication.

66. When dividing signed numbers, care must be taken not to divide by zero. Why?

CHALLENGE

Simplify:

67. $[-|-10|(6-11)] \div [(8-13)(11-10)]$

68. $[(14-20)(-5-9)] \div [-(-12)(-8+7)]$

69. $\left(-\frac{5}{6}-\frac{1}{2}\right)\left(-\frac{2}{3}+\frac{1}{6}\right) \div \left(\frac{1}{3}-\frac{3}{4}\right)$

70. $\left(-\frac{1}{3}-\frac{1}{4}\right)\left(-\frac{1}{3}+\frac{1}{6}\right) \div \left(\frac{1}{3}-\frac{3}{4}\right)$

71. $(-0.82-1.28)(1.84-2.12) \div [3.14+(-3.56)]$

GROUP WORK

72. Determine the temperature on the first of each month over a 12-month period in one city in Alaska, one city in Canada, and one city in Hawaii. Find the average monthly changes in temperature for each city. Find the averages for the temperatures on the first of the month for each city. Make a chart or graph from these data. Be sure no two groups choose the same cities. Compare the results with your classmates.

MAINTAIN YOUR SKILLS

Simplify.

73. $16 \div 4 + 8 - 5$

74. $75 \div 3 \cdot 5 + 3 - (5 - 2)$

75. $14 - 3^2 + 8 - 3 \cdot 2 + 11$

76. $(17 - 3 \cdot 5)^3 + [16 - (19 - 2 \cdot 8)]$

77. Find the volume of a cylinder that has a radius of 8 in. and a height of 24 in. (Let $\pi \approx 3.14$.)

78. Find the volume of a cone that has a radius of 12 in. and a height of 9 in. (Let $\pi \approx 3.14$.)

79. An underground gasoline storage tank is a cylinder that is 72 in. in diameter and 18 ft long. If there are 231 in.^3 in a gallon, how many gallons of gasoline will the tank hold? Round the answer to the nearest gallon.

80. A swimming pool is to be dug and the dirt hauled away. The pool is to be 27 ft long, 16 ft wide, and 6 ft deep. How many cubic yards of dirt must be removed?

81. To remove the dirt from the swimming pool in Exercise 80, trucks that can haul 8 yd^3 per load are used. How many truckloads will there be?

82. A real estate broker sells a lot that measures 88.75 ft by 180 ft. The sale price is \$2 per square foot. If the broker's commission is 8%, how much does she make?

8.6 Order of Operations: A Review

HOW & WHY

OBJECTIVE Do any combination of operations with signed numbers.

The order of operations for signed numbers is the same as that for whole numbers, fractions, and decimals.

OBJECTIVE

Do any combination of operations with signed numbers.

To evaluate expressions with more than one operation

Step 1. Parentheses: Do the operations within grouping symbols first (parentheses, fraction bar, and so on), in the order given in steps 2, 3, and 4.
Step 2. Exponents: Do the operations indicated by exponents.
Step 3. Multiply and **divide**: Do only multiplication and division as they appear from left to right.
Step 4. Add and **subtract**: Do addition and subtraction as they appear from left to right.

EXAMPLES A–G

DIRECTIONS: Perform the indicated operations.

STRATEGY: Follow the order of operations.

A. Perform the indicated operations: $-63 + (-21) \div 7$

$$\begin{aligned} -63 + (-21) \div 7 &= -63 + (-3) \\ &= -66 \end{aligned}$$

B. Perform the indicated operations: $(-15)(-3) - 44 \div (-11)$

$$\begin{aligned} (-15)(-3) - 44 \div (-11) &= 45 - (-4) \\ &= 45 + 4 \qquad \text{Add the opposite of } -4. \\ &= 49 \end{aligned}$$

C. Perform the indicated operations: $12 \div (-0.16) + 3(-1.45)$

$$\begin{aligned} 12 \div (-0.16) + 3(-1.45) &= -75 + (-4.35) \\ &= -79.35 \end{aligned}$$

D. Perform the indicated operations: $10 - \left(\frac{3}{5}\right)(-20)$

$$\begin{aligned} 10 - \left(\frac{3}{5}\right)(-20) &= 10 - (-12) \\ &= 10 + 12 \qquad \text{Add the opposite of } -12. \\ &= 22 \end{aligned}$$

E. Perform the indicated operations: $3(-5)^2 - 3^2 + 7(-4)^2$

$$\begin{aligned} 3(-5)^2 - 3^2 + 7(-4)^2 &= 3(25) - 9 + 7(16) \\ &= 75 - 9 + 112 \\ &= 178 \end{aligned}$$

WARM-UPS A–G

A. Perform the indicated operations: $-56 + (-32) \div 4$

B. Perform the indicated operations: $(-8)(5) - 72 \div (-12)$

C. Perform the indicated operations: $9 \div (-0.15) + 6(-2.15)$

D. Perform the indicated operations: $15 - \left(\frac{7}{12}\right)(-36)$

E. Perform the indicated operations: $(-6)(-3)^2 + 43 - (-5)^2$

Answers to Warm-Ups

A. −64 **B.** −34 **C.** −72.9 **D.** 36 **E.** −36

F. Perform the indicated operations:
$(-28)(14) - 225 \div (-3)$

G. How many degrees Celsius is $-4°F$?

CALCULATOR EXAMPLE

F. Perform the indicated operations: $(-18)(23) - (-84) \div (-7)$

The result is -426.

G. Hilda keeps the thermostat on her furnace set at 68°F. Her pen pal in Germany says that her thermostat is set at 20°C. They wonder whether the two temperatures are equal. Hilda finds the formula, given below, for changing degrees Celsius to degrees Fahrenheit. Use this formula to find out whether 68°F is equal to 20°C.

$$C = \frac{5}{9}(F - 32)$$

STRATEGY: To find out whether 68°F = 20°C, substitute 68 for F in the formula.

$$C = \frac{5}{9}(F - 32)$$

$$C = \frac{5}{9}(68 - 32)$$

$$C = \frac{5}{9}(36)$$

$$C = 20$$

Therefore, 68°F equals 20°C.

Answers to Warm-Ups

F. -317

G. The Celsius temperature is $-20°C$.

EXERCISES 8.6

OBJECTIVE Do any combination of operations with signed numbers.

Perform the indicated operations.

A.

1. $2(-7) - 10$

2. $13 + 4(-5)$

3. $(-2)(-4) + 11$

4. $15 + (-3)(-5)$

5. $3(-6) + 12$

6. $(-5)(6) + 19$

7. $-7 + 3(-3)$

8. $-14 + (-6)2$

9. $(-3)8 \div 4$

10. $(-8)6 \div 3$

11. $(-4)8 \div (-4)$

12. $(-7)12 \div (-6)$

13. $(-8) \div 4(-2)$

14. $(-18) \div 3(2)$

15. $(-3)^2 + (-2)^2$

16. $6^2 - 4^2$

17. $(11 - 3) + (9 - 6)$

18. $(4 - 9) + (5 - 7)$

19. $(3 - 5)(6 - 10)$

20. $(8 - 5)(11 - 15)$

21. $(-3)^2 + 4(-2)$

22. $(-2)^2 - 4(-2)$

23. $-5 + (6 - 8) - 5(3)$

24. $-7 + (5 - 11) - 4(2)$

B.

25. $(-13)(-2) + (-16)2$

26. $(-16)(-5) + (-14)5$

27. $(9-7)(-2 - 5) + (15 - 9)(2 + 7)$

28. $(10 - 15)(-4 - 3) + (12 - 7)(3 + 2)$

29. $7(-11 + 5) - 44 \div (-11)$

30. $8(-7 + 4) - 54 \div (-9)$

31. $18(-2) \div (-6) - 14$

32. $(-4)(-9) \div (-12) + 11$

33. $-120 \div (-20) - (9 - 11)$

34. $-135 \div (-15) - (12 - 17)$

35. $-2^3 - (-2)^3$

36. $-4^3 - 4^3$

37. $-35 \div 7(-5) - 7^2$

38. $-28 \div 7(-4) - 7^2$

39. $2^2(5 - 4)(7 - 3)^2$

40. $3^2(8 - 6)(6 - 8)^2$

41. $(9 - 13) + (-5)(-2) - (-2)5 - 3^3$

42. $(8 - 11) - (7)(-3) + (-4)(3) - 2^2$

43. $(-3)(-2)(-3) - (-4)(-3) - (3)(-5)$

44. $(-5)(-6)(-1) - (-3)(-6) - (-5)(2)$

45. $(-1)(-6)^2(-2) - (-3)^2(-2)^3$

46. $(-1)(-3)^3(-4) - (-4)^2(-2)^2$

C.

47. Find the sum of the product of 12 and -4 and the product of -3 and -12.

48. Find the difference of the product of 3 and 9 and the product of -8 and 3.

Exercises 49–52 refer to the following table, which shows temperatures a satellite recorded during a 5-day period at one spot on the surface of Mars.

Temperatures on the Surface of Mars

	Day 1	Day 2	Day 3	Day 4	Day 5
5:00 A.M.	−92°C	−88°C	−115°C	−103°C	−74°C
9:00 A.M.	−57°C	−49°C	−86°C	−93°C	−64°C
1:00 P.M.	−52°C	−33°C	−46°C	−48°C	−10°C
6:00 P.M.	−45°C	−90°C	−102°C	−36°C	−42°C
11:00 P.M.	−107°C	−105°C	−105°C	−98°C	−90°C

49. What is the average temperature recorded during day 5?

50. What is the average temperature recorded at 6:00 P.M.?

51. What was the average high temperature recorded for the 5 days?

52. What was the average low temperature recorded for the 5 days?

Simplify.

53. $[-3 + (-6)]^2 - [-8 - 2(-3)]^2$

54. $[-5(-9) - (-6)^2]^2 + [(-8)(-1)^3 + 2]^2$

55. $[46 - 3(-4)^2]^3 - [-7(1)^3 + (-5)(-8)]$

56. $[30 - (-5)^2]^2 - [-8(-2) - (-2)(-4)]^2$

57. $-15 - \dfrac{8^2 - (-4)}{3^2 + 3}$

58. $-22 + \dfrac{9^2 - 6}{6^2 - 11}$

59. $\dfrac{12(8 - 24)}{5^2 - 3^2} \div (-12)$

60. $\dfrac{15(12 - 45)}{6^2 - 5^2} \div (-9)$

61. $-8|125 - 321| - 21^2 + 8(-7)$

62. $-9|482 - 632| - 17^2 + 9(-9)$

63. $-6(8^2 - 9^2)^2 - (-7)20$

64. $-5(6^2 - 7^2)^2 - (-8)19$

65. Find the difference of the quotient of 28 and −7 and the product of −4 and −3.

66. Find the sum of the product of −3 and 7 and the quotient of −15 and −5.

67. Keshia buys a TV for \$95 down and \$47 per month for 15 months. What is the total price she pays for the TV? (*Hint*: When a deferred payment plan is used, the total cost of the article is the down payment plus the total of the monthly payments.)

68. The E-Z Chair Company advertises recliners for \$40 down and \$17 per month for 24 months. What is the total cost of the recliner?

69. During a "blue light" special, Kmart sold 24 fishing poles at a loss of \$3 per pole. During the remainder of the day, the store sold 9 poles at a profit of \$7 per pole. Express the profit or loss on the sale of fishing poles as a signed number.

70. Fly America sells 40 seats on Flight 402 at a loss of \$52 per seat (−\$52). Fly America also sells 67 seats at a profit of \$78 per seat. Express the profit or loss on the sale of the seats as a signed number.

Exercises 71–75 refer to the chapter application.

71. For each continent, calculate the average of the highest and lowest point. Which continent has the largest average and which has the smallest average?

Exercises 72–74. The following table gives the altitudes of selected cities around the world.

Altitudes of Cities

City	Altitude (ft)	City	Altitude (ft)
Athens, Greece	300	Mexico City, Mexico	7347
Bangkok, Thailand	0	New Delhi, India	770
Berlin, Germany	110	Quito, Ecuador	9222
Bogotá, Colombia	8660	Rome, Italy	95
Jakarta, Indonesia	26	Tehran, Iran	5937
Jerusalem, Israel	2500	Tokyo, Japan	30

72. Find a group of five cities whose average altitude is less than 100 ft.

73. Find a group of three cities whose average altitude is approximately 350 ft.

74. Find four cities with an average altitude of approximately 7000 ft.

75. The treasurer of a local club records the following transactions during one month:

Opening Balance	\$4756
Deposit	\$345
Check # 34	\$212
Check # 35	\$1218
New check cost	\$15
Deposit	\$98
National Dues Paid	\$450
Electric bill	\$78

What is the balance at the end of the month?

76. The Chicago Bears made the following plays during a quarter of a game:

3 plays lost 8 yd each
8 plays lost 5 yd each
1 quarterback sack lost 23 yd
1 pass for 85 yd
5 plays gained 3 yd each
2 plays gained 12 yd each
1 fumble lost 7 yd
2 passes for 10 yd each

Determine the average movement per play during this quarter. Round to the nearest tenth.

77. Consider traveling north and east as positive values and traveling south and west as negative values. A certain trip requires going 81 miles north followed by 67 miles west. The next day the trip requires traveling 213 miles south followed by 107 miles west. The last day of the trip requires traveling 210 miles north and 83 miles east. What is the net result of this trip? Determine your position at the end of the trip in relation to your starting point.

78. Try this game on your friends. Have them pick a number. Tell them to double it, then add 20 to that number, divide the sum by 4, subtract 5 from that quotient, square the difference, and multiply the square by 4. They should now have the square of the original number. Write a mathematical representation of this riddle.

STATE YOUR UNDERSTANDING

Locate the error in Exercises 79 and 80. Indicate why each is not correct. Determine the correct answer.

79. $$\begin{aligned} 2[3 + 5(-4)] &= 2[8(-4)] \\ &= 2[-32] \\ &= -64 \end{aligned}$$

80. $$\begin{aligned} 3 - [5 - 2(6 - 4^2)^3] &= 3 - [5 - 2(6 - 16)^3] \\ &= 3 - [5 - 2(-10)^3] \\ &= 3 - [5 - (-20)^3] \\ &= 3 - [5 - (-8000)] \\ &= 3 - [5 + 8000] \\ &= 3 - 8005 \\ &= -8002 \end{aligned}$$

81. Is there ever a case when exponents are not computed first? If so, give an example.

CHALLENGE

Simplify.

82. $\dfrac{3^2 - 5(-2)^2 + 8 + [4 - 3(-3)]}{4 - 3(-2)^3 - 18}$

83. $\dfrac{(5 - 9)^2 + (-6 + 8)^2 - (14 - 6)^2}{[3 - 4(7) + 3^3]^2}$

84. $\dfrac{3(4 - 7)^2 + 2(5 - 8)^3 - 18}{(6 - 9)^2 + 6}$

GROUP WORK

85. Engage the entire class in a game of KRYPTO. This card game consists of 41 cards numbered from −20 to 20. Each group gets four cards. A card is chosen at random from the remaining cards and the number is put on the board. Each group must find a way—using addition, subtraction, multiplication, and/or division—to combine their given cards to equal the number on the board. Operations may be used more than once.

MAINTAIN YOUR SKILLS

Add.

86. $(-17.2) + (-18.6) + (-2.7) + 9.1$

87. $(28.31) + (-8.14) + (-21.26) + (-16)$

Subtract:

88. $48 - (-136)$

89. $-62.7 - (-78.8)$

Multiply.

90. $(-36)(84)(-21)$

91. $(-62)(-22)(-30)$

Divide.

92. $(-800) \div (-32)$

93. $(-25{,}781) \div 3.5$

94. The four Zapple brothers form a company. The first year the company loses \$5832 (−\$5832). The brothers share equally in the loss. Represent each brother's loss as a signed number.

95. AVI Biopharma stock recorded the following gains and losses for the week:

Monday	loss	0.34
Tuesday	loss	0.54
Wednesday	gain	1.32
Thursday	gain	0.67
Friday	loss	0.672

Use signed numbers to find out whether the stock gains or loses for the week?

8.7 Solving Equations

VOCABULARY

Recall that the **coefficient** of the variable is the number that is multiplied times the variable.

OBJECTIVE

Solve equations of the form $ax + b = c$ or $ax - b = c$, where a, b, and c are signed numbers.

HOW & WHY

OBJECTIVE Solve equations of the form $ax + b = c$ or $ax - b = c$, where a, b, and c are signed numbers.

Note: Before starting this section, you may want to review Getting Ready for Algebra sections in earlier chapters.

The solutions of equations that are of the form $ax + b = c$ and $ax - b = c$, using signed numbers, involve two operations to isolate the variable. To isolate the variable is to get an equation in which the variable is the only symbol on a particular side of the equation.

To find the solution of an equation of the form $ax + b = c$ or $ax - b = c$

1. Add (subtract) the constant to (from) each side of the equation.
2. Divide both sides by the coefficient of the variable.

EXAMPLES A–D

DIRECTIONS: Solve.

STRATEGY: First, add or subtract the constant to or from both sides of the equation. Second, divide both sides of the equation by the coefficient of the variable.

A. Solve: $-6x + 23 = -7$

$$-6x + 23 = -7 \quad \text{Original equation.}$$
$$-6x + 23 - 23 = -7 - 23 \quad \text{Subtract.}$$
$$-6x = -30$$
$$\frac{-6x}{-6} = \frac{-30}{-6} \quad \text{Divide.}$$
$$x = 5$$

CHECK: Substitute 5 for x in the original equation.

$$-6(5) + 23 = -7$$
$$-30 + 23 = -7$$
$$-7 = -7$$

The solution is $x = 5$.

B. Solve: $-43 = -14x - 71$

$$-43 = -14x - 71 \quad \text{Original equation.}$$
$$-43 + 71 = -14x - 71 + 71 \quad \text{Add.}$$
$$28 = -14x$$
$$\frac{28}{-14} = \frac{-14x}{-14} \quad \text{Divide.}$$
$$-2 = x$$

WARM-UPS A–D

A. Solve: $-7x + 12 = -23$

B. Solve: $37 = -13x - 15$

Answers to Warm-Ups

A. $x = 5$ **B.** $x = -4$

CHECK: Substitute -2 in the original equation.

$$-43 = -14(-2) - 71$$
$$-43 = 28 - 71$$
$$-43 = 28 + (-71)$$
$$-43 = -43$$

The solution is $x = -2$.

C. $6x - 45 = -87$

C. Solve: $7x - 32 = -88$

$$7x - 32 = -88 \quad \text{Original equation.}$$
$$+32 \quad +32 \quad \text{Add 32 to both sides.}$$
$$7x = -56$$
$$\frac{7x}{7} = \frac{-56}{7} \quad \text{Divide both sides by 7.}$$
$$x = -8 \quad \text{The check is left for the student.}$$

The solution is $x = -8$.

D. Using the formula $s = v + gt$, find the initial velocity (v) in feet per second of a sky diver if, after a time (t) of 5 seconds, she reaches a speed (s) of 180 feet per second, and $g = 32$ feet per second squared.

D. Using the formula $s = v + gt$, find the initial velocity (v) in feet per second of a sky diver if, after a time (t) of 3.5 seconds, she reaches a speed (s) of 125 feet per second, and $g = 32$ feet per second squared.

STRATEGY: Substitute into the formula and solve the resulting equation.

Formula:

$s = v + gt$

Substitute:

$$\frac{125\text{ ft}}{1\text{ sec}} = v + \left(\frac{32\text{ ft}}{\text{sec}^2}\right)(3.5\text{ sec})$$

$$\frac{125\text{ ft}}{1\text{ sec}} = v + \frac{112\text{ ft}}{1\text{ sec}}$$ One factor of seconds divides out.

$$\frac{125\text{ ft}}{1\text{ sec}} - \frac{112\text{ ft}}{1\text{ sec}} = v + \frac{112\text{ ft}}{1\text{ sec}} - \frac{112\text{ ft}}{1\text{ sec}}$$ Subtract 112 feet per second from both sides.

$$\frac{13\text{ ft}}{1\text{ sec}} = v$$ The check is left for the student.

* The sky diver's initial velocity was 13 feet per second.

Answers to Warm-Ups

C. $x = -7$

D. The initial velocity is 20 feet per second.

Name ________________ Class ________________ Date ________________

EXERCISES 8.7

OBJECTIVE Solve equations of the form $ax + b = c$ or $ax - b = c$, where a, b, and c, are signed numbers.

Solve.

A.

1. $-3x + 25 = 4$

2. $-4y + 11 = -9$

3. $-6 + 3x = 9$

4. $-11 + 5y = 14$

5. $4y - 9 = -29$

6. $3x - 13 = -43$

7. $2a - 11 = 3$

8. $5a + 17 = 17$

9. $-5x + 12 = -23$

10. $-11y - 32 = -65$

11. $4x - 12 = 28$

12. $9y - 14 = 4$

B.

13. $-14 = 2x - 8$

14. $26 = 3x - 4$

15. $-40 = 5x - 10$

16. $-30 = -5x - 10$

17. $-6 = -8x - 6$

18. $9 = -5x + 9$

19. $-10 = -4x + 2$

20. $20 = -8x + 4$

21. $-14y - 1 = -99$

22. $-16x + 5 = -27$

23. $-3 = -8a - 3$

24. $-12 = 5b + 18$

C.

25. $-0.6x - 0.15 = 0.15$

26. $-1.05y + 5.08 = 1.72$

27. $0.03x + 2.3 = 1.55$

28. $0.02x - 2.4 = 1.22$

29. $-135x - 674 = 1486$

30. $94y + 307 = -257$

31. $-102y + 6 = 414$

32. $-63c + 22 = 400$

33. Find the time (t) it takes a free-falling sky diver to reach a speed (s) of 147 feet per second if $v = 19$ and $g = 32$. The formula is $s = v + gt$.

34. Find the time it takes a free-falling sky diver to reach a speed (s) of 211 feet per second if $v = 19$ and $g = 32$. The formula is $s = v + gt$.

35. If 98 is added to six times some number, the sum is 266. What is the number?

36. If 73 is added to 11 times a number, the sum is -158. What is the number?

37. The difference of 15 times a number and 181 is -61. What is the number?

38. The difference of 24 times a number and 32 is -248. What is the number?

39. A formula for distance traveled is $2d = t^2a + 2v$, where d represents distance, v represents initial velocity, t represents time, and a represents acceleration. Find a if $d = 244$, $v = -20$, and $t = 4$.

40. Use the formula in Exercise 39 to find a if $d = 240$, $v = -35$, and $t = 5$.

41. The formula for the balance of a loan (D) is $D = B - NP$, where P represents the monthly payment, N represents the number of payments, and B represents the money borrowed. Find N when $D = \$575$, $B = \$925$, and $P = \$25$.

42. Use the formula in Exercise 41 to determine the monthly payment (P) if $D = \$820$, $B = \$1020$, and $N = 5$.

Exercises 43–45 refer to the chapter application. Use negative numbers to represent feet below sea level.

43. The high point of Australia is 12,558 ft more than four times the lowest point of one of the other continents. Write an algebraic equation that describes this relationship. Which continent's lowest point fits the description?

44. The lowest point of Antarctica is 2183 ft less than 12 times the lowest point of one of the other continents. Write an algebraic equation that describes this relationship. Which continent's lowest point fits this description?

45. The Mariana Trench in the Pacific Ocean is about 2000 ft deeper than twice one of the other oceans' deepest parts. Write an algebraic equation that describes this relationship. Which ocean's deepest part fits this description? See Exercise 63, Section 8.1.

Solve.

46. $5x + 12 + (-9) = 18$

47. $8z - 12 + (-6) = 38$

48. $-3b - 12 + (-4) = 11 + (-6)$

49. $-5z - 15 + 6 = -21 - 18$

STATE YOUR UNDERSTANDING

50. Explain what it means to solve an equation.

51. Explain how to solve the equation $-3x + 10 = 4$.

CHALLENGE

Solve.

52. $8x - 9 = 3x + 6$

53. $7x + 14 = 3x - 2$

54. $9x + 16 = 7x - 12$

55. $10x + 16 = 5x + 6$

GROUP WORK

56. Bring in your last two electricity bills. Develop a formula for determining how your bills are computed. Share your group's synopsis with the class. Try to predict your next month's bill.

CHAPTER 8

KEY CONCEPTS

SECTION 8.1 OPPOSITES AND ABSOLUTE VALUE

Definitions and Concepts	Examples
Positive numbers are greater than zero.	Positive numbers: $4, 7.31, \frac{5}{9}$
Negative numbers are written with a dash (−) and are less than zero.	Negative numbers: $-4, -7.31, -\frac{5}{9}$
The opposite of a signed number is the number that is the same distance from zero on a number line but has the opposite sign.	Opposites: 4 and −4, 7.31 and −7.31, $\frac{5}{9}$ and $-\frac{5}{9}$
The absolute value of a signed number is its distance from zero on a number line.	$\lvert -5 \rvert = 5$ $\lvert 5 \rvert = 5$

SECTION 8.2 ADDING SIGNED NUMBERS

Definitions and Concepts	Examples
To add signed numbers: • If the signs are alike, add their absolute values and use the common sign.	$4 + 9 = 13$ $(-4) + (-9) = -13$
• If the signs are unlike, subtract the smaller absolute value from the larger absolute value. The sum has the sign of the number with the larger absolute value.	$-4 + 9 = 9 - 4 = 5$ $4 + (-9) = -(9 - 4) = -5$

SECTION 8.3 SUBTRACTING SIGNED NUMBERS

Definitions and Concepts	Examples
To subtract signed numbers: • Rewrite as an addition problem by adding the opposite of the number to be subtracted. • Find the sum.	$5 - 8 = 5 + (-8) = -3$ $-5 - 8 = (-5) + (-8) = -13$ $5 - (-8) = 5 + 8 = 13$

SECTION 8.4 MULTIPLYING SIGNED NUMBERS

Definitions and Concepts	Examples
To multiply signed numbers: • Find the product of the absolute values.	$3(-9) = -27$
• If there are an even number of negative factors, the product is positive.	$(-3)(-9) = 27$ $(-2)(-2)(-2)(-2) = 16$
• If there are an odd number of negative factors, the product is negative.	$(-4)(-3)(-2) = -24$

SECTION 8.5 DIVIDING SIGNED NUMBERS

Definitions and Concepts	Examples
To divide signed numbers: • Find the quotient of the absolute values. • If the signs are alike, the quotient is positive. • If the signs are unlike, the quotient is negative.	$-12 \div 6 = -2$ $(-12) \div (-6) = 2$ $12 \div (-6) = -2$

SECTION 8.6 ORDER OF OPERATIONS: A REVIEW

Definitions and Concepts	Examples
The order of operations for signed numbers is the same as that for whole numbers: • Parentheses • Exponents • Multiplication/Division • Addition/Subtraction	$(-5)(-2)^3 + (-24) \div (6 - 8)$ $(-5)(-2)^3 + (-24) \div (-2)$ $(-5)(-8) + (-24) \div (-2)$ $40 + 12$ 52

SECTION 8.7 SOLVING EQUATIONS

Definitions and Concepts	Examples
To solve equations of the form $ax + b = c$ or $ax - b = c$: • Add (or subtract) the constant to both sides. • Divide both sides by the coefficient of the variable.	$3x + 8 = -6$ $3x + 8 - 8 = -6 - 8$ $3x = -14$ $\frac{3x}{3} = \frac{-14}{3}$ $x = -\frac{14}{3}$

CHAPTER 8

REVIEW EXERCISES

Section 8.1

Find the opposite of the signed number.

1. -39 **2.** 57 **3.** -0.91 **4.** -0.134

Find the absolute value of the signed number.

5. $|-16.5|$ **6.** $|-386|$ **7.** $|71|$ **8.** $|3.03|$

9. Find the opposite of $|-6.4|$.

10. If 93 miles north is represented by $+93$ miles, how would you represent 93 miles south?

Section 8.2

Add.

11. $-75 + (-23)$ **12.** $-75 + 23$ **13.** $75 + (-23)$

14. $65 + (-45) + (-82)$ **15.** $-7.8 + (-5.3) + 9.9$

16. $-24 + 65 + (-17) + 31$ **17.** $-6.8 + (-4.3) + 7.12 + 3.45$

18. $-\frac{7}{12} + \frac{7}{15} + \left(-\frac{7}{10}\right)$

19. The Miami Dolphins made the following consecutive plays during a recent football game: 9-yd gain, 5-yd loss, 6-yd loss, and a 12-yd gain. Did they get a first down?

20. Intel stock has the following changes in one week: up 0.78, down 1.34, down 2.78, up 3.12, and down 0.15. What is the net change for the week?

Section 8.3

Subtract.

21. $19 - (-3)$ **22.** $-45 - 81$ **23.** $16 - (-75)$

24. $-134 - (-134)$ **25.** $-4.56 - 3.25$ **26.** $-4.56 - (-3.25)$

27. Find the difference between -127 and -156.

28. Subtract -56 from -45.

29. At the beginning of the month, Maria's bank account had a balance of \$562.75. At the end of the month, the account was overdrawn by \$123.15 (−\$123.15). If there were no deposits during the month, what was the total amount of the checks Maria wrote?

30. Microsoft stock opened the day at 69.74 and closed the day at 67.82. Did the stock gain or lose for the day? Express the gain or loss as a signed number.

Section 8.4

Multiply.

31. $-6(11)$ **32.** $5(-28)$ **33.** $-1.2(3.4)$

34. $7.4(-5.1)$ **35.** $-3(-17)$ **36.** $-7(-21)$

37. $-4.03(-2.1)$ **38.** $(-1)(-4)(-6)(5)(-2)$

39. Kroger's promotes a gallon of milk as a loss leader. If Kroger loses 45¢ per gallon, what will be the total loss if the store sells 632 gallons? Express the loss as a signed number.

40. Pedro owns 723.5 shares of Pfizer. If the stock loses \$0.32 (−\$0.32) a share, what is Pedro's loss, expressed as a signed number?

Section 8.5

Divide.

41. $-18 \div 6$ **42.** $153 \div (-3)$ **43.** $-45 \div (-9)$

44. $-4.14 \div (-1.2)$ **45.** $-2448 \div 153$ **46.** $-8342 \div (-97)$

47. Find the quotient of -84.3 and 1.5.

48. Divide -712 by -32.

49. A share of UPS stock loses 15.5 points in 4 days. Express the average daily loss as a signed number.

50. Albertson's grocery store lost \$240.50 on the sale of Wheaties as a loss leader. If the store sold 650 boxes of the cereal during the sale, what is the loss per box, expressed as a signed number?

Section 8.6

Perform the indicated operations.

51. $5(-9) - 11$ **52.** $-3(17) + 45$ **53.** $-18 + (-4)(5) - 12$

54. $-84 \div 4(7)$ **55.** $-72 \div (-12)(3) + 6(-7)$

56. $(7 - 4)(4)(-3) + 6(7 - 9)$

57. $(-4)^2(-1)(-1) + (-6)(-3) - 17$

58. $(-1)(-3^2)(-4) + 4(-5) - (-18 - 5)$

59. Find the difference of the quotient of 71 and -2.5 and the product of 3.2 and -2.4.

60. A local airline sells 65 seats for \$324 each and 81 seats for \$211 each. If the break-even point for the airline is \$256 a seat, express the profit or loss for the airline for this flight as a signed number.

Section 8.7

Solve.

61. $7x + 25 = -10$

62. $-6x + 21 = -21$

63. $71 - 5x = -54$

64. $-55 = 3x + 41$

65. $-43 = 7x - 43$

66. $12y - 9 = -45$

67. $78a + 124 = -890$

68. $-55b + 241 = -144$

69. A formula for relating degrees Fahrenheit (F) and degrees Celsius (C) is $9C = 5F - 160$. Find the degrees Fahrenheit that is equal to -22°C.

70. Using the formula in Exercise 69, find the degrees Fahrenheit that is equal to -8°C.

Name ______________________ Class ______________________ Date ____________

CHAPTER 8

TRUE / FALSE CONCEPT REVIEW

ANSWERS

Check your understanding of the language of basic mathematics. Tell whether each of the following statements is true (always true) or false (not always true). For those statements that you judge to be false, revise them to make them true.

1. Negative numbers are found to the left of zero on the number line. **1.** ____________

2. The opposite of a signed number is always positive. **2.** ____________

3. The absolute value of a number is always positive. **3.** ____________

4. The opposite of a signed number is the same distance from zero as the number on the number line but in the opposite direction. **4.** ____________

5. The sum of two signed numbers is always positive or negative. **5.** ____________

6. The sum of a positive signed number and a negative signed number is always positive. **6.** ____________

7. To find the sum of a positive signed number and a negative signed number, subtract their absolute values and use the sign of the number with the larger absolute value. **7.** ____________

8. To subtract two signed numbers, add their absolute values. **8.** ____________

9. If a negative number is subtracted from a positive number, the difference is always positive. **9.** ____________

10. The product of two negative numbers is never negative. **10.** ____________

11. The sign of the product of a positive number and a negative number depends on which number has the larger absolute value. **11.** ____________

12. The sign of the quotient, when dividing two signed numbers, is the same as the sign obtained when multiplying the two numbers. **12.** ____________

13. The order of operations for signed numbers is the same as the order of operations for positive numbers. **13.** ____________

14. Subtracting a number from both sides of an equation results in an equation that has the same solution as the original equation. **14.** ____________

Name ______________________ Class ______________________ Date ____________

CHAPTER 8

TEST

ANSWERS

Perform the indicated operations.

1. $-32 + (-19) + 39 + (-21)$

2. $(45 - 52)(-16 + 21)$

3. $\left(-\frac{3}{8}\right) \div \left(\frac{3}{10}\right)$

4. $\left(-\frac{7}{15}\right) - \left(-\frac{3}{5}\right)$

5. $(-11 - 5) - (5 - 22) + (-6)$

6. $-5.78 + 6.93$

7. **a.** $-(-17)$
b. $|-33|$

8. $-65 - (-32)$

9. $(-18 + 6) \div 3 \cdot 4 - (-7)(-2)(-1)$

10. $-110 \div (-55)$

11. $(-6)(-8)(2)$

12. $(|-7|)(-3)(-1)(-1)$

13. $-63.2 - 45.7$

14. $(-2)^2(-2)^2 + 4^2 \div (2)(3)$

15. $21.84 \div (-0.7)$

16. $\left(-\frac{1}{3}\right) + \frac{5}{6} + \left(-\frac{1}{2}\right) + \left(-\frac{1}{6}\right)$

17. $-112 \div (-8)$

18. $-56 - 24$

19. $(-7)(4 - 13)(-2) - 5(-2 - 6)$

1. ____________
2. ____________
3. ____________
4. ____________
5. ____________
6. ____________
7. ____________
8. ____________
9. ____________
10. ____________
11. ____________
12. ____________
13. ____________
14. ____________
15. ____________
16. ____________
17. ____________
18. ____________
19. ____________

 Name ______________________ Class ______________________ Date ______________

20. $\left(-\frac{3}{8}\right)\left(\frac{12}{15}\right)$ 20. ______________

21. $-45 + (-23)$ 21. ______________

22. $6(-7) + 37$ 22. ______________

23. $-16 = 5x + 14$ 23. ______________

24. $7x - 32 = 17$ 24. ______________

25. $24 - 6a = 45$ 25. ______________

26. Ms. Rosier loses an average of 1.05 lb per week (−1.05 lb) during her 16-week diet. Express Ms. Rosier's total weight loss during the 16 weeks as a signed number. 26. ______________

27. The temperature in Chicago ranges from a high of 12°F to a low of −9°F within a 24-hr period. What is the drop in temperature expressed as a signed number? 27. ______________

28. A stock on the New York Stock Exchange opens at 17.65 on Monday. It records the following changes during the week: Monday, +0.37; Tuesday, −0.67; Wednesday, +1.23; Thursday, −0.87; Friday, +0.26. What is its closing price on Friday? 28. ______________

29. What Fahrenheit temperature is equal to a reading of −10°C? Use the formula $F = \frac{9}{5}C + 32$. 29. ______________

30. Find the average of −11, −15, 23, −19, 10, and −12. 30. ______________

CHAPTER 8

GROUP PROJECT (4 weeks)

OPTIONAL

On three consecutive Mondays, locate the final scores for each of the three major professional golf tours in the United States: the Professional Golf Association, PGA; the Ladies Professional Golf Association, LPGA; and the Senior Professional Golf Association, Senior PGA. These scores can usually be found on the summary page in the sports section of the daily newspaper.

a. Record the scores, against par, for the 30 top finishers and ties on each tour. Display the data using bar graphs for week 1, line graphs for week 2, and pictorial graphs for week 3. Which type of graph best displays the data?

b. Calculate the average score, against par, for each tour for each week. When finding the average, if there is a remainder and if it is half of or more than the divisor, round up; otherwise, round down. Now average the average scores for each tour. Which tour scored the best? Why?

c. What is the difference between the best and worst scores on each tour for each week?

d. What is the average amount of money earned by the player whose scores were recorded on each tour for each week? Which tour pays the best?

e. How much did the winner on each tour earn per stroke under par in the second week of your data? Compare the results. Is this a good way to compare the earnings on the tour? If not, why not?

APPENDIX A

CALCULATORS

The wide availability and economical price of current hand-held calculators make them ideal for doing time-consuming arithmetic operations. Even people who are very good at math use calculators under certain circumstances (for instance, when balancing their checkbooks). You are encouraged to use a calculator as you work through this text. Learning the proper and appropriate use of a calculator is a vital skill for today's math students.

As with all new skills, your instructor will give you guidance as to where and when to use it. Calculators are especially useful in the following instances:

- For doing the fundamental operations of arithmetic (addition, subtraction, multiplication, and division)
- For finding powers or square roots of numbers
- For evaluating complicated arithmetic expressions
- For checking solutions to equations

Several different kinds of calculators are available.

- A basic 4-function calculator will add, subtract, multiply and divide. Sometimes these calculators also have a square root key. These calculators are not powerful enough to do all of the math in this text, and they are not recommended for math students at this level.
- A scientific calculator generally has about eight rows of keys on it and is usually labeled "scientific." Look for keys labeled "sin," "tan," and "log." Scientific calculators also have power keys and parenthesis keys, and the order of operations is built into them. These calculators are recommended for math students at this level.
- A graphing calculator also has about eight rows of keys, but it has a large, nearly square display screen. These calculators are very powerful, and you may be required to purchase them in later math courses. However, you will not need all that power to be successful in this course, and they are significantly more expensive than scientific calculators.

We will assume that you are operating a scientific calculator. (Some of the keystrokes are different on graphing calculators, so if you are using one of these calculators, please consult your owner's manual.) Study the following table to discover how the basic keys are used.

Expression	Key Strokes	Display
$144 \div 3 - 7$	[144] [÷] [3] [−] [7] [=]	41.
$3(2) + 4(5)$	[3] [×] [2] [+] [4] [×] [5] [=]	26.
$13^2 - 2(12 + 10)$	[13] [x^2] [−] [2] [×] [(] [12] [+] [10] [)] [=]	125.
$\frac{28 + 42}{10}$	[(] [28] [+] [42] [)] [÷] [10] [=]	7.
	or	
	[28] [+] [42] [=] [÷] [10] [=]	7.
$\frac{288}{6 + 12}$	[288] [÷] [(] [6] [+] [12] [)] [=]	16.
$19^2 - 3^5$	[19] [x^2] [−] [3] [x^y] [5] [=]	118.
$2\frac{1}{3} + \frac{5}{6}$	[2] [$a^b/_c$] [1] [$a^b/_c$] [3] [+] [5] [$a^b/_c$] [6] [=]	3 ⌋ 1 ⌋ 6

Notice that the calculator does calculations when you hit the [=] or ENTER key. The calculator automatically uses the order of operations when you enter more than one operation before hitting [=] or ENTER. Notice that if you begin a sequence with an operation sign, the calculator automatically uses the number currently displayed as part of the calculation. There are three operations that require only one number: squaring a number, square rooting a number, and taking the opposite of a number. In each case, enter the number first and then hit the appropriate operation key. Be especially careful with fractions. Remember that when there is addition or subtraction inside a fraction, the fraction bar acts as a grouping symbol. But the only way to convey this to your calculator is by using the grouping symbols [(] and [)]. Notice that the fraction key is used between the numerator and denominator of a fraction and also between the whole number and fractional part of a mixed number. It automatically calculates the common denominator when necessary.

Model Problem Solving

Practice the following problems until you can get the results shown.

	Answers
a. $47 + \frac{525}{105}$	52
b. $\frac{45 + 525}{38}$	15
c. $\frac{648}{17 + 15}$	20.25
d. $\frac{140 - 5(6)}{11}$	10
e. $\frac{3870}{9(7) + 23}$	45
f. $\frac{5(73) + 130}{33}$	15
g. $100 - 2^5$	68
h. $100 - (-2)^5$	132
i. $4\frac{2}{7} - 3\frac{3}{5}$	$\frac{24}{35}$

PRIME FACTORS OF NUMBERS 1 THROUGH 100

	Prime Factors		Prime Factors		Prime Factors		Prime Factors
1	none	26	$2 \cdot 13$	51	$3 \cdot 17$	76	$2^2 \cdot 19$
2	2	27	3^3	52	$2^2 \cdot 13$	77	$7 \cdot 11$
3	3	28	$2^2 \cdot 7$	53	53	78	$2 \cdot 3 \cdot 13$
4	2^2	29	29	54	$2 \cdot 3^3$	79	79
5	5	30	$2 \cdot 3 \cdot 5$	55	$5 \cdot 11$	80	$2^4 \cdot 5$
6	$2 \cdot 3$	31	31	56	$2^3 \cdot 7$	81	3^4
7	7	32	2^5	57	$3 \cdot 19$	82	$2 \cdot 41$
8	2^3	33	$3 \cdot 11$	58	$2 \cdot 29$	83	83
9	3^2	34	$2 \cdot 17$	59	59	84	$2^2 \cdot 3 \cdot 7$
10	$2 \cdot 5$	35	$5 \cdot 7$	60	$2^2 \cdot 3 \cdot 5$	85	$5 \cdot 17$
11	11	36	$2^2 \cdot 3^2$	61	61	86	$2 \cdot 43$
12	$2^2 \cdot 3$	37	37	62	$2 \cdot 31$	87	$3 \cdot 29$
13	13	38	$2 \cdot 19$	63	$3^2 \cdot 7$	88	$2^3 \cdot 11$
14	$2 \cdot 7$	39	$3 \cdot 13$	64	2^6	89	89
15	$3 \cdot 5$	40	$2^3 \cdot 5$	65	$5 \cdot 13$	90	$2 \cdot 3^2 \cdot 5$
16	2^4	41	41	66	$2 \cdot 3 \cdot 11$	91	$7 \cdot 13$
17	17	42	$2 \cdot 3 \cdot 7$	67	67	92	$2^2 \cdot 23$
18	$2 \cdot 3^2$	43	43	68	$2^2 \cdot 17$	93	$3 \cdot 31$
19	19	44	$2^2 \cdot 11$	69	$3 \cdot 23$	94	$2 \cdot 47$
20	$2^2 \cdot 5$	45	$3^2 \cdot 5$	70	$2 \cdot 5 \cdot 7$	95	$5 \cdot 19$
21	$3 \cdot 7$	46	$2 \cdot 23$	71	71	96	$2^5 \cdot 3$
22	$2 \cdot 11$	47	47	72	$2^3 \cdot 3^2$	97	97
23	23	48	$2^4 \cdot 3$	73	73	98	$2 \cdot 7^2$
24	$2^3 \cdot 3$	49	7^2	74	$2 \cdot 37$	99	$3^2 \cdot 11$
25	5^2	50	$2 \cdot 5^2$	75	$3 \cdot 5^2$	100	$2^2 \cdot 5^2$

APPENDIX C

SQUARES AND SQUARE ROOTS (0 TO 199)

n	n^2	$\sqrt{n}$	n	n^2	$\sqrt{n}$	n	n^2	$\sqrt{n}$	n	n^2	$\sqrt{n}$
0	0	0.000	50	2,500	7.071	100	10,000	10.000	150	22,500	12.247
1	1	1.000	51	2,601	7.141	101	10,201	10.050	151	22,801	12.288
2	4	1.414	52	2,704	7.211	102	10,404	10.100	152	23,104	12.329
3	9	1.732	53	2,809	7.280	103	10,609	10.149	153	23,409	12.369
4	16	2.000	54	2,916	7.348	104	10,816	10.198	154	23,716	12.410
5	25	2.236	55	3,025	7.416	105	11,025	10.247	155	24,025	12.450
6	36	2.449	56	3,136	7.483	106	11,236	10.296	156	24,336	12.490
7	49	2.646	57	3,249	7.550	107	11,449	10.344	157	24,649	12.530
8	64	2.828	58	3,346	7.616	108	11,664	10.392	158	24,964	12.570
9	81	3.000	59	3,481	7.681	109	11,881	10.440	159	25,281	12.610
10	100	3.162	60	3,600	7.746	110	12,100	10.488	160	25,600	12.649
11	121	3.317	61	3,721	7.810	111	12,321	10.536	161	25,921	12.689
12	144	3.464	62	3,844	7.874	112	12,544	10.583	162	26,244	12.728
13	169	3.606	63	3,969	7.937	113	12,769	10.630	163	26,569	12.767
14	196	3.742	64	4,096	8.000	114	12,996	10.677	164	26,896	12.806
15	225	3.873	65	4,225	8.062	115	13,225	10.724	165	27,225	12.845
16	256	4.000	66	4,356	8.124	116	13,456	10.770	166	27,556	12.884
17	289	4.123	67	4,489	8.185	117	13,689	10.817	167	27,889	12.923
18	324	4.243	68	4,624	8.246	118	13,924	10.863	168	28,224	12.961
19	361	4.359	69	4,761	8.307	119	14,161	10.909	169	28,561	13.000
20	400	4.472	70	4,900	8.367	120	14,400	10.954	170	28,900	13.038
21	441	4.583	71	5,041	8.426	121	14,641	11.000	171	29,241	13.077
22	484	4.690	72	5,184	8.485	122	14,884	11.045	172	29,584	13.115
23	529	4.796	73	5,329	8.544	123	15,129	11.091	173	29,929	13.153
24	576	4.899	74	5,476	8.602	124	15,376	11.136	174	30,276	13.191
25	625	5.000	75	5,625	8.660	125	15,625	11.180	175	30,625	13.229
26	676	5.099	76	5,776	8.718	126	15,876	11.225	176	30,976	13.266
27	729	5.196	77	5,929	8.775	127	16,129	11.269	177	31,329	13.304
28	784	5.292	78	6,084	8.832	128	16,384	11.314	178	31,684	13.342
29	841	5.385	79	6,241	8.888	129	16,641	11.358	179	32,041	13.379
30	900	5.477	80	6,400	8.944	130	16,900	11.402	180	32,400	13.416
31	961	5.568	81	6,561	9.000	131	17,161	11.446	181	32,761	13.454
32	1,024	5.657	82	6,724	9.055	132	17,424	11.489	182	33,124	13.491
33	1,089	5.745	83	6,889	9.110	133	17,689	11.533	183	33,489	13.528
34	1,156	5.831	84	7,056	9.165	134	17,956	11.576	184	33,856	13.565
35	1,225	5.916	85	7,225	9.220	135	18,225	11.619	185	34,225	13.601
36	1,296	6.000	86	7,396	9.274	136	18,496	11.662	186	34,596	13.638
37	1,369	6.083	87	7,569	9.327	137	18,769	11.705	187	34,969	13.675
38	1,444	6.164	88	7,744	9.381	138	19,044	11.747	188	35,344	13.711
39	1,521	6.245	89	7,921	9.434	139	19,321	11.790	189	35,721	13.748
40	1,600	6.325	90	8,100	9.487	140	19,600	11.832	190	36,100	13.784
41	1,681	6.403	91	8,281	9.539	141	19,881	11.874	191	36,481	13.820
42	1,764	6.481	92	8,464	9.592	142	20,164	11.916	192	36,864	13.856
43	1,849	6.557	93	8,649	9.644	143	20,449	11.958	193	37,249	13.892
44	1,936	6.633	94	8,836	9.659	144	20,736	12.000	194	37,636	13.928
45	2,025	6.708	95	9,025	9.747	145	21,025	12.042	195	38,025	13.964
46	2,116	6.782	96	9,216	9.798	146	21,316	12.083	196	38,416	14.000
47	2,209	6.856	97	9,409	9.849	147	21,609	12.124	197	38,809	14.036
48	2,304	6.928	98	9,604	9.899	148	21,904	12.166	198	39,204	14.071
49	2,401	7.000	99	9,801	9.950	149	22,201	12.207	199	39,601	14.107
n	n^2	$\sqrt{n}$	n	n^2	$\sqrt{n}$	n	n^2	$\sqrt{n}$	n	n^2	$\sqrt{n}$

CHAPTERS 1–4

MIDTERM EXAMINATION

ANSWERS

1. Write the place value of the digit 6 in 569,234. **1.** ________

2. Add:

$$\begin{array}{r} 3,678 \\ 91 \\ 341 \\ 89,612 \\ +\quad 297 \\ \hline \end{array}$$

2. ________

3. Write the word name for 345,804. **3.** ________

4. Subtract:

$$\begin{array}{r} 8934 \\ -5877 \\ \hline \end{array}$$

4. ________

5. Add: $843 + 16,867 + 45 + 8$ **5.** ________

6. Multiply: (532)(78) **6.** ________

7. Estimate the product and multiply:

$$\begin{array}{r} 803 \\ \times 906 \\ \hline \end{array}$$

7. ________

8. Divide: $56\overline{)9352}$ **8.** ________

9. Divide: $23\overline{)83,672}$ **9.** ________

10. Perform the indicated operations: $8 - 3 \cdot 2 + 15 \div 3$ **10.** ________

11. Find the sum of the quotient of 72 and 9 and the product of 33 and 2. **11.** ________

12. Find the average of 452, 149, 86, and 101. **12.** ________

13. Find the average, median, and mode of 22, 34, 34, 67, 98, 67, 52, and 34. **13.** ________

14. The graph shows the number of Honda vehicles sold at a local dealership by model.

a. Which model has the highest sales? **14. a.** ________

b. How many more Accords are sold than Passports? **b.** ________

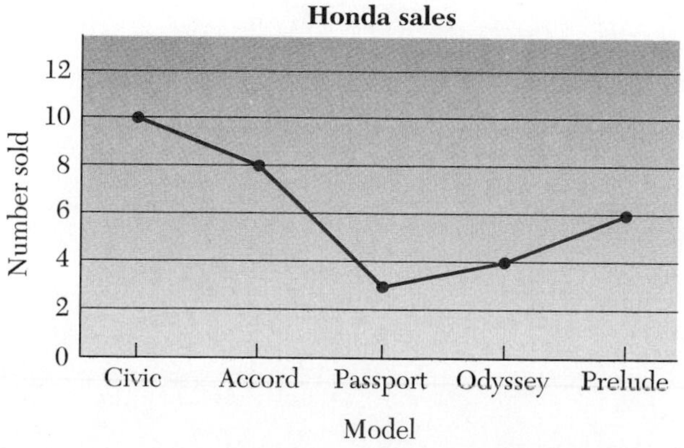

15. Write the least common multiple (LCM) of 12, 15, and 18. 15. ____________

16. What is the greatest common factor (GCF) of 60, 75, and 90? 16. ____________

17. Is 107 a prime number or a composite number? 17. ____________

18. Is 396 a multiple of 9? 18. ____________

19. List the first five multiples of 23. 19. ____________

20. List all the factors of 304. 20. ____________

21. Write the prime factorization of 304. 21. ____________

22. Change to a mixed number: $\frac{67}{8}$ 22. ____________

23. Change to an improper fraction: $17\frac{4}{7}$ 23. ____________

24. Which of these fractions are proper? $\frac{3}{4}, \frac{7}{6}, \frac{8}{8}, \frac{7}{8}, \frac{9}{8}, \frac{5}{4}, \frac{4}{4}$ 24. ____________

25. List these fractions from the smallest to the largest: $\frac{5}{9}, \frac{5}{8}, \frac{7}{12}, \frac{2}{3}$ 25. ____________

26. Simplify: $\frac{144}{216}$ 26. ____________

27. Multiply and simplify: $\frac{4}{15} \cdot \frac{5}{8} \cdot \frac{6}{7}$ 27. ____________

28. Multiply: $\left(4\frac{3}{4}\right)\left(11\frac{5}{8}\right)$. Write the answer as a mixed number. 28. ____________

29. Divide and simplify: $\frac{21}{15} \div \frac{14}{25}$ 29. ____________

30. What is the reciprocal of $4\frac{1}{3}$? 30. ____________

31. Add: $\frac{4}{15} + \frac{7}{12}$ 31. ____________

32. Add: $6\frac{2}{3}$ $+11\frac{7}{8}$ 32. ____________

33. Subtract: $11 - 7\frac{3}{5}$ 33. ____________

34. Subtract: $34\frac{4}{9}$ $-26\frac{4}{5}$ 34. ____________

35. Find the average of $4\frac{5}{6}$, $6\frac{2}{3}$, and $11\frac{3}{4}$. 35. ____________

36. Perform the indicated operations: $\frac{4}{5} - \frac{1}{2} \cdot \frac{5}{6} \div \frac{5}{6}$ 36. ____________

37. Write the place value name for "seventy thousand and three hundred two thousandths." 37. ____________

38. Round 2.07499 to the nearest tenth, hundredth, and thousandth. 38. ____________

39. Write 0.14 as a fraction. 39. ____________

40. List the following numbers from smallest to largest: 1.034, 1.109, 1.044, 1.094. 1.02, 1.07 40. ____________

41. Add: $34.6 + 2.009 + 0.34 + 2.054 + 16.9$ 41. ____________

42. Subtract: $\begin{array}{r} 3.7 \\ -1.956 \\ \hline \end{array}$ 42. ____________

43. Multiply: $\begin{array}{r} 4.025 \\ \times \quad 3.6 \\ \hline \end{array}$ 43. ____________

44. Multiply: $0.012 \times 10{,}000$ 44. ____________

45. Divide: $0.012 \div 10{,}000$ 45. ____________

46. Write 0.00371 in scientific notation. 46. ____________

47. Divide: $2.35\overline{)31.96}$ 47. ____________

48. Change $\frac{7}{13}$ to a decimal rounded to the nearest thousandth. 48. ____________

49. Find the average and median of 4.5, 6.9, 8.3, 9.5, 3.1, 10.3, and 2.2. 49. ____________

50. Perform the indicated operations: $(5.5)^2 - 2.3(4.1) + 11.8 - 9.3 \div 0.3$ 50. ____________

51. Find the perimeter of a trapezoid with bases of 34 ft and 29 ft and sides of 7 ft and 5 ft. 51. ____________

52. Find the area of a rectangle that is $3\frac{3}{4}$ ft wide and $7\frac{2}{5}$ ft long. 52. ____________

53. On Thursday, October 4, the following counts of chinook salmon going over the dams were recorded: Bonneville, 1577; The Dalles, 1589; John Day, 1854; McNary, 1361; Ice Harbor, 295; Little Goose, 274. How many chinook salmon were counted? 53. ____________

54. Recently, GE declared a stock dividend of 16¢ a share. Maria owns 453.24 shares of GE. To the nearest cent, what was Maria's dividend? 54. ____________

55. Houng earned $632.80 for 35 hours of work last week. What is Houng's hourly wage? 55. ____________

Name ______________________ Class ______________________ Date __________

CHAPTERS 1–8

FINAL EXAMINATION

ANSWERS

1. Add: $\frac{7}{8} + \frac{9}{16}$ — 1. ________

2. Write the LCM (least common multiple) of 14, 18, and 21. — 2. ________

3. Subtract: $\begin{array}{r} 73.56 \\ -39.67 \\ \hline \end{array}$ — 3. ________

4. Add: $19.04 + 0.073 + 0.74$ — 4. ________

5. Divide: $\frac{5}{6} \div \frac{5}{7}$ — 5. ________

6. Multiply: $2\frac{4}{9} \cdot 21$ — 6. ________

7. Which of these number is a prime number? 99, 199, 299, 699 — 7. ________

8. Divide: $0.58\overline{)56.75}$. Round the answer to the nearest hundredth. — 8. ________

9. Subtract: $14 - 7\frac{7}{10}$ — 9. ________

10. Multiply: (0.00721)(100,000) — 10. ________

11. Round to the nearest hundredth: 82.04499 — 11. ________

12. Multiply: (8.3)(7.02) — 12. ________

13. Write as a fraction and simplify: 86% — 13. ________

14. Add: $\begin{array}{r} 9\frac{11}{15} \\ + \frac{5}{6} \\ \hline \end{array}$ — 14. ________

15. Solve the proportion: $\frac{7}{12} = \frac{x}{4}$ — 15. ________

16. What is the place value of the 6 in 23.0067? — 16. ________

17. List these fractions from the smallest to the largest: $\frac{3}{4}, \frac{4}{5}, \frac{7}{10}$ — 17. ________

18. Change to percent: $\frac{17}{20}$ — 18. ________

19. Divide: $27\overline{)659}$ — 19. ________

 Name ______________ Class ______________ Date ______________

20. Write as an approximate decimal to the nearest thousandth: $\frac{11}{23}$ **20.** ______________

21. Write the place value name for "six thousand and fifteen thousandths." **21.** ______________

22. Divide: $0.016\overline{)61}$ **22.** ______________

23. Write as a decimal: $11\frac{3}{8}\%$ **23.** ______________

24. Forty-six percent of what number is 16.1? **24.** ______________

25. Write as a percent: 4.52 **25.** ______________

26. Write as a decimal: $\frac{53}{200}$ **26.** ______________

27. Eighty-one percent of 45 is what number? **27.** ______________

28. Multiply: (0.17)(1.2)(0.15) **28.** ______________

29. If 6 books cost $117.48, how much would 15 books cost at the same rate? **29.** ______________

30. A CD player is priced at $480. It is on sale for $345. What is the percent of discount based on the original price? **30.** ______________

31. List the first five multiples of 41. **31.** ______________

32. Write the word name for 9045.045. **32.** ______________

33. Is the following proportion true or false? $\frac{1.9}{22} = \frac{5.8}{59}$ **33.** ______________

34. Write the prime factorization of 420. **34.** ______________

35. Simplify: $\frac{105}{180}$ **35.** ______________

36. Change to a fraction and simplify: 0.945 **36.** ______________

37. Change to a mixed number: $\frac{255}{7}$ **37.** ______________

38. Multiply and simplify: $\frac{5}{28} \cdot \frac{35}{20}$ **38.** ______________

39. Change to a fraction: 0.075 **39.** ______________

40. At a service station, 29 of 50 drivers come in for a "fill-up." What percent of the drivers wants a full tank of gas? **40.** ______________

41. Is 733 a multiple of 3? **41.** ______________

42. Subtract: $14\frac{3}{5} - \frac{11}{15}$ 42. ____________

43. Change to an improper fraction: $13\frac{7}{9}$ 43. ____________

44. List the following decimals from smallest to largest: 2.32, 2.332, 2.299, 2.322 44. ____________

45. Find the greatest common factor (GCF) of 102, 170, 340, and 408. 45. ____________

46. Divide: $674.81 \div 10{,}000$ 46. ____________

47. Divide: $6\frac{3}{8} \div 4\frac{5}{16}$ 47. ____________

48. Write a ratio to compare 5 in. to 5 ft (using common units) and simplify. 48. ____________

49. Mildred calculates that she pays $1.21 for gas and oil to drive 5 miles. In addition, she pays 30¢ for maintenance for each 5 miles she travels. How much will it cost her to drive 6000 miles? 49. ____________

50. Twenty percent of a family's income is spent on food, 8% on transportation, 35% on housing, 10% on heat and utilities, 7% on insurance, and the rest on miscellaneous expenses. If the family's income is $2100 per month, how much is spent on miscellaneous expenses? 50. ____________

51. The sales tax on a $55 purchase is $4.75. What is the sales tax rate, to the nearest tenth of a percent? 51. ____________

52. Perform the indicated operations: $32 - 5 \cdot 6 + 45 \div 9$ 52. ____________

53. Find the average of $\frac{7}{15}$, $3\frac{3}{4}$, and $5\frac{5}{6}$. 53. ____________

54. Perform the indicated operations: $7.8 - 2.1(3.6) \div 4$ 54. ____________

55. Add:

$$\begin{array}{r} 4\text{ hr } 35\text{ min } 16\text{ sec} \\ +2\text{ hr } 42\text{ min } 55\text{ sec} \\ \hline \end{array}$$

55. ____________

56. Subtract:

$$\begin{array}{r} 7\text{ m } 27\text{ cm} \\ -4\text{ m } 35\text{ cm} \\ \hline \end{array}$$

56. ____________

57. Convert 5¢ per gram to dollars per kilogram. 57. ____________

58. Find the perimeter of a trapezoid with bases of 45.7 ft and 52.8 ft and sides of 13.7 ft and 15.3 ft. 58. ____________

59. Find the area of a triangle with base 5.7 m and height 4.7 m. 59. ____________

60. Find the area of the following geometric figure (let $\pi \approx 3.14$). 60. ____________

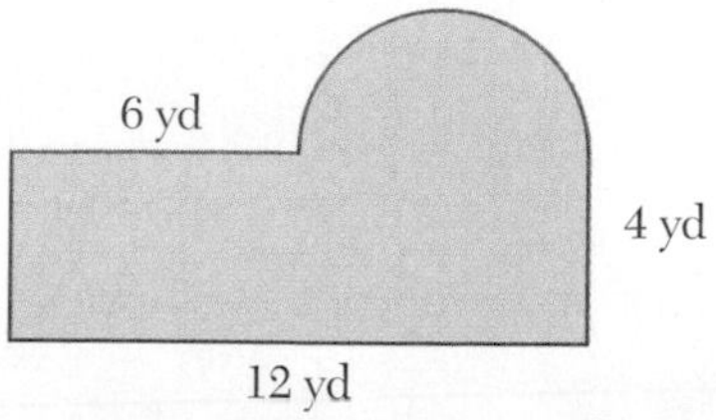

61. Find the volume of a box with length $2\frac{1}{4}$ ft, width 18 in., and height 8 in. (in cubic inches). 61. ______

62. Find the square root of 12.84 to the nearest hundredth. 62. ______

63. Find the hypotenuse of a right triangle with legs of 14 cm and 17 cm. Find to the nearest tenth of a centimeter. 63. ______

64. Add: $(-34) + (-23) + 41$ 64. ______

65. Subtract: $(-45) - (-72)$ 65. ______

66. Multiply: $(3)(-4)(-1)(-6)$ 66. ______

67. Divide: $(-125) \div (-25)$ 67. ______

68. Perform the indicated operations: $(-6 - 4)(-4) \div (-5) - (-10)$ 68. ______

69. Solve: $12a + 98 = 2$ 69. ______

70. Find the Fahrenheit temperature that is equivalent to 45°C. Use the formula $9C = 5F - 160$. 70. ______

ANSWERS

CHAPTER 1

SECTION 1.1

1. seven hundred eighty-one **3.** eight hundred ninety **5.** seven thousand, twenty **7.** 57 **9.** 7500 **11.** 100,000,000 **13.** thirty-six thousand, four hundred fifty **15.** three hundred six thousand, four hundred fifty **17.** forty-five million **19.** 243,700 **21.** 22,570 **23.** 17,000,000,000 **25.** < **27.** > **29.** < **31.** < **33.** 740 **35.** 1700

	Number	Ten	Hundred	Thousand	Ten Thousand
37.	304,285	304,290	304,300	304,000	300,000
39.	8,657,896	8,657,900	8,657,900	8,658,000	8,660,000

41. The percentage who exercise regularly is 40%.
43. The under $15,000 category has the highest percentage of nonexercisers.
45. As income level goes up, so does the percentage of regular exercisers.
47. Substance abusers, families with children, children, the mentally ill, and single women all increased over the 9-year period. Single men, military veterans, and the employed all decreased.
49. Single men, single women, and children made up 87% of the homeless in 1988. In 1997, the percentage was the same. **51.** 560,353,730
53. thirteen thousand, one hundred forty-eight dollars
55. < **57.** 1000 **59.** 81,630,000
61. 63,700; 63,800. Rounding the second time changes the tens digit to 5, and rounding to the hundreds place from 63,750 results in a different answer. If told to simply round to the nearest hundred, the first method is correct.
63. Kimo wrote "Eighteen thousand, four hundred sixty-five" on the check.
65. Ducks Unlimited estimated that three hundred eighty-nine thousand, five hundred ducks spent the winter at the refuge.
67. The place value name for the bid is $36,407.
69. To the nearest thousand dollars, the value of the Intel shares is $847,000.
71. The metal industries released 327,000,000 pounds of toxic material.
73. The per capita income in Maine is twenty thousand, five hundred twenty-seven dollars.
75. Maine has the smallest per capita income.
77. To the nearest million miles, the distance from Earth to the sun is 93,000,000 miles.

79.

Month	Number of Marriages	Month	Number of Marriages
June	242,000	December	184,000
May	241,000	April	172,000
August	239,000	November	171,000
July	235,000	February	155,000
October	231,000	March	118,000
September	225,000	January	110,000

81. The rivers, in order of increasing length, are: Yukon, Colorado, Arkansas, Rio Grande, Missouri, Mississippi.
83. Yes, the motor vehicle department's estimate was correct.
85. Jif has fewer of the following nutrients: fat, sodium, carbohydrates, and sugars.
87. The 1999–2000 payroll figures seem to be rounded to the nearest hundred.
89. The top payroll for the 1999–2000 season is less than the top payroll for the 2000–2001 season.
91. Base 10 is a good name for our number system because each place value in the system is 10 times the previous place and one-tenth the succeeding place.
93. Rounding a number is a method of calculating an approximation of that number. The purpose is to get an idea of the value of the number without listing digits that do not add to our understanding. The number 87,452 rounds to 87,000, to the nearest thousand, because 452 is less than halfway between 0 and 1000. Rounded to the nearest hundred, it is 87,500, because 52 is over halfway between 0 and 100.
95. three trillion, four hundred fifty-six billion, seven hundred nine million, two hundred thirty thousand
97. 5 **99.** 0

SECTION 1.2

1. 97 **3.** 979 **5.** 707 **7.** 1 **9.** 5737 **11.** 6571 **13.** 17,500 **15.** 201 **17.** 437 **19.** 423 **21.** 10 **23.** 175 **25.** 163 **27.** 4500 **29.** 1100 **31.** 10,000 **33.** 300 **35.** 4800 **37.** 19,000; 19,918 **39.** 161,000; 160,498 **41.** 5000; 4686 **43.** 34,000; 33,633 **45.** 41,000; 40,883 **47.** 36 cm **49.** 221 in. **51.** 40 ft **53.** 134 in.
55. The total number of Fords, Toyotas, and Lexuses sold is 3561. **57.** 163 more Hondas than Fords are sold.

59. The total number of the three best-selling cars sold is 5320.
61. For the week, 3589 salmon went through the ladder. Tuesday's count was 363 more than Saturday's count.
63. The estimated number of arrests for property crimes is 1,611,832. **65.** Sasha consumes 520 calories.
67. Sasha could have eaten 1180 calories for breakfast and lunch. **69.** A total of 15,136 trees can be harvested.
71. Fong's Grocery still owes $13,100.
73. The median family income was probably rounded to the nearest hundred. The San Francisco median income was $9300 higher than Seattle's median income.
75. On average, 530 more husbands killed their wives than wives killed their husbands.
77. On average, 209 people killed a sibling.
79. 64,422 more General Motors vehicles were sold than Ford vehicles.

81.

	June Sales
General Motors	457,000
Ford	392,000
Daimler Chrysler	245,000

83. The total number of HIV infections in Asian regions is 6,900,000. **85.** The perimeter is 160 ft.
87. Blanche needs 184 in. of lace.
89. Explanations for subtracting that are aimed at 6-year-olds are usually based on physical objects. For example, $15 - 9 = 6$ because when 9 circles (or pencils or apples or whatever) are removed from 15 circles, 6 circles are left. Count the circles that are empty.

○○○●●
○○○●●
●●●●●

91. A sum is the result of adding numbers. The sum of 8, 8, and 2 is 18. Mathematically, we write $8 + 8 + 2 = 18$.
93. seven hundred thousand, nine hundred
95. The dollar sales for the nine cars is $165,991. The Accord's sales were $26,585 more than the Civic's sales.
97. A = 7, B = 2, C = 3, D = 1

GETTING READY FOR ALGEBRA

1. $x = 12$ **3.** $x = 23$ **5.** $z = 14$ **7.** $c = 39$
9. $a = 151$ **11.** $x = 14$ **13.** $y = 58$
15. $k = 168$ **17.** $x = 11$ **19.** $w = 116$
21. The markup is $348.
23. The length of the garage is 9 meters.
25. Let S represent the EPA City rating of the Saturn and I represent the EPA City rating of the Impreza: $I + 4 = S$. The Impreza has a rating of 24 mpg.
27. Let B represent the total dollars budgeted in a category, S represent the dollars spent in a category, and R represent the dollars not yet spent in a category: $S + R = B$.

SECTION 1.3

1. 310 **3.** 132 **5.** 268 **7.** 511 **9.** 477
11. 0 **13.** 2640 **15.** thousands **17.** 3177
19. 4218 **21.** 1596 **23.** 3588 **25.** 153,900
27. 38,684 **29.** 29,070 **31.** 21,100
33. 1800 **35.** 16,800 **37.** 7200 **39.** 58,400
41. 15,000; 15,600 **43.** 35,000; 38,471
45. 20,000; 18,525 **47.** 320,000; 328,396
49. 168 square yd **51.** 225 square ft
53. 1150 square cm **55.** 56 square ft
57. 37,352 square cm **59.** 180 square m
61. 432 square ft **63.** 120 square ft **65.** 279,265
67. 2,100,000; 2,375,592
69. The Rotary Club estimates it will sell 4230 dozen roses.
71. The gross receipts from the sale of Mustang Coupes are $629,496.
73. The estimated gross receipts from all models are $1,340,000. The actual gross receipts are $1,273,000, rounded to the nearest thousand.
75. Washington County grew by 24,516 people during 1997.
77. The comptroller realized $28,585 from the sale of the shares. **79.** There are 6 feet in a fathom.
81. There are about 18,228 ft in a league.
83. It is not possible to be 20,000 leagues under the sea. The author was taking literary license.
85. Sirius is about 47,040,000,000,000 miles from Earth.
87. The term paper could have up to 64,800 dots.
89. A RAM of 128 Kb is 131,072 bytes.
91. A tablespoon of olive oil has 126 calories from fat.
93. To the nearest thousand gallons, in a 31-day month the water usage is 9,934,000 gallons.
95. The shovels cost Mr. Garcia $3900. The gross income from the sale of the shovels is $6825. The profit from the sale of the shovels is $2925.
97. If the Rams doubled their payroll, they would rank first in the 2000–2001 season.
99. Explanations for multiplication that are aimed at 8-year-olds are usually based on physical objects. For example, $3(8) = 24$ because 3 groups, each of which has 8 circles (or pencils or apples or whatever), contain a total of 24 circles. Count the total number of circles.

○○○○ ○○○○ ○○○○
○○○○ ○○○○ ○○○○

101. A product is the result of multiplying numbers. The product of 11 and 8 is 88. Mathematically, we write $11 \times 8 = 88$.
103. To the nearest thousand, the crop is worth $23,439,000.
105. A = 3, B = 2, C = 4, D = 5, E = 9

SECTION 1.4

1. 7 **3.** 5 **5.** 107 **7.** 61 **9.** 16 **11.** 40
13. 71 **15.** 2 R 5 **17.** divisor **19.** 4062
21. 32 **23.** 26 **25.** 96 **27.** 51
29. 173 R 12 **31.** 14 R 17 **33.** 58 **35.** 1020
37. 10 **39.** 90 **41.** 58 **43.** 56
45. 370; 362 **47.** 420; 439 **49.** 400; 433 R 41
51. 1900; 1790 R 414 **53.** 18,000; 18,911 R 133

55. The estimated taxes paid per return in week 2 is \$4700. **57.** The taxes paid per return in week 3 is \$4100, rounded to the nearest hundred. **59.** 210; 175 R 333
61. The survey finds that 128 trees per acre are ready to harvest.
63. It takes the company 520 hours to process the beans.
65. You would need to spend \$2,055,000 per day.
67. The population density of China was about 338 people per square mile.
69. The population density of the United States was about 72 people per square mile.
71. There are about three pets per household in the United States.
73. Each California representative represents about 625,380 people.
75. The gross state product per person in Florida was about \$26,210.
77. There are about 113 calories per serving.
79. Juan can take 8 capsules per day.
81. The Rams' payroll was approximately \$2,743,000 per game.
83. Jerry Rice averaged almost 20 yards per reception.
85. The remainder, after division, is the amount left over after all possible groups of the appropriate size have been formed. When 46 objects are separated into groups of 9 (you can make 5 groups of 9), there is 1 object remaining. So $46 \div 9 = 5$ R 1.
87. The company will get the greater gross return in France. The difference in gross receipts is \$7125.
89. $A = 2, B = 7, C = 9$

GETTING READY FOR ALGEBRA

1. $x = 5$ **3.** $c = 18$ **5.** $x = 4$ **7.** $b = 46$
9. $x = 12$ **11.** $y = 312$ **13.** $x = 24$
15. $b = 24{,}408$ **17.** $x = 61$ **19.** $w = 782$
21. The width of the garden plot is 17 ft.
23. He sells 2340 lb of crab.
25. The wholesale cost of one set is \$310.
27. Let L represent the low temperature in July and H represent the high temperature in January: $2H = L$. The high temperature is 30°F.

SECTION 1.5

1. 12^8 **3.** 81 **5.** 8 **7.** 121
9. base; exponent; power or value **11.** 216
13. 324 **15.** 10,000 **17.** 729 **19.** 6561
21. 3700 **23.** 110,000 **25.** 17 **27.** 73
29. exponent **31.** 36,300,000 **33.** 231
35. 35,910,000 **37.** 302 **39.** 131,000,000,000
41. 820 **43.** 10^{12} **45.** 50,625
47. 387,420,489 **49.** 47,160,000,000,000
51. 7900 **53.** The size of Salvador's lot is 12,000 ft^2.
55. The college's operating budget is \$73,000,000.
57. There are approximately 25,500,000,000,000 miles from Earth to Alpha Centauri.
59. The distance, 6 trillion miles, can also be written as 6,000,000,000,000 and as 6×10^{12}.
61. There are 2^4, or 16, bacteria after 4 hours.
63. The number of bacteria will exceed 1000 during the tenth hour.
65. The surface area of the Pacific Ocean is about 64,200,000 square miles. **67.** A gigabyte is 10^9 bytes.
69. The Steelers' payroll was $54{,}613 \times 10^3$.
71. 25^6 is larger than the Cardinals' payroll.
73. The expression 4^{10} represents the product of 10 fours—that is, $4 \times 4 \times 4 \times 4 \times 4 \times 4 \times 4 \times 4 \times 4 \times 4$.
75. Mitchell's grandparents will deposit \$531,441 on his twelfth birthday. They will have deposited a total of \$797,160. **77.** 9736

SECTION 1.6

1. 53 **3.** 0 **5.** 34 **7.** 23 **9.** 36 **11.** 39
13. 30 **15.** 43 **17.** 31 **19.** 7 **21.** 29
23. 106 **25.** 24 **27.** 58 **29.** 145
31. 161 **33.** 45
35. There were 595 more mallards and canvasbacks than teals and woodducks.
37. Four times the number of woodducks added to the number of mallards would be 165 more than the number of teals and canvasbacks. **39.** 58 **41.** 436
43. The Hair Barn used supplies costing \$1130 for the month.
45. The trucker's average weekly income is \$1225 per week. His yearly income is \$61,250.
47. Marla consumes 1150 calories for breakfast.
49. Sally's total charge is \$108.
51. Clay and Connie have \$42 left on their certificate.
53. June's target training rate is 120 bpm.
55. Your target training rate decreases as you age because the MHR decreases. **57.** The result is \$84,631,550.
59. Using the order of operations, multiplication takes precedence over addition, so $3 + 5 \cdot 4 = 3 + 20$, or 23.
61. 196

GETTING READY FOR ALGEBRA

1. $x = 7$ **3.** $y = 33$ **5.** $x = 6$ **7.** $c = 32$
9. $x = 2$ **11.** $c = 12$ **13.** $a = 196$
15. $b = 21$ **17.** Remy bought seven tickets.
19. Rana made six arrangements.
21. $15m = C$; Jessica can purchase 200 minutes from AT&T.
23. $9m + 696 = C$; Jessica can purchase 256 minutes from Pace.

SECTION 1.7

1. 6 **3.** 9 **5.** 12 **7.** 8 **9.** 4 **11.** 7
13. 12 **15.** 30 **17.** 40 **19.** 31 **21.** 128
23. 145 **25.** 117 **27.** 54 **29.** 56 **31.** 26
33. 41 **35.** 91 **37.** 116 **39.** 82 **41.** 17
43. 3 **45.** no mode **47.** 38 **49.** 32 and 47

51. no mode
53. The average number of ducks per species is 675. The median number of ducks per species is 740.
55. The golfer has an average score of 84, a median score of 81, and a mode score of 78.
57. The average gas mileage for the cars is 30 mpg, the median mileage is 31 mpg, and the mode mileage is 31 mpg.
59. The average weight of the football team is 201 lb.
61. The average cost per mile to build the new lines is $63 million.
63. The average cost of the top five coffee makers is about $52. The third-, fourth-, and fifth-ranked coffee makers all retail for less than the average price of $52.
65. The average population was about 3,686,220 over the 10-year period.

67. Population by Year in South Carolina

Year	Population	Year	Population
1990	3,486,703	1996	3,726,121
1991	3,526,606	1997	3,766,024
1992	3,566,509	1998	3,805,927
1993	3,606,412	1999	3,845,830
1994	3,646,315	2000	3,885,733
1995	3,686,218		

69. The average yearly attendance for the 3 years was about 10,533,333.
71. The average assets for the top three banks were $314 billion.
73. The average assets for the top five banks were about $293 billion.
75. The mean number of unique visitors for the top five web sites is 8,651,000.
77. The median number of unique visitors for the top five web sites is 9,933,000.
79. Four web sites are above the mean and six are below it.
81. Half of the houses on Jupiter Island cost $4 million or more and half cost $4 million or less.
83. In 1900, about half of all women getting married for the first time were 22 years or younger. In 1996, about half of all women getting married for the first time were 25 years or younger. This means that women are waiting longer to get married. The 22-, 23-, and 24-year-old women are now below the average of those getting married for the first time.
85. The average payroll was $87,162,000.
87. The average payroll of the Patriots and Jaguars was about $55,100,000. The difference in the payrolls was about $3 million, so add half of that ($1,500,000) to the Jaguars' payroll. The calculated average payroll was $55,110,950.
89. The average of 2, 4, 5, 5, and 9 is their sum, 25, divided by 5, the number of numbers. So, the mean is 5. The average gives one possible measure of the center of the group.

SECTION 1.8

1. 10–11 **3.** 250
5. The total number of calls made is 1425. **7.** 40
9. full-size
11. The number of cars in for repair is 400. **13.** 1996
15. The increase in production is 12,500 units.
17. The average production is 20,500 units.
19. The total paid for paint and lumber is $45,000.
21. Steel casting costs $15,000 less than plastic.
23. The company will pay $40,000 for doubling production.

25.

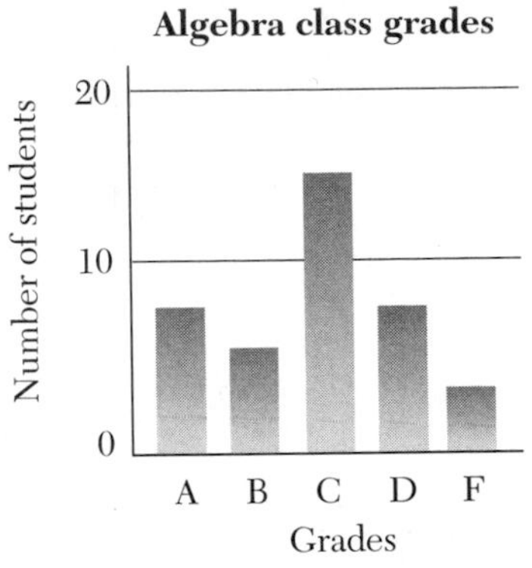

27.

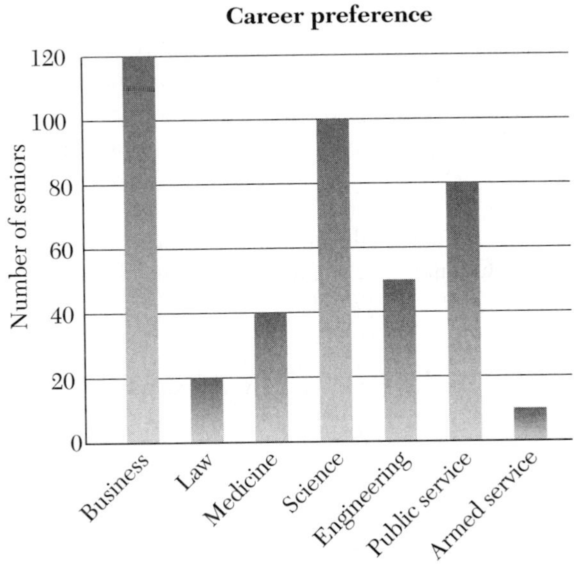

29.

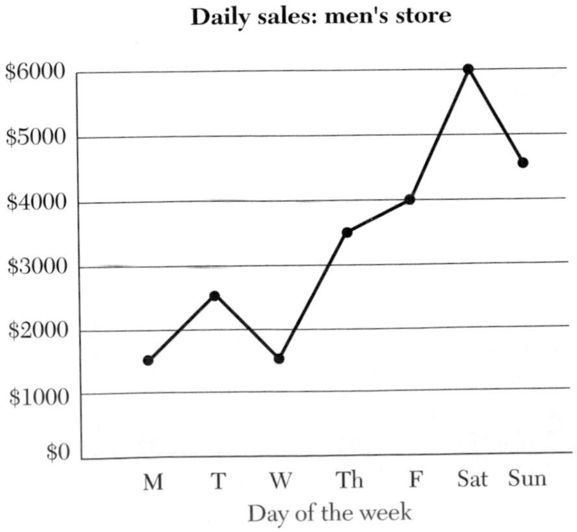

31.

Smith family income

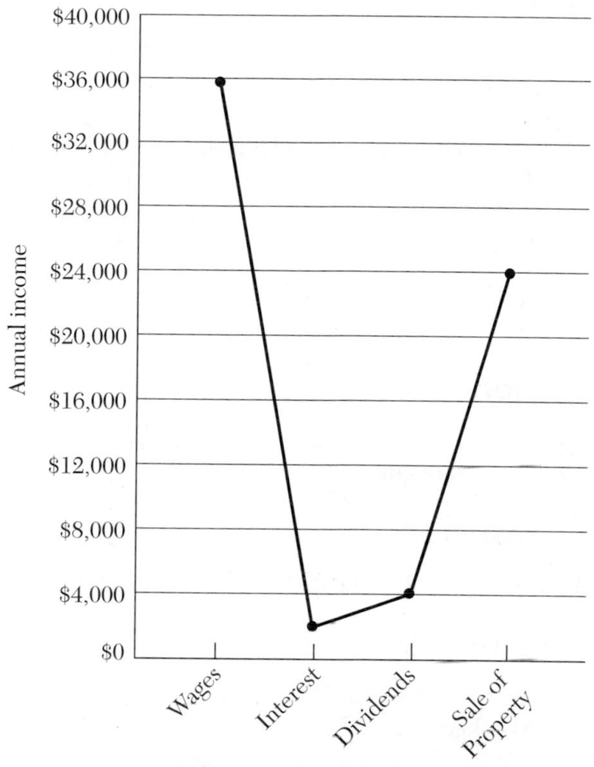

33.

Average cost of a three-bedroom house

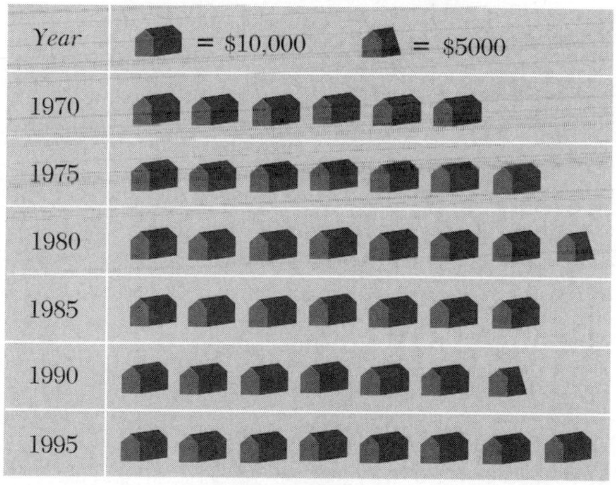

35.

Oil production in barrels

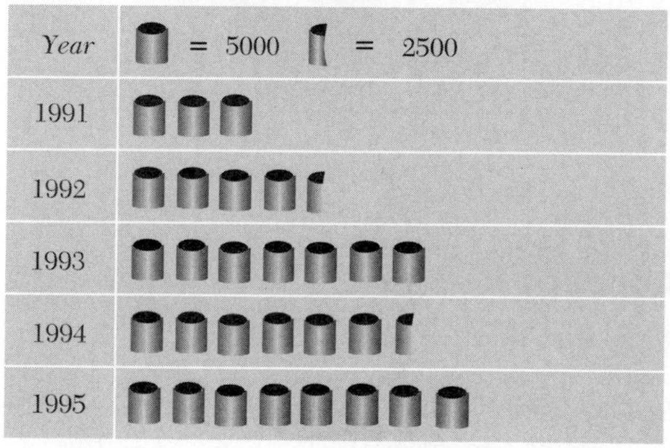

37.

Maximum retirement benefits

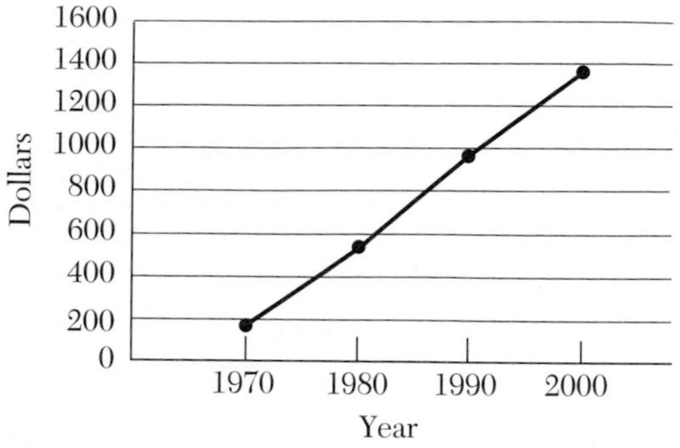

39.

Median price of a single-family home in Miami

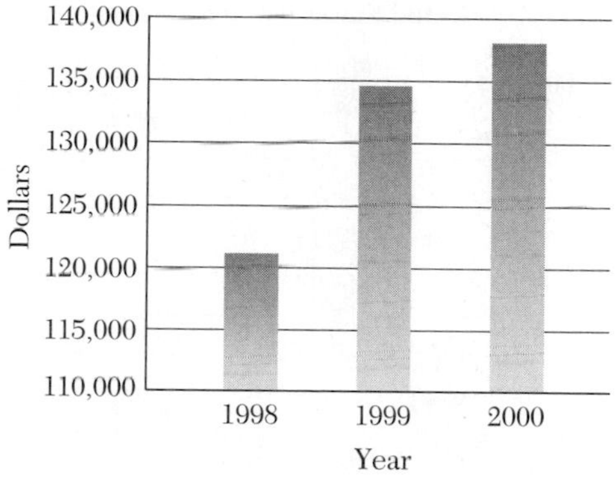

41. General Motors; Chrysler

43. The total number of U.S. workers is 675,000.

45.

Attendance at Utaki Park

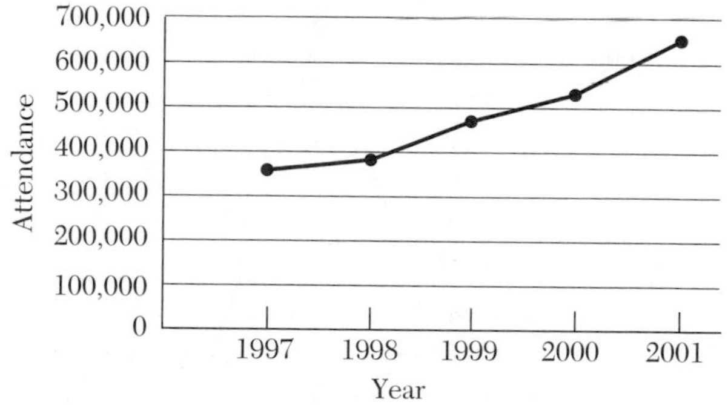

47.

Overnight camping at Lizard Lake State Park

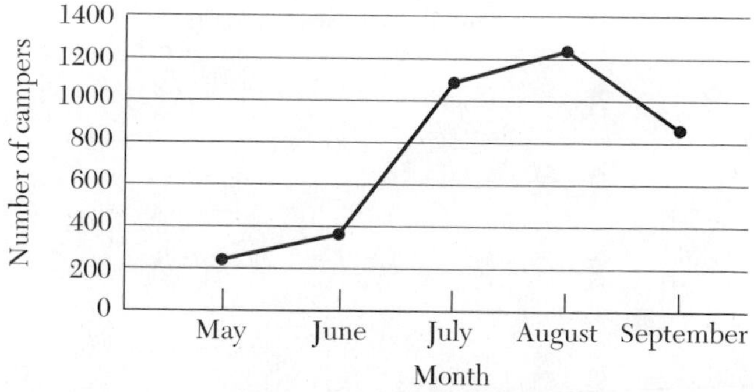

49.

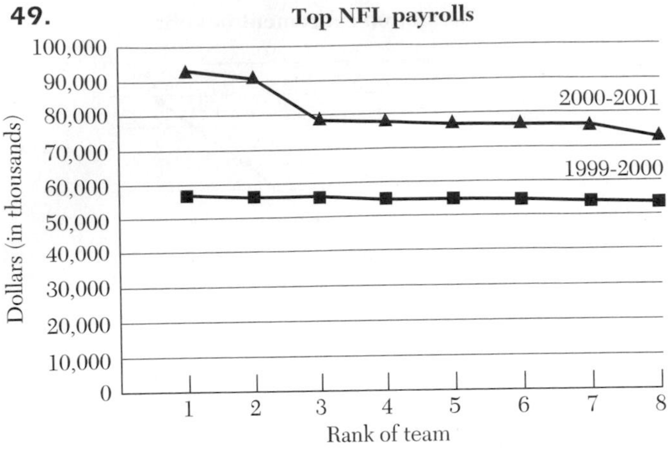

51.

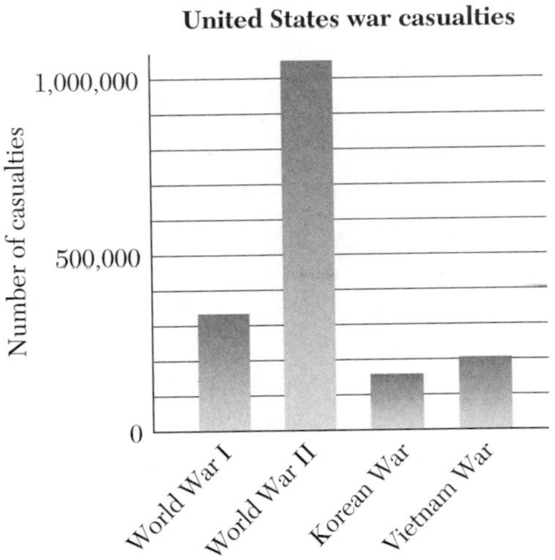

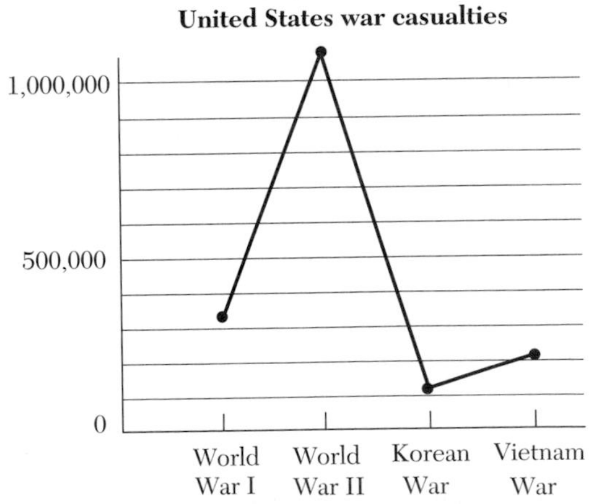

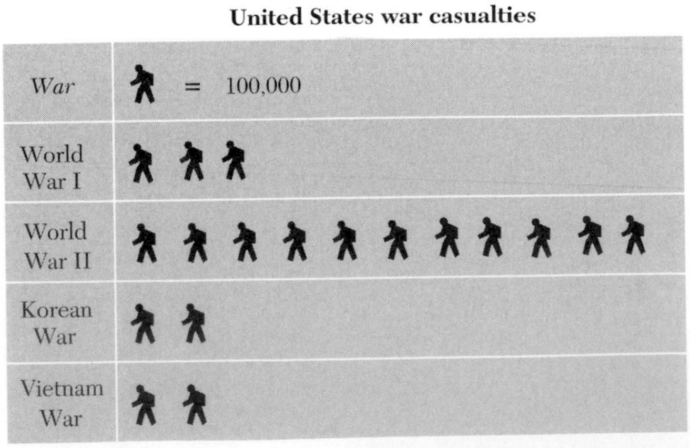

CHAPTER 1 Review Exercises

Section 1.1

1. four hundred fifty-one thousand, eight hundred ten
3. 90,337 **5.** < **7.** <
9. 183,660, 183,700, 184,000, 180,000
11. There are 74 more people in the under-15 age group.
13. There are 383 more people in the under-15 age group.

Section 1.2

15. 400; 477 **17.** 1700; 1764 **19.** 500; 431
21. 1000; 1492 **23.** 102 in.

Section 1.3

25. 12,000; 11,774 **27.** 80,000; 76,779
29. 605 square cm

Section 1.4

31. 29; 32 **33.** 11,000; 10,717 R 29

Section 1.5

35. 1331 **37.** 18,000 **39.** 516,000,000
41. $1,600,000,000

Section 1.6

43. 1 **45.** 41

Section 1.7

47. 180, 159, no mode **49.** 4388, 3572, 8312

Section 1.8

51. The highest temperature is recorded at 3 P.M. and the lowest at 4 A.M.

53.

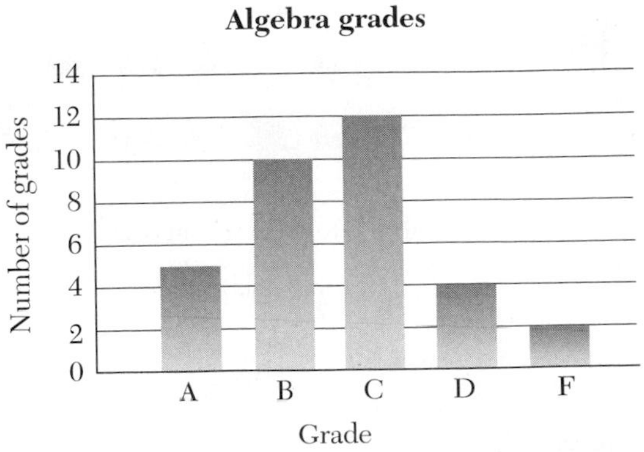

CHAPTER 1 True/False Concept Review

1. false; to write a billion takes 10 digits. **2.** true
3. true **4.** false; seven is less than 23.
5. false; 2567 > 2566 **6.** true **7.** true **8.** true
9. false; the sum is 87. **10.** true **11.** true
12. false; the product is 45. **13.** true **14.** true

15. true **16.** false; a number multiplied by 0 is 0.
17. true **18.** false; the quotient is 26. **19.** true
20. true **21.** false; division by zero is undefined.
22. false; the value is 49. **23.** true **24.** true
25. false; the product is 450,000. **26.** true
27. false; in $(3 + 1)^2$, the addition is done first.
28. false; in $(11 - 7) \cdot 9$, subtraction is done first.
29. true **30.** true **31.** true **32.** true
33. false; the median of 34, 54, 14, 44, 67, 81, and 90 is 54.
34. true

CHAPTER 1 Test

1. 124 **2.** 2765 **3.** 13 **4.** 25,286 **5.** >
6. 8,300,000 **7.** 362,956 **8.** 560,072 **9.** 1412
10. 138,300 **11.** 20,000 **12.** 200,000
13. 729 **14.** 6839 **15.** 92 ft **16.** 21,000
17. 8256 **18.** three thousand, six **19.** 49
20. 273,212 **21.** 806 **22.** 675,940,000
23. 1500; 1512 R 23 **24.** 29 **25.** 204 square cm
26. 514 **27.** average, 579; median, 633; mode, 674
28. It will take the secretary 140 minutes, or 2 hours and 20 minutes, to type 15 pages.
29. Each person won $3,146,500. Each person will receive $157,325 per year.
30. Chevrolet **31.** 700 **32.** 100
33. Divisions A and B have the same number and the most employees.
34. There are 225 more employees in Division A.
35. There are a total of 1380 employees.

CHAPTER 2

SECTION 2.1

1. no **3.** yes **5.** yes **7.** no **9.** yes
11. no **13.** yes **15.** yes **17.** no
19. 2, 3, 5 **21.** 2 **23.** 3, 5 **25.** 3, 5
27. 3, 5 **29.** yes **31.** no **33.** yes **35.** yes
37. no **39.** yes **41.** no **43.** no **45.** yes
47. 6, 10 **49.** 9 **51.** none **53.** 6, 10
55. 6, 9, 10 **57.** Yes, 1221 is divisible by 3.
59. The band can march in rows of 5 but not rows of 3 or 10 with the same number in each row.
61. Lucia can divide the class into groups of 5 or 6 because 120 is divisible by 5 and 6. Groups of 9 are not possible because 120 is not divisible by 9.
63. Yes, each person will receive 96 recordings.
65. There are 15 animals in the pen.
67. There is a total of 24 elephants and riders in the act.
69. A number is divisible by 5 when the remainder after dividing by 5 is 0. For example, 115 is divisible by 5 because $115 \div 5 = 23$ R 0. However, 116 is not divisible by 5 because $116 \div 5 = 23$ R 1.
71. In the divisibility test for 2, we look at the last digit. If the last digit is even, then the number is divisible by 2. In the divisibility test for 3, we look at the sum of the digits. If the sum is divisible by 3, then the number is divisible by 3.
73. yes **75.** yes; a number divisible by both 3 and 10 is divisible by 30. **81.** 5,060,000; 5,100,000
83. 21 **85.** 196 cm^2 **87.** 13, 26, 39, 52, 65, and 78
89. 1231, 2462, 3693, 4924, 6155, and 7386

SECTION 2.2

1. 3, 6, 9, 12, 15 **3.** 17, 34, 51, 68, 85
5. 12, 24, 36, 48, 60 **7.** 25, 50, 75, 100, 125
9. 50, 100, 150, 200, 250 **11.** 56, 112, 168, 224, 280
13. 68, 136, 204, 272, 340 **15.** 83, 166, 249, 332, 415
17. 157, 314, 471, 628, 785
19. 345, 690, 1035, 1380, 1725 **21.** yes **23.** yes
25. no **27.** no **29.** no **31.** yes **33.** yes
35. yes **37.** yes **39.** multiple of 6 and 9
41. multiple of 6 and 15 **43.** multiple of 6 and 15
45. multiple of 13 **47.** multiple of 13 and 19
49. No, Jean's car will not be selected because 16 is not a multiple of 5.
51. The students should work problems 4, 8, 12, 16, 20, 24, 28, 32, 36, 40, 44, 48, and 50.
53. Because goats have 4 feet, the number of goats' feet in the pen must be a multiple of 4. But 26 is not a multiple of 4, so Katy counted incorrectly.
55. The team should check bottles with the numbers 510, 525, 540, 555, 570, 585, 600, 615, 630, 645, 660, 675, and 690.
57. Yes, the gear is in its original position because 260 is a multiple of 20.
59. Yes, the number of in-line skating injuries is a multiple of the number of skateboard injuries. There were approximately three times as many injuries from in-line skating as from skateboarding.
61. Using 2, 3, 4, and 10 with the Pythagorean triple 3, 4, 5, we have 6, 8, 10; 9, 12, 15; 12, 16, 20; and 30, 40, 50. Verification: $6^2 + 8^2 = 10^2$ $(36 + 64 = 100)$; $9^2 + 12^2 = 15^2$ $(81 + 144 = 225)$; $12^2 + 16^2 = 20^2$ $(144 + 256 = 400)$; and $30^2 + 40^2 = 50^2$ $(900 + 1600 = 2500)$.
63. Using 2, 3, and 10 with the Pythagorean triple 5, 12, 13, we have 10, 24, 26; 15, 36, 39; and 50, 120, 130. Verification: $10^2 + 24^2 = 26^2$ $(100 + 576 = 676)$; $15^2 + 36^2 = 39^2$ $(225 + 1296 = 1521)$; $50^2 + 120^2 = 130^2$ $(2500 + 14{,}400 = 16{,}900)$.
65. The different words help us to describe the process we are using. Saying "9 is a factor of 135" means that 9 times some whole number is 135. The process is multiplication. The statement "9 is a divisor of 135" means that 9 divides 135 evenly. The process is division.
67. yes **69.** 5814 **73.** 1, 2, 3, 4, 6, 12
75. 1, 3, 5, 15 **77.** 1, 17 **79.** 6 **81.** 17
83. No, 12 is divisible by both 3 and 6 but not by 18.

SECTION 2.3

1. $1 \cdot 16, 2 \cdot 8, 4 \cdot 4$ **3.** $1 \cdot 19$ **5.** $1 \cdot 20, 2 \cdot 10, 4 \cdot 5$
7. $1 \cdot 46, 2 \cdot 23$ **9.** $1 \cdot 49, 7 \cdot 7$ **11.** $1 \cdot 95, 5 \cdot 19$

13. $1 \cdot 100, 2 \cdot 50, 4 \cdot 25, 5 \cdot 20, 10 \cdot 10$
15. $1 \cdot 112, 2 \cdot 56, 4 \cdot 28, 7 \cdot 16, 8 \cdot 14$
17. $1 \cdot 115, 5 \cdot 23$
19. $1 \cdot 444, 2 \cdot 222, 3 \cdot 148, 4 \cdot 111, 6 \cdot 74, 12 \cdot 37$
21. 1, 2, 4, 8, 16 **23.** 1, 23
25. 1, 2, 3, 5, 6, 10, 15, 30 **27.** 1, 3, 17, 51
29. 1, 5, 13, 65
31. 1, 2, 3, 4, 6, 8, 9, 12, 18, 24, 36, 72
33. 1, 2, 3, 4, 6, 7, 12, 14, 21, 28, 42, 84
35. 1, 2, 3, 6, 17, 34, 51, 102 **37.** 1, 2, 61, 122
39. 1, 2, 3, 6, 9, 18, 27, 54, 81, 162 **41.** 8 **43.** 2
45. 15 **47.** 24 **49.** 26 **51.** 8 **53.** 6
55. 14 **57.** 16 **59.** 4 **61.** $1 \cdot 445, 5 \cdot 89$
63. $1 \cdot 652, 2 \cdot 326, 4 \cdot 163$
65. $1 \cdot 680, 2 \cdot 340, 4 \cdot 170, 5 \cdot 136, 8 \cdot 85, 10 \cdot 68, 17 \cdot 40, 20 \cdot 34$

67.

Number of Programs	Length of Each
1	120 min
2	60 min
3	40 min
4	30 min
5	24 min
6	20 min
8	15 min
10	12 min
12	10 min
15	8 min
20	6 min
24	5 min
30	4 min
40	3 min
60	2 min
120	1 min

69. To provide 1 adult for every 4 infants, the school needs 6 adults. To provide 1 adult for every 3 infants, it needs 8 adults.
71. The band can make rectangles of 2 by 25 or 10 by 5.
73. We know that the number of band members is not a multiple of 4 or 5. In fact, the number of members is 1 more than a multiple of 4 and also 1 more than a multiple of 5. The number of members must be in the list 5, 9, 13, 17, 21, 25, 29, 33, ... It must also be in the list 6, 11, 16, 21, 26, 31, ... The smallest number in both lists is 21.
75. The divisors of 6 are 1, 2, 3, and 6. The sum $1 + 2 + 3 = 6$. Therefore, 6 is a perfect number.
77. The divisors of 28 are 1, 2, 4, 7, 14, and 28. Because $1 + 2 + 4 + 7 + 14 = 28$, 28 is a perfect number.
79. The difference between a factor and a divisor is the operation we use. If we are multiplying, we generally use the word *factor*. If we are dividing, we generally use the word *divisor*.
81. The largest factor is 991. **85.** 2499
87. 38,016 **89.** 33 **91.** 27 R 3
93. There is enough wire to wire 41 speakers, with 16 feet left over.

SECTION 2.4

1. composite **3.** prime **5.** composite
7. prime **9.** composite **11.** prime
13. composite **15.** prime **17.** composite
19. prime **21.** composite **23.** composite
25. prime **27.** composite **29.** prime
31. composite **33.** prime **35.** prime
37. composite **39.** prime **41.** composite
43. composite **45.** composite **47.** prime
49. prime **51.** composite **53.** composite
55. The last prime year in the 20th century was 1999. The first prime year in the 21st century is 2003.
57. 1903 is not a prime number.
59. The number of impulses was 59.
61. Only one arrangement is possible because 19 is a prime number. **63.** $M_3 = 2^3 - 1 = 7$; $M_5 = 2^5 - 1 = 31$; and $M_7 = 2^7 - 1 = 127$
65. $M_3 \times (2^{3-1}) = 7 \times (2^2) = 7 \times 4 = 28$
67. A prime number has only two factors: itself and 1. The number 7 is prime. A composite number has three or more factors: itself, 1, and at least one other factor. The number 9 is composite because it has three factors: 1, 3, and 9. A composite number is composed of a number or numbers other than itself and 1.
69. There is no limit to the number of factors a composite number can have. If you start multiplying counting numbers together, no matter where you stop, the product will be a composite number (with a lot of factors!).
71. composite **75.** Yes; the final quotient is 5.
77. Yes; the final quotient is 29.
79. No, the quotient of 1029 and 3 is 343, which is not divisible by 3. So the final quotient is 343.
81. No, the quotient of 2880 and 5 is 576, which is not divisible by 5. So the final quotient is 576.
83. Yes; the final quotient is 11.

SECTION 2.5

1. $2^2 \cdot 3$ **3.** $2 \cdot 3^2$ **5.** $2 \cdot 11$ **7.** 5^2 **9.** $2^2 \cdot 7$
11. 2^5 **13.** $2^3 \cdot 5$ **15.** $2^2 \cdot 17$ **17.** $2^2 \cdot 19$
19. $2^3 \cdot 11$ **21.** $7 \cdot 13$ **23.** $2 \cdot 3 \cdot 17$ **25.** $3 \cdot 5 \cdot 7$
27. prime **29.** $2 \cdot 7 \cdot 11$ **31.** $2 \cdot 3^4$ **33.** $2^2 \cdot 3^2 \cdot 5$
35. $2 \cdot 97$ **37.** $2 \cdot 3 \cdot 53$ **39.** $2^2 \cdot 3^2 \cdot 11$ **41.** 8
43. 15 **45.** 4 **47.** 1 **49.** 8 **51.** 23
53. 6 **55.** 14 **57.** 70 **59.** 29 **61.** $3 \cdot 101$
63. $17 \cdot 19$ **65.** $2^3 \cdot 3 \cdot 19$ **67.** prime **69.** 5^4
71. $2 \cdot 3 \cdot 7 \cdot 29$ **73.** Answers will vary.
75. Answers will vary.
77. The numbers 119 and 143 are relatively prime because they have no common factors, as we can see from the prime factorization of each: $119 = 7 \cdot 17$ and $143 = 11 \cdot 13$.
79. The number 97 is relatively prime to 180.
81. A number is written in prime-factored form when (1) it is a prime number or (2) it is written as a product and every factor in the product is a prime number. For example, $18 = 2 \cdot 3^2$ is in prime-factored form, whereas $18 = 3 \cdot 6$ and $18 = 2 \cdot 9$ are not. **83.** $2 \cdot 3 \cdot 5 \cdot 73$ **85.** $2^8 \cdot 7$

87. The perimeter is 56 inches. **89.** yes
91. no **93.** yes
95. Answers will vary but may include 15, 30, 45, and 60.

SECTION 2.6

1. 10 **3.** 9 **5.** 14 **7.** 14 **9.** 18 **11.** 36
13. 12 **15.** 36 **17.** 24 **19.** 10 **21.** 60
23. 48 **25.** 40 **27.** 36 **29.** 72 **31.** 48
33. 180 **35.** 60 **37.** 48 **39.** 120 **41.** 72
43. 204 **45.** 560 **47.** 24 **49.** 240
51. The two gears will return to their original positions in one turn of the 36-tooth gear because 9 is a divisor of 36.
53. The two gears will return to their original positions in two turns of the 36-tooth gear.
55. The least price they could pay is $100, which is the LCM of 20, 50, and 5.
57. To find the LCM of 20, 24, and 45, first find the prime factorization of each number. Then build the LCM by using each different prime factor to the highest power that occurs in any single number. **59.** 720 **61.** 1296
65. 1, 2, 4, 5, 10, 20, 23, 46, 92, 115, 230, 460
67. 1, 2, 4, 5, 7, 8, 10, 14, 16, 20, 28, 35, 40, 56, 70, 80, 112, 140, 280, 560 **69.** yes **71.** no
73. The average attendance was 1350 people.

CHAPTER 2 Review Exercises
Section 2.1

1. 6, 36, 636 **3.** 15, 255, 525 **5.** 36, 63
7. 444, 666, 888 **9.** 645 **11.** 7, 9 **13.** 6, 7, 9
15. 4, 5, 10 **17.** 6, 15
19. Yes, the profit can be evenly divided because 2060 is divisible by 5.

Section 2.2

21. 8, 16, 24, 32, 40 **23.** 51, 102, 153, 204, 255
25. 85, 170, 255, 340, 425 **27.** 122, 244, 366, 488, 610
29. 141, 282, 423, 564, 705 **31.** multiple of 6 and 9
33. multiple of 6 and 15 **35.** multiple of 15
37. multiple of 12 and 16
39. Yes, the prediction is accurate because 605 is the 11th multiple of 55.

Section 2.3

41. $1 \cdot 15, 3 \cdot 5$ **43.** $1 \cdot 38, 2 \cdot 19$
45. $1 \cdot 236, 2 \cdot 118, 4 \cdot 59$ **47.** 1, 3, 11, 33
49. 1, 2, 3, 6, 7, 14, 21, 42 **51.** 1, 2, 3, 6, 13, 26, 39, 78
53. 3 **55.** 21 **57.** 15
59. The possible energy costs are $1590 for 1 person, $795 for 2 people, $530 for 3 people, $318 for 5 people, $265 for 6 people, $159 for 10 people.

Section 2.4

61. prime **63.** prime **65.** composite
67. prime **69.** composite **71.** prime
73. prime **75.** composite **77.** prime
79. The prime number year after 2003 is 2011.

Section 2.5

81. 3^3 **83.** $2 \cdot 19$ **85.** $2^2 \cdot 13$ **87.** $2 \cdot 3 \cdot 37$
89. $2^2 \cdot 3^2 \cdot 7$ **91.** 6 **93.** 30 **95.** 8 **97.** 5
99. $2003 = 2003$; $2004 = 2^2 \cdot 3 \cdot 167$; $2005 = 5 \cdot 401$; and so forth

Section 2.6

101. 24 **103.** 60 **105.** 144 **107.** 100
109. 210 **111.** 24 **113.** 72 **115.** 180
117. 840
119. Possible answers include 2 and 49; 14 and 49; 2 and 98; and 14 and 98. Other answers are possible.

CHAPTER 2 True/False Concept Review

1. false; not all multiples of 3 end with the digit 3. For example, 15 is a multiple of 3.
2. true **3.** true **4.** true **5.** true **6.** true
7. false; only one multiple of 300 is also a factor of 300—itself.
8. false; the square of 25 is 625. Twice 25 is 50.
9. false; not all natural numbers ending in 6 are divisible by 6. For example, 16 is not divisible by 6. **10.** true
11. false; not all natural numbers ending in 9 are divisible by 3. For example, 29 is not divisible by 3.
12. false; the GCF of 6 and 8 is 2.
13. false; the GCF of three numbers is a divisor of each of the numbers. **14.** true **15.** true
16. false; all prime numbers except 2 are odd.
17. true
18. false; every prime number has exactly two factors.
19. true **20.** true
21. false; the least common multiple (LCM) of three different prime numbers is the product of the three numbers.
22. false; the greatest common factor (GCF) of three different prime numbers is 1.
23. false; the largest divisor of the least common multiple (LCM) of three numbers is not necessarily the largest of the three numbers. For example, the LCM of 2, 4, and 6 is 12. The largest divisor of 12 is 12, not 6.
24. true **25.** true

CHAPTER 2 Test

1. yes **2.** 1, 2, 4, 7, 8, 14, 16, 28, 56, 112 **3.** yes
4. 4 **5.** 96 **6.** $1 \cdot 75, 3 \cdot 25, 5 \cdot 15$ **7.** $2^3 \cdot 5 \cdot 7$
8. 252 **9.** 6 **10.** 208, 221, 234, 247 **11.** no
12. prime **13.** composite **14.** $5 \cdot 11^2$
15. 504 **16.** 2 **17.** 299 **18.** 360
19. No, a set of prime factors has only one product.
20. Any two of the following sets of numbers: 2, 3, 7; or 2, 3, 14; or 2, 3, 21; or 2, 3, 42; or 2, 7, 42; or 3, 7, 14; or 6, 7, 42; or 14, 21, 42

CHAPTER 3

SECTION 3.1

1. $\frac{5}{7}$ **3.** $\frac{4}{7}$ **5.** $\frac{4}{5}$ **7.** $\frac{4}{3}$ **9.** $\frac{8}{5}$ **11.** $\frac{13}{9}$

13. Proper fractions: $\frac{3}{7}, \frac{4}{7}, \frac{5}{7}, \frac{6}{7}$; improper fractions: $\frac{7}{7}, \frac{8}{7}, \frac{9}{7}$

15. Proper fractions: none; improper fractions: $\frac{17}{16}, \frac{18}{17}, \frac{29}{19}, \frac{30}{21}, \frac{23}{23}$

17. Proper fractions: $\frac{10}{11}, \frac{3}{5}$; improper fractions: $\frac{7}{4}, \frac{13}{13}, \frac{20}{19}$

19. Proper fractions: $\frac{9}{10}, \frac{102}{103}$; improper fractions: $\frac{5}{5}, \frac{18}{18}, \frac{147}{147}$ **21.** $42\frac{3}{5}$ **23.** $30\frac{1}{7}$ **25.** $12\frac{4}{9}$

27. $7\frac{7}{13}$ **29.** $14\frac{21}{22}$ **31.** $35\frac{1}{12}$ **33.** $\frac{41}{9}$

35. $\frac{12}{1}$ **37.** $\frac{31}{4}$ **39.** $\frac{89}{3}$ **41.** $\frac{76}{5}$ **43.** $\frac{155}{8}$

45. The fraction $\frac{0}{\boxed{1}}$ is a proper fraction. Any whole number except 0 can be put into the denominator.

47. The error occurred because both numerator and denominator were multiplied by 2: $2\frac{2}{5} = \frac{12}{5}$.

49. $\frac{7}{10}$ **51.** $\frac{11}{8}$

53. 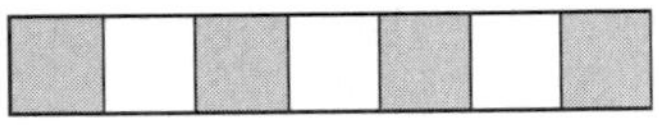

55. Of the class, $\frac{13}{21}$ are women.

57. The mixed numbers $11\frac{1}{2}$ and $6\frac{1}{8}$ can be written as improper fractions because they represent numbers larger than 1. The dimensions of nonstandard mail are longer than $\frac{23}{2}$ in., taller than $\frac{49}{8}$ in., and/or thicker than $\frac{1}{4}$ in.

59. The tank is $\frac{3}{8}$ full.

61. The number 4 is closest to the mark.

63. The company needs 2495 sections.

65. Seven in. is $\frac{7}{12}$ of a foot, so 5′ 7″ is $5\frac{7}{12}$ ft.

67. The cup contains $\frac{5}{8}$ cup of oil.

69. She can use four $\frac{1}{2}$-c measures of sugar.

71. To change $\frac{34}{5}$ to a mixed number, divide 34 by 5. Write the remainder, 4, over the divisor, 5, to form the fractional part. So, $\frac{34}{5} = 6\frac{4}{5}$. To change $7\frac{3}{8}$ to an improper fraction, multiply the denominator, 8, times the whole number 7. Add the product to 3 to get 59. Write this sum over the denominator. So, $7\frac{3}{8} = \frac{59}{8}$.

73. A mixed number has a value of 1 or more. A proper fraction has a value less than 1.

75. $\frac{144}{9}; \frac{2304}{144}$

77. There will be 87 pieces of rope, 5 pieces for each child. There will be 12 pieces left over.

83. 39 **85.** 15 **87.** composite **89.** $3^2 \cdot 31$

91. Bonnie averaged 201 miles per tank and 50 miles per gallon.

SECTION 3.2

1. $\frac{1}{2}$ **3.** $\frac{2}{3}$ **5.** $\frac{2}{5}$ **7.** $\frac{4}{5}$ **9.** $\frac{3}{4}$ **11.** $\frac{3}{5}$

13. $\frac{4}{5}$ **15.** $\frac{5}{3}$ **17.** $\frac{7}{9}$ **19.** $\frac{4}{5}$ **21.** 13

23. $\frac{8}{3}$ **25.** $\frac{1}{3}$ **27.** $\frac{1}{3}$ **29.** $\frac{3}{4}$ **31.** $\frac{29}{36}$

33. $\frac{2}{3}$ **35.** $\frac{11}{15}$ **37.** $\frac{3}{4}$ **39.** $\frac{9}{16}$ **41.** $\frac{3}{4}$

43. 2 **45.** $\frac{5}{7}$ **47.** $\frac{16}{21}$ **49.** $\frac{3}{4}$ **51.** $\frac{26}{33}$

53. $\frac{5}{6}$ **55.** $\frac{4}{5}$ **57.** $\frac{4}{5}$

59. Four inches is $\frac{1}{3}$ of a foot, so 5′ 4″ is $5\frac{1}{3}$ feet.

61. Toronto's goalie saved $\frac{15}{16}$ of the shots on goal. The opposing team scored 2 points.

63. $\frac{2}{3}$ of the plugs are fouled.

65. She has $\frac{1}{5}$ of her shift left.

67. The DA prosecutes $\frac{11}{16}$ of his cases successfully.

69. Of the elk, $\frac{17}{26}$ are cows. **71.** $\frac{7}{36}$

73. Find the prime factorization of both numerator and denominator. Then eliminate the common factors. The product of the remaining factors is the simplified fraction: $\frac{\cancel{3} \cdot \cancel{5} \cdot \cancel{5} \cdot 7}{\cancel{3} \cdot 3 \cdot \cancel{5} \cdot \cancel{5} \cdot 5} = \frac{7}{15}$.

75. Yes, all simplify to $\frac{5}{12}$. **79.** 616 **81.** 990

83. 3552 **85.** 18 **87.** 16

SECTION 3.3

1. $\frac{3}{16}$ **3.** $\frac{15}{28}$ **5.** $\frac{7}{4}$ or $1\frac{3}{4}$ **7.** $\frac{1}{5}$ **9.** $\frac{1}{4}$

11. $\frac{1}{3}$ **13.** $\frac{3}{2}$ **15.** 1 **17.** $\frac{5}{9}$ **19.** 0

21. $\frac{10}{3}$ **23.** $\frac{1}{8}$ **25.** none **27.** $\frac{14}{33}$ **29.** 8

31. $\frac{27}{28}$ **33.** $\frac{2}{3}$ **35.** 3 **37.** $\frac{9}{25}$ **39.** 1

41. $\frac{64}{15}$ or $4\frac{4}{15}$ **43.** $\frac{3}{4}$ **45.** $\frac{3}{20}$ **47.** $\frac{10}{11}$

49. 1 **51.** $\frac{7}{12}$ **53.** $\frac{3}{10}$ **55.** $\frac{9}{4}$ or $2\frac{1}{4}$

57. $\frac{8}{3}$ or $2\frac{2}{3}$

59. $\frac{2}{3}\cdot\frac{\boxed{5}}{16} = \frac{5}{24}$ because $\frac{2}{3}\cdot\frac{\boxed{5}}{16} = \frac{5}{24}$ can be simplified to $\frac{1}{3}\cdot\frac{\boxed{5}}{8}$; then we multiply 1 times 5 to get 5.

61. The error comes from inverting $\frac{4}{5}$ before multiplying. We invert only in division: $\frac{3}{8}\cdot\frac{4}{5} = \frac{12}{40} = \frac{3}{10}$.

63. There are 8 combinations of four children with a boy as firstborn, and 8 combinations with a girl as firstborn.

65. One combination has a firstborn boy followed by three girls.

67. When it is three-fourths full, the container holds $\frac{21}{20}$, or $1\frac{1}{20}$, gallons.

69. Becky might save 41 gallons per year.

71. **a.** Hannah needs 140 cookies. **b.** Six recipes will make 144 cookies. **c.** She needs 3 tsp of salt. **d.** She needs $4\frac{1}{2}$ c of brown sugar.

73. **a.** Sally would have to make $\frac{3}{8}$ of the recipe **b.** Sally needs $1\frac{1}{8}$ lb of cheese.

75. Multiply $\frac{35}{24}$ by $\frac{14}{40}$ (the reciprocal of $\frac{40}{14}$). Then simplify: $\frac{35}{24} \div \frac{40}{14} = \frac{49}{96}$. **77.** $\frac{9}{10}$ **83.** $3\frac{3}{5}$ **85.** $1\frac{13}{16}$

87. $\frac{9}{2}$ **89.** $\frac{16}{5}$ **91.** Kevin can haul 15 scoops.

SECTION 3.4

1. $\frac{5}{6}$ **3.** $2\frac{5}{8}$ **5.** 8 **7.** $22\frac{1}{2}$ **9.** 24

11. $4\frac{1}{7}$ **13.** $5\frac{11}{16}$ **15.** 0 **17.** $37\frac{1}{2}$

19. $16\frac{1}{2}$ **21.** $35\frac{1}{5}$ **23.** 105 **25.** $4\frac{1}{2}$

27. $\frac{3}{4}$ **29.** 2 **31.** $\frac{13}{18}$ **33.** $1\frac{1}{3}$ **35.** $\frac{2}{3}$

37. $\frac{7}{30}$ **39.** $1\frac{1}{4}$ **41.** $14\frac{1}{2}$ **43.** $8\frac{1}{28}$

45. $17\frac{7}{16}$ **47.** $4\frac{2}{3}$ **49.** 186 **51.** $\frac{1}{2}$

53. Mixed numbers cannot be multiplied by multiplying the whole number parts and the fractional parts separately. Change them to fractions first: $\frac{5}{3}\cdot\frac{3}{2} = \frac{5}{2} = 2\frac{1}{2}$.

55. The interior of the shed has $6253\frac{1}{2}$ in.^{2} of space.

57. 1115 steelhead were counted.

59. The iron content is 22 parts per million.

61. Karla can make 192 glasses of juice.

63. The less-efficient car emits 3000 lb of CO_2.

65. **a.** Hannah needs $22\frac{1}{2}$ c of flour. **b.** She needs 15 c of sugar.

67. **a.** Quan can make $\frac{2}{3}$ of the recipe. **b.** Quan should use 4 tbs of butter. **c.** Quan should use 1 tsp of vanilla.

69. When we divide by $1\frac{1}{2}$, we are dividing by $\frac{3}{2}$. Dividing by $\frac{3}{2}$ is the same as multiplying by $\frac{2}{3}$. Because $\frac{2}{3}$ is less than 1, the multiplication gives us a smaller number.

71. 1230 **75.** $8\frac{1}{8}$ **77.** $1\frac{2}{15}$ **79.** $13\frac{1}{5}$

81. 88 **83.** No, because 234,572 is not a multiple of 3.

GETTING READY FOR ALGEBRA

1. $x = \frac{3}{8}$ **3.** $y = \frac{10}{9}$ or $y = 1\frac{1}{9}$ **5.** $z = \frac{15}{16}$

7. $x = \frac{17}{8}$ or $x = 2\frac{1}{8}$ **9.** $a = \frac{1}{4}$

11. $b = \frac{3}{2}$ or $b = 1\frac{1}{2}$ **13.** $z = 2$ **15.** $a = \frac{11}{30}$

17. The distance is $\frac{3}{2}$ m or $1\frac{1}{2}$ m.

19. 180 lb of tin were recycled.

SECTION 3.5

1. $\frac{6}{8}, \frac{9}{12}, \frac{12}{16}, \frac{15}{20}$ **3.** $\frac{14}{16}, \frac{21}{24}, \frac{28}{32}, \frac{35}{40}$

5. $\frac{8}{18}, \frac{12}{27}, \frac{16}{36}, \frac{20}{45}$ **7.** $\frac{18}{22}, \frac{27}{33}, \frac{36}{44}, \frac{45}{55}$

9. $\frac{14}{6}, \frac{21}{9}, \frac{28}{12}, \frac{35}{15}$ **11.** 8 **13.** 28 **15.** 28

17. 27 **19.** 15 **21.** 8 **23.** 38 **25.** 16

27. 57 **29.** 104 **31.** $\frac{3}{7}, \frac{4}{7}, \frac{5}{7}$ **33.** $\frac{3}{8}, \frac{1}{2}, \frac{3}{4}$

35. $\frac{1}{3}, \frac{3}{8}, \frac{1}{2}$ **37.** true **39.** false **41.** true

43. $\frac{2}{3}, \frac{3}{4}, \frac{4}{5}$ **45.** $\frac{17}{30}, \frac{3}{5}, \frac{2}{3}, \frac{14}{15}$ **47.** $\frac{11}{24}, \frac{17}{36}, \frac{35}{72}$

49. $\frac{3}{7}, \frac{13}{28}, \frac{17}{35}$ **51.** $3\frac{2}{3}, 3\frac{7}{9}, 3\frac{5}{6}$ **53.** true

55. true **57.** true **59.** LCM = 24; $\frac{12}{24}, \frac{16}{24}, \frac{4}{24}, \frac{15}{24}$

61. From smallest to largest, the strengths are $\frac{3}{32}, \frac{1}{8}, \frac{1}{4}, \frac{5}{16}, \frac{3}{8}, \frac{1}{2}$, and $\frac{9}{16}$.
63. More of the population is from the ethnic group Fula.
65. African Americans are a larger portion of the state's population. **67.** Chang's measurement is heaviest.
69. a. Pecan chocolate chip cookies give them the bigger profit. **b.** Three-fourths of a cup is more than she needs.
71. To simplify a fraction is to find a fraction with a smaller numerator and smaller denominator that is equivalent to the original fraction. To build a fraction is to find a fraction with a larger numerator and larger denominator that is equivalent to the original.
73. $\frac{50}{70}, \frac{65}{91}, \frac{115}{161}, \frac{560}{784}, \frac{2905}{4067}$ **77.** 8 **79.** 30
81. 32 **83.** $2^3 \cdot 3^2$
85. The pharmacist should give her 56 capsules.

SECTION 3.6

1. $\frac{6}{7}$ **3.** $\frac{1}{2}$ **5.** 2 **7.** $\frac{4}{5}$ **9.** $\frac{10}{13}$ **11.** $1\frac{1}{4}$

13. $\frac{1}{2}$ **15.** $\frac{5}{16}$ **17.** $\frac{3}{5}$ **19.** $\frac{3}{4}$ **21.** $\frac{23}{24}$

23. $\frac{5}{12}$ **25.** $\frac{13}{16}$ **27.** $\frac{19}{30}$ **29.** $\frac{4}{5}$ **31.** $\frac{14}{15}$

33. $1\frac{1}{12}$ **35.** $1\frac{2}{5}$ **37.** $1\frac{5}{16}$ **39.** $2\frac{5}{8}$

41. $2\frac{5}{81}$ **43.** $2\frac{19}{90}$ **45.** $\frac{3}{5}$ **47.** $\frac{7}{15}$

49. $1\frac{13}{144}$

51. Contributions and tuition and fees were $\frac{149}{200}$ of the total revenues for the school.
53. Asia and Africa contain more than $\frac{19}{20}$ of the world's Muslims.
55. When adding fractions, the parts of the unit must be the same size. So we must keep the common denominator in the result and not add the denominators: $\frac{1}{5} + \frac{2}{5} = \frac{3}{5}$.

57. The recipe makes $3\frac{3}{4}$ gal of punch.

59. Jonnie needs a bolt that is $1\frac{13}{16}$ in. long.

61. The length of the pin is $\frac{3}{4}$ in.

63. The length of the rod is $3\frac{3}{8}$ in.

65. a. The three recipes require $1\frac{1}{4}$ tsp of baking soda.

b. The order requires $3\frac{3}{4}$ tsp of baking soda.

c. The three recipes require 2 tsp of salt.
d. The order requires 3 tsp of salt.
67. Each denominator indicates how many parts a unit has been divided into. Unless the units are the same size, it would be like adding apples and oranges. You cannot add units of different sizes.

69. $\frac{4}{7}$

71. Jim consumes 4 g of fat, which is $\frac{6}{55}$ of the number of calories.

73. $10\frac{5}{12}$ **75.** $19\frac{1}{15}$ **77.** $6\frac{11}{16}$ **79.** 78

81. The average class size is 40.

SECTION 3.7

1. $3\frac{4}{5}$ **3.** 8 **5.** $8\frac{1}{6}$ **7.** $6\frac{2}{9}$ **9.** $6\frac{1}{14}$

11. $15\frac{2}{15}$ **13.** $15\frac{1}{16}$ **15.** 11 **17.** $26\frac{5}{8}$

19. $15\frac{11}{24}$ **21.** $10\frac{1}{6}$ **23.** $22\frac{1}{80}$ **25.** $49\frac{5}{6}$

27. $719\frac{31}{36}$ **29.** $86\frac{11}{30}$ **31.** $55\frac{11}{12}$ **33.** $118\frac{1}{45}$

35. $103\frac{11}{70}$ **37.** $50\frac{3}{5}$ **39.** $12\frac{14}{15}$ **41.** $64\frac{3}{8}$

43. Nancy needs $98\frac{1}{2}$ in. of molding for the frame.

45. Dave's driveway is $267\frac{3}{4}$ in. long.

47. Michelle worked $23\frac{3}{4}$ hours.

49. The perimeter is $43\frac{1}{2}$ in.

51. The total rainfall for the three cities is $92\frac{5}{24}$ in.

53. $61\frac{27}{80}$ mi of road need to be resurfaced. It will cost about $920,000.

55. a. One recipe will make $\frac{6}{5}$, or $1\frac{1}{5}$, boxes.

b. The recipe will make $\frac{6}{5}$, or $1\frac{1}{5}$, boxes.

57. I prefer the methods of this section rather than changing to improper fractions first because (a) it avoids fractions with large numerators and (b) it requires fewer steps.

59. Yes, the statement is true. **63.** 84 **65.** 424

67. 15 **69.** $\frac{6}{11}$

71. It will take 186 lb of fresh prunes.

SECTION 3.8

1. $\frac{2}{7}$ **3.** $\frac{1}{4}$ **5.** $\frac{1}{3}$ **7.** $\frac{1}{2}$ **9.** $\frac{7}{16}$ **11.** $\frac{7}{45}$

13. $\frac{1}{6}$ **15.** $\frac{5}{18}$ **17.** $\frac{1}{4}$ **19.** $\frac{9}{20}$ **21.** $\frac{1}{24}$

23. $\frac{7}{48}$ **25.** $\frac{1}{21}$ **27.** $\frac{1}{16}$ **29.** $\frac{1}{18}$ **31.** $\frac{13}{24}$

33. $\frac{2}{5}$ **35.** $\frac{5}{36}$ **37.** $\frac{2}{75}$ **39.** $\frac{17}{48}$ **41.** $\frac{1}{36}$

43. $\frac{7}{30}$ **45.** $\frac{5}{12}$

47. The board will be $1\frac{1}{8}$ in. thick.

49. The relish has less fat by $\frac{1}{63}$ of a gram.

51. In that year, $\frac{12}{25}$ of the population did not hold stock.

53. In 1999, about $\frac{1}{10}$ of the total sales was neither CDs nor cassettes.

55. About $\frac{1}{4}$ of the weight lost on a reduced calorie diet is water.

57. The difference in diameters is $\frac{11}{16}$ in.

59. She can use $\frac{1}{2}$ c less 2 tbs.

61. They are not equal because the denominators (parts of the unit), 4 and 2, are not the same. We cannot get the result by subtracting the numerators and the denominators.

We must first find the common denominator. So, $\frac{3}{4} - \frac{1}{2} = \frac{3}{4} - \frac{2}{4} = \frac{1}{4}$.

63. $\frac{218}{625}$ **67.** 12,200 **69.** $6\frac{2}{5}$ **71.** $5\frac{3}{7}$

73. $27\frac{7}{10}$ **75.** It will take them 8 days to build the wall.

SECTION 3.9

1. $3\frac{2}{5}$ **3.** $2\frac{1}{6}$ **5.** $5\frac{1}{8}$ **7.** $6\frac{4}{7}$ **9.** $68\frac{1}{8}$

11. $2\frac{1}{2}$ **13.** $11\frac{3}{4}$ **15.** $155\frac{1}{9}$ **17.** $6\frac{11}{12}$

19. $15\frac{2}{5}$ **21.** $56\frac{1}{3}$ **23.** $25\frac{9}{20}$ **25.** $21\frac{3}{4}$

27. $15\frac{47}{48}$ **29.** $28\frac{1}{3}$ **31.** $2\frac{7}{32}$ **33.** $7\frac{11}{36}$

35. $74\frac{7}{8}$ **37.** $36\frac{17}{120}$ **39.** $27\frac{31}{36}$ **41.** $1\frac{3}{20}$

43. $\frac{5}{6}$

45. If 1 is borrowed from 16 in order to subtract the fractional part, $\frac{1}{4}$, then the whole-number part is $15 - 13$, or 2. So, $16 - 13\frac{1}{4} = (15 + \frac{4}{4}) - (13 + \frac{1}{4}) = 2\frac{3}{4}$.

47. She trims $1\frac{1}{2}$ lb. **49.** He has $18\frac{9}{20}$ tons left.

51. Nancy has $2\frac{1}{3}$ ft of the remnant left over.

53. Amber is charged for $12\frac{7}{8}$ extra pounds.

55. They have $17\frac{7}{10}$ mi to go.

57. The Williams family recycles $381\frac{3}{10}$ more pounds per year.

59. For a 6-year period, $77\frac{7}{8}$ more inches of rain fall on Hartville.

61. First change the fractions to equivalent fractions with a common denominator. Thus, $\frac{1}{3} = \frac{8}{24}$ and $\frac{5}{8} = \frac{15}{24}$. Because $\frac{15}{24}$ is the larger fraction, we must borrow 1 from the 4 and add it to $\frac{8}{24}$. We have $4\frac{8}{24} = 3 + 1\frac{8}{24} = 3\frac{32}{24}$. Now we can subtract: $3\frac{32}{24} - 2\frac{15}{24} = 1\frac{17}{24}$.

63. No, they are not equal.

65. The snail's net distance is $3\frac{2}{3}$ ft. It will take 6 days to gain over 20 ft.

67. 80 **69.** 28 **71.** 12 **73.** $\frac{8}{15}$

75. Karla averaged 28 miles per gallon.

GETTING READY FOR ALGEBRA

1. $a = \frac{1}{2}$ **3.** $c = \frac{5}{8}$ **5.** $x = \frac{11}{72}$ **7.** $y = 1\frac{38}{63}$

9. $a = 1\frac{11}{40}$ **11.** $c = 4\frac{1}{6}$ **13.** $x = 3\frac{5}{36}$

15. $3\frac{1}{6} = w$ **17.** $a = 36\frac{4}{9}$ **19.** $c = 21\frac{2}{21}$

21. The height 10 years ago was $42\frac{5}{8}$ ft.

23. She bought 46 lb of nails.

SECTION 3.10

1. $\frac{1}{3}$ **3.** $\frac{1}{15}$ **5.** $\frac{1}{2}$ **7.** $\frac{3}{4}$ **9.** 0 **11.** $\frac{1}{2}$

13. $\frac{11}{12}$ **15.** $\frac{3}{8}$ **17.** $\frac{7}{9}$ **19.** $\frac{5}{6}$ **21.** $\frac{5}{6}$

23. 0 **25.** $1\frac{13}{64}$ **27.** $\frac{1}{6}$ **29.** $\frac{1}{2}$ **31.** $\frac{2}{5}$

33. $\frac{4}{11}$ **35.** $\frac{13}{45}$ **37.** 4 **39.** $1\frac{1}{16}$ **41.** $\frac{3}{8}$

43. $3\frac{4}{9}$ **45.** $7\frac{1}{5}$ **47.** $\frac{5}{12}$ **49.** 1 **51.** $6\frac{1}{54}$

53. $4\frac{1}{12}$

55. The average length is $33\frac{2}{3}$ in.

57. The mothers need 55 cups of cereal, $1\frac{1}{4}$ cups of butter, $33\frac{3}{4}$ cups of chocolate chips, $27\frac{1}{2}$ cups of marshmallows, 40 cups of pretzels, and 10 cups of raisins.

59. The class average is $\frac{7}{10}$ of the problems correct.

61. There are $108\frac{1}{4}$ oz of seafood in the carton. The average cost is 48¢ per ounce.

63. a. The average amount is $2\frac{1}{12}$ c of flour.

b. The average amount is $1\frac{1}{6}$ c of sugar.

65. 1. Do operations in parentheses first, following steps 2, 3, and 4. **2.** Do exponents next. **3.** Do multiplication and division as they occur. **4.** Do addition and subtraction as they occur. The order is the same as for whole numbers.

67. $2\frac{57}{1000}$ **69.** The total amount paid is $19,444.

71. $6\frac{1}{9}$ **73.** $4\frac{4}{27}$ **75.** $\frac{1}{7}$

77. $2 \cdot 5^2 \cdot 13$ **79.** The tabletop is $1\frac{1}{4}$ in. thick.

CHAPTER 3 Review Exercises
Section 3.1

1. $\frac{3}{5}$ **3.** $\frac{9}{10}$ **5.** Proper fractions: $\frac{11}{12}, \frac{3}{20}$

7. $6\frac{5}{12}$ **9.** $114\frac{2}{3}$ **11.** $\frac{103}{12}$ **13.** $\frac{77}{6}$ **15.** $\frac{17}{1}$

17. The wholesaler can pack $2684\frac{19}{24}$ cases of beans.

Section 3.2

19. $\frac{5}{7}$ **21.** $\frac{4}{7}$ **23.** $\frac{3}{5}$ **25.** 17 **27.** $\frac{1}{3}$

29. $\frac{22}{23}$ **31.** $\frac{1}{5}$ **33.** $\frac{2}{3}$

35. He has $\frac{3}{5}$ of his shift left.

Section 3.3

37. $\frac{2}{25}$ **39.** $\frac{18}{55}$ **41.** 1 **43.** $\frac{8}{3}$ **45.** $\frac{3}{8}$

47. $\frac{3}{4}$ **49.** $\frac{21}{50}$

51. Rent accounts for $\frac{1}{7}$ of their income.

Section 3.4

53. $1\frac{5}{16}$ **55.** 12 **57.** $3\frac{1}{3}$ **59.** $18\frac{3}{4}$ **61.** 2

63. $1\frac{1}{8}$ **65.** $3\frac{3}{5}$ **67.** $28\frac{1}{5}$

69. She harvests $8362\frac{1}{2}$ bushels.

Section 3.5

71. $\frac{4}{6}, \frac{6}{9}, \frac{10}{15}, \frac{16}{24}$ **73.** $\frac{6}{28}, \frac{9}{42}, \frac{15}{70}, \frac{24}{112}$ **75.** 18

77. 120 **79.** $\frac{1}{2}, \frac{3}{5}, \frac{7}{10}$ **81.** $\frac{1}{5}, \frac{2}{9}, \frac{3}{11}$

83. $\frac{8}{25}, \frac{59}{100}, \frac{3}{5}, \frac{31}{50}$ **85.** true **87.** false

89. The smallest is $\frac{7}{16}$ ton and the largest is $\frac{3}{4}$ ton.

Section 3.6

91. $\frac{7}{11}$ **93.** $\frac{2}{3}$ **95.** $\frac{5}{8}$ **97.** $\frac{3}{5}$ **99.** $\frac{7}{15}$

101. $1\frac{5}{8}$ **103.** The bamboo grew $1\frac{1}{8}$ in.

Section 3.7

105. $6\frac{5}{14}$ **107.** $15\frac{13}{15}$ **109.** $13\frac{5}{24}$

111. $25\frac{43}{80}$ **113.** 80 **115.** $60\frac{7}{12}$

117. Russ's mileage was $11\frac{13}{24}$ miles.

Section 3.8

119. $\frac{1}{2}$ **121.** $\frac{1}{2}$ **123.** $\frac{13}{30}$ **125.** $\frac{1}{30}$

127. $\frac{19}{75}$ **129.** $\frac{3}{40}$ **131.** She has $\frac{5}{12}$ oz left.

Section 3.9

133. $118\frac{1}{6}$ **135.** $15\frac{1}{5}$ **137.** $1\frac{5}{8}$ **139.** $6\frac{7}{48}$

141. $6\frac{13}{24}$ **143.** $5\frac{25}{32}$

145. a. During the year, $3\frac{5}{8}$ in. more rain falls in Westport.

b. In a 10-yr period, $277\frac{1}{12}$ in. more rain falls in Salem.

Section 3.10

147. 1 **149.** $1\frac{7}{12}$ **151.** $\frac{1}{2}$ **153.** $1\frac{1}{64}$

155. $\frac{15}{32}$ **157.** $1\frac{13}{30}$

159. The class average was $\frac{7}{10}$ of the problems correct.

CHAPTER 3 True/False Concept Review

1. false; use more than one unit, with one partially shaded and the others totally shaded. **2.** true
3. false; the numerator always equals the denominator, so the fraction is improper. **4.** true **5.** true
6. false; every improper fraction can be changed to a mixed number or a whole number. **7.** true **8.** true
9. true **10.** true **11.** true **12.** true
13. false; unlike fractions have different denominators.
14. false; add the whole number and add the fractions.
15. true **16.** true **17.** true
18. false; the product of two fractions is sometimes smaller than either of the fractions.

CHAPTER 3 Test

1. $15\frac{1}{4}$ **2.** $1\frac{5}{24}$ **3.** $\frac{79}{9}$ **4.** $\frac{3}{8}, \frac{2}{5}, \frac{3}{7}$ **5.** $\frac{17}{1}$

6. 35 **7.** $9\frac{2}{15}$ **8.** $17\frac{23}{40}$ **9.** $\frac{3}{4}$ **10.** $\frac{2}{3}$

11. $6\frac{4}{5}$ **12.** $\frac{5}{8}$ **13.** $\frac{1}{2}$ **14.** $\frac{1}{3}$ **15.** $\frac{12}{35}$

16. $6\frac{13}{20}$ **17.** $\frac{3}{8}$ **18.** $\frac{5}{7}$ **19.** $\frac{8}{21}$ **20.** $3\frac{1}{5}$

21. $\frac{7}{8}, \frac{7}{9}, \frac{8}{9}$ **22.** $3\frac{3}{4}$ **23.** $8\frac{13}{40}$ **24.** $\frac{7}{12}$

25. $5\frac{5}{12}$ **26.** $\frac{5}{7}$ **27.** $1\frac{15}{16}$ **28.** $\frac{9}{50}$

29. false **30.** $\frac{5}{5}, \frac{6}{6}, \frac{7}{7}$

31. There are 22 truckloads of hay in the rail car.
32. She needs to make 25 pounds of candy.

CHAPTERS 1–3 Cumulative Review

1. six thousand, ninety-one **3.** 1,000,310
5. 654,800 **7.** > **9.** 134,199 **11.** 1593
13. 154 cm **15.** 24,893 **17.** 257 **19.** 7^8
21. 321,000 **23.** 32 **25.** 45,406, 39,445
27. Phil Mickelson had the lowest score at 64.
29. Week 4 had the highest sales.
31. Holli spent $3404 on redecorating.
33. The Portland Symphony generated $262,124.
35. 2 **37.** 2, 3 **39.** 140, 280, 420, 560, 700
41. no **43.** $1 \cdot 495, 3 \cdot 165, 5 \cdot 99, 9 \cdot 55, 11 \cdot 45, 15 \cdot 33$
45. 1, 3, 5, 11, 15, 33, 55, 165 **47.** 15 **49.** prime

51. $2 \cdot 3 \cdot 5^2 \cdot 79$ **53.** 210 **55.** $\frac{9}{5}$ **57.** $3\frac{3}{8}$

59. $\frac{43}{6}$ **61.** $\frac{11}{17}$ **63.** $\frac{3}{5}$ **65.** $\frac{25}{24}$ **67.** $18\frac{17}{18}$

69. $1\frac{15}{28}$ **71.** $\frac{12}{14}, \frac{18}{21}, \frac{24}{28}, \frac{30}{35}$ **73.** 27

75. false **77.** $\frac{10}{11}$ **79.** $\frac{11}{105}$ **81.** $4\frac{41}{60}$

83. $\frac{31}{72}$ **85.** average, $\frac{55}{108}$; median, $\frac{35}{72}$

87. $16\frac{1}{32}$ in.^{2}

89. The Perez family spends $\frac{7}{12}$ of their income on these items.

CHAPTER 4

SECTION 4.1

1. seventeen hundredths
3. two hundred sixty-seven thousandths
5. four and five ten-thousandths
7. six and forty-seven hundredths
9. 0.16 **11.** 0.113 **13.** 2.019 **15.** 0.0032
17. seven hundred three thousandths
19. seventy and three hundredths
21. eighteen and two hundred five ten-thousandths
23. sixty and nine thousandths **25.** 0.012
27. 800.042 **29.** 505.005 **31.** 23.23

	Unit	Tenth	Hundredth
33. 23.777	24	23.8	23.78
35. 639.875	640	639.9	639.88
37. 0.8156	1	0.8	0.82

39. $33.54 **41.** $398.61

	Ten	Hundredth	Thousandth
43. 27.6647	30	27.66	27.665
45. 245.2454	250	245.25	245.245
47. 0.7728	0	0.77	0.773

49. $23 **51.** $1129
53. The word name is one and three tenths inches.
55. To the nearest tenth of an inch, the precipitation in week 1 is 1.1 inches.
57. Ngo writes "Sixty-four and seventy-nine hundredths dollars."
59. The position of the arrow to the nearest hundredth is 1.62.

61. The position of the arrow to the nearest tenth is 1.6.
63. five hundred sixty-seven and nine thousand twenty-three ten-thousandths
65. The account value to the nearest cent is \$1617.37.
67. 213.1101
69. The population density of Oman is about 31 people per square mile.
71. The number appears to have been rounded to the nearest hundredth.

		Hundred	Hundredth	Ten-thousandth
73.	982.678456	1000	982.68	982.6785
75.	78,042.38875	78,000	78,042.39	78,042.3888

77. The speed was rounded to the thousandths place.
79. The speed is 763 mph to the nearest mile per hour and 763.0 to the nearest tenth.
81. In 43.29, the numeral 4 is in the tens place, which is two places to the left of the decimal point. This means there are four tens in the number. In 18.64, the 4 is in the hundredths place, which is two places to the right of the decimal point. This means there are four hundredths in the number.
83. The rounded value, 8.283, is greater than the original value: $8.283 > 8.28282828$ or $8.28282828 < 8.283$.
85. The place value name is 0.0000215 **89.** 55
91. 425 **93.** true
95. 4110, 4117, 4118, 4121, 4123
97. Jose must get 84 problems correct.

SECTION 4.2

1. $\frac{53}{100}$ **3.** $\frac{3}{4}$ **5.** $\frac{213}{1000}$ **7.** $\frac{41}{50}$ **9.** $\frac{18}{25}$
11. $5\frac{63}{100}$ **13.** $\frac{7}{8}$ **15.** $5\frac{9}{25}$ **17.** $7\frac{123}{250}$
19. $\frac{4}{5}$ **21.** 0.1, 0.6, 0.7 **23.** 0.15, 0.17, 0.61
25. 4.159, 4.16, 4.161 **27.** false **29.** true
31. 0.052, 0.053, 0.053003, 0.053016, 0.0533
33. 0.88579, 0.88799, 0.888, 0.8881
35. 32.006, 32.059, 32.06, 32.095
37. false **39.** true
41. As a reduced fraction, the probability is $\frac{1}{8}$.
43. The best bid, lowest, for the restaurant is 98.35¢ made by Circle K Meats.
45. As a reduced fraction, the number of inches of precipitation in the fourth week is $\frac{3}{50}$ inch.
47. The least precipitation is in week 4. The most precipitation is in week 3.
49. $\frac{217}{400}$ **51.** $65\frac{3}{400}$ **53.** Hoa needs less soap.
55. 5.00088, 5.0009, 5.00091, 5.001, 5.00101
57. 37.59, 37.79, 37.90, 37.95, 37.96, 38.01, 38.15, 38.25
59. Maria should choose the $\frac{6}{7}$ yard.
61. Cape Verde has fewer people per square mile.
63. The fat grams are $7\frac{19}{20}$, $8\frac{9}{10}$, and $10\frac{17}{20}$. The least amount of fat is in the frozen plain potatoes.
65. Tim Hardaway made $\frac{167}{200}$ of his free throws. He missed $\frac{33}{200}$ of his free throws.
67. The table, sorted by averages from highest to lowest, is

Top Ten PBA Averages, Central Region, 2000

Bowler	Number of Games	Average
Bob Learn	46	220.26
Ted Hannah	69	219.04
Sean Quinn	46	215.97
Jim Johnson	56	210.03
Jeff Zaffio	42	209.21
Chris Matthews	32	208.87
Jeff Wellis	33	208.37
Pat Cumbrowski	54	208.01
Mike Eaton	42	207.88
Paul Ashby	56	207.85

69. $\frac{11}{25}, \frac{101}{250}, \frac{1011}{25,000}$ **73.** 13,676 **75.** 2898
77. $\frac{11}{8}$ or $1\frac{3}{8}$ **79.** $\frac{3}{32}$
81. Pedro counted 1577 vehicles.

SECTION 4.3

1. 0.9 **3.** 3.6 **5.** 9 **7.** 27.43 **9.** 3
11. 23.332 **13.** 23.4 **15.** 2.697 **17.** 9.0797
19. 103.482 **21.** 90.1044 **23.** 831.267
25. 28.596 **27.** 0.3 **29.** 2.1 **31.** 5.1
33. 6.48 **35.** 20.87 **37.** 0.266 **39.** 1.564
41. 3.999 **43.** 67.044 **45.** 6.94 **47.** 0.217
49. 0.15 **51.** 0.3 **53.** 6 **55.** 0.08
57. 2.3 **59.** 0.04; 0.04019 **61.** 0.18; 0.1714
63. 10; 7.2189 **65.** 1.3; 1.2635 **67.** 0.3; 0.2719
69. Manuel bought 49.1 gallons of gas on his trip.
71. 603.3
73. Total sales for the 10-year period were \$193.7 billion.
75. \$115.9 billion more was invested in the 1990s than in the 1980s. **77.** Jack has \$0.44 remaining.
79. The total cost of the groceries is \$30.85.
81. The total length of the Seikan and the Dai-shimizu is 76.44 km.
83. The period from 2005 to 2010 is projected as the largest increase, 3.7 million.
85. The combined height is 75 ft.
87. The distance is 6.59375 in.

89. Sera is 0.778 second faster than Muthoni.
91. The relay race was completed in 38.95 sec.
93. The fourth sprinter must have a time of 13.16 sec or less.
95. When subtracting fractions, we must first get common denominators and then subtract the numerators. When subtracting decimals, we align decimal points. This results in columns of like place value, which are then subtracted. The numbers in each column have the same place value, which means that if they were written as fractions, they would have common denominators. Thus, aligning the decimal points results in subtracting numbers with common denominators.
97. A total of 23 8.75s must be added.
99. The missing number is 0.1877.
101. The difference is 1.5725. **105.** 660,824
107. 26,656 **109.** $\frac{2}{3}$ **111.** $23\frac{5}{8}$
113. The Candy Basket will need 10,125 pieces of chocolate peanut clusters to fill the order.

GETTING READY FOR ALGEBRA

1. $x = 11.1$ **3.** $y = 13.83$ **5.** $t = 0.484$
7. $x = 12.51$ **9.** $w = 7.2$ **11.** $t = 8.45$
13. $a = 3.84$ **15.** $x = 16.17$ **17.** $a = 17.7$
19. $s = 7.273$ **21.** $c = 596.36$
23. Two years ago the price of the water heater was $427.73.
25. The markup is $158.66.
27. Let C represent the cost of the groceries; then $19 + C = $43. The shopper can spend $24 on groceries.

SECTION 4.4

1. 4.5 **3.** 9 **5.** 0.27 **7.** 0.2 **9.** 0.042
11. 0.064 **13.** four **15.** 0.05364 **17.** 6.67
19. 2.5092 **21.** 2.6988 **23.** 22.797
25. 0.4784 **27.** 24 **29.** 0.0035 **31.** 1.6
33. 0.0018 **35.** 20; 22.631 **37.** 0.028; 0.023852
39. 0.16; 0.167232 **41.** 0.004; 0.004455
43. Grant purchased a total of 97.7 gallons of gas.
45. To the nearest cent, Grant paid $33.65 for the fifth fill-up.
47. Grant paid the least for his fill-up at $1.699 per gallon.
49. 450; 413.7936 **51.** 8000; 7980.21
53. 2820.270 **55.** 4804.71
57. Joanne pays $680.06 for 14.75 yards.
59. It costs $186 to rent the midsize car.
61. It costs $10.56 less to rent the compact car.
63. Store 3 is selling the freezer-refrigerator for the least total cost. **65.** Steve burns 1662.6 calories per week.
67. The total linear feet of steel is 744.975 ft.
69. The total weight of red meat consumed is 528 lb in 1970, 505.6 lb in 1980, and 449.2 lb in 1990. One reason is the greater awareness of potential heart problems associated with the consumption of large quantities of red meat.
71. The number of gallons of water used by each type of toilet is as follows: pre-1970, 1,149,005 gallons; 1970s, 731,185 gallons; low-flow, 323,810.5 gallons. A total of 825,194.5 gallons of water is saved by using the low-flow model. **73.** The Gregory estate owes $18,906 in taxes.
75. The dive was awarded 66.99 points.
77. The number of decimal places in the product is the sum of the number of decimal places in the two factors. So the product of 5.73 and 4.2 will have three decimal places.
79. The smallest whole number is 84.
81. The missing number is 0.0216. **85.** 750,000
87. 150 **89.** 803,000 **91.** 38 **93.** 42,100

SECTION 4.5

1. 0.367 **3.** 821 **5.** 628.33 **7.** 1.359
9. 79.13 **11.** 8700 **13.** right **15.** 12,563
17. 0.395 **19.** 36,900 **21.** 0.90775
23. 293,000 **25.** 13.6794 **27.** 0.005819
29. 4.2×10^5 **31.** 3.5×10^{-4} **33.** 4.1953×10^3
35. 70,000 **37.** 0.008 **39.** 83,500
41. 8.2×10^4 **43.** 7.2×10^{-6} **45.** 8.21×10^{-11}
47. 9.75002×10^2 **49.** 0.00002912
51. 7,110,000,000 **53.** 0.0000000311 **55.** 567.23
57. The total cost of the shoes is $2768.
59. The total cost of Ms. James's land is $110,000.
61. The land area of Earth is 5.2×10^7 square miles.
63. The length of a red light ray is 7.2×10^{-8} cm.
65. The speed of light is approximately 11,160,000 miles per minute.
67. The shortest wavelength of visible light is 0.00004 cm.
69. The total energy use of the four families is 1,245,600,000 BTUs.
71. The per capita consumption of milk in Cabot Cove in 1995 was 27.6 gallons.
73. They are multiplying the batting average by 1000.
75. The mathematically calculated average is 0.424.
77. The approximate length of 1 parsec is 1.92×10^{13} miles.
79. 6.5×10^6 **83.** 108 **85.** 28 R 4 **87.** 2416
89. 4820
91. The width of the rectangle is 18 in., or 1.5 ft.

SECTION 4.6

1. 0.7 **3.** 6.6 **5.** 183.1 **7.** 2020
9. whole number **11.** 13.4 **13.** 0.52 **15.** 0.5
17. 7.6 **19.** 15.189 **21.** 0.098 **23.** 13
25. 1.9 **27.** 420 **29.** 100; 97.334
31. 0.017; 0.016 **33.** 14; 14.444 **35.** 80.87
37. To the nearest tenth of a cent, the unit price of oranges is $1.473 per pound.
39. To the nearest tenth of a cent, the unit price of rib steak is $2.969 per pound.
41. To the nearest cent, the cost of 7 lb of ham is $24.42.
43. The total cost is $27.08. **45.** 190; 191.29
47. 5.8; 6.21
49. To the nearest mile, June got 30 miles per gallon.
51. To the nearest foot, 288 feet of cable are on the spool.
53. She worked 9.75 hours each day.
55. The cost of pouring each cubic foot of concrete is $54.

57. Denmark has the smallest area and Norway has the smallest population.
59. Denmark, the country with the largest population density, is almost 10 times as crowded as Norway.
61. To the nearest hundredth, the ERA is 1.35.
63. The fraction is $\frac{7}{10}$; it means that the runner is successful 7 out of 10 times.
65. The decimal point in a quotient is positioned directly above the decimal point in the dividend when dividing by a whole number. When dividing by a decimal, move the decimal point in the divisor to make it a whole number. Move the decimal point in the dividend the same number of places. Now the decimal point in the quotient is positioned above the relocated decimal in the dividend. This works because a division problem may be thought of as a fraction $\left(\frac{\text{dividend}}{\text{divisor}}\right)$, and moving the decimals in both the dividend and the divisor the same number of places to the right is the same as multiplying a fraction by 1 in the form of some power of 10 over itself.
67.

Division Procedure

Operation on Decimals	Procedure	Example
Division	If the divisor is not a whole number, move the decimal point to the right until it is a whole number. Move the decimal point in the dividend the same number of places to the right. Now place the decimal point in the quotient directly above the decimal point in the dividend. Now divide.	$0.25\overline{)3.5675}$ $\begin{array}{r} 14.27 \\ 25\overline{)356.75} \\ \underline{25} \\ 106 \\ \underline{100} \\ 67 \\ \underline{50} \\ 175 \\ \underline{175} \\ 0 \end{array}$

71. $\frac{5}{6}$ **73.** $\frac{52}{11}$ **75.** $17\frac{11}{12}$ **77.** 68 **79.** $\frac{3}{5}$

GETTING READY FOR ALGEBRA

1. $x = 7$ **3.** $y = 308.5$ **5.** $t = 0.23$ **7.** $m = 0.06$
9. $q = 562.5$ **11.** $h = 28.125$ **13.** $y = 2.66$
15. $c = 0.2028$ **17.** $x = 1.04$ **19.** $s = 0.02415$
21. $z = 85.7616$ **23.** There are 23 servings.
25. The current is 7.5 amp.
27. The length of the rectangle is 18.4 ft.
29. Let S represent the number of students; then $20S = 3500$. The instructor has 175 students in her classes.

SECTION 4.7

1. 0.75 **3.** 0.375 **5.** 0.6875 **7.** 3.85
9. 11.488

		Tenth	Hundredth
11.	$\frac{5}{7}$	0.7	0.71
13.	$\frac{9}{11}$	0.8	0.82
15.	$\frac{7}{13}$	0.5	0.54
17.	$\frac{8}{15}$	0.5	0.53
19.	$5\frac{13}{18}$	5.7	5.72

21. $0.\overline{72}$ **23.** $0.41\overline{6}$

		Hundredth	Thousandth
25.	$\frac{17}{53}$	0.32	0.321
27.	$\frac{29}{65}$	0.45	0.446

29. $0.\overline{384615}$ **31.** $0.\overline{857142}$
33. The micrometer reading will be 1.625 in.
35. The measurements are 0.375 in., 1.25 in., and 0.5 in.

		Hundredths	Thousandths	Ten-thousandths
37.	$\frac{42}{47}$	0.89	0.894	0.8936
39.	$6\frac{42}{165}$	6.25	6.255	6.2545

41. The cost of the fabric is $\$4\frac{108}{125}$, or \$4.864 per yard. It is easier to use decimals because you can avoid using fractions like $\frac{608}{125}$.
43. Texas has the winning jump.
45. On June 1, there were 15.45 hr of daylight in Portland.
47. Takahashi's winning time was $2\frac{697}{1800}$ hr, or about 2.39 hr.
49. The larger is 0.0036.
51. The fraction is less than 0.1. Yes, the decimal equivalent is 0.015375.
55. 14 **57.** 13 **59.** 406 **61.** $\frac{67}{96}$
63. The average sale price of a house was \$257,682.

SECTION 4.8

1. 1 **3.** 0.05 **5.** 0.3 **7.** 16.3 **9.** 0.2
11. 0.08 **13.** 14.58 **15.** 130 **17.** 21.3
19. 26.264 **21.** 5.5 **23.** 8.1 **25.** 3
27. 8.4 **29.** 3.1 **31.** 6.62, 6.75
33. 9.9625, 10.11 **35.** 13.9, 11.8
37. 0.6225, 0.637 **39.** 0.64774, 0.5421

41. Elmer spent \$16.11.
43. The average closing price of Oracle was \$16.815.
45. 52.89 **47.** 0.361 **49.** 27.8825
51. The Adams family will make monthly payments of \$58.23. **53.** Tim had an average of 6.7 yd per carry.
55. The average high temperature for the cities was 19.3°C.
57. The average range of temperatures for the cities was 13.0°C. Ottawa had a range of 13°C.
59. The median of the low temperatures was 5.5°C.
61. The Yankees averaged 3.8 runs per game. The Mets averaged 3.2 runs per game.
63. In the problem $0.3(5.1)^2 + 8.3 \div 5$, we divide 8.3 by 5 and add the result to $0.3(5.1)^2$. We do this because the order of operations requires that we divide before adding. In the problem $[0.3(5.1)^2 + 8.3] \div 5$, the entire quantity $[0.3(5.1)^2 + 8.3]$ is divided by 5. The insertion of the brackets has changed the order of operations so that we add before dividing because the addition is inside grouping symbols.
65. $3.62 \div (0.02 + 72.3 \cdot 0.2) = 0.25$
67. $(1.4^2 - 0.8)^2 = 1.3456$ **71.** 0.78125
73. 1.88 **75.** $\frac{91}{125}$ **77.** $4\frac{16}{25}$
79. The clerk should price the VCR at \$289.45.

Section 4.4

33. 16.238 **35.** 0.0123046 **37.** 0.14; 0.148296
39. 0.0019 **41.** Pauline will pay \$1243.31 for 23.75 m.

Section 4.5

43. 0.007693 **45.** 73,210 **47.** 7.8×10^{-3}
49. 4.2×10^{-6} **51.** 70,000,000 **53.** 0.0808
55. The average price of a pair of shoes is \$97.35.

Section 4.6

57. 0.037 **59.** 177.1875 **61.** 0.84
63. 70; 76.00
65. To the nearest cent, the average donation was \$2525.45.

Section 4.7

67. 0.5625 **69.** 17.296 **71.** 0.61
73. $0.\overline{692307}$
75. The value of the Microsoft share is 63.22.

Section 4.8

77. 2.75 **79.** 26.498 **81.** 51.449
83. 7.912, 8.09 **85.** 0.519, 0.56135
87. David spent \$26.50.

GETTING READY FOR ALGEBRA

1. $x = 8.16$ **3.** $x = 3$ **5.** $0.1 = t$ **7.** $x = 741$
9. $x = 0.32$ **11.** $m = 63.1$ **13.** $y = 23.56$
15. $p = 4.375$ **17.** $24.3 = x$ **19.** $h = 26.8$
21. $c = 6.275$ **23.** The Celsius temperature is 120°C.
25. Gina has made 15 payments.
27. Let H represent the hours of labor; then $36H + 137.50 = 749.50$. She put in 17 hours of labor on the repair.

CHAPTER 4 Review Exercises
Section 4.1

1. five and eighteen hundredths
3. ten and thirty-two thousandths
5. 17.05 **7.** 400.04

	Tenth	Hundredth	Thousandth
9. 34.7648	34.8	34.76	34.765
11. 0.467215	0.5	0.47	0.467

Section 4.2

13. $\frac{16}{25}$ **15.** $\frac{8}{3125}$ **17.** 0.89, 0.95, 1.01
19. 0.515, 5.007, 5.012, 5.101 **21.** true

Section 4.3

23. 29.288 **25.** 6.5131 **27.** 160; 165.2
29. 32 in. **31.** Hilda's take-home pay is \$2340.24.

CHAPTER 4 True/False Concept Review

1. false; the word name is seven hundred nine thousandths.
2. true **3.** false; the expanded form is $\frac{8}{10} + \frac{5}{100}$.
4. true
5. false; the number on the left is smaller because 732,687 is less than 740,000. **6.** true **7.** true **8.** true
9. false; the numeral in the tenths place is 7 followed by a 4, so it rounds to 356.7. **10.** true
11. false; $9.7 - 0.2 = 9.5$, as we can see when the decimal points are aligned.
12. false; not if we leave out the extra zeros in a product such as $0.5(3.02) = 1.51$. **13.** true
14. false; the decimal moves to the left, making the dividend smaller.
15. false; move the decimal left when the exponent is negative. **16.** true
17. false; most fractions have no exact terminating decimal form. For example, the fractions $\frac{1}{3}, \frac{1}{6}, \frac{1}{7}$, and $\frac{1}{9}$ have no exact terminating decimal equivalents.
18. false; the rounded value is 0.78. **19.** true
20. true

CHAPTER 4 Test

1. 4.030 **2.** 0.6299, 0.6307, 0.638, 0.6389, 0.642
3. fifty-six and seventy-two thousandths
4. 42.572 **5.** 0.088 **6.** 63.00 **7.** 72.163
8. $12\frac{21}{40}$ **9.** 3.41×10^{-6} **10.** 0.74 **11.** 35,000
12. 21.39 **13.** 6.907 **14.** 0.000745
15. 413.085 **16.** 0.523 **17.** 6.392×10^{-4}
18. 31.667 **19.** 0.062928 **20.** 0.00748
21. The average Sunday offering was \$3801.25.
22. 4953.653 **23.** Grant pays \$15.41.
24. Michael Jordan had played in 683 games.
25. The player's slugging percentage is 765.
26. Jerry lost 0.03 lb more than Harold.

CHAPTER 5

SECTION 5.1

1. $\frac{1}{7}$ **3.** $\frac{1}{3}$ **5.** $\frac{8}{9}$ **7.** $\frac{1}{2}$ **9.** $\frac{9}{10}$ **11.** $\frac{10}{9}$
13. $\frac{13 \text{ cars}}{6 \text{ families}}$ **15.** $\frac{55 \text{ miles}}{1 \text{ hour}}$ **17.** $\frac{21 \text{ miles}}{1 \text{ gallon}}$
19. $\frac{23 \text{ trees}}{2 \text{ rows}}$ **21.** $\frac{2 \text{ trees}}{7 \text{ ft}}$ **23.** $\frac{2 \text{ books}}{5 \text{ students}}$
25. $\frac{43 \text{ students}}{2 \text{ rooms}}$ **27.** $\frac{15 \text{ pies}}{2 \text{ sales}}$
29. 38 students per class **31.** 4 ft per sec
33. 66 yd per billboard **35.** 1250 miles per quart
37. 2.6 children per family **39.** 83.3 ft per sec
41. 268.8 lb per mi^2 **43.** 741.7 gal per hr
45. **a.** The price per rose of the special is \$9.98.
b. This is a savings of \$2.90 over buying three separate roses.
47. **a.** The small bag costs 16.5¢ per ounce, and the large bag costs 11.65¢ per ounce. **b.** Jerry can save \$2.91 by buying the large bag instead of 5 small bags.
49. **a.** The ratio of compact spaces to larger spaces is $\frac{3}{4}$.
b. The ratio of compact spaces to total spaces is $\frac{3}{7}$.
51. The ratio of sale price to regular price is $\frac{2}{3}$.
53. The population density of Dryton City is about 97.6 people per square mile. **55.** Answers will vary.
57. **a.** The ratio of laundry use to toilet use is $\frac{7}{20}$.
b. The ratio of bath and shower use to dishwashing use is $\frac{16}{3}$.
59. A total of 1.25 mg of lead is enough to pollute 25 L of water.
61. The 14-oz can is 6.36¢ per ounce and the 1-lb can is 6.25¢ per ounce, so the 1-lb can is the better buy.
63. The 5-lb sale bag is the better buy, at a cost of 99¢ per pound.
65. The 16-oz can has a smaller unit price of 8¢ per oz.
67. The retrieve rate is $\frac{105 \text{ in.}}{5 \text{ turns}} = 21$ in. per turn.
69. The average number of credits per student is $\frac{4020 \text{ credits}}{645 \text{ students}} \approx 6$ credits per student. The FTE for the term is $\frac{4020 \text{ credits}}{15 \text{ credits}} = 268$ FTE.
71. An ant is about 0.4 in. long and a horse is about 6 ft long. The ratio of ant to horse is $\frac{0.4}{72} \approx 0.0056$. The reduction rate is approximately 0.0038. The rates are off mathematically. Because the ant-to-horse rate is larger, an ant would appear even larger than a horse to the reduced kids.
73. A ratio compares two measurements. The ratio of \$5 to \$1000 is $\frac{5}{1000}$ or $\frac{1}{200}$. A rate is a ratio that compares unlike measurements. The rate of 350 cents per 20 oz is $\frac{350 \text{ cents}}{20 \text{ oz}}$, or $\frac{35 \text{ cents}}{2 \text{ oz}}$. A unit rate compares unlike measurements, where the second measurement is 1 unit. The unit rate of 350 cents to 20 oz is 17.5¢ per ounce.
75. Examples of 2-to-1 ratios are eyes to people, hands to people, and legs to people. Examples of 3-to-1 ratios are triplets to mothers, petals to trillium flowers, and leaves to poison ivy stems.
79. $\frac{1}{14}$ **81.** 0.195 **83.** 0.525 **85.** $4\frac{8}{15}$
87. $\frac{94}{125}$

SECTION 5.2

1. true **3.** false **5.** true **7.** true **9.** false
11. false **13.** true **15.** $a = 8$ **17.** $c = 6$
19. $y = 6$ **21.** $c = 10$ **23.** $x = 18$ **25.** $p = \frac{1}{2}$
27. $x = 10.5$ **29.** $z = 4.4$ **31.** $w = 9\frac{1}{3}$
33. $a = 12.8$ **35.** $c = 0.6$ **37.** $b = \frac{3}{8}$
39. $x = 0.35$ **41.** $w = 0.05$ **43.** $y = 216$
45. $t = 8$ **47.** $w \approx 1.4$ **49.** $y \approx 54.4$
51. $a \approx 0.22$ **53.** $c \approx 0.95$
55. The value in the box must be 10, so the cross products are the same.
57. The error is multiplying straight across instead of finding cross products. Multiplying straight across goes with fraction multiplication, but cross multiplication goes with proportions.
59. Fran's minimum allowable level of HDL is about 48.
61. Westridge should have about 34 teachers.
63. **a.** The number of men who suffer from migraines in the group is 102. **b.** The number of migraine sufferers is 245. **c.** Of 350 migraine sufferers, 70 migraine headaches may be related to diet.

65. The proportion is true if the butterfly has a wingspan of about 1.9 in. This is realistic.

67. $3.5(y) = \left(\frac{1}{4}\right)7$ — To solve the proportion, first cross multiply.

$3.5y = \frac{7}{4}$

$y = \frac{7}{4} \div 3.5$ — Rewrite as division. We can simplify using decimals or fractions.

If we choose fractions, we get $y = \frac{7}{4} \div \frac{7}{2}$, or $y = \frac{1}{2}$.

If we choose decimals, we get $y = 1.75 \div 3.5$, or $y = 0.5$.

69. $a = 10$ **71.** $w \approx 8.809$ **75.** 6.8265 **77.** 48,350 **79.** 25.375 **81.** 2.001 **83.** Quan pays $22.64.

SECTION 5.3

1. 6 **3.** 4 **5.** $\frac{6}{4} = \frac{15}{h}$ **7.** 35 **9.** 15 **11.** $\frac{35}{15} = \frac{h}{24}$ **13.** 8 **15.** 48 **17.** $\frac{8}{48} = \frac{T}{288}$ **19.** T **21.** $\frac{3}{65} = \frac{T}{910}$

23. The school will need 4 additional teachers.

25. $\frac{1.5}{30} = \frac{14}{x}$ **27.** Idamae must knit 100 rows.

29. Los Angeles would have had 2,792,048 cars, buses, and trucks.

31. The family would spend $9000 on food.

33. Fifty pounds of fertilizer will cover 2500 ft^2.

35. $\frac{12}{15} = \frac{x}{30}$ **37.** Hazel will earn $127.50.

39. Mike's onions cost $1.55.

41. It will take 17 hr to go 935 mi.

43. The second room will cost $350.90.

45. Ida needs 0.4 cm^3 (0.4 cc) of the drug.

47. A pound of gumdrops costs $4.80.

49. Betty needs 12 lb of cashews and 28 lb of peanuts.

51. Debra needs 42 quarts of blue paint.

53. Mario needs 70 lb of ground round for the meatballs.

55. The computer costs £845 (British pounds).

57. The comparable price for a 60-month battery is $59.75.

59. 12 m^3 of air must have less than 2820 mg of ozone.

61. The store must reduce the price of each box by 10¢ in order to beat the warehouse outlet's price.

63. Assuming a Cheerio is about 0.5 in., the set designer must make it 11 ft in diameter.

65. A proportion is composed of two equal ratios or rates. The speed of a vehicle can be measured in different units: $\frac{60 \text{ mi}}{1 \text{ hr}} = \frac{88 \text{ ft}}{1 \text{ sec}}$. Food recipes can be cut or expanded by keeping the ingredient measures in proportion. Prices for many items, especially food and clothing, are proportional to the number purchased.

67. From a consumer's viewpoint, it is better if there is a price break when buying larger quantities. Many products have a lower unit price when bought in bulk or larger packages.

69. The drive shaft speed is 1120 revolutions per minute.

73. 37.41; 37.415 **75.** $0.00872 < 0.011$

77. The gas costs $19.03. **79.** $\frac{173}{200}$ **81.** 0.559

CHAPTER 5 Review Exercises
Section 5.1

1. $\frac{1}{5}$ **3.** $\frac{6}{5}$ **5.** $\frac{4}{5}$ **7.** $\frac{2}{3}$ **9.** $\frac{9 \text{ people}}{10 \text{ chairs}}$ **11.** $\frac{1 \text{ videotape}}{7 \text{ DVDs}}$ **13.** $\frac{8 \text{ lb}}{3 \text{ ft}}$ **15.** $\frac{85 \text{ people}}{3 \text{ committees}}$

17. 25 mph **19.** 9¢ per pound of potatoes

21. 37.5 mpg **23.** 16.1 gal per minute

25. No, the first part of the country has a rate of 3.5 TVs per household, and the second part of the country has a rate of about 3.3 TVs per household.

Section 5.2

27. true **29.** false **31.** true **33.** $r = 9$ **35.** $t = 4$ **37.** $f = \frac{1}{2}$ **39.** $r = 10\frac{2}{3}$ **41.** $t = 5\frac{1}{3}$ **43.** $a \approx 4.2$ **45.** $c \approx 38.4$

47. The store brand must be cheaper than $5.07 for a 25-load box.

Section 5.3

49. Merle will study 37.5 hr per week.

51. Juan must work 125 hr to pay for his tuition.

53. Larry will make $43.75.

CHAPTER 5 True/False Concept Review

1. true

2. false; a ratio is a comparison of two numbers or measures written as a fraction.

3. false; $\frac{18 \text{ miles}}{1 \text{ gallon}} = \frac{54 \text{ miles}}{3 \text{ gallons}}$ or $\frac{18 \text{ miles}}{1 \text{ hour}} = \frac{54 \text{ miles}}{3 \text{ hours}}$

4. true **5.** true

6. false; in a proportion, two ratios are *stated* to be equal but they may not be equal.

7. false; like measures have the same units. Three feet and 1 yard are *equivalent* measures.

8. false; ratios that are called rates compare unlike units.

9. true

10. false; the table should look like this:

	First Tree	Second Tree
Height	18	x
Shadow	17	25

CHAPTER 5 Test

1. $\frac{4}{5}$ **2.** Ken would answer 45 questions correctly.
3. $w = 0.9$ **4.** false **5.** $y = 6.5$ **6.** true
7. Mary should be paid $84.24. **8.** $\frac{1}{9}$
9. The price of 10 cans of peas is $8.20.
10. One would expect 17.5 lb of bones.
11. $x = 0.625$ **12.** They will catch 24 fish.
13. Jennie will need 10 gal of gas. **14.** $a = 5.3$
15. yes **16.** The tree will cast a 10.5-ft shadow.
17. The population density is 37.2 people per square mile.
18. $y = 1.65$ **19.** The crew could complete 26 jobs.
20. There are 30 females in the class.

CHAPTERS 1–5 Cumulative Review

1. 2104 **3.** 3134 **5.** 11,248 **7.** 8845
9. 128 **11.** 49
13. The sophomore class had the most students taking algebra.
15.

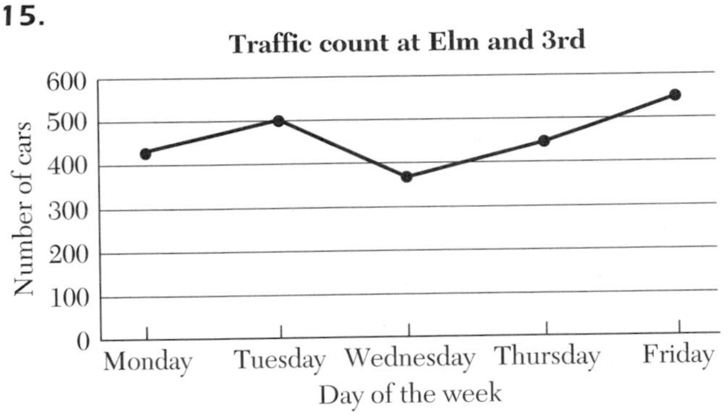

17. It is divisible by 6, 9, and 10.
19. No, 24 is not a multiple of 678.
21. The factors of 190 are 1, 2, 5, 10, 19, 38, 95, and 190.
23. The number 303 is composite.
25. The LCM is 7308 and the GCF is 4.
27. $\frac{103}{9}$ **29.** $\frac{16}{19}$ **31.** $\frac{3}{5}$ **33.** $\frac{9}{16}$ **35.** $38\frac{2}{5}$
37. $\frac{29}{20}$ or $1\frac{9}{20}$ **39.** $\frac{43}{90}$ **41.** $8\frac{1}{60}$ **43.** $\frac{29}{56}$
45. $\frac{41}{50}$ **47.** Jamie must get 32 problems correct.
49. The machinist must cut off $\frac{7}{32}$ in.
51. five hundred twelve hundred-thousandths
53. 4000.0004 **55.** $\frac{7}{200}$
57. 0.009, 0.042, 0.091, 0.2, 0.62 **59.** 65.665
61. 127.47 **63.** Jose has $2.10 left.
65. 0.024566 **67.** 319.5 **69.** 0.056341
71. 3.609×10^2
73. The cleaning solvent will last about 75 days.
75. 8.85 **77.** 1.224
79. average, 13.72; median, 12.78; mode, 12.11
81. Andy spends $96.80 on supplies. **83.** $\frac{1}{3}$
85. 18.2 mi per gal **87.** false **89.** $x = 45$
91. $c \approx 149.0$ **93.** the unknown, x **95.** 12 ft
97. The second flagpole is 18 ft tall.
99. Greg will earn $91.50 if he sells $610 worth of clothes.

CHAPTER 6

SECTION 6.1

1. 22% **3.** 72% **5.** 140% **7.** 39% **9.** 7%
11. 66% **13.** 44% **15.** 45% **17.** 170%
19. 100% **21.** 150% **23.** 42.5% **25.** 80%
27. 93.75% **29.** $83\frac{1}{3}\%$ **31.** $66\frac{2}{3}\%$ **33.** 12
35. The percent of eligible voters who exercised their right to vote was 82%.
37. $31\frac{1}{4}\%$ **39.** 460% **41.** $53\frac{1}{6}\%$
43. The telephone tax is 8%.
45. The annual interest rate is 5.65%.
47. The tax rate is 2.75%.
49. In 1968, 55% of property crime arrests were juveniles.
51. In 2000, 56% of bachelor's degree recipients were women.
53. Carol spends 67% of her money on the outfit.
55. The percent paid in interest was 15%.
57. Percent is the amount per hundred. It is the numerator of a fraction with a denominator of 100. It is the number of hundredths in a decimal.
59. $\frac{101}{500}$; 20.2% **63.** 2398.5 **65.** 176,300
67. 1460 **69.** 1
71. Ms. Henderson earned $1989.90 during the month.

SECTION 6.2

1. 67% **3.** 651% **5.** 5% **7.** 139%
9. 1700% **11.** 0.35% **13.** 73.2% **15.** 45%
17. 600% **19.** 41.6% **21.** 7.31% **23.** 8000%
25. 1781% **27.** 0.034% **29.** 710%
31. 3.45% **33.** 0.047% **35.** 20.54%
37. $81.16\overline{6}\%$, or $81\frac{1}{6}\%$ **39.** 10.05%
41. The tax on income is 27%.
43. The percent of items sold is 46%.
45. The percent of contestants who withdraw is 17.8%.
47. The price of a snow blower increased by a factor of 137%.
49. The swim team wins 71.5% of its meets.
51. Barry Bonds's batting average was 30%, which means he got a hit 30% of the times he was at bat.

53. In 2006, the number of physical therapy jobs will be 170.4% of the number of jobs in 1996.
55. The percent of income paid for both Social Security and Medicare is 7.65%.
57. The sale price is 89% of the original price. The store will advertise "11% off."
59. Percent means parts out of 100, so a number written as a percent represents a part of 100. For example, 36% means 36 parts out of 100. Another way to represent 36 out of 100 is to write 0.36, or 36 hundredths. The two forms differ by the position of the decimal point and the presence of the % symbol. When a number is in its percent form, the % symbol is used and the decimal point is two places farther to the right than in the decimal form of the number.
61. 1,800,000% **63.** 42.5%; $42\frac{5}{9}\%$ **65.** 0.1245
67. 0.1743 **69.** 0.346 **71.** 3.873 **73.** 2.6472

SECTION 6.3

1. 0.19 **3.** 0.53 **5.** 0.73 **7.** 0.0641
9. 5.03 **11.** 2.165 **13.** 0.00012 **15.** 2.13
17. 0.0019 **19.** 23.4 **21.** 0.00037 **23.** 1
25. 2.12 **27.** 0.005 **29.** 0.00625 **31.** 0.3775
33. 0.0000014 **35.** 0.00875 **37.** 1.83554
39. 4.755 **41.** The employees will get a raise of 0.0825.
43. The Credit Union will use 0.0975 to compute the interest.
45. Unemployment is down by a factor of 0.0035.
47. 0.005 **49.** 0.623 **51.** 0.038
53. The river pollution due to agricultural runoff is 0.6 of the total pollution.
55. In 1993, Americans had a cholesterol rate that is a factor of 0.04 lower than in 1981.
57. Answers will vary.
59. One yard is about 0.914 meter.
61. The Petronas Towers are about 1.02 times the height of the Sears Tower.
63. Roger Clemens won 0.87 of the games where he was involved in the decision.
65. Carol will use 0.45 to calculate the savings.
67. The decimal will be greater than 1 when the % is greater than 100%. For example, 235% written as a decimal is 2.35 and 3.4% written as a decimal is 0.034.
69. 0.6, 0.569 **71.** 0.875 **73.** 1.1875 **75.** 1.8
77. $\frac{143}{200}$ **79.** Greg's take-home pay is $206.55.

SECTION 6.4

1. 77% **3.** 74% **5.** 52% **7.** 50% **9.** 44%
11. 15% **13.** 130% **15.** 137.5% **17.** 4.3%
19. 65% **21.** 460% **23.** $66\frac{2}{3}\%$ **25.** $383\frac{1}{3}\%$
27. $283\frac{1}{3}\%$ **29.** $12\frac{1}{4}\%$, or 12.25% **31.** 70.6%
33. 55.6% **35.** 84.6%
37. 1139.1% **39.** 9.5%
41. Kobe Bryant made 95% of his free throws.
43. The voter turnout was 57.1%.
45. The soccer club won 76.5% of their games.
47. 15.75% **49.** 2.86% **51.** 657.58%
53. Miquel is taking 162.5% of the recommended daily consumption.
55. The current number of post offices is $\frac{28}{77}$, or about 36.4% of the number in 1900.
57. St. Croix is $\frac{3}{5}$, or 60%, of the area of the Virgin Islands.
59. In 1995, about 57.8% of the days were smoggy. In 1998, about 45.8% of the days were smoggy. Possible reasons include better emission controls on cars and trucks and more pollution controls on factories.
61. The percent taken off the original price is about 48%.
63. 2.3% **67.** $\frac{33}{250}$ **69.** $16\frac{12}{25}$ **71.** $\frac{211}{2000}$
73. 6.96261 **75.** 0.129

SECTION 6.5

1. $\frac{3}{25}$ **3.** $\frac{9}{20}$ **5.** $\frac{13}{10}$, or $1\frac{3}{10}$ **7.** 5 **9.** $\frac{4}{5}$
11. $\frac{18}{25}$ **13.** $\frac{3}{4}$ **15.** $\frac{7}{20}$ **17.** $\frac{3}{2}$, or $1\frac{1}{2}$
19. $\frac{67}{100}$ **21.** $\frac{71}{200}$ **23.** $\frac{17}{250}$ **25.** $\frac{61}{200}$
27. $\frac{1}{400}$ **29.** $\frac{3}{40}$ **31.** $\frac{2}{225}$ **33.** $\frac{21}{40}$
35. $\frac{89}{180}$ **37.** $\frac{3}{80}$ **39.** $\frac{133}{60}$, or $2\frac{13}{60}$
41. George and Ethel paid $\frac{6}{25}$ of their income in taxes.
43. The monthly mortgage payment is $\frac{8}{25}$ of their monthly income.
45. Grant Hill made $\frac{21}{25}$ of his shots. **47.** $\frac{19}{125}$
49. $\frac{2}{45}$ **51.** $\frac{29}{4000}$
53. It represents $\frac{27}{25}$, or $1\frac{2}{25}$, of last year's enrollment.
55. The fraction of residents between the ages of 25 and 40 is $\frac{7}{40}$.
57. The fraction of Mediterranean fruit flies that are eliminated is $\frac{22}{25}$.

59. In Vietnam, $\frac{47}{50}$ of the population is literate. It means that 47 out of every 50 people are literate.

61. In 1998 in Iceland, $\frac{8}{25}$ of the population used the Internet. It means that 8 of every 25 people in Iceland use the Internet.

63. Juanita pays $\frac{18}{25}$ of the original price.

65. Sales can be described by either fractions or percents. For example, you might hear about a sale of $\frac{1}{3}$ off or of 33% off. Common statistics are also given in either form. You might hear that $\frac{1}{5}$ of the residents of a town are over 65 years old or that 20% of the residents are over 65 years old. It is important to note that in such circumstances fractions are generally approximations of the actual statistics. Percent forms of statistics may be approximations but are generally considered more accurate than fractional forms.

67. $\frac{1}{400,000}$ **69.** $1\frac{10,001}{20,000}$ **73.** 0.6921875

75. $\frac{26}{125}$ **77.** 500.7% **79.** 1.128

81. $\frac{2953 \text{ gallons}}{20 \text{ people}}$

SECTION 6.6

1. 0.1; 10% **3.** $\frac{3}{4}$; 75% **5.** $1\frac{9}{20}$; 1.45

7. $\frac{1}{1000}$; 0.1% **9.** 2.25; 225% **11.** $\frac{11}{200}$; 0.055

13. $\frac{1}{200}$; 0.005 **15.** $\frac{5}{8}$; 0.625 **17.** $\frac{43}{50}$; 86%

19. $\frac{2}{25}$; 8% **21.** $\frac{1}{4}$; 0.25 **23.** $\frac{2}{5}$; 0.4

25. $\frac{1}{8}$; 12.5%, or $12\frac{1}{2}$%

27. The tires are 25% off the regular price.

29. Super Value Grocery offers the best deal.

31. Vendor B offers the best deal and falls within the boss's authorized amount.

33. Evan had the best batting average.

35. Peru has the higher literacy rate.

37. The TV sells for $\frac{2}{5}$ off the regular price.

39. Melinda gets the best deal from Machines Etc.

41. Fractions are commonly used to describe units with subdivisions that are not powers of 10. For example, inches are subdivided into fourths and eighths, so it is common to have measures of $14\frac{3}{8}$ inches. Rates are usually stated using decimals forms. For example, unit pricing is generally given in tenths of a cent, as in Cheerios costing 19.4¢ per ounce. Statistics are most often given using percent forms. For example, 52% of total orchestra expenses are salaries of the musicians.

45. $y = 41.4$ **47.** $x = 3.735$ **49.** $B = 200$

51. $A \approx 9.5$

53. The Bacons should collect $33,200 in insurance money.

SECTION 6.7

1. 34 **3.** 42 **5.** 200% **7.** 50% **9.** 40

11. 44.1 **13.** 80% **15.** 100 **17.** 2

19. 100 **21.** 7.5 **23.** 1.2% **25.** 150

27. 31.35 **29.** 15.5 **31.** 325 **33.** 76.8%

35. 11.836 **37.** 62.5% **39.** 205 **41.** 14.9%

43. 403.36 **45.** 227.265 **47.** 162.1%

49. $\frac{171}{200}$ **51.** $53\frac{4}{15}$ **53.** 55.6%

55. The problem is in equating a 40¢ profit with a 40% profit. A 40¢ profit means that the company makes 40¢ on each part sold. A 40% profit means that the company makes 40% of the cost on each part. But 40% of 30¢ is only 12¢. So a 40% profit would mean selling the parts for 42¢. The company is making a much higher profit than that. The 40¢ profit is actually 133% of the cost.

57. $\frac{17}{75}$ **61.** $x = 45$ **63.** $a = 0.25$, or $a = \frac{1}{4}$

65. $w = 20.2$ **67.** The number of miles is 115.

69. The number of inches needed is $20\frac{1}{2}$.

SECTION 6.8

1. Sarah pays $17.82 in sales taxes.

3. Joan's yearly salary was $17,545.

5. Frank must sell $1355 worth of subscriptions.

7. Debbi's total salary was $509.24.

9. The price of the TV was $578.

11. Ilga owes $19,882.80 on the new van.

13. John's score was 93%.

15. The annual income from rent is 9.3%.

17. Adams High School won 68% of the games.

19. The percent off is about 14%.

21. Gravel is 50% of the mixture.

23. The diesel engine is down 1% less time than the gasoline engine.

25. The percent of increased mileage is 8%.

27. By 2005, it is predicted, there will be about 444,000 human services workers.

29. One hamburger contains 630 calories.

31. Answers will vary.

33. The percent of calories from fat is as follows: light mayonnaise, 90%; cocktail peanuts, 82%; Wheat Thins, 33%; cream sandwich cookies, 23%.

35. The percent of non-Caucasians in Head Start is 67%.

37. Development costs more than taxes.

39. The amount spent on salaries was $960,000.

41.

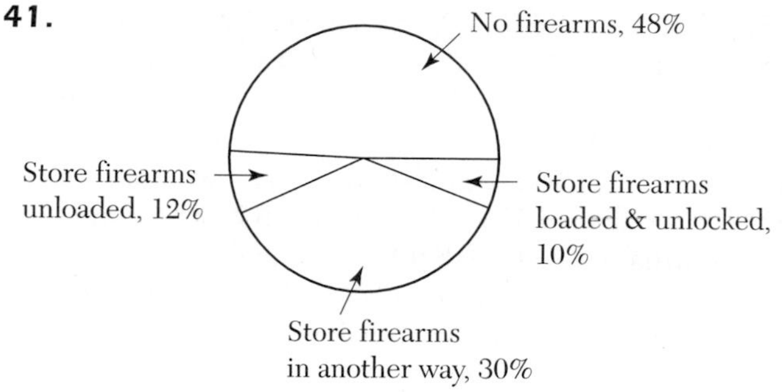

43. In 1900, the U.S. population was about 75.7 million people.
45. In 1999, the U.S. population was about 271.6 million people.
47. In 1990, there were about 255,328 minority residents in Cleveland.
49. *R* is more than 100% and *A* is greater than *B*.
51. Carol's baby increased his weight by 215% during the first year.
53. Nonfood items accounted for 30.4% of the cost. Meat products accounted for 23.5%.
55. Sally receives $7560 and Rita receives $5040.
57. Joe pays a total of $17,508 for the car.
59. The engine develops 113.75 horsepower.
61. Peter attended 90% of the classes.
63. The score on the algebra test is 85%.

SECTION 6.9

1. The percent of markup based on the cost is 25%.
3. The selling price of the tool is $13.34.
5. The TV is on sale for $402.71.
7. Maria earned $655.65 last week.
9. Matthew's total sales for the week were $5187.50.
11. The rebate on the canopy is $26.74.
13. The property tax rate is 1.2%.
15. Hazel contributes $585 to her retirement fund and $60.94 to the United Way.
17. The CEO will earn a bonus of $7071.75.
19. The rate of depreciation was 14%.
21. The associates receive $6336, $5148, $4752, and $3564, respectively.
23. Ms. James earned commissions of $1177.30 for the week.
25. The cost of the hutch to the store was $795.
27. The taxes on the house would be $431.20.
29. The average wage at Swift Copier will now be $9.89 per hour. 31. The profit for the store is $89.
33. The average number of hours per week decreased about 21% over the last 100 years.
35. The number of database administrators/computer support specialists/computer scientists is expected to increase by about 117% over the 10-year period.
37. On April 13, 2000, the Dow Jones closed at 10,924.07 points. 39. The New Balances are sold at 15% off.
41. The final price of the blazer is $80.40. The savings is 52% off the original price.
43. You pay $893.75 for the golf clubs.
45. You will save $11.44 on the purchase.
47. A 10% increase in $100 is a $10 increase and results in $110. Now a 10% decrease from $110 is a decrease of $11 and results in $99. The second 10% is taken on a larger value.
49. The selling price of the lounge chair is $104.62.
51. The percent of savings with the diesel engine is 42%.
53. 2917 55. 4912 57. 19,096 59. 259
61. Geri earned $435.16 last week.

CHAPTER 6 Review Exercises
Section 6.1

1. 65% 3. 82% 5. $57\frac{1}{7}\%$

Section 6.2

7. 43.8% 9. 30.9% 11. 0.017%
13. The Seattle Mariners won 83.9% of their first 31 games.

Section 6.3

15. 0.67 17. 0.0747 19. 0.000625
21. The decimal used to compute the interest is 0.19785.

Section 6.4

23. 91% 25. 504.6875% 27. 13.5%
29. The Little League team won about 70.37% of their games.

Section 6.5

31. $\frac{12}{25}$ 33. $\frac{39}{500}$ 35. $\frac{43}{200}$

37. The Bears defense was on the field $\frac{31}{50}$ of the game.

Section 6.6

39. 0.36 41. $\frac{18}{25}$ 43. $\frac{3}{200}$ 45. 3.275

Section 6.7

47. 32.9 49. 21.25% 51. 52.48 53. 8.8%

Section 6.8

55. Melinda's yearly salary was $13,432.98.
57. Omni Plastics currently employs 330.
59. The Trailblazers lost 68% of their last 25 games.
61. Most students received a B grade.

Section 6.9

63. The golf club bag is marked up 80% based on the cost.
65. The cost of the suit to the consumer is $405.54.
67. The property tax rate is about 2.9%.
69. The New Balance shoes are sold at about 28% off.

CHAPTER 6 True/False Concept Review

1. true **2.** true
3. false; change the fraction to a decimal and then change the decimal to a percent.
4. false; move the decimal point two places to the right and add a percent symbol.
5. true **6.** false; the base unit is always 100. **7.** true
8. true **9.** false; the proportion is $\frac{A}{B} = \frac{X}{100}$.
10. true **11.** true **12.** false; $7\frac{1}{2} = 750\%$
13. false; 0.009% = 0.00009 **14.** false; B = 28,000
15. true **16.** true
17. false; it is the same as a decrease of 27.75%.
18. false; her salary is 99% of what it was Monday.
19. true **20.** true **21.** true
22. false; $\frac{1}{2}\% = 0.005$

CHAPTER 6 Test

1. The percent of discount on the computer is 24.3%.
2. 7.436% **3.** The population is 21% Hispanic.
4. 71.7% **5.** 68.75% **6.** 61 **7.** $\frac{12}{35}$
8. 791% **9.** 4.836 **10.** 372.7%
11. $\frac{68}{25}$, or $2\frac{18}{25}$ **12.** 0.4371
13. The Siekas family spends $1001.05 per month on food.
14. 0.8125 **15.** 81.25% **16.** $\frac{78}{125}$ **17.** 62.4%
18. $\frac{37}{200}$ **19.** 0.185 **20.** 65.6% **21.** 0.00619
22. The selling price of the tire is $61.11.
23. The sale price of the Hilfiger jacket is $104.68.
24. The percent of increase in population is 30.3%.
25. The percent of B grades is 23.4%.
26. Gas prices decreased 14%.
27. The suit sells for $558.72.
28. The percent of calories from fat is 47.4%.
29. The number of students who took the final exam is 62.
30. The percent of increase in the price of the radial tire was 12.1%.

CHAPTER 7

SECTION 7.1

1. feet or meters **3.** inches or millimeters
5. inches or centimeters **7.** feet or meters
9. feet or meters **11.** inches or centimeters
13. feet or meters **15.** yards, feet, or meters
17. 60 in. **19.** 32,000 m **21.** 42 ft
23. 0.082 km **25.** 600,000 cm **27.** 0.19 mi
29. 2.25 ft **31.** 2100.58 cm **33.** 74.57 mi
35. 1.51 ft **37.** 78 ft **39.** 4 in. **41.** 200 mm
43. 16 ft **45.** 2600 mi **47.** 7 ft 5 in.
49. 1.75 mi **51.** 24,075 mm **53.** 6 ft 7 in.
55. 12,512 mi **57.** 19 ft 7 in. **59.** 8 yd 3 in.
61. Martha should buy 18 ft 8 in. of molding.
63. Rosa can make 6 napkins; she will have 20 cm of fabric left over. **65.** Answers will vary.
67. Nine lanes will require eight dividers for a total of 400 m.
69. Figures for the Amazon and the Congo Rivers appear to be rounded because of the two and three terminal zeros. This implies they are probably estimates.
71. The Amazon River is about twice as long as the São Francisco River.
73. The yard is 20 ft wide and 28 ft long.
75. The main walkway is 4 ft wide, and the others are 2 ft wide.
77. Only like measures may be added or subtracted. Add (subtract) the numbers and keep the common measure.
81. 0.0000478 **83.** $\frac{9}{35}$ **85.** $17\frac{21}{40}$ **87.** 24.96
89. 95

SECTION 7.2

1. 35 qt **3.** 8 mL **5.** 4 gal **7.** 24 c
9. 4.5 gal **11.** 414 mL **13.** 59 c **15.** 0 qt
17. 112 c **19.** 132.48 L **21.** 20 g **23.** 20 lb
25. 45 kg **27.** 8000 g **29.** 7 g **31.** 208 lb
33. 4728 oz **35.** 650 g **37.** 0.036 g
39. 2.63 oz **41.** 0°C **43.** 5°C **45.** 72 hr
47. 95°F **49.** 70 days **51.** 37.8°C **53.** 8.9°C
55. 87.8°F **57.** 17,520 hr **59.** 0.01 day
61. 21 lb 6 oz **63.** 8 days 15 hr 40 min
65. 4 lb 5 oz **67.** Normal body temperature is 37°C.
69. The patient gets 840 mg of medication in a week.
71. The total time for the team is 3 min 20.2 sec.
73. Each child gets 0.2 L, or 200 mL, of 7-Up.
75. Seven bags contain 2086 g of potato chips.
77.

Kilograms	54	58	63	69	76	85	97	130
Pounds	119	128	139	152	168	187	214	287

79. The Jahre Viking weighs about 674 million kg fully loaded. **81.** He needs 11 servings of cereal and milk to get sufficient protein.
83. Because grams measure weight and inches measure length, it is not possible to add these measures.
89. 848 **91.** 41.94 **93.** $45\frac{7}{20}$ **95.** 153.45
97. 4820 m

SECTION 7.3

1. 19 in. **3.** 28 m **5.** 34 mm **7.** 26 km
9. 83 mm **11.** 170 m **13.** 90 ft **15.** 100 cm

17. 72 yd **19.** 168 m **21.** 12.56 in.
23. 37.68 mm **25.** 18.84 km **27.** 49.13 cm
29. 61.70 yd **31.** 37.70 m **33.** 54 cm
35. 350 mm **37.** 105.13 ft
39. The five frames require 16 ft 8 in. of molding.
41. It will take Hazel 1 hr 48 min to bind the rug.
43. The inside dimensions of the frame are 26 in. by 20 in.
45. They need 117 ft of fence.
47. The football player ran 19,376 yd, which is about 11 mi.
49. Marylane needs 72 ft of lumber for the project.
51. The total cost of the lumber is $112.41.
53. Yolande needs about 170 inches of lace for the second row. **55.** The perimeter of the patio is 60 ft.
57. The center line divides the yard into identical but mirror-imaged halves.

59.

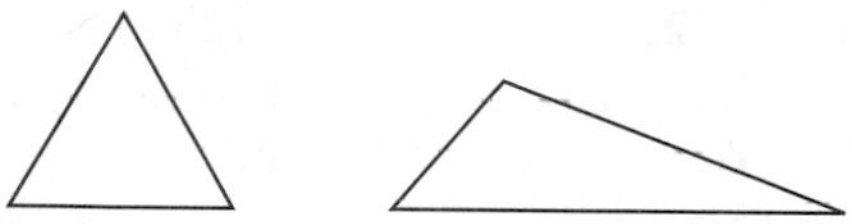

Symmetrical Not symmetrical

61. Perimeter is the distance around a flat surface. To find the perimeter of the figure, add the lengths of the four sides. If the figure were a park, the sides could be measured in miles or kilometers. If the figure were a room, the sides could be measured in feet or meters. If it were a piece of paper, the sides could be measured in inches or centimeters.
63. A square has all four sides equal. This doesn't have to be true for a rectangle. **67.** 950,000 **69.** 385
71. 3004 **73.** 5 ft 5 in. **75.** $51

SECTION 7.4

1. 16 km^2 **3.** 20 yd^2 **5.** 55 ft^2 **7.** 52 m^2
9. $78\frac{4}{7}$ cm^2 **11.** 132 km^2 **13.** 1652 yd^2
15. 1792 $in.^2$ **17.** 396 cm^2 **19.** 135 $in.^2$
21. 630 m^2 **23.** 120.7016 $in.^2$ **25.** 432 $in.^2$
27. 36 ft^2 **29.** 27,878,400 ft^2 **31.** 9 ft^2
33. 13 yd^2 **35.** 40.5 ft^2 **37.** 0.43 mi^2
39. 0.39 $in.^2$ **41.** 3.59 yd^2 **43.** 13,935.46 cm^2
45. 101 ft^2 **47.** 6550 m^2 **49.** 5775 $in.^2$
51. 280.13 cm^2 **53.** No, she needs more than 3 gal.
55. April and Larry need about 151 ft^2 of tile.
57. Debbie needs 360 oz of weed killer.
59. The doors require 36 ft^2 of glass.
61. The gable requires 162 ft^2 of sheathing.
63. The table can be covered with 1332 $in.^2$ of pad.
65. Sheila should buy about 3 yd of material.
67. The region can be covered with 6 bags of seed.
69. The den can be paneled with 6 sheets of paneling.
71. The patio has an area of 326 ft^2.
73. Area is usually measured in square units. For example, square feet and square meters can be used. Acres can also be used to measure land area.
75. Cover the area with as many whole squares as possible. For the leftover area, estimate half or quarter squares. Then find the total of all the squares.

77. 3100 ft^2 will be walkway and the remaining 20,900 ft^2 will be lawn. **81.** 457,900 **83.** 46 **85.** 17
87. average 56; median 18

89.

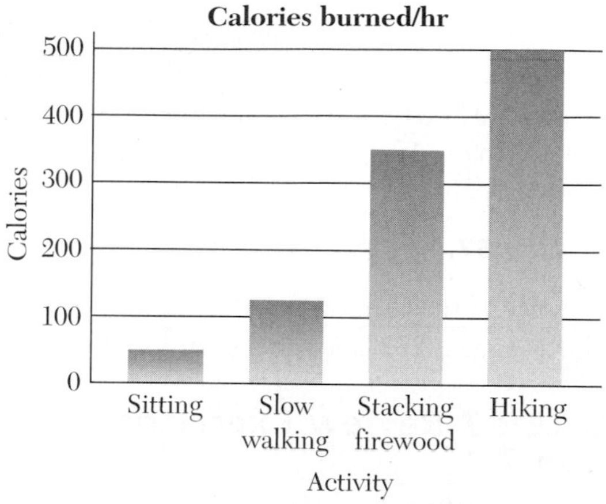

SECTION 7.5

1. 800 m^3 **3.** 250 ft^3 **5.** 120 cm^3 **7.** 50.24 ft^3
9. 31,500 mL **11.** 48 ft^3 **13.** 3052.08 $in.^3$
15. 3060 $in.^3$ **17.** 43,800 cm^3 **19.** 1728 $in.^3$
21. 27 ft^3 **23.** 46,656 $in.^3$ **25.** 4 ft^3 **27.** 8 yd^3
29. 24,192 $in.^3$ **31.** 147,197,952,000 ft^3
33. 0.09 ft^3 **35.** 4.17 km^3 **37.** 0.28 ft^3
39. 5880 $in.^3$ **41.** 5520 cm^3 **43.** 174 ft^3
45. 20,900 mm^3
47. The flower boxes need three bags of potting soil.
49. The total volume of the silos is 330,000 ft^3.
51. The slab is 187,776 $in.^3 \approx 109$ $ft^3 \approx 4$ yd^3 of concrete.
53. The cement contractor needs 127 ft of lumber for the form. **55.** The trash can has a 154-$in.^2$ base.
57. Norma needs 10 yd^3 of mushroom compost.
59. Lucy can fit 12 balls into the box.
61. The volume of an object is the amount of space it will hold or displace. Volume is a consideration when trying to fit a birthday gift into a box, deciding how many servings of orange juice are in a container, or filling up the tank in your car with gasoline.
63. In the formula $V = \ell wh$, ℓw represents the area of the base. Therefore, we can rewrite the formula as $V = \ell wh = bh$.
65. The hole for the pool is 200 m^3. Lining the pool requires about 28.7 m^3 of concrete. **67.** 512 **69.** 99
71. 169 **73.** 3600 **75.** 72 $in.^2$, or 0.5 ft^2

SECTION 7.6

1. 5 **3.** 13 **5.** 15 **7.** 20 **9.** $\frac{9}{7}$ **11.** 0.8
13. 11.83 **15.** 27 **17.** 3.27 **19.** 0.68
21. 105 **23.** $b = 8$ **25.** $c = 25$ **27.** $a = 20$
29. $b = 16$ **31.** $b \approx 21.79$ **33.** $c \approx 12.37$
35. $b \approx 48.99$ **37.** $b \approx 3.32$
39. The hypotenuse is 100 cm.
41. The side of the square is 12 ft.
43. The side of the square is 17 m.
45. The top of the ladder is about 11.62 ft up the house.

47. It is about 127.28 ft from 1st base to 3rd base.
49. Harry needs 12 ft of wire for the supports.
51. A square root of a number is a number that, when squared, gives back the original number. To find $\sqrt{289}$, we look for the number whose square is 289. Because $20^2 = 400$, we know that it is less than 20. Working backward, we have $18^2 = 324$ (still too big but close) and $17^2 = 289$. Therefore, $\sqrt{289} = 17$.
53. The rod will fit in the second box but not in the first one.
55. 6750 **57.** $16\frac{1}{5}$ **59.** $8\frac{5}{8}$
61. The second tree is 9.375 ft tall.
63. The average is 22.6 and the median is 29.

CHAPTER 7 Review Exercises
Section 7.1

1. 60 cm **3.** 3 yd 14 in. **5.** 1 ft 4 in.
7. 10,560 ft **9.** 381 mm

Section 7.2

11. 54 gal **13.** 21 oz **15.** 10 hr 20 min
17. 1.5 lb **19.** 15°C

Section 7.3

21. $P = 100$ cm **23.** $P = 41$ in. **25.** $P = 38$ yd
27. The perimeter is 260 m.
29. The backyard requires 174 ft of fencing.

Section 7.4

31. $A = 1968$ mm^2 **33.** $A = 120$ cm^2
35. $A = 63$ yd^2
37. The area of the square is 250,000 mm^2.
39. One coat must cover 206 ft^2, so two coats must cover 412 ft^2; she does not have enough paint.

Section 7.5

41. $V = 4050$ cm^3 **43.** $V \approx 502{,}654.82$ mm^3
45. $V = 288$ m^3 **47.** The volume is about 2930.67 in.^{3}.
49. There are 46,656 in.^{3} in a cubic yard.

Section 7.6

51. 17 **53.** $\frac{7}{9}$, or about 0.78 **55.** $b = 15$
57. $b \approx 10.39$
59. The top of the ladder is about 9.54 ft high.

CHAPTER 7 True/False Concept Review

1. false; metric measurements are the most commonly used in the world. **2.** true
3. false; a liter is a measure of capacity or volume.
4. true
5. false; volume is the measure of the inside of a solid.
6. false; the volume of a cube is $V = s^3$. **7.** true
8. true **9.** true
10. false; in a right triangle, the longest side is the hypotenuse. **11.** false; $\sqrt{4} = 2$ **12.** true
13. false; a trapezoid has four sides. **14.** true
15. true **16.** true
17. false; volume can also be measured in gallons and liters.
18. false; the sum of the squares of the lengths of the legs of a right triangle is equal to the square of the hypotenuse.
19. false; 1 ft^2 = 144 in.^{2}
20. false; the prefix kilo- means 1000.

CHAPTER 7 Test

1. 7454 mm **2.** 136 cm **3.** 1728 in.^{3}
4. The room can be covered by 80 ft^2 of flooring.
5. 14.142 **6.** 22 cm **7.** 30 in.^{2}
8. 1 gal 3 qt 1 pt
9. In the English system, there are pounds, ounces, and tons. In the metric system, there are grams, milligrams, and kilograms. **10.** 288 in.^{2}
11. The windows require 50 ft of molding.
12. 264 m^3 **13.** 48 cm^2 **14.** The volume is 20 ft^3.
15. Each bar will weigh 114 g.
16. The dog runs will cost $132. **17.** 13 mm
18. 140.52 km^2 **19.** Each player runs 1920 yd.
20. Area is a measure of two-dimensional space, whereas volume is a measure of three-dimensional space.
21. Jared's average bench-press is 193 lb.
22. Li needs about 251.2 in. of lace. She must buy 7 yd.
23. The box has a larger volume than the bag, so all the dog food will fit in the box.
24. Clayton and Sam are about 10.44 km apart.
25. Neela needs more than 400 in.^{2} of wrapping paper.

CHAPTERS 1–7 Cumulative Review

1. six thousand thirty-five **3.** 309, 375, 381, 384, 390
5. 436 **7.** 7755 **9.** 740
11. average, 35; median, 34; mode, 46
13. Dental hygienists have the lowest average salary.
15. Sylvia collected pledges totaling $1155.
17. Millie earned $559.
19. 1, 2, 3, 6, 9, 18, 27, 54, 81, 162
21. 23, 46, 69, 92, 115 **23.** It is divisible by 2 and 3.
25. $2 \cdot 3 \cdot 5 \cdot 23$ **27.** $\frac{3}{7}$ **29.** $25\frac{1}{2}$ **31.** $\frac{8}{9}$
33. $\frac{17}{10}$, or $1\frac{7}{10}$ **35.** $\frac{31}{140}$ **37.** $1\frac{3}{4}$
39. The number of students who passed the mathematics competency exam is 141.
41. ten and forty-seven thousandths
43. 1.046, 1.307, 1.3491, 1.35, 1.4005 **45.** 95.78
47. 3.45×10^{-5} **49.** 13.45 **51.** 0.6875
53. 22.32
55. Bill should cut 2.125 inches from each end of the board.

57. The ratio of the females to the total class is $\frac{11}{20}$.
59. A = 108
61. The tree will cast a shadow of 18.25 ft. 63. 0.058
65. 63.5% 67. 34.375% 69. $\frac{341}{500}$ 71. 7.119
73. 48.5 75. Helga earned $517.20 last week.
77. 8 yd 2 ft 6 in. 79. 650 mL 81. 24,076.8 ft
83. 1.59 kg
85. The temperature measured in Celsius is 43.1°C.
87. 108 in. 89. The area of the triangle is 17.5 ft^2.
91. 452 $in.^2$ 93. The volume is about 1017.36 cm^3.
95. The carpet will cost Trudy $6175. 97. 6.08
99. The hypotenuse is 45 ft in length.

CHAPTER 8

SECTION 8.1

1. 4 3. −5 5. 2.3 7. $-\frac{3}{5}$ 9. $\frac{6}{7}$
11. −61 13. 42 15. 3.78 17. $-\frac{17}{5}$
19. −0.55 21. −113.8 23. 0.0123 25. 4
27. 24 29. 3.17 31. $\frac{5}{11}$ 33. $\frac{3}{11}$
35. 32 or −32 37. 0.0065 39. 355
41. $\frac{11}{3}$ 43. $\frac{33}{7}$ 45. 0 47. 0.341
49. $-\frac{5}{17}$ 51. −78
53. The opposite of a gain of 1.54 is −1.54.
55. The opposite of a reading of 14°C is −14°C.
57. The opposite of 1915 B.C. is A.D. 1915.
59. As signed numbers, the distance up the mountain is +1415 ft and the distance down the mountain is −1143 ft.
61. Eighty miles south would be represented by −80 miles.

63.

Ocean	Deepest Part	(ft)	Average Depth (ft)
Pacific	Mariana Trench	−35,840	−12,925
Atlantic	Puerto Rico Trench	−28,232	−11,730
Indian	Java Trench	−23,376	−12,598
Arctic	Eurasia Basin	−17,881	−3407
Mediterranean	Ionian Basin	−16,896	−4926

65. 24 67. −14
69. On the first play, the Seahawks made −8 yd. The opposite of the loss of 8 yd is +8 yd.
71. The distances driven by the Golden family expressed as signed numbers are +137 miles, +212 miles, −98 miles, and −38 miles.
73. The positive real numbers and zero each have an absolute value that is the number itself. This is because the absolute value of a number is its distance (distance is positive) from zero, and the positive numbers are all a positive distance from zero. 75. 7 77. positive 81. 52
83. 22 85. 7173 mL 87. 20 pt
89. Two aspirin tablets contain 154.3 g.

SECTION 8.2

1. 2 3. 2 5. −15 7. −14 9. 0
11. −12 13. −46 15. −14 17. −26
19. 7 21. −6 23. −108 25. 0 27. 27
29. −10 31. −7 33. −8.3 35. 2.8
37. $-\frac{7}{12}$ 39. $-\frac{7}{6}$ 41. −121 43. −189
45. −30 47. 81 49. −757 51. −114
53. The sum of the temperatures recorded at 9:00 A.M. is −349°C.
55. The sum of the temperatures recorded at 11:00 P.M. is −505°C.
57. The net change in weight the cargo master should report is −108 lb.
59. The altitude of the airplane at 4 P.M. is 29,870 ft.
61. The inventory at the end of the month is 30,370 volumes.
63. No, the Bills were 1 yd short of a first down.
65. a. The net change in Nordstrom stock for the week was up 1.76.
b. The closing price of Nordstrom stock is 74.59.
67. The net gain for the month is $5077.
69. Adding signed numbers can be thought of as moving along a number line according to the absolute value and sign of the numbers. Starting at zero, move as many spaces as the absolute value of the first number. Move right if the number is positive and left if it is negative. Then move according to the second number. The sign of the sum is determined by which side of zero you end up on. So if you have moved right for more spaces than you have moved left, the sum will be positive. And if you have moved left more spaces than you have moved right, the sum will be negative. Symbolically, compare the absolute values of the numbers being added. The sign of the one with the largest absolute value (corresponding to the most spaces moved in a particular direction) is the sign of the answer.
71. −124 73. 32 77. 267 79. 199
81. The circumference is 106.76 cm.
83. The perimeter is 195 cm.
85. The cost of pouring the walk is $12,348.

SECTION 8.3

1. 4 3. −12 5. −5 7. 3 9. −20
11. −9 13. 32 15. −34 17. −28 19. 1
21. 0 23. −18 25. −80 27. 118
29. −119 31. 11 33. −182 35. 204
37. 0 39. 144 41. 173 43. −9.98
45. −1.44 47. −56.14 49. 116 51. −687
53. The change in temperature is 85°C.
55. Joe wrote checks totaling $796.32.
57. The range of altitude for each continent is Africa, 19,852 ft; Antarctica, 25,191 ft; Asia, 30,340 ft; Australia, 7362 ft; Europe, 18,602 ft; North America, 20,602 ft; and

South America, 22,965 ft. The smallest range is in Australia. It means it is a rather flat continent.
59. The depth of the ocean floor at Mauna Kea is −19,680 feet, or 19,680 feet below sea level.
61. The difference in temperatures recorded at 11:00 P.M. on day 3 and day 5 is −15°C.
63. The difference between the highest and lowest temperatures recorded during the 5 days is 105°C.
65. Thomas's bank balance is −\$6.41, or \$6.41 overdrawn.
67. Janna wrote checks totaling \$477.14.
69. Gina's sales total −\$164.45, or a loss of \$164.45.
71. The difference in coal imports between 1980 and 1970 is −0.46 quadrillion BTU. The difference in coal imports between 1990 and 1980 is −0.31 quadrillion BTU. The difference in coal imports between 1990 and 1970 is −0.77 quadrillion BTU.
73. Starting at zero, move 12 units to the right, positive. Now if we were adding −8, we would move 8 units to the left of +12, but because we are subtracting, we reverse the direction and move an additional 8 units to the right, ending at +20.
75. The first problem is addition and the second is subtraction. The answer to the two problems is the same, −3.
77. −12.68 **79.** −41 **83.** 546
85. 1,731,084
87. The area of the square is 278.89 m^2.
89. The area of the triangle is 56.16 m^2.
91. The number of tiles needed is 896.

SECTION 8.4

1. −8 **3.** −10 **5.** −80 **7.** −77 **9.** 11
11. −306 **13.** −255 **15.** −9 **17.** −350
19. $-\frac{1}{4}$ **21.** 3 **23.** 28 **25.** 99 **27.** 60
29. −11 **31.** 210 **33.** 5.4 **35.** 24.2
37. $\frac{1}{6}$ **39.** 1.645 **41.** −2520 **43.** 465
45. 7.14 **47.** $-\frac{3}{10}$ **49.** −14.035 **51.** 0.5423
53. 0 **55.** 21 **57.** −240 **59.** $\frac{1}{3}$
61. The Celsius measure is −15°C.
63. Ms. Riles's loss, expressed as a signed number, is −20.8 lb.
65. The Dow Jones loss over 12 days is −20.88.
67. −49 **69.** −42 **71.** 108
73. Safeway's loss, expressed as a signed number, is −\$65.62.
75. Winn Dixie's loss, expressed as a signed number, is −\$31.38.
77. Safeway's loss, expressed as a signed number, is −\$250.79.
79. The trader paid \$6257.80 for the stock. She received \$4321.50 when she sold the stock. She lost \$1936.30, or −\$1936.30, on the sale of the stock. **81.** Asia
83. One of the ways to interpret the opposite of a number is to consider it as −1 times the number. So, in the expression -3^2, we have two operations, multiplying by −1 and squaring. The rules for order of operations require that we do the exponent before multiplying. So, $-3^2 = -1(3)^2 = -1(9) = -9$. In the expression $(-3)^2$, we have the same two operations but the parentheses require that we multiply first, so −3 is squared or $(-3)^2 = 9$.
85. −70 **87.** 56 **91.** 68 **93.** 268
95. The piecework wage is \$2.60 per piece.
97. The tip of the hand moves 9.42 in.
99. The cost per square foot for the rental is \$0.5625 per square foot per year.

SECTION 8.5

1. −2 **3.** −4 **5.** −3 **7.** −8 **9.** 8
11. −6 **13.** −2.02 **15.** −35 **17.** −3
19. −1.2 **21.** 2 **23.** 3 **25.** 7 **27.** 6
29. −5 **31.** 7 **33.** 8 **35.** 4.04 **37.** $\frac{1}{2}$
39. 5 **41.** −45 **43.** 116 **45.** −4.5
47. 0 **49.** −0.0026 **51.** −452 **53.** 16
55. Each member's share of the loss is −\$125.65.
57. Mr. Harkness lost an average of −4.6 lb per week.
59. The average daily loss for the stock is −3.82 points.
61. Answers will vary. **63.** The continent is Australia.
65. The division rules are the same as a consequence of the fact that division is the inverse of multiplication. Every division fact can be rewritten as a multiplication fact. So, we have: $a \div b = a$ (reciprocal of b).
67. −10 **69.** $-\frac{8}{5}$ **71.** −1.4 **73.** 7
75. 18
77. The volume of the cylinder is about 4823.04 in.^3.
79. The tank will hold approximately 3805 gallons.
81. There will be 12 truckloads.

SECTION 8.6

1. −24 **3.** 19 **5.** −6 **7.** −16 **9.** −6
11. 8 **13.** 4 **15.** 13 **17.** 11 **19.** 8
21. 1 **23.** −22 **25.** −6 **27.** 40
29. −38 **31.** −8 **33.** 8 **35.** 0 **37.** −24
39. 64 **41.** −11 **43.** −15 **45.** 144
47. −12
49. The average temperature recorded during day 5 is −56°C.
51. The average high temperature recorded for the 5 days is −34°C.
53. 77 **55.** −41 **57.** $-\frac{62}{3}$ **59.** 1
61. −2065 **63.** −1594 **65.** −16
67. Keshia pays a total of \$800 for the TV set.
69. Kmart has a loss on the fishing poles of −\$9.
71. The average of the highest and lowest points are: Africa, 9414 ft; Antarctica, 4268.5 ft; Asia, 13,858 ft; Australia, 3629 ft; Europe, 9209 ft; North America, 10,019 ft;

and South America, 11,351.5 ft. The largest average is in Asia. The smallest average is in Australia.
73. The three cities are Athens, Bangkok, and New Delhi.
75. The balance at the end of the month is $3226.
77. The net result of the trip is −13 miles. At the end of the trip, you are 78 miles north and 91 miles west of the starting point.
79. The problem is worked according to the order of operations. So we work inside the brackets first, 3 + 5(−4). This is where the error has been made. We must multiply first, so

$$\begin{aligned} 2[3 + 5(-4)] &= 2[3 + (-20)] \\ &= 2[-17] \\ &= -34 \end{aligned}$$

81. Yes, if a different operation is inside parentheses. For example, in $(3 - 6)^2$, the subtraction is done first. So, $(3 - 6)^2 = (-3)^2 = 9$.
83. −11 **87.** −17.09 **89.** 16.1 **91.** −40,920
93. −7366
95. The stock gains 0.438 point for the week.

SECTION 8.7

1. $x = 7$ **3.** $x = 5$ **5.** $y = -5$ **7.** $a = 7$
9. $x = 7$ **11.** $x = 10$ **13.** $x = -3$
15. $x = -6$ **17.** $x = 0$ **19.** $x = 3$ **21.** $y = 7$
23. $a = 0$ **25.** $x = -0.5$ **27.** $x = -25$
29. $x = -16$ **31.** $y = -4$
33. It will take the sky diver 4 seconds to reach a falling speed of 147 feet per second. **35.** The number is 28.
37. The number is 8. **39.** The acceleration is 33.
41. There will be 14 payments.
43. The equation is $7310 - 4L = 12{,}558$, where L represents the lowest point. Asia's lowest point fits the description.
45. The equation is $-35{,}840 - 2D = -2000$, where D represents the deepest part. The Ionian Basin's deepest part fits this description.
47. $z = 7$ **49.** $z = 6$
51.

$$\begin{aligned} -3x + 10 &= 4 \\ -10 \quad & -10 \quad \text{Add } -10 \text{ to both sides.} \\ -3x &= -6 \quad \text{Simplify.} \\ \frac{-3x}{-3} &= \frac{-6}{-3} \quad \text{Divide both sides by } -3. \\ x &= 2 \quad \text{Simplify.} \end{aligned}$$

53. $x = -4$ **55.** $x = -2$

CHAPTER 8 Review Exercises
Section 8.1

1. 39 **3.** 0.91 **5.** 16.5 **7.** 71 **9.** −6.4

Section 8.2

11. −98 **13.** 52 **15.** −3.2 **17.** −0.53
19. Yes, the Dolphins gained 10 yd.

Section 8.3

21. 22 **23.** 91 **25.** −7.81 **27.** 29
29. Maria wrote checks totaling $685.90.

Section 8.4

31. −66 **33.** −4.08 **35.** 51 **37.** 8.463
39. Kroger's loss, expressed as a signed number, is −$284.40.

Section 8.5

41. −3 **43.** 5 **45.** −16 **47.** −56.2
49. The average daily loss of the stock is −3.875 points.

Section 8.6

51. −56 **53.** −50 **55.** −24 **57.** 17
59. −20.72

Section 8.7

61. $x = -5$ **63.** $x = 25$ **65.** $x = 0$
67. $a = -13$
69. The Fahrenheit temperature is −7.6°F.

CHAPTER 8 True/False Concept Review

1. true
2. false; the opposite of a positive number is negative.
3. false; the absolute value of a nonzero number is always positive. **4.** true
5. false; the sum of two signed numbers can be positive, negative, or zero.
6. false; the sum of a positive signed number and a negative signed number may be positive, negative, or zero.
7. true
8. false; to subtract two signed numbers, add the opposite of the number to be subtracted. **9.** true **10.** true
11. false; the sign of the product of a positive number and a negative number is negative. **12.** true **13.** true
14. true

CHAPTER 8 Test

1. −33 **2.** −35 **3.** $-\frac{5}{4}$ **4.** $\frac{2}{15}$ **5.** −5
6. 1.15 **7. a.** 17 **b.** 33 **8.** −33 **9.** −2
10. 2 **11.** 96 **12.** −21 **13.** −108.9
14. 40 **15.** −31.2 **16.** $-\frac{1}{6}$ **17.** 14
18. −80 **19.** −86 **20.** $-\frac{3}{10}$ **21.** −68
22. −5 **23.** $x = -6$ **24.** $x = 7$ **25.** $a = -3.5$
26. Ms. Rosier lost −16.8 lb.
27. The drop in temperature is −21°F.
28. The closing on Friday is 17.97.
29. The Fahrenheit temperature is 14°F.
30. The average is −4.

CHAPTERS 1–4 Midterm Exam

1. ten thousand **2.** 94,019
3. three hundred forty-five thousand, eight hundred four.

4. 3057 **5.** 17,763 **6.** 41,496
7. 720,000; 727,518 **8.** 167 **9.** 3637 R 21
10. 7 **11.** 74 **12.** 197
13. average, 51; median, 43; mode, 34
14. a. Civic has the highest sales.
b. Five more Accords are sold than Passports.
15. 180 **16.** 15 **17.** prime number **18.** yes
19. 23, 46, 69, 92, 115
20. 1, 2, 4, 8, 16, 19, 38, 76, 152, 304
21. $2 \cdot 2 \cdot 2 \cdot 2 \cdot 19 = 2^4 \cdot 19$ **22.** $8\frac{3}{8}$ **23.** $\frac{123}{7}$
24. $\frac{3}{4}, \frac{7}{8}$ **25.** $\frac{5}{9}, \frac{7}{12}, \frac{5}{8}, \frac{2}{3}$ **26.** $\frac{2}{3}$ **27.** $\frac{1}{7}$
28. $55\frac{7}{32}$ **29.** $2\frac{1}{2}$ **30.** $\frac{3}{13}$ **31.** $\frac{17}{20}$
32. $18\frac{13}{24}$ **33.** $3\frac{2}{5}$ **34.** $7\frac{29}{45}$ **35.** $7\frac{3}{4}$
36. $\frac{3}{10}$ **37.** 70,000.302 **38.** 2.1, 2.07, 2.075
39. $\frac{7}{50}$ **40.** 1.02, 1.034, 1.044, 1.07, 1.094, 1.109
41. 55.903 **42.** 1.744 **43.** 14.49 **44.** 120
45. 0.0000012 **46.** 3.71×10^{-3} **47.** 13.6
48. 0.538 **49.** average, 6.4; median, 6.9
50. 1.62 **51.** The perimeter is 75 ft.
52. The area is $27\frac{3}{4}$ ft^2.
53. 6950 chinook were counted.
54. Maria received $72.52 as a dividend.
55. Houng earns $18.08 per hour.

CHAPTERS 1–8 Final Exam

1. $\frac{23}{16}$, or $1\frac{7}{16}$ **2.** 126 **3.** 33.89 **4.** 19.853
5. $\frac{7}{6}$, or $1\frac{1}{6}$ **6.** $51\frac{1}{3}$ **7.** 199 **8.** 97.84
9. $6\frac{3}{10}$ **10.** 721 **11.** 82.04 **12.** 58.266
13. $\frac{43}{50}$ **14.** $10\frac{17}{30}$ **15.** $x = \frac{7}{3}$, or $2\frac{1}{3}$
16. thousandths **17.** $\frac{7}{10}, \frac{3}{4}, \frac{4}{5}$ **18.** 85%
19. 24 R 11 **20.** 0.478 **21.** 6000.015
22. 3812.5 **23.** 0.11375 **24.** 35 **25.** 452%
26. 0.265 **27.** 36.45 **28.** 0.0306
29. $239.70 **30.** 28.125%
31. 41, 82, 123, 164, 205
32. nine thousand forty-five and forty-five thousandths
33. false **34.** $2 \cdot 2 \cdot 3 \cdot 5 \cdot 7 = 2^2 \cdot 3 \cdot 5 \cdot 7$
35. $\frac{7}{12}$ **36.** $\frac{189}{200}$ **37.** $36\frac{3}{7}$ **38.** $\frac{5}{16}$
39. $\frac{3}{40}$ **40.** 58% **41.** no **42.** $13\frac{13}{15}$
43. $\frac{124}{9}$ **44.** 2.299, 2.32, 2.322, 2.332 **45.** 34
46. 0.067481 **47.** $1\frac{11}{23}$ **48.** $\frac{1}{12}$ **49.** $1812
50. $420 **51.** 8.6% **52.** 7 **53.** $3\frac{7}{20}$
54. 5.91 **55.** 7 hr 18 min 11 sec **56.** 2 m 92 cm
57. $50 per kilogram **58.** The perimeter is 127.5 ft.
59. 13.395 m^2 **60.** 62.13 yd^2 **61.** 3888 in.^{3}
62. 3.58
63. The hypotenuse of the triangle is about 22.0 cm.
64. −16 **65.** 27 **66.** −72 **67.** 5 **68.** 2
69. $a = -8$
70. The Fahrenheit temperature is 113°F.

INDEX

A